Student Solutions Manual

Laurel Technical Services

Beginning & Intermediate Algebra

SECOND EDITION

K. Elayn Martin-Gay

Upper Saddle River, NJ 07458

Executive Editor: Karin E. Wagner
Project Manager: Mary Beckwith
Assistant Managing Editor: John Matthews
Production Editor: James Buckley
Supplement Cover Manager: Paul Gourhan
Supplement Cover Designer: PM Workshop Inc.
Manufacturing Buyer: Lisa McDowell

Upper Saddle River, NJ 07458

Printed in the United States of America

10 9 8 7 6 5

ISBN 0-13-017338-X

Prentice-Hall International (UK) Limited, London
Prentice-Hall of Australia Pty. Limited, Sydney
Prentice-Hall Canada, Inc., Toronto
Prentice-Hall Hispanoamericana, S.A., Mexico City
Prentice-Hall of India Private Limited, New Delhi
Pearson Education Asia Pte. Ltd., Singapore
Prentice-Hall of Japan, Inc., Tokyo
Editora Prentice-Hall do Brazil, Ltda., Rio de Janeiro

Table of Contents

Chapter 1

Exercise Set 1.2

1. $4 < 10$ since 4 is to the left of 10 on the number line.

3. $7 > 3$ since 7 is to the right of 3 on the number line.

5. $6.26 = 6.26$

7. $0 < 7$ since 0 is to the left of 7 on the number line.

9. $32 < 212$ since 32 is to the left of 212 on the number line.

11. $44{,}300 > 34{,}611$ since 44,300 is to the right of 34,611 on the number line.

13 True, since $11 = 11$ is true.

15. False, since 10 is to the left of 11 on the number line.

17. $3+8 \geq 3(8)$
$11 \geq 24$
False, since neither $11 > 24$ nor $11 = 24$ is true.

19. True, since 7 is to the right of 0 on the number line.

21. $30 \leq 45$

23. $8 < 12$

25. $5 \geq 4$

27. $15 \neq -2$

29. The integer 535 represents 535 feet. The integer −8 represents 8 feet below sea level.

31. The integer −433,853 represents a decrease in attendance of 433,853.

33. The integer 350 represents a deposit of \$350. The integer −126 represents a withdrawal of \$126.

35. 1988 because this bar is the shortest.

37. 1988, 1989, 1990, 1991 because the heights of these bars are less than 5000.

39. In 1993 there were 6068 communities with curbside plastic recycling and in 1990 there were 2649 communities with curbside plastic recycling. 6068 is to the right of 2649 on the number line.
$6068 \geq 2649$

41. whole, integers, rational, real

43. integers, rational, real

45. natural, whole, integers, rational, real

47. rational, real

49. irrational, real

51. False; for example, $\frac{1}{4}$ is a rational number but not an integer.

53. True; the set of natural numbers is $\{1, 2, 3, 4, 5, \ldots\}$.

55. True; 0 corresponds to a point on the number line.

57. True; the set of whole numbers is $\{0, 1, 2, 3, 4, \ldots\}$.

59. False; the set of irrational numbers is the set of all numbers that correspond to points on the number line but that are not rational numbers.

61. $-10 > -100$ since −10 is to the right of −100 on the number line.

63. $32 > 5.2$ since 32 is to the right of 5.2 on the number line.

65. $\frac{18}{3} < \frac{24}{3}$ since $\frac{18}{3} = 6$ and $\frac{24}{3} = 8$, $6 < 8$.

67. $-51 < -50$ since -51 is to the left of -50 on the number line.

69. $|-5| > -4$ since $|-5| = 5$ and 5 is to the right of -4 on the number line.

71. $|-1| = |1|$ since $|-1| = 1$ and $|1| = 1$.

73. $|-2| < |-3|$ since $|-2| = 2$ and $|-3| = 3$ and 2 is to the left of 3 on the number line.

75. $|0| < |-8|$ since $|0| = 0$ and $|-8| = 8$ and 0 is to the left of 8 on the number line.

77. $-0.04 > -26.7$ since -0.04 is to the right of -26.7 on the number line.

79. The sun is brighter, since $-26.7 < -0.04$.

81. The sun, since on the number line -26.7 is to the left of all other numbers listed, and therefore, -26.7 is smaller than all other numbers listed.

83. $20 \leq 25$

85. $6 > 0$

87. $-12 < -10$

89. Answers may vary.

Section 1.3

Mental Math

1. $\frac{3}{8}$

2. $\frac{1}{4}$

3. $\frac{5}{7}$

4. $\frac{2}{5}$

5. numerator; denominator

6. $\frac{11}{7}$

Exercise Set 1.3

1. $33 = 3 \cdot 11$

3. $98 = 2 \cdot 49 = 2 \cdot 7 \cdot 7$

5. $20 = 2 \cdot 10 = 2 \cdot 2 \cdot 5$

7. $75 = 3 \cdot 25 = 3 \cdot 5 \cdot 5$

9. $45 = 3 \cdot 15 = 3 \cdot 3 \cdot 5$

11. $\frac{2}{4} = \frac{2}{2 \cdot 2} = \frac{1}{2}$

13. $\frac{10}{15} = \frac{2 \cdot 5}{3 \cdot 5} = \frac{2}{3}$

15. $\frac{3}{7}$ is already in lowest terms.

17. $\frac{18}{30} = \frac{2 \cdot 3 \cdot 3}{2 \cdot 3 \cdot 5} = \frac{3}{5}$

19. $\frac{1}{2} \cdot \frac{3}{4} = \frac{1 \cdot 3}{2 \cdot 2 \cdot 2} = \frac{3}{8}$

21. $\frac{2}{3} \cdot \frac{3}{4} = \frac{2 \cdot 3}{3 \cdot 2 \cdot 2} = \frac{1}{2}$

23. $\frac{1}{2} \div \frac{7}{12} = \frac{1}{2} \cdot \frac{12}{7} = \frac{1 \cdot 2 \cdot 2 \cdot 3}{2 \cdot 7} = \frac{6}{7}$

25. $\frac{3}{4} \div \frac{1}{20} = \frac{3}{4} \cdot \frac{20}{1} = \frac{3 \cdot 2 \cdot 2 \cdot 5}{2 \cdot 2} = 15$

27. $\frac{7}{10} \cdot \frac{5}{21} = \frac{7 \cdot 5}{2 \cdot 5 \cdot 3 \cdot 7} = \frac{1}{6}$

29. $2\frac{7}{9}\cdot\frac{1}{3}=\frac{25}{9}\cdot\frac{1}{3}=\frac{5\cdot 5\cdot 1}{3\cdot 3\cdot 3}=\frac{25}{27}$

31. $l\cdot w=\frac{11}{12}\cdot\frac{3}{5}=\frac{11\cdot 3}{2\cdot 2\cdot 3\cdot 5}=\frac{11}{20}$

The area is $\frac{11}{20}$ square mile.

33. $\frac{4}{5}-\frac{1}{5}=\frac{4-1}{5}=\frac{3}{5}$

35. $\frac{4}{5}+\frac{1}{5}=\frac{4+1}{5}=\frac{5}{5}=1$

37. $\frac{17}{21}-\frac{10}{21}=\frac{17-10}{21}=\frac{7}{21}=\frac{7}{7\cdot 3}=\frac{1}{3}$

39. $\frac{23}{105}+\frac{4}{105}=\frac{23+4}{105}=\frac{27}{105}=\frac{3\cdot 3\cdot 3}{3\cdot 5\cdot 7}=\frac{9}{35}$

41. $\frac{7}{10}=\frac{7}{10}\cdot\frac{3}{3}=\frac{7\cdot 3}{10\cdot 3}=\frac{21}{30}$

43. $\frac{2}{9}=\frac{2}{9}\cdot\frac{2}{2}=\frac{2\cdot 2}{9\cdot 2}=\frac{4}{18}$

45. $\frac{4}{5}=\frac{4}{5}\cdot\frac{4}{4}=\frac{4\cdot 4}{5\cdot 4}=\frac{16}{20}$

47. $\frac{2}{3}+\frac{3}{7}$ LCD = 21

$\frac{2}{3}\cdot\frac{7}{7}=\frac{14}{21}$

$\frac{3}{7}\cdot\frac{3}{3}=\frac{9}{21}$

$\frac{14}{21}+\frac{9}{21}=\frac{14+9}{21}=\frac{23}{21}$

49. $2\frac{13}{15}-1\frac{1}{5}=\frac{43}{15}-\frac{6}{5}$ LCD = 15

$\frac{6}{5}\cdot\frac{3}{3}=\frac{18}{15}$

$\frac{43}{15}-\frac{18}{15}=\frac{43-18}{15}=\frac{25}{15}=1\frac{2}{3}$

51. $\frac{5}{22}-\frac{5}{33}$ LCD = 66

$\frac{5}{22}\cdot\frac{3}{3}=\frac{15}{66}$

$\frac{5}{33}\cdot\frac{2}{2}=\frac{10}{66}$

$\frac{15}{66}-\frac{10}{66}=\frac{15-10}{66}=\frac{5}{66}$

53. $\frac{12}{5}-1$ LCD = 5

$1\cdot\frac{5}{5}=\frac{5}{5}$

$\frac{12}{5}-\frac{5}{5}=\frac{12-5}{5}=\frac{7}{5}$

55. $1-\frac{3}{10}-\frac{5}{10}=\frac{10}{10}-\frac{3}{10}-\frac{5}{10}$

$=\frac{10-3-5}{10}$

$=\frac{2}{10}$

$=\frac{1}{5}$

57. $1-\frac{1}{4}-\frac{3}{8}=\frac{8}{8}-\frac{2}{8}-\frac{3}{8}=\frac{8-2-3}{8}=\frac{3}{8}$

59. $1-\frac{1}{2}-\frac{1}{6}-\frac{2}{9}=\frac{18}{18}-\frac{9}{18}-\frac{3}{18}-\frac{4}{18}$

$=\frac{18-9-3-4}{18}$

$=\frac{2}{18}$

$=\frac{1}{9}$

61. $\frac{10}{21}+\frac{5}{21}=\frac{10+5}{21}=\frac{15}{21}=\frac{3\cdot 5}{3\cdot 7}=\frac{5}{7}$

63. $\frac{10}{3}-\frac{5}{21}$ LCD = 21
$\frac{10}{3}\cdot\frac{7}{7}=\frac{70}{21}$
$\frac{70}{21}-\frac{5}{21}=\frac{70-5}{21}=\frac{65}{21}$

65. $\frac{2}{3}\cdot\frac{3}{5}=\frac{2\cdot 3}{3\cdot 5}=\frac{2}{5}$

67. $\frac{3}{4}\div\frac{7}{12}=\frac{3}{4}\cdot\frac{12}{7}=\frac{3\cdot 2\cdot 2\cdot 3}{2\cdot 2\cdot 7}=\frac{9}{7}$

69. $\frac{5}{12}+\frac{4}{12}=\frac{5+4}{12}=\frac{9}{12}=\frac{3\cdot 3}{2\cdot 2\cdot 3}=\frac{3}{4}$

71. $5+\frac{2}{3}$ LCD = 3
$\frac{5}{1}\cdot\frac{3}{3}=\frac{15}{3}$
$\frac{15}{3}+\frac{2}{3}=\frac{15+2}{3}=\frac{17}{3}$

73. $\frac{7}{8}\div 3\frac{1}{4}=\frac{7}{8}\div\frac{13}{4}=\frac{7}{8}\cdot\frac{4}{13}=\frac{7\cdot 2\cdot 2}{2\cdot 2\cdot 2\cdot 13}=\frac{7}{26}$

75. $\frac{7}{18}\div\frac{14}{36}=\frac{7}{18}\cdot\frac{36}{14}=\frac{7\cdot 2\cdot 2\cdot 3\cdot 3}{2\cdot 3\cdot 3\cdot 2\cdot 7}=1$

77. $\frac{23}{105}-\frac{2}{105}=\frac{23-2}{105}=\frac{21}{105}=\frac{3\cdot 7}{3\cdot 5\cdot 7}=\frac{1}{5}$

79. $1\frac{1}{2}+3\frac{2}{3}=\frac{3}{2}+\frac{11}{3}$ LCD = 6
$\frac{3}{2}\cdot\frac{3}{3}=\frac{9}{6}$
$\frac{11}{3}\cdot\frac{2}{2}=\frac{22}{6}$
$\frac{9}{6}+\frac{22}{6}=\frac{31}{6}=5\frac{1}{6}$

81. $\frac{2}{3}-\frac{5}{9}+\frac{5}{6}$ LCD = 18
$\frac{2}{3}\cdot\frac{6}{6}=\frac{12}{18}$
$\frac{5}{9}\cdot\frac{2}{2}=\frac{10}{18}$
$\frac{5}{6}\cdot\frac{3}{3}=\frac{15}{18}$
$\frac{12}{18}-\frac{10}{18}+\frac{15}{18}=\frac{12-10+15}{18}=\frac{17}{18}$

83. $5+4\frac{1}{8}+15\frac{3}{4}+10\frac{1}{2}+15\frac{3}{4}+4\frac{1}{8}$
$=5+\frac{33}{8}+\frac{63}{4}+\frac{21}{2}+\frac{63}{4}+\frac{33}{8}$
$=\frac{40}{8}+\frac{33}{8}+\frac{126}{8}+\frac{84}{8}+\frac{126}{8}+\frac{33}{8}$
$=\frac{40+33+126+84+126+33}{8}$
$=\frac{442}{8}$
$=55\frac{1}{4}$ feet

85. $80-74\frac{17}{25}=\frac{2000}{25}-\frac{1867}{25}$
$=\frac{2000-1867}{25}$
$=\frac{133}{25}$
$=5\frac{8}{25}$ meters

87. Answers may vary.

89. $5\frac{1}{2}-2\frac{1}{8}=\frac{11}{2}-\frac{17}{8}$
$=\frac{44}{8}-\frac{17}{8}$
$=\frac{44-17}{8}$
$=\frac{27}{8}$
$=3\frac{3}{8}$ miles

91. $\frac{3}{4}$

93. $1-\frac{3}{4}-\frac{1}{200}=\frac{200}{200}-\frac{150}{200}-\frac{1}{200}$
$=\frac{200-150-1}{200}$
$=\frac{49}{200}$

Section 1.4

Calculator Explorations

1. $5^3 = 125$

3. $9^5 = 59{,}049$

5. $2(20-5) = 30$

7. $24(862-455) + 89 = 9857$

Mental Math

1. multiply

2. add

3. subtract

4. divide

Exercise Set 1.4

1. $3^5 = 3\cdot3\cdot3\cdot3\cdot3 = 243$

3. $3^3 = 3\cdot3\cdot3 = 27$

5. $1^5 = 1\cdot1\cdot1\cdot1\cdot1 = 1$

7. $5^1 = 5$

9. $\left(\frac{1}{5}\right)^3=\frac{1}{5}\cdot\frac{1}{5}\cdot\frac{1}{5}=\frac{1}{125}$

11. $\left(\frac{2}{3}\right)^4=\frac{2}{3}\cdot\frac{2}{3}\cdot\frac{2}{3}\cdot\frac{2}{3}=\frac{16}{81}$

13. $7^2 = 7\cdot7 = 49$

15. $4^2 = 4\cdot4 = 16$

17. $(1.2)^2 = (1.2)(1.2) = 1.44$

19. $5 + 6\cdot2 = 5 + 12 = 17$

21. $4\cdot8 - 6\cdot2 = 32 - 12 = 20$

23. $2(8-3) = 2(5) = 10$

25. $2+(5-2)+4^2 = 2+3+4^2$
$= 2+3+16$
$= 5+16$
$= 21$

27. $5\cdot3^2 = 5\cdot9 = 45$

29. $\frac{1}{4}\cdot\frac{2}{3}-\frac{1}{6}=\frac{2}{12}-\frac{1}{6}=\frac{1}{6}-\frac{1}{6}=0$

31. $\frac{6-4}{9-2}=\frac{2}{7}$

33. $2[5+2(8-3)] = 2[5+2(5)]$
$= 2[5+10]$
$= 2[15]$
$= 30$

35. $\frac{19-3\cdot5}{6-4}=\frac{19-15}{2}=\frac{4}{2}=2$

37. $\frac{|6-2|+3}{8+2\cdot5}=\frac{4+3}{8+10}=\frac{7}{18}$

39. $\dfrac{3+3(5+3)}{3^2+1}=\dfrac{3+3(8)}{9+1}=\dfrac{3+24}{10}=\dfrac{27}{10}$

41. $\dfrac{6+|8-2|+3^2}{18-3}=\dfrac{6+|6|+3^2}{15}$
$=\dfrac{6+6+9}{15}$
$=\dfrac{21}{15}$
$=\dfrac{3\cdot 7}{3\cdot 5}$
$=\dfrac{7}{5}$

43. No, since in the absence of grouping symbols we always perform multiplications or divisions before additions or subtractions in any expression. Thus, in the expression $2+3\cdot 5$, we should first find the product $3\cdot 5$ and then add 2.
$2+3\cdot 5=2+15=17$

45. a. $(6+2)\cdot(5+3)=8\cdot 8=64$

b. $(6+2)\cdot 5+3=8\cdot 5+3$
$=40+3$
$=43$

c. $6+2\cdot 5+3=6+10+3=19$

d. $6+2\cdot(5+3)=6+2\cdot 8=6+16=22$

47. Replace y with 3.
$3y=3(3)=9$

49. Replace x with 1 and z with 5.
$\dfrac{z}{5x}=\dfrac{5}{5(1)}=\dfrac{5}{5}=1$

51. Replace x with 1.
$3x-2=3(1)-2=3-2=1$

53. Replace x with 1 and y with 3.
$|2x+3y|=|2(1)+3(3)|=|2+9|=|11|=11$

55. Replace y with 3.
$5y^2=5\cdot 3^2=5\cdot 9=45$

57. If $x=12$, $y=8$, and $z=4$, then
$\dfrac{x}{z}+3y=\dfrac{12}{4}+3(8)=3+24=27.$

59. If $x=12$ and $y=8$, then
$x^2-3y+x=(12)^2-3(8)+12$
$=144-24+12$
$=132$

61. If $x=12$, $y=8$, and $z=4$, then
$\dfrac{x^2+z}{y^2+2z}=\dfrac{(12)^2+4}{(8)^2+2(4)}=\dfrac{144+4}{64+8}$
$=\dfrac{148}{72}=\dfrac{37}{18}.$

63. We evaluate the expression $16t^2$ for each value of t.
When $t=1$, we have
$16t^2=16\cdot 1^2=16\cdot 1=16.$
When $t=2$, we have
$16t^2=16\cdot 2^2=16\cdot 4=64.$
When $t=3$, we have
$16t^2=16\cdot 3^2=16\cdot 9=144.$
When $t=4$, we have
$16t^2=16\cdot 4^2=16\cdot 16=256.$

Time t (in seconds)	Distance $16t^2$ (in feet)
1	16
2	64
3	144
4	256

65. $3x-6=9$
$3(5)-6\overset{?}{=}9$
$15-6\overset{?}{=}9$
$9=9$
Solution

67. $2x+6=5x-1$
$2(0)+6 \stackrel{?}{=} 5(0)-1$
$0+6 \stackrel{?}{=} 0-1$
$6 \neq -1$
Not a solution

69. $2x-5=5$
$2(8)-5 \, 0 \, 5$
$16-5 \, 0 \, 5$
$11 \neq 5$
Not a solution

71. $x+6=x+6$
$2+6 \stackrel{?}{=} 2+6$
$8=8$
Solution

73. $x=5x+15$
$0 \stackrel{?}{=} 5(0)+15$
$0 \stackrel{?}{=} 0+15$
$0 \neq 15$
Not a solution

75. $x+15$

77. $x-5$

79. $3x \mid 22$

81. $1+2=9 \div 3$

83. $3 \neq 4 \div 2$

85. $5+x=20$

87. $13-3x=13$

89. $\frac{12}{x}=\frac{1}{2}$

91. Answers may vary.

93. $2l+2w=2(8)+2(6)$
$=16+12$
$=28$
The perimeter of the rectangle is 28 meters.

95. $lw=(120)(100)$
$=12{,}000$
The area of the lot is 12,000 square feet.

97. $\frac{I}{PT}=\frac{126.75}{(650)(3)}=\frac{126.75}{1950}=0.065$
The interest rate is 6.5%.

99. If $m=228$ minutes, then the monthly bill is
$4.95+0.10m=4.95+0.10(228)$
$=4.95+22.8$
$=\$27.75.$

Section 1.5

Mental Math

1. negative

2. positive

3. 0

4. negative

5. negative

6. 0

Exercise Set 1.5

1. $6+3=9$

3. $-6+(-8)=-14$

5. $8+(-7)=1$

7. $-14+2=-12$

9. $-2+(-3)=-5$

11. $-9+(-3)=-12$

13. $-7+3=-4$

15. $10+(-3)=7$

17. $5+(-7)=-2$

19. $-16+16=0$

21. $27+(-46)=-19$

23. $-18+49=31$

25. $-33 + (-14) = -47$

27. $6.3 + (-8.4) = -2.1$

29. $|-8| + (-16) = 8 + (-16) = -8$

31. $117 + (-79) = 38$

33. $-9.6 + (-3.5) = -13.1$

35. $-\frac{3}{8} + \frac{5}{8} = \frac{2}{8} = \frac{1}{4}$

37. $-\frac{7}{16} + \frac{1}{4} = -\frac{7}{16} + \frac{4 \cdot 1}{4 \cdot 4} = -\frac{7}{16} + \frac{4}{16} = -\frac{3}{16}$

39.
$$\begin{aligned} -\frac{7}{10} + \left(-\frac{3}{5}\right) &= -\frac{7}{10} + \left(-\frac{2 \cdot 3}{2 \cdot 5}\right) \\ &= -\frac{7}{10} + \left(-\frac{6}{10}\right) \\ &= -\frac{13}{10} \end{aligned}$$

41. $-15 + 9 + (-2) = -6 + (-2) = -8$

43. $-21 + (-16) + (-22) = -37 + (-22) = -59$

45. $-23 + 16 + (-2) = -7 + (-2) = -9$

47. $|5 + (-10)| = |-5| = 5$

49. $6 + (-4) + 9 = 2 + 9 = 11$

51.
$$\begin{aligned} [-17 + (-4)] + [-12 + 15] &= [-21] + [3] \\ &= -18 \end{aligned}$$

53. $|9 + (-12)| + |-16| = |-3| + |-16| = 3 + 16 = 19$

55.
$$\begin{aligned} &-1.3 + [0.5 + (-0.3) + 0.4] \\ &= -1.3 + (0.2 + 0.4) \\ &= -1.3 + 0.6 \\ &= -0.7 \end{aligned}$$

57. Tuesday, because the longest bar in the positive direction is for Tuesday.

59. 7°; the longest bar in the positive direction represents 7°.

61. $\frac{(-4°) + 3° + 7° + (-2°) + 1°}{5} = \frac{5°}{5} = 1°$

63. $-15° + 9° = -6°$

65. $-1312 + 658 = -654$ feet

67.
$$\begin{aligned} &-41.1 + (-126.7) + (-51.0) \\ &= -167.8 + (-51.0) \\ &= -218.8 \text{ million dollars} \end{aligned}$$

69. $-6 + (-7) + (-4) = -13 + (-4) = -17$

71. The opposite of 6 is -6.

73. The opposite of -2 is 2.

75. The opposite of 0 is 0.

77. Since $|-6| = 6$, the opposite of $|-6|$ is -6.

79. Answers may vary.

81. Since $|-2| = 2$, $-|-2| = -2$.

83. $-|0| = -0 = 0$

85. Since $\left|-\frac{2}{3}\right| = \frac{2}{3}$, $-\left|-\frac{2}{3}\right| = -\frac{2}{3}$.

87. Answers may vary.

89. $-a$ is a negative number.

91. $a + a$ is a positive number.

93.
$$\begin{aligned} x + 9 &= 5 \\ -4 + 9 &\stackrel{?}{=} 5 \\ 5 &= 5 \end{aligned}$$
Solution

95.
$$\begin{aligned} y + (-3) &= -7 \\ -1 + (-3) &\stackrel{?}{=} -7 \\ -4 &\neq -7 \end{aligned}$$
Not a solution

Exercise Set 1.6

1. $-6-4=-6+(-4)=-10$

3. $4-9=4+(-9)=-5$

5. $16-(-3)=16+(3)=19$

7. $\frac{1}{2}-\frac{1}{3}=\frac{1}{2}+\left(-\frac{1}{3}\right)$
$=\frac{3\cdot 1}{3\cdot 2}+\left(-\frac{2\cdot 1}{2\cdot 3}\right)$
$=\frac{3}{6}+\left(-\frac{2}{6}\right)$
$=\frac{1}{6}$

9. $-16-(-18)=-16+(18)=2$

11. $-6-5=-6+(-5)=-11$

13. $7-(-4)=7+(4)=11$

15. $-6-(-11)=-6+(11)=5$

17. $16-(-21)=16+(21)=37$

19. $9.7-16.1=9.7+(-16.1)=-6.4$

21. $-44-27=-44+(-27)=-71$

23. $-21-(-21)=-21+(21)=0$

25. $-2.6-(-6.7)=-2.6+(6.7)=4.1$

27. $-\frac{3}{11}-\left(-\frac{5}{11}\right)=-\frac{3}{11}+\left(\frac{5}{11}\right)=\frac{2}{11}$

29. $-\frac{1}{6}-\frac{3}{4}=-\frac{1}{6}+\left(-\frac{3}{4}\right)$
$=-\frac{4\cdot 1}{4\cdot 6}+\left(-\frac{6\cdot 3}{6\cdot 4}\right)$
$=-\frac{4}{24}+\left(-\frac{18}{24}\right)$
$=-\frac{22}{24}$
$=-\frac{2\cdot 11}{2\cdot 12}$
$=-\frac{11}{12}$

31. $8.3-(-0.62)=8.3+(0.62)=8.92$

33. $8-(-5)=8+5=13$

35. $-6-(-1)=-6+1=-5$

37. $7-8=7+(-8)=-1$

39. $-8-15=-8+(-15)=-23$

41. Answers may vary.

43. $-10-(-8)+(-4)-20$
$=-10+8+(-4)+(-20)$
$=-2+(-4)+(-20)$
$=-6+(-20)$
$=-26$

45. $5-9+(-4)-8-8$
$=5+(-9)+(-4)+(-8)+(-8)$
$=-4+(-4)+(-8)+(-8)$
$=-8+(-8)+(-8)$
$=-16+(-8)$
$=-24$

47. $-6-(2-11)=-6-[2+(-11)]$
$=-6-(-9)$
$=-6+9$
$=3$

49. $3^3-8\cdot 9=27-8\cdot 9$
$=27-72$
$=27+(-72)$
$=-45$

51. $2-3(8-6)=2-3[8+(-6)]$
$=2-3(2)$
$=2-6$
$=2+(-6)$
$=-4$

53. $(3-6)+4^2=[3+(-6)]+4^2$
$=(-3)+4^2$
$=-3+16$
$=13$

55. $-2+[(8-11)-(-2-9)]$
$=-2+[(8+(-11))-(-2+(-9))]$
$=-2+[(-3)-(-11)]$
$=-2+[-3+11]$
$=-2+[8]$
$=6$

57. $|-3|+2^2+[-4-(-6)]=3+2^2+[-4+6]$
$=3+2^2+[2]$
$=3+4+2$
$=7+2$
$=9$

59. Replace x with -5 and y with 4.
$x-y=-5-4=-5+(-4)=-9$

61. Replace x with -5, y with 4, and t with 10.
$|x|+2t-8y=|-5|+2(10)-8(4)$
$=5+20-32$
$=5+20+(-32)$
$=25+(-32)$
$=-7$

63. Replace x with -5 and y with 4.
$\frac{9-x}{y+6}=\frac{9-(-5)}{4+6}=\frac{9+5}{4+6}=\frac{14}{10}=\frac{2\cdot 7}{2\cdot 5}=\frac{7}{5}$

65. Replace x with -5 and y with 4.
$y^2-x=4^2-(-5)$
$=16-(-5)$
$=16+5$
$=21$

67. Replace x with -5 and t with 10.
$\frac{|x-(-10)|}{2t}=\frac{|-5-(-10)|}{2(10)}$
$=\frac{|-5+10|}{20}$
$=\frac{|5|}{20}$
$=\frac{5}{20}$
$=\frac{5}{4\cdot 5}$
$=\frac{1}{4}$

69. January, –22°; the longest bar in the negative direction is –22° and represents January.

71. $6°-18°=6°+(-18°)=-12°$

73. $30°-(-19°)=30°+19°=49°$

75. $44°-(-56°)=44°+56°=100°$
The temperature fell 100°.

77. $2-5-20=2+(-5)+(-20)$
$=-3+(-20)$
$=-23$
The total loss is 23 yards.

79. $-322-62=-322+(-62)=-384$
Aristotal was born in 384 B.C.

81. $-1\frac{5}{8}-\frac{3}{4}=-1\frac{5}{8}+\left(-\frac{3}{4}\right)$
$=-\frac{13}{8}+\left(-\frac{3}{4}\right)$
$=-\frac{13}{8}+\left(-\frac{6}{8}\right)$
$=-\frac{19}{8}$
$=-2\frac{3}{8}$

The overall change is $-2\frac{3}{8}$ points.

83. $22{,}834 - (-131) = 22{,}834 + 131$
$= 22{,}965$
Mount Aconcagua is 22,965 feet higher than Valdes Peninsula.

85. These angles are supplementary, so their sum is 180°.
$y = 180° - 50° = 130°$

87. These angles are complementary, so their sum is 90°.
$x = 90° - 60° = 30°$

89.
$$\begin{aligned} x - 9 &= 5 \\ -4 - 9 &\stackrel{?}{=} 5 \\ -4 + (-9) &\stackrel{?}{=} 5 \\ -13 &\neq 5 \end{aligned}$$
Not a solution

91.
$$\begin{aligned} -x + 6 &= -x - 1 \\ -(-2) + 6 &\stackrel{?}{=} -(-2) - 1 \\ 2 + 6 &\stackrel{?}{=} 2 - 1 \\ 2 + 6 &\stackrel{?}{=} 2 + (-1) \\ 8 &\neq 1 \end{aligned}$$
Not a solution

93.
$$\begin{aligned} -x - 13 &= -15 \\ -2 - 13 &\stackrel{?}{=} -15 \\ -2 + (-13) &\stackrel{?}{=} -15 \\ -15 &= -15 \end{aligned}$$
Solution

95. True

97. False

99. $56{,}875 + (-87{,}262)$
Negative since $87{,}262 > 56{,}875$
$56{,}875 - 87{,}262 = -30{,}387$

Section 1.7

Calculator Explorations

1. $-38(26 - 27) = 38$

3. $134 + 25(68 - 91) = -441$

5. $\frac{-50(294)}{175 - 265} = 163.\overline{3}$
(Note: a calculator will display 163.33333..., where the number of 3s to the right of the decimal point depends on the width of the calculator screen.)

7. $9^5 - 4550 = 54{,}499$

9. $(-125)^2 = 15{,}625$

Mental Math

1. positive

2. positive

3. negative

4. negative

5. positive

6. negative

Exercise Set 1.7

1. $-6(4) = -24$

3. $2(-1) = -2$

5. $-5(-10) = 50$

7. $-3 \cdot 4 = -12$

9. $-6(-7) = 42$

11. $2(-9) = -18$

13. $-\frac{1}{2}\left(-\frac{3}{5}\right) = \frac{1 \cdot 3}{2 \cdot 5} = \frac{3}{10}$

15. $-\frac{3}{4}\left(-\frac{8}{9}\right) = \frac{3 \cdot 8}{4 \cdot 9} = \frac{24}{36} = \frac{2 \cdot 12}{3 \cdot 12} = \frac{2}{3}$

17. $5(-1.4) = -7$

19. $-0.2(-0.7) = 0.14$

21. $-10(80) = -800$

23. $4(-7) = -28$

25. $(-5)(-5) = 25$

27. $\frac{2}{3}\left(-\frac{4}{9}\right) = -\frac{2 \cdot 4}{3 \cdot 9} = -\frac{8}{27}$

29. $-11(11) = -121$

31. $-\frac{20}{25}\left(\frac{5}{16}\right) = -\frac{20 \cdot 5}{25 \cdot 16}$
$= -\frac{100}{400}$
$= -\frac{4 \cdot 5 \cdot 5}{5 \cdot 5 \cdot 4 \cdot 4}$
$= -\frac{1}{4}$

33. $-2.1(-0.4) = 0.84$

35. $(-1)(2)(-3)(-5) = (-2)(-3)(-5)$
$= (6)(-5)$
$= -30$

37. $(2)(-1)(-3)(5)(3) = (-2)(-3)(5)(3)$
$= (6)(5)(3)$
$= (30)(3)$
$= 90$

39. True

41. False

43. $(-2)^4 = (-2)(-2)(-2)(-2) = 16$

45. $-1^5 = -(1 \cdot 1 \cdot 1 \cdot 1 \cdot 1) = -1$

47. $(-5)^2 = (-5)(-5) = 25$

49. $-7^2 = -(7 \cdot 7) = -49$

51. The reciprocal of 9 is $\frac{1}{9}$ since $9 \cdot \frac{1}{9} = 1$.

53. The reciprocal of $\frac{2}{3}$ is $\frac{3}{2}$ since $\frac{2}{3} \cdot \frac{3}{2} = 1$.

55. The reciprocal of -14 is $-\frac{1}{14}$ since $(-14)\left(-\frac{1}{14}\right) = 1$.

57. The reciprocal of $-\frac{3}{11}$ is $-\frac{11}{3}$ since $\left(-\frac{3}{11}\right)\left(-\frac{11}{3}\right) = 1$.

59. The reciprocal of 0.2 is $\frac{1}{0.2}$ since $(0.2)\left(\frac{1}{0.2}\right) = 1$.

61. The reciprocal of $\frac{1}{-6.3}$ is -6.3 since $\left(\frac{1}{-6.3}\right)(-6.3) = 1$.

63. $\frac{18}{-2} = 18 \cdot -\frac{1}{2} = -9$

65. $\frac{-16}{-4} = -16 \cdot -\frac{1}{4} = 4$

67. $\frac{-48}{12} = -48 \cdot \frac{1}{12} = -4$

69. $\frac{0}{-4} = 0 \cdot -\frac{1}{4} = 0$

71. $-\frac{15}{3} = -15 \cdot \frac{1}{3} = -5$

73. $\frac{5}{0}$ is undefined

75. $\frac{-12}{-4} = -12 \cdot -\frac{1}{4} = 3$

77. $\frac{30}{-2} = 30 \cdot -\frac{1}{2} = -15$

79. $\frac{6}{7} \div \left(-\frac{1}{3}\right) = \frac{6}{7} \cdot \left(-\frac{3}{1}\right) = -\frac{18}{7}$

81. $-\frac{5}{9} \div \left(-\frac{3}{4}\right) = -\frac{5}{9} \cdot \left(-\frac{4}{3}\right) = \frac{20}{27}$

83. $-\frac{4}{9} \div \frac{4}{9} = -\frac{4}{9} \cdot \frac{9}{4} = -1$

85. $\frac{-9(-3)}{-6} = \frac{27}{-6} = -\frac{9}{2}$

87. $\frac{12}{9-12} = \frac{12}{-3} = -4$

89. $\frac{-6^2+4}{-2} = \frac{-36+4}{-2} = \frac{-32}{-2} = 16$

91. $\frac{8+(-4)^2}{4-12} = \frac{8+16}{4-12} = \frac{24}{-8} = -3$

93. $\frac{22+(3)(-2)}{-5-2} = \frac{22-6}{-5-2} = \frac{16}{-7} = -\frac{16}{7}$

95. $\frac{-3-5^2}{2(-7)} = \frac{-3-25}{-14} = \frac{-28}{-14} = 2$

97. $\frac{6-2(-3)}{4-3(-2)} = \frac{6+6}{4+6} = \frac{12}{10} = \frac{6}{5}$

99. $\frac{-3-2(-9)}{-15-3(-4)} = \frac{-3+18}{-15+12} = \frac{15}{-3} = -5$

101.
$$\begin{aligned}\frac{|5-9|+|10-15|}{|2(-3)|} &= \frac{|-4|+|-5|}{|-6|}\\ &= \frac{4+5}{6}\\ &= \frac{9}{6}\\ &= \frac{3}{2}\end{aligned}$$

103. If $x = -5$ and $y = -3$, then
$$\begin{aligned}3x+2y &= 3(-5)+2(-3)\\ &= -15+(-6)\\ &= -21.\end{aligned}$$

105. If $x = -5$ and $y = -3$, then
$$\begin{aligned}2x^2-y^2 &= 2(-5)^2-(-3)^2\\ &= 2(25)-(9)\\ &= 50+(-9)\\ &= 41.\end{aligned}$$

107. If $x = -5$ and $y = -3$, then
$$\begin{aligned}x^3+3y &= (-5)^3+3(-3)\\ &= -125+(-9)\\ &= -134.\end{aligned}$$

109. If $x = -5$ and $y = -3$, then
$$\begin{aligned}\frac{2x-5}{y-2} &= \frac{2(-5)-5}{-3-2}\\ &= \frac{-10+(-5)}{-3+(-2)}\\ &= \frac{-15}{-5}\\ &= 3.\end{aligned}$$

111. If $x = -5$ and $y = -3$, then
$$\frac{6-y}{x-4} = \frac{6-(-3)}{-5-4} = \frac{6+3}{-5+(-4)} = \frac{9}{-9} = -1.$$

113. 4(–\$124.5 million) = –\$498 million

115. Answers may vary.

117. 1, –1

119. Positive

121. Not possible to determine

123. Negative

125. $-2+\frac{-15}{3} = -2+(-5) = -7$

127. $2[-5+(-3)] = 2[-8] = -16$

129.
$$\begin{aligned}-5x &= -35\\ -5(7) &\stackrel{?}{=} -35\\ -35 &= -35\end{aligned}$$
Solution

131.
$$\begin{aligned}\frac{x}{-10} &= 2\\ \frac{-20}{-10} &\stackrel{?}{=} 2\\ 2 &= 2\end{aligned}$$
Solution

133.
$$\begin{aligned}-3x-5 &= -20\\ -3(5)-5 &\stackrel{?}{=} -20\\ -15-5 &\stackrel{?}{=} -20\\ -15+(-5) &\stackrel{?}{=} -20\\ -20 &= -20\end{aligned}$$
Solution

Exercise Set 1.8

1. $x + 16 = 16 + x$

3. $-4 \cdot y = y \cdot (-4)$

5. $xy = yx$

7. $2x + 13 = 13 + 2x$

9. $(xy) \cdot z = x \cdot (yz)$

11. $2 + (a + b) = (2 + a) + b$

13. $4 \cdot (ab) = (4a) \cdot b$

15. $(a + b) + c = a + (b + c)$

17. $8 + (9 + b) = (8 + 9) + b = 17 + b$

19. $4(6y) = (4 \cdot 6)y = 24y$

21. $\frac{1}{5}(5y) = \left(\frac{1}{5} \cdot 5\right)y = 1 \cdot y = y$

23. $$\begin{aligned}(13 + a) + 13 &= 13 + (a + 13)\\ &= 13 + (13 + a)\\ &= (13 + 13) + a\\ &= 26 + a\end{aligned}$$

25. $-9(8x) = (-9 \cdot 8)x = -72x$

27. $\frac{3}{4}\left(\frac{4}{3}s\right) = \left(\frac{3}{4} \cdot \frac{4}{3}\right)s = 1 \cdot s = s$

29. Answers may vary.

31. $4(x + y) = 4(x) + 4(y) = 4x + 4y$

33. $9(x - 6) = 9(x) - 9(6) = 9x - 54$

35. $2(3x + 5) = 2(3x) + 2(5) = 6x + 10$

37. $7(4x - 3) = 7(4x) - 7(3) = 28x - 21$

39. $3(6 + x) = 3(6) + 3(x) = 18 + 3x$

41. $-2(y - z) = -2(y) - (-2)(z) = -2y + 2z$

43. $-7(3y + 5) = -7(3y) + (-7)(5) = -21y - 35$

45. $$\begin{aligned}5(x + 4m + 2) &= 5(x) + 5(4m) + 5(2)\\ &= 5x + 20m + 10\end{aligned}$$

47. $$\begin{aligned}-4(1 - 2m + n) &= -4(1) - (-4)(2m) + (-4)(n)\\ &= -4 + 8m - 4n\end{aligned}$$

49. $$\begin{aligned}-(5x + 2) &= -1(5x + 2)\\ &= (-1)(5x) + (-1)(2)\\ &= -5x - 2\end{aligned}$$

51. $$\begin{aligned}-(r - 3 - 7p) &= -1(r - 3 - 7p)\\ &= (-1)(r) - (-1)(3) - (-1)(7p)\\ &= -r + 3 + 7p\end{aligned}$$

53. $$\begin{aligned}\frac{1}{2}(6x + 8) &= \frac{1}{2}(6x) + \frac{1}{2}(8)\\ &= \left(\frac{1}{2} \cdot 6\right)x + \left(\frac{1}{2} \cdot 8\right)\\ &= 3x + 4\end{aligned}$$

55. $$\begin{aligned}-\frac{1}{3}(3x - 9y) &= -\frac{1}{3}(3x) - \left(-\frac{1}{3}\right)(9y)\\ &= \left(-\frac{1}{3} \cdot 3\right)x - \left(-\frac{1}{3} \cdot 9\right)y\\ &= (-1)x - (-3)y\\ &= -x + 3y\end{aligned}$$

57. $$\begin{aligned}3(2r + 5) - 7 &= 3(2r) + 3(5) - 7\\ &= 6r + 15 - 7\\ &= 6r + 8\end{aligned}$$

59. $$\begin{aligned}-9(4x + 8) + 2 &= -9(4x) + (-9)(8) + 2\\ &= -36x - 72 + 2\\ &= -36x - 70\end{aligned}$$

61. $$\begin{aligned}-4(4x + 5) - 5 &= -4(4x) + (-4)(5) - 5\\ &= -16x - 20 - 5\\ &= -16x - 25\end{aligned}$$

63. $4 \cdot 1 + 4 \cdot y = 4(1 + y)$

65. $11x + 11y = 11(x + y)$

67. $(-1) \cdot 5 + (-1) \cdot x = -1(5 + x)$

69. $30a + 30b = 30(a + b)$

71. $3 \cdot 5 = 5 \cdot 3$
The commutative property of multiplication

73. $2 + (x + 5) = (2 + x) + 5$
The associative property of addition

75. $9(3 + 7) = 9 \cdot 3 + 9 \cdot 7$
The distributive property

77. $(4 \cdot y) \cdot 9 = 4 \cdot (y \cdot 9)$
The associative property of multiplication

79. $0 + 6 = 6$
The identity property of addition

81. $-4(y + 7) = -4 \cdot y + (-4) \cdot 7$
The distributive property

83. $-4 \cdot (8 \cdot 3) = (8 \cdot -4) \cdot 3$
The associative and commutative properties of multiplication

85.

Expression	Opposite	Reciprocal
8	-8	$\frac{1}{8}$

87.

Expression	Opposite	Reciprocal
x	$-x$	$\frac{1}{x}$

89.

Expression	Opposite	Reciprocal
$2x$	$-2x$	$\frac{1}{2x}$

91. These are not commutative. The result is different (hopefully better) if the order is "studying for the test" then "taking a test" rather than the order "taking a test" then "studying for the test."

93. These are commutative. The result will be the same (both shoes on) regardless of which shoe is put on first.

95. Answers may vary.

Exercise Set 1.9

1. The number of teenagers expected to use the Internet in 1999 is about 7.8 million.

3. From the graph, the greatest increase in the heights of successive bars is from 1999 to 2002. The greatest increase is in 2002.

5. Look for the shortest bar, which is the bar representing the PGA/LPGA tours. The PGA/LPGA tours spent the least amount of money on advertising.

7. Major League Baseball and the NBA each spent over $10,000,000 on advertising.

9. To approximate the amount spent, move from the right edge of the NBA bar vertically downward to the dollar axis. The NBA spent about $15 million on advertising.

11. France because it represents the tallest bar.

13. France, U.S., Spain, Italy; These bars are all taller than 30.

15. 33 million; Go to the top of the Italy bar, then move horizontally left until the vertical axis is reached. Approximate the number.

17. *Pocahontas* generated approximately 142 million dollars, or $142,000,000.

19. *Snow White and the Seven Dwarfs* is the film that generated the most income before 1990 because it is the longest bar of those prior to 1990.

21. Answers may vary.

23. Find the highest point of the line graph, then move vertically downward to the year axis. The highest percent of arson fires started by juveniles occurred in 1994.

25. 1989 and 1990 appear to be the same because the line connecting them is horizontal.

27. We locate the year 1997 along the year axis and move vertically upward until the line is reached. From this point on the line, we move horizontally to the left until the percent axis is reached. We find that approximately 54% of the arson fires were started by juveniles in 1997.

29. The greatest increase in the percent of arson fires started by juveniles occurred in 1994. Notice that the line graph is steepest between the years 1993 and 1994.

31. The pulse rate was approximately 59 beats per minute 5 minutes before lighting a cigarette.

33. 5 minutes before lighting a cigarette, the pulse rate was approximately 59 beats per minute. 10 minutes after lighting a cigarette, the pulse rate had increased to about 85 beats per minute.
$85 - 59 = 26$
The pulse rate increased by 26 beats per minute between 5 minutes before and 10 minutes after lighting a cigarette.

35. Locate 1991 along the year axis. Move vertically up until the line is reached. From this point on the line, move horizontally to the left until the vertical axis is reached. Read the number of students.
In 1991, there were 20 students per computer.

37. The greatest decrease in number of students per computer occured in 1985. Notice that the line graph is steepest between the years 1984 and 1985.

39. Answers may vary.

41. The height of the bar representing the number of women in the workforce in 1890 is 4 million.

43. The height of the bar representing the number of men in the workforce in 1990 is 69 million.

45. The heights of the bars representing the number of women in the workforce first exceeded 20 million in 1960.

47. 69 million – 57 million = 12 million
The difference in the number of men and women in the workforce in 1990 is 12 million.

49. New Orleans, Louisiana is located at latitude 30° north and longitude 90° west.

51. Answers may vary.

Chapter 1 Review Exercises

1. $8 < 10$ since 8 is to the left of 10 on the number line.

2. $7 > 2$ since 7 is to the right of 2 on the number line.

3. $-4 > -5$ since -4 is to the right of -5 on the number line.

4. $\frac{12}{2} > -8$ since $\frac{12}{2} = 6$ and 6 is to the right of -8 on the number line.

5. $|-7| < |-8|$ since $|-7| = 7$ and $|-8| = 8$ and 7 is to the left of 8 on the number line.

6. $|-9| > -9$ since $|-9| = 9$ and 9 is to the right of -9 on the number line.

7. $-|-1| = -1$ since $|-1| = 1$ and $-|-1| = -1$.

8. $|-14| = -(-14)$ since $|-14| = 14$ and $-(-14) = 14$.

9. $1.2 > 1.02$ since 1.2 is to the right of 1.02 on the number line.

10. $-\frac{3}{2} < -\frac{3}{4}$ since $-\frac{3}{2}$ is to the left of $-\frac{3}{4}$ on the number line.

11. $4 \geq -3$

12. $6 \neq 5$

13. $0.03 < 0.3$

14. $50 > 40$ since 50 is to the right of 40 on the number line.

15. a. $\{1, 3\}$

b. $\{0, 1, 3\}$

c. $\{-6, 0, 1, 3\}$

d. $\{-6, 0, 1, 1\frac{1}{2}, 3, 9.62\}$

e. $\{\pi\}$

f. $\{-6, 0, 1, 1\frac{1}{2}, 3, \pi, 9.62\}$

16. a. $\{2, 5\}$

b. $\{2, 5\}$

c. $\{-3, 2, 5\}$

d. $\{-3, -1.6, 2, 5, \frac{11}{2}, 15.1\}$

e. $\{\sqrt{5}, 2\pi\}$

f. $\{-3, -1.6, 2, 5, \frac{11}{2}, 15.1, \sqrt{5}, 2\pi\}$

17. Friday because −4 is the negative number with the largest absolute value.

18. Wednesday because +5 is the largest positive number.

19. $36 = 2 \cdot 18 = 2 \cdot 2 \cdot 9 = 2 \cdot 2 \cdot 3 \cdot 3$

20. $120 = 2 \cdot 60$
$= 2 \cdot 2 \cdot 30$
$= 2 \cdot 2 \cdot 2 \cdot 15$
$= 2 \cdot 2 \cdot 2 \cdot 3 \cdot 5$

21. $\frac{8}{15} \cdot \frac{27}{30} = \frac{2 \cdot 2 \cdot 2 \cdot 3 \cdot 3 \cdot 3}{3 \cdot 5 \cdot 2 \cdot 3 \cdot 5} = \frac{12}{25}$

22. $\frac{7}{8} \div \frac{21}{32} = \frac{7}{8} \cdot \frac{32}{21} = \frac{7 \cdot 2 \cdot 2 \cdot 2 \cdot 2 \cdot 2}{2 \cdot 2 \cdot 2 \cdot 3 \cdot 7} = \frac{4}{3}$

23. $\frac{7}{15} + \frac{5}{6}$ LCD = 30
$\frac{7}{15} \cdot \frac{2}{2} = \frac{14}{30}$
$\frac{5}{6} \cdot \frac{5}{5} = \frac{25}{30}$
$\frac{14}{30} + \frac{25}{30} = \frac{14 + 25}{30} = \frac{39}{30} = \frac{3 \cdot 13}{2 \cdot 3 \cdot 5} = \frac{13}{10}$

24. $\frac{3}{4} - \frac{3}{20}$ LCD = 20
$\frac{3}{4} \cdot \frac{5}{5} = \frac{15}{20}$
$\frac{15}{20} - \frac{3}{20} = \frac{15 - 3}{20} = \frac{12}{20} = \frac{2 \cdot 2 \cdot 3}{2 \cdot 2 \cdot 5} = \frac{3}{5}$

25. $2\frac{3}{4} + 6\frac{5}{8} = \frac{11}{4} + \frac{53}{8}$ LCD = 8
$\frac{11}{4} \cdot \frac{2}{2} = \frac{22}{8}$
$\frac{22}{8} + \frac{53}{8} = \frac{22 + 53}{8} = \frac{75}{8} = 9\frac{3}{8}$

26. $7\frac{1}{6} - 2\frac{2}{3} = \frac{43}{6} - \frac{8}{3}$ LCD = 6
$\frac{8}{3} \cdot \frac{2}{2} = \frac{16}{6}$
$\frac{43}{6} - \frac{16}{6} = \frac{43 - 16}{6} = \frac{27}{6} = \frac{9}{2} = 4\frac{1}{2}$

27. $5 \div \frac{1}{3} = 5 \cdot \frac{3}{1} = 15$

28. $2 \cdot 8\frac{3}{4} = 2 \cdot \frac{35}{4} = \frac{70}{4} = \frac{35}{2} = 17\frac{1}{2}$

29. $1-\frac{1}{6}-\frac{1}{4}=1\cdot\frac{12}{12}-\frac{1}{6}\cdot\frac{2}{2}-\frac{1}{4}\cdot\frac{3}{3}$
$=\frac{12}{12}-\frac{2}{12}-\frac{3}{12}$
$=\frac{12-2-3}{12}$
$=\frac{7}{12}$

30. Area $=1\frac{1}{3}\cdot\frac{7}{8}=\frac{4}{3}\cdot\frac{7}{8}=\frac{28}{24}=\frac{7}{6}=1\frac{1}{6}$

The area is $1\frac{1}{6}$ meters.

Perimeter $=\frac{7}{8}+\frac{7}{8}+1\frac{1}{3}+1\frac{1}{3}$
$=\frac{7}{8}+\frac{7}{8}+\frac{4}{3}+\frac{4}{3}$
$=\frac{7}{8}\cdot\frac{3}{3}+\frac{7}{8}\cdot\frac{3}{3}+\frac{4}{3}\cdot\frac{8}{8}+\frac{4}{3}\cdot\frac{8}{8}$
$=\frac{21}{24}+\frac{21}{24}+\frac{32}{24}+\frac{32}{24}$
$=\frac{21+21+32+32}{24}$
$=\frac{106}{24}$
$=\frac{53}{12}$
$=4\frac{5}{12}$

The perimeter is $4\frac{5}{12}$ meters.

31. Area $=\frac{3}{11}\cdot\frac{3}{11}+\frac{5}{11}\cdot\frac{5}{11}$
$=\frac{9}{121}+\frac{25}{121}$
$=\frac{9+25}{121}$
$=\frac{34}{121}$

The area is $\frac{34}{121}$ square inch.

Perimeter
$=\frac{5}{11}+\frac{5}{11}+\frac{3}{11}+\frac{2}{11}+\frac{3}{11}+\frac{3}{11}+\frac{5}{11}$
$=\frac{5+5+3+2+3+3+5}{11}$
$=\frac{26}{11}$
$=2\frac{4}{11}$

The perimeter is $2\frac{4}{11}$ inches.

32. $7\frac{1}{2}-6\frac{1}{8}=\frac{15}{2}-\frac{49}{8}$
$=\frac{15}{2}\cdot\frac{4}{4}-\frac{49}{8}$
$=\frac{60}{8}-\frac{49}{8}$
$=\frac{60-49}{8}$
$=\frac{11}{8}$
$=1\frac{3}{8}$

She needs to cut a piece $1\frac{3}{8}$ feet long.

33. $1\frac{1}{8}+1\frac{13}{16}=\frac{9}{8}+\frac{29}{16}$
$=\frac{18}{16}+\frac{29}{16}$
$=\frac{18+29}{16}$
$=\frac{47}{16}$
$=2\frac{15}{16}$

The total weight is $2\frac{15}{16}$ pounds.

34. $1\frac{1}{2}+1\frac{11}{16}+1\frac{3}{4}+1\frac{5}{8}+\frac{11}{16}+1\frac{1}{8}$
$=\frac{3}{2}+\frac{27}{16}+\frac{7}{4}+\frac{13}{8}+\frac{11}{16}+\frac{9}{8}$
$=\frac{24}{16}+\frac{27}{16}+\frac{28}{16}+\frac{26}{16}+\frac{11}{16}+\frac{18}{16}$
$=\frac{24+27+28+26+11+18}{16}$
$=\frac{134}{16}$
$=\frac{67}{8}$
$=8\frac{3}{8}$

The total weight is $8\frac{3}{8}$ pounds.

35. $2\frac{15}{16}+8\frac{3}{8}=\frac{47}{16}+\frac{67}{8}$
$=\frac{47}{16}+\frac{134}{16}$
$=\frac{47+134}{16}$
$=\frac{181}{16}$
$=11\frac{5}{16}$

The combined weight is $11\frac{5}{16}$ pounds.

36. Jioke weighed the most because $1\frac{13}{16}$ is farther to the right on the number line than any of the other numbers.

37. Odera weighed the least because $\frac{11}{16}$ is farther to the left on the number line than any of the other numbers.

38. $1\frac{13}{16}-\frac{11}{16}=\frac{29}{16}-\frac{11}{16}$
$=\frac{29-11}{16}$
$=\frac{18}{16}$
$=\frac{9}{8}$
$=1\frac{1}{8}$

The heaviest baby weighed $1\frac{1}{8}$ pounds more than the lightest baby.

39. $5\frac{1}{2}-1\frac{5}{8}=\frac{11}{2}-\frac{13}{8}$
$=\frac{44}{8}-\frac{13}{8}$
$=\frac{44-13}{8}$
$=\frac{31}{8}$
$=3\frac{7}{8}$

She had gained $3\frac{7}{8}$ pounds.

40. $4\frac{5}{32}-1\frac{1}{8}=\frac{133}{32}-\frac{9}{8}$
$=\frac{133}{32}-\frac{36}{32}$
$=\frac{133-36}{32}$
$=\frac{97}{32}$
$=3\frac{1}{32}$

He had gained $3\frac{1}{32}$ pounds.

41. $2^4=2\cdot 2\cdot 2\cdot 2=16$

42. $5^2 = 5 \cdot 5 = 25$

43. $\left(\frac{2}{7}\right)^2 = \frac{2}{7} \cdot \frac{2}{7} = \frac{4}{49}$

44. $\left(\frac{3}{4}\right)^3 = \frac{3}{4} \cdot \frac{3}{4} \cdot \frac{3}{4} = \frac{27}{64}$

45. $6 \cdot 3^2 + 2 \cdot 8 = 6 \cdot 9 + 2 \cdot 8 = 54 + 16 = 70$

46. $68 - 5 \cdot 2^3 = 68 - 5 \cdot 8 = 68 - 40 = 28$

47.
$$\begin{aligned} 3(1 + 2 \cdot 5) + 4 &= 3(1 + 10) + 4 \\ &= 3(11) + 4 \\ &= 33 + 4 \\ &= 37 \end{aligned}$$

48.
$$\begin{aligned} 8 + 3(2 \cdot 6 - 1) &= 8 + 3(12 - 1) \\ &= 8 + 3(11) \\ &= 8 + 33 \\ &= 41 \end{aligned}$$

49.
$$\begin{aligned} \frac{4 + |6 - 2| + 8^2}{4 + 6 \cdot 4} &= \frac{4 + 4 + 64}{4 + 24} \\ &= \frac{72}{28} \\ &= \frac{2 \cdot 2 \cdot 2 \cdot 3 \cdot 3}{2 \cdot 2 \cdot 7} \\ &= \frac{18}{7} \end{aligned}$$

50.
$$\begin{aligned} 5[3(2 + 5) - 5] &= 5[3(7) - 5] \\ &= 5(21 - 5) \\ &= 5(16) \\ &= 80 \end{aligned}$$

51. $20 - 12 = 2 \cdot 4$

52. $\frac{9}{2} > -5$

53. If $x = 6$ and $y = 2$, then
$2x + 3y = 2(6) + 3(2) = 12 + 6 = 18.$

54. If $x = 6$, $y = 2$, and $z = 8$, then
$$\begin{aligned} x(y + 2z) &= 6[2 + 2(8)] \\ &= 6(2 + 16) \\ &= 6(18) \\ &= 108. \end{aligned}$$

55. If $x = 6$, $y = 2$, and $z = 8$, then
$$\frac{x}{y} + \frac{z}{2y} = \frac{6}{2} + \frac{8}{2(2)} = 3 + \frac{8}{4} = 3 + 2 = 5.$$

56. If $x = 6$ and $y = 2$, then
$$\begin{aligned} x^2 - 3y^2 &= 6^2 - 3(2)^2 \\ &= 36 - 3(4) \\ &= 36 - 12 \\ &= 24. \end{aligned}$$

57. $180 - 37 - 80 = 63°$

58.
$$\begin{aligned} 7x - 3 &= 18 \\ 7(3) - 3 &\stackrel{?}{=} 18 \\ 21 - 3 &\stackrel{?}{=} 18 \\ 18 &= 18 \end{aligned}$$
Solution

59.
$$\begin{aligned} 3x^2 + 4 &= x - 1 \\ 3(1)^2 + 4 &\stackrel{?}{=} 1 - 1 \\ 3(1) + 4 &\stackrel{?}{=} 0 \\ 3 + 4 &\stackrel{?}{=} 0 \\ 7 &\neq 0 \end{aligned}$$
Not a solution

60. The opposite of -9 is 9.

61. The opposite of $\frac{2}{3}$ is $-\frac{2}{3}$.

62. The opposite of $|-2|$ is -2 since $|-2| = 2$.

63. The opposites of $-|-7|$ is 7 since $-|-7| = -7$.

64. $-15 + 4 = -11$

65. $-6 + (-11) = -17$

66. $\frac{1}{16}+\left(-\frac{1}{4}\right)=\frac{1}{16}+\left(-\frac{1}{4}\cdot\frac{4}{4}\right)$
$=\frac{1}{16}+\left(-\frac{4}{16}\right)$
$=\frac{1+(-4)}{16}$
$=-\frac{3}{16}$

67. $-8+|-3|=-8+3=-5$

68. $-4.6+(-9.3)=-13.9$

69. $-2.8+6.7=3.9$

70. $-282+728=446$
Your elevation is 446 feet.

71. $6-20=6+(-20)=-14$

72. $-3.1-8.4=-3.1+(-8.4)=-11.5$

73. $-6-(-11)=-6+11=5$

74. $4-15=4+(-15)=-11$

75. $-21-16+3(8-2)=-21-16+3(6)$
$=-21-16+18$
$=-21+(-16)+18$
$=-37+18$
$=-19$

76. $\frac{11-(-9)+6(8-2)}{2+3\cdot 4}=\frac{11+9+6(6)}{2+12}$
$=\frac{11+9+36}{14}$
$=\frac{56}{14}$
$=4$

77. If $x=3$, $y=-6$, and $z=-9$, then
$2x^2-y+z=2(3)^2-(-6)+(-9)$
$=2(9)-(-6)+(-9)$
$=18-(-6)+(-9)$
$=18+6+(-9)$
$=24+(-9)$
$=15.$

78. If $x=3$ and $y=-6$, then
$\frac{y-x+5x}{2x}=\frac{-6-3+5(3)}{2(3)}$
$=\frac{-6-3+15}{6}$
$=\frac{-6+(-3)+15}{6}$
$=\frac{-9+15}{6}$
$=\frac{6}{6}$
$=1.$

79. $50+1-2+5+1-4=50+1+(-2)+5+1+(-4)$
$=51+(-2)+5+1+(-4)$
$=49+5+1+(-4)$
$=54+1+(-4)$
$=55+(-4)$
$=\$51$

80. If E is 412 and I is 536, then
$E-I=412-536=412+(-536)=-124.$
$-\$124$ billion

81. The reciprocal of -6 is $-\frac{1}{6}$ since $(-6)\left(-\frac{1}{6}\right)=1.$

82. The reciprocal of $\frac{3}{5}$ is $\frac{5}{3}$ since $\left(\frac{3}{5}\right)\left(\frac{5}{3}\right)=1.$

83. $6(-8)=-48$

84. $(-2)(-14)=28$

85. $\frac{-18}{-6}=3$

86. $\frac{42}{-3}=-14$

87. $-3(-6)(-2)=18(-2)=-36$

88. $(-4)(-3)(0)(-6)=(12)(0)(-6)=(0)(-6)=0$

89. $\frac{4(-3)+(-8)}{2+(-2)} = \frac{-12+(-8)}{2+(-2)}$
$= \frac{-20}{0}$
Undefined—division by zero is not defined.

90. $\frac{3(-2)^2-5}{-14} = \frac{3(4)-5}{-14}$
$= \frac{12-5}{-14}$
$= \frac{7}{-14}$
$= -\frac{1}{2}$

91. $-\frac{6}{0}$ is undefined.

92. $\frac{0}{-2} = 0$

93. $\frac{-9+(-7)+1}{3} = \frac{-16+1}{3} = \frac{-15}{3} = -5$

94. $\frac{-1+0+(-3)+0}{4} = \frac{-1+(-3)}{4} = \frac{-4}{4} = -1$

95. $-6+5=5+(-6)$
The commutative property of addition

96. $6 \cdot 1 = 6$
The multiplicative identity property

97. $3(8-5) = 3 \cdot 8 + 3(-5)$
The distributive property

98. $4+(-4)=0$
The additive inverse property

99. $2+(3+9)=(2+3)+9$
The associative property of addition

100. $2 \cdot 8 = 8 \cdot 2$
The commutative property of multiplication

101. $6(8+5) = 6 \cdot 8 + 6 \cdot 5$
The distributive property

102. $(3 \cdot 8) \cdot 4 = 3 \cdot (8 \cdot 4)$
The associative property of multiplication

103. $4 \cdot \frac{1}{4} = 1$
Multiplicative inverse

104. $8+0=8$
The additive identity property

105. $4(8+3) = 4(3+8)$
The commutative property of addition

106. Locate 1999 along the year axis. Move vertically up until the line is reached. From this point on the line, move horizontally to the left until the vertical axis is reached. Read the number of subscribers in 1999. There were 76 million subscribers.

107. Subtract the number of subscribers in 1998 from those in 1999.
$76 - 69 = 76 + (-69) = 7$
There was an increase of 7 million subscribers in 1999.

108. The greatest number of subscribers is represented by the highest point on the graph. In 1999 the number of subscribers was the greatest.

109. The trend shows that the number of subscribers is increasing.

110. This is represented by the longest bar. San Francisco had the greatest increase in annual rent. It was 8.3%.

111. This is represented by the shortest bar. Miami had the least increase in annual rent. It was 1.8%.

112. Choose the areas whose bars are shorter than 3.2%. Detroit, Los Angeles/Orange County, Washington/Baltimore, Philadelphia, and Miami had rental increases below the national average.

113. Choose the areas whose bars are longer than 3.2%. New York, Cleveland, Houston, Chicago, Boston, Dallas/Ft. Worth, and San Francisco had rental increases above the national average.

Chapter 1 Test

1. $|-7| > 5$

2. $(9 + 5) \geq 4$

3. $-13 + 8 = -5$

4. $-13 - (-2) = -13 + 2 = -11$

5. $6 \cdot 3 - 8 \cdot 4 = 18 - 32 = 18 + (-32) = -14$

6. $(13)(-3) = -39$

7. $(-6)(-2) = 12$

8. $\dfrac{|-16|}{-8} = \dfrac{16}{-8} = -2$

9. $\dfrac{-8}{0}$
Undefined—division by zero is not defined.

10. $\dfrac{|-6|+2}{5-6} = \dfrac{6+2}{5-6} = \dfrac{8}{5+(-6)} = \dfrac{8}{-1} = -8$

11.
$$\begin{aligned}\frac{1}{2} - \frac{5}{6} &= \frac{1}{2} + \left(-\frac{5}{6}\right) \\ &= \frac{1}{2} \cdot \frac{3}{3} + \left(-\frac{5}{6}\right) \\ &= \frac{3}{6} + \left(-\frac{5}{6}\right) \\ &= \frac{3+(-5)}{6} \\ &= -\frac{2}{6} \\ &= -\frac{2}{2 \cdot 3} \\ &= -\frac{1}{3}\end{aligned}$$

12.
$$\begin{aligned}-1\frac{1}{8} + 5\frac{3}{4} &= -\frac{9}{8} + \frac{23}{4} \\ &= -\frac{9}{8} + \frac{23}{4} \cdot \frac{2}{2} \\ &= -\frac{9}{8} + \frac{46}{8} \\ &= \frac{-9+46}{8} \\ &= \frac{37}{8} \\ &= 4\frac{5}{8}\end{aligned}$$

13.
$$\begin{aligned}-\frac{3}{5} + \frac{15}{8} &= -\frac{3}{5} \cdot \frac{8}{8} + \frac{15}{8} \cdot \frac{5}{5} \\ &= -\frac{24}{40} + \frac{75}{40} \\ &= \frac{-24+75}{40} \\ &= \frac{51}{40}\end{aligned}$$

14.
$$\begin{aligned}3(-4)^2 - 80 &= 3(16) - 80 \\ &= 48 - 80 \\ &= 48 + (-80) \\ &= -32\end{aligned}$$

15.
$$\begin{aligned}6[5 + 2(3-8) - 3] &= 6[5 + 2(3 + (-8)) - 3] \\ &= 6[5 + 2(-5) - 3] \\ &= 6[5 + (-10) - 3] \\ &= 6[5 + (-10) + (-3)] \\ &= 6[-5 + (-3)] \\ &= 6[-8] \\ &= -48\end{aligned}$$

16. $\dfrac{-12 + 3 \cdot 8}{4} = \dfrac{-12 + 24}{4} = \dfrac{12}{4} = 3$

17. $\dfrac{(-2)(0)(-3)}{-6} = \dfrac{(0)(-3)}{-6} = \dfrac{0}{-6} = 0$

18. $-3 > -7$ since -3 is to the right of -7 on the number line.

19. $4 > -8$ since 4 is to the right of -8 on the number line.

20. $|-3| > 2$ since $|-3| = 3$ and 3 is to the right of 2 on the number line.

21. $|-2| = -1-(-3)$ since $|-2| = 2$ and $-1-(-3) = -1+3 = 2$.

22. $2221 < 10{,}993$ since 2221 is to the left of 10,993 on the number line.

23. **a.** $\{1, 7\}$

b. $\{0, 1, 7\}$

c. $\{-5, -1, 0, 1, 7\}$

d. $\left\{-5, -1, 0, \frac{1}{4}, 1, 7, 11.6\right\}$

e. $\{\sqrt{7}, 3\pi\}$

f. $\left\{-5, -1, 0, \frac{1}{4}, 1, 7, 11.6, \sqrt{7}, 3\pi\right\}$

24. If $x = 6$ and $y = -2$, then $x^2 + y^2 = (6)^2 + (-2)^2 = 36 + 4 = 40$.

25. If $x = 6$, $y = -2$, and $z = -3$, then $x + yz = 6 + (-2)(-3) = 6 + 6 = 12$.

26. If $x = 6$ and $y = -2$, then

$$\begin{aligned} 2 + 3x - y &= 2 + 3(6) - (-2) \\ &= 2 + 18 + 2 \\ &= 20 + 2 \\ &= 22. \end{aligned}$$

27. If $x = 6$, $y = -2$, and $z = -3$, then

$$\begin{aligned} \frac{y+z-1}{x} &= \frac{-2+(-3)-1}{6} \\ &= \frac{-2+(-3)+(-1)}{6} \\ &= \frac{-5+(-1)}{6} \\ &= \frac{-6}{6} \\ &= -1. \end{aligned}$$

28. $8 + (9 + 3) = (8 + 9) + 3$
The associative property of addition

29. $6 \cdot 8 = 8 \cdot 6$
The commutative property of multiplication

30. $-6(2 + 4) = -6 \cdot 2 + (-6) \cdot 4$
The distributive property

31. $\frac{1}{6}(6) = 1$
Multiplicative inverse

32. The opposite of -9 is 9.

33. The reciprocal of $-\frac{1}{3}$ is -3 since $-\frac{1}{3}(-3) = 1$.

34. Second down, because -10 is the negative number with the greatest absolute value.

35. Yes; $5 + (-10) + (-2) + 29 = 22$ which is the number of yards needed for a touchdown.

36. $-14 + 31 = 17°$

37.

$$\begin{aligned} 356 + 460 + (-166) &= 816 + (-166) \\ &= 650 \end{aligned}$$

The total net income for these three years was \$650 million.

38. $280(-1.5) = -420$
She lost \$420.

39. Locate 1993 along the year axis. Move vertically up until the line is reached. From this point on the line, move horizontally to the left until the vertical axis is reached. Read the amount. In 1993, the revenue was \$8 billion.

40. Locate 1997 along the year axis. Move vertically up until the line is reached. From this point on the line, move horizontally to the left until the vertical axis is reached. Read the amount. In 1997, the revenue was \$25 billion.

41. Subtract the revenue from 1993 from that of 1995.
$13.5 - 8 = 13.5 + (-8) = 5.5$
The increase in revenue from 1993 to 1995 was $5.5 billion.

42. 1996, since the increase from 1995 to 1996 is greater than any other year.

43. This is represented by the longest bar. Indiana was the top steel producer with 25.2 million tons.

44. This is represented by the shortest bar. Texas produced the least (of the top steel producing states) with 5 million tons.

45. Move to the right edge of the Ohio bar, then move vertically downward to the horizontal axis. Read the amount produced. Ohio produced 16 million tons of raw steel.

46. Subtract the amount produced in Texas from that produced in Pennsylvania.
$8 - 5 = 8 + (-5) = 3$
3 million tons more steel was produced in Pennsylvania than in Texas.

Chapter 2

Section 2.1

Mental Math

1. The numerical coefficient of $-7y$ is -7.

2. The numerical coefficient of $3x$ is 3.

3. The numerical coefficient of x is 1 since x is $1x$.

4. The numerical coefficient of $-y$ is -1 since $-y$ is $-1y$.

5. The numerical coefficient of $17x^2y$ is 17.

6. The numerical coefficient of $1.2xyz$ is 1.2.

7. Like terms, since the variable and its exponent match.

8. Unlike terms, since the exponents on x are not the same.

9. Unlike terms, since the exponents on z are not the same.

10. Like terms, since each variable and its exponent match.

11. Like terms, since $wz = zw$ by the commutative property.

12. Unlike terms, since the variables do not match.

Exercise Set 2.1

1. $7y + 8y = (7 + 8)y = 15y$

3. $8w - w + 6w = (8 - 1 + 6)w = 13w$

5. $$\begin{aligned} 3b - 5 - 10b - 4 &= 3b - 10b - 5 - 4 \\ &= (3 - 10)b + (-5 - 4) \\ &= -7b - 9 \end{aligned}$$

7. $$\begin{aligned} m - 4m + 2m - 6 &= (1 - 4 + 2)m - 6 \\ &= -m - 6 \end{aligned}$$

9. $5(y - 4) = 5y - 20$

11. $7(d - 3) + 10 = 7d - 21 + 10 = 7d - 11$

13. $-(3x - 2y + 1) = -3x + 2y - 1$

15. $$\begin{aligned} 5(x + 2) - (3x - 4) &= 5x + 10 - 3x + 4 \\ &= 5x - 3x + 10 + 4 \\ &= 2x + 14 \end{aligned}$$

17. Answers may vary.

19. $$\begin{aligned} (6x + 7) + (4x - 10) &= 6x + 7 + 4x - 10 \\ &= 6x + 4x + 7 - 10 \\ &= 10x - 3 \end{aligned}$$

21. $$\begin{aligned} (3x - 8) - (7x + 1) &= 3x - 8 - 7x - 1 \\ &= 3x - 7x - 8 - 1 \\ &= -4x - 9 \end{aligned}$$

23. $7x^2 + 8x^2 - 10x^2 = 5x^2$

25. $$\begin{aligned} 6x - 5x + x - 3 + 2x &= 6x - 5x + x + 2x - 3 \\ &= 4x - 3 \end{aligned}$$

27. $$\begin{aligned} -5 + 8(x - 6) &= -5 + 8x - 48 \\ &= 8x - 5 - 48 \\ &= 8x - 53 \end{aligned}$$

29. $$\begin{aligned} 5g - 3 - 5 - 5g &= 5g - 5g - 3 - 5 \\ &= -8 \end{aligned}$$

31. $$\begin{aligned} 6.2x - 4 + x - 1.2 &= 6.2x + x - 4 - 1.2 \\ &= 7.2x - 5.2 \end{aligned}$$

33. $2k - k - 6 = k - 6$

35. $$\begin{aligned} 0.5(m + 2) + 0.4m &= 0.5m + 1 + 0.4m \\ &= 0.5m + 0.4m + 1 \\ &= 0.9m + 1 \end{aligned}$$

37. $-4(3y - 4) = -12y + 16$

39. $$\begin{aligned} 3(2x - 5) - 5(x - 4) &= 6x - 15 - 5x + 20 \\ &= 6x - 5x - 15 + 20 \\ &= x + 5 \end{aligned}$$

41. $3.4m - 4 - 3.4m - 7 = 3.4m - 3.4m - 4 - 7$
$= -11$

43. $6x + 0.5 - 4.3x - 0.4x + 3$
$= 6x - 4.3x - 0.4x + 0.5 + 3$
$= 1.3x + 3.5$

45. $-2(3x - 4) + 7x - 6 = -6x + 8 + 7x - 6$
$= -6x + 7x + 8 - 6$
$= x + 2$

47. $-9x + 4x + 18 - 10x = -9x + 4x - 10x + 18$
$= -15x + 18$

49. $5k - (3k - 10) = 5k - 3k + 10 = 2k + 10$

51. $(3x + 4) - (6x - 1) = 3x + 4 - 6x + 1$
$= 3x - 6x + 4 + 1$
$= -3x + 5$

53.

twice a number	decreased by	4
$2x$	$-$	4

$2x - 4$

55.

three-fourths of a number	increased by	12
$\frac{3}{4}x$	+	12

$\frac{3}{4}x + 12$

57.

the sum of 5 times a number and –2	added to	7 times the number
$5x + (-2)$	+	$7x$

$5x + (-2) + 7x = -2 + 12x$

59. Subtract $5m - 6$ from $m - 9$.
$(m - 9) - (5m - 6) = m - 9 - 5m + 6$
$= -4m - 3$

61.

eight	times	the sum of a number and 6
8	$\cdot$	$(x + 6)$

$8(x + 6)$

63.

double a number	minus	the sum of the number and 10
$2x$	$-$	$(x + 10)$

$2x - (x + 10) = 2x - x - 10 = x - 10$

65.

seven	multiplied by	the quotient of a number and 6
7	$\cdot$	$\frac{x}{6}$

$7 \cdot \frac{x}{6} = \frac{7x}{6}$

67.

the sum of			
2	3 times a number	–9	4 times the number
2 +	$3x$	+ (–9) +	$4x$

$2 + 3x - 9 + 4x = 7x - 7$

69. $2(5x) + 2(4x - 1) = 10x + 8x - 2 = 18x - 2$
$(18x - 2)$ feet

71. Balanced

73. Balanced

75. $12 \cdot (x + 2) + (3x - 1) = 12x + 24 + 3x - 1$
$= 12x + 3x + 24 - 1$
$= 15x + 23$

$(15x + 23)$ inches

77. $\frac{5}{8} \Rightarrow \frac{8}{5}$

79. $2 \Rightarrow \frac{1}{2}$

81. $-\frac{1}{9} \Rightarrow -9$

83. $\frac{3x}{3} = \left(\frac{3}{3}\right)x = (1)x = x$

85. $-5\left(-\frac{1}{5}y\right) = \left(-5 \cdot -\frac{1}{5}\right)y = (1)y = y$

87. $\frac{3}{5}\left(\frac{5}{3}x\right) = \left(\frac{3}{5} \cdot \frac{5}{3}\right)x = (1)x = x$

89. $5b^2c^3 + 8b^3c^2 - 7b^3c^2 = 5b^2c^3 + b^3c^2$

91.
$$\begin{aligned} & 3x - (2x^2 - 6x) + 7x^2 \\ &= 3x - 2x^2 + 6x + 7x^2 \\ &= -2x^2 + 7x^2 + 3x + 6x \\ &= 5x^2 + 9x \end{aligned}$$

93.
$$\begin{aligned} & -(2x^2y + 3z) + 3z - 5x^2y \\ &= -2x^2y - 3z + 3z - 5x^2y \\ &= -2x^2y - 5x^2y - 3z + 3z \\ &= -7x^2y \end{aligned}$$

Section 2.2

Mental Math

1. $x + 4 = 6$
$x = 2$

2. $x + 7 = 10$
$x = 3$

3. $n + 18 = 30$
$n = 12$

4. $z + 22 = 40$
$z = 18$

5. $b - 11 = 6$
$b = 17$

6. $d - 16 = 5$
$d = 21$

7. $3a = 27$
$a = 9$

8. $9c = 54$
$c = 6$

9. $5b = 10$
$b = 2$

10. $7t = 14$
$t = 2$

11. $6x = -30$
$x = -5$

12. $8r = -64$
$r = -8$

Exercise Set 2.2

1.
$$\begin{aligned} x + 7 &= 10 \\ x + 7 - 7 &= 10 - 7 \\ x &= 3 \end{aligned}$$

Check:
$$\begin{aligned} x + 7 &= 10 \\ 3 + 7 &\stackrel{?}{=} 10 \\ 10 &= 10 \end{aligned}$$

The solution is 3.

3.
$$\begin{aligned} x - 2 &= -4 \\ x - 2 + 2 &= -4 + 2 \\ x &= -2 \end{aligned}$$

Check:
$$\begin{aligned} x - 2 &= -4 \\ -2 - 2 &\stackrel{?}{=} -4 \\ -4 &= -4 \end{aligned}$$

The solution is –2.

5.
$$\begin{aligned} 3 + x &= -11 \\ 3 + x - 3 &= -11 - 3 \\ x &= -14 \end{aligned}$$

Check:
$$\begin{aligned} 3 + x &= -11 \\ 3 + (-14) &\stackrel{?}{=} -11 \\ -11 &= -11 \end{aligned}$$

The solution is –14.

7. $r - 8.6 = -8.1$
$r - 8.6 + 8.6 = -8.1 + 8.6$
$r = 0.5$

Check: $r - 8.6 = -8.1$
$0.5 - 8.6 \stackrel{?}{=} -8.1$
$-8.1 = -8.1$

The solution is 0.5.

9. $8x = 7x - 3$
$8x - 7x = 7x - 3 - 7x$
$x = -3$

Check: $8x = 7x - 3$
$8(-3) \stackrel{?}{=} 7(-3) - 3$
$-24 \stackrel{?}{=} -21 - 3$
$-24 = -24$

The solution is –3.

11. $5b - 0.7 = 6b$
$5b - 0.7 - 5b = 6b - 5b$
$-0.7 = b$

Check: $5b - 0.7 = 6b$
$5(-0.7) - 0.7 \stackrel{?}{=} 6(-0.7)$
$-3.5 - 0.7 \stackrel{?}{=} -4.2$
$-4.2 = -4.2$

The solution is –0.7.

13. $7x - 3 = 6x$
$7x - 3 - 6x = 6x - 6x$
$x - 3 = 0$
$x - 3 + 3 = 0 + 3$
$x = 3$

Check: $7x - 3 = 6x$
$7(3) - 3 \stackrel{?}{=} 6(3)$
$21 - 3 \stackrel{?}{=} 18$
$18 = 18$

The solution is 3.

15. $3x - 6 = 2x + 5$
$3x - 6 - 2x = 2x + 5 - 2x$
$x - 6 = 5$
$x - 6 + 6 = 5 + 6$
$x = 11$
The solution is 11.

17. $3t - t - 7 = t - 7$
$2t - 7 = t - 7$
$2t - 7 - t = t - 7 - t$
$t - 7 = -7$
$t - 7 + 7 = -7 + 7$
$t = 0$
The solution is 0.

19. $7x + 2x = 8x - 3$
$9x = 8x - 3$
$9x - 8x = 8x - 3 - 8x$
$x = -3$
The solution is –3.

21. $-2(x + 1) + 3x = 14$
$-2x - 2 + 3x = 14$
$-2 + x = 14$
$-2 + x + 2 = 14 + 2$
$x = 16$
The solution is 16.

23. $-5x = 20$
$\frac{-5x}{-5} = \frac{20}{-5}$
$x = -4$
The solution is –4.

25. $3x = 0$
$\frac{3x}{3} = \frac{0}{3}$
$x = 0$
The solution is 0.

27. $-x = -12$
$\frac{-x}{-1} = \frac{-12}{-1}$
$x = 12$
The solution is 12.

29. $3x + 2x = 50$
$5x = 50$
$\frac{5x}{5} = \frac{50}{5}$
$x = 10$
The solution is 10.

31.
$$\frac{2}{3}x = -8$$
$$\frac{3}{2}\left(\frac{2}{3}x\right) = \frac{3}{2}(-8)$$
$$x = \frac{-24}{2}$$
$$x = -12$$
The solution is −12.

33.
$$\frac{1}{6}d = \frac{1}{2}$$
$$\frac{6}{1}\left(\frac{1}{6}d\right) = \frac{6}{1}\left(\frac{1}{2}\right)$$
$$d = \frac{6}{2}$$
$$d = 3$$
The solution is 3.

35.
$$\frac{a}{-2} = 1$$
$$-2\left(\frac{a}{-2}\right) = -2(1)$$
$$a = -2$$
The solution is −2.

37.
$$\frac{k}{7} = 0$$
$$7\left(\frac{k}{7}\right) = 7(0)$$
$$k = 0$$
The solution is 0.

39. Answers may vary.

41.
$$2x - 4 = 16$$
$$2x - 4 + 4 = 16 + 4$$
$$2x = 20$$
$$\frac{2x}{2} = \frac{20}{2}$$
$$x = 10$$

Check:
$$2x - 4 = 16$$
$$2(10) - 4 \stackrel{?}{=} 16$$
$$20 - 4 \stackrel{?}{=} 16$$
$$16 = 16$$
The solution is 10.

43.
$$-x + 2 = 22$$
$$-x + 2 - 2 = 22 - 2$$
$$-x = 20$$
$$\frac{-x}{-1} = \frac{20}{-1}$$
$$x = -20$$
Check:
$$-x + 2 = 22$$
$$-(-20) + 2 \stackrel{?}{=} 22$$
$$20 + 2 \stackrel{?}{=} 22$$
$$22 = 22$$
The solution is −20.

45.
$$6a + 3 = 3$$
$$6a + 3 - 3 = 3 - 3$$
$$6a = 0$$
$$\frac{6a}{6} = \frac{0}{6}$$
$$a = 0$$

Check:
$$6a + 3 = 3$$
$$6(0) + 3 \stackrel{?}{=} 3$$
$$0 + 3 \stackrel{?}{=} 3$$
$$3 = 3$$
The solution is 0.

47.
$$6x + 10 = -20$$
$$6x + 10 - 10 = -20 - 10$$
$$6x = -30$$
$$\frac{6x}{6} = \frac{-30}{6}$$
$$x = -5$$
Check:
$$6x + 10 = -20$$
$$6(-5) + 10 \stackrel{?}{=} -20$$
$$-30 + 10 \stackrel{?}{=} -20$$
$$-20 = -20$$
The solution is −5.

49.
$$5-0.3k=5$$
$$5-0.3k-5=5-5$$
$$-0.3k=0$$
$$\frac{-0.3k}{-0.3}=\frac{0}{-0.3}$$
$$k=0$$

Check: $5-0.3k=5$
$$5-0.3(0)\stackrel{?}{=}5$$
$$5-0\stackrel{?}{=}5$$
$$5=5$$

The solution is 0.

51.
$$-2x+\frac{1}{2}=\frac{7}{2}$$
$$-2x+\frac{1}{2}-\frac{1}{2}=\frac{7}{2}-\frac{1}{2}$$
$$-2x=\frac{6}{2}$$
$$-2x=3$$
$$\frac{-2x}{-2}=\frac{3}{-2}$$
$$x=-\frac{3}{2}$$

Check:
$$-2x+\frac{1}{2}=\frac{7}{2}$$
$$-2\left(-\frac{3}{2}\right)+\frac{1}{2}\stackrel{?}{=}\frac{7}{2}$$
$$3+\frac{1}{2}\stackrel{?}{=}\frac{7}{2}$$
$$\frac{6}{2}+\frac{1}{2}\stackrel{?}{=}\frac{7}{2}$$
$$\frac{7}{2}=\frac{7}{2}$$

The solution is $-\frac{3}{2}$.

53.
$$\frac{x}{3}+2=-5$$
$$\frac{x}{3}+2-2=-5-2$$
$$\frac{x}{3}=-7$$
$$3\left(\frac{x}{3}\right)=3(-7)$$
$$x=-21$$

Check:
$$\frac{x}{3}+2=-5$$
$$\frac{-21}{3}+2\stackrel{?}{=}-5$$
$$-7+2\stackrel{?}{=}-5$$
$$-5=-5$$

The solution is −21.

55.
$$10=2x-1$$
$$10+1=2x-1+1$$
$$11=2x$$
$$\frac{11}{2}=\frac{2x}{2}$$
$$\frac{11}{2}=x$$

Check: $10=2x-1$
$$10\stackrel{?}{=}2\left(\frac{11}{2}\right)-1$$
$$10\stackrel{?}{=}11-1$$
$$10=10$$

The solution is $\frac{11}{2}$.

57.
$$6z-8-z+3=0$$
$$5z-5=0$$
$$5z-5+5=0+5$$
$$5z=5$$
$$\frac{5z}{5}=\frac{5}{5}$$
$$z=1$$

Check:
$$6z-8-z+3=0$$
$$6(1)-8-1+3\stackrel{?}{=}0$$
$$6-8-1+3\stackrel{?}{=}0$$
$$0=0$$

The solution is 1.

59. $10 - 3x - 6 - 9x = 7$

$$\begin{aligned} 4 - 12x &= 7 \\ 4 - 12x - 4 &= 7 - 4 \\ -12x &= 3 \\ \frac{-12x}{-12} &= \frac{3}{-12} \\ x &= -\frac{1}{4} \end{aligned}$$

Check:

$$\begin{aligned} 10 - 3x - 6 - 9x &= 7 \\ 10 - 3\left(-\frac{1}{4}\right) - 6 - 9\left(-\frac{1}{4}\right) &\stackrel{?}{=} 7 \\ 10 + \frac{3}{4} - 6 + \frac{9}{4} &\stackrel{?}{=} 7 \\ 10 - 6 + \frac{3}{4} + \frac{9}{4} &\stackrel{?}{=} 7 \\ 4 + \frac{12}{4} &\stackrel{?}{=} 7 \\ 4 + 3 &\stackrel{?}{=} 7 \\ 7 &= 7 \end{aligned}$$

The solution is $-\frac{1}{4}$.

61. $\frac{5}{6}x = 10$

$$\begin{aligned} \frac{6}{5}\left(\frac{5}{6}x\right) &= \frac{6}{5}(10) \\ x &= \frac{60}{5} \\ x &= 12 \end{aligned}$$

Check:

$$\begin{aligned} \frac{5}{6}x &= 10 \\ \frac{5}{6}(12) &\stackrel{?}{=} 10 \\ \frac{60}{6} &\stackrel{?}{=} 10 \\ 10 &= 10 \end{aligned}$$

The solution is 12.

63.

$$\begin{aligned} 1 &= 0.4x - 0.6x - 5 \\ 1 &= -0.2x - 5 \\ 1 + 5 &= -0.2x - 5 + 5 \\ 6 &= -0.2x \\ \frac{6}{-0.2} &= \frac{-0.2x}{-0.2} \\ -30 &= x \end{aligned}$$

Check:

$$\begin{aligned} 1 &= 0.4x - 0.6x - 5 \\ 1 &\stackrel{?}{=} 0.4(-30) - 0.6(-30) - 5 \\ 1 &\stackrel{?}{=} -12 + 18 - 5 \\ 1 &= 1 \end{aligned}$$

The solution is −30.

65.

$$\begin{aligned} z - 5z &= 7z - 9 - z \\ -4z &= 6z - 9 \\ -4z - 6z &= 6z - 9 - 6z \\ -10z &= -9 \\ \frac{-10z}{-10} &= \frac{-9}{-10} \\ z &= \frac{9}{10} \end{aligned}$$

Check:

$$\begin{aligned} z - 5z &= 7z - 9 - z \\ \frac{9}{10} - 5\left(\frac{9}{10}\right) &\stackrel{?}{=} 7\left(\frac{9}{10}\right) - 9 - \frac{9}{10} \\ \frac{9}{10} - \frac{45}{10} &\stackrel{?}{=} \frac{63}{10} - 9 - \frac{9}{10} \\ -\frac{36}{10} &\stackrel{?}{=} \frac{54}{10} - 9 \\ -\frac{36}{10} &\stackrel{?}{=} \frac{54}{10} - \frac{90}{10} \\ -\frac{36}{10} &= -\frac{36}{10} \end{aligned}$$

The solution is $\frac{9}{10}$.

67.
$$0.4x - 0.6x - 5 = 1$$
$$-0.2x - 5 = 1$$
$$-0.2x - 5 + 5 = 1 + 5$$
$$-0.2x = 6$$
$$\frac{-0.2x}{-0.2} = \frac{6}{-0.2}$$
$$x = -30$$

Check:
$$0.4x - 0.6x - 5 = 1$$
$$0.4(-30) - 0.6(-30) - 5 \stackrel{?}{=} 1$$
$$-12 + 18 - 5 \stackrel{?}{=} 1$$
$$1 = 1$$

The solution is –30.

69.
$$6 - 2x + 8 = 10$$
$$14 - 2x = 10$$
$$14 - 14 - 2x = 10 - 14$$
$$-2x = -4$$
$$\frac{-2x}{-2} = \frac{-4}{-2}$$
$$x = 2$$

Check:
$$6 - 2x + 8 = 10$$
$$6 - 2(2) + 8 \stackrel{?}{=} 10$$
$$6 - 4 + 8 \stackrel{?}{=} 10$$
$$10 = 10$$

The solution is 2.

71.
$$-3a + 6 + 5a = 7a - 8a$$
$$2a + 6 = -a$$
$$2a - 2a + 6 = -a - 2a$$
$$6 = -3a$$
$$\frac{6}{-3} = \frac{-3a}{-3}$$
$$-2 = a$$

Check:
$$-3a + 6 + 5a = 7a - 8a$$
$$-3(-2) + 6 + 5(-2) \stackrel{?}{=} 7(-2) - 8(-2)$$
$$6 + 6 - 10 \stackrel{?}{=} -14 + 16$$
$$2 = 2$$

The solution is –2.

73.
$$20 = -3(2x + 1) + 7x$$
$$20 = -6x - 3 + 7x$$
$$20 = x - 3$$
$$20 + 3 = x - 3 + 3$$
$$23 = x$$

Check:
$$20 = -3(2x + 1) + 7x$$
$$20 \stackrel{?}{=} -3[2(23) + 1] + 7(23)$$
$$20 \stackrel{?}{=} -3(46 + 1) + 161$$
$$20 \stackrel{?}{=} -3(47) + 161$$
$$20 \stackrel{?}{=} -141 + 161$$
$$20 = 20$$

The solution is 23.

75. If the sum of two numbers is 20 and one number is p, we find the other number by subtracting p from 20. The other number is $20 - p$.

77. Since the original board is 10 feet long and once piece is x feet long, we find the length of the other piece by subtracting x from 10. Thus, the length of the other piece is $(10 - x)$ feet.

79. Since the sum of two angles is 180° and one angle measures $x°$, we find the measure of the supplement by subtracting $x°$ from 180°. Thus, the supplement measures $(180 - x)°$.

81. Since Charles received n votes and April received 284 more votes than Charles, we find the number of votes April received by adding 284 to n. Thus, April received $(n + 284)$ votes.

83. If the length of the Verrazano-Narrows Bridge is m feet and the Golden Gate Bridge is 60 feet shorter than the Verrazano-Narrows Bridge, we find the length of the Golden Gate Bridge by subtracting 60 from m. Thus, the length of the Golden Gate Bridge is $(m - 60)$ feet.

85. Since the sum of the three angles is 180°, we find the measure of the third angle by subtracting the measures of the first two angles, $x°$ and $(2x + 7)°$, from 180°.

$$180° - x° - (2x+7)° = (180 - x - 2x - 7)° = (173 - 3x)°$$

The third angle measures $(173 - 3x)°$.

87.
$$200 + 150 + 400 + x = 1000$$
$$750 + x = 1000$$
$$750 - 750 + x = 1000 - 750$$
$$x = 250$$
The nurse must give the patient 250 milliliters more fluid.

89. Answers may vary.

91. If x = the first even integer, then $x + 2$ = the second consecutive even integer, $x + 4$ = the third consecutive even integer, and $x + 6$ = the fourth consecutive even integer. Their sum is
$$x + (x+2) + (x+4) + (x+6)$$
$$= x + x + 2 + x + 4 + x + 6$$
$$= x + x + x + x + 2 + 4 + 6$$
$$= 4x + 12.$$

93. If x = the first integer, then $x + 1$ = the second integer. The sum of 20 and the second integer is
$20 + (x + 1) = 20 + x + 1 = x + 21.$

95.
$$-3.6x = 10.62$$
$$\frac{-3.6x}{-3.6} = \frac{10.62}{-3.6}$$
$$x = -2.95$$

97.
$$7x - 5.06 = -4.92$$
$$7x - 5.06 + 5.06 = -4.92 + 5.06$$
$$7x = 0.14$$
$$\frac{7x}{7} = \frac{0.14}{7}$$
$$x = 0.02$$

99. $5x + 2(x - 6) = 5x + 2x - 12 = 7x - 12$

101. $6(2z + 4) + 20 = 12z + 24 + 20 = 12z + 44$

103. $-(x - 1) + x = -x + 1 + x = 1$

Section 2.3

Calculator Explorations

1. Equation: $2x = 48 + 6x$
Possible solution: $x = -12$
Left side: $2(-12) = -24$
Right side: $48 + 6(-12) = -24$
Solution, since the left side equals the right side.

3. Equation: $5x - 2.6 = 2(x + 0.8)$
Possible solution: $x = 4.4$
Left side: $5(4.4) - 2.6 = 19.4$
Right side: $2(4.4 + 0.8) = 10.4$
Not a solution, since the left side does not equal the right side.

5. Equation: $\frac{564x}{4} = 200x - 11(649)$
Possible solution: $x = 121$
Left side: $\frac{564(121)}{4} = 17{,}061$
Right side: $200(121) - 11(649) = 17{,}061$
Solution, since the left side equals the right side.

Exercise Set 2.3

1.
$$-2(3x - 4) = 2x$$
$$-6x + 8 = 2x$$
$$-6x - 2x + 8 = 2x - 2x$$
$$-8x + 8 = 0$$
$$-8x + 8 - 8 = 0 - 8$$
$$-8x = -8$$
$$\frac{-8x}{-8} = \frac{-8}{-8}$$
$$x = 1$$

3.
$$\begin{aligned} 4(2n-1) &= (6n+4)+1 \\ 8n-4 &= 6n+4+1 \\ 8n-4 &= 6n+5 \\ 8n-6n-4 &= 6n-6n+5 \\ 2n-4 &= 5 \\ 2n-4+4 &= 5+4 \\ 2n &= 9 \\ \frac{2n}{2} &= \frac{9}{2} \\ n &= \frac{9}{2} \end{aligned}$$

5.
$$\begin{aligned} 5(2x-1)-2(3x) &= 1 \\ 10x-5-6x &= 1 \\ 4x-5 &= 1 \\ 4x-5+5 &= 1+5 \\ 4x &= 6 \\ \frac{4x}{4} &= \frac{6}{4} \\ x &= \frac{3}{2} \end{aligned}$$

7.
$$\begin{aligned} 6(x-3)+10 &= -8 \\ 6x-18+10 &= -8 \\ 6x-8 &= -8 \\ 6x-8+8 &= -8+8 \\ 6x &= 0 \\ \frac{6x}{6} &= \frac{0}{6} \\ x &= 0 \end{aligned}$$

9.
$$\begin{aligned} \frac{3}{4}x-\frac{1}{2} &= 1 \\ 4\left(\frac{3}{4}x-\frac{1}{2}\right) &= 4\cdot 1 \\ 4\left(\frac{3}{4}x\right)-4\left(\frac{1}{2}\right) &= 4 \\ 3x-2 &= 4 \\ 3x-2+2 &= 4+2 \\ 3x &= 6 \\ \frac{3x}{3} &= \frac{6}{3} \\ x &= 2 \end{aligned}$$

11.
$$\begin{aligned} x+\frac{5}{4} &= \frac{3}{4}x \\ 4\left(x+\frac{5}{4}\right) &= 4\left(\frac{3}{4}x\right) \\ 4(x)+4\left(\frac{5}{4}\right) &= 4\left(\frac{3}{4}x\right) \\ 4x+5 &= 3x \\ 4x-3x+5 &= 3x-3x \\ x+5 &= 0 \\ x+5-5 &= 0-5 \\ x &= -5 \end{aligned}$$

13.
$$\begin{aligned} \frac{x}{2}-1 &= \frac{x}{5}+2 \\ 10\left(\frac{x}{2}-1\right) &= 10\left(\frac{x}{5}+2\right) \\ 10\left(\frac{x}{2}\right)-10\cdot 1 &= 10\left(\frac{x}{5}\right)+10\cdot 2 \\ 5x-10 &= 2x+20 \\ 5x-2x-10 &= 2x-2x+20 \\ 3x-10 &= 20 \\ 3x-10+10 &= 20+10 \\ 3x &= 30 \\ \frac{3x}{3} &= \frac{30}{3} \\ x &= 10 \end{aligned}$$

15.
$$\begin{aligned} \frac{6(3-z)}{5} &= -z \\ 5\cdot\frac{6(3-z)}{5} &= 5(-z) \\ 6(3-z) &= -5z \\ 18-6z &= -5z \\ 18-6z+6z &= -5z+6z \\ 18 &= z \end{aligned}$$

17.
$$\frac{2(x+1)}{4} = 3x-2$$
$$4\cdot\frac{2(x+1)}{4} = 4(3x-2)$$
$$2(x+1) = 12x-8$$
$$2x+2 = 12x-8$$
$$2x-2x+2 = 12x-2x-8$$
$$2 = 10x-8$$
$$2+8 = 10x-8+8$$
$$10 = 10x$$
$$\frac{10}{10} = \frac{10x}{10}$$
$$1 = x$$

19.
$$0.50x+0.15(70) = 0.25(142)$$
$$50x+15(70) = 25(142)$$
$$50x+1050 = 3550$$
$$50x+1050-1050 = 3550-1050$$
$$50x = 2500$$
$$\frac{50x}{50} = \frac{2500}{50}$$
$$x = 50$$

21.
$$0.12(y-6)+0.06y = 0.08y-0.07(10)$$
$$12(y-6)+6y = 8y-7(10)$$
$$12y-72+6y = 8y-70$$
$$18y-72 = 8y-70$$
$$18y-72-8y = 8y-70-8y$$
$$10y-72 = -70$$
$$10y-72+72 = -70+72$$
$$10y = 2$$
$$\frac{10y}{10} = \frac{2}{10}$$
$$y = 0.2$$

23.
$$5x-5 = 2(x+1)+3x-7$$
$$5x-5 = 2x+2+3x-7$$
$$5x-5 = 5x-5$$
$$5x-5x-5 = 5x-5x-5$$
$$-5 = -5$$
$$-5+5 = -5+5$$
$$0 = 0$$
All real numbers

25.
$$\frac{x}{4}+1 = \frac{x}{4}$$
$$4\left(\frac{x}{4}+1\right) = 4\left(\frac{x}{4}\right)$$
$$4\left(\frac{x}{4}\right)+4(1) = 4\left(\frac{x}{4}\right)$$
$$x+4 = x$$
$$x-x+4 = x-x$$
$$4 = 0$$
No solution

27.
$$3x-7 = 3(x+1)$$
$$3x-7 = 3x+3$$
$$3x-3x-7 = 3x-3x+3$$
$$-7 = 3$$
No solution

29. Answers may vary.

31. Answers may vary.

33.
$$4x+3 = 2x+11$$
$$4x-2x+3 = 2x-2x+11$$
$$2x+3 = 11$$
$$2x+3-3 = 11-3$$
$$2x = 8$$
$$\frac{2x}{2} = \frac{8}{2}$$
$$x = 4$$

35.
$$-2y-10 = 5y+18$$
$$-2y+2y-10 = 5y+2y+18$$
$$-10 = 7y+18$$
$$-10-18 = 7y+18-18$$
$$-28 = 7y$$
$$\frac{-28}{7} = \frac{7y}{7}$$
$$-4 = y$$

37.
$$0.6x-0.1 = 0.5x+0.2$$
$$6x-1 = 5x+2$$
$$6x-5x-1 = 5x-5x+2$$
$$x-1 = 2$$
$$x-1+1 = 2+1$$
$$x = 3$$

39.
$$\begin{aligned} 2y+2&=y\\ 2y-2y+2&=y-2y\\ 2&=-y\\ \frac{2}{-1}&=\frac{-y}{-1}\\ -2&=y \end{aligned}$$

41.
$$\begin{aligned} 3(5c-1)-2&=13c+3\\ 15c-3-2&=13c+3\\ 15c-5&=13c+3\\ 15c-13c-5&=13c-13c+3\\ 2c-5&=3\\ 2c-5+5&=3+5\\ 2c&=8\\ \frac{2c}{2}&=\frac{8}{2}\\ c&=4 \end{aligned}$$

43.
$$\begin{aligned} x+\frac{7}{6}&=2x-\frac{7}{6}\\ 6\left(x+\frac{7}{6}\right)&=6\left(2x-\frac{7}{6}\right)\\ 6(x)+6\left(\frac{7}{6}\right)&=6(2x)+6\left(-\frac{7}{6}\right)\\ 6x+7&=12x-7\\ 6x-12x+7&=12x-12x-7\\ -6x+7&=-7\\ -6x+7-7&=-7-7\\ -6x&=-14\\ \frac{-6x}{-6}&=\frac{-14}{-6}\\ x&=\frac{14}{6}\\ x&=\frac{7}{3} \end{aligned}$$

45.
$$\begin{aligned} 2(x-5)&=7+2x\\ 2x-10&=7+2x\\ 2x-2x-10&=7+2x-2x\\ -10&=7 \end{aligned}$$
No solution

47.
$$\begin{aligned} \frac{2(z+3)}{3}&=5-z\\ 3\cdot\frac{2(z+3)}{3}&=3(5-z)\\ 2(z+3)&=3(5-z)\\ 2z+6&=15-3z\\ 2z+3z+6&=15-3z+3z\\ 5z+6&=15\\ 5z+6-6&=15-6\\ 5z&=9\\ \frac{5z}{5}&=\frac{9}{5}\\ z&=\frac{9}{5} \end{aligned}$$

49.
$$\begin{aligned} \frac{4(y-1)}{5}&=-3y\\ 5\cdot\frac{4(y-1)}{5}&=5(-3y)\\ 4(y-1)&=5(-3y)\\ 4y-4&=-15y\\ 4y-4y-4&=-15y-4y\\ -4&=-19y\\ \frac{-4}{-19}&=\frac{-19y}{-19}\\ \frac{4}{19}&=y \end{aligned}$$

51.
$$\begin{aligned} 8-2(a-1)&=7+a\\ 8-2a+2&=7+a\\ -2a+10&=7+a\\ -2a-a+10&=7+a-a\\ -3a+10&=7\\ -3a+10-10&=7-10\\ -3a&=-3\\ \frac{-3a}{-3}&=\frac{-3}{-3}\\ a&=1 \end{aligned}$$

53.
$$\begin{aligned} 2(x+3)-5&=5x-3(1+x)\\ 2x+6-5&=5x-3-3x\\ 2x+1&=2x-3\\ 2x-2x+1&=2x-2x-3\\ 1&=-3 \end{aligned}$$
No solution

55. $\frac{5x-7}{3} = x$

$3 \cdot \frac{5x-7}{3} = 3(x)$

$5x - 7 = 3x$

$5x - 5x - 7 = 3x - 5x$

$-7 = -2x$

$\frac{-7}{-2} = \frac{-2x}{-2}$

$\frac{7}{2} = x$

57. $\frac{9+5v}{2} = 2v - 4$

$2 \cdot \frac{9+5v}{2} = 2(2v-4)$

$9 + 5v = 4v - 8$

$9 + 5v - 4v = 4v - 4v - 8$

$9 + v = -8$

$9 - 9 + v = -8 - 9$

$v = -17$

59. $-3(t-5) + 2t = 5t - 4$

$-3t + 15 + 2t = 5t - 4$

$-t + 15 = 5t - 4$

$-t + t + 15 = 5t + t - 4$

$15 = 6t - 4$

$15 + 4 = 6t - 4 + 4$

$19 = 6t$

$\frac{19}{6} = \frac{6t}{6}$

$\frac{19}{6} = t$

61. $0.02(6t-3) = 0.12(t-2) + 0.18$

$2(6t-3) = 12(t-2) + 18$

$12t - 6 = 12t - 24 + 18$

$12t - 6 = 12t - 6$

$12t - 6 - 12t = 12t - 6 - 12t$

$6 = 6$

All real numbers

63. $0.06 - 0.01(x+1) = -0.02(2-x)$

$6 - 1(x+1) = -2(2-x)$

$6 - x - 1 = -4 + 2x$

$5 - x = -4 + 2x$

$5 - x + x = -4 + 2x + x$

$5 = -4 + 3x$

$5 + 4 = -4 + 4 + 3x$

$9 = 3x$

$\frac{9}{3} = \frac{3x}{3}$

$3 = x$

65. $\frac{3(x-5)}{2} = \frac{2(x+5)}{3}$

$6 \cdot \frac{3(x-5)}{2} = 6 \cdot \frac{2(x+5)}{3}$

$3 \cdot \frac{3(x-5)}{1} = 2 \cdot \frac{2(x+5)}{1}$

$9(x-5) = 4(x+5)$

$9x - 45 = 4x + 20$

$9x - 4x - 45 = 4x - 4x + 20$

$5x - 45 = 20$

$5x - 45 + 45 = 20 + 45$

$5x = 65$

$\frac{5x}{5} = \frac{65}{5}$

$x = 13$

67.

$$\begin{aligned} 1000(7x-10) &= 50(412+100x) \\ 7000x-10,000 &= 20,600+5000x \\ 7000x-5000x-10,000 &= 20,600+5000x-5000x \\ 2000x-10,000 &= 20,600 \\ 2000x-10,000+10,000 &= 20,600+10,000 \\ 2000x &= 30,600 \\ \frac{2000x}{2000} &= \frac{30,600}{2000} \\ x &= 15.3 \end{aligned}$$

69.

$$\begin{aligned} 0.035x+5.112 &= 0.010x+5.107 \\ 35x+5112 &= 10x+5107 \\ 35x-10x+5112 &= 10x-10x+5107 \\ 25x+5112 &= 5107 \\ 25x+5112-5112 &= 5107-5112 \\ 25x &= -5 \\ \frac{25x}{25} &= \frac{-5}{25} \\ x &= -0.2 \end{aligned}$$

71. Let x represent the number.

Equation: $2x+\frac{1}{5}=3x-\frac{4}{5}$

Solution:

$$\begin{aligned} 2x+\frac{1}{5} &= 3x-\frac{4}{5} \\ 5\left(2x+\frac{1}{5}\right) &= 5\left(3x-\frac{4}{5}\right) \\ 5(2x)+5\left(\frac{1}{5}\right) &= 5(3x)+5\left(-\frac{4}{5}\right) \\ 10x+1 &= 15x-4 \\ 10x-15x+1 &= 15x-15x-4 \\ -5x+1 &= -4 \\ -5x+1-1 &= -4-1 \\ -5x &= -5 \\ \frac{-5x}{-5} &= \frac{-5}{-5} \\ x &= 1 \end{aligned}$$

The number is 1.

73. Let x represent the number.
Equation: $2x + 7 = x + 6$
Solution:
$$2x+7=x+6$$
$$2x-x+7=x-x+6$$
$$x+7=6$$
$$x+7-7=6-7$$
$$x=-1$$
The number is −1.

75. Let x represent the number.
Equation: $3x - 6 = 2x + 8$
Solution:
$$3x-6=2x+8$$
$$3x-2x-6=2x-2x+8$$
$$x-6=8$$
$$x-6+6=8+6$$
$$x=14$$
The number is 14.

77. Let x represent the number.
Equation: $\frac{1}{3}x = \frac{5}{6}$
Solution:
$$\frac{1}{3}x=\frac{5}{6}$$
$$\frac{3}{1}\cdot\frac{1}{3}x=\frac{5}{6}\cdot\frac{3}{1}$$
$$x=\frac{5}{2}$$
The number is $\frac{5}{2}$.

79. Let x represent the number.
Equation: $x - 4 = 2x$
Solution:
$$x-4=2x$$
$$x-x-4=2x-x$$
$$-4=x$$
The number is −4.

81. Let x represent the number.
Equation: $\frac{x}{4}+\frac{1}{2}=\frac{3}{4}$
Solution:
$$\frac{x}{4}+\frac{1}{2}=\frac{3}{4}$$
$$4\left(\frac{x}{4}+\frac{1}{2}\right)=4\left(\frac{3}{4}\right)$$
$$4\left(\frac{x}{4}\right)+4\left(\frac{1}{2}\right)=4\left(\frac{3}{4}\right)$$
$$x+2=3$$
$$x+2-2=3-2$$
$$x=1$$
The number is 1.

83.
$$x+x+x+2x+2x=28$$
$$7x=28$$
$$\frac{7x}{7}=\frac{28}{7}$$
$$x=4$$
$$2x=8$$
The lengths of the sides are 2 cm, 2 cm, 2 cm, 4 cm and 4 cm.

85. Let x represent the number.
Equation: $10 - 5x = 3x$
Solution:
$$10-5x=3x$$
$$10-5x+5x=3x+5x$$
$$10=8x$$
$$\frac{10}{8}=\frac{8x}{8}$$
$$\frac{5}{4}=x$$
The number is $\frac{5}{4}$.

87. Midway; it has the tallest bar.

89.
$$x+55=2x-90$$
$$x-x+55=2x-x-90$$
$$55=x-90$$
$$55+90=x-90+90$$
$$145=x$$
145 cities, towns, or villages are named Five Points.

91. $|2^3 - 3^2| - |5 - 7| = |8 - 9| - |5 - 7|$
$= |-1| - |-2|$
$= 1 - 2$
$= -1$

93. $\frac{5}{4+3\cdot 7} = \frac{5}{4+21} = \frac{5}{25} = \frac{1}{5}$

95. The perimeter is the sum of the lengths of the sides.
$x + (2x - 3) + (3x - 5) = 6x - 8$
The perimeter is $(6x - 8)$ meters.

97.
$$x(x-3) = x^2 + 5x + 7$$
$$x^2 - 3x = x^2 + 5x + 7$$
$$x^2 - x^2 - 3x = x^2 - x^2 + 5x + 7$$
$$-3x = 5x + 7$$
$$-3x - 5x = 5x - 5x + 7$$
$$-8x = 7$$
$$\frac{-8x}{-8} = \frac{7}{-8}$$
$$x = -\frac{7}{8}$$

99.
$$2z(z+6) = 2z^2 + 12z - 8$$
$$2z^2 + 12z = 2z^2 + 12z - 8$$
$$2z^2 - 2z^2 + 12z = 2z^2 - 2z^2 + 12z - 8$$
$$12z = 12z - 8$$
$$12z - 12z = 12z - 12z - 8$$
$$0 = -8$$
No solution

101.
$$n(3+n) = n^2 + 4n$$
$$3n + n^2 = n^2 + 4n$$
$$3n + n^2 - n^2 = n^2 - n^2 + 4n$$
$$3n = 4n$$
$$3n - 3n = 4n - 3n$$
$$0 = n$$

Exercise Set 2.4

1. Let x represent the number.
$$2x + \frac{1}{5} = 3x - \frac{4}{5}$$
$$5\left(2x + \frac{1}{5}\right) = 5\left(3x - \frac{4}{5}\right)$$
$$10x + 1 = 15x - 4$$
$$10x + 1 - 10x = 15x - 4 - 10x$$
$$1 = 5x - 4$$
$$1 + 4 = 5x - 4 + 4$$
$$5 = 5x$$
$$\frac{5}{5} = \frac{5x}{5}$$
$$1 = x$$
The number is 1.

3. Let x represent the number.
$$2(x-8) = 3(x+3)$$
$$2x - 16 = 3x + 9$$
$$2x - 16 - 2x = 3x + 9 - 2x$$
$$-16 = x + 9$$
$$-16 - 9 = x + 9 - 9$$
$$-25 = x$$
The number is -25.

5. Let x represent the number.
$$2x \cdot 3 = 5x - \frac{3}{4}$$
$$6x = 5x - \frac{3}{4}$$
$$6x - 5x = 5x - \frac{3}{4} - 5x$$
$$x = -\frac{3}{4}$$
The number is $-\frac{3}{4}$.

7. Let x represent the number.
$$3(x+5) = 2x - 1$$
$$3x + 15 = 2x - 1$$
$$3x + 15 - 2x = 2x - 1 - 2x$$
$$x + 15 = -1$$
$$x + 15 - 15 = -1 - 15$$
$$x = -16$$
The number is -16.

9. Let x = salary of the governor of Nebraska. Then $2x$ = salary of the governor of Washington.

$$x+2x=195{,}000$$
$$3x=195{,}000$$
$$\frac{3x}{3}=\frac{195{,}000}{3}$$
$$x=65{,}000$$

The salary of the governor of Nebraska is \$65,000. The salary of the governor of Washington is $2 \cdot 65{,}000 = \$130{,}000$.

11. If x = length of the first piece, then $2x$ = length of the second piece, and $5x$ = length of the third piece.

$$x+2x+5x=40$$
$$8x=40$$
$$\frac{8x}{8}=\frac{40}{8}$$
$$x=5$$

The first piece is 5 inches long, the second piece is $2 \cdot 5 = 10$ inches long, and the third piece is $5 \cdot 5 = 25$ inches long.

13. The cost of renting the car is equal to the daily rental charge plus \$0.29 per mile. Let x = number of miles.

$$2 \cdot 24.95+0.29x=100$$
$$49.90+0.29x=100$$
$$49.90+0.29x-49.90=100-49.90$$
$$0.29x=50.10$$
$$\frac{0.29x}{0.29}=\frac{50.10}{0.29}$$
$$x=172$$

You can drive 172 whole miles on a budget of \$100.

15. Let x = measure of each of the two equal angles, then $2x + 30$ = measure of the third angle.

$$x+x+2x+30=180$$
$$4x+30=180$$
$$4x+30-30=180-30$$
$$4x=150$$
$$\frac{4x}{4}=\frac{150}{4}$$
$$x=37.5$$

The angles measure 37.5°, 37.5°, and 105°.

17. Let x = number of votes for Randall. Then $x + 13{,}288$ = number of votes for Brown.

$$x+x+13{,}288=119{,}436$$
$$2x+13{,}288=119{,}436$$
$$2x+13{,}288-13{,}288=119{,}436-13{,}288$$
$$2x=106{,}148$$
$$\frac{2x}{2}=\frac{106{,}148}{2}$$
$$x=53{,}074$$

$x + 13{,}288 = 66{,}362$

Brown: 66,362 votes

Randall: 53,074 votes

19.

$$x+3x=180$$
$$4x=180$$
$$\frac{4x}{4}=\frac{180}{4}$$
$$x=45$$

$3x = 135$

45° and 135°

21. Let x = code for Belgium. Then $x + 1$ = code for France, and $x + 2$ = code for Spain.

$$x+(x+1)+(x+2)=99$$
$$3x+3=99$$
$$3x+3-3=99-3$$
$$3x=96$$
$$\frac{3x}{3}=\frac{96}{3}$$
$$x=32$$

$x + 1 = 33$

$x + 2 = 34$

Belgium: 32

France: 33

Spain: 34

23. Let x = growth rate of human toenails.

$$4x=0.8$$
$$\frac{4x}{4}=\frac{0.8}{4}$$
$$x=0.2$$

Human toenails grow at a rate of 0.2 inch per year.

25. Let x = height of the probe. Then $2x - 19$ = diameter of the probe.

$$x + (2x - 19) = 83$$
$$x + 2x - 19 = 83$$
$$3x - 19 = 83$$
$$3x - 19 + 19 = 83 + 19$$
$$3x = 102$$
$$\frac{3x}{3} = \frac{102}{3}$$
$$x = 34$$

The probe is 34 inches tall and has a diameter of $2(34) - 19 = 49$ inches.

27. Let x = Knicks' score. Then $x + 1$ = Spurs' score.

$$x + x + 1 = 155$$
$$2x + 1 = 155$$
$$2x + 1 - 1 = 155 - 1$$
$$2x = 154$$
$$\frac{2x}{2} = \frac{154}{2}$$
$$x = 77$$

$x + 1 = 78$

Spurs: 78

Knicks: 77

29. Let x = number of rotations

$$360x = 900$$
$$\frac{360x}{360} = \frac{900}{360}$$
$$x = 2\frac{1}{2}$$

There were $2\frac{1}{2}$ rotations.

31. Let x = number of Democratic governors. Then $x + 14$ = number of Republican governors.

$$x + x + 14 = 50 - 2$$
$$2x + 14 = 48$$
$$2x + 14 - 14 = 48 - 14$$
$$2x = 34$$
$$\frac{2x}{2} = \frac{34}{2}$$
$$x = 17$$

$x + 14 = 31$

17 Democratic governors

31 Republican governors

33. If x = length of the shorter piece, then $2x + 2$ = length of the longer piece.

$$x + (2x + 2) = 17$$
$$x + 2x + 2 = 17$$
$$3x + 2 = 17$$
$$3x + 2 - 2 = 17 - 2$$
$$3x = 15$$
$$\frac{3x}{3} = \frac{15}{3}$$
$$x = 5$$

The shorter piece is 5 feet long. The longer piece is 12 feet long.

35. Let x = smallest angle.

$$x + (x + 2) + (x + 4) = 180$$
$$3x + 6 = 180$$
$$3x + 6 - 6 = 180 - 6$$
$$3x = 174$$
$$\frac{3x}{3} = \frac{174}{3}$$
$$x = 58$$

$x + 2 = 60$

$x + 4 = 62$

The angles are 58°, 60°, and 62°.

37. Let x = number of miles.

$$34 + 0.20x = 104$$
$$34 + 0.20x - 34 = 104 - 34$$
$$0.20x = 70$$
$$\frac{0.20x}{0.20} = \frac{70}{0.20}$$
$$x = 350$$

You drove 350 miles.

39. Answers may vary.

41. Texas and Florida

43. Let x = amount spent by Pennsylvania.
Then $2x - 8.1$ = amount spent by Hawaii.

$$x + 2x - 8.1 = 60.9$$
$$3x - 8.1 = 60.9$$
$$3x - 8.1 + 8.1 = 60.9 + 8.1$$
$$3x = 69.0$$
$$\frac{3x}{3} = \frac{69.0}{3}$$
$$x = 23.0$$

$2x - 8.1 = 37.9$

Hawaii: $37.9 million

Pennsylvania: $23 million

45. Let x = width. Then $1.6x$ = length.

$$2x + 2(1.6x) = 78$$
$$2x + 3.2x = 78$$
$$5.2x = 78$$
$$x = 15$$

$1.6x = 24$

The dimensions are 15 feet by 24 feet.

47. Answers may vary.

49. Answers may vary.

51. c

53. $-2 + (-8) = -2 - 8 = -10$

55. $-11 + 2 = -9$

57. $-12 - 3 = -15$

59. $5(-x) = x + 60$

61. $50 - (x + 9) = 0$

Exercise Set 2.5

1.
$$A = bh$$
$$45 = 15 \cdot h$$
$$\frac{45}{15} = \frac{15 \cdot h}{15}$$
$$3 = h$$

3.
$$S = 4lw + 2wh$$
$$102 = 4(7)(3) + 2(3)h$$
$$102 = 84 + 6h$$
$$102 - 84 = 84 + 6h - 84$$
$$18 = 6h$$
$$\frac{18}{6} = \frac{6h}{6}$$
$$3 = h$$

5.
$$A = \frac{1}{2}h(B + b)$$
$$180 = \frac{1}{2}h(11 + 7)$$
$$180 = \frac{1}{2}h(18)$$
$$180 = 9h$$
$$\frac{180}{9} = \frac{9h}{9}$$
$$20 = h$$

7.
$$P = a + b + c$$
$$30 = 8 + 10 + c$$
$$30 = 18 + c$$
$$30 - 18 = 18 + c - 18$$
$$12 = c$$

9.
$$C = 2\pi r$$
$$15.7 = 2\pi r$$
$$\frac{15.7}{2\pi} = \frac{2\pi r}{2\pi}$$
$$\frac{15.7}{2(3.14)} = r$$
$$2.5 = r$$

11.
$$I = PRT$$
$$3750 = (25,000)(0.05)T$$
$$3750 = 1250T$$
$$\frac{3750}{1250} = \frac{1250T}{1250}$$
$$3 = T$$

13.
$$V = \frac{1}{3}\pi r^2 h$$
$$565.2 = \frac{1}{3}\pi(6^2)h$$
$$565.2 = \frac{1}{3}\pi(36)h$$
$$565.2 = 12\pi h$$
$$\frac{565.2}{12\pi} = \frac{12\pi h}{12\pi}$$
$$15.0 = h$$

15.
$$f = 5gh$$
$$\frac{f}{5g} = \frac{5gh}{5g}$$
$$\frac{f}{5g} = h$$

17.
$$V = LWH$$
$$\frac{V}{LH} = \frac{LWH}{LH}$$
$$\frac{V}{LH} = W$$

19.
$$3x + y = 7$$
$$3x - 3x + y = 7 - 3x$$
$$y = 7 - 3x$$

21.
$$A = P + PRT$$
$$A - P = P - P + PRT$$
$$A - P = PRT$$
$$\frac{A-P}{PT} = \frac{PRT}{PT}$$
$$\frac{A-P}{PT} = R$$

23.
$$V = \frac{1}{3}Ah$$
$$3(V) = 3\left(\frac{1}{3}Ah\right)$$
$$3V = Ah$$
$$\frac{3V}{h} = \frac{Ah}{h}$$
$$\frac{3V}{h} = A$$

25.
$$P = a + b + c$$
$$P - b = a + b - b + c$$
$$P - b = a + c$$
$$P - b - c = a + c - c$$
$$P - b - c = a$$

27.
$$S = 2\pi rh + 2\pi r^2$$
$$S - 2\pi r^2 = 2\pi rh + 2\pi r^2 - 2\pi r^2$$
$$S - 2\pi r^2 = 2\pi rh$$
$$\frac{S - 2\pi r^2}{2\pi r} = \frac{2\pi rh}{2\pi r}$$
$$\frac{S - 2\pi r^2}{2\pi r} = h$$

29.
$$A = lw$$
$$52,400 = 400 \cdot w$$
$$\frac{52,400}{400} = \frac{400 \cdot w}{400}$$
$$131 = w$$
The width of the sign is 131 feet.

31.
$$d = rt$$
$$375 = 50 \cdot t$$
$$\frac{375}{50} = \frac{50 \cdot t}{50}$$
$$7.5 = t$$
It would take 7.5 hours.

33. Use the formula $C = \frac{5}{9}(F - 32)$ with $F = 14$.
$$C = \frac{5}{9}(14 - 32) = \frac{5}{9}(-18) = -10$$
14°F is the same as –10°C.

35. $V = lwh$
$V = (8)(3)(6)$
$V = 144$
Since the tank has a volume of 144 cubic feet, and each piranha requires 1.5 cubic feet, the tank can hold $\frac{144}{1.5} = 96$ piranhas.

37. $d = rt$
$d = 55(2.5)$
$d = 137.5$
The cities are 137.5 miles apart.

39. $A = \frac{1}{2}h(B+b)$

$A = \frac{1}{2}(60)(130+70) = \frac{1}{2}(60)(200) = 6000$

Since the area of the lawn is 6000 square feet and since each bag covers 4000 square feet, two bags must be purchased.

41. $d = rt$

$25{,}000 = 4000 \cdot t$

$\frac{25{,}000}{4000} = \frac{4000 \cdot t}{4000}$

$6.25 = t$

It will take 6.25 hours.

43. $V = lwh$

$V = (10)(8)(10)$

$V = 800$

The minimum volume of the box must be 800 cubic feet.

45. Use the formula for the area of a circle, $A = \pi r^2$, to solve for the number of square inches of pizza purchased in each case. For one 16-inch pizza use $r = 8$.

$A = \pi r^2 = \pi\left(8^2\right) = 64\pi$ square inches. For two 10-inch pizzas use $r = 5$ and multiply the result by 2.

$A = 2(\pi r^2) = 2\pi(5^2) = 2\pi(25) = 50\pi$ square inches

One 16-inch pizza gives more pizza for the price.

47. $d = rt$

$42.8 = 552t$

$\frac{42.8}{552} = \frac{552t}{552}$

$0.077536 \approx t$

$(0.077536 \text{ hours})\left(\frac{60 \text{ minutes}}{\text{hour}}\right)$

≈ 4.65 minutes

The test run would last approximately 4.65 minutes.

49. $d = rt$

$135 = 60 \cdot t$

$\frac{135}{60} = \frac{60 \cdot t}{60}$

$2.25 = t$

It will take 2.25 hours.

51. Use $F = \frac{9}{5}C + 32$ with $C = -78.5$.

$F = \frac{9}{5}C + 32$

$F = \frac{9}{5}(-78.5) + 32$

$= -141.3 + 32$

$= -109.3$

-78.5°C is the same as -109.3°F.

53. Use $d = rt$ with $d = 93{,}000{,}000$ and $r = 186{,}000$.

$d = rt$

$93{,}000{,}000 = 186{,}000 \cdot t$

$\frac{93{,}000{,}000}{186{,}000} = \frac{186{,}000 \cdot t}{186{,}000}$

$500 = t$

It takes 500 seconds or $8\frac{1}{3}$ minutes.

55. Use the formula for the volume of a sphere, $V = \frac{4}{3}\pi r^3$, with $r = 2000$.

$V = \frac{4}{3}\pi(2000)^3$

$= \frac{4}{3}\pi(8{,}000{,}000{,}000)$

$= 33{,}510{,}321{,}638$

The volume of the fireball was 33,510,321,638 cubic miles.

57. Use the formula $d = rt$ with $d = 25{,}120$ and $r = 270{,}000$.

$d = rt$

$25{,}120 = 270{,}000 \cdot t$

$\frac{25{,}120}{270{,}000} = \frac{270{,}000 \cdot t}{270{,}000}$

$0.093 \approx t$

Thus, it takes approximately 0.093 second for a bolt of lightning to travel around the world once. In one second it can travel around the world $\frac{1}{0.093} \approx 10.7$ times.

59. $\frac{20 \text{ miles}}{1 \text{ hour}} = \frac{20 \text{ miles}}{1 \text{ hour}} \cdot \frac{5280 \text{ feet}}{1 \text{ mile}} \cdot \frac{1 \text{ hour}}{3600 \text{ sec}}$
$= \frac{88}{3}$ feet per second

Use $d = rt$ with $d = 1300$ feet and $r = \frac{88}{3}$ feet per second.

$$1300 = \frac{88}{3}t$$
$$\frac{3}{88}(1300) = \left(\frac{3}{88}\right)\left(\frac{88}{3}t\right)$$
$$44.3 \approx t$$

It took about 44.3 seconds.

61. Use the formula $C = \frac{5}{9}(F - 32)$ to find when $C = F$.

$$C = \frac{5}{9}(F-32)$$
$$F = \frac{5}{9}(F-32)$$
$$9(F) = 9\left[\frac{5}{9}(F-32)\right]$$
$$9F = 5(F-32)$$
$$9F = 5F - 160$$
$$9F - 5F = 5F - 160 - 5F$$
$$4F = -160$$
$$\frac{4F}{4} = \frac{-160}{4}$$
$$F = -40$$

–40°F is the same as –40°C.

63. $V = (2L)(2W)(2H) = 8LWH$
The volume is multiplied by 8.

65. $\frac{9}{x+5}$

67. $3(x + 4)$

69. $2(10 + 4x)$

71. $3(x - 12)$

Section 2.6

Mental Math

1. not correct

2. not correct

3. correct

4. correct

Exercise Set 2.6

1. $120\% = 1.2$

3. $22.5\% = 0.225$

5. $0.12\% = 0.0012$

7. $0.75 = 75\%$

9. $2 = 200\%$

11. $\frac{1}{8} = 0.125 = 12.5\%$

13. 38%

15. $38\% + 16\% = 54\%$

17. 38% of 360° = 0.38(360°) = 136.8°

19. Answers may vary.

21. 4%

23. 4% of 360° = 0.04(360°) = 14.4°

25. 37% of 135,000 = 0.37(135,000) = 49,950

27. Let x = unknown number.
$x = 0.16(70)$
$x = 11.2$

29. Let x = unknown percent.
$28.6 = x(52)$
$0.55 = x$
55%

31. Let x = unknown number.
$45 = 0.25x$
$180 = x$

33. $0.23(20) = 4.6$

35. Let x = unknown number.
$40 = 0.80x$
$50 = x$

37. Let x = unknown percent.
$144 = x(480)$
$0.3 = x$
30%

39. Decrease = \$156(0.25) = \$39
Sale price = \$156 − \$39 = \$117

41. To find how much shorter the men's record throw is than the women's, we find 3.7% of 252 feet.
$3.7\% \text{ of } 252 = 0.037(252)$
$= 9.324$
The men's record throw is 9.324 feet less than the women's.
$252 - 9.324 = 243$
The men's record is 243 feet.

43. 55.40% of those surveyed have used over-the-counter drugs to combat the common cold.

45. Since 23.70% of those surveyed have used over-the-counter drugs for allergies, we find 23.70% of 230.
$23.70\% \text{ of } 230 = 0.237(230) = 54.51$
54 people used over-the-counter drugs for allergies.

47. No, because many people have used over-the-counter drugs for more than one of the categories listed.

49. Since 26% of men doze off, we find 26% of 121.
$26\% \text{ of } 121 = 0.26(121) = 31.46$
We would expect 31 of the men to have dozed off.

51. The percent of total number of women from each service is the ratio of the number of women of the service to the total.
Navy: $\frac{50,287}{186,697} \approx 27\%$
Marines: $\frac{9696}{186,697} \approx 5\%$
Air Force: $\frac{64,427}{186,697} \approx 35\%$
Coast Guard: $\frac{3879}{186,697} \approx 2\%$
Total:
$31\% + 27\% + 5\% + 35\% + 2\% = 100\%$

53. $\frac{1800}{1,200,000} = 0.0015 = 0.15\%$

55. $0.44x = 10.4$
$\frac{0.44x}{0.44} = \frac{10.4}{0.44}$
$x = 23.6$
23.6 million

57. Increase = $70 - 40 = 30$
$30 = x(40)$
$0.75 = x$
75% increase

59. Increase = $15.65 - 12.62 = 3.03$
$3.03 = x(12.62)$
$0.24 = x$
24% increase; No

61. Increase = $5000 - 3 = 4997$
$4997 = x(3)$
$1665.67 = x$
166,567% increase

63. 16%

65. $24\% \text{ of } 50 = 0.24(50) = 12$

67. Answers may vary.

69. $\frac{34,611 - 38,831}{38,831} = \frac{-4220}{38,831} = -0.109$
There was an 11% decrease.

71. $\frac{23}{300} \approx 0.077 = 7.7\%$
About 7.7% of the daily recommended carbohydrate intake is contained in one serving.

73. Find the ratio of the calories from fat to the total calories. Each serving contains 6 grams of fat.
$6 \cdot 9 = 54$ calories
$\frac{54}{280} \approx 0.193 = 19.3\%$
About 19.3% of the calories in each serving come from fat.

75. Answers may vary.

77. $2a + b - c = 2(5) + (-1) - 3 = 10 - 1 - 3 = 6$

79. $4ab - 3bc = 4(-5)(-8) - 3(-8)(2)$
$= 160 + 48 = 208$

81. $n^2 - m^2 = (-3)^2 - (-8)^2 = 9 - 64 = -55$

Section 2.7

Mental Math

1. $5x > 10$
$x > 2$

2. $4x < 20$
$x < 5$

3. $2x \geq 16$
$x \geq 8$

4. $9x \leq 63$
$x \leq 7$

Exercise Set 2.7

1. $\{x|x < -3\}$

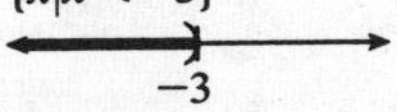

−3

$(-\infty, -3)$

3. $\{x|x \geq 0.3\}$

0.3

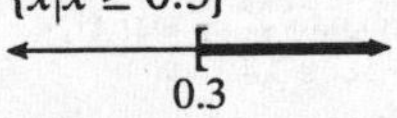

$[0.3, \infty)$

5. $\{x|5 < x\}$

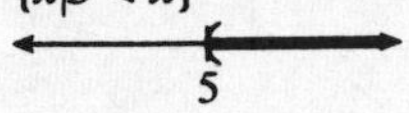

5

$(5, \infty)$

7. $\{x|-2 < x < 5\}$

−2 5

$(-2, 5)$

9. $\{x|5 > x > -1\}$

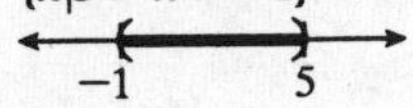

−1 5

$(-1, 5)$

11. $2x < -6$
$\frac{2x}{2} < \frac{-6}{2}$
$x < -3$

−3

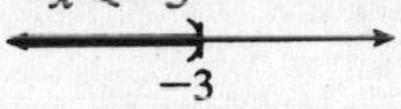

$(-\infty, -3)$

13. $x - 2 \geq -7$
$x - 2 + 2 \geq -7 + 2$
$x \geq -5$

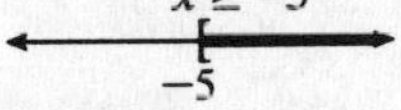

−5

$[-5, \infty)$

15. $-8x \leq 16$
$\frac{-8x}{-8} \geq \frac{16}{-8}$
$x \geq -2$

−2

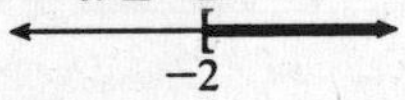

$[-2, \infty)$

17. $15 + 2x \geq 4x - 7$
$15 + 2x - 2x \geq 4x - 7 - 2x$
$15 \geq 2x - 7$
$15 + 7 \geq 2x - 7 + 7$
$22 \geq 2x$
$\frac{22}{2} \geq \frac{2x}{2}$
$11 \geq x$

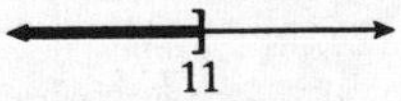

11

$(-\infty, 11]$

19.
$$\frac{3x}{4} \ge 2$$
$$4\left(\frac{3x}{4}\right) \ge 4(2)$$
$$3x \ge 8$$
$$\frac{3x}{3} \ge \frac{8}{3}$$
$$x \ge \frac{8}{3}$$

$\frac{8}{3}$

$\left[\frac{8}{3}, \infty\right)$

21.
$$3(x-5) < 2(2x-1)$$
$$3x - 15 < 4x - 2$$
$$3x - 15 - 3x < 4x - 2 - 3x$$
$$-15 < x - 2$$
$$-15 + 2 < x - 2 + 2$$
$$-13 < x$$

-13

$(-13, \infty)$

23.
$$\frac{1}{2} + \frac{2}{3} \ge \frac{x}{6}$$
$$6\left(\frac{1}{2} + \frac{2}{3}\right) \ge 6\left(\frac{x}{6}\right)$$
$$3 + 4 \ge x$$
$$7 \ge x$$

7

$(-\infty, 7]$

25.
$$4(x-1) \ge 4x - 8$$
$$4x - 4 \ge 4x - 8$$
$$4x - 4 - 4x \ge 4x - 8 - 4x$$
$$-4 \ge -8$$

This is true for all values of x, so the inequality is true for all real numbers.

0

$(-\infty, \infty)$

27.
$$7x < 7(x-2)$$
$$7x < 7x - 14$$
$$7x - 7x < 7x - 14 - 7x$$
$$0 < -14$$

This is a false statement for all values of x, so the inequality has no solution.

0

$\varnothing$

29. Answers may vary.

31. $\varnothing$; Answers may vary.

33.
$$5x + 3 > 2 + 4x$$
$$x + 3 > 2$$
$$x > -1$$

-1

$(-1, \infty)$

35.
$$8x - 7 \le 7x - 5$$
$$x - 7 \le -5$$
$$x \le 2$$

2

$(-\infty, 2]$

37.
$$5x > 10$$
$$x > 2$$

2

$(2, \infty)$

39.
$$-4x \le 32$$
$$x \ge -8$$

-8

$[-8, \infty)$

41.
$$-2x + 7 \ge 9$$
$$-2x \ge 2$$
$$x \le -1$$

-1

$(-\infty, -1]$

43. $4(2x+1)>4$
$8x+4>4$
$8x>0$
$x>0$

(0, ∞)

45. $\dfrac{x+7}{5}>1$
$x+7>5$
$x>-2$

(–2, ∞)

47. $-6x+2\ge 2(5-x)$
$-6x+2\ge 10-2x$
$-4x+2\ge 10$
$-4x\ge 8$
$x\le -2$

(–∞, –2]

49. $4(3x-1)\le 5(2x-4)$
$12x-4\le 10x-20$
$2x-4\le -20$
$2x\le -16$
$x\le -8$

(–∞, –8]

51. $\dfrac{-5x+11}{2}\le 7$
$-5x+11\le 14$
$-5x\le 3$
$x\ge -\dfrac{3}{5}$

$\left[-\dfrac{3}{5},\infty\right)$

53. $8x-16\le 10x+2$
$-16\le 2x+2$
$-18\le 2x$
$-9\le x$

[–9, ∞)

55. $2(x-3)>70$
$2x-6>70$
$2x>76$
$x>38$

(38, ∞)

57. $-5x+4\le -4(x-1)$
$-5x+4\le -4x+4$
$-5x\le -4x$
$0\le x$

[0, ∞)

59. $\dfrac{1}{4}(x-7)\ge x+2$
$x-7\ge 4x+8$
$x-15\ge 4x$
$-15\ge 3x$
$-5\ge x$

(–∞, –5]

61. $\dfrac{2}{3}(x+2)<\dfrac{1}{5}(2x+7)$
$2(x+2)<\dfrac{3}{5}(2x+7)$
$10(x+2)<3(2x+7)$
$10x+20<6x+21$
$10x<6x+1$
$4x<1$
$x<\dfrac{1}{4}$

$\left(-\infty,\dfrac{1}{4}\right)$

63. $4(x-6)+2x-4 \geq 3(x-7)+10x$
$4x-24+2x-4 \geq 3x-21+10x$
$6x-28 \geq 13x-21$
$6x \geq 13x+7$
$-7x \geq 7$
$x \leq -1$

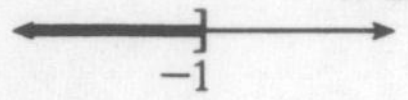

-1

$(-\infty, -1]$

65. $\frac{5x+1}{7} - \frac{2x-6}{4} \geq -4$
$5x+1-\frac{7(2x-6)}{4} \geq -28$
$4(5x+1)-7(2x-6) \geq -112$
$20x+4-14x+42 \geq -112$
$6x+46 \geq -112$
$6x \geq -158$
$x \geq -\frac{79}{3}$

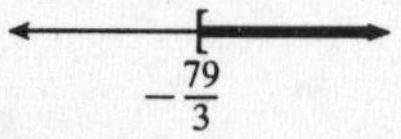

$-\frac{79}{3}$

$\left[-\frac{79}{3}, \infty\right)$

67. $\frac{-x+2}{2} - \frac{1-5x}{8} < -1$
$4(-x+2)-(1-5x) < -8$
$-4x+8-1+5x < -8$
$7+x < -8$
$x < -15$

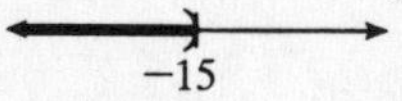

-15

$(-\infty, -15)$

69. $x < 200$ recommended
$200 \leq x \leq 240$ borderline
$x > 240$ high

71. Let x = number.
$2x+6 > -14$
$2x+6-6 > -14-6$
$2x > -20$
$\frac{2x}{2} > \frac{-20}{2}$
$x > -10$

73. Let x = number of people.
$50+34x \leq 3000$
$50+34x-50 \leq 3000-50$
$34x \leq 2950$
$\frac{34x}{34} \leq \frac{2950}{34}$
$x \leq 86.7$
They can invite a maximum of 86 people.

75. $P = 2L + 2W$
$2x+2(15) \leq 100$
$2x+30 \leq 100$
$2x+30-30 \leq 100-30$
$2x \leq 70$
$\frac{2x}{2} \leq \frac{70}{2}$
$x \leq 35$
The maximum length is 35 centimeters.

77.

Principal	Rate	Time	Interest
10,000	0.11	1	1100
5,000	x	1	$5000x$

$1100+5000x \geq 1600$
$1100-1100+5000x \geq 1600-1100$
$5000x \geq 500$
$\frac{5000x}{5000} \geq \frac{500}{5000}$
$x \geq 0.10$
$x \geq 10\%$

79. Let x = score of 3rd game.
$\frac{146+201+x}{3} \geq 180$
$3\left(\frac{146+201+x}{3}\right) \geq (3)180$
$146+201+x \geq 540$
$347+x \geq 540$
$347-347+x \geq 540-347$
$x \geq 193$

81. $(2)^3 = (2)(2)(2) = 8$

83. $(1)^{12} = 1$

85. $\left(\frac{4}{7}\right)^2 = \frac{4}{7}\cdot\frac{4}{7} = \frac{4\cdot4}{7\cdot7} = \frac{16}{49}$

87. 51 million; point on graph is (1995, 51).

89. 1996; the increase from 1995 to 1996 is the greatest.

91.
$$\begin{aligned} x(x+4) &> x^2-2x+6 \\ x^2+4x &> x^2-2x+6 \\ x^2-x^2+4x &> x^2-x^2-2x+6 \\ 4x &> -2x+6 \\ 4x+2x &> -2x+2x+6 \\ 6x &> 6 \\ \frac{6x}{6} &> \frac{6}{6} \\ x &> 1 \end{aligned}$$

1

93.
$$\begin{aligned} x^2+6x-10 &< x(x-10) \\ x^2+6x-10 &< x^2-10x \\ x^2-x^2+6x-10 &< x^2-x^2-10x \\ 6x-10 &< -10x \\ 6x-6x-10 &< -10x-6x \\ -10 &< -16x \\ \frac{-10}{-16} &> \frac{-16x}{-16} \\ \frac{10}{16} &> x \\ \frac{5}{8} &> x \end{aligned}$$

$\frac{5}{8}$

95.
$$\begin{aligned} x(2x-3) &\le 2x^2-5x \\ 2x^2-3x &\le 2x^2-5x \\ 2x^2-2x^2-3x &\le 2x^2-2x^2-5x \\ -3x &\le -5x \\ -3x+5x &\le -5x+5x \\ 2x &\le 0 \\ \frac{2x}{2} &\le \frac{0}{2} \\ x &\le 0 \end{aligned}$$

0

Chapter 2 Review Exercises

1. $5x - x + 2x = 4x + 2x = 6x$

2. $0.2z - 4.6x - 7.4z = -4.6x - 7.2z$

3. $\frac{1}{2}x + 3 + \frac{7}{2}x - 5 = \frac{8}{2}x - 2 = 4x - 2$

4. $\frac{4}{5}y + 1 + \frac{6}{5}y + 2 = \frac{10}{5}y + 3 = 2y + 3$

5. $2(n-4) + n - 10 = 2n - 8 + n - 10 = 3n - 18$

6. $3(w+2) - (12-w) = 3w + 6 - 12 + w$
$= 4w - 6$

7. $x + 5 - (7x - 2) = x + 5 - 7x + 2 = -6x + 7$

8. $y - 0.7 - (1.4y - 3) = y - 0.7 - 1.4y + 3$
$= -0.4y + 2.3$

9. Let x = number.
$3x - 7$

10. Let x = number.
$2(x + 2.8) + 3x$

11.
$$\begin{aligned} 8x+4 &= 9x \\ 8x-8x+4 &= 9x-8x \\ 4 &= x \end{aligned}$$

12.
$$\begin{aligned} 5y-3 &= 6y \\ 5y-5y-3 &= 6y-5y \\ -3 &= y \end{aligned}$$

13.
$$\begin{aligned} 3x-5&=4x+1\\ 3x-3x-5&=4x-3x+1\\ -5&=x+1\\ -5-1&=x+1-1\\ -6&=x \end{aligned}$$

14.
$$\begin{aligned} 2x-6&=x-6\\ 2x-6+6&=x-6+6\\ 2x&=x\\ 2x-x&=x-x\\ x&=0 \end{aligned}$$

15.
$$\begin{aligned} 4(x+3)&=3(1+x)\\ 4x+12&=3+3x\\ 4x-3x+12&=3+3x-3x\\ x+12&=3\\ x+12-12&=3-12\\ x&=-9 \end{aligned}$$

16.
$$\begin{aligned} 6(3+n)&=5(n-1)\\ 18+6n&=5n-5\\ 18+6n-5n&=5n-5n-5\\ 18+n&=-5\\ 18-18+n&=-5-18\\ n&=-23 \end{aligned}$$

17. $x-5+\underline{5}=3+\underline{5}$

18. $x+9-\underline{9}=-2-\underline{9}$

19. Let x = number.
Then $10-x$ = other number.

20. $(x-5)$ inches

21. $180-(x+5)=180-x-5=175-x$ or $(175-x)°$

22.
$$\begin{aligned} \frac{3}{4}x&=-9\\ \frac{4}{3}\left(\frac{3}{4}x\right)&=\frac{4}{3}(-9)\\ x&=-12 \end{aligned}$$

23.
$$\begin{aligned} \frac{x}{6}&=\frac{2}{3}\\ 6\cdot\frac{x}{6}&=6\cdot\frac{2}{3}\\ x&=4 \end{aligned}$$

24.
$$\begin{aligned} -3x+1&=19\\ -3x+1-1&=19-1\\ -3x&=18\\ \frac{-3x}{-3}&=\frac{18}{-3}\\ x&=-6 \end{aligned}$$

25.
$$\begin{aligned} 5x+25&=20\\ 5x+25-25&=20-25\\ 5x&=-5\\ \frac{5x}{5}&=\frac{-5}{5}\\ x&=-1 \end{aligned}$$

26.
$$\begin{aligned} 5x+x&=9+4x-1+6\\ 6x&=4x+14\\ -4x+6x&=-4x+4x+14\\ 2x&=14\\ \frac{2x}{2}&=\frac{14}{2}\\ x&=7 \end{aligned}$$

27.
$$\begin{aligned} -y+4y&=7-y-3-8\\ 3y&=-y-4\\ y+3y&=y-y-4\\ 4y&=-4\\ \frac{4y}{4}&=-\frac{4}{4}\\ y&=-1 \end{aligned}$$

28. $x+(x+2)+(x+4)=3x+6$

29.
$$\begin{aligned} \frac{2}{7}x-\frac{5}{7}&=1\\ 2x-5&=7\\ 2x&=12\\ \frac{2x}{2}&=\frac{12}{2}\\ x&=6 \end{aligned}$$

30.
$$\frac{5}{3}x+4=\frac{2}{3}x$$
$$\frac{5}{3}x-\frac{5}{3}x+4=\frac{2}{3}x-\frac{5}{3}x$$
$$4=-\frac{3}{3}x$$
$$4=-x$$
$$\frac{4}{-1}=\frac{-x}{-1}$$
$$-4=x$$

31.
$$-(5x+1)=-7x+3$$
$$-5x-1=-7x+3$$
$$-5x+7x-1=-7x+7x+3$$
$$2x-1=3$$
$$2x-1+1=3+1$$
$$2x=4$$
$$\frac{2x}{2}=\frac{4}{2}$$
$$x=2$$

32.
$$-4(2x+1)=-5x+5$$
$$-8x-4=-5x+5$$
$$-8x+5x-4=-5x+5x+5$$
$$-3x-4=5$$
$$-3x-4+4=5+4$$
$$-3x=9$$
$$\frac{-3x}{-3}=\frac{9}{-3}$$
$$x=-3$$

33.
$$-6(2x-5)=-3(9+4x)$$
$$-12x+30=-27-12x$$
$$-12x+12x+30=-27-12x+12x$$
$$30=-27$$
No solution

34.
$$3(8y-1)=6(5+4y)$$
$$24y-3=30+24y$$
$$24y-24y-3=30+24y-24y$$
$$-3=30$$
No solution

35.
$$\frac{3(2-z)}{5}=z$$
$$5\cdot\frac{3(2-z)}{5}=5\cdot z$$
$$3(2-z)=5z$$
$$6-3z=5z$$
$$6-3z+3z=5z+3z$$
$$6=8z$$
$$\frac{6}{8}=\frac{8z}{8}$$
$$\frac{6}{8}=z$$
$$z=\frac{3}{4}$$

36.
$$\frac{4(n+2)}{5}=-n$$
$$5\left[\frac{4(n+2)}{5}\right]=5(-n)$$
$$4(n+2)=-5n$$
$$4n+8=-5n$$
$$4n-4n+8=-5n-4n$$
$$8=-9n$$
$$\frac{8}{-9}=\frac{-9n}{-9}$$
$$-\frac{8}{9}=n$$

37.
$$5(2n-3)-1=4(6+2n)$$
$$10n-15-1=24+8n$$
$$10n-16=24+8n$$
$$10n-16+16=24+16+8n$$
$$10n=40+8n$$
$$10n-8n=40+8n-8n$$
$$2n=40$$
$$\frac{2n}{2}=\frac{40}{2}$$
$$n=20$$

38.
$$\begin{aligned}-2(4y-3)+4&=3(5-y)\\-8y+6+4&=15-3y\\-8y+10&=15-3y\\-8y+3y+10&=15-3y+3y\\-5y+10&=15\\-5y+10-10&=15-10\\-5y&=5\\\frac{-5y}{-5}&=\frac{5}{-5}\\y&=-1\end{aligned}$$

39.
$$\begin{aligned}9z-z+1&=6(z-1)+7\\8z+1&=6z-6+7\\8z+1&=6z+1\\8z-6z+1&=6z-6z+1\\2z+1&=1\\2z+1-1&=1-1\\2z&=0\\\frac{2z}{2}&=\frac{0}{2}\\z&=0\end{aligned}$$

40.
$$\begin{aligned}5t-3-t&=3(t+4)-15\\4t-3&=3t+12-15\\4t-3&=3t-3\\4t-3t-3&=3t-3t-3\\t-3&=-3\\t-3+3&=-3+3\\t&=0\end{aligned}$$

41.
$$\begin{aligned}-n+10&=2(3n-5)\\-n+10&=6n-10\\-n+n+10&=6n+n-10\\10&=7n-10\\10+10&=7n-10+10\\20&=7n\\\frac{20}{7}&=\frac{7n}{7}\\\frac{20}{7}&=n\end{aligned}$$

42.
$$\begin{aligned}-9-5a&=3(6a-1)\\-9-5a&=18a-3\\-9-5a+5a&=18a+5a-3\\-9&=23a-3\\-9+3&=23a-3+3\\-6&=23a\\\frac{-6}{23}&=\frac{23a}{23}\\-\frac{6}{23}&=a\end{aligned}$$

43.
$$\begin{aligned}\frac{5(c+1)}{6}&=2c-3\\5(c+1)&=6(2c-3)\\5c+5&=12c-18\\5c-5c+5&=12c-5c-18\\5&=7c-18\\5+18&=7c-18+18\\23&=7c\\\frac{23}{7}&=\frac{7c}{7}\\\frac{23}{7}&=c\end{aligned}$$

44.
$$\begin{aligned}\frac{2(8-a)}{3}&=4-4a\\3\left[\frac{2(8-a)}{3}\right]&=3(4-4a)\\2(8-a)&=12-12a\\16-2a&=12-12a\\16-2a+12a&=12-12a+12a\\16+10a&=12\\16-16+10a&=12-16\\10a&=-4\\\frac{10a}{10}&=\frac{-4}{10}\\a&=-\frac{4}{10}=-\frac{2}{5}\end{aligned}$$

45.
$$\begin{aligned}
200(70x-3560) &= -179(150x-19{,}300)\\
14{,}000x-712{,}000 &= -26{,}850x+3{,}454{,}700\\
14{,}000x+26{,}850x-712{,}000 &= -26{,}850x+26{,}850x+3{,}454{,}700\\
40{,}850x-712{,}000 &= 3{,}454{,}700\\
40{,}850x-712{,}000+712{,}000 &= 3{,}454{,}700+712{,}000\\
40{,}850x &= 4{,}166{,}700\\
\frac{40{,}850x}{40{,}850} &= \frac{4{,}166{,}700}{40{,}850}\\
x &= 102
\end{aligned}$$

46.
$$\begin{aligned}
1.72y-0.04y &= 0.42\\
1.68y &= 0.42\\
\frac{1.68y}{1.68} &= \frac{0.42}{1.68}\\
y &= 0.25
\end{aligned}$$

47. Let x = number.
$$\begin{aligned}
\frac{x}{3} &= x-2\\
x &= 3(x-2)\\
x &= 3x-6\\
x-x &= 3x-x-6\\
0 &= 2x-6\\
0+6 &= 2x-6+6\\
6 &= 2x\\
\frac{6}{2} &= \frac{2x}{2}\\
3 &= x
\end{aligned}$$
The number is 3.

48. Let x = number.
$$\begin{aligned}
2(x+6) &= -x\\
2x+12 &= -x\\
2x-2x+12 &= -x-2x\\
12 &= -3x\\
\frac{12}{-3} &= \frac{-3x}{-3}\\
-4 &= x
\end{aligned}$$
The number is –4.

49. Let x = side of base.
Then $68 + 3x$ = height.
$$\begin{aligned}
68+3x+x &= 1380\\
68+4x &= 1380\\
68-68+4x &= 1380-68\\
4x &= 1312\\
\frac{4x}{4} &= \frac{1312}{4}\\
x &= 328
\end{aligned}$$
Thus, the height = 68 + 3(328) = 1052 feet.

50. If x = the length of the shorter piece, then $2x$ = the length of the longer piece.
$$\begin{aligned}
x+2x &= 12\\
3x &= 12\\
\frac{3x}{3} &= \frac{12}{3}\\
x &= 4
\end{aligned}$$
$2x = 8$
The lengths of the pieces are 4 feet and 8 feet.

51. Let x = smaller area code.
Then $34 + 3x$ = the larger area code.
$$\begin{aligned}
34+3x+x &= 1262\\
34+4x &= 1262\\
34-34+4x &= 1262-34\\
4x &= 1228\\
\frac{4x}{4} &= \frac{1228}{4}\\
x &= 307
\end{aligned}$$
$34 + 3x = 955$
The area codes are 307 and 955.

52. Let x = smallest integer. Then $(x + 2)$ and $(x + 4)$ are the other two integers.

$$x+(x+2)+(x+4)=-114$$
$$3x+6=-114$$
$$3x=-120$$
$$x=-40$$

$x+2=-38$
$x+4=-36$
The integers are –40, –38, and –36.

53.
$$P=2l+2w$$
$$46=2(14)+2w$$
$$46=28+2w$$
$$46-28=28-28+2w$$
$$18=2w$$
$$\frac{18}{2}=\frac{2w}{2}$$
$$9=w$$

54.
$$V=lwh$$
$$192=8(6)h$$
$$192=48h$$
$$\frac{192}{48}=\frac{48h}{48}$$
$$4=h$$

55.
$$y=mx+b$$
$$y-b=mx+b-b$$
$$y-b=mx$$
$$\frac{y-b}{x}=\frac{mx}{x}$$
$$\frac{y-b}{x}=m$$

56.
$$r=vst-9$$
$$r+9=vst-9+9$$
$$r+9=vst$$
$$\frac{r+9}{vt}=\frac{vst}{vt}$$
$$\frac{r+9}{vt}=s$$

57.
$$2y-5x=7$$
$$2y-5x+5x=7+5x$$
$$2y=7+5x$$
$$2y-7=7-7+5x$$
$$2y-7=5x$$
$$\frac{2y-7}{5}=\frac{5x}{5}$$
$$\frac{2y-7}{5}=x$$

58.
$$3x-6y=-2$$
$$3x-3x-6y=-2-3x$$
$$-6y=-2-3x$$
$$\frac{-6y}{-6}=\frac{-2-3x}{-6}$$
$$y=\frac{-2-3x}{-6}$$
$$y=\frac{-1(-2-3x)}{-1(-6)}$$
$$y=\frac{2+3x}{6}$$

59.
$$C=\pi D$$
$$\frac{C}{D}=\frac{\pi D}{D}$$
$$\frac{C}{D}=\pi$$

60.
$$C=2\pi r$$
$$\frac{C}{2r}=\frac{2\pi r}{2r}$$
$$\frac{C}{2r}=\pi$$

61. Use the formula for the volume of a rectangular box and substitute $V = 900$, $l = 20$, and $h = 3$. Solve for w.
$$V=lwh$$
$$900=(20)(w)(3)$$
$$900=60w$$
$$\frac{900}{60}=\frac{60w}{60}$$
$$15=w$$
The width of the pool is 15 meters

62. $C = \frac{5}{9}(F - 32)$

$C = \frac{5}{9}(104 - 32)$

$C = \frac{5}{9}(72)$

$C = 40$

104°F is the same as 40°C.

63. Use the distance formula and substitute $D = 10{,}000$ and $R = 125$. Solve for T.

$D = RT$

$10{,}000 = 125T$

$\frac{10{,}000}{125} = \frac{125T}{125}$

$80 = T$

It will take 80 minutes, or 1 hour 20 minutes, to finish the race.

64. $0.12(250) = 30$

65. $1.10(85) = 93.5$

66. Let x = percent.

$9 = x(45)$

$\frac{9}{45} = \frac{45x}{45}$

$0.2 = x$

20%

67. Let x = percent.

$59.5 = x(85)$

$\frac{59.5}{85} = \frac{85x}{85}$

$0.7 = x$

70%

68. Let x = unknown number.

$137.5 = 1.25x$

$\frac{137.5}{1.25} = \frac{1.25x}{1.25}$

$110 = x$

69. Let x = unknown number.

$768 = 0.6x$

$\frac{768}{0.6} = \frac{0.6x}{0.6}$

$1280 = x$

70. We would expect 0.126(50,000) = 6300 homes to be phoneless.

71. 6%; the height of the 'Nap' bar is at 6%.

72. Eat from the Minibar; because this is the tallest bar.

73. We would expect 0.40(300) = 120 travelers to relax by watching TV.

74. No; some business travelers may have chosen more than one category.

75. $\frac{210 - 180}{210} = \frac{30}{210} = 0.143$

The number of employees decreased 14.3%.

76. $3(x - 5) > -(x + 3)$

$3x - 15 > -x - 3$

$4x - 15 > -3$

$4x > 12$

$x > 3$

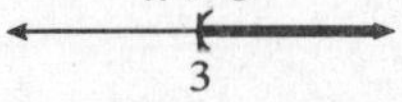

$(3, \infty)$

77. $-2(x + 7) \geq 3(x + 2)$

$-2x - 14 \geq 3x + 6$

$-14 \geq 5x + 6$

$-20 \geq 5x$

$-4 \geq x$

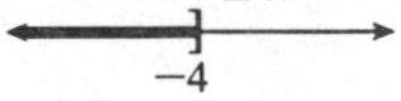

$(-\infty, -4]$

78. $4x - (5 + 2x) < 3x - 1$

$4x - 5 - 2x < 3x - 1$

$2x - 5 < 3x - 1$

$-5 < x - 1$

$-4 < x$

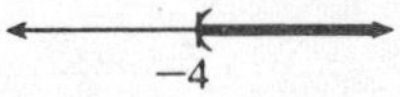

$(-4, \infty)$

79. $3(x-8)<7x+2(5-x)$
$3x-24<7x+10-2x$
$3x-24<5x+10$
$-24<2x+10$
$-34<2x$
$-17<x$

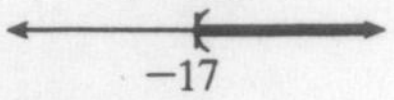

$(-17, \infty)$

80. $24 \geq 6x-2(3x-5)+2x$
$24 \geq 6x-6x+10+2x$
$24 \geq 10+2x$
$14 \geq 2x$
$7 \geq x$

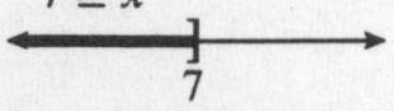

$(-\infty, 7]$

81. $48+x \geq 5(2x+4)-2x$
$48+x \geq 10x+20-2x$
$48+x \geq 8x+20$
$28+x \geq 8x$
$28 \geq 7x$
$4 \geq x$

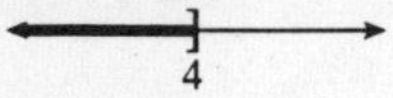

$(-\infty, 4]$

82. $\frac{x}{3}+\frac{1}{2}>\frac{2}{3}$
$2x+3>4$
$2x>1$
$x>\frac{1}{2}$

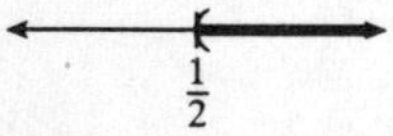

$\left(\frac{1}{2}, \infty\right)$

83. $x+\frac{3}{4}<\frac{-x}{2}+\frac{9}{4}$
$4x+3<-2x+9$
$6x+3<9$
$6x<6$
$x<1$

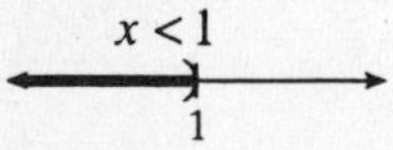

$(-\infty, 1)$

84. $\frac{x-5}{2} \leq \frac{3}{8}(2x+6)$
$4(x-5) \leq 3(2x+6)$
$4x-20 \leq 6x+18$
$4x \leq 6x+38$
$-2x \leq 38$
$x \geq -19$

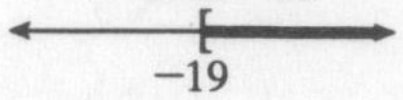

$[-19, \infty)$

85. $\frac{3(x-2)}{5}>\frac{-5(x-2)}{3}$
$3(x-2)>\frac{-25(x-2)}{3}$
$9(x-2)>-25(x-2)$
$9x-18>-25x+50$
$34x-18>50$
$34x>68$
$x>2$

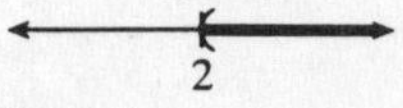

$(2, \infty)$

86. Let x = Tina's sales.
$175+0.05x \geq 300$
$17{,}500+5x \geq 30{,}000$
$5x \geq 12{,}500$
$\frac{5x}{5} \geq \frac{12{,}500}{5}$
$x \geq 2500$
Her minimum sales must be \$2500.

87. Let x = score on next round.
$\frac{76+82+79+x}{4}<80$
$4\left[\frac{76+82+79+x}{4}\right]<4(80)$
$76+82+79+x<320$
$237+x<320$
$237-237+x<320-237$
$x<83$
Her next score must be less than 83.

Chapter 2 Test

1. $2y - 6 - y - 4 = y - 10$

2. $2.7x + 6.1 + 3.2x - 4.9 = 5.9x + 1.2$

3. $$\begin{aligned} 4(x-2) - 3(2x-6) &= 4x - 8 - 6x + 18 \\ &= -2x + 10 \end{aligned}$$

4. $$\begin{aligned} -5(y+1) + 2(3-5y) &= -5y - 5 + 6 - 10y \\ &= -15y + 1 \end{aligned}$$

5. $$\begin{aligned} -\frac{4}{5}x &= 4 \\ -\frac{5}{4}\left(-\frac{4}{5}x\right) &= -\frac{5}{4}(4) \\ x &= -5 \end{aligned}$$

6. $$\begin{aligned} 4(n-5) &= -(4-2n) \\ 4n - 20 &= -4 + 2n \\ 4n - 2n - 20 &= -4 + 2n - 2n \\ 2n - 20 &= -4 \\ 2n - 20 + 20 &= -4 + 20 \\ 2n &= 16 \\ \frac{2n}{2} &= \frac{16}{2} \\ n &= 8 \end{aligned}$$

7. $$\begin{aligned} 5y - 7 + y &= -(y + 3y) \\ 5y - 7 + y &= -y - 3y \\ 6y - 7 &= -4y \\ 6y - 6y - 7 &= -4y - 6y \\ -7 &= -10y \\ \frac{-7}{-10} &= \frac{-10y}{-10} \\ \frac{-7}{-10} &= y \\ \frac{7}{10} &= y \end{aligned}$$

8. $$\begin{aligned} 4z + 1 - z &= 1 + z \\ 3z + 1 &= 1 + z \\ 3z - z + 1 &= 1 + z - z \\ 2z + 1 &= 1 \\ 2z + 1 - 1 &= 1 - 1 \\ 2z &= 0 \\ \frac{2z}{2} &= \frac{0}{2} \\ z &= 0 \end{aligned}$$

9. $$\begin{aligned} \frac{2(x+6)}{3} &= x - 5 \\ 3\left[\frac{2(x+6)}{3}\right] &= 3(x-5) \\ 2(x+6) &= 3(x-5) \\ 2x + 12 &= 3x - 15 \\ 2x - 2x + 12 &= 3x - 2x - 15 \\ 12 &= x - 15 \\ 12 + 15 &= x - 15 + 15 \\ 27 &= x \end{aligned}$$

10. $$\begin{aligned} \frac{4(y-1)}{5} &= 2y + 3 \\ 5\left[\frac{4(y-1)}{5}\right] &= 5(2y+3) \\ 4(y-1) &= 5(2y+3) \\ 4y - 4 &= 10y + 15 \\ 4y - 10y - 4 &= 10y - 10y + 15 \\ -6y - 4 &= 15 \\ -6y - 4 + 4 &= 15 + 4 \\ -6y &= 19 \\ \frac{-6y}{-6} &= \frac{19}{-6} \\ y &= -\frac{19}{6} \end{aligned}$$

11. $\frac{1}{2} - x + \frac{3}{2} = x - 4$

$$-x + \frac{4}{2} = x - 4$$
$$-x + 2 = x - 4$$
$$-x + x + 2 = x + x - 4$$
$$2 = 2x - 4$$
$$2 + 4 = 2x - 4 + 4$$
$$6 = 2x$$
$$\frac{6}{2} = \frac{2x}{2}$$
$$3 = x$$

12. $\frac{1}{3}(y + 3) = 4y$

$$3\left[\frac{1}{3}(y+3)\right] = 3(4y)$$
$$y + 3 = 12y$$
$$y - y + 3 = 12y - y$$
$$3 = 11y$$
$$\frac{3}{11} = \frac{11y}{11}$$
$$\frac{3}{11} = y$$

13. $-0.3(x - 4) + x = 0.5(3 - x)$

$$-3(x - 4) + 10x = 5(3 - x)$$
$$-3x + 12 + 10x = 15 - 5x$$
$$12 + 7x = 15 - 5x$$
$$12 + 7x + 5x = 15 - 5x + 5x$$
$$12 + 12x = 15$$
$$12 - 12 + 12x = 15 - 12$$
$$12x = 3$$
$$\frac{12x}{12} = \frac{3}{12}$$
$$x = 0.25$$

14. $-4(a + 1) - 3a = -7(2a - 3)$

$$-4a - 4 - 3a = -14a + 21$$
$$-7a - 4 = -14a + 21$$
$$-7a + 14a - 4 = -14a + 14a + 21$$
$$7a - 4 = 21$$
$$7a - 4 + 4 = 21 + 4$$
$$7a = 25$$
$$\frac{7a}{7} = \frac{25}{7}$$
$$a = \frac{25}{7}$$

15. Let x = number.

$$x + \frac{2}{3}x = 35$$
$$\frac{3}{3}x + \frac{2}{3}x = 35$$
$$\frac{5}{3}x = 35$$
$$\frac{3}{5}\left(\frac{5}{3}x\right) = \frac{3}{5}(35)$$
$$x = 21$$

16. Area of deck

$A = lw$

$A = (20 \text{ ft})(35 \text{ ft})$

$A = 700$ sq ft

Two coats are needed.

Twice the Area = 2(700 sq ft)

Twice the Area = 1400 sq ft

$$\frac{1400 \text{ sq ft}}{200 \text{ sq ft / gal}}$$

7 gallons

17.

Principal · rate · time = interest				
Amoxil	x	0.10	1	$0.10x$
IBM	$2x$	0.12	1	$0.24x$

$$0.10x + 0.24x = 2890$$
$$0.34x = 2890$$
$$x = 8500$$

$2x = 17{,}000$

Amoxil: $8500

IBM: $17,000

18.

	rate · time = distance		
1st train	50	x	$50x$
2nd train	64	x	$64x$

$$50x + 64x = 285$$
$$114x = 285$$
$$\frac{114x}{114} = \frac{285}{114}$$
$$x = 2\frac{1}{2} \text{ hours}$$

19.
$$y = mx + b$$
$$-14 = -2x - 2$$
$$-12 = -2x$$
$$6 = x$$

20.
$$V = \pi r^2 h$$
$$\frac{V}{\pi r^2} = \frac{\pi r^2 h}{\pi r^2}$$
$$\frac{V}{\pi r^2} = h$$

21.
$$3x - 4y = 10$$
$$3x - 3x - 4y = 10 - 3x$$
$$-4y = 10 - 3x$$
$$\frac{-4y}{-4} = \frac{10 - 3x}{-4}$$
$$y = \frac{10 - 3x}{-4}$$
$$y = \frac{-1(10 - 3x)}{-1(-4)}$$
$$y = \frac{-10 + 3x}{4}$$
$$y = \frac{3x - 10}{4}$$

22.
$$3x - 5 > 7x + 3$$
$$3x - 7x - 5 > 7x - 7x + 3$$
$$-4x - 5 > 3$$
$$-4x - 5 + 5 > 3 + 5$$
$$-4x > 8$$
$$\frac{-4x}{-4} < \frac{8}{-4}$$
$$x < -2$$
$(-\infty, -2)$

−2

23.
$$x + 6 > 4x - 6$$
$$x - 4x + 6 > 4x - 4x - 6$$
$$-3x + 6 > -6$$
$$-3x + 6 - 6 > -6 - 6$$
$$-3x > -12$$
$$\frac{-3x}{-3} < \frac{-12}{-3}$$
$$x < 4$$
$(-\infty, 4)$

4

24.
$$\frac{2(5x+1)}{3} > 2$$
$$3\left[\frac{2(5x+1)}{3}\right] > 3(2)$$
$$2(5x+1) > 3(2)$$
$$10x + 2 > 6$$
$$10x + 2 - 2 > 6 - 2$$
$$10x > 4$$
$$\frac{10x}{10} > \frac{4}{10}$$
$$x > \frac{4}{10}$$
$$x > \frac{2}{5}$$

$\left(\frac{2}{5}, \infty\right)$

$\frac{2}{5}$

25.
$$\frac{x}{5}-\frac{3}{10}\le\frac{1}{2}$$
$$10\left(\frac{x}{5}-\frac{3}{10}\right)\le 10\left(\frac{1}{2}\right)$$
$$2x-3\le 5$$
$$2x-3+3\le 5+3$$
$$2x\le 8$$
$$\frac{2x}{2}\le\frac{8}{2}$$
$$x\le 4$$

4

$(-\infty, 4]$

26. 81.3%

27. $0.047(126.2) = \$5.9314$ billion

28. $0.067(360) = 24.12°$

29. 17%; the bar for 1997, e-mail extends to 17%.

30. $37\% - 24\% = 13\%$; the lengths of the bars for computers for 1997 and 1994 are 37% and 24% respectively.

31.
$$\begin{aligned}26\% \text{ of } 23{,}000 &= 0.26(23{,}000)\\ &= 5980\end{aligned}$$

Chapter 3

Section 3.1

Mental Math

1. Point A is (5, 2).
2. Point B is (2, 5).
3. Point C is (3, –1).
4. Point D is (–1, 3).
5. Point E is (–5, –2).
6. Point F is (–3, 5).
7. Point G is (–1, 0).
8. Point H is (0, –3).

Exercise Set 3.1

1. quadrant I

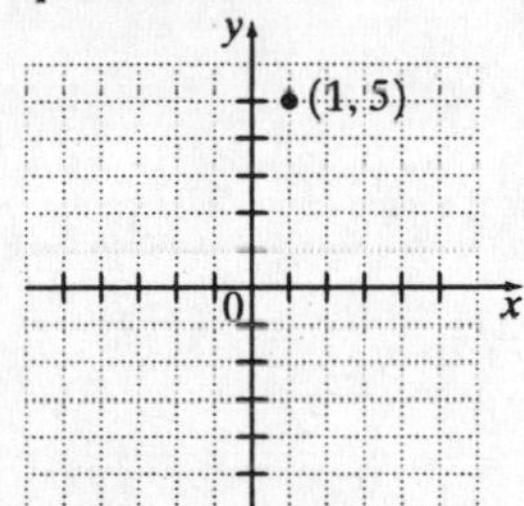

3. no quadrant, x-axis

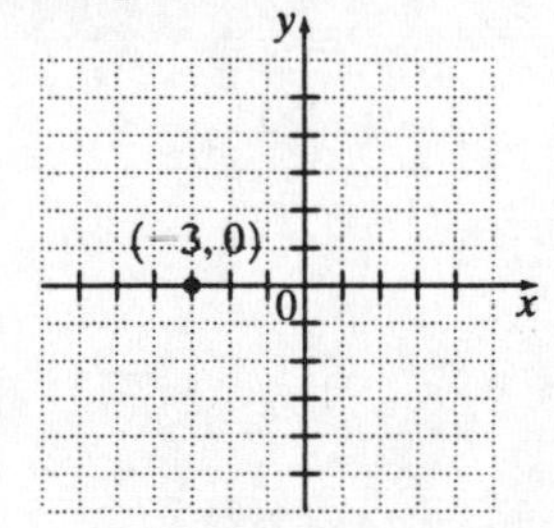

5. quadrant IV

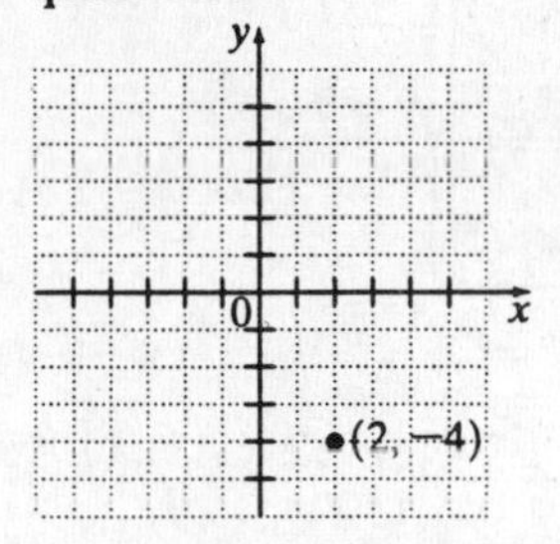

7. no quadrant, x-axis

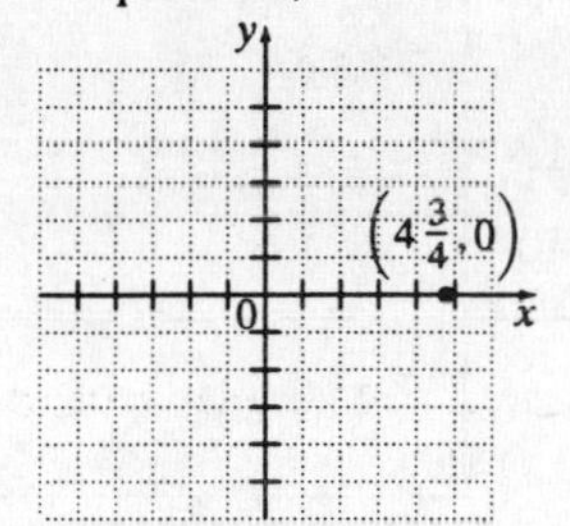

9. no quadrant, origin

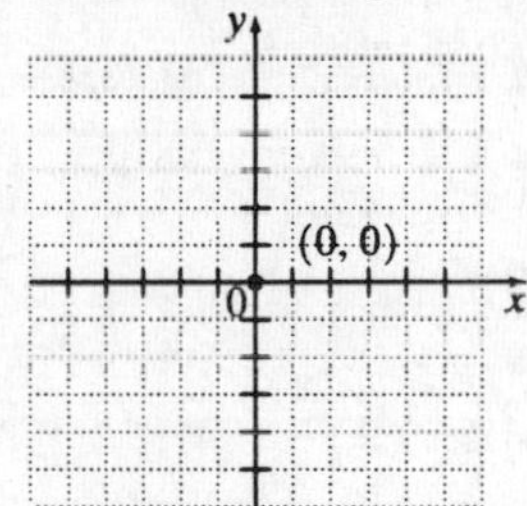

11. no quadrant, y-axis

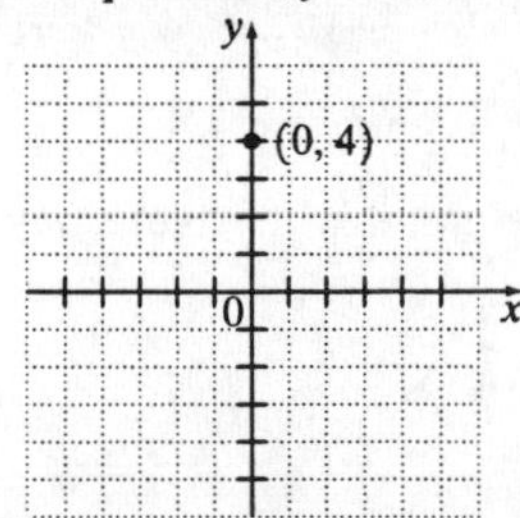

13. A: (0, 0); B: $\left(3\frac{1}{2},\ 0\right)$; C: (3, 2); D: (–1, 3); E: (–2, –2); F: (0, –1); G: (2, –1)

15. $2(4) + 2(9) = 26$ units

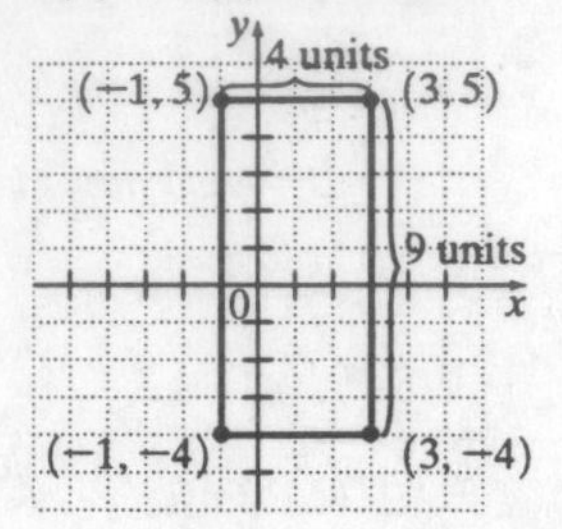

17. a. (1991, 1.14), (1992, 1.13), (1993, 1.11), (1994, 1.11), (1995, 1.15), (1996, 1.23), (1997, 1.23), (1998, 1.06), (1999, 1.17)

b.

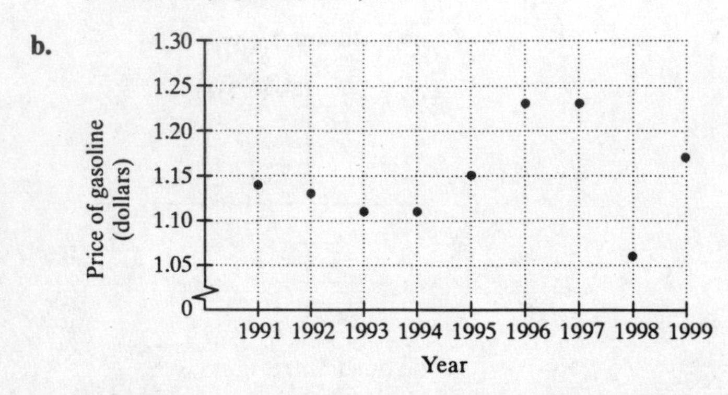

19. a. (1994, 578), (1995, 613), (1996, 654), (1997, 675), (1998, 717)

b.

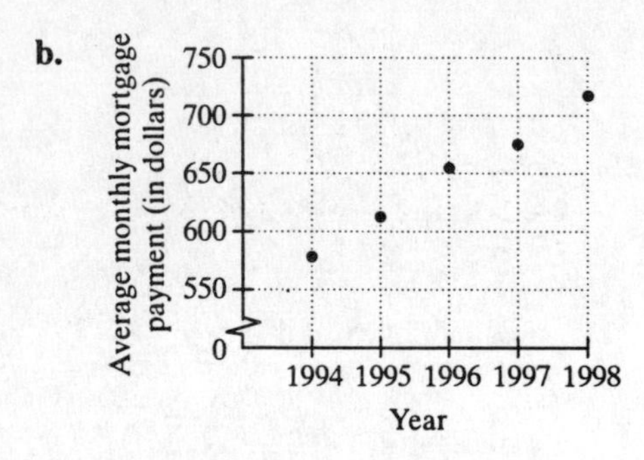

c. Average monthly mortgage payment increases each year.

21. $2x + y = 7$

(3, 1) $2(3) + 1 = 7$

$6 + 1 = 7$

yes $7 = 7$ True

$2x + y = 7$

(7, 0) $2(7) + 0 = 7$

$14 + 0 = 7$

no $14 = 7$ False

$2x + y = 7$

(0, 7) $2(0) + 7 = 7$

$0 + 7 = 7$

yes $7 = 7$ True

23. $y = -5x$

(−1, −5) $-5 = -5(-1)$

no $-5 = 5$ False

$y = -5x$

(0, 0) $0 = -5(0)$

yes $0 = 0$ True

$y = -5x$

(2, −10) $-10 = -5(2)$

yes $-10 = -10$ True

25. $x = 5$

(4, 5) no $4 = 5$ False

(5, 4) yes $5 = 5$ True

(5, 0) yes $5 = 5$ True

27. $x + 2y = 9$

(5, 2) $5 + 2(2) = 9$

$5 + 4 = 9$

yes $9 = 9$ True

$x + 2y = 9$

(0, 9) $0 + 2(9) = 9$

no $18 = 9$ False

29. $2x - y = 11$

(3, −4) $2(3) - (-4) = 11$

$6 + 4 = 11$

no $10 = 11$ False

$2x - y = 11$

(9, 8) $2(9) - 8 = 11$

$18 - 8 = 11$

no $10 = 11$ False

31. $x = \frac{1}{3}y$

(0, 0) $0 = \frac{1}{3}(0)$

yes $0 = 0$ True

$x = \frac{1}{3}y$

(3, 9) $3 = \frac{1}{3}(9)$

yes $3 = 3$ True

33. $y = -2$

(−2, −2) yes $-2 = -2$ True

$y = -2$

(5, −2) yes $-2 = -2$ True

35. $x - 4y = 4$; (, −2)

Let $y = -2$ and solve for x.

$x - 4(-2) = 4$

$x + 8 = 4$

$x = -4$

(−4, −2)

$x - 4y = 4$; (4,)

Let $x = 4$ and solve for y.

$4 - 4y = 4$

$-4y = 0$

$y = 0$

(4, 0)

37. $3x + y = 9; (0,\ \ )$
Let $x = 0$ and solve for y.
$3(0) + y = 9$
$0 + y = 9$
$y = 9$
$(0, 9)$

$3x + y = 9; (\ \ , 0)$
Let $y = 0$ and solve for x.
$3x + 0 = 9$
$3x = 9$
$x = 3$
$(3, 0)$

39. $y = -7; (11,\ \ )$
When $x = 11$, $y = -7$.
$(11, -7)$

$y = -7; (\ \ , -7)$
Let $y = -7$ in the equation.
$-7 = -7$
This is an identity; it is true for all x.
$(2, -7)$ is an example.

41. $x + 3y = 6$
First row:
Let $x = 0$ and solve for y.
$0 + 3y = 6$
$3y = 6$
$y = 2$
Put 2 in the first row of the table under y.
Second row:
Let $y = 0$ and solve for x.
$x + 3(0) = 6$
$x + 0 = 6$
$x = 6$
Put 6 in the second row of the table under x.
Third row:
Let $y = 1$ and solve for x.
$x + 3(1) = 6$
$x + 3 = 6$
$x = 3$
Put 3 in the third row of the table under x.

x	y
0	2
6	0
3	1

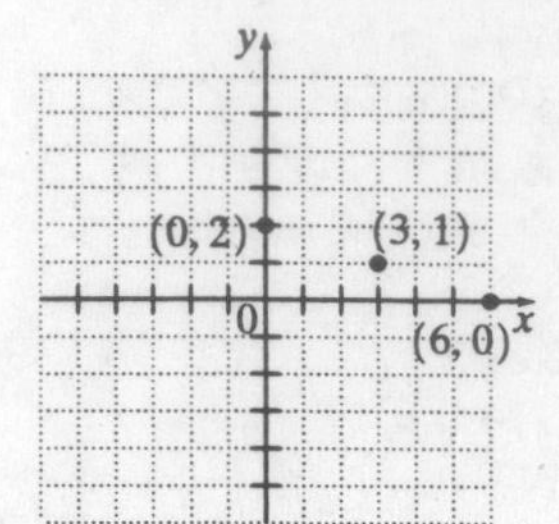

43. $2x - y = 12$
First row:
Let $x = 0$ and solve for y.
$2(0) - y = 12$
$0 - y = 12$
$-y = 12$
$y = -12$
Put -12 in the first row of the table under y.
Second row:
Let $y = -2$ and solve for x.
$2x - (-2) = 12$
$2x + 2 = 12$
$2x = 10$
$x = 5$
Put 5 in the second row of the table under x.
Third row:
Let $x = -3$ and solve for y.
$2(-3) - y = 12$
$-6 - y = 12$
$-y = 18$
$y = -18$
Put -18 in the third row of the table under y.

x	y
0	-12
5	-2
-3	-18

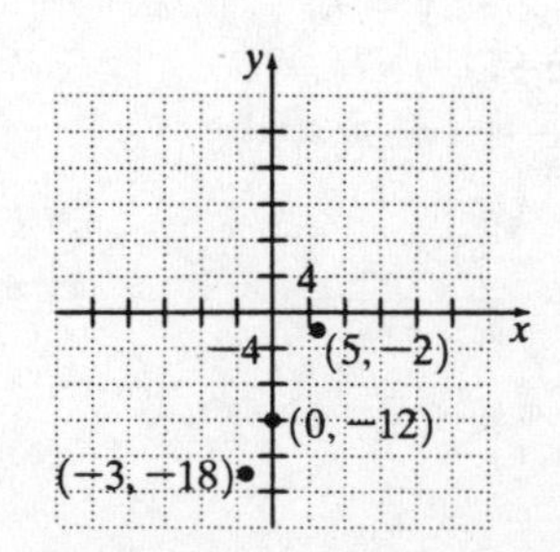

45. $2x + 7y = 5$
First row:
Let $x = 0$ and solve for y.

$$2(0) + 7y = 5$$
$$0 + 7y = 5$$
$$7y = 5$$
$$y = \frac{5}{7}$$

Put $\frac{5}{7}$ in the first row of the table under y.
Second row:
Let $y = 0$ and solve for x.

$$2x + 7(0) = 5$$
$$2x + 0 = 5$$
$$2x = 5$$
$$x = \frac{5}{2}$$

Put $\frac{5}{2}$ in the second row of the table under x.
Third row:
Let $y = 1$ and solve for x.

$$2x + 7(1) = 5$$
$$2x + 7 = 5$$
$$2x = -2$$
$$x = -1$$

Put −1 in the third row of the table under x.

x	y
0	$\frac{5}{7}$
$\frac{5}{2}$	0
−1	1

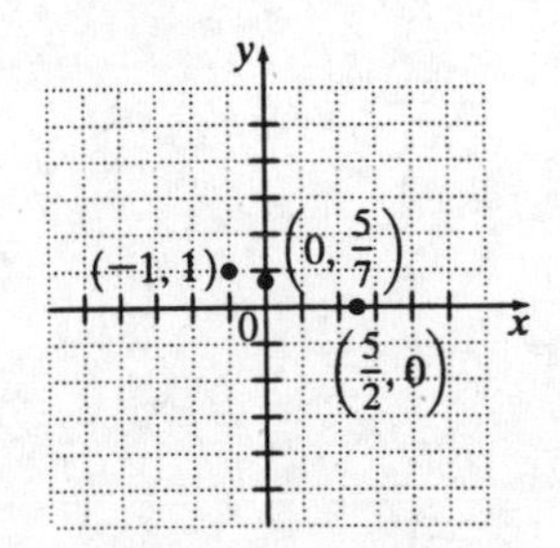

47. $x = 3$
Regardless of the value of y, the value of x is 3.

x	y
3	0
3	−0.5
3	$\frac{1}{4}$

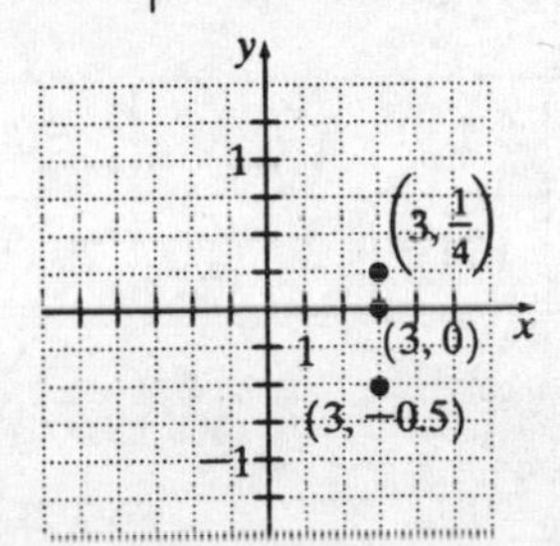

49. $x = -5y$
First row:
Let $y = 0$ and solve for x.
$x = 5(0)$
$x = 0$
Put 0 in the first row of the table under x.
Second row:
Let $y = 1$ and solve for x.
$x = -5(1)$
$x = -5$
Put −5 in the second row of the table under x.
Third row:
Let $x = 10$ and solve for y.
$10 = -5y$
$y = -2$
Put −2 in the third row of the table under y.

x	y
0	0
−5	1
10	−2

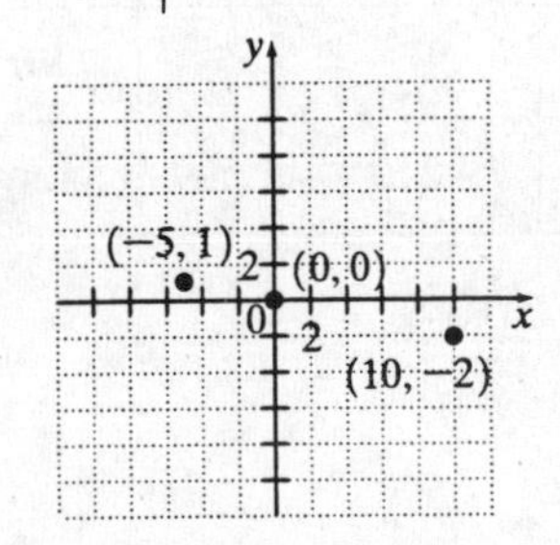

51. Answers may vary.

53. a. To complete the table, substitute each value of x into $y = 80x + 5000$ and solve for y.

$$y = 80(100) + 5000 = 13{,}000$$
$$y = 80(200) + 5000 = 21{,}000$$
$$y = 80(300) + 5000 = 29{,}000$$

x	100	200	300
y	13,000	21,000	29,000

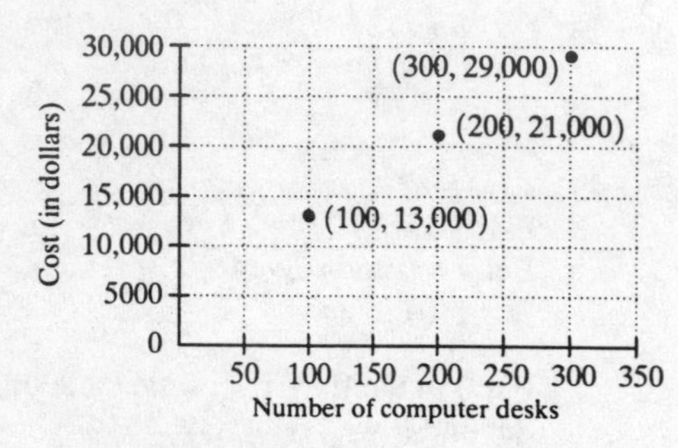

b. Let $y = 8600$ in the equation and solve for x.

$$y = 80x + 5000$$
$$8600 = 80x + 5000$$
$$3600 = 80x$$
$$45 = x$$

45 desks can be produced for \$8600.

55. a. To complete the table, substitute each value of x into $y = 0.364x + 21.939$ and solve for y.

$$y = 0.364(20) + 21.939 = 29.219$$
$$y = 0.364(65) + 21.939 = 45.599$$
$$y = 0.364(90) + 21.939 = 54.699$$

x	20	65	90
y	29.219	45.599	54.699

b. Let $y = 50$ in the equation and solve for x.

$$50 = 0.364x + 21.939$$
$$50 - 21.939 = 0.364x + 21.939 - 21.939$$
$$28.061 = 0.364x$$
$$\frac{28.061}{0.364} = \frac{0.364x}{0.364}$$
$$77 \approx x$$

Since x represents the number of years after 1900, the year is $1900 + 77 = 1977$. In 1977 the population density was approximately 50 people per square mile.

57. (5,670)
5 corresponds to the year 1995.
670 corresponds to the number of Target stores.
In 1995, there were 670 Target stores.

59. Year 6: $736 - 670 = 66$ stores
Year 7: $796 - 736 = 60$ stores
Year 8: $851 - 796 = 55$ stores

61. The graphs of (a, b) and (b, a) are the same when $a = b$.

63. $(+, -)$; quadrant IV

65. $(-, y)$; quadrant II or III

67.
$$x + y = 5$$
$$x - x + y = 5 - x$$
$$y = 5 - x$$

69.
$$2x + 4y = 5$$
$$2x - 2x + 4y = 5 - 2x$$
$$4y = 5 - 2x$$
$$\frac{4y}{4} = \frac{5 - 2x}{4}$$
$$y = \frac{5}{4} - \frac{1}{2}x$$
$$y = -\frac{1}{2}x + \frac{5}{4}$$

71.
$$10x = -5y$$
$$\frac{10x}{-5} = \frac{-5y}{-5}$$
$$-2x = y$$
$$y = -2x$$

73.
$$x - 3y = 6$$
$$x - 3y - x = 6 - x$$
$$-3y = 6 - x$$
$$\frac{-3y}{-3} = \frac{6-x}{-3}$$
$$y = -2 + \frac{1}{3}x$$
$$y = \frac{1}{3}x - 2$$

Section 3.2

Graphing Calculator Explorations

1.

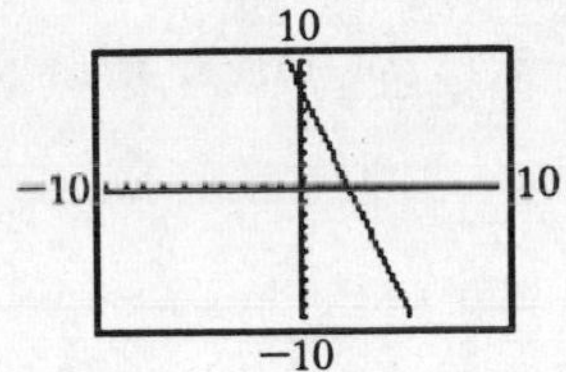

3.

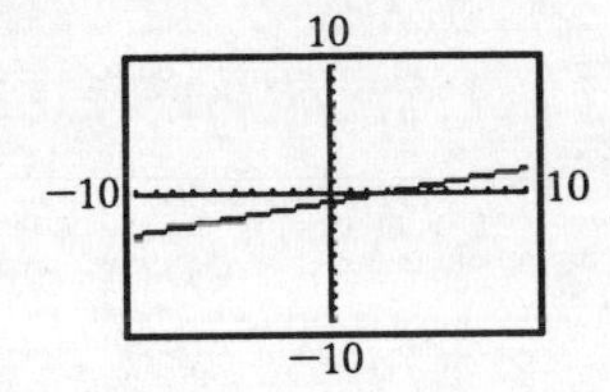

5.

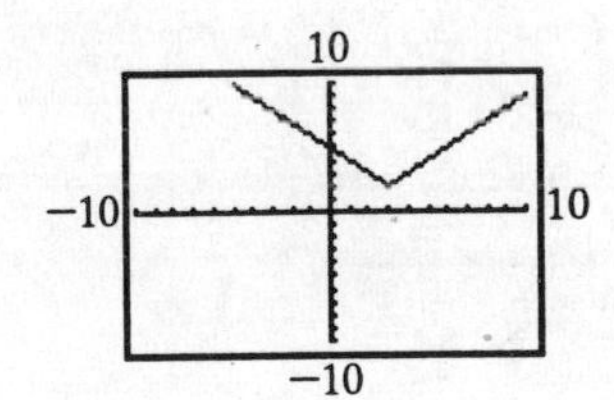

7.

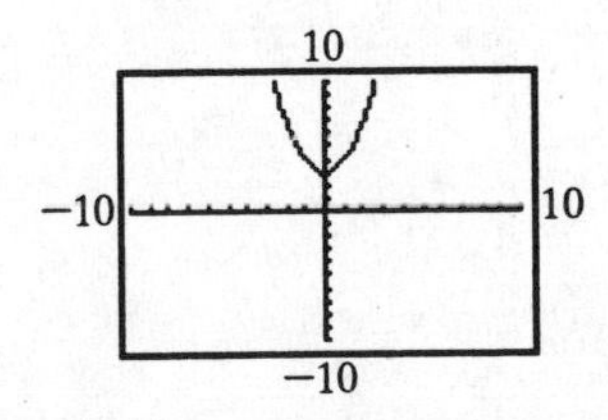

Exercise Set 3.2

1. Yes; it can be written in the form $Ax + By = C$.
$$-x = 3y + 10$$
$$-x - 3y = 10$$
$$(-1)x + (-3)y = 10$$

3. Yes; it can be written in the form $Ax + By = C$.
$$x = y$$
$$x - y = 0$$
$$x + (-1)y = 0$$

5. No; x is squared.

7. Yes; it can be written in the form $Ax + By = C$.
$$y = -1$$
$$0x + y = -1$$

9. $x + y = 4$
Find 3 points.

x	y
1	3
0	4
−2	6

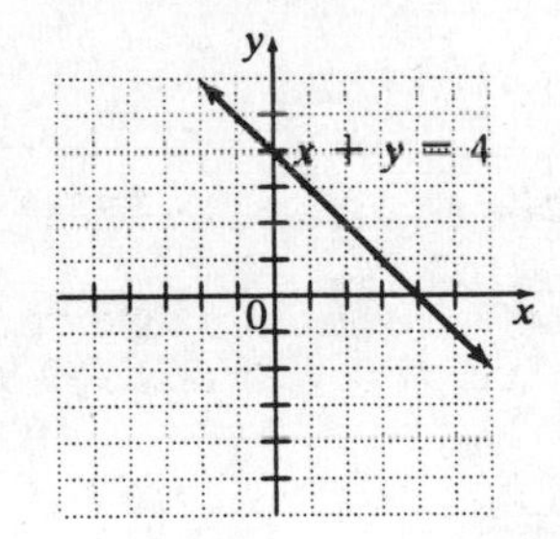

11. $x - y = -2$
Find 3 points.

x	y
0	2
–2	0
1	3

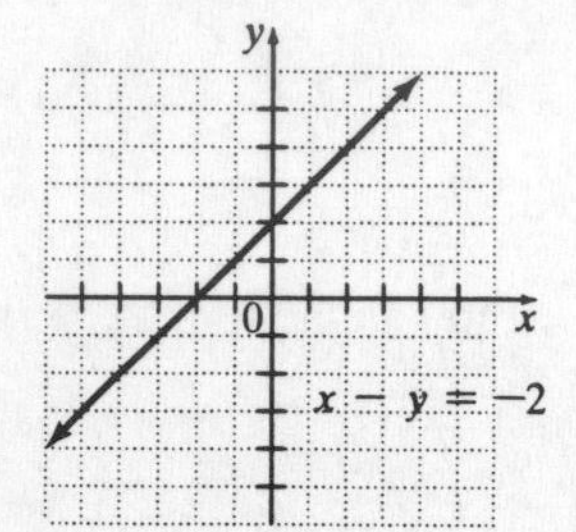

13. $x - 2y = 6$
Find 3 points.

x	y
0	–3
6	0
4	–1

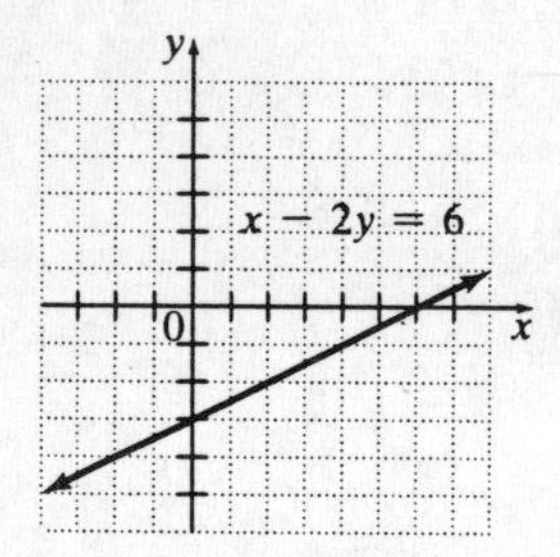

15. $y = 6x + 3$
Find 3 points.

x	y
0	3
–1	–3
1	9

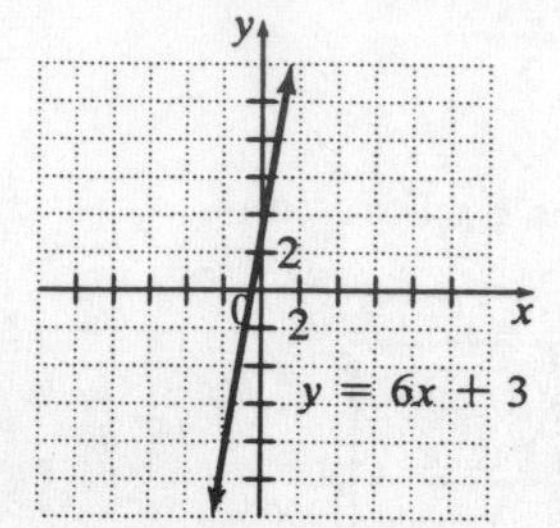

17. $x - 2y = -6$
Find 3 points.

x	y
0	3
–6	0
–4	1

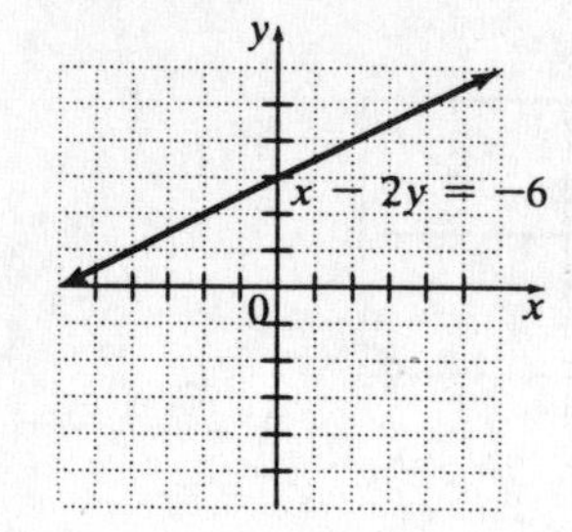

19. $y = 6x$
Find 3 points.

x	y
0	0
1	6
–1	–6

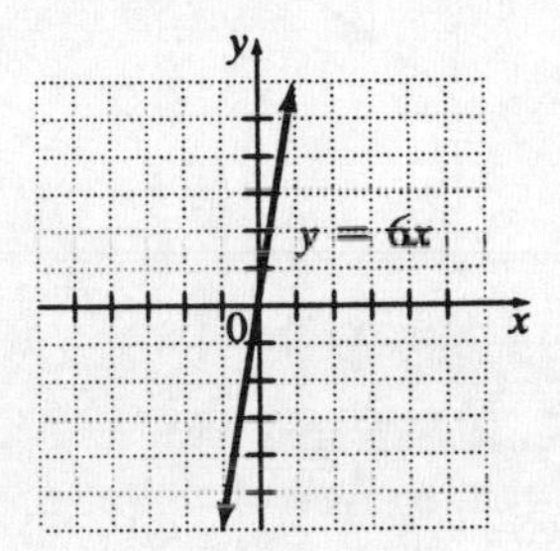

21. Find 3 points for each line.
$y = 5x$

x	y
0	0
1	5
–1	–5

$y = 5x + 4$

x	y
0	4
–1	–1
–2	–6

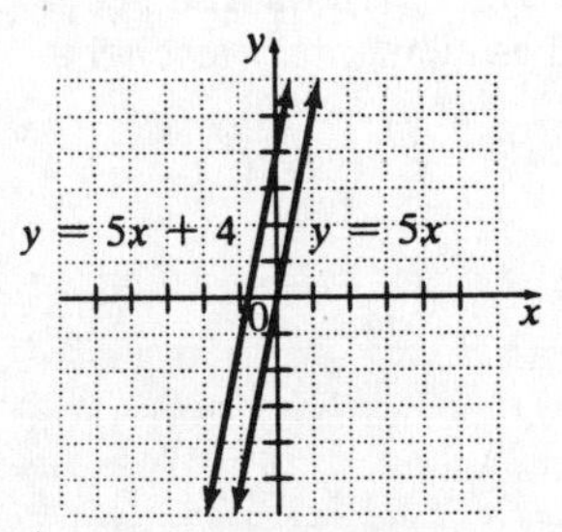

Answers may vary.

23. Find 3 points for each line.
$y = -2x$

x	y
2	–4
0	0
–2	4

$y = -2x - 3$

x	y
–2	1
–1	–1
0	–3

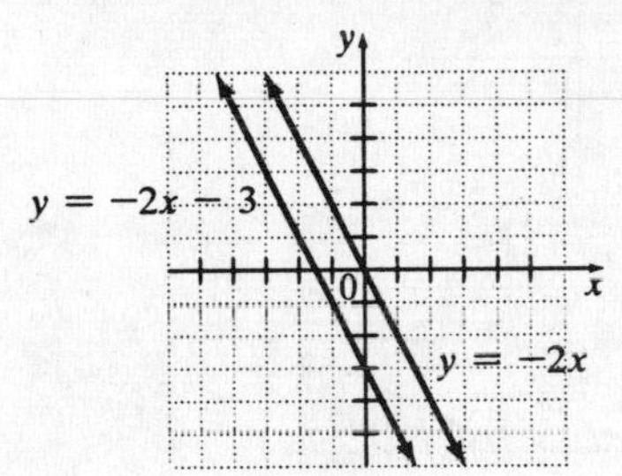

Answers may vary.

25. Find three points for each line.

$y = \frac{1}{2}x$

x	y
0	0
–2	–1
2	1

$y = \frac{1}{2}x + 2$

x	y
0	2
2	3
–2	1

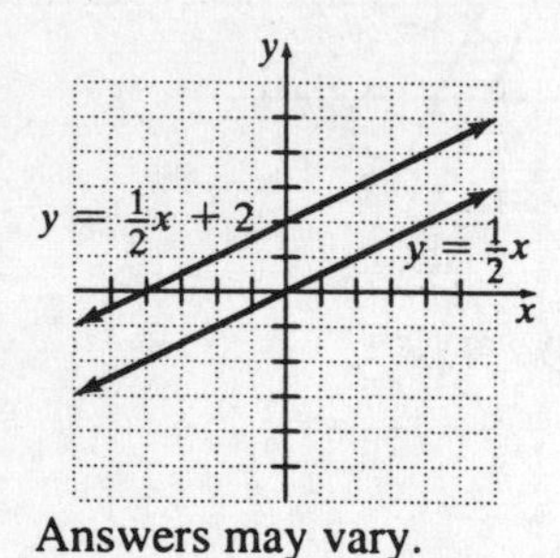

Answers may vary.

27. The graph of $y = 5x + 5$ crosses the y-axis at 5. Thus, its graph is c.

29. The graph of $y = 5x - 1$ crosses the y-axis at –1. Thus, its graph is d.

31. $y = 5x + 15$

x	y
0	15
4	35
10	65

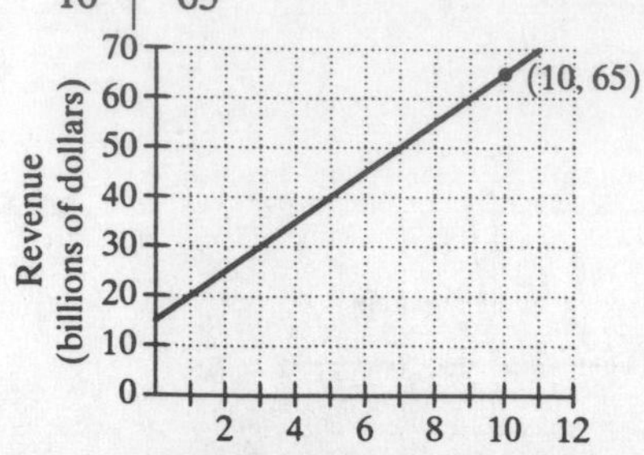

In 2006 ($x = 10$), the revenue will be 65 billion dollars.

33. $y = 20x + 1539$

x	y
0	1539
5	1639
8	1699

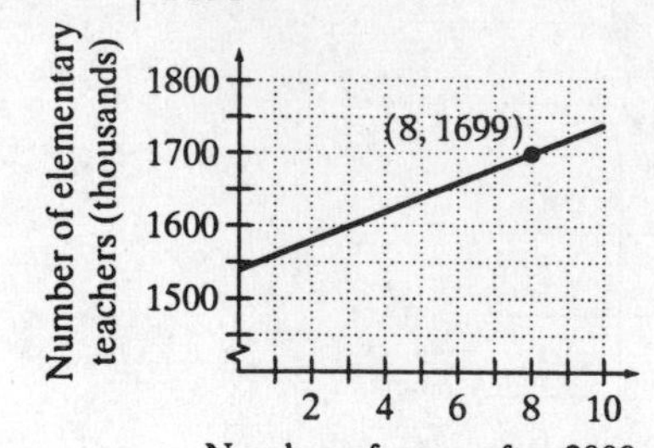

In 2008 ($x = 8$), 1699 thousand people will be employed as elementary teachers.

35. Linear; it is written in the form $Ax + By = C$.
$x + y = 3$
Find 3 points.

x	y
0	3
1	2
3	0

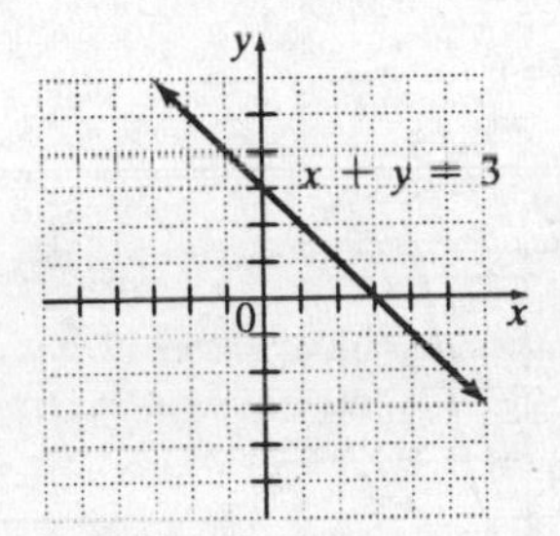

37. Linear; it is written in the form $y = mx + b$.
$y = 4x$
Find 3 points.

x	y
0	0
1	4
−1	−4

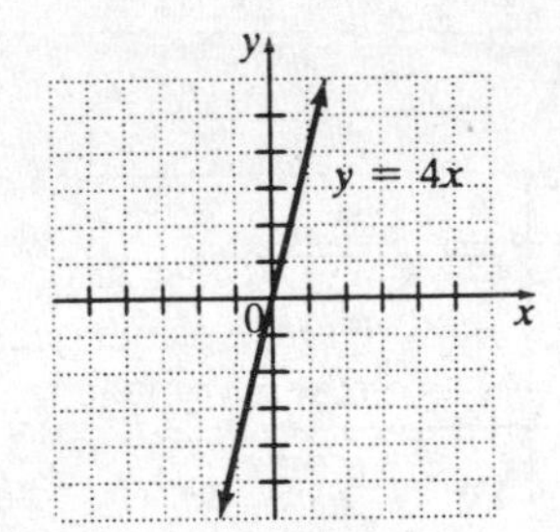

39. Linear; it is written in the form $y = mx + b$.
$y = 4x - 2$
Find 3 points.

x	y
0	−2
1	2
−1	−6

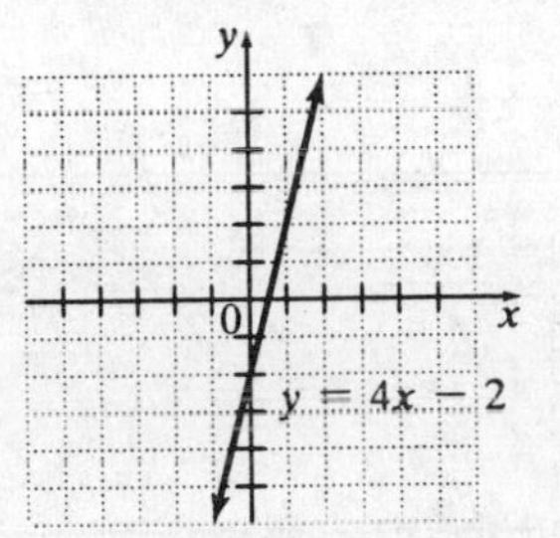

41. Not linear; x is inside absolute value.
$y = |x| + 3$
Find several points.

x	y
−2	5
−1	4
0	3
1	4
2	5

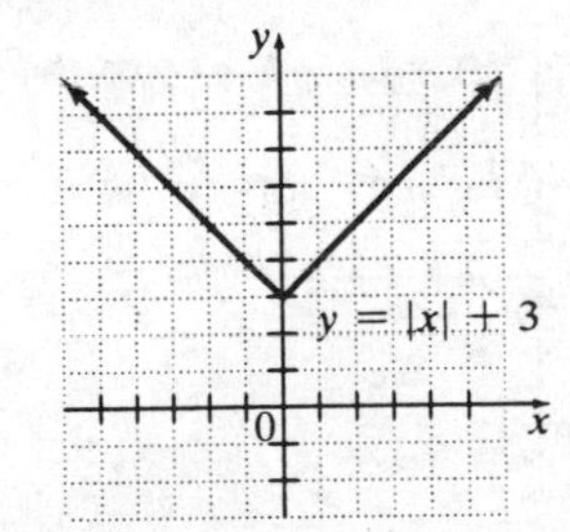

43. Linear; it is written in the form $Ax + By = C$.
$2x - y = 5$
Find 3 points.

x	y
0	–5
1	–3
2	–1

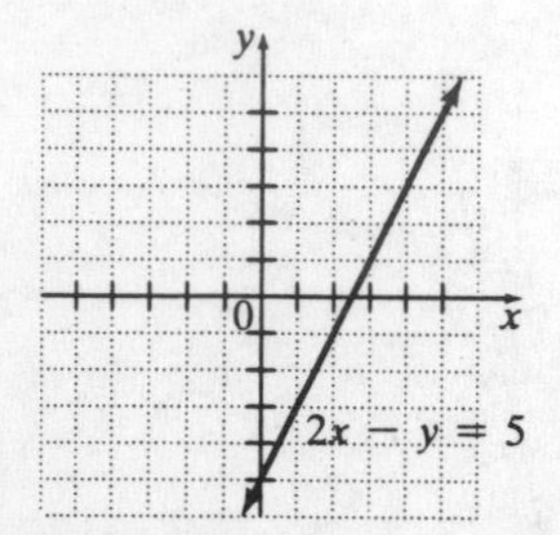

45. Not linear; the x is squared.
$y = 2x^2$
Find several points.

x	y
–1	2
$-\frac{1}{2}$	$\frac{1}{2}$
0	0
$\frac{1}{2}$	$\frac{1}{2}$
1	2

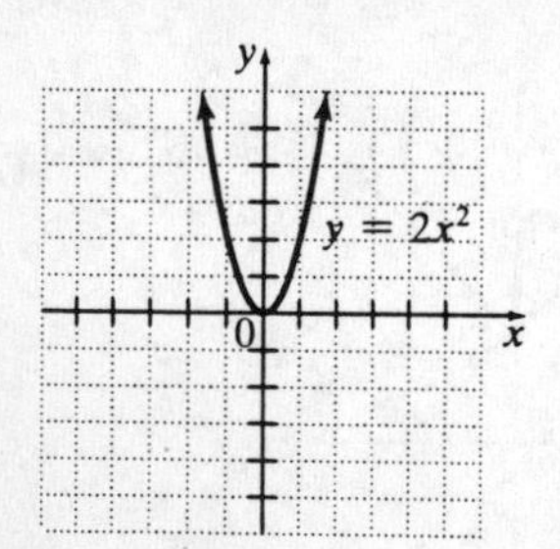

47. Not linear; the x is squared.
$y = x^2 - 3$
Find several points.

x	y
–2	1
–1	–2
0	–3
1	–2
2	1

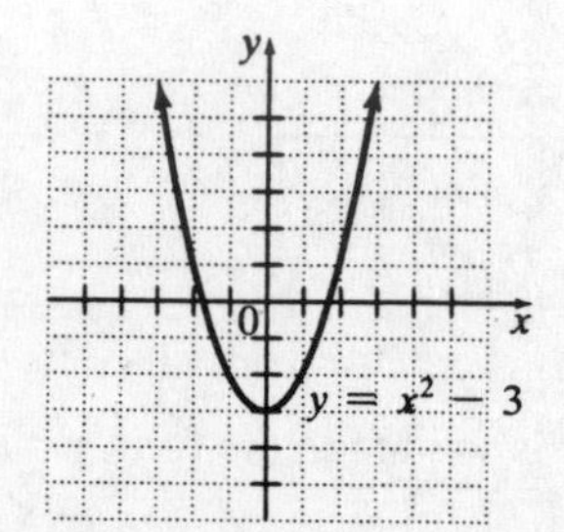

49. Linear; it is written in the form $y = mx + b$.
$y = -2x$
Find 3 points.

x	y
–1	2
0	0
1	–2

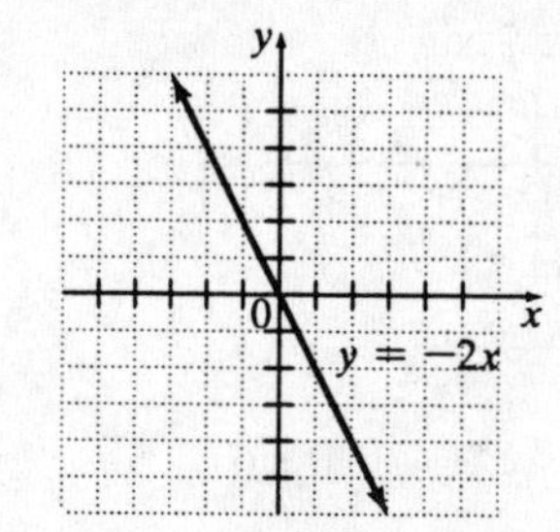

51. Linear; it is written in the form $y = mx + b$.
$y = -2x + 3$
Find 3 points.

x	y
0	3
1	1
2	−1

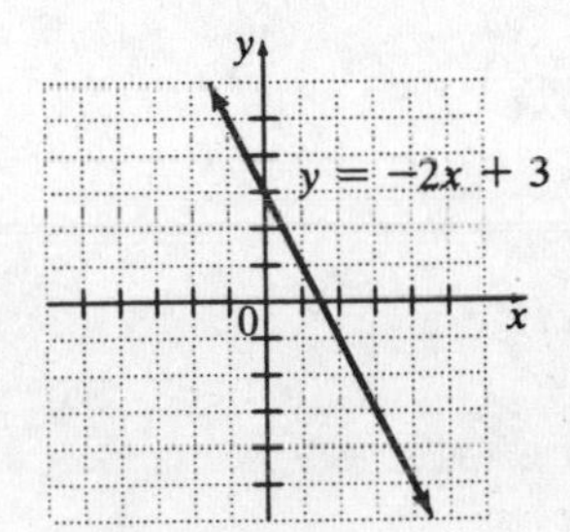

53. Not linear; x is inside absolute value.
$y = |x + 2|$
Find several points.

x	y
−4	2
−3	1
−2	0
−1	1
0	2
1	3

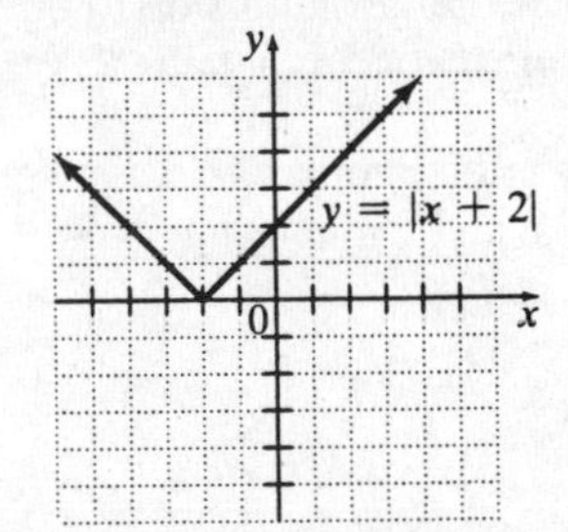

55. Not linear; the x is cubed.
$y = x^3$
Let $x = -3, -2, -1, 0, 1, 2$.

x	y
−3	−27
−2	−8
−1	−1
0	0
1	1
2	8

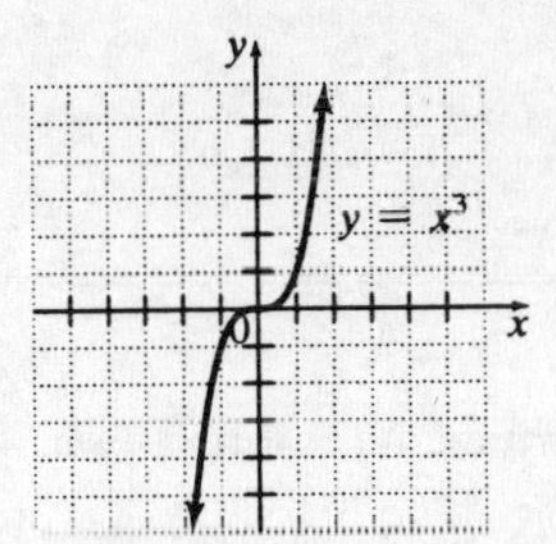

57. Not linear; the x is inside absolute value.
$y = -|x|$
Find several points.

x	y
−2	−2
−1	−1
0	0
1	−1
2	−2
3	−3

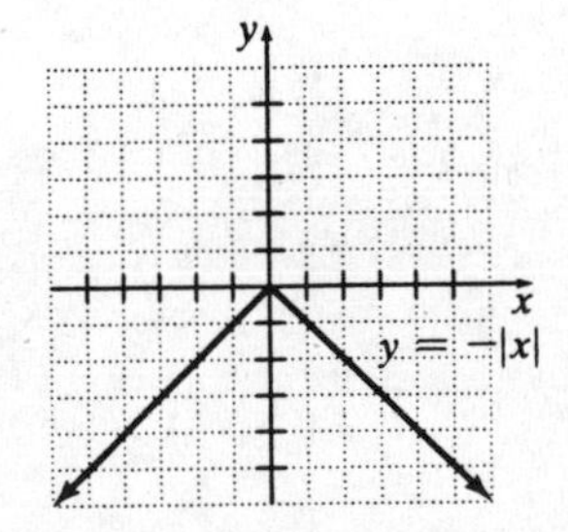

59. Linear; it is written in the form $y = mx + b$.

$y = \frac{1}{3}x - 1$

Find 3 points.

x	y
–3	–2
0	–1
3	0

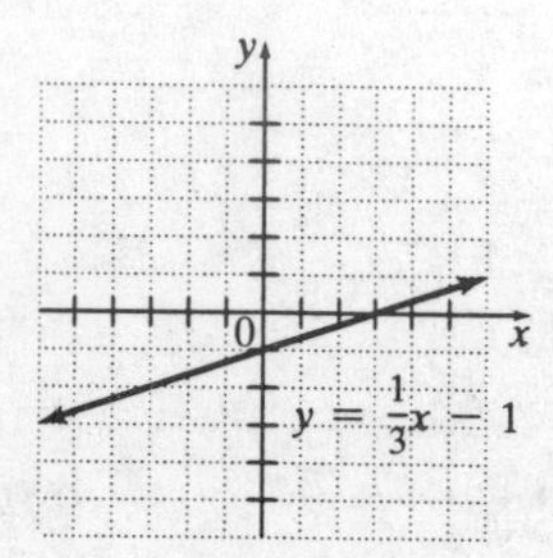

61. Linear; it is written in the form $y = mx + b$.

$y = -\frac{3}{2}x + 1$

Find 3 points.

x	y
–2	4
0	1
2	–2

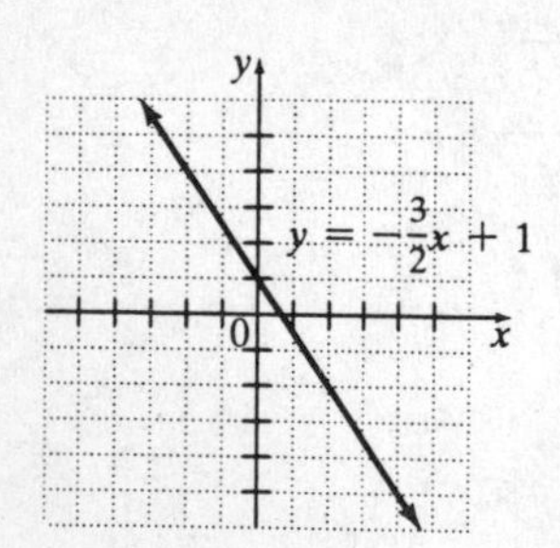

63. a. The equation $y = 2x + 6$ is linear; it is in the form $y = mx + b$. Find 3 points.

x	y
0	6
1	8
2	10

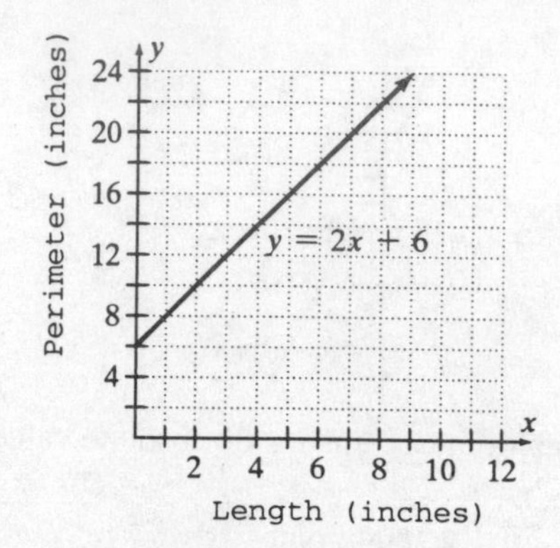

b. From the graph, we see that when $x = 4$, $y = 14$. Thus, the perimeter is 14 inches.

65. 1991: in April it rose to $4.25.

67. Answers may vary.

69. B; The height of the graph is 40 until the beginning of the fall semester in late August. Then the height drops to 0 because Moe quit his job. It stays at 0 until the middle of December when the Christmas break begins. At this point, the height jumps up to 60 when Moe begins working again.

71. C; The height of the graph varies between a minimum of 10 and a maximum of 30.

73. Answers may vary.

75. $y = 3x + 5$
This is a linear equation. To graph it, find 3 points.

x	y
0	5
–1	2
–2	–1

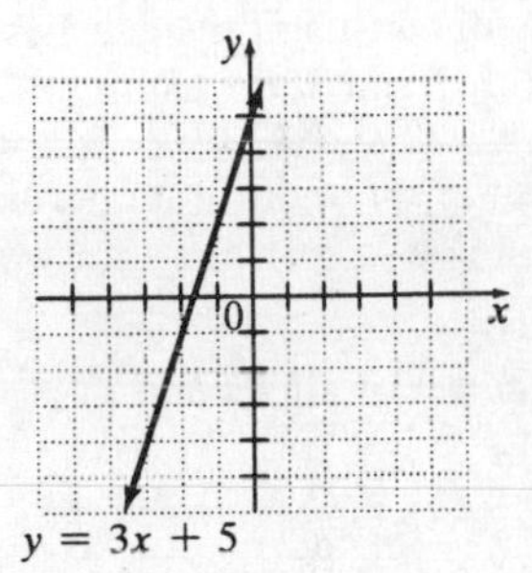

77. $y = x^2 + 2$
This equation is not linear since the x is squared. To graph it, find several points.

x	y
–2	6
–1	3
0	2
1	3
2	6

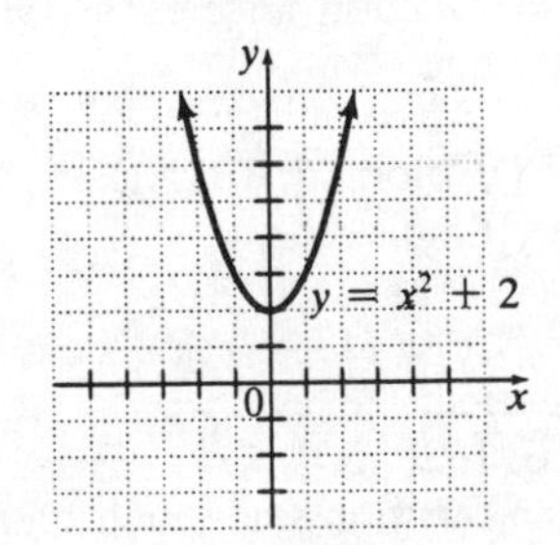

79.

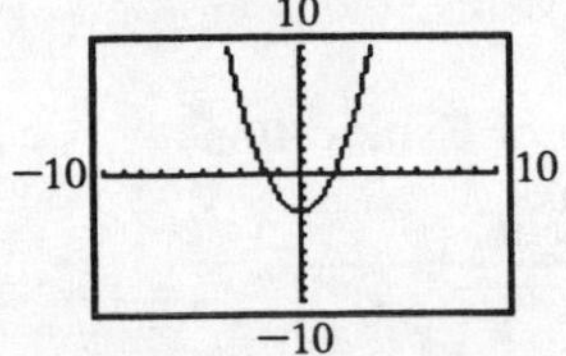

81.

10
–10
10
–10

83. (4, –1)

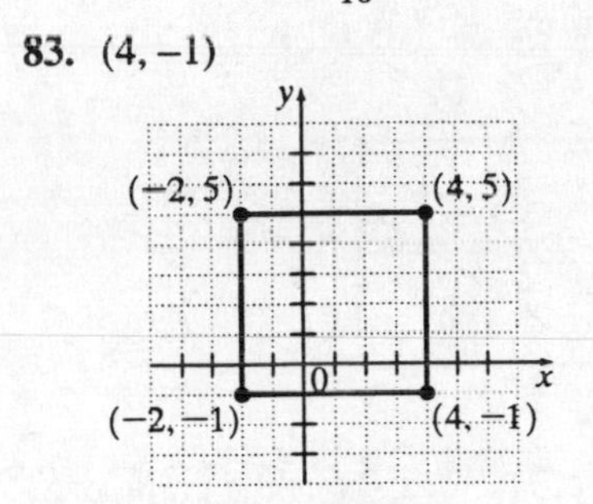

85.
$$\begin{aligned} 3(x-2)+5x &= 6x-16 \\ 3x-6+5x &= 6x-16 \\ 8x-6 &= 6x-16 \\ 2x-6 &= -16 \\ 2x &= -10 \\ x &= -5 \end{aligned}$$

87.
$$\begin{aligned} 3x+\frac{2}{5} &= \frac{1}{10} \\ 3x &= \frac{1}{10}-\frac{2}{5} \\ 3x &= \frac{1}{10}-\frac{4}{10} \\ 3x &= -\frac{3}{10} \\ x &= -\frac{1}{10} \end{aligned}$$

Section 3.3

Graphing Calculator Explorations

1. 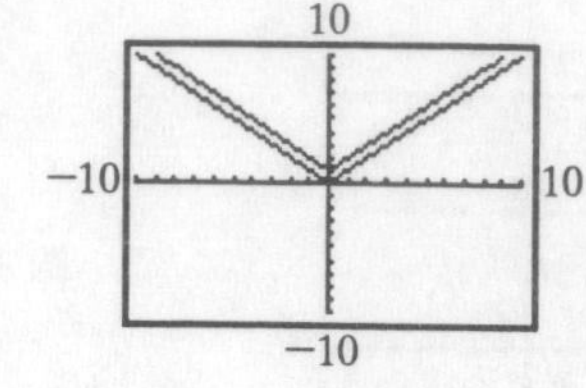

The graph of $g(x) = |x| + 1$ is the graph of $f(x) = |x|$ moved 1 unit upward.

3. 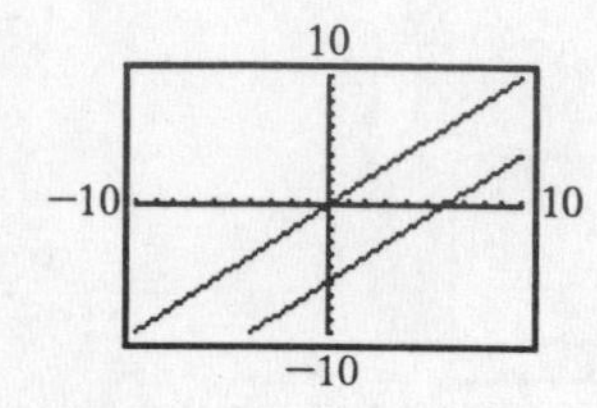

The graph of $H(x) = x - 6$ is the graph of $f(x) = x$ moved 6 units downward.

5. 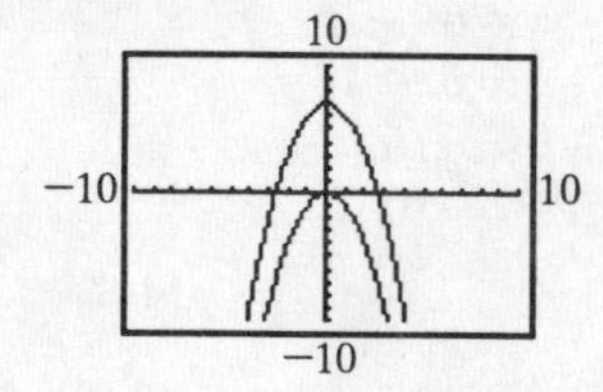

The graph of $F(x) = -x^2 + 7$ is the graph of $f(x) = -x^2$ moved 7 units upward.

Exercise Set 3.3

1. The domain is the set of all first coordinates, $\{-1, 0, -2, 5\}$.
The range is the set of all second coordinates, $\{7, 6, 2\}$.
The relation is a function since each x-value is assigned only one y-value.

3. The domain is the set of all first coordinates, $\{-2, 6, 7\}$.
The range is the set of all second coordinates, $\{4, -3, -8\}$.
The relation is not a function since -2 is paired with both 4 and -3.

5. The domain is the set of all first coordinates, $\{1\}$.
The range is the set of all second coordinates, $\{1, 2, 3, 4\}$.
The relation is not a function since 1 is paired with both 1 and 2, for example.

7. The domain is the set of all first coordinates, $\left\{\frac{3}{2}, 0\right\}$.
The range is the set of all second coordinates, $\left\{\frac{1}{2}, -7, \frac{4}{5}\right\}$.
The relation is not a function since $\frac{3}{2} = 1\frac{1}{2}$ is paired with both $\frac{1}{2}$ and -7.

9. The domain is the set of all first coordinates, $\{-3, 0, 3\}$.
The range is the set of all second coordinates, $\{-3, 0, 3\}$.
The relation is a function since each x-value is assigned only one y-value.

11. The relation is $\{(-1, 2), (1, 1), (2, 1), (3, 1)\}$.
The domain is $\{-1, 1, 2, 3\}$.
The range is $\{2, 1\}$.
The relation is a function since each x-value is assigned only one y-value.

13. The relation is {(Colorado, 6), (Alaska, 1), (Delaware, 1), (Illinois, 20), (Connecticut, 6), (Texas, 30)}.
The domain is {Colorado, Alaska, Delaware, Illinois, Connecticut, Texas}.
The range is {6, 1, 20, 30}.
The relation is a function since each state is assigned only one number.

15. The relation is {(32°, 0°), (104°, 40°), (212°, 100°), (50°, 10°)}.
The domain is {32°, 104°, 212°, 50°}.
The range is {0°, 40°, 10°, 100°}.
The relation is a function since each temperature in degrees Fahrenheit is assigned only one temperature in degrees Celsius.

17. The relation is {(2, 0), (−1, 0), (5, 0), (100, 0)}.
The domain is {2, −1, 5, 100}.
The range is {0}.
The relation is a function since each x-value is assigned only one y-value.

19. The relation is a function since each algebra student has only one grade average.

21. Answers may vary.

23. The graph is the graph of a function since each vertical line intersects the graph in only one point.

25. The graph is not the graph of a function since the vertical line $x = 0$ intersects the graph in two points.

27. The graph is the graph of a function since each vertical line intersects the graph in only one point.

29. To approximate the time of the sunrise on September 1, we find the mark on the horizontal axis that corresponds to September 1. From this mark, we move vertically upward until the graph is reached. From that point on the graph, we move horizontally to the left until the vertical axis is reached. We are about $\frac{1}{3}$ of the way from 5 A.M. to 6 A.M., so the time is approximately 5:20 A.M.

31. Answers may vary.

33.

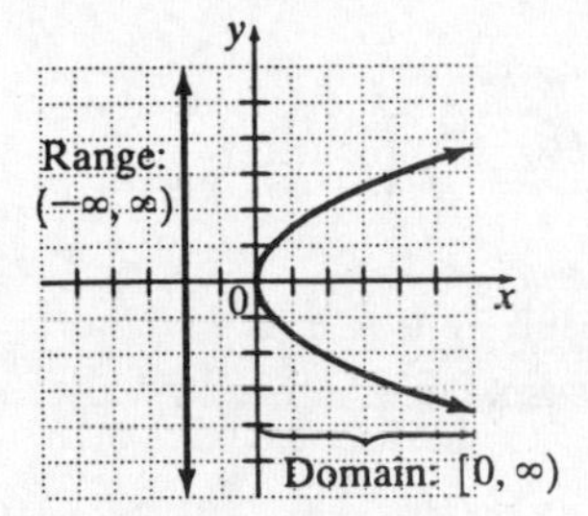

Domain = [0, ∞)
Range = All reals
The relation is not a function since the vertical line $x = 1$ intersects the graph in more than one point.

35.

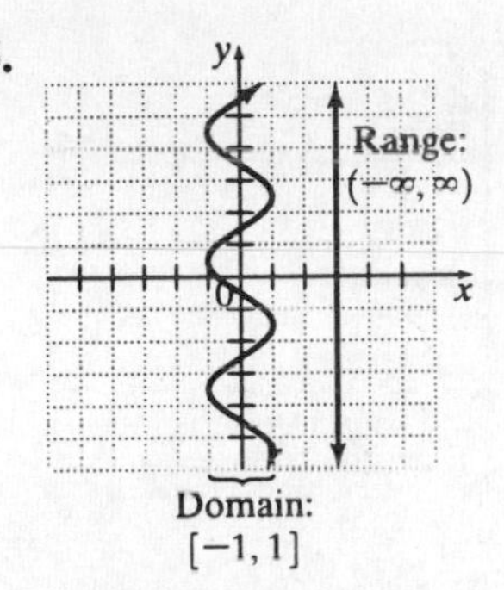

Domain = [−1, 1]
Range = All reals
The relation is not a function since the vertical line $x = 0$ intersects the graph in more than one point.

37.

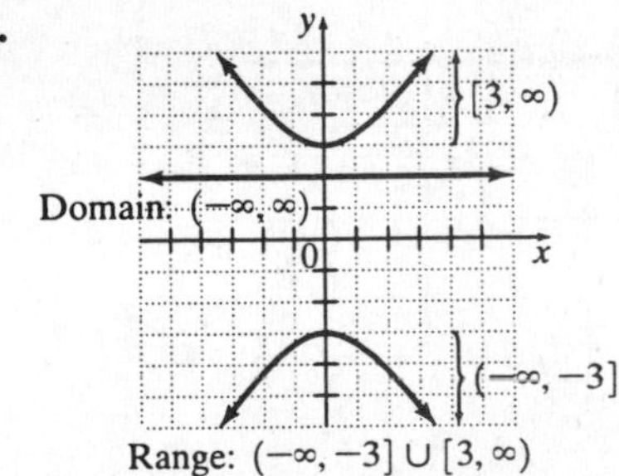

Domain = (−∞, ∞)
Range = (−∞, −3] ∪ [3, ∞)
The relation is not a function since the vertical line $x = 3$ intersects the graph in more than one point.

39.

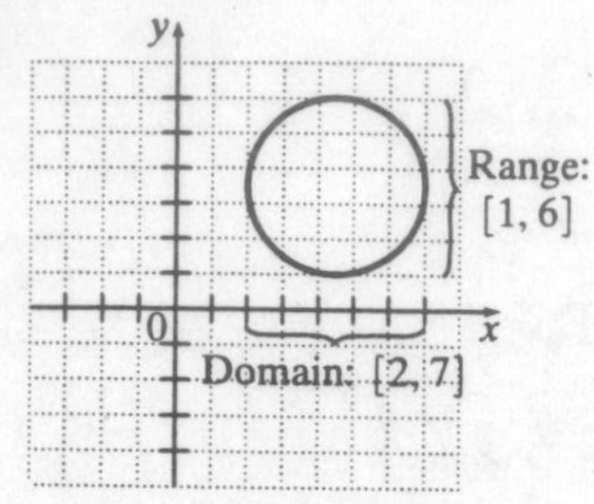

Domain = [2, 7]
Range = [1, 6]
The relation is not a function since the vertical line $x = 4$ intersects the graph in more than one point.

41.

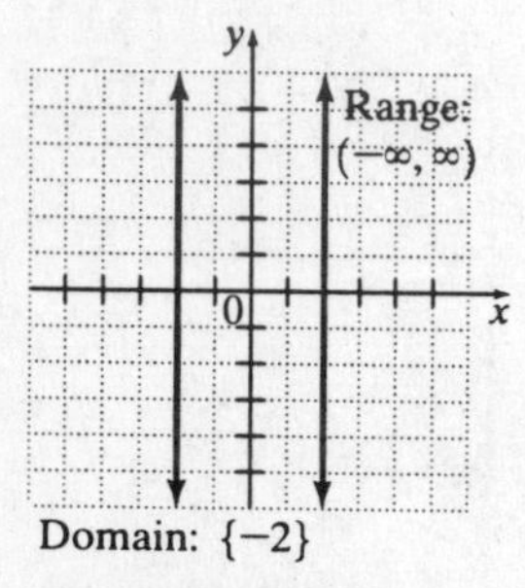

Domain = {−2}
Range = (−∞, ∞)
The relation is not a function since the vertical line $x = -2$ touches the graph in more than one point.

43.

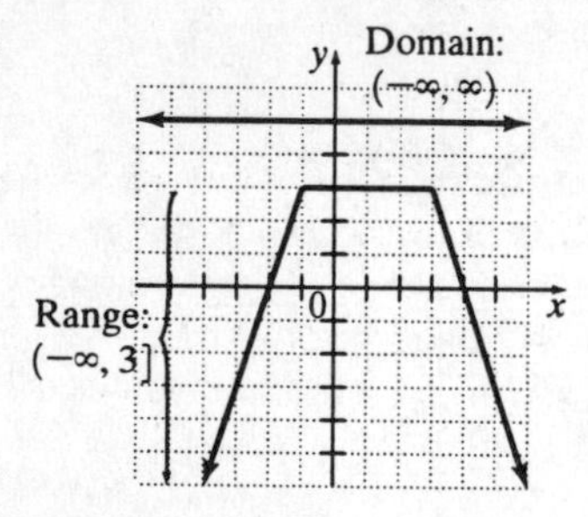

Domain = (−∞, ∞)
Range = (−∞, 3]
The relation is a function since each vertical line intersects the graph in exactly one point.

45. Answers may vary.

47. $y = x + 1$
For each x-value substituted into the equation, the addition performed gives a single result, so only one y-value will be associated with each x-value. Thus, $y = x + 1$ is a function.

49. $x = 2y^2$
If $y = -2$, then $x = 8$, and if $y = 2$, then $x = 8$. So the one x-value 8 corresponds to two y-values −2 and 2. Thus, $x = 2y^2$ is not a function.

51. $y - x = 7$
The equation $y - x = 7$ is equivalent to the equation $y = x + 7$, obtained by adding x to each side of the original equation. For each x-value substituted into $y = x + 7$, the addition performed gives a single result, so only one y-value will be associated with each x-value. Thus, $y - x = 7$ is a function.

53. $y = \frac{1}{x}$
For each non-zero x-value substituted into the equation, the division performed gives a single result, so only one y-value will be associated with each x-value. Thus, $y = \frac{1}{x}$ is a function.

55. $f(x) = 3x + 3$
$f(4) = 3(4) + 3 = 12 + 3 = 15$

57. $h(x) = 5x^2 - 7$
$$\begin{aligned} h(-3) &= 5(-3)^2 - 7 \\ &= 5(9) - 7 \\ &= 45 - 7 \\ &= 38 \end{aligned}$$

59. $g(x) = 4x^2 - 6x + 3$
$$\begin{aligned} g(2) &= 4(2)^2 - 6(2) + 3 \\ &= 4(4) - 12 + 3 \\ &= 16 - 12 + 3 \\ &= 7 \end{aligned}$$

61. $g(x) = 4x^2 - 6x + 3$

$$\begin{aligned} g(0) &= 4(0)^2 - 6(0) + 3 \\ &= 4(0) - 0 + 3 \\ &= 0 - 0 + 3 \\ &= 3 \end{aligned}$$

63. $f(x) = \frac{1}{2}x$

a. $f(0) = \frac{1}{2}(0) - 0$

b. $f(2) = \frac{1}{2}(2) = 1$

c. $f(-2) = \frac{1}{2}(-2) = -1$

65. $g(x) = 2x^2 + 4$

a.
$$\begin{aligned} g(-11) &= 2(-11)^2 + 4 \\ &= 2(121) + 4 \\ &= 242 + 4 \\ &= 246 \end{aligned}$$

b.
$$\begin{aligned} g(-1) &= 2(-1)^2 + 4 \\ &= 2(1) + 4 \\ &= 2 + 4 \\ &= 6 \end{aligned}$$

c.
$$\begin{aligned} g\left(\frac{1}{2}\right) &= 2\left(\frac{1}{2}\right)^2 + 4 \\ &= 2\left(\frac{1}{4}\right) + 4 \\ &= \frac{1}{2} + \frac{8}{2} \\ &= \frac{9}{2} \end{aligned}$$

67. $f(x) = -5$

a. $f(2) = -5$

b. $f(0) = -5$

c. $f(606) = -5$

69. $f(x) = 1.3x^2 - 2.6x + 5.1$

a.
$$\begin{aligned} f(2) &= 1.3(2)^2 - 2.6(2) + 5.1 \\ &= 1.3(4) - 5.2 + 5.1 \\ &= 5.2 - 5.2 + 5.1 \\ &= 5.1 \end{aligned}$$

b.
$$\begin{aligned} f(-2) &= 1.3(-2)^2 - 2.6(-2) + 5.1 \\ &= 1.3(4) + 5.2 + 5.1 \\ &= 5.2 + 5.2 + 5.1 \\ &= 15.5 \end{aligned}$$

c.
$$\begin{aligned} f(3.1) &= 1.3(3.1)^2 - 2.6(3.1) + 5.1 \\ &= 1.3(9.61) - 8.06 + 5.1 \\ &= 12.493 - 8.06 + 5.1 \\ &= 9.533 \end{aligned}$$

71. $f(1) = -10$
The input value 1 is x, and the output value -10 is y, so the ordered pair is $(1, -10)$.

73. $f(-1)$ is the y-coordinate of the point on the graph having x-coordinate -1. We see that $(-1, -2)$ is a point on the graph, so $f(-1) = -2$.

75. The values of x such that $f(x) = -5$ are the x-coordinates of the points on the graph having y-coordinate -5. There are two points on the graph having y-coordinate -5, $(-4, -5)$ and $(0, -5)$. Thus, the values of x are -4 and 0.

77. A function may have an infinite number of x-intercepts. An x-intercept is a point on the graph having y-coordinate 0. The definition of a function does not restrict the number of x-values that can be associated with a single y-value.

79. a. On the graph, 0 corresponds to the year 1993, so 1 corresponds to the year 1994. In 1994, approximately $13.4 billion was spent.

b. To approximate the amount spent in 1994 using the function $f(x) = 1.882x + 11.79$, we find $f(1)$, since $x = 1$ corresponds to the year 1994.
$f(1) = 1.882(1) + 11.79 = 13.672$
Thus, in 1994, approximately $13.672 billion was spent on research and development.

81. To predict the amount of money that will be spent in the year 2005, we use $f(x) = 1.882x + 11.79$ and find $f(12)$, since year 12 corresponds to 2005.
$f(12) = 1.882(12) + 11.79 = 34.374$
We predict that in the year 2005, the Pharmaceutical Manufacturers Association will spend $34.374 billion on research and development.

83. In order to rewrite the equation $y = x + 7$ using function notation, we replace the y in the equation with $f(x)$.
$f(x) = x + 7$

85. To find the area of the circle whose radius is 5 centimeters, we use the function $A(r) = \pi r^2$ and find $A(5)$.
$A(5) = \pi(5)^2 = 25\pi$
The area of the circle is 25π square centimeters.

87. To find the volume of a cube whose side is 14 inches, we use the function $V(x) = x^3$ and find $V(14)$.
$V(14) = (14)^3 = 2744$
The volume of the cube is 2744 cubic inches.

89. To find the height of a woman whose femur measures 46 centimeters, we use the function $H(f) = 2.59f + 47.24$ and find $H(46)$.
$H(46) = 2.59(46) + 47.24 = 166.38$
The height of the woman is 166.38 centimeters.

91. To find the dosage for a dog that weighs 30 pounds, we use the function $D(x) = \frac{136}{25}x$ and find $D(30)$.
$D(30) = \frac{136}{25}(30) = \frac{816}{5} = 163.2$
A 30-pound dog should be given 163.2 milligrams of Ivermectin to prevent heartworms.

93. a. To find $C(2)$, we replace x with 2 in the function $C(x) = 1.7x + 88$.
$C(2) = 1.7(2) + 88 = 91.4$
$x = 2$ corresponds to the year 1997, since x is the number of years since 1995. Thus, the per capita consumption of poultry was 91.4 pounds in 1997.

b. To predict the per capita consumption of poultry in the United States in 2006, we use the function $C(x) = 1.7x + 88$ and find $C(11)$, since $x = 11$ corresponds to the year 2006.
$C(11) = 1.7(11) + 88 = 106.7$
We predict that per capita consumption of poultry in 2006 will be 106.7 pounds.

95. $x - y = -5$

x	0	–5	1
y	5	0	6

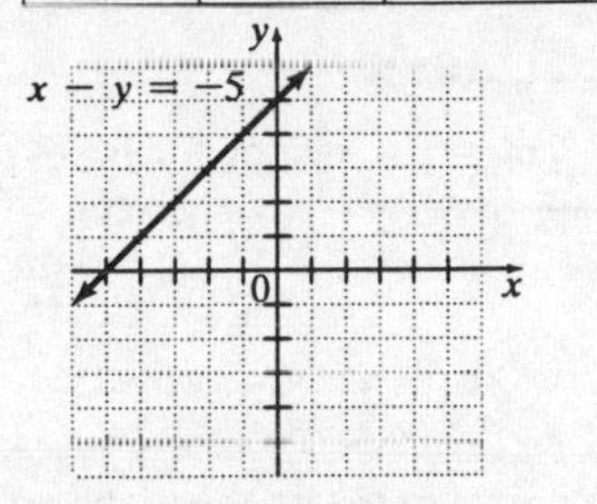

97. $7x + 4y = 8$

x	0	$\frac{8}{7}$	$\frac{12}{7}$
y	2	0	–1

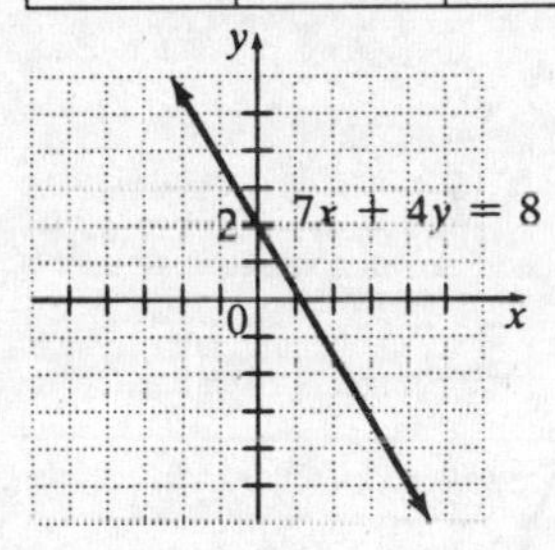

99. $y = 6x$

x	0	0	–1
y	0	0	–6

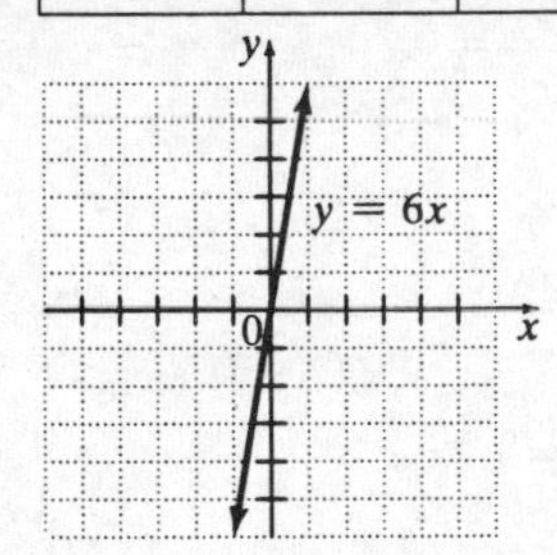

101.

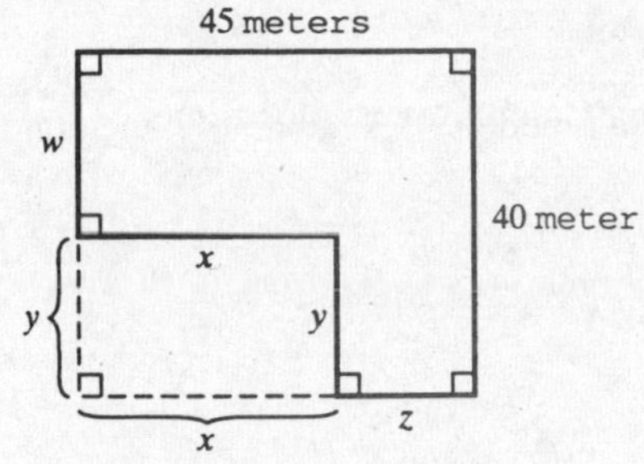

Let w, x, y and z be the lengths of the sides of the figure as in the diagram. Then the perimeter P of the figure is
$P = 40 + 45 + w + x + y + z.$
Note that if the sides labeled w and y were placed end-to-end, the resulting line segment would have length equal to the length of the right side of the figure, and if the sides labeled x and z were placed end-to-end, the resulting line segment would have length equal to the length of the upper side of the figure. Thus, $w + y = 40$ and $x + z = 45$.
Then

$$\begin{aligned} P &= 40 + 45 + w + x + y + z \\ &= 85 + (w + y) + (x + z) \\ &= 85 + 40 + 45 \\ &= 170. \end{aligned}$$

The perimeter of the figure is 170 meters.

103. $g(x) = -3x + 12$

a. $g(s) = -3(s) + 12 = -3s + 12$

b. $g(r) = -3(r) + 12 = -3r + 12$

105. $f(x) = x^2 - 12$

a. $f(12) = (12)^2 - 12 = 144 - 12 = 132$

b. $f(a) = (a)^2 - 12 = a^2 - 12$

Section 3.4

Graphing Calculator Explorations

1. $x = 3.5y$

$$\frac{x}{3.5} = \frac{3.5y}{3.5}$$

$$y = \frac{x}{3.5}$$

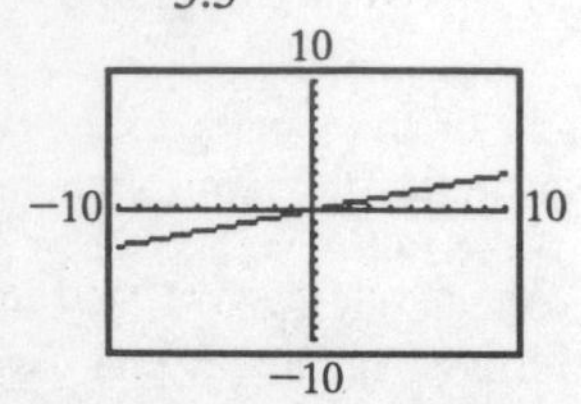

3. $5.78x + 2.31y = 10.98$

$$2.31y = -5.78x + 10.98$$

$$\frac{2.31y}{2.31} = \frac{-5.78}{2.31}x + \frac{10.98}{2.31}$$

$$y = -\frac{5.78}{2.31}x + \frac{10.98}{2.31}$$

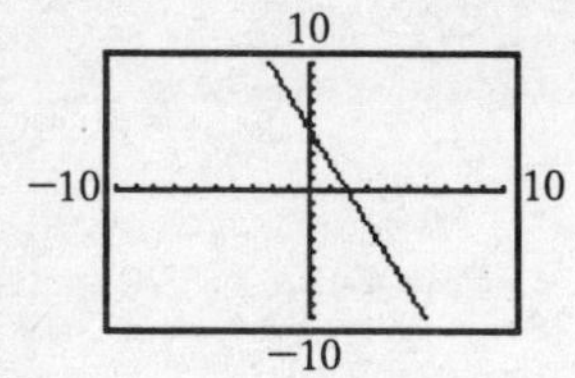

5. $y - |x| = 3.78$

$$y = |x| + 3.78$$

7. $y - 5.6x^2 = 7.7x + 1.5$

$$y = 5.6x^2 + 7.7x + 1.5$$

Exercise Set 3.4

1. $f(x) = -2x$

Plot three points to obtain the graph.

x	y
0	0
–1	2
1	–2

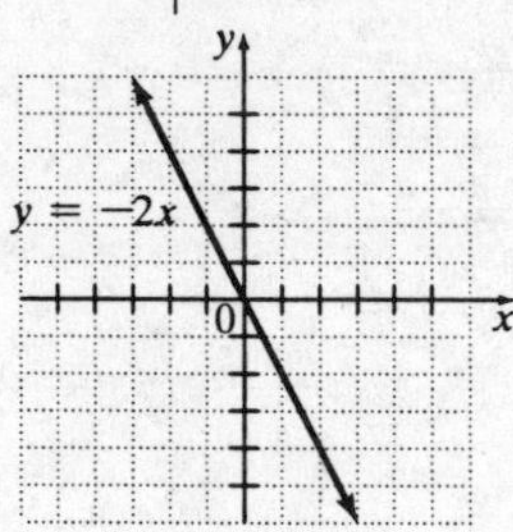

3. $f(x) = -2x + 3$

Plot three points to obtain the graph.

x	y
0	3
1	1
–1	5

5. $f(x) = \frac{1}{2}x$

Plot three points to obtain the graph.

x	y
0	0
2	1
–2	–1

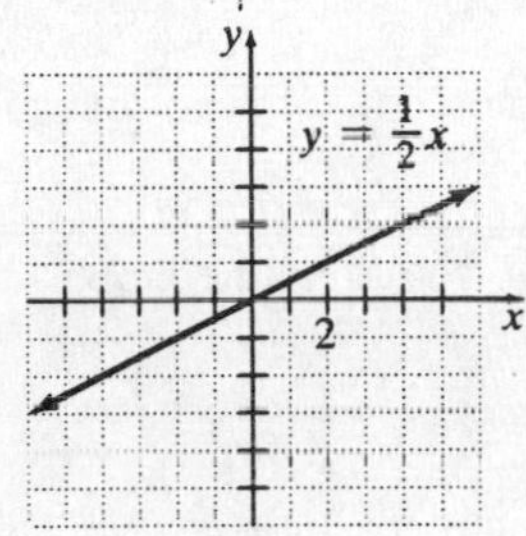

7. $f(x) = \frac{1}{2}x - 4$

Plot three points to obtain the graph.

x	y
0	–4
2	–3
4	–2

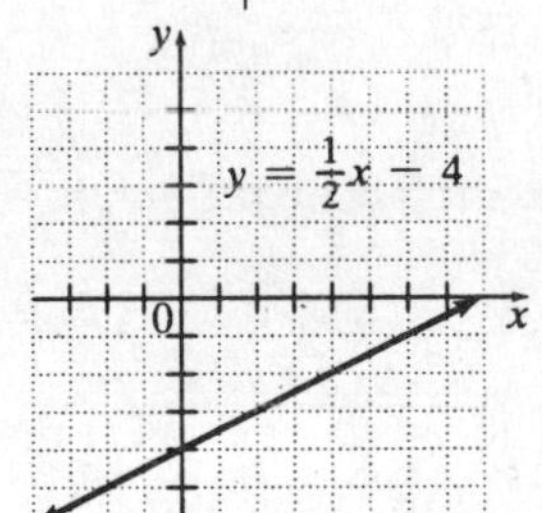

9. The graph of $f(x) = 5x - 3$ is the same as the graph of $f(x) = 5x$ shifted downward 3 units. This graph is C.

11. The graph of $f(x) = 5x + 1$ is the same as the graph of $f(x) = 5x$ shifted upward 1 unit. This graph is D.

13. $x - y = 3$

To find the x-intercept, let $y = 0$ and solve for x.

$x - 0 = 3$

$x = 3$

$(3, 0)$

To find the y-intercept, let $x = 0$ and solve for y.

$0 - y = 3$

$-y = 3$

$y = -3$

$(0, -3)$

To find a third ordered pair to check our work, let $x = 2$ and solve for y.

$2 - y = 3$

$-y = 1$

$y = -1$

$(2, -1)$

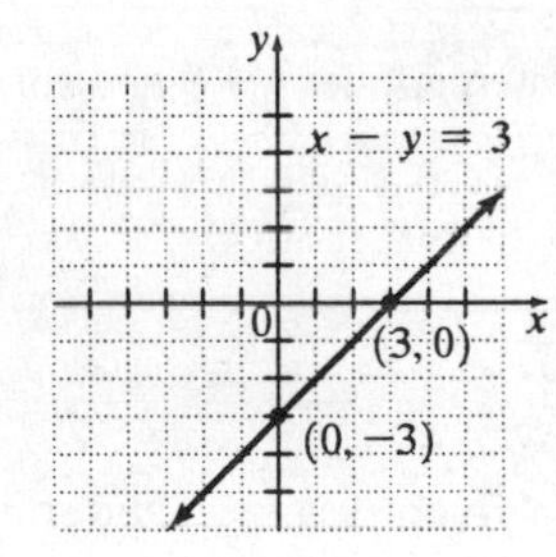

15. $x = 5y$
To find the x-intercept, let $y = 0$ and solve for x.
$x = 5(0)$
$x = 0$
$(0, 0)$
The x-intercept is also the y-intercept.
We find two other ordered pairs to be able to graph the line and check our work.
Let $x = 5$ and solve for y.
$5 = 5y$
$1 = y$
$(5, 1)$
Let $x = -5$ and solve for y.
$-5 = 5y$
$-1 = y$
$(-5, -1)$

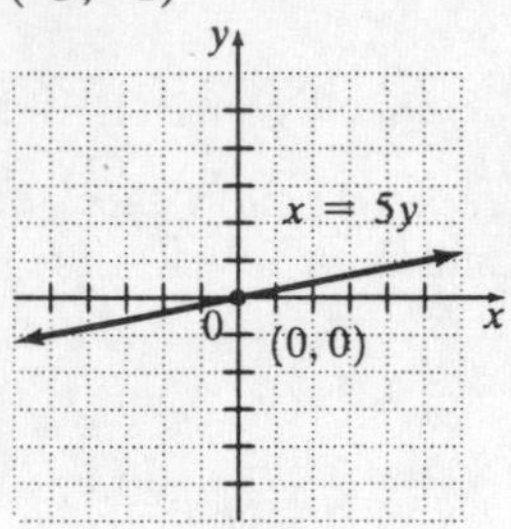

17. $-x + 2y = 6$
To find the x-intercept, let $y = 0$ and solve for x.
$-x + 2(0) = 6$
$-x + 0 = 6$
$-x = 6$
$x = -6$
$(-6, 0)$
To find the y-intercept, let $x = 0$ and solve for y.
$-(0) + 2y = 6$
$0 + 2y = 6$
$2y = 6$
$y = 3$
$(0, 3)$
To find a third ordered pair to check our work, let $x = -2$ and solve for y.
$-(-2) + 2y = 6$
$2 + 2y = 6$
$2y = 4$
$y = 2$
$(-2, 2)$

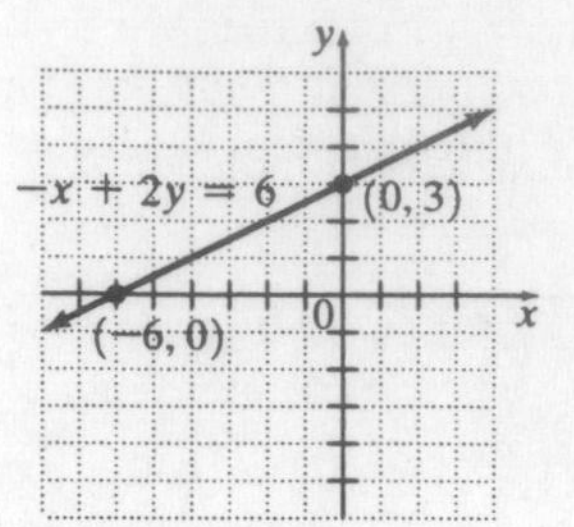

19. $2x - 4y = 8$
To find the x-intercept, let $y = 0$ and solve for x.
$2x - 4(0) = 8$
$2x - 0 = 8$
$2x = 8$
$x = 4$
$(4, 0)$
To find the y-intercept, let $x = 0$ and solve for y.
$2(0) - 4y = 8$
$0 - 4y = 8$
$-4y = 8$
$y = -2$
$(0, -2)$
To find a third ordered pair to check our work, let $x = 2$ and solve for y.
$2(2) - 4y = 8$
$4 - 4y = 8$
$-4y = 4$
$y = -1$
$(2, -1)$

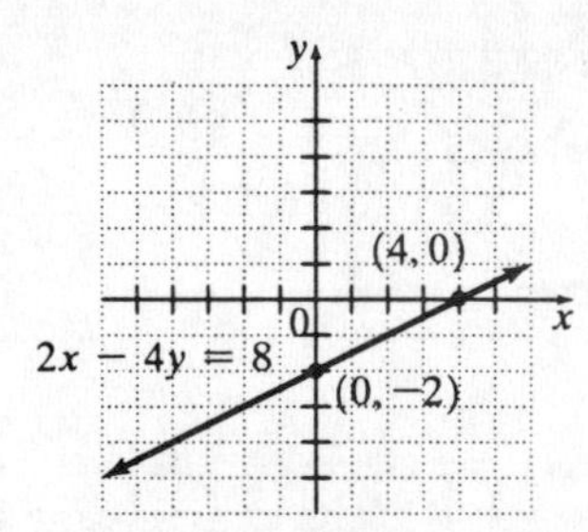

21. Answers may vary.

23. The equation $x = -1$ is of the form $x = c$, where c is a real number. Thus, its graph is a vertical line with x-intercept $(-1, 0)$.

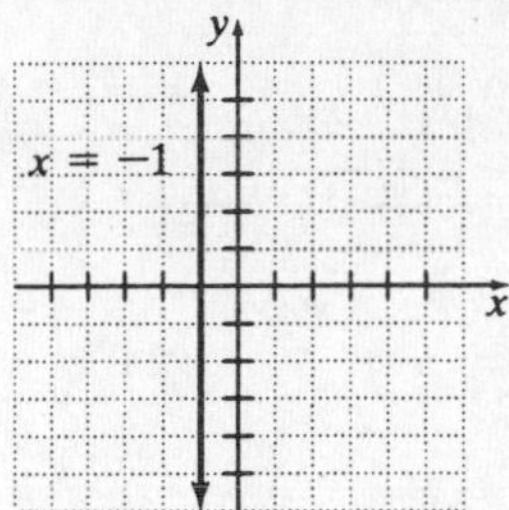

25. The equation $y = 0$ is of the form $y = c$, where c is a real number. Thus, its graph is a horizontal line with y-intercept $(0, 0)$.

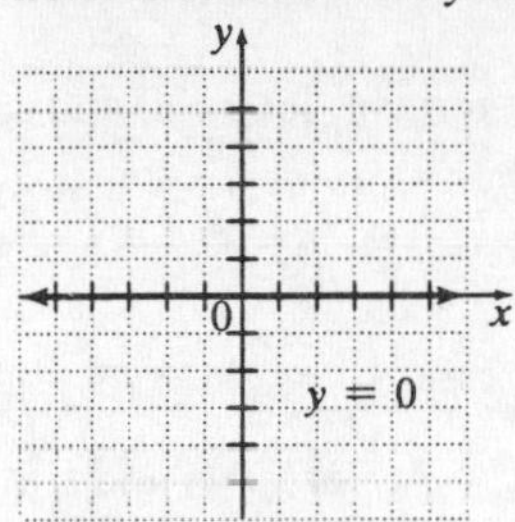

27. The equation $y + 7 = 0$ can be written in the form $y = c$, where c is a real number.

$y + 7 = 0$
$y = -7$

Thus, its graph is a horizontal line with y-intercept $(0, -7)$.

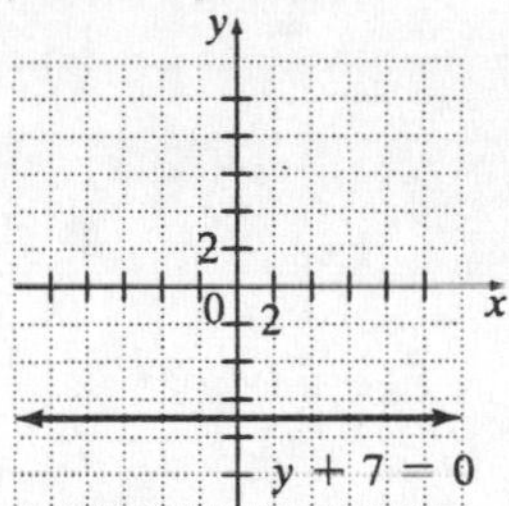

29. The graph of $y = 2$ is a horizontal line with y-intercept $(0, 2)$. This describes graph C.

31. The equation $x - 2 = 0$ can be written in the form $x = 2$. Thus, its graph is a vertical line with x-intercept $(2, 0)$. This describes graph A.

33. The line $x = 0$ is a vertical line on which every point is a y-intercept. However, no other vertical line has a y-intercept.

In Exercises 35–59, only one method of graphing lines will be illustrated in each problem. Other methods may be used.

35. $x + 2y = 8$
Find 3 points.

x	y
−2	5
0	4
2	3

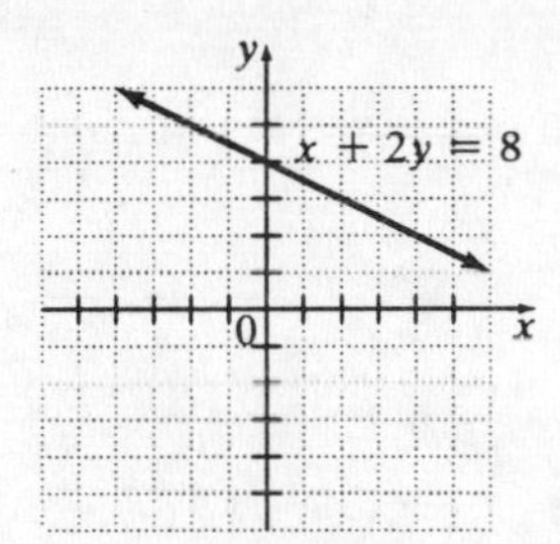

37. $f(x) = \frac{3}{4}x + 2$ or $y = \frac{3}{4}x + 2$
Find 3 points.

x	y
−4	−1
0	2
4	5

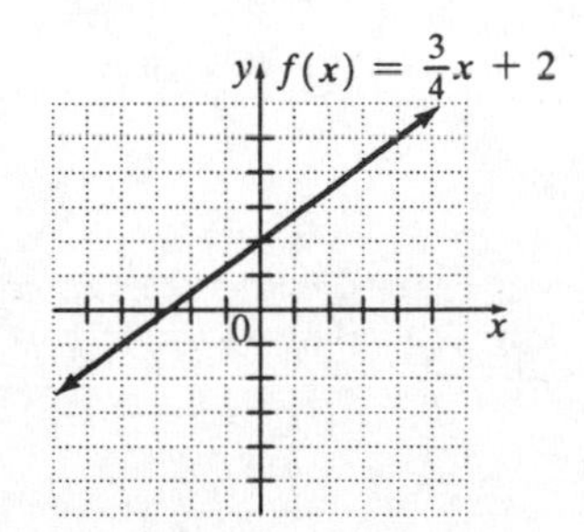

39. $x = -3$
The equation is of the form $x = c$, where c is a real number. Thus, its graph is a vertical line with x-intercept $(-3, 0)$.

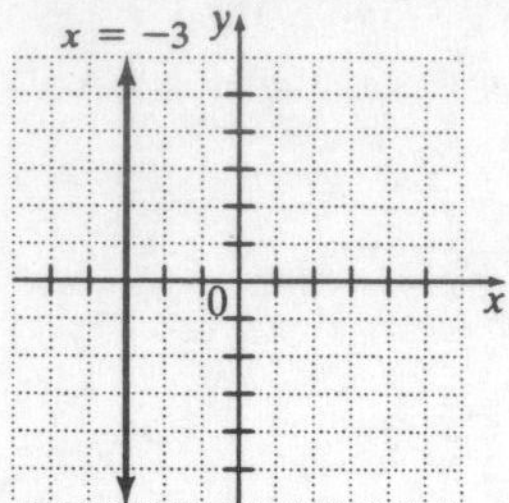

41. $3x + 5y = 7$
Find 3 points.

x	y
0	$\frac{7}{5}$
$\frac{7}{3}$	0
$\frac{2}{3}$	1

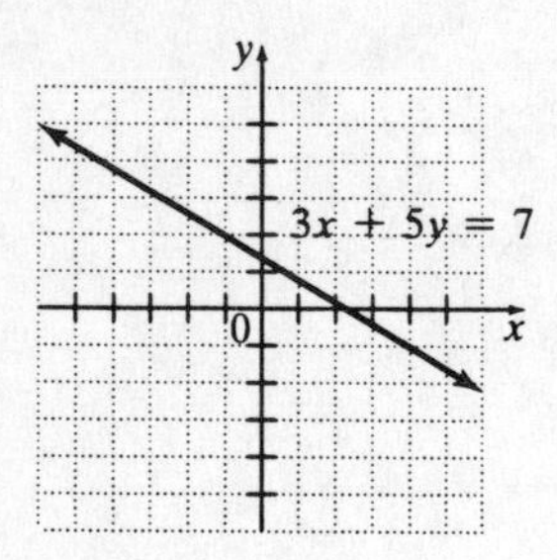

43. $f(x) = x$ or $y = x$
Find 3 points.

x	y
−1	−1
0	0
1	1

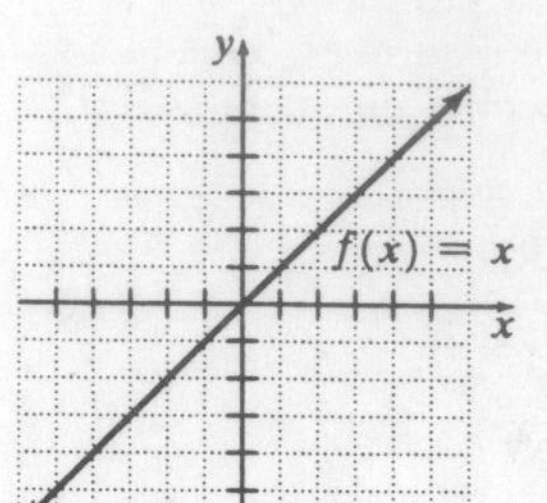

45. $x + 8y = 8$
To find the x-intercept, let $y = 0$ and solve for x.

$$x + 8(0) = 8$$
$$x = 8$$

(8, 0)
To find the y-intercept, let $x = 0$ and solve for y.

$$0 + 8y = 8$$
$$8y = 8$$
$$y = 1$$

(0, 1)
To find a third point, let $x = 4$ and solve for y.

$$4 + 8y = 8$$
$$8y = 4$$
$$y = \frac{4}{8} = \frac{1}{2}$$

$$\left(4, \frac{1}{2}\right)$$

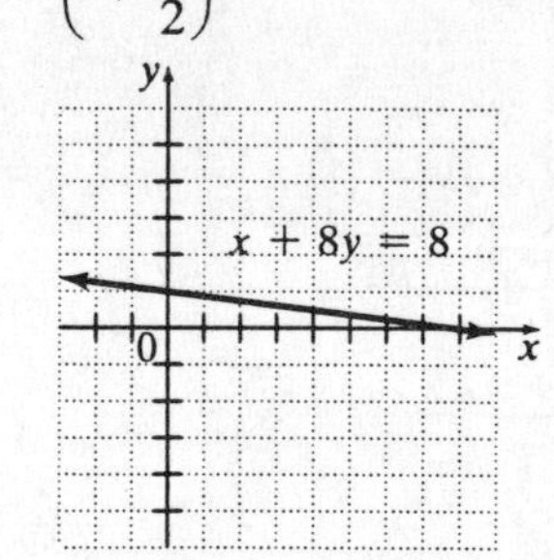

47. $5 = 6x - y$
Find 3 points.

x	y
0	$\frac{5}{6}$
–5	0
1	1

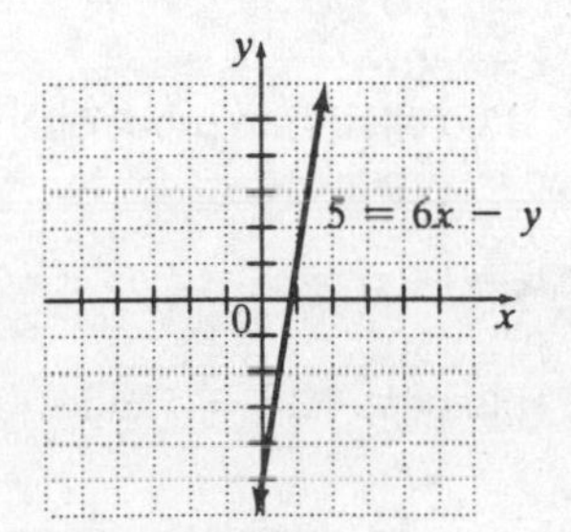

49. $-x + 10y = 11$
Find 3 points.

x	y
0	$\frac{11}{10}$
–11	0
9	2

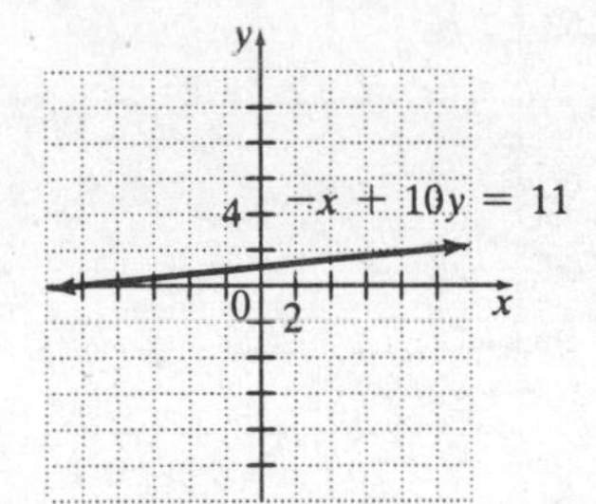

51. $y = 1$
The equation is of the form $y = c$, where c is a real number. Thus, its graph is a horizontal line with y-intercept (0, 1).

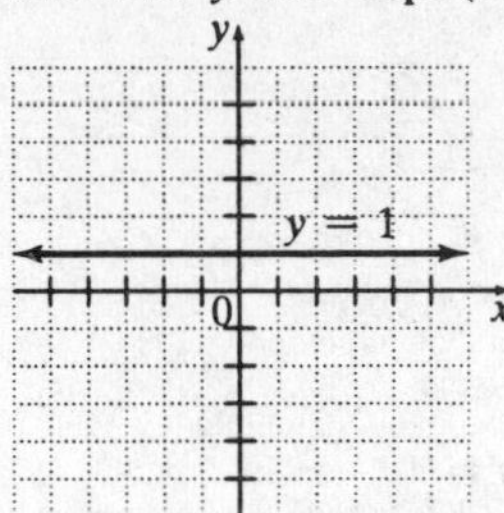

53. $f(x) = \frac{1}{2}x$ or $y = \frac{1}{2}x$
Find 3 points.

x	y
–2	–1
0	0
2	1

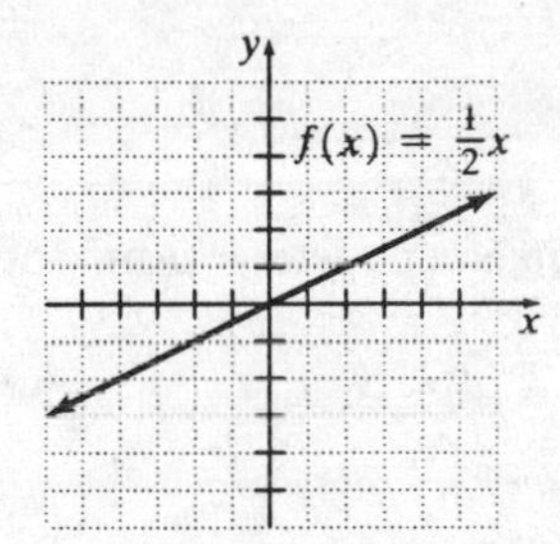

55. $x + 3 = 0$
The equation $x + 3 = 0$ can be rewritten in the form $x = -3$. Thus, its graph is a vertical line with x-intercept (–3, 0).

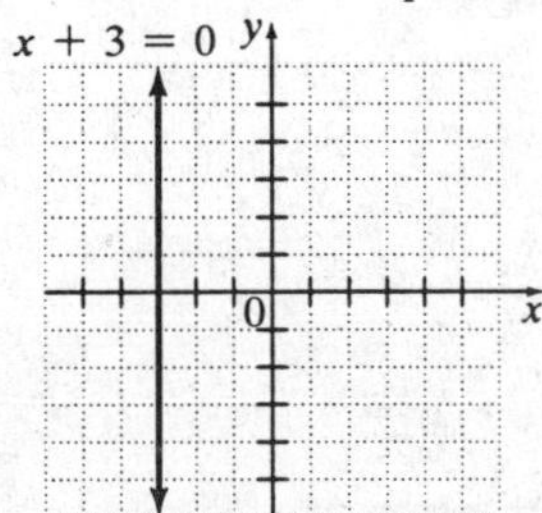

57. $f(x) = 4x - \frac{1}{3}$ or $y = 4x - \frac{1}{3}$

Find 3 points.

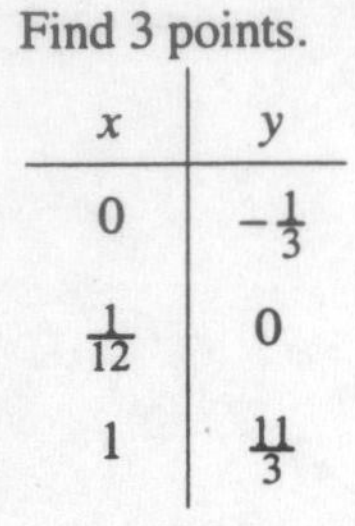

x	y
0	$-\frac{1}{3}$
$\frac{1}{12}$	0
1	$\frac{11}{3}$

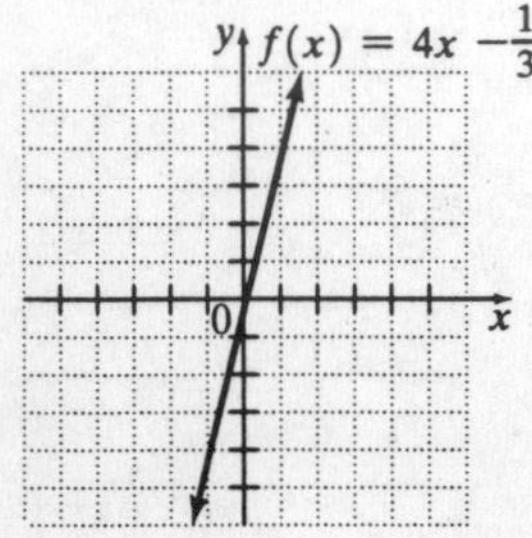

59. $2x + 3y = 6$

To find the x-intercept, let $y = 0$ and solve for x.

$$2x + 3(0) = 6$$
$$2x = 6$$
$$x = 3$$

(3, 0)

To find the y-intercept, let $x = 0$ and solve for y.

$$2(0) + 3y = 6$$
$$3y = 6$$
$$y = 2$$

(0, 2)

To find a third point, let $x = -3$ and solve for y.

$$2(-3) + 3y = 6$$
$$-6 + 3y = 6$$
$$3y = 12$$
$$y = 4$$

(–3, 4)

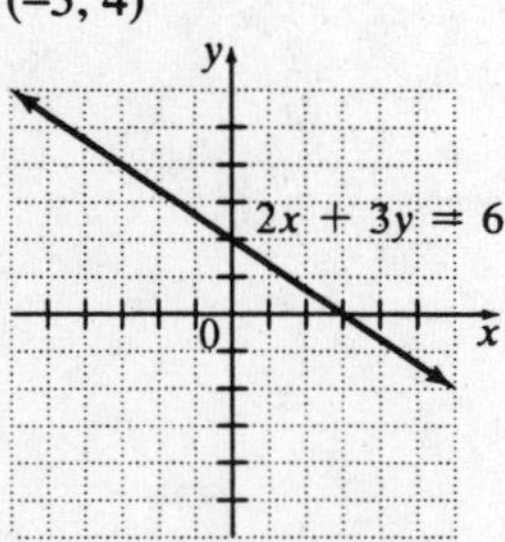

61. $2x + 3y = 1500$

a. $2(0) + 3y = 1500$
$$3y = 1500$$
$$y = 500$$

(0, 500); If no tables are produced, 500 chairs can be produced.

b. $2x + 3(0) = 1500$
$$2x = 1500$$
$$x = 750$$

(750, 0); If no chairs are produced, 750 tables can be produced.

c. $2(50) + 3y = 1500$
$$100 + 3y = 1500$$
$$3y = 1400$$
$$y \approx 466.7$$

If 50 tables are produced, the company can make a maximum of 466 chairs.

63. $C(x) = 0.2x + 24$

a. $C(200) = 0.2(200) + 24$
$$= 40 + 24$$
$$= 64$$

The cost of driving the car 200 miles is $64.

b.

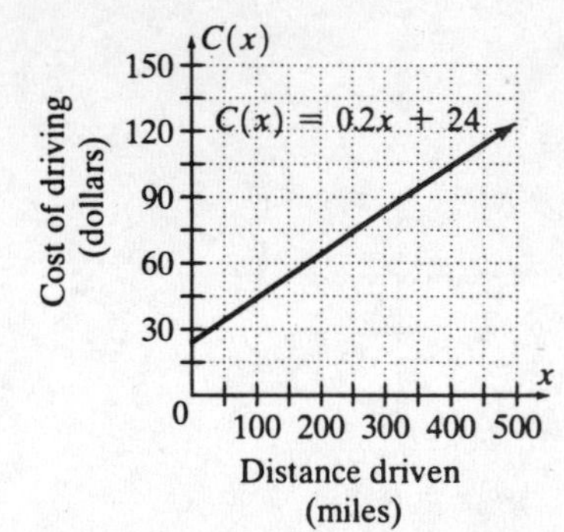

c. The line moves upward from left to right.

65. $f(x) = 72.9x + 785.2$

a. We find $f(20)$ since 2010 is 20 years after 1990.

$$f(20) = 72.9(20) + 785.2 = 1458 + 785.2 = 2243.2$$

The estimated yearly cost in 2010 is $2243.20.

b. We replace $f(x)$ with 2000 and solve for x.

$$2000 = 72.9x + 785.2$$
$$1214.8 = 72.9x$$
$$16.66 \approx x$$
$$1990 + 17 = 2007$$

The yearly cost will first exceed $2000 in the year 2007.

c. Answers may vary.

67. $y = 1.2x + 23.6$

a. Since the equation $y = 1.2x + 23.6$ is written in the form $y = mx + b$, we know the graph of the equation crosses the y-axis at $b = 23.6$. Thus, the y-intercept is $(0, 23.6)$.

b. In 1995 (0 years after 1995), U.S. farm expenses for livestock feed were 23.6 billion dollars.

69. The line $x = 5$ is a vertical line that intersects the x-axis at $(5, 0)$. A line parallel to it will also be vertical. The equation of the vertical line that intersects the x-axis at $(1, 0)$ is $x = 1$.

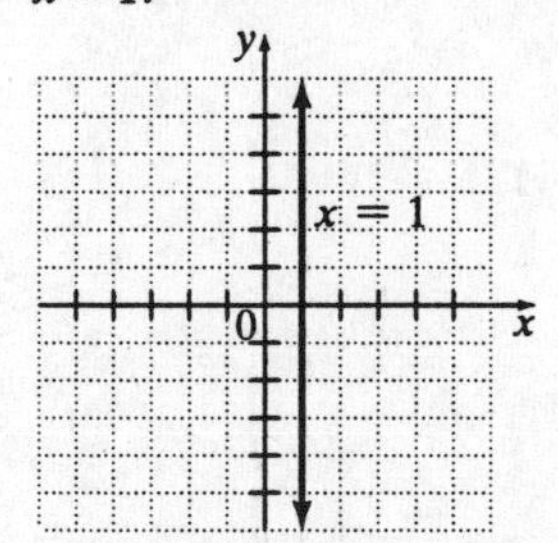

71.

73. To graph $-x + 2y = 6$ with a graphing calculator, we must first solve the equation for y.

$$-x + 2y = 6$$
$$2y = x + 6$$
$$y = \frac{1}{2}x + 3$$

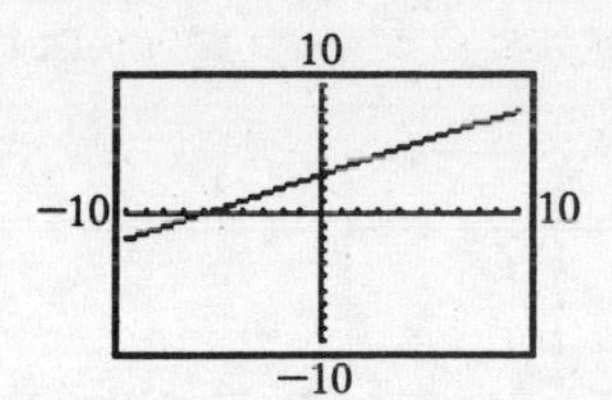

75. $\frac{-6-3}{2-8} = \frac{-9}{-6} = \frac{3}{2}$

77. $\frac{-8-(-2)}{-3-(-2)} = \frac{-8+2}{-3+2} = \frac{-6}{-1} = 6$

79. $\frac{0-6}{5-0} = -\frac{6}{5}$

Section 3.5

Graphing Calculator Explorations

1.

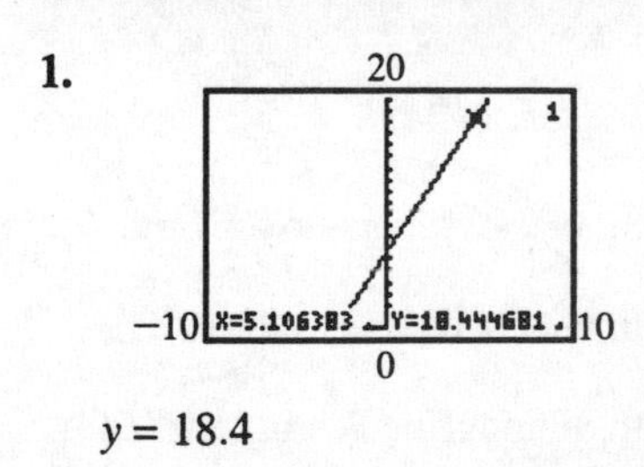

$y = 18.4$

3. 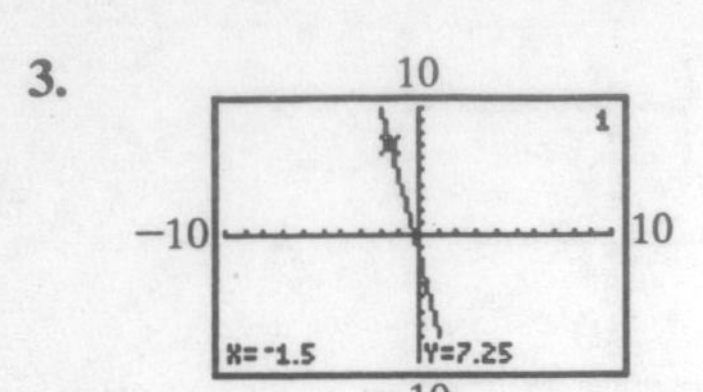

$x = -1.5$

5. 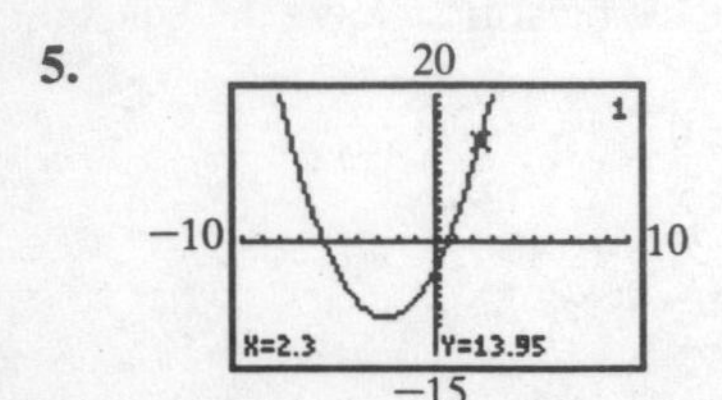

$y = 14.0$

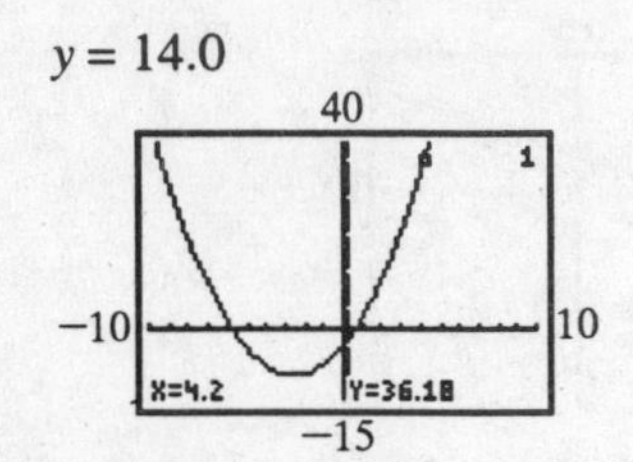

$x = 4.2$

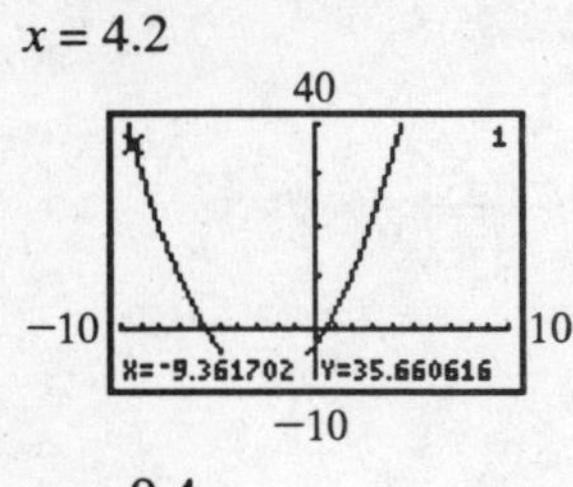

$x = -9.4$

Mental Math

1. A line with $m = \frac{7}{6}$ slants upward.
2. A line with $m = -3$ slants downward.
3. A line with $m = 0$ slants horizontally.
4. A line with m undefined slants vertically.

Exercise Set 3.5

1. $m = \frac{y_2 - y_1}{x_2 - x_1} = \frac{11-2}{8-3} = \frac{9}{5}$

3. $m = \frac{y_2 - y_1}{x_2 - x_1} = \frac{8-1}{1-3} = \frac{7}{-2} = -\frac{7}{2}$

5. $m = \frac{y_2 - y_1}{x_2 - x_1} = \frac{3-8}{4-(-2)} = \frac{-5}{6} = -\frac{5}{6}$

7. $m = \frac{y_2 - y_1}{x_2 - x_1} = \frac{-4-(-6)}{4-(-2)} = \frac{2}{6} = \frac{1}{3}$

9. $m = \frac{y_2 - y_1}{x_2 - x_1} = \frac{11-(-1)}{-12-(-3)} = \frac{12}{-9} = -\frac{4}{3}$

11. $m = \frac{y_2 - y_1}{x_2 - x_1} = \frac{5-5}{3-(-2)} = \frac{0}{5} = 0$

13. $m = \frac{y_2 - y_1}{x_2 - x_1} = \frac{-5-1}{-1-(-1)} = \frac{-6}{0}$
undefined slope

15. $m = \frac{y_2 - y_1}{x_2 - x_1} = \frac{0-6}{-3-0} = \frac{-6}{-3} = 2$

17. $m = \frac{y_2 - y_1}{x_2 - x_1}$
$= \frac{4-2}{-3-(-1)}$
$= \frac{2}{-3+1}$
$= \frac{2}{-2}$
$= -1$

19. $\frac{\text{rise}}{\text{run}} = \frac{6}{10} = \frac{3}{5}$

21. $\frac{\text{rise}}{\text{run}} = \frac{2}{16} = 0.125$
12.5%

23. $\frac{\text{rise}}{\text{run}} = \frac{2580}{6450} = 0.40$
40%

25. $\frac{\text{rise}}{\text{run}} = \frac{0.25}{12} \approx 0.02$

27. (1999, 99), (2002, 144)
$m = \frac{144-99}{2002-1999} = \frac{45}{3} = 15$
Every one year there are/should be 15 million more internet users.

29. (5000, 1800), (20,000, 7200)
$m = \frac{7200-1800}{20,000-5000} = \frac{5400}{15,000} = 0.36$
It costs \$0.36 per 1 mile to own and operate a compact car.

31. l_1 has a negative slope; l_2 has a positive slope.
l_2 has the greater slope.

33. l_1 has a negative slope; l_2 has a 0 slope.
l_2 has the greater slope.

35. l_2 has a steeper positive slope than l_1.
l_2 has the greater slope.

37. **a.** $m \text{ for } l_1 = \frac{-2-4}{2-(-1)} = \frac{-6}{3} = -2$

$m \text{ for } l_2 = \frac{2-6}{-4-(-8)} = \frac{-4}{4} = -1$

$m \text{ for } l_3 = \frac{-4-0}{0-(-6)} = \frac{-4}{6} = -\frac{2}{3}$

b. lesser

39. $f(x) = -2x + 6$
$y = -2x + 6$
$m = -2, b = 6$

41. $-5x + y = 10$
$y = 5x + 10$
$m = 5, b = 10$

43. $-3x - 4y = 6$
$-4y = 3x + 6$
$y = -\frac{3}{4}x - \frac{3}{2}$
$m = -\frac{3}{4},\ b = -\frac{3}{2}$

45. $f(x) = -\frac{1}{4}x$
$y = -\frac{1}{4}x$
$m = -\frac{1}{4},\ b = 0$

47. $f(x) = 2x - 3$
$y = 2x - 3$
$m = 2, b = -3$
D

49. $f(x) = -2x - 3$
$y = -2x - 3$
$m = -2, b = -3$
C

51. $y = -2$
$m = 0$

53. $x = 4$
Slope is undefined.

55. $y - 7 = 0$
$y = 7$
$m = 0$

57. Answers may vary.

59. $f(x) = x + 2$
$y = x + 2$
$m = 1, b = 2$

61. $4x - 7y = 28$
$-7y = -4x + 28$
$y = \frac{4}{7}x - 4$
$m = \frac{4}{7},\ b = -4$

63. $2y - 7 = x$
$2y = x + 7$
$y = \frac{1}{2}x + \frac{7}{2}$
$m = \frac{1}{2},\ b = \frac{7}{2}$

65. $x = 7$
Slope is undefined.
There is no *y*-intercept.

67. $f(x) = \frac{1}{7}x$

$y = \frac{1}{7}x$

$m = \frac{1}{7}, \ b = 0$

69. $x - 7 = 0$
$x = 7$
Slope is undefined.
There is no *y*-intercept.

71. $2y + 4 = -7$
$2y = -11$
$y = -\frac{11}{2}$
$m = 0, \ b = -\frac{11}{2}$

73. $f(x) = 5x - 6 \quad g(x) = 5x + 2$
$y = 5x - 6 \quad y = 5x + 2$
$m = 5 \quad m = 5$
Parallel, since they have the same slope.

75. $2x - y = -10 \quad 2x + 4y = 2$
$-y = -2x - 10 \quad 4y = -2x + 2$
$y = 2x + 10 \quad y = -\frac{1}{2}x + \frac{1}{2}$
$m = 2 \quad m = -\frac{1}{2}$
Perpendicular, since the product of their slopes is –1.

77. $x + 4y = 7 \quad 2x - 5y = 0$
$4y = -x + 7 \quad -5y = -2x$
$y = -\frac{1}{4}x + \frac{7}{4} \quad y = \frac{2}{5}x$
$m = -\frac{1}{4} \quad m = \frac{2}{5}$
Neither, since their slopes are not equal, nor does the product of the slopes equal –1.

79. Answers may vary.

81. $y = 1054.7x + 23{,}285.9$

a. $x = 1996 - 1991 = 5$
$y = 1054.7(5) + 23{,}285.9$
$y = 28{,}559.4$
The income is $28,559.40.

b. $m = 1054.7$; the annual average income increases $1054.70 each year.

c. $b = 23{,}285.9$; at year $x = 0$, or 1991, the annual average income was $23,285.90.

83. $-76x + 10y = 1130$
$10y = 76x + 1130$
$y = 7.6x + 113$

a. $m = 7.6$; $b = 113$

b. The number of people employed as paralegals increases by 7.6 thousand each year.

c. There were 113 thousand paralegals employed in 1996.

85. $f(x) = 72.9x + 785.2$
$y = 72.9x + 785.2$

a. $m = 72.9$; the yearly cost of tuition increases by $72.90 each year.

b. $b = 785.2$; the yearly cost of tuition in 1990 was $785.20.

87. $f(x) = x$
$m = 1$
The slope of a parallel line is 1.

89. $f(x) = x$
$m = 1$
The slope of a perpendicular line is –1.

91. $-3x + 4y = 10$
$4y = 3x + 10$
$y = \frac{3}{4}x + \frac{5}{2}$
A parallel line has the same slope.
$m = \frac{3}{4}$

93. a. $B = (6, 20)$

b. $C = (10, 13)$

c. $m = \frac{13-20}{10-6} = -\frac{7}{4}$, or -1.75 yards per second

d. $F = (22, 2)$
$G = (26, 8)$
$m = \frac{8-2}{26-22} = \frac{6}{4} = \frac{3}{2}$,
or 1.5 yards per second

95. $-4x + 2y = 5$
$2x - y = 7$

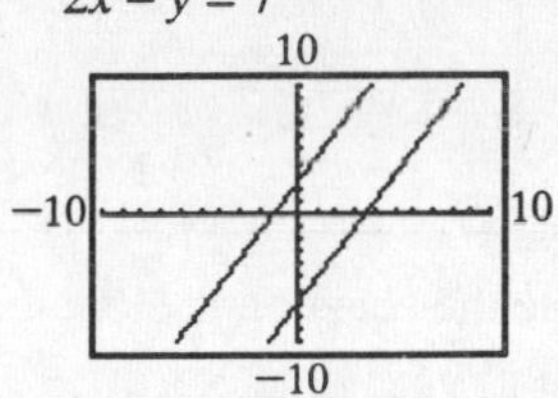

97. a. $y = \frac{1}{2}x + 1$
$y = x + 1$
$y = 2x + 1$

10
−10
10
−10

b. $y = -\frac{1}{2}x + 1$
$y = -x + 1$
$y = -2x + 1$

10
−10
10
−10

c. True

99.
$$y - 0 = -3[x - (-10)]$$
$$y = -3(x + 10)$$
$$y = -3x - 30$$

101.
$$y - 9 = -8[x - (-4)]$$
$$y - 9 = -8(x + 4)$$
$$y - 9 = -8x - 32$$
$$y = -8x - 23$$

Section 3.6

Mental Math

1. $m = -4$, $b = 12$

2. $m = \frac{2}{3}$, $b = -\frac{7}{2}$

3. $m = 5$, $b = 0$

4. $m = -1$, $b = 0$

5. $m = \frac{1}{2}$, $b = 6$

6. $m = -\frac{2}{3}$, $b = 5$

7. Parallel; both have slope $m = 12$.

8. Parallel; both have slope $m = -5$.

9. Neither; the slopes of the lines, -9 and $\frac{3}{2}$, are not equal and the product of the slopes, $-9 \cdot \frac{3}{2} = \frac{-27}{2}$, is not equal to -1.

10. Neither; the slopes of the lines, 2 and $\frac{1}{2}$, are not equal and the product of the slopes, $2 \cdot \frac{1}{2} = 1$, is not equal to -1.

Exercise Set 3.6

1. $m = -1; b = 1$
$y = mx + b$
$y = (-1)x + 1$
$y = -x + 1$

3. $m = 2;\ b = \frac{3}{4}$
$y = mx + b$
$y = 2x + \frac{3}{4}$

5. $m = \frac{2}{7};\ b = 0$
$y = mx + b$
$y = \frac{2}{7}x + 0$
$y = \frac{2}{7}x$

7. $y = 5x$
The equation $y = 5x$ is equivalent to $y = 5x + 0$, from which we see that the slope of the line is 5 and the y-intercept is (0, 0). To graph the line, we first plot the y-intercept (0, 0). To find another point, we use the slope $5 = \frac{5}{1}$. Starting at (0, 0), we move 5 units up and then 1 unit to the right, arriving at the point (1, 5).

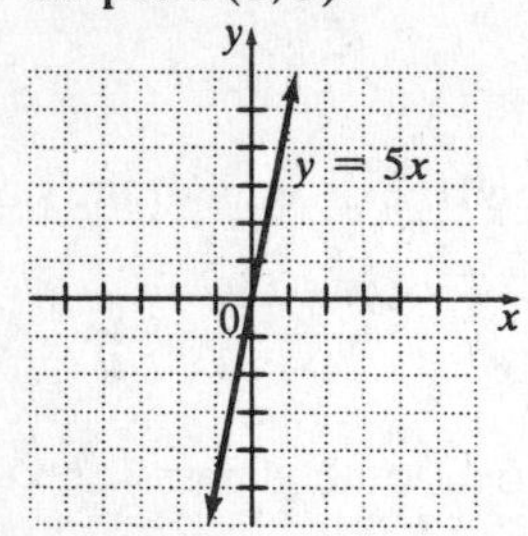

9. $x + y = 7$
First, write the equation in slope-intercept form by solving for y.
$y = -x + 7$

Thus, the slope of the line is -1 and the y-intercept is (0, 7). To graph the line, we first plot the y-intercept (0, 7). To find another point, we use the slope $-1 = \frac{-1}{1}$.
Starting at the point (0, 7), we move 1 unit down (we move down because the numerator has a negative value) and then 1 unit to the right, arriving at the point (1, 6).

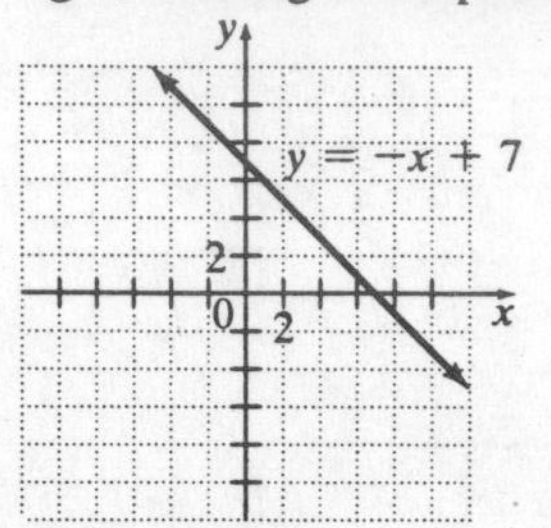

11. $-3x + 2y = 3$
First, write the equation in slope-intercept form by solving for y.
$2y = 3x + 3$
$y = \frac{3}{2}x + \frac{3}{2}$

Thus the slope of the line is $\frac{3}{2}$ and the y-intercept is $\left(0, \frac{3}{2}\right)$. To graph the line, we first plot the y-intercept $\left(0, \frac{3}{2}\right)$. To find another point, we use the slope $\frac{3}{2}$. Starting at $\left(0, \frac{3}{2}\right)$, we move 3 units up and then 2 units to the right, arriving at the point $\left(2, \frac{9}{2}\right)$.

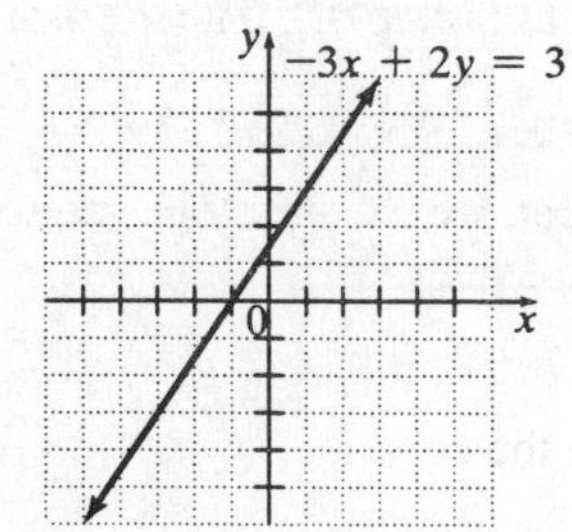

13. $m = 3;\ (x_1,\ y_1) = (1,\ 2)$
$y - y_1 = m(x - x_1)$
$y - 2 = 3(x - 1)$
$y - 2 = 3x - 3$
$y = 3x - 1$

15. $m = -2;\ (x_1,\ y_1) = (1,\ -3)$
$y - y_1 = m(x - x_1)$
$y - (-3) = -2(x - 1)$
$y + 3 = -2x + 2$
$y = -2x - 1$

17. $m = \frac{1}{2};(x_1,\ y_1) = (-6,\ 2)$
$y - y_1 = m(x - x_1)$
$y - 2 = \frac{1}{2}[x - (-6)]$
$y - 2 = \frac{1}{2}(x + 6)$
$y - 2 = \frac{1}{2}x + 3$
$y = \frac{1}{2}x + 5$

19. $m = -\frac{9}{10};\ (x_1,\ y_1) = (-3,\ 0)$
$y - y_1 = m(x - x_1)$
$y - 0 = -\frac{9}{10}[x - (-3)]$
$y = -\frac{9}{10}(x + 3)$
$y = -\frac{9}{10}x - \frac{27}{10}$

21. The y-intercept of the line is (0, 3). To go from (0, 3) to the other point indicated, we go 2 units down and 1 unit to the right. Thus, the slope of the line is $\frac{-2}{1} = -2$. Since we know the slope and the y-intercept, we use the slope-intercept form to write the equation.
$y = mx + b$
$y = -2x + 3$
Then we rewrite the equation in standard form.
$2x + y = 3$

23. The two points on the line that are marked are (–2, 1) and (4, 5). To go from (–2, 1) to (4, 5), we move 4 units up and then 6 units to the right. Thus, the slope is $\frac{4}{6} = \frac{2}{3}$. Since we know the slope and a point, we use the point-slope form to write the equation and then simplify to write it in standard form.
$y - y_1 = m(x - x_1)$
$y - 1 = \frac{2}{3}(x + 2)$
$3y - 3 = 2x + 4$
$2x - 3y = -7$

In each of Exercises 25–31, we are given two points and asked to find an equation. To accomplish this, we first find the slope, using the formula $m = \frac{y_2 - y_1}{x_2 - x_1}$, and then use the slope and one of the points to write the equation, using the point-slope form $y - y_1 = m(x - x_1)$. To write the equation using function notation, we must first solve the equation for y and then replace y with $f(x)$.

25. The points are (2, 0) and (4, 6).
$m = \frac{y_2 - y_1}{x_2 - x_1} = \frac{6 - 0}{4 - 2} = \frac{6}{2} = 3$
$y - y_1 = m(x - x_1)$
$y - 0 = 3(x - 2)$
$y = 3x - 6$
$f(x) = 3x - 6$

27. The points are (–2, 5) and (–6, 13).
$m = \frac{y_2 - y_1}{x_2 - x_1} = \frac{13 - 5}{-6 - (-2)} = \frac{8}{-4} = -2$
$y - 5 = -2(x - (-2))$
$y - 5 = -2(x + 2)$
$y - 5 = -2x - 4$
$y = -2x + 1$
$f(x) = -2x + 1$

29. The points are (–2, –4) and (–4, –3).

$$m = \frac{y_2 - y_1}{x_2 - x_1} = \frac{-3-(-4)}{-4-(-2)} = \frac{1}{-2} = -\frac{1}{2}$$

$$y - y_1 = m(x - x_1)$$
$$y - (-4) = -\frac{1}{2}(x - (-2))$$
$$y + 4 = -\frac{1}{2}(x + 2)$$
$$y + 4 = -\frac{1}{2}x - 1$$
$$y = -\frac{1}{2}x - 5$$
$$f(x) = -\frac{1}{2}x - 5$$

31. The points are (–3, –8) and (–6, –9).

$$m = \frac{y_2 - y_1}{x_2 - x_1} = \frac{-9-(-8)}{-6-(-3)} = \frac{-1}{-3} = \frac{1}{3}$$

$$y - y_1 = m(x - x_1)$$
$$y - (-8) = \frac{1}{3}(x - (-3))$$
$$y + 8 = \frac{1}{3}(x + 3)$$
$$y + 8 = \frac{1}{3}x + 1$$
$$y = \frac{1}{3}x - 7$$
$$f(x) = \frac{1}{3}x - 7$$

33. Answers may vary.

35. $f(0)$ is the y-coordinate of the point on the graph having x-coordinate 0. This point is (0, –2). Thus, $f(0) = -2$.

37. $f(2)$ is the y-coordinate of the point on the graph having x-coordinate 2. This point is (2, 2). Thus, $f(2) = 2$.

39. The value of x such that $f(x) = -6$ is the x-coordinate of the point having y-coordinate –6. This point is (–2, –6). Thus, $f(-2) = -6$, so $x = -2$.

41. Since the slope is 0, the line is horizontal and has an equation of the form $y = c$. Since the line passes through (–2, –4), the equation is $y = -4$.

43. A vertical line has an equation of the form $x = c$. Since the line passes through the point (4, 7), the equation is $x = 4$.

45. A horizontal line has an equation of the form $y = c$. Since the line passes through the point (0, 5), the equation is $y = 5$.

47. Since the line is parallel to $f(x) = 4x - 2$ it has the same slope as $f(x) = 4x - 2$, which is $m = 4$. Since we know the slope of the line and that it passes through (3, 8), we use the point-slope form to write the equation.

$$y - y_1 = m(x - x_1)$$
$$y - 8 = 4(x - 3)$$
$$y - 8 = 4x - 12$$
$$y = 4x - 4$$
$$f(x) = 4x - 4$$

49. Since the line is perpendicular to $3y = x - 6$, we first find the slope of $3y = x - 6$ by rewriting it in slope-intercept form.

$$3y = x - 6$$
$$y = \frac{1}{3}x - 2$$

Since the slope of this line is $\frac{1}{3}$, the slope of any line perpendicular to it has slope $m = -\frac{1}{\frac{1}{3}} = -3$. Since we know the slope of the line and that it passes through (2, –5), we use the point-slope form to write the equation.

$$y - y_1 = m(x - x_1)$$
$$y - (-5) = -3(x - 2)$$
$$y + 5 = -3x + 6$$
$$y = -3x + 1$$
$$f(x) = -3x + 1$$

51. Since the line is parallel to $3x + 2y = 5$, it has the same slope. We find the slope of $3x + 2y = 5$ by rewriting the equation in slope-intercept form.

$$3x + 2y = 5$$
$$2y = -3x + 5$$
$$y = -\frac{3}{2}x + \frac{5}{2}$$

The line has slope $m = -\frac{3}{2}$ and passes through $(-2, -3)$. We use the point-slope form to write the equation

$$y - y_1 = m(x - x_1)$$
$$y - (-3) = -\frac{3}{2}(x - (-2))$$
$$y + 3 = -\frac{3}{2}(x + 2)$$
$$y + 3 = -\frac{3}{2}x - 3$$
$$y = -\frac{3}{2}x - 6$$
$$f(x) = -\frac{3}{2}x - 6$$

53. Since we know the slope and a point, we use point-slope form.

$$y - y_1 = m(x - x_1)$$
$$y - 3 = 2(x - (-2))$$
$$y - 3 = 2(x + 2)$$
$$y - 3 = 2x + 4$$
$$-2x + y = 7$$
$$2x - y = -7$$

55. Since we know two points, we first find the slope and then use the point-slope form.

$$m = \frac{y_2 - y_1}{x_2 - x_1} = \frac{2 - 6}{5 - 1} = \frac{-4}{4} = -1$$
$$y - y_1 = m(x - x_1)$$
$$y - 6 = -1(x - 1)$$
$$y - 6 = -x + 1$$
$$y = -x + 7$$
$$f(x) = -x + 7$$

57. Since we know the slope and y-intercept, we use slope-intercept form.

$$y = mx + b$$
$$y = -\frac{1}{2}x + 11$$
$$2y = -x + 22$$
$$x + 2y = 22$$

59. Since we know two points, one of which is the y-intercept, we first find the slope and then use the slope-intercept form.

$$m = \frac{y_2 - y_1}{x_2 - x_1} = \frac{-6 - (-4)}{0 - (-7)} = \frac{-2}{7}$$
$$y = mx + b$$
$$y = -\frac{2}{7}x - 6$$
$$7y = -2x - 42$$
$$2x + 7y = -42$$

61. Since we know the slope and a point, we use the point-slope form.

$$y - y_1 = m(x - x_1)$$
$$y - 0 = -\frac{4}{3}(x - (-5))$$
$$3y = -4(x + 5)$$
$$3y = -4x - 20$$
$$4x + 3y = -20$$

63. A vertical line has an equation of the form $x = c$. Since the line passes through $(-2, -10)$, the equation is $x = -2$.

65. We begin by finding the slope of $2x + 4y = 9$, since the desired line, being parallel to this line, has the same slope. To find the slope of the line, we rewrite $2x + 4y = 9$ in slope-intercept form.

$$2x + 4y = 9$$
$$4y = -2x + 9$$
$$y = -\frac{1}{2}x + \frac{9}{4}$$

Thus, the slope of the line is $-\frac{1}{2}$ and it passes through (6, –2). We now use the point-slope form to write the equation.

$$y - y_1 = m(x - x_1)$$
$$y - (-2) = -\frac{1}{2}(x - 6)$$
$$2(y + 2) = -x + 6$$
$$2y + 4 = -x + 6$$
$$x + 2y = 2$$

67. Since the slope of the line is 0, the line is horizontal. The equation of the horizontal line passing through (–9, 12) is $y = 12$.

69. We begin by finding the slope of $8x - y = 9$, since the desired line, being parallel to this line, has the same slope. To find the slope of $8x - y = 9$, we rewrite the equation in slope-intercept form.

$$8x - y = 9$$
$$8x - 9 = y$$
$$y = 8x - 9$$

Thus, the slope of the line is 8 and it passes through (6, 1). We now use the point-slope form to write the equation.

$$y - y_1 = m(x - x_1)$$
$$y - 1 = 8(x - 6)$$
$$y - 1 = 8x - 48$$
$$47 = 8x - y$$
$$8x - y = 47$$

71. Since $y = 9$ is a horizontal line, any line perpendicular to it is vertical. The equation of the vertical line passing through (5, –6) is $x = 5$.

73. Since we know two points, we first find the slope and then use the point-slope form.

$$m = \frac{y_2 - y_1}{x_2 - x_1} = \frac{-5 - (-8)}{-6 - 2} = \frac{3}{-8} = -\frac{3}{8}$$
$$y - y_1 = m(x - x_1)$$
$$y - (-8) = -\frac{3}{8}(x - 2)$$
$$y + 8 = -\frac{3}{8}x + \frac{3}{4}$$
$$y = -\frac{3}{8}x + \frac{3}{4} - \frac{32}{4}$$
$$y = -\frac{3}{8}x - \frac{29}{4}$$
$$f(x) = -\frac{3}{8}x - \frac{29}{4}$$

75. a. The ordered pairs are (0, 3280) and (2, 4760). We use the two ordered pairs to find the slope of the line through the two points. Then since we will know the slope and the y-intercept (0, 3280), we will write the equation using the slope-intercept form.

$$m = \frac{y_2 - y_1}{x_2 - x_1} = \frac{4760 - 3280}{2 - 0} = \frac{1480}{2} = 740$$
$$y = mx + b$$
$$y = 740x + 3280$$

The equation $y = 740x + 3280$ describes the relationship between time and number of electric-powered vehicles, where x is the number of years after 1996.

b. In 2005, $x = 9$.

$y = 740(9) + 3280 = 9940$

We predict that 9940 electric-powered vehicles will be in use in 2005.

77. a. We begin by writing two ordered pairs (x, P) from the information given: (1, 30,000) and (4, 66,000). We use the ordered pairs to write a linear equation by first finding the slope of the line and then using the slope and one ordered pair in the point-slope form.

$$m = \frac{y_2 - y_1}{x_2 - x_1} = \frac{66{,}000 - 30{,}000}{4 - 1} = 12{,}000$$

$$y - y_1 = m(x - x_1)$$

$$y - 30{,}000 = 12{,}000(x - 1)$$

$$y = 12{,}000x + 18{,}000$$

To write this in function notation, we substitute $P(x)$ for y.

$P(x) = 12{,}000x + 18{,}000$

The linear function

$P(x) = 12{,}000x + 18{,}000$ expresses profit as a function of time.

b. $P(7) = 12{,}000(7) + 18{,}000$

$= 102{,}000$

We predict that the company's profits at the end of the seventh year will be \$102,000.

c. To predict when the company's profit should reach \$126,000, we replace $P(x)$ with 126,000 in the equation and solve for x.

$$126{,}000 = 12{,}000x + 18{,}000$$

$$108{,}000 = 12{,}000x$$

$$9 = x$$

The profit should reach \$126,000 at the end of the 9th year.

79. a. The line passes through the points (0, 109,000) and (4, 128,000). We first use these ordered pairs to find the slope of the line and then use the slope and y-intercept (0, 109,000) to write the equation in the slope-intercept form.

$$m = \frac{y_2 - y_1}{x_2 - x_1} = \frac{128{,}400 - 109{,}900}{4 - 0}$$

$$= \frac{18{,}500}{4} = 4625$$

$$y = mx + b$$

$$y = 4625x + 109{,}900$$

The linear equation

$y = 4625x + 109{,}900$ models the median existing home price in terms of the number of years after 1994.

b. Since 2008 is 14 years after 1994, we let $x = 14$ in the equation.

$y = 4625(14) + 109{,}900 = 174{,}650$

We predict that the median existing home price for the year 2008 will be \$174,650.

81. a. The line passes through (0, 225) and (10, 391). We first use these ordered pairs to find the slope of the line and then use the slope and y-intercept (0, 225) to write the equation in the slope-intercept form.

$$m = \frac{y_2 - y_1}{x_2 - x_1} = \frac{391 - 225}{10 - 0} = 16.6$$

$$y = mx + b$$

$$y = 16.6x + 225$$

The linear equation $y = 16.6x + 225$ models the number of people (in thousands) employed as medical assistants x years after 1996.

b. Since 2004 is 8 years after 1996, we let $x = 8$ in the equation.

$y = 16.6(8) + 225 = 357.8$

We estimate that 357.8 thousand people will be employed as medical assistants in the year 2004.

83. The function from Exercise 55 is $f(x) = -x + 7$.

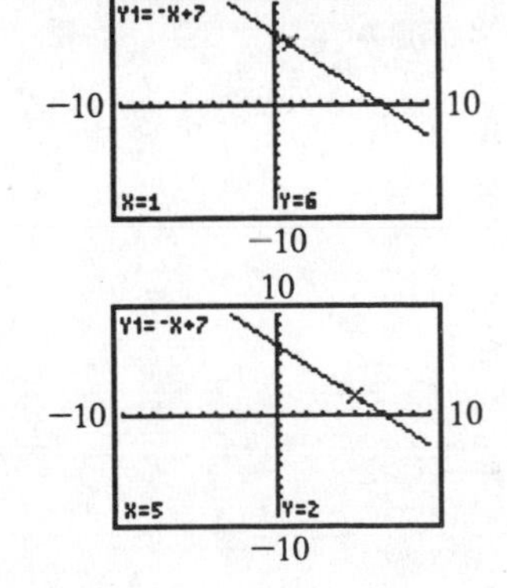

85. The equation from Exercise 61 is $4x + 3y = -20$. To graph the line, we must first solve for x.

$$4x+3y=-20$$
$$3y=-4x-20$$
$$y=-\frac{4}{3}x-\frac{20}{3}$$

Y1=-(4/3)X-20/3, window $[-10, 10]$ by $[-10, 10]$, X=-5 Y=0

We see that the line has a negative slope since it goes down from left to right.

87.
$$2x-7\le 21$$
$$2x\le 28$$
$$x\le 14$$
$(-\infty, 14]$

14

89.
$$5(x-2)\ge 3(x-1)$$
$$5x-10\ge 3x-3$$
$$2x\ge 7$$
$$x\ge \frac{7}{2}$$
$\left[\frac{7}{2}, \infty\right)$

$\frac{7}{2}$

91.
$$\frac{x}{2}+\frac{1}{4}<\frac{1}{8}$$
$$8\left(\frac{x}{2}+\frac{1}{4}\right)<8\left(\frac{1}{8}\right)$$
$$4x+2<1$$
$$4x<-1$$
$$x<-\frac{1}{4}$$
$\left(-\infty, -\frac{1}{4}\right)$

$-\frac{1}{4}$

93. First, the midpoint of the segment with endpoints $(3, -1)$ and $(-5, 1)$ is $(-1, 0)$. The slope of this segment is

$$m=\frac{y_2-y_1}{x_2-x_1}=\frac{1-(-1)}{-5-3}=\frac{1+1}{-8}=\frac{2}{-8}=-\frac{1}{4}.$$

A line perpendicular to this segment must have slope 4. Finally, the equation of the perpendicular bisector is given by

$$y-y_1=m(x-x_1)$$
$$y-0=4(x-(-1))$$
$$y=4(x+1)$$
$$y=4x+4$$
$$-4x+y=4.$$

95. First, the midpoint of the segment with endpoints $(-2, 6)$ and $(-22, -4)$ is $(-12, 1)$. The slope of this segment is

$$m=\frac{y_2-y_1}{x_2-x_1}=\frac{-4-6}{-22-(-2)}=\frac{-10}{-22+2}=\frac{-10}{-20}=\frac{1}{2}.$$

A line perpendicular to this segment must have slope -2. Finally, the equation of the perpendicular bisector is given by

$$y-y_1=m(x-x_1)$$
$$y-1=-2(x-(-12))$$
$$y-1=-2(x+12)$$
$$y-1=-2x-24$$
$$2x+y=-23.$$

97. First, the midpoint of the segment with endpoints $(2, 3)$ and $(-4, 7)$ is $(-1, 5)$. The slope of this segment is

$$m=\frac{y_2-y_1}{x_2-x_1}=\frac{3-7}{2-(-4)}=\frac{-4}{2+4}=\frac{-4}{6}=-\frac{2}{3}.$$

A line perpendicular to this segment must have slope $\frac{3}{2}$. Finally, the equation of the perpendicular bisector is given by

$$y-y_1=m(x-x_1)$$
$$y-5=\frac{3}{2}(x-(-1))$$
$$2(y-5)=2\left[\frac{3}{2}(x+1)\right]$$
$$2y-10=3(x+1)$$
$$2y-10=3x+3$$
$$-13=3x-2y$$
$$3x-2y=-13$$

Exercise Set 3.7

1. $x < 2$
   Graph $x = 2$ as a dashed line, because the inequality symbol is $<$.
   Test: (0, 0)
   $0 < 2$ True
   Shade the half-plane that contains (0, 0).

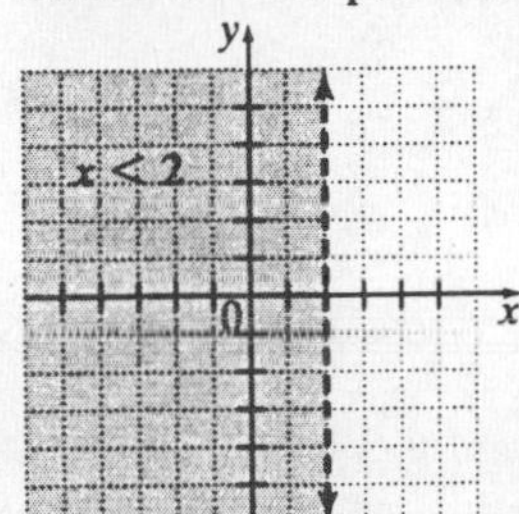

3. $x - y \geq 7$
   Graph $x - y = 7$ as a solid line, because the inequality symbol is $\geq$.
   Test: (0, 0)
   $0 - 0 \geq 7$
   $0 \geq 7$ False
   Shade the half-plane that does not contain (0, 0).

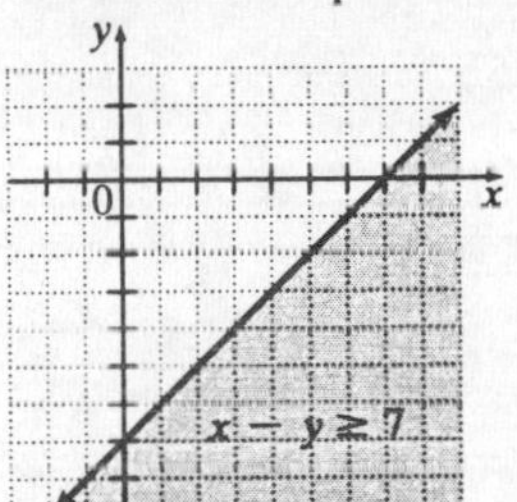

5. $3x + y > 6$
   Graph $3x + y = 6$ as a dashed line, because the inequality symbol is $>$.
   Test: (0, 0)
   $3(0) + 0 > 6$
   $0 > 6$ False
   Shade the half-plane that does not contain (0, 0).

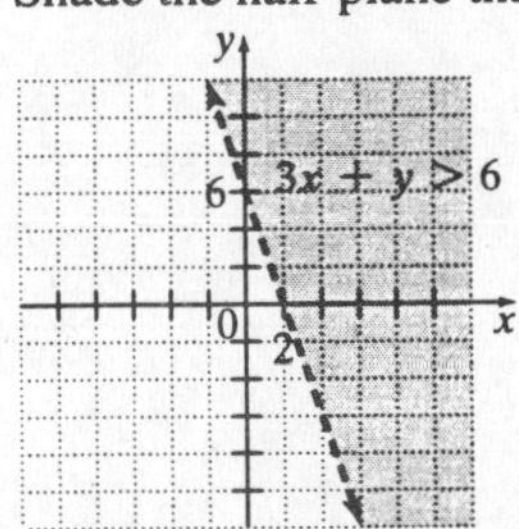

7. $y \leq -2x$
   Graph $y = -2x$ as a solid line because the inequality symbol is $\leq$.
   Test: (1, 1)
   $1 \leq -2(1)$
   $1 \leq -2$ False
   Shade the half-plane that does not contain (1, 1).

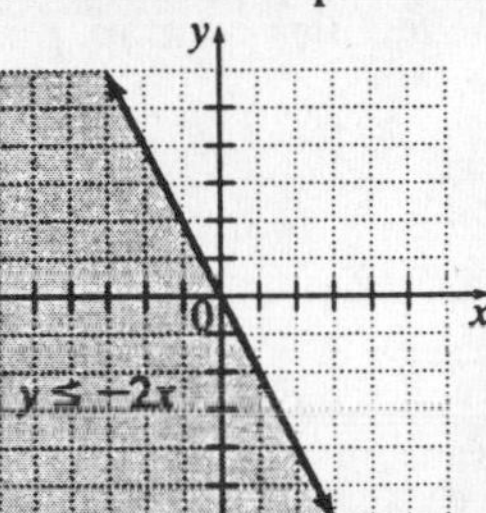

9. $2x + 4y \geq 8$
   Graph $2x + 4y = 8$ as a solid line because the inequality symbol is $\geq$.
   Test: (0, 0)
   $2(0) + 4(0) \geq 8$
   $0 \geq 8$ False
   Shade the half-plane that does not contain (0, 0).

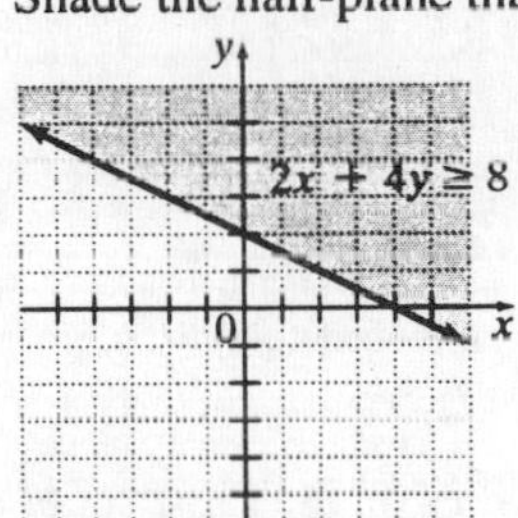

11. $5x + 3y > -15$
    Graph $5x + 3y = -15$ as a dashed line because the inequality symbol is $>$.
    Test: (0, 0)
    $5(0) + 3(0) > -15$
    $0 > -15$ True
    Shade the half-plane that contains (0, 0).

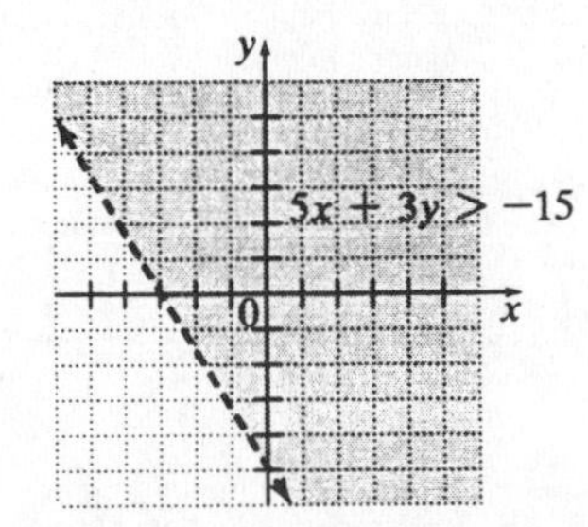

13. Answers may vary. A dashed boundary line should be used when the inequality contains either < or >.

15. $x \geq 3$ and $y \leq -2$
Graph both inequalities on the same set of axes. The intersection is the darker region where the graphs of the two inequalities overlap.

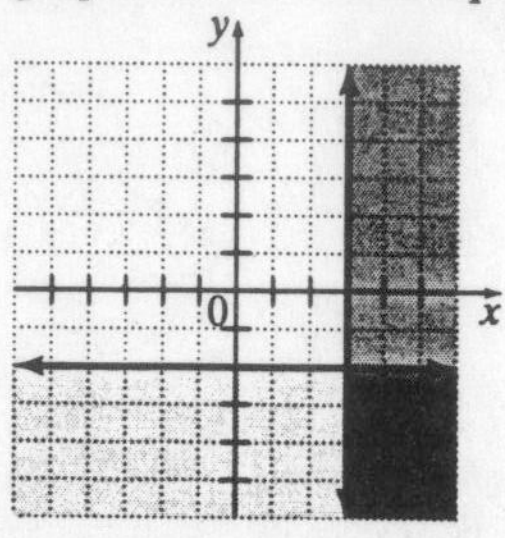

17. $x \leq -2$ or $y \geq 4$
Graph both inequalities on the same set of axes. The union is the entire shaded region.

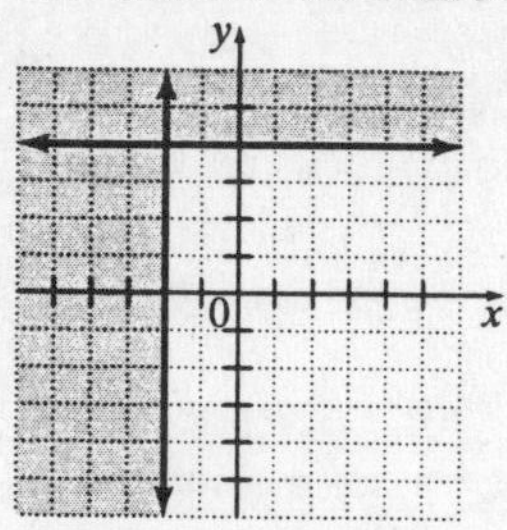

19. $x - y < 3$ and $x > 4$
Graph both inequalities on the same set of axes. The intersection is the darker region where the graphs of the two inequalities overlap.

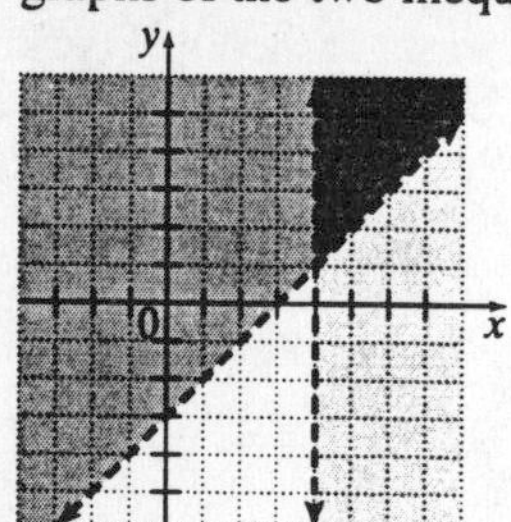

21. $x + y \leq 3$ or $x - y \geq 5$
Graph both inequalities on the same set of axes. The union is the entire shaded region.

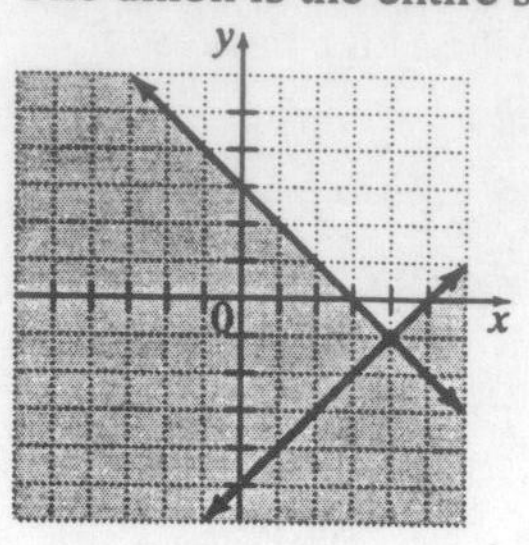

23. $y \geq -2$

25. $x - 6y < 12$

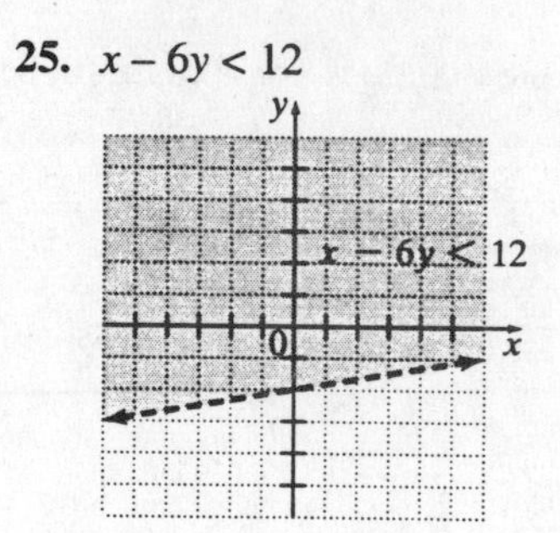

27. $x > 5$

29. $-2x + y \leq 4$

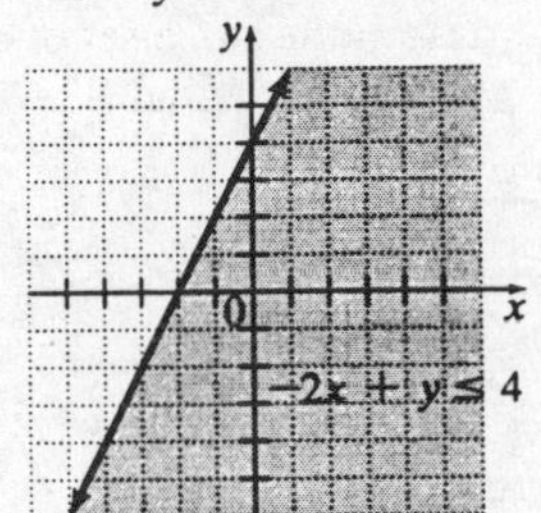

31. $x - 3y < 0$

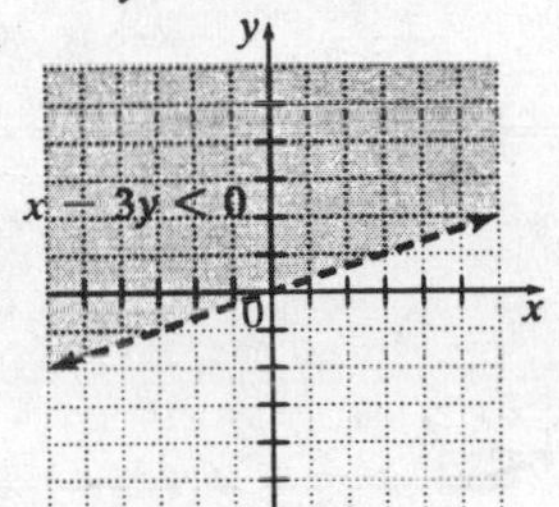

33. $3x - 2y \leq 12$

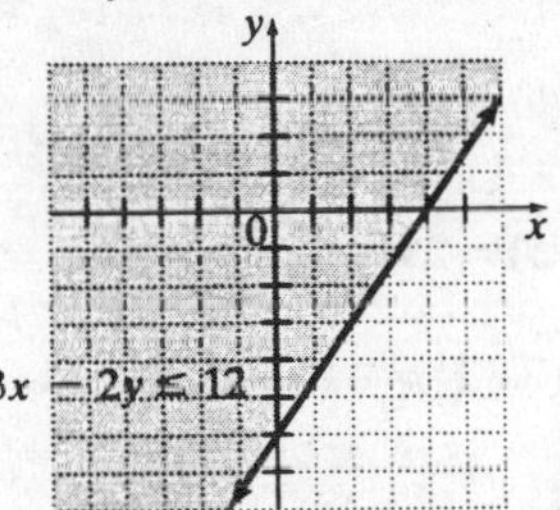

35. $x - y \geq 2$ or $y < 5$

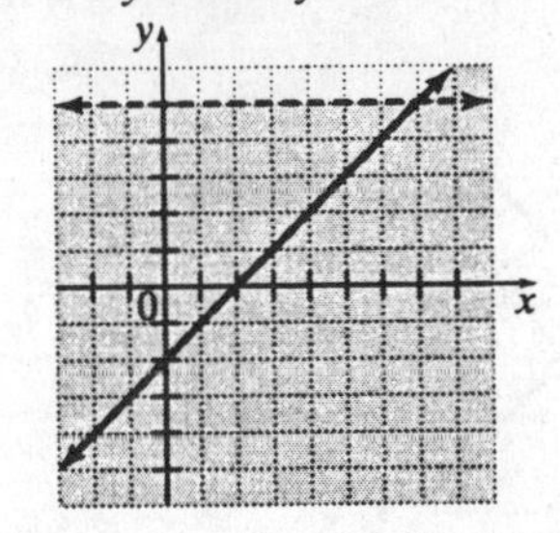

37. $x + y \leq 1$ and $y \leq -1$

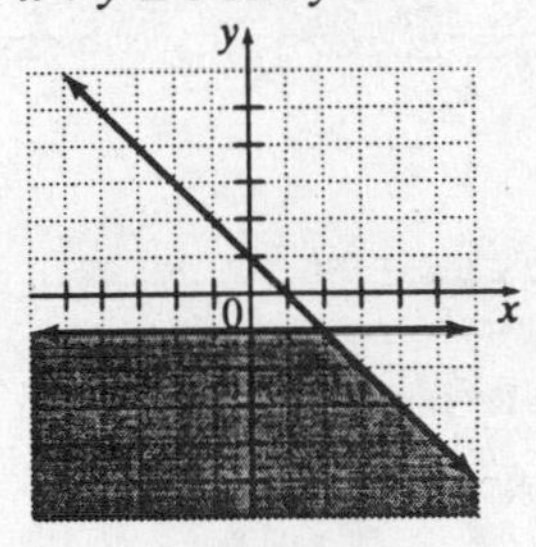

39. $2x + y > 4$ or $x \geq 1$

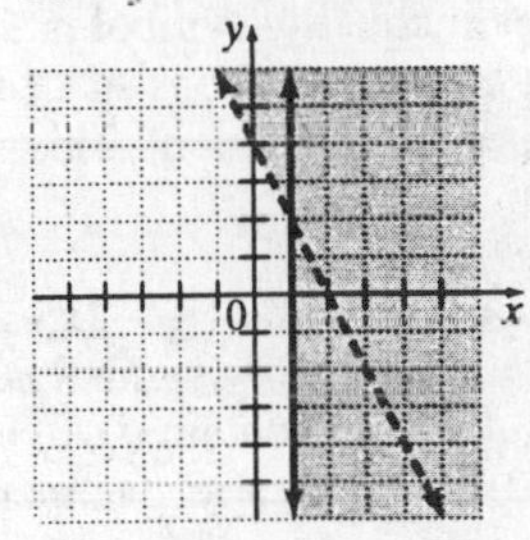

41. $x \geq -2$ and $x \leq 1$

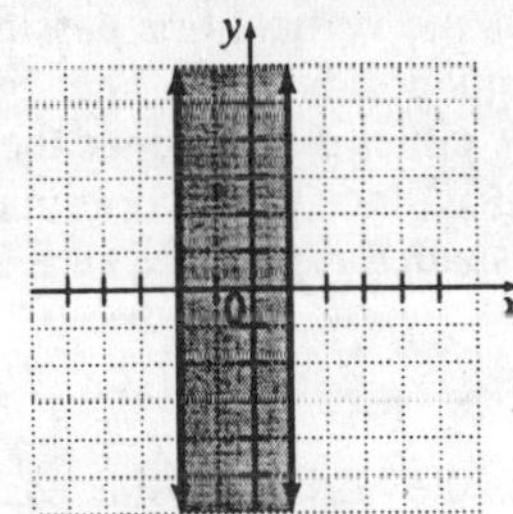

43. $x + y \leq 0$ or $3x - 6y \geq 12$

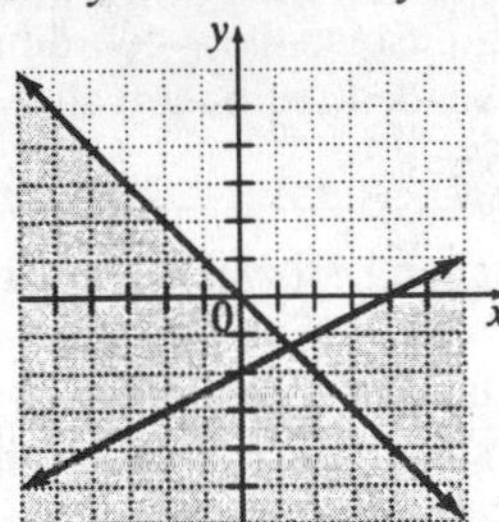

45. $2x - y > 3$ and $x \geq 0$

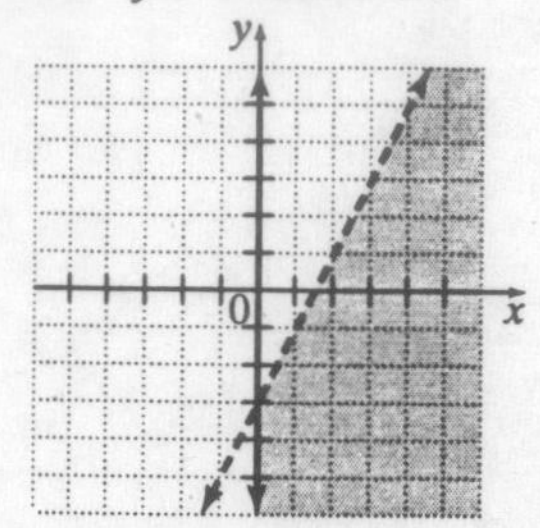

47. The boundary line of the graph of $y \leq 2x + 3$ is a solid line since the inequality symbol is $\leq$. The half-plane that contains (0, 0) is shaded since $0 \leq 2(0) + 3$ is a true statement. This describes graph D.

49. The boundary line of the graph of $y > 2x + 3$ is a dashed line since the inequality symbol is >. The half-plane that does not contain (0, 0) is shaded since $0 > 2(0) + 3$ is a false statement. This describes graph A.

51. The boundary line is the vertical line passing through (2, 0); its equation is $x = 2$. The inequality symbol is either $\leq$ or $\geq$ since the boundary line is solid. Since the x-coordinates of the points in the shaded region are all ≥ 2, the inequality is
$x \geq 2$.

53. The boundary line is the horizontal line passing through (0, –3); its equation is $y = -3$. The inequality symbol is either $\leq$ or $\geq$ since the boundary line is solid. Since the y-coordinates of the points in the shaded region are all ≤ -3, the inequality is $y \leq -3$.

55. The boundary line is the horizontal line through (0, 4); its equation is $y = 4$. The inequality is either < or > since the boundary line is dashed. Since the y-coordinates of the points are all ≥ 4, the inequality is $y \geq 4$.

57. The boundary line is the vertical line passing through (1, 0); its equation is $x = 1$. The inequality is either < or > since the boundary line is dashed. Since the x-coordinates of the points in the shaded region are all < 1, the inequality is $x < 1$.

59. The inequalities are $0 \leq x \leq 20$ and $y \geq 10$.

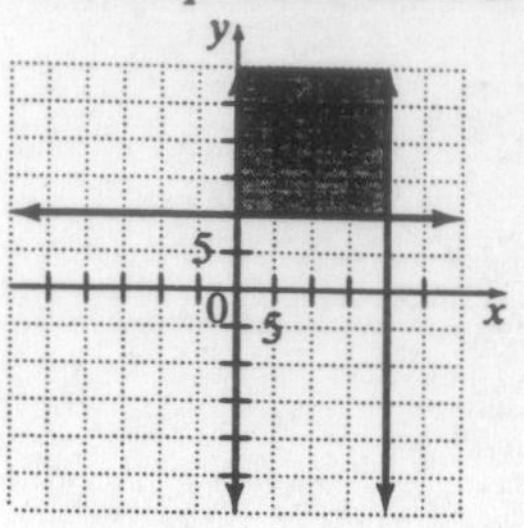

61. $\begin{cases} x \geq 0 \\ y \geq 0 \\ 2x + 4y \leq 40 \end{cases}$

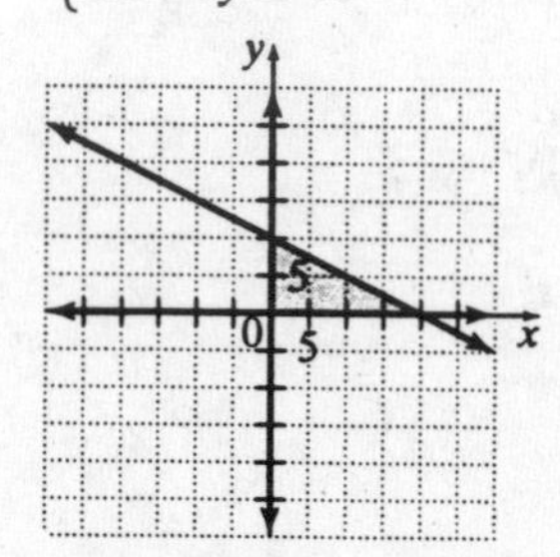

63. $3^2 = 3 \cdot 3 = 9$

65. $(-5)^2 = (-5)(-5) = 25$

67. $-2^4 = -(2)(2)(2)(2) = -16$

69. $\left(\frac{2}{7}\right)^2 = \frac{2^2}{7^2} = \frac{4}{49}$

71.

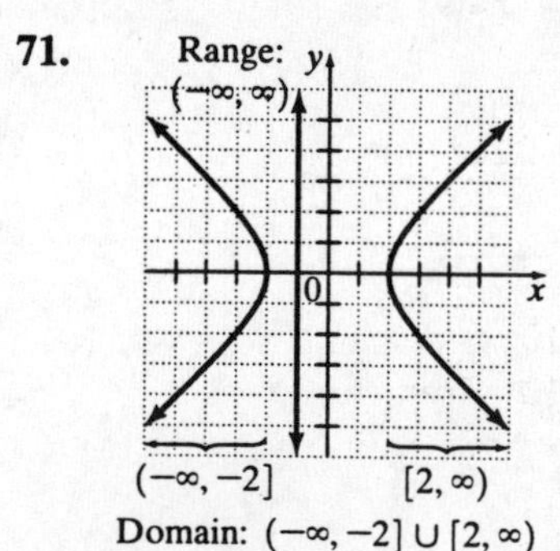

Domain: $(-\infty, 2] \cup [2, \infty)$
Range: $(-\infty, \infty)$
This relation is not a function, since many vertical lines intersect the graph in more than one point.

Chapter 3 Review Exercises

1. $A(2, -1)$, quadrant IV
$B(-2, 1)$, quadrant II
$C(0, 3)$, y-axis
$D(-3, -5)$, quadrant III

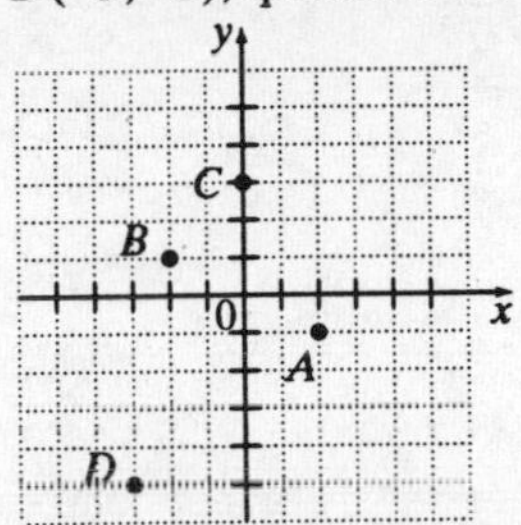

2. $A(-3, 4)$, quadrant II
$B(4, -3)$, quadrant IV
$C(-2, 0)$, x-axis
$D(-4, 1)$, quadrant II

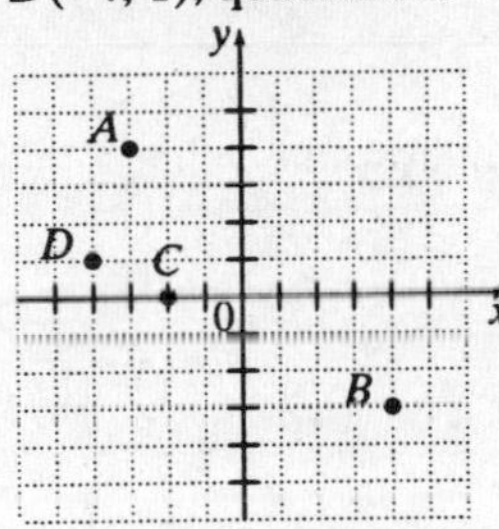

3. $7x - 8y = 56$
(0, 56); no
$7(0) - 8(56) = 56$
$-448 = 56$ False
(8, 0); yes
$7(8) - 8(0) = 56$
$56 = 56$ True

4. $-2x + 5y = 10$
(−5, 0); yes
$-2(-5) + 5(0) = 10$
$10 = 10$ True
(1, 1); no
$-2(1) + 5(1) = 10$
$-2 + 5 = 10$
$3 = 10$ False

5. a. (8.00, 1), (7.50, 10), (6.50, 25), (5.00, 50), (2.00, 100)

b.

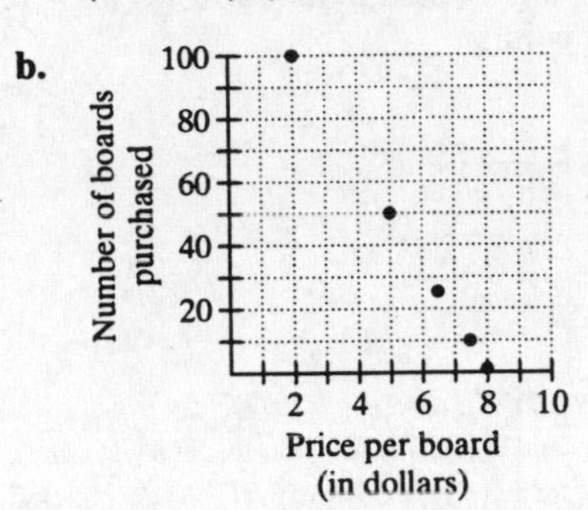

6. a. (1992, 46), (1993, 45), (1994, 47), (1995, 49), (1996, 51), (1997, 53)

b.

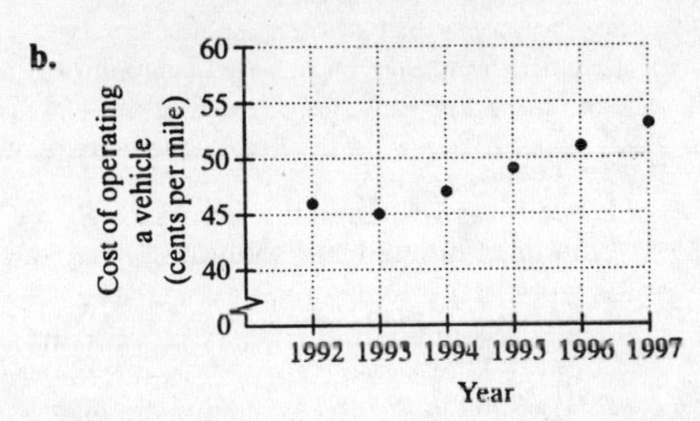

7. $y = 3x$
Linear; it is written in the form $y = mx + b$.
Find 3 points:

x	y
−1	−3
0	0
1	3

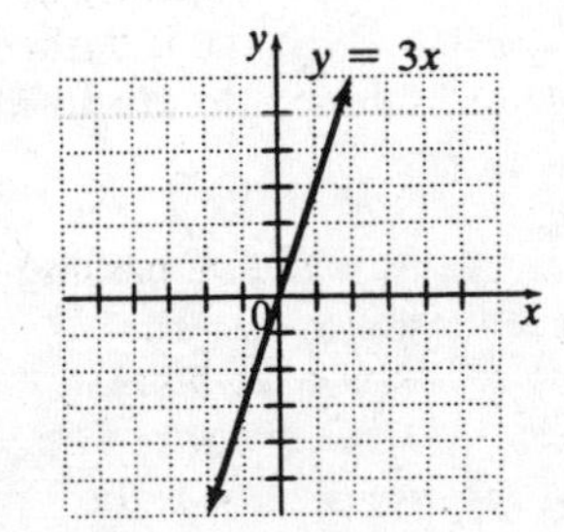

8. $y = 5x$
Linear; it is written in the form $y = mx + b$.
Find 3 points:

x	y
–1	–5
0	0
1	5

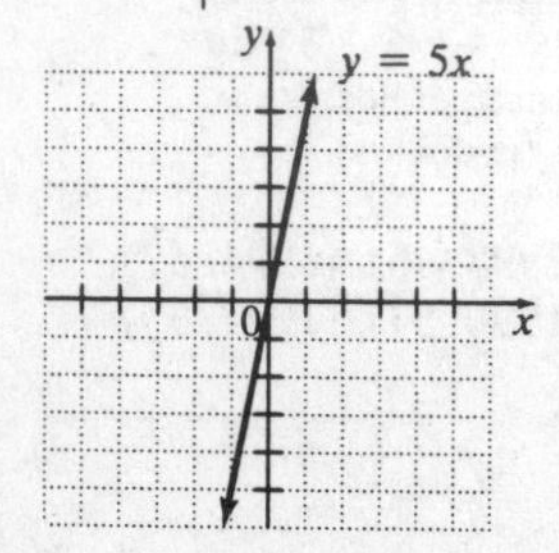

9. $3x - y = 4$
Linear; it is written in the from $Ax + By = C$.
Find three ordered pair solutions.

x	y
0	–4
1	–1
2	2

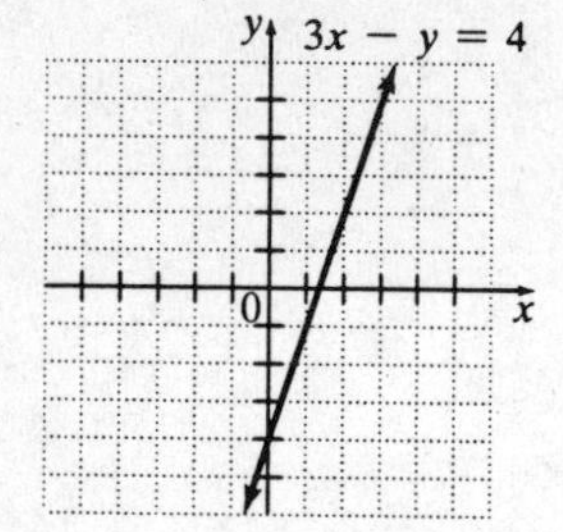

10. $x - 3y = 2$
Linear; it is written in the form $Ax + By = C$.
Find three ordered pair solutions.

x	y
–4	–2
–1	–1
2	0

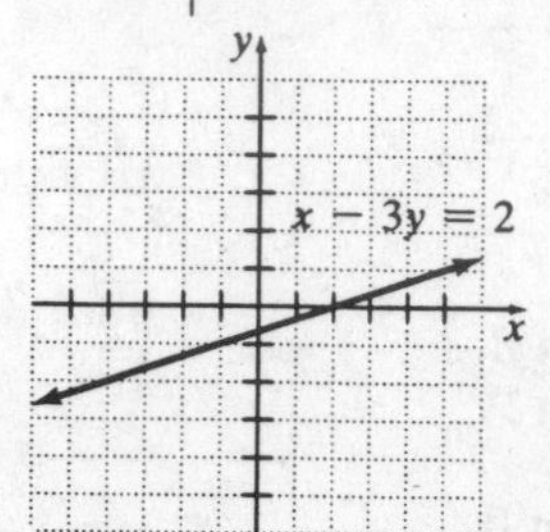

11. $y = |x| + 4$
Not linear; the x is in absolute value.

x	y
–3	7
–2	6
–1	5
0	4
1	5
2	6
3	7

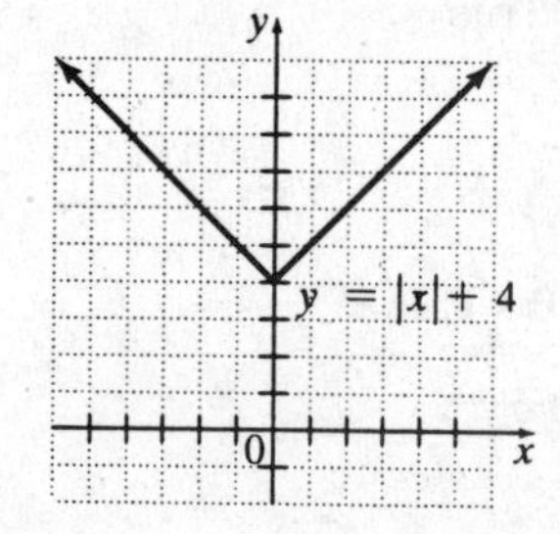

12. $y = x^2 + 4$
Not linear; the x is squared.

x	y
−3	13
−2	8
−1	5
0	4
1	5
2	8
3	13

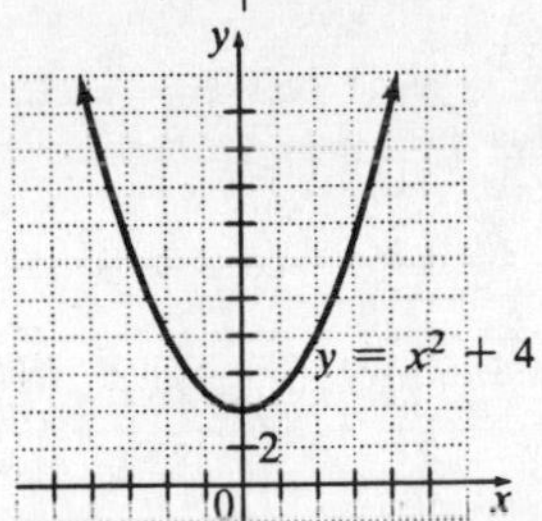

13. $y = -\frac{1}{2}x + 2$
Linear; it is written in the form $y = mx + b$.
Find three ordered pair solutions.

x	y
−2	3
0	2
2	1

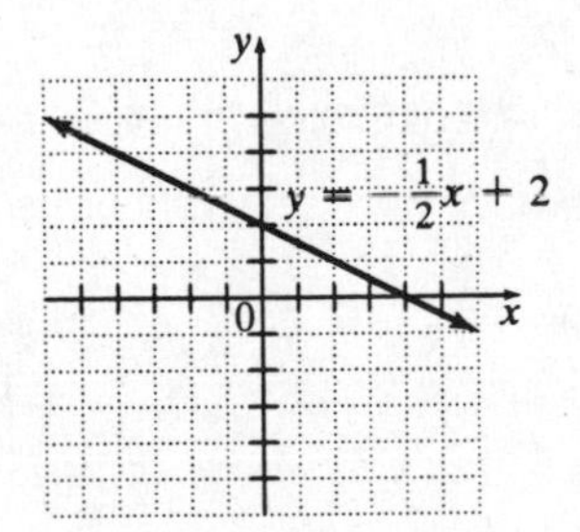

14. $y = -x + 5$
Linear; it is written in the form $y = mx + b$.
Find three ordered pair solutions.

x	y
−2	7
0	5
2	3

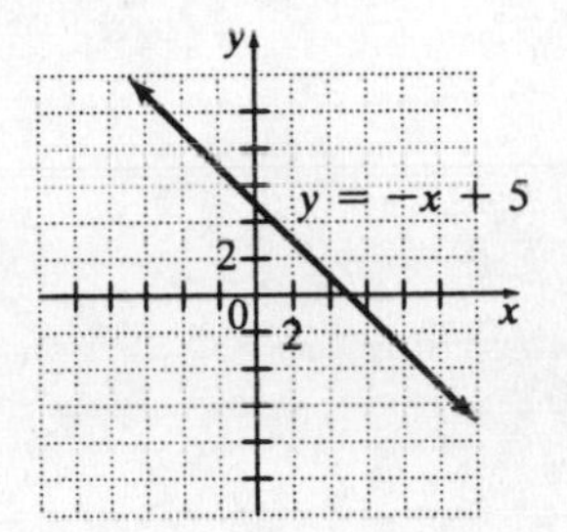

15. $y = 2x - 1$
Linear; it is written in the from $y = mx + b$.
Find three ordered pair solutions.

x	y
−1	−3
0	−1
1	1

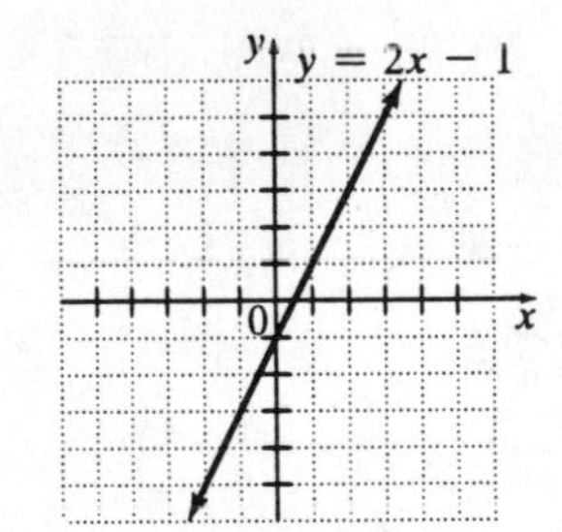

16. $y = \frac{1}{3}x + 1$
Linear; it is written in the form $y = mx + b$.
Find three ordered pair solutions.

x	y
−3	0
0	1
3	2

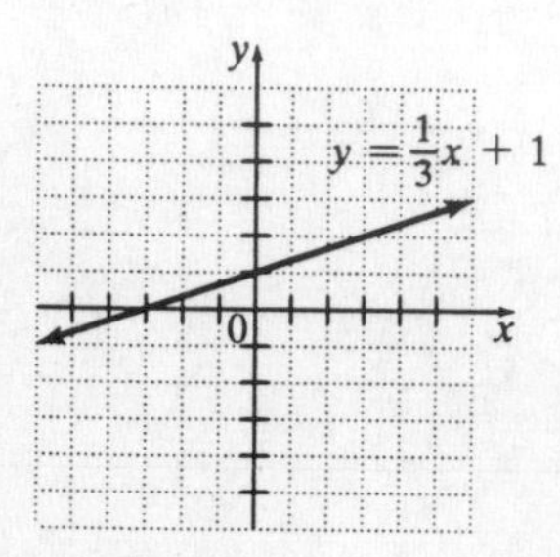

17. $y = -1.36x$
Linear; it is written in the form $y = mx + b$.
Find three ordered pair solutions.

x	y
−3	4.08
0	0
3	−4.08

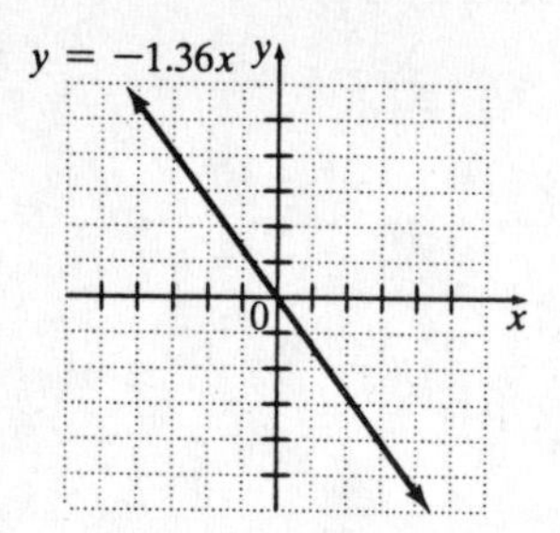

18. $y = 2.1x + 5.9$
Linear; it is written in the form $y = mx + b$.
Find three ordered pair solutions.

x	y
−2	1.7
−1	3.8
0	5.9

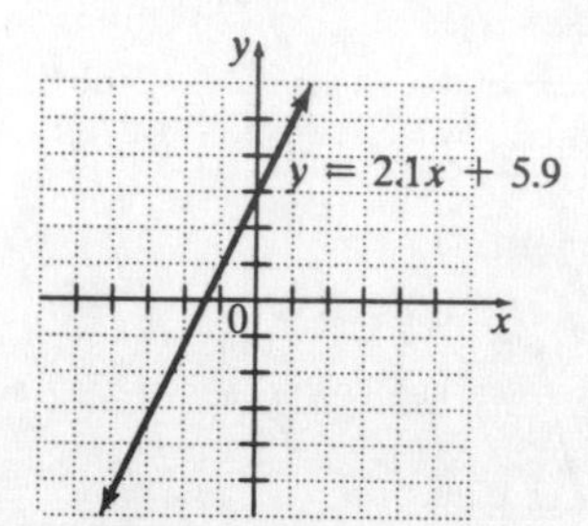

19. The domain is the set of all first coordinates, $\left\{-\frac{1}{2}, 6, 0, 25\right\}$.
The range is the set of all second coordinates $\left\{\frac{3}{4}, -12, 25\right\}$.
The relation is a function since each x-coordinate is paired with only one y-coordinate.

20. The domain is the set of all first coordinates, $\left\{\frac{3}{4}, -12, 25\right\}$.
The range is the set of all second coordinates, $\left\{-\frac{1}{2}, 6, 0, 25\right\}$.
The relation is not a function since the x-coordinate $\frac{3}{4} = 0.75$ is paired with both $-\frac{1}{2}$ and 6.

21. The relation is {(2, 2), (2, 4), (4, 5), (6, 5), (8, 6)}.
The domain is the set of all first coordinates, {2, 4, 6, 8}.
The range is the set of all second coordinates, {2, 4, 5, 6}.
The relation is not a function because the x-coordinate 2 is paired with both 2 and 4.

22. The relation is {(triangle, 3), (square, 4), (rectangle, 4), (parallelogram, 4)}.
The domain is the set of all first coordinates, {triangle, square, rectangle, parallelogram}.
The range is the set of all second coordinates, {3, 4}.
The relation is a function since each x-coordinate is paired with only one y-coordinate.

23.

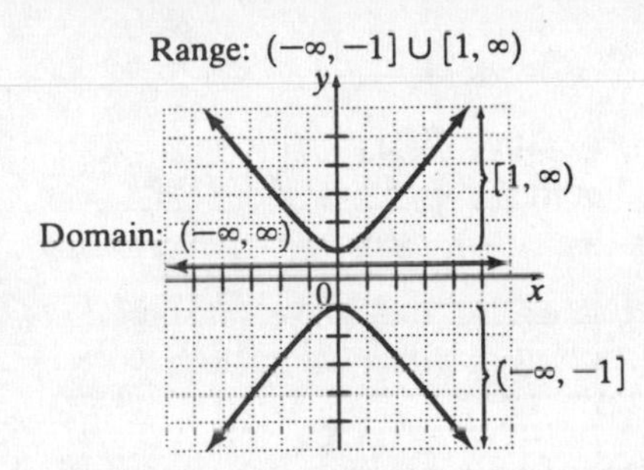

Domain: $(-\infty, \infty)$
Range: $(-\infty, -1] \cup [1, \infty)$
Not a function; the vertical line $x = 0$ intersects the graph in more than one point.

24.

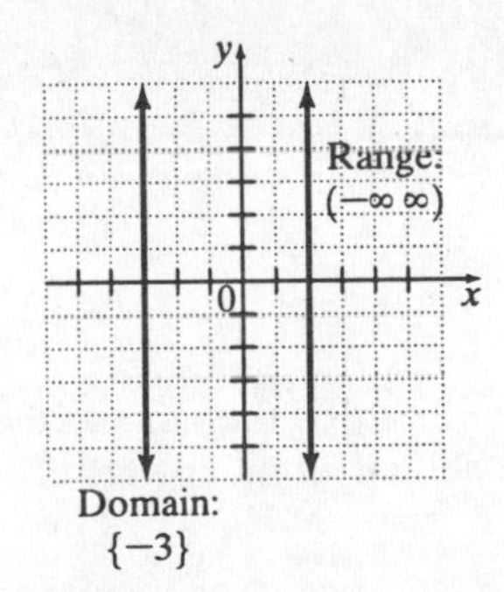

Domain: $\{-3\}$
Range: $(-\infty, \infty)$
Not a function; the vertical line $x = -3$ intersects the graph in more than one point.

25.

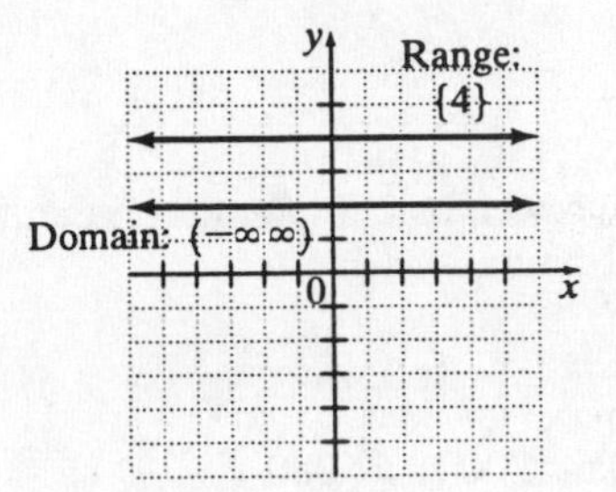

Domain: $(-\infty, \infty)$
Range: $\{4\}$
Function; no vertical line intersects the graph in more than one point.

26.

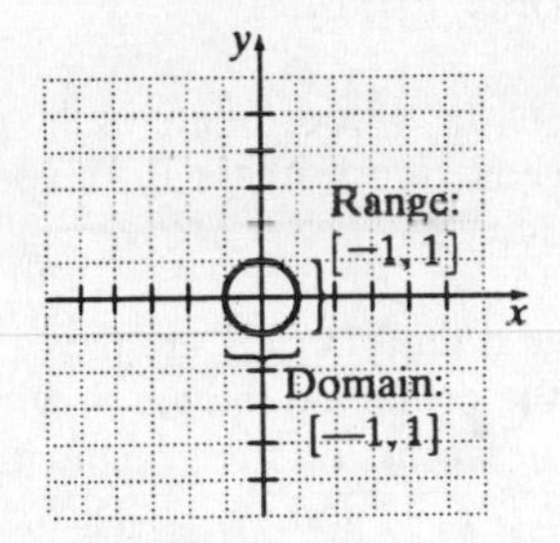

Domain: $[-1, 1]$
Range: $[-1, 1]$
Not a function; the vertical line $x = 0$ intersects the graph in more than one point.

27. $f(x) = x - 5$
$f(2) = 2 - 5 = -3$

28. $g(x) = -3x$
$g(0) = -3(0) = 0$

29. $g(x) = -3x$
$g(-6) = -3(-6) = 18$

30. $h(x) = 2x^2 - 6x + 1$
$h(-1) = 2(-1)^2 - 6(-1) + 1$
$= 2(1) + 6 + 1$
$= 9$

31. $h(x) = 2x^2 - 6x + 1$
$h(1) = 2(1)^2 - 6(1) + 1 = 2 - 6 + 1 = -3$

32. $f(x) = x - 5$
$f(5) = 5 - 5 = 0$

33. Find $J(150)$.
$J(x) = 2.54x$
$J(150) = 2.54(150) = 381$
A person who weighs 250 pounds on Earth weights 381 pounds on Jupiter.

34. Find $J(2000)$.
$J(x) = 2.54x$
$J(2000) = 2.54(2000) = 5080$
The probe that weighs 2000 pounds on Earth weighs 5080 pounds on Jupiter.

35. The value of $f(-1)$ is the y-coordinate of the point on the graph that has x-coordinate -1. The point is $(-1, 0)$. Thus, $f(-1) = 0$.

36. The value of $f(1)$ is the y-coordinate of the point on the graph that has x-coordinate 1. The point is $(1, 2)$. Thus, $f(1) = 2$.

37. The values of x such that $f(x) = 1$ are the x-coordinates of the points on the graph with y-coordinate 1. The points on the graph with y-coordinate 1 are $(-2, 1)$ and $(4, 1)$. Thus, the values of x are -2 and 4.

38. The values of x such that $f(x) = -1$ are the x-coordinates of the points on the graph with y-coordinate -1. The points on the graph with y-coordinate -1 are $(0, -1)$ and $(2, -1)$. Thus, the values of x are 0 and 2.

39. $f(x) = x$
Find 3 points.

x	y
-1	-1
0	0
1	1

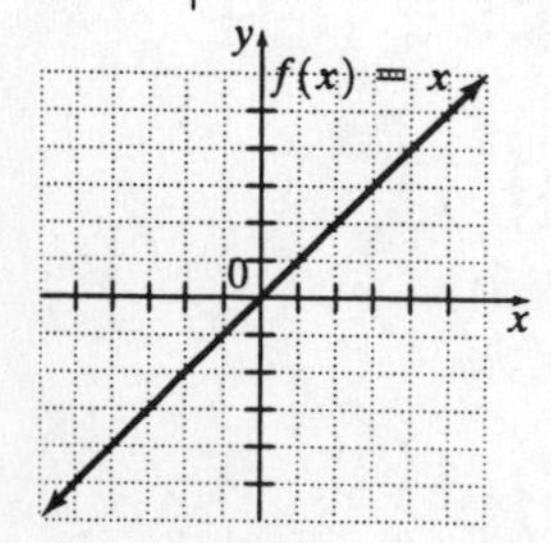

40. $f(x) = -\frac{1}{3}x$
Find 3 points.

x	y
-3	1
0	0
3	-1

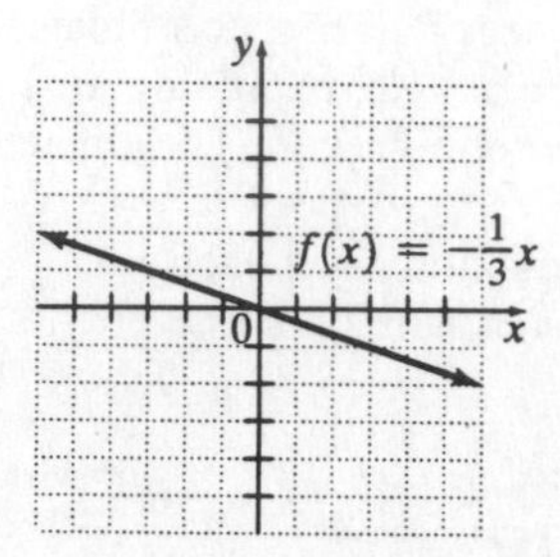

41. $g(x) = 4x - 1$ or $y = 4x - 1$
Find 3 points.

x	y
-1	3
0	-1
1	-5

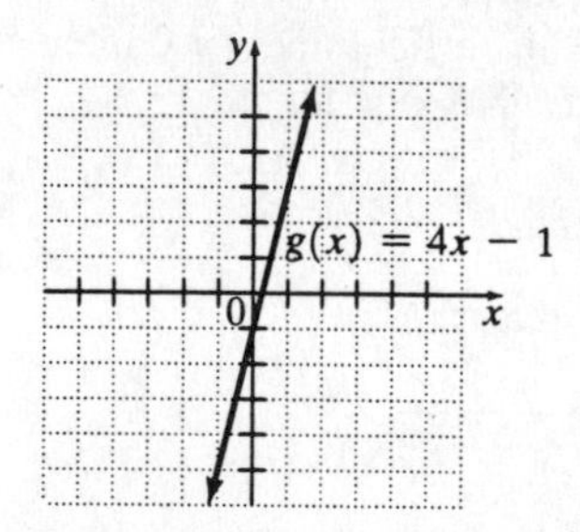

42. The graph of $f(x) = 3x + 1$ is the same as the graph of $f(x) = 3x$ shifted 1 unit upward. This describes graph C.

43. The graph of $f(x) = 3x - 2$ is the same as the graph of $f(x) = 3x$ shifted 2 units downward. This describes graph A.

44. The graph of $f(x) = 3x + 2$ is the same as the graph of $f(x) = 3x$ shifted 2 units upward. This describes graph B.

45. The graph of $f(x) = 3x - 5$ is the same as the graph of $f(x) = 3x$ shifted 5 units downward. This describes graph D.

46. $4x + 5y = 20$

Let $x = 0$.	Let $y = 0$.
$4(0) + 5y = 20$	$4x + 5(0) = 20$
$y = 4$	$x = 5$
$(0, 4)$	$(5, 0)$

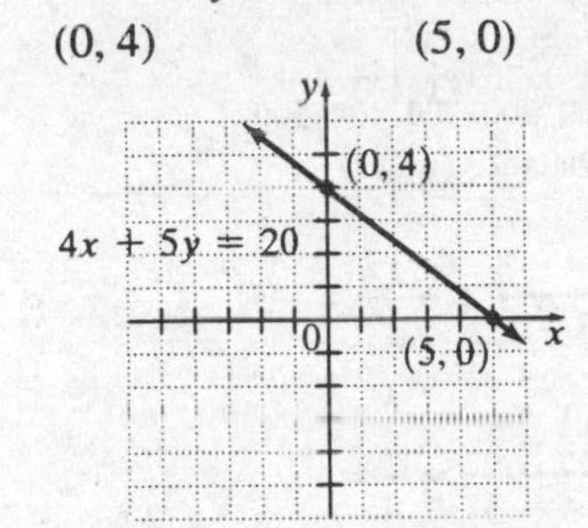

47. $3x - 2y = -9$

Let $x = 0$.	Let $y = 0$.
$3(0) - 2y = -9$	$3x - 2(0) = -9$
$y = \frac{9}{2}$	$x = -3$
$\left(0, \frac{9}{2}\right)$	$(-3, 0)$

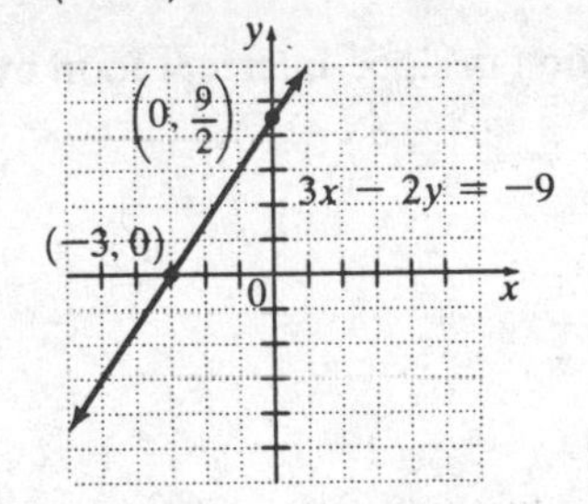

48. $4x - y = 3$

Let $x = 0$.	Let $y = 0$.
$4(0) - y = 3$	$4x - 0 = 3$
$y = -3$	$x = \frac{3}{4}$
$(0, -3)$	$\left(\frac{3}{4}, 0\right)$

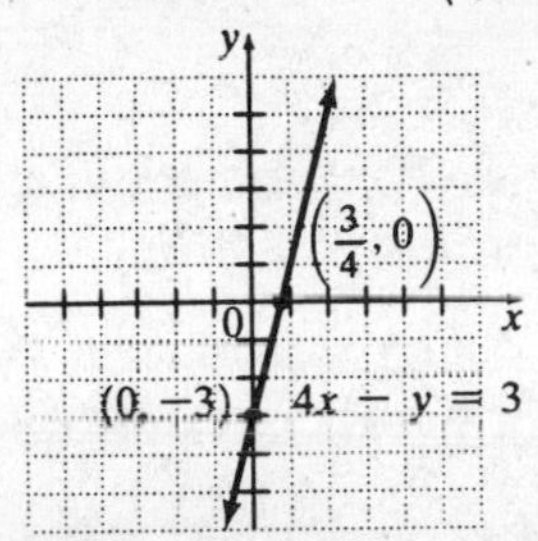

49. $2x + 6y = 9$

Let $x = 0$.	Let $y = 0$.
$2(0) + 6y = 9$	$2x + 6(0) = 9$
$y = \frac{3}{2}$	$x = \frac{9}{2}$
$\left(0, \frac{3}{2}\right)$	$\left(\frac{9}{2}, 0\right)$

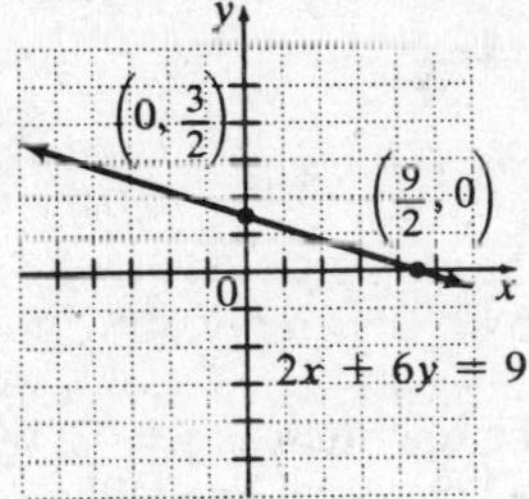

50. $y = 5$

Let $x = 0$.	Let $y = 0$.
$y = 5$	$0 = 5$
$(0, 5)$	no x-intercept

The graph is the horizontal line with y-intercept $(0, 5)$.

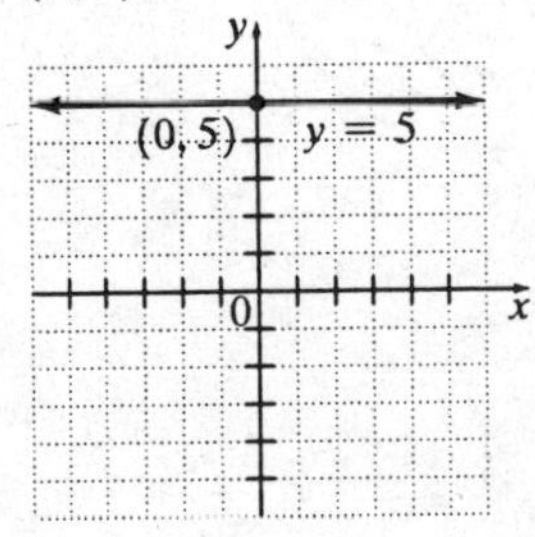

51. $x = -2$
Let $x = 0$. $0 = -2$ no y-intercept
Let $y = 0$. $x = -2$ $(-2, 0)$
The graph is the vertical line with x-intercept $(-2, 0)$.

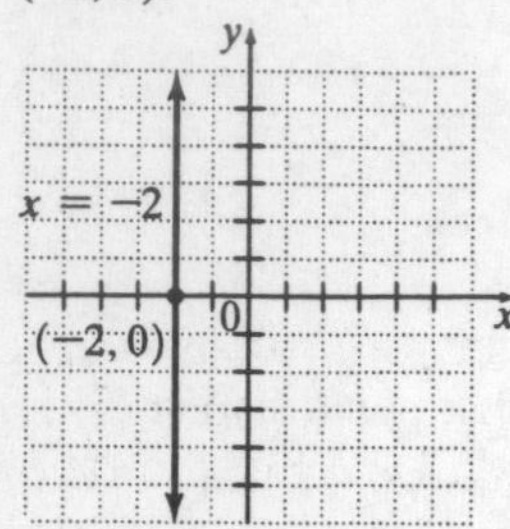

52. $x - 2 = 0$
The equation can be rewritten as $x = 2$, which is the equation of a vertical line with x-intercept $(2, 0)$.

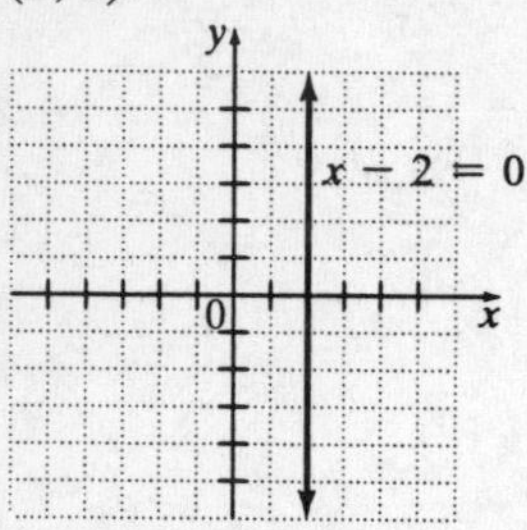

53. $y + 3 = 0$
The equation can be rewritten as $y = -3$, which is the equation of a horizontal line with y-intercept $(0, -3)$.

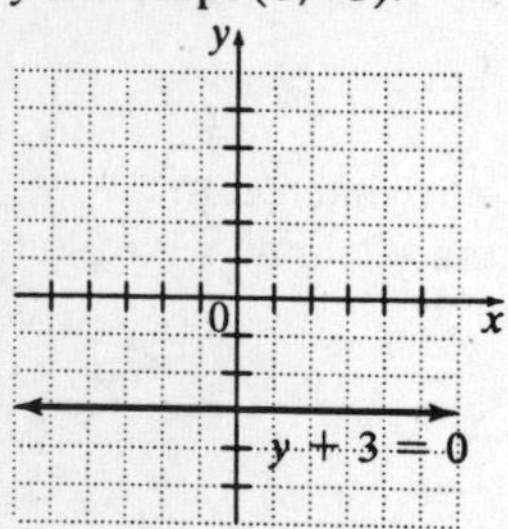

54. $C(x) = 0.3x + 42$

a. Find $C(150)$.
$$C(150) = 0.3(150) + 42 = 45 + 42 = 87$$
The cost of renting a minivan for one day and driving it 150 miles is \$87.

b.

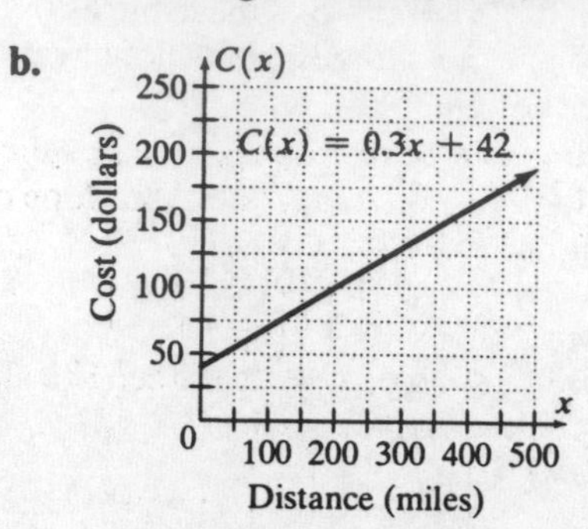

55. $m = \dfrac{y_2 - y_1}{x_2 - x_1} = \dfrac{-4 - 8}{6 - 2} = \dfrac{-12}{4} = -3$

56. $m = \dfrac{y_2 - y_1}{x_2 - x_1} = \dfrac{13 - 9}{5 - (-3)} = \dfrac{4}{8} = \dfrac{1}{2}$

57. $m = \dfrac{y_2 - y_1}{x_2 - x_1} = \dfrac{6 - (-4)}{-3 - (-7)} = \dfrac{10}{4} = \dfrac{5}{2}$

58. $m = \dfrac{y_2 - y_1}{x_2 - x_1} = \dfrac{7 - (-2)}{-5 - 7} = \dfrac{9}{-12} = -\dfrac{3}{4}$

59. Rewrite the equation in slope-intercept form by solving for y.
$$6x - 15y = 20$$
$$6x - 20 = 15y$$
$$y = \frac{2}{5}x - \frac{4}{3}$$
$m = \frac{2}{5}$, $b = -\frac{4}{3}$

60. Rewrite the equation in slope-intercept form by solving for y.

$$4x + 14y = 21$$
$$14y = -4x + 21$$
$$y = -\frac{2}{7}x + \frac{3}{2}$$
$$m = -\frac{2}{7},\ b = \frac{3}{2}$$

61. $y - 3 = 0$
The equation can be rewritten as $y = 3$, which is the equation of a horizontal line. The slope of a horizontal line is 0.

62. $x = -5$
This is the equation of a vertical line. The slope of a vertical line is undefined.

63. $m = \frac{y_2 - y_1}{x_2 - x_1} = \frac{42.92 - 39.2}{1998 - 1995} = \frac{3.72}{3} = 1.24$
Each year, 1.24 million more persons have a bachelors degree or higher.

64. $m = \frac{y_2 - y_1}{x_2 - x_1} = \frac{859 - 805}{1997 - 1995} = \frac{54}{2} = 27$
Each year, 27 million more people go on vacations.

65. l_2 has the greater slope, since both lines have a positive slope but l_2 is steeper.

66. l_1 has the greater slope, since its slope is 0 and the slope of l_2 is negative.

67. $C(x) = 0.3x + 42$
The function is written in slope-intercept form where $m = 0.3$ and $b = 42$.

a. The slope is 0.3; the cost increases by $0.30 for each additional mile driven.

b. The y-intercept is (0, 42); the cost for 0 miles driven is $42.

68. $f(x) = -2x + 6 \qquad g(x) = 2x - 1$
$m = -2 \qquad m = 2$
Neither; the slopes are not the same and their product is not -1.

69. $-x + 3y = 2 \qquad 6x - 18y = 3$

$$3y = x + 2 \qquad -18y = -6x + 3$$
$$y = \frac{1}{3}x + \frac{2}{3} \qquad y = \frac{1}{3}x - \frac{1}{6}$$
$$m = \frac{1}{3},\ b = \frac{2}{3} \qquad m = \frac{1}{3},\ b = -\frac{1}{6}$$

Parallel. Since their slopes are equal and their y-intercepts are different.

70. $y = -x + 1$
$m = -1, b = 1$
To graph the equation, first plot the y-intercept (0, 1). To find another point, start at (0, 1) and move 1 unit downward and 1 unit to the right since the slope is $-1 = \frac{-1}{1}$.

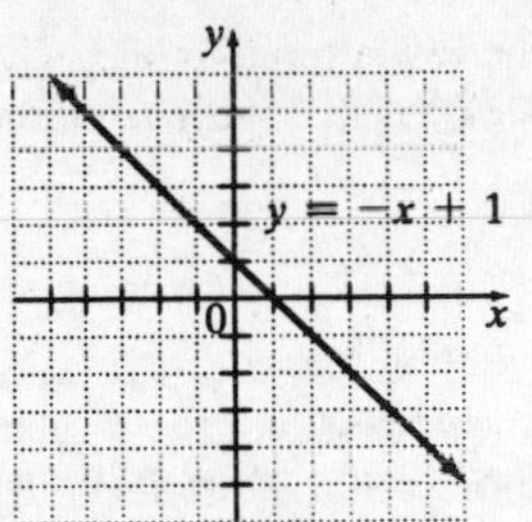

71. $y = 4x - 3$
$m = 4, b = -3$
To graph the equation, first plot the y-intercept (0, −3). To find another point, start at (0, −3) and move 4 units upward and 1 unit to the right, since the slope is $4 = \frac{4}{1}$.

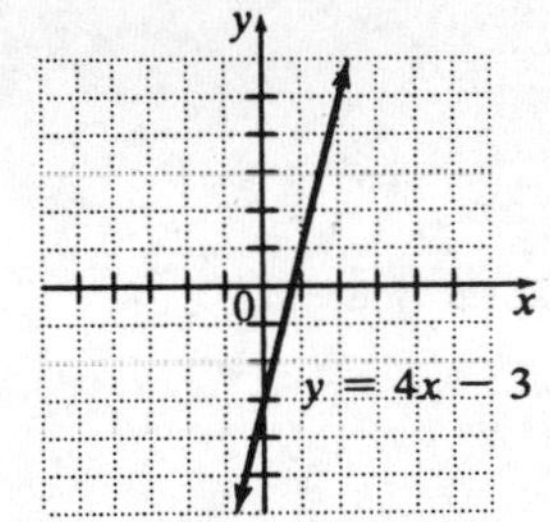

72. Rewrite the equation in slope-intercept form.

$$3x - y = 6$$
$$-y = -3x + 6$$
$$y = 3x - 6$$

$m = 3, b = -6$

To graph the equation, first plot the y-intercept $(0, -6)$. To find another point, start at $(0, -6)$ and move 3 units upward and 1 unit to the right since the slope is $3 = \frac{3}{1}$.

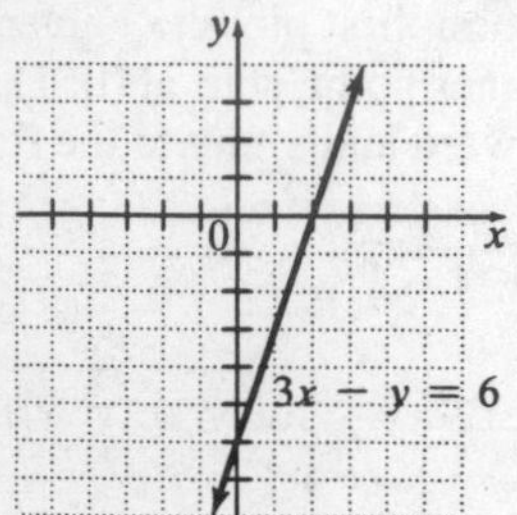

73. $y = -5x$

$m = -5, b = 0$

To graph the equation, first plot the y-intercept $(0, 0)$. To find another point, start at $(0, 0)$ and move 5 units downward and 1 unit to the right, since the slope is $-5 = \frac{-5}{1}$.

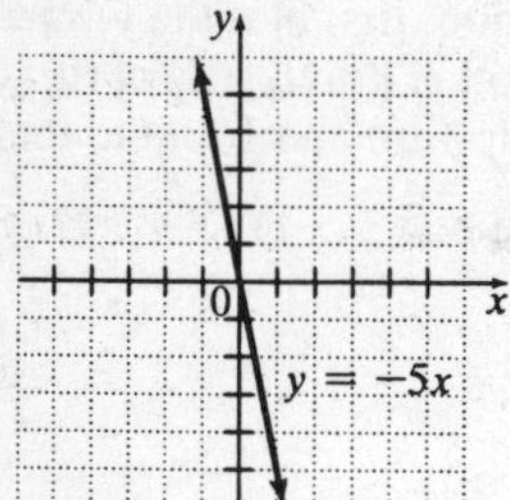

74. Horizontal lines have slope = 0.
The y-intercept is -1.
$y = -1$

75. Vertical lines have undefined slope.
The x-intercept is -2.
$x = -2$

76. The slope is undefined; the line is vertical.
The x-intercept is -4.
$x = -4$

77. The line is horizontal with y-intercept = 5.
$y = 5$

78. Since we are given the slope and a point that the line passes through, we use the point-slope form to write the equation.

$$y - y_1 = m(x - x_1)$$
$$y - 5 = 3(x - (-3))$$
$$y - 5 = 3(x + 3)$$
$$y - 5 = 3x + 9$$
$$-3x + y = 14$$
$$3x - y = -14$$

79. Since we are given the slope and a point that the line passes through, we use the point-slope form to write the equation.

$$y - y_1 = m(x - x_1)$$
$$y - (-2) = 2(x - 5)$$
$$y + 2 = 2x - 10$$
$$-2x + y = -12$$
$$2x - y = 12$$

80. Since we are given two points, we first use the points to find the slope and then use the slope and one point to write the equation with the point-slope form.

$$m = \frac{y_2 - y_1}{x_2 - x_1} = \frac{-2 - (-1)}{-4 - (-6)} = \frac{-2 + 1}{-4 + 6} = -\frac{1}{2}$$
$$y - y_1 = m(x - x_1)$$
$$y - (-1) = -\frac{1}{2}(x - (-6))$$
$$2(y + 1) = -(x + 6)$$
$$2y + 2 = -x - 6$$
$$x + 2y = -8$$

81. Since we are given two points, we first use the points to find the slope and then use the slope and one point to write the equation with the point-slope form.

$$m = \frac{y_2 - y_1}{x_2 - x_1} = \frac{-8 - 3}{-4 - (-5)} = \frac{-8 - 3}{-4 + 5} = \frac{-11}{1} = -11$$
$$y - y_1 = m(x - x_1)$$
$$y - 3 = -11(x - (-5))$$
$$y - 3 = -11(x + 5)$$
$$y - 3 = -11x - 55$$
$$11x + y = -52$$

82. The graph of $x = 4$ is a vertical line. A line perpendicular to $x = 4$ is horizontal. The equation of the horizontal line passing through (–2, 3) is $y = 3$.

83. The graph of $y = 8$ is a horizontal line. A line parallel to it will also be horizontal. The equation of the horizontal line passing through (–2, –5) is $y = -5$.

84. Since we know the slope and the *y*-intercept, we use the slope-intercept form.

$$y = mx + b$$
$$y = -\frac{2}{3}x + 4$$
$$f(x) = -\frac{2}{3}x + 4$$

85. Since we know the slope and the *y*-intercept, we use the slope-intercept form.

$$y = mx + b$$
$$y = -x - 2$$
$$f(x) = -x - 2$$

86. First find the slope of $6x + 3y = 5$ by rewriting the equation in slope-intercept form. The desired line, being parallel to it, has the same slope.

$$6x + 3y = 5$$
$$3y = -6x + 5$$
$$y = -2x + \frac{5}{3}$$

The slope is –2. Now use the slope and the point (2, –6) to write the equation of the desired line.

$$y - y_1 = m(x - x_1)$$
$$y - (-6) = -2(x - 2)$$
$$y + 6 = -2x + 4$$
$$y = -2x - 2$$
$$f(x) = -2x - 2$$

87. First find the slope of $3x + 2y = 8$ by rewriting the equation in slope-intercept form. The desired line, being parallel to it, has the same slope.

$$3x + 2y = 8$$
$$2y = -3x + 8$$
$$y = -\frac{3}{2}x + 4$$

The slope is $-\frac{3}{2}$. Now use the slope and the point (–4, –2) to write the equation of the desired line.

$$y - y_1 = m(x - x_1)$$
$$y - (-2) = -\frac{3}{2}(x - (-4))$$
$$y + 2 = -\frac{3}{2}(x + 4)$$
$$y + 2 = -\frac{3}{2}x - 6$$
$$y = -\frac{3}{2}x - 8$$
$$f(x) = -\frac{3}{2}x - 8$$

88. First find the slope of $4x + 3y = 5$ by rewriting the equation in slope-intercept form.

$$4x + 3y = 5$$
$$3y = -4x + 5$$
$$y = -\frac{4}{3}x + \frac{5}{3}$$

Thus, the slope of this line is $-\frac{4}{3}$. The slope of the desired line is $\frac{3}{4}$, the negative reciprocal of $-\frac{4}{3}$, since the lines are perpendicular. Use the slope $\frac{3}{4}$ and the point (–6, –1) to write the equation.

$$y - y_1 = m(x - x_1)$$
$$y - (-1) = \frac{3}{4}(x - (-6))$$
$$4(y + 1) = 3(x + 6)$$
$$4y + 4 = 3x + 18$$
$$4y = 3x + 14$$
$$y = \frac{3}{4}x + \frac{7}{2}$$
$$f(x) = \frac{3}{4}x + \frac{7}{2}$$

89. First find the slope of $2x - 3y = 6$ by rewriting the equation in slope-intercept form.

$$2x - 3y = 6$$
$$3y = 2x - 6$$
$$y = \frac{2}{3}x - 2$$

Thus, the slope of this line is $\frac{2}{3}$. The slope of the desired line is $-\frac{3}{2}$, the negative reciprocal of $\frac{2}{3}$, since the lines are perpendicular. Use the slope $-\frac{3}{2}$ and the point (−4, 5) to write the equation.

$$y - y_1 = m(x - x_1)$$
$$y - 5 = -\frac{3}{2}(x - (-4))$$
$$y - 5 = -\frac{3}{2}(x + 4)$$
$$y - 5 = -\frac{3}{2}x - 6$$
$$y = -\frac{3}{2}x - 1$$
$$f(x) = -\frac{3}{2}x - 1$$

90. a. Use ordered pairs (0, 42) and (3, 58). First find the slope.

$$m = \frac{y_2 - y_1}{x_2 - x_1} = \frac{58 - 42}{3 - 0} = \frac{16}{3}$$

Now use the slope $\frac{16}{3}$ and y-intercept (0, 42) to write the equation.

$$y = mx + b$$
$$y = \frac{16}{3}x + 42$$

The linear equation $y = \frac{16}{3}x + 42$ models the number of U.S. paging subscribers (in millions) in terms of the number of years x past 1996.

b. Let $x = 11$ in the equation since 2007 is 11 years after 1996.

$$y = \frac{16}{3}(11) + 42 \approx 100.7$$

There will be about 101 million subscribers in the year 2007.

91. a. Use ordered pairs (0, 43) and (22, 60). First find the slope.

$$m = \frac{y_2 - y_1}{x_2 - x_1} = \frac{60 - 43}{22 - 0} = \frac{17}{22}$$

Now use the slope $\frac{17}{22}$ and the y-intercept (0, 43) to write the equation.

$$y = mx + b$$
$$y = \frac{17}{22}x + 43$$

The linear equation $y = \frac{17}{22}x + 43$ models the number of people (in millions) reporting arthritis in terms of the number of years x past 1998.

b. Let $x = 12$ in the equation since 2010 is 12 years after 1998.

$$y = \frac{17}{22}(12) + 43 \approx 52.3$$

There will be about 52 million people reporting arthritis in 2010.

92. $3x + y > 4$

Graph $3x + y = 4$ as a dashed line since the inequality symbol is >.

Test (0, 0).

$3(0) = 0 > 4$ False

Shade the half-plane that does not contain (0, 0).

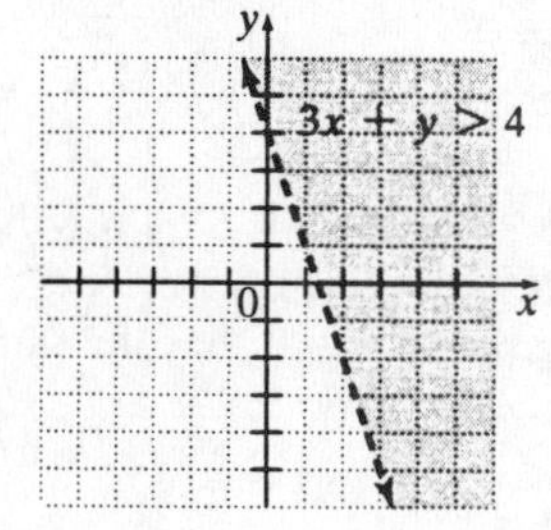

93. $\frac{1}{2}x - y < 2$

Graph the line $\frac{1}{2}x - y = 2$ as a dashed line since the inequality symbol is <.

Test (0, 0).

$\frac{1}{2}(0) - 0 < 2$ True

Shade the half-plane containing (0, 0).

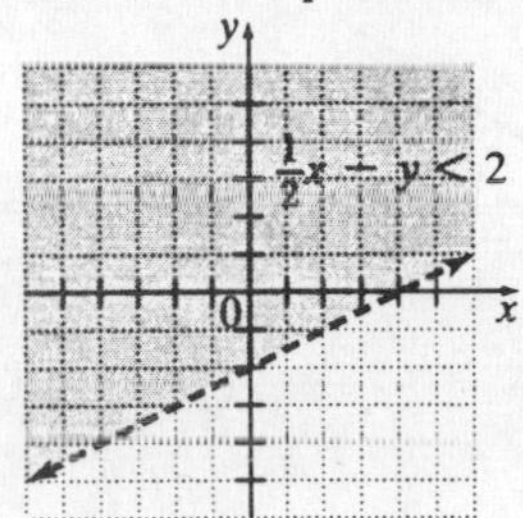

94. $5x - 2y \le 9$

Graph the line $5x - 2y = 9$ as a solid line since the inequality symbol is ≤.

Test (0, 0).

$5(0) - 2(0) \le 9$ True

Shade the half-plane containing (0, 0).

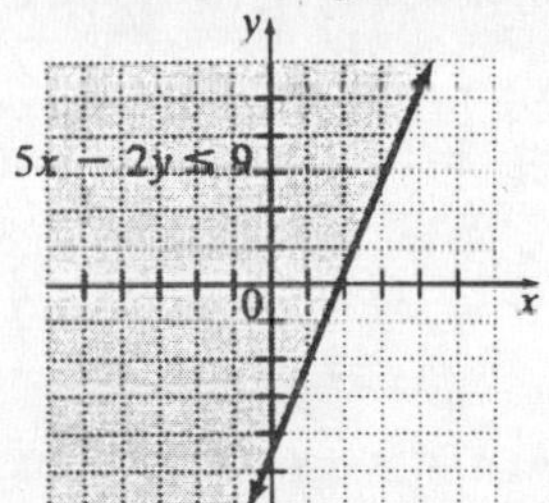

95. $3y \ge x$

Graph the line $3y = x$ as a solid line since the inequality symbol is ≥.

Test (2, 2).

$3(2) \ge 2$ True

Shade the half-plane containing (2, 2).

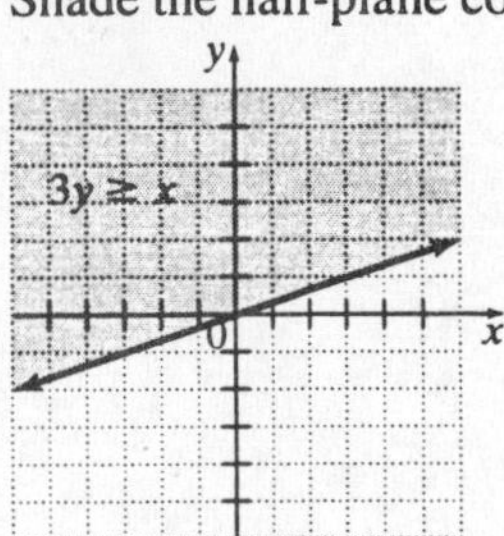

96. $y < 1$

Graph $y = 1$ as a dashed line since the inequality symbol is <.

Test (0, 0).

$0 < 1$ True

Shade the half-plane containing (0, 0).

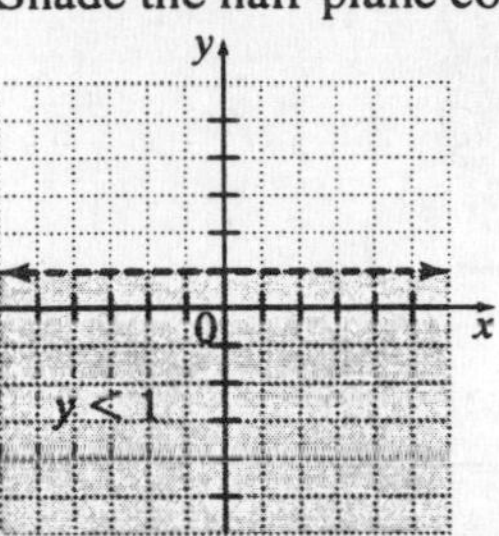

97. $x > -2$

Graph $x = -2$ as a dashed line since the inequality symbol is >.

Test (0, 0).

$0 > -2$ True

Shade the half-plane containing (0, 0).

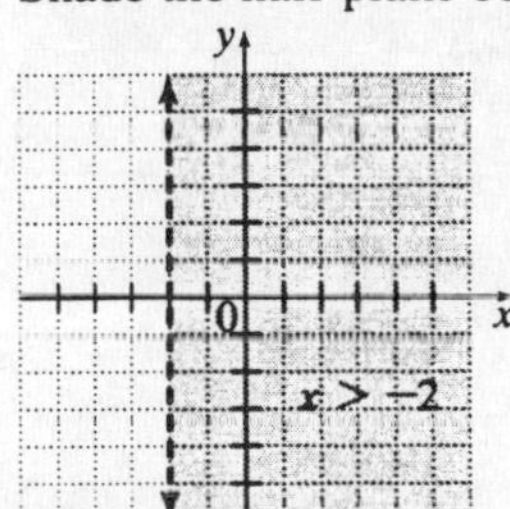

98. $y > 2x + 3$ or $x \le -3$

Graph both inequalities on the same set of axes, the union is the entire shaded area.

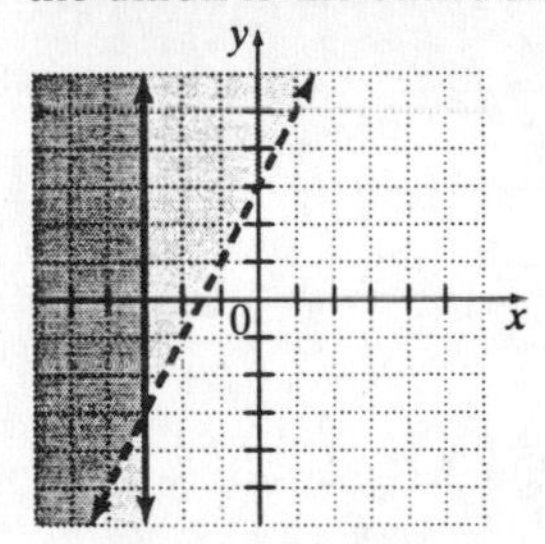

99. $2x < 3y + 8$ and $y \geq -2$
Graph both inequalities on the same set of axes; the intersection is the darker region where the shaded half-planes overlap.

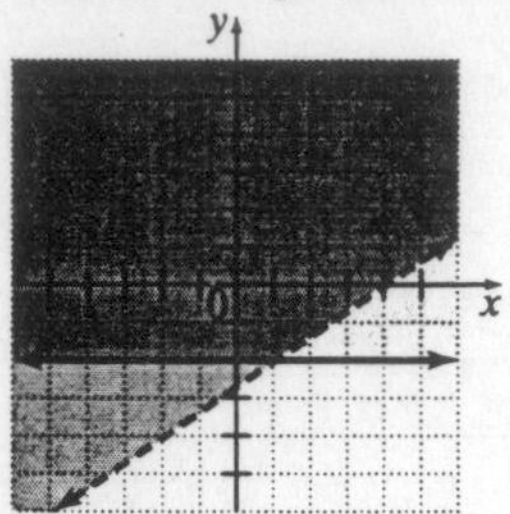

Chapter 3 Test

1.

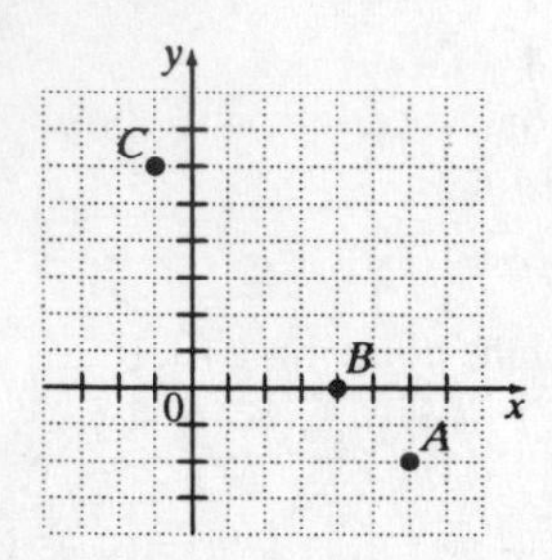

A is in quadrant IV.
B is on the *x*-axis, no quadrant.
C is in quadrant II.

2. $2y - 3x = 12$
Let $x = -6$.
$$2y - 3(-6) = 12$$
$$2y + 18 = 12$$
$$2y = -6$$
$$y = -3$$
$(-6, -3)$

3. $2x - 3y = -6$
Find intercepts.

Let $x = 0$.	Let $y = 0$.
$2(0) - 3y = -6$	$2x - 3(0) = -6$
$-3y - 6$	$2x = -6$
$y = 2$	$x = -3$
$(0, 2)$	$(-3, 0)$

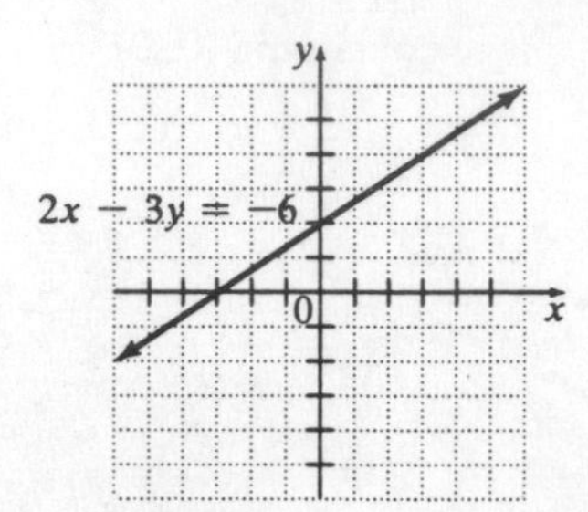

4. $4x + 6y = 7$
Rewrite the equation in slope-intercept form by solving for *y*.
$$4x + 6y = 7$$
$$6y = -4x + 7$$
$$y = -\frac{2}{3}x + \frac{7}{6}$$

Plot the *y*-intercept $\left(0, \frac{7}{6}\right)$. Then, from $\left(0, \frac{7}{6}\right)$ move 2 units downward and 3 units to the right to find another point.

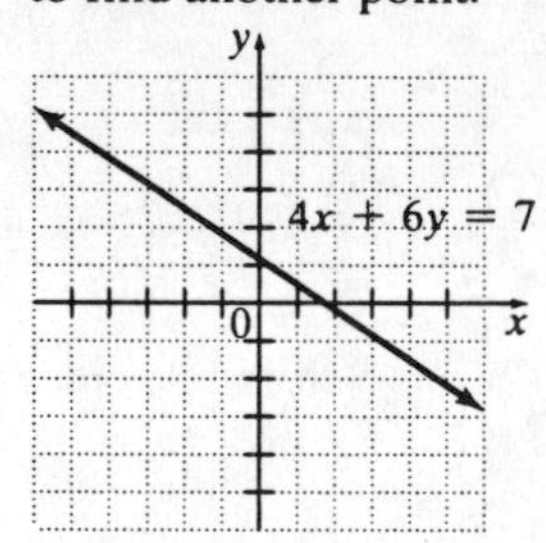

5. $y = \frac{2}{3}x$

Find 3 points.

x	y
–3	–2
0	0
3	2

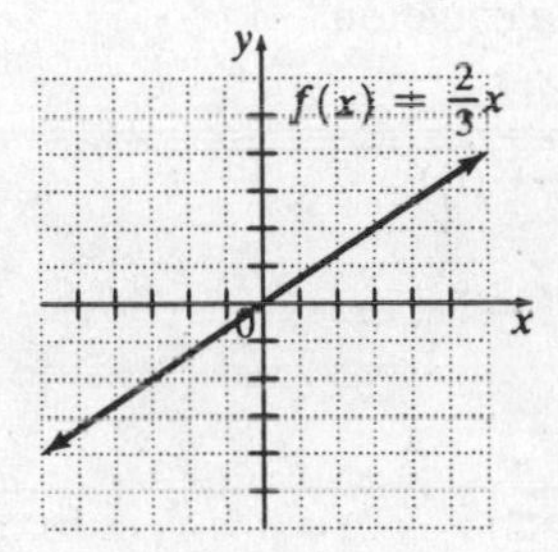

6. $y = -3$

This is a horizontal line with y-intercept $(0, -3)$.

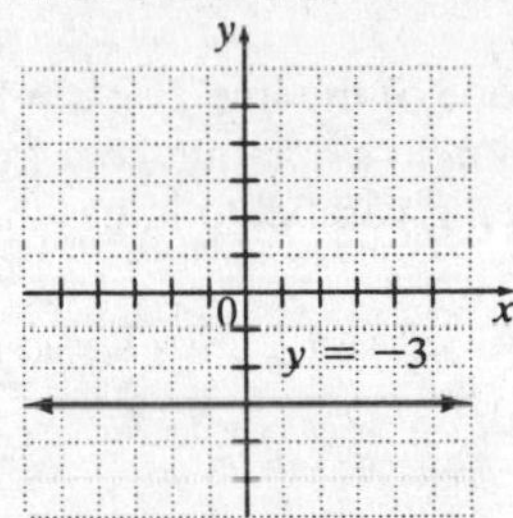

7. $m = \frac{10-(-8)}{-7-5} = \frac{18}{-12} = -\frac{3}{2}$

8. $3x + 12y = 8$

$12y = -3x + 8$

$y = -\frac{1}{4}x + \frac{2}{3}$

$m = -\frac{1}{4}$ and $b = \frac{2}{3}$

9. $f(x) = (x-1)^2$

x	$f(x)$
–2	9
–1	4
0	1
1	0
2	1
3	4
4	9

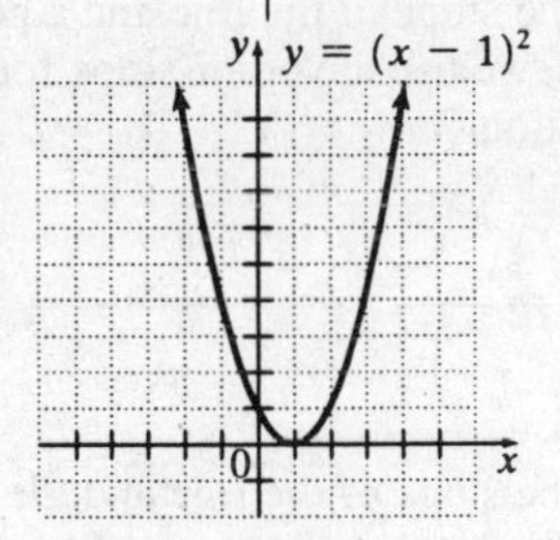

10. $g(x) = |x| + 2$

x	$g(x)$
–3	5
–2	4
–1	3
0	2
1	3
2	4
3	5

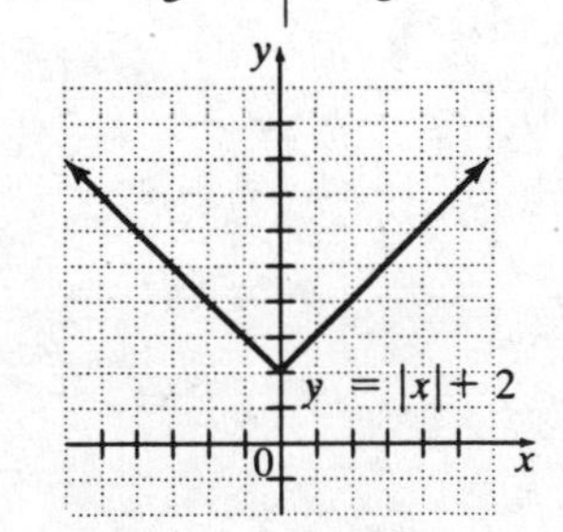

11. A horizontal line has an equation of the form $y = c$. Since the desired line passes through (2, –8), the equation is $y = -8$.

12. A vertical line has an equation of the form $x = c$. Since the desired line passes through (–4, –3), the equation is $x = -4$.

13. Since $x = 5$ is a vertical line, the desired line, being perpendicular to it, is horizontal. The equation of a horizontal line has the form $y = c$. Since the line passes through (3, –2), the equation is $y = -2$.

14. Since we know the slope of the line and a point it passes through, we use the point-slope form to write the equation.

$$y - y_1 = m(x - x_1)$$
$$y - (-1) = -3(x - 4)$$
$$y + 1 = -3x + 12$$
$$3x + y = 11$$

15. Since we know the slope of the line and the y-intercept, we use the slope-intercept form to write the equation.

$$y = mx + b$$
$$y = 5x + (-2)$$
$$-5x + y = -2$$
$$5x - y = 2$$

16. Since we know two points, we first find the slope and then use the slope and one point to write the equation using the point-slope form.

$$m = \frac{y_2 - y_1}{x_2 - x_1} = \frac{-3-(-2)}{6-4} = \frac{-3+2}{6-4} = \frac{-1}{2} = -\frac{1}{2}$$
$$y - y_1 = m(x - x_1)$$
$$y - (-2) = -\frac{1}{2}(x - 4)$$
$$2(y + 2) = -(x - 4)$$
$$2y + 4 = -x + 4$$
$$2y = -x$$
$$y = -\frac{1}{2}x$$
$$f(x) = -\frac{1}{2}x$$

17. First we find the slope of $3x - y = 4$ by rewriting the equation in slope-intercept form.

$$3x - y = 4$$
$$-y = -3x + 4$$
$$y = 3x - 4$$

The slope of $3x - y = 4$ is 3. Thus, the slope of a line perpendicular to it is $-\frac{1}{3}$, the negative reciprocal of 3. Use the slope $-\frac{1}{3}$ and the point (–1, 2) to write the equation.

$$y - y_1 = m(x - x_1)$$
$$y - 2 = -\frac{1}{3}(x - (-1))$$
$$3(y - 2) = -(x + 1)$$
$$3y - 6 = -x - 1$$
$$3y = -x + 5$$
$$y = -\frac{1}{3}x + \frac{5}{3}$$
$$f(x) = -\frac{1}{3}x + \frac{5}{3}$$

18. First we find the slope of the line $2y + x = 3$ by rewriting the equation in slope-intercept form. The desired line, being parallel to it, has the same slope.

$$2y + x = 3$$
$$2y = -x + 3$$
$$y = -\frac{1}{2}x + \frac{3}{2}$$

Now use the slope $-\frac{1}{2}$ and the point (3, –2) to write the equation.

$$y - y_1 = m(x - x_1)$$
$$y - (-2) = -\frac{1}{2}(x - 3)$$
$$2(y + 2) = -(x - 3)$$
$$2y + 4 = -x + 3$$
$$2y = -x - 1$$
$$y = -\frac{1}{2}x - \frac{1}{2}$$
$$f(x) = -\frac{1}{2}x - \frac{1}{2}$$

19. Find the slope of L_1 by rewriting the equation in slope-intercept form.

$2x - 5y = 8$

$-5y = -2x - 8$

$5y = 2x - 8$

$y = \frac{2}{5}x - \frac{8}{5}$

Thus, the slope of L_1 is $\frac{2}{5}$.

Find the slope of L_2 using the two points.

$m = \frac{y_2 - y_1}{x_2 - x_1} = \frac{-1-4}{-1-1} = \frac{-5}{-2} = \frac{5}{2}$

The slope of L_2 is $\frac{5}{2}$.

$\frac{2}{5} \neq \frac{5}{2}$

$\frac{2}{5} \cdot \frac{5}{2} = 1 \neq -1$

The slopes are not equal, and their product is not -1.

Therefore, lines L_1 and L_2 are neither parallel nor perpendicular.

20. $x \leq -4$

Graph the line $x = -4$ as a solid line since the inequality symbol is $\leq$.

Test (0, 0).

$0 \leq -4$ False

Shade the half-plane that does not contain (0, 0).

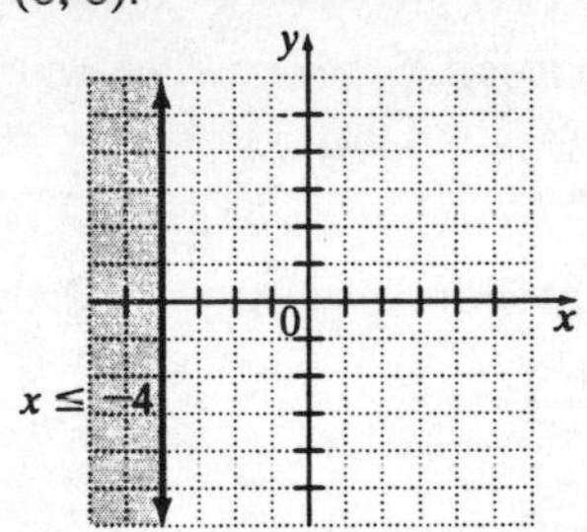

21. $y > -2$

Graph the line $y = -2$ as a dashed line since the inequality symbol is >.

Test (0, 0).

$0 > -2$ True

Shade the half-plane containing (0, 0).

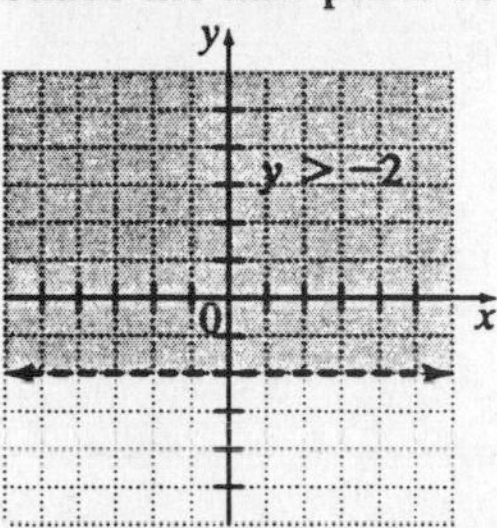

22. $2x - y > 5$

Graph the line $2x - y = 5$ as a dashed line since the inequality symbol is >.

Test (0, 0).

$2(0) - 0 > 5$ False

Shade the half-plane that does not contain (0, 0).

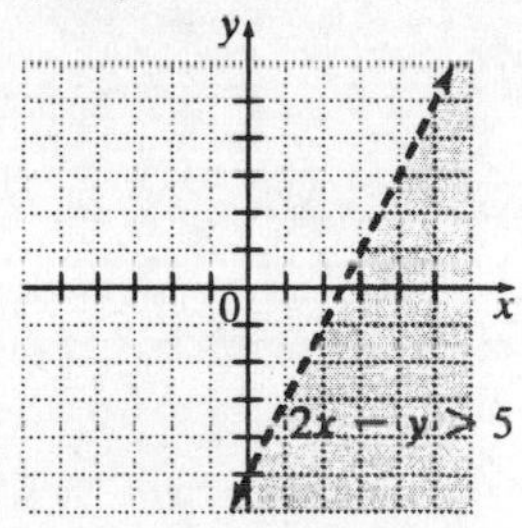

23. $2x + 4y < 6$ and $y \leq -4$

Graph both inequalities on the same set of axes. The intersection is the darkest region where the half-planes overlap.

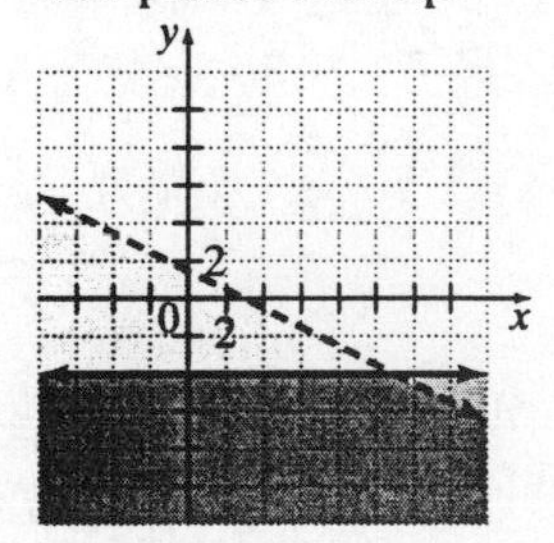

24.

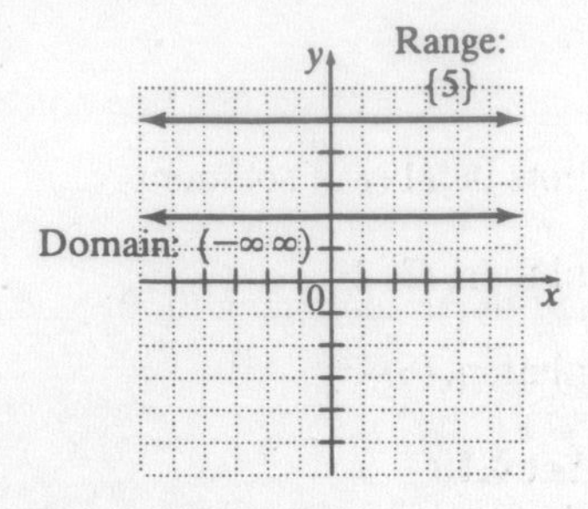

Domain: $(-\infty, \infty)$
Range: $\{5\}$
Function; no vertical line intersects the graph in more than one point.

25.

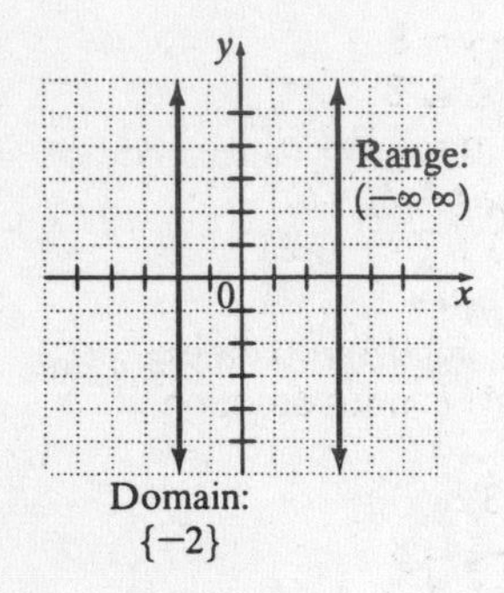

Domain: $\{-2\}$
Range: $(-\infty, \infty)$
Not a function; the line $x = -2$ intersects the graph in more than one point.

26.

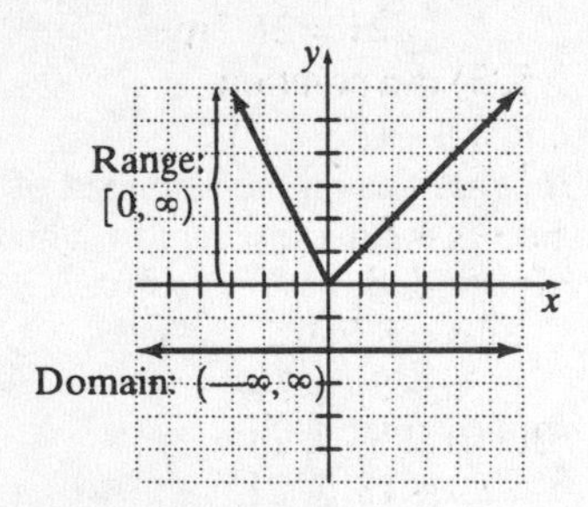

Domain: $(-\infty, \infty)$
Range: $[0, \infty)$
Function; no vertical line intersects the graph in more than one point.

27.

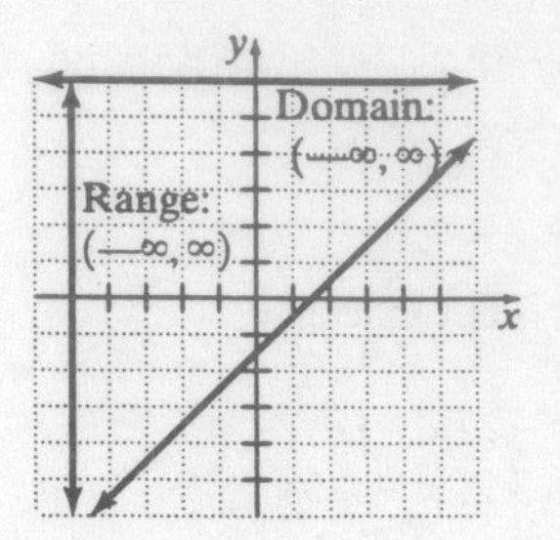

Domain: $(-\infty, \infty)$
Range: $(-\infty, \infty)$
Function; no vertical line intersects the graph in more than one point.

28. $f(x) = 732x + 21,428$

a. Let $x = 2$ since 1998 is 2 years after 1996.
$f(2) = 732(2) + 21,428 = 22,892$
The average earnings were $22,892.

b. Let $x = 9$ since 2005 is 9 years after 1996.
$f(9) = 732(9) + 21,428 = 28,016$
The average earning will be $28,016.

c. We went to find x such that $f(x) = 30,000$.
$$732x + 21,428 = 30,000$$
$$732x = 8572$$
$$x = \frac{8572}{732} \approx 11.71$$
The average earnings will first exceed $30,000 in about year 12 or 2008.

d. The slope is 732; the yearly earnings for high school graduates increases by $732 per year.

e. The y-intercept is (0, 21,428); the yearly earning for a high school graduate in 1996 was $21,428.

29. The line passes through (1996, 1332) and (1998, 1474).
$$m = \frac{y_2 - y_1}{x_2 - x_1} = \frac{1474 - 1332}{1998 - 1996} = \frac{142}{2} = 71.$$
Each year, 71 million more movie tickets are sold.

Chapter 4

Section 4.1

Calculator Explorations

1. $\begin{cases} y = -2.68x + 1.21 \\ y = 5.22x - 1.68 \end{cases}$

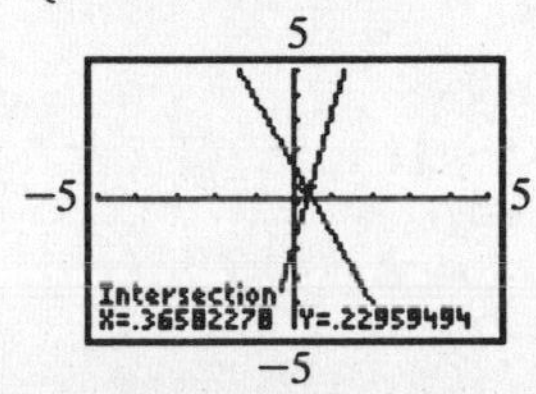

(0.37, 0.23)

3. $\begin{cases} 4.3x - 2.9y = 5.6 \\ 8.1x + 7.6y = -14.1 \end{cases}$

First, solve each equation for *y*.

$\begin{cases} y = \dfrac{4.3x - 5.6}{2.9} \\ y = \dfrac{-8.1x - 14.1}{7.6} \end{cases}$

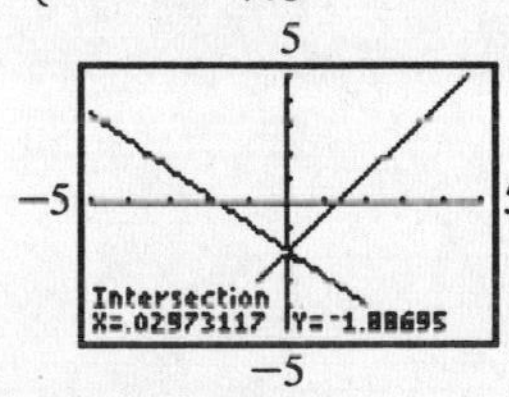

(0.03, −1.89)

Mental Math

1. 1 solution, (−1, 3)
2. no solution
3. infinite number of solutions
4. 1 solution, (3, 4)
5. no solution
6. infinite number of solutions
7. 1 solution, (3, 2)
8. 1 solution, (0 −3)

Exercise Set 4.1

1. $\begin{cases} x + y = 8 \\ 3x + 2y = 21 \end{cases}$

   a. (2, 4):
   $x + y = 8$
   $2 + 4 \stackrel{?}{=} 8$
   $6 = 8$ False
   $3x + 2y = 21$
   $3(2) + 2(4) \stackrel{?}{=} 21$
   $6 + 8 \stackrel{?}{=} 21$
   $14 = 21$ False
   (2, 4) is not a solution.

   b. (5, 3):
   $x + y = 8$
   $5 + 3 \stackrel{?}{=} 8$
   $8 = 8$ True
   $3x + 2y = 21$
   $3(5) + 2(3) \stackrel{?}{=} 21$
   $15 + 6 \stackrel{?}{=} 21$
   $21 = 21$ True
   (5, 3) is a solution.

   c. (1, 9):
   $x + y = 8$
   $1 + 9 \stackrel{?}{=} 8$
   $10 = 8$ False
   $3x + 2y = 21$
   $3(1) + 2(9) \stackrel{?}{=} 21$
   $3 + 18 \stackrel{?}{=} 21$
   $21 = 21$ True
   (1, 9) is not a solution.

3. $\begin{cases} 3x - y = 5 \\ x + 2y = 11 \end{cases}$

a. (2, –1):
$$\begin{aligned} 3x - y &= 5 \\ 3(2) - (-1) &\stackrel{?}{=} 5 \\ 6 + 1 &\stackrel{?}{=} 5 \\ 7 &= 5 \text{ False} \\ x + 2y &= 11 \\ 2 + 2(-1) &\stackrel{?}{=} 11 \\ 2 - 2 &\stackrel{?}{=} 11 \\ 0 &= 11 \text{ False} \end{aligned}$$
(2, –1) is not a solution.

b. (3, 4):
$$\begin{aligned} 3x - y &= 5 \\ 3(3) - 4 &\stackrel{?}{=} 5 \\ 9 - 4 &\stackrel{?}{=} 5 \\ 5 &= 5 \text{ True} \\ x + 2y &= 11 \\ 3 + 2(4) &\stackrel{?}{=} 11 \\ 3 + 8 &\stackrel{?}{=} 11 \\ 11 &= 11 \text{ True} \end{aligned}$$
(3, 4) is a solution.

c. (0, –5):
$$\begin{aligned} 3x - y &= 5 \\ 3(0) - (-5) &\stackrel{?}{=} 5 \\ 0 + 5 &\stackrel{?}{=} 5 \\ 5 &= 5 \text{ True} \\ x + 2y &= 11 \\ 0 + 2(-5) &\stackrel{?}{=} 11 \\ -10 &= 11 \text{ False} \end{aligned}$$
(0, –5) is not a solution.

5. $\begin{cases} 2y = 4x \\ 2x - y = 0 \end{cases}$

a. (–3, –6):
$$\begin{aligned} 2y &= 4x \\ 2(-6) &\stackrel{?}{=} 4(-3) \\ -12 &= -12 \text{ True} \\ 2x - y &= 0 \\ 2(-3) - (-6) &\stackrel{?}{=} 0 \\ -6 + 6 &\stackrel{?}{=} 0 \\ 0 &= 0 \text{ True} \end{aligned}$$
(–3, –6) is a solution.

b. (0, 0):
$$\begin{aligned} 2y &= 4x \\ 2(0) &\stackrel{?}{=} 4(0) \\ 0 &= 0 \text{ True} \\ 2x - y &= 0 \\ 2(0) - 0 &\stackrel{?}{=} 0 \\ 0 &= 0 \text{ True} \end{aligned}$$
(0, 0) is a solution.

c. (1, 2):
$$\begin{aligned} 2y &= 4x \\ 2(2) &\stackrel{?}{=} 4(1) \\ 4 &= 4 \text{ True} \\ 2x - y &= 0 \\ 2(1) - 2 &\stackrel{?}{=} 0 \\ 2 - 2 &\stackrel{?}{=} 0 \\ 0 &= 0 \text{ True} \end{aligned}$$
(1, 2) is a solution.

7. Answers may vary.

9. $\begin{cases} y = x + 1 \\ y = 2x - 1 \end{cases}$

Graph both linear equations on the same set of axes.

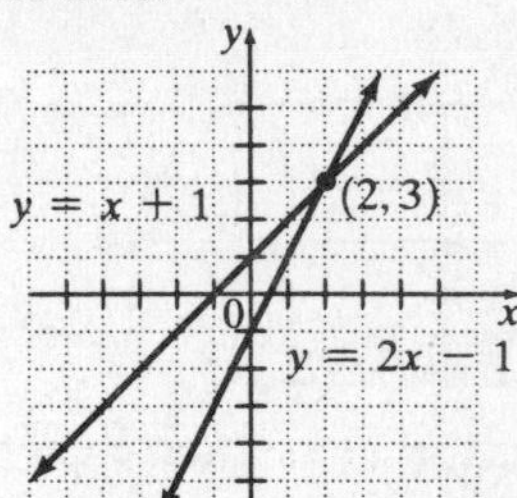

The solution is the intersection point of the two lines, (2, 3). Since this system has only one solution, it is consistent and independent.

11. $\begin{cases} 2x + y = 0 \\ 3x + y = 1 \end{cases}$

Graph both linear equations on the same set of axes.

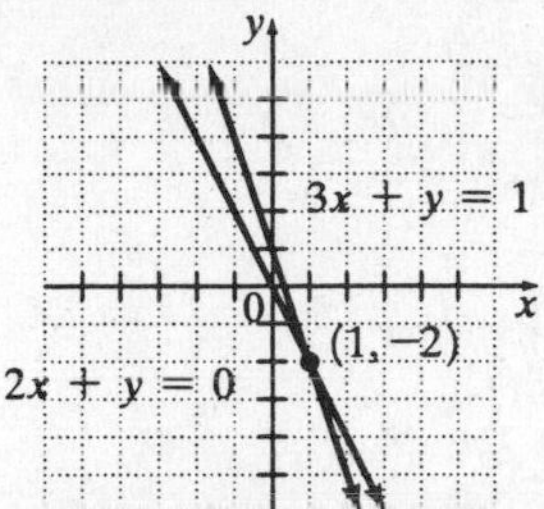

The solution is the intersection point of the two lines, (1, –2). Since this system has only one solution, it is consistent and independent.

13. $\begin{cases} y = -x - 1 \\ y = 2x + 5 \end{cases}$

Graph both linear equations on the same set of axes.

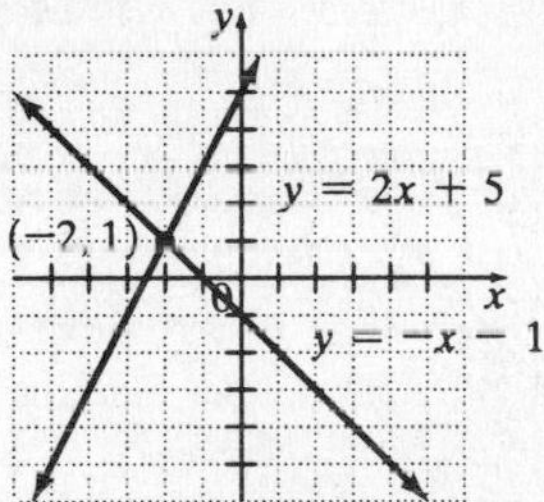

The solution is the intersection point of the two lines, (–2, 1). Since this system has only one solution, it is consistent and independent.

15. $\begin{cases} 2x - y = 6 \\ y = 2 \end{cases}$

Graph both linear equations on the same set of axes.

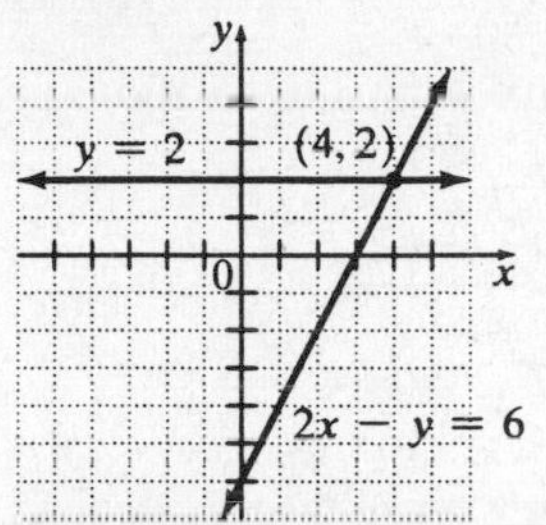

The solution is the intersection point of the two lines, (4, 2). Since this system has only one solution, it is consistent and independent.

17. $\begin{cases} x + y = 5 \\ x + y = 6 \end{cases}$

Graph both linear equations on the same set of axes.

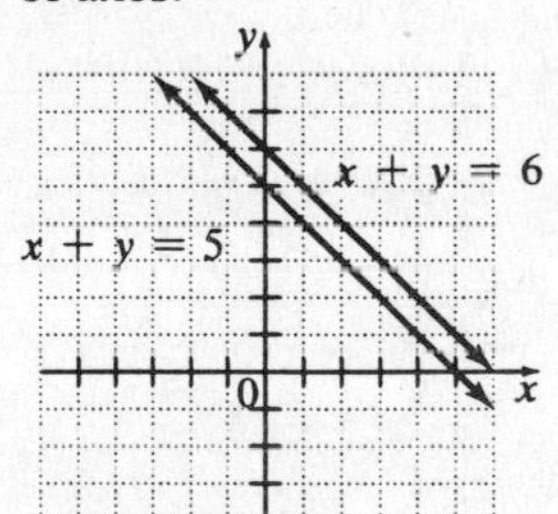

Since the lines are parallel, the system has no solution. The system is inconsistent and independent.

19. $\begin{cases} y-3x=-2 \\ 6x-2y=4 \end{cases}$

Graph both linear equations on the same set of axes.

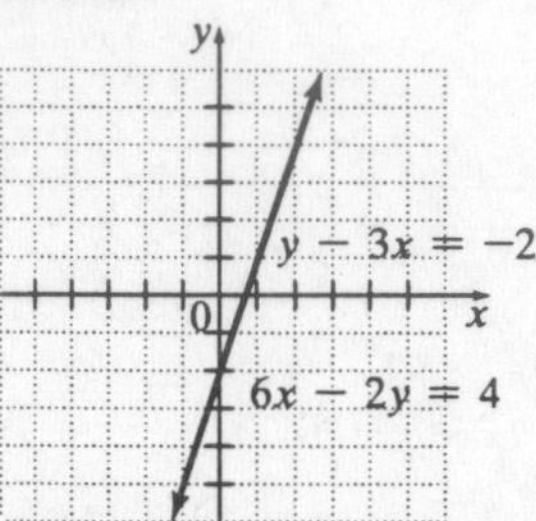

Since the graphs of the equationss are the same line, there are infinitely many solutions. This system is consistent and dependent.

21. $\begin{cases} x-2y=2 \\ 3x+2y=-2 \end{cases}$

Graph both linear equations on the same set of axes.

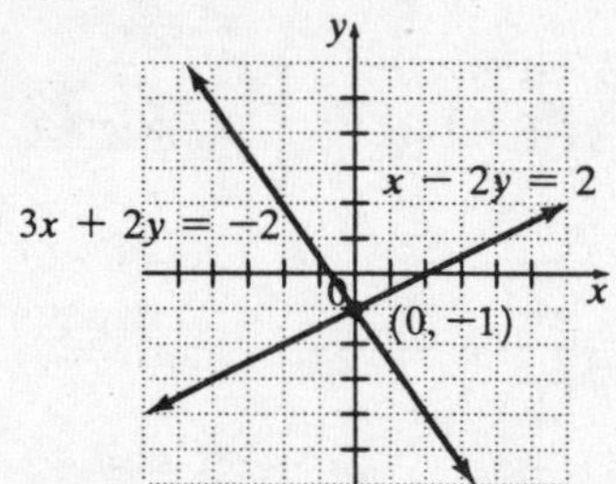

The solution is the intersection point of the two lines, (0, –1). Since this system has only one solution, it is consistent and independent.

23. $\begin{cases} \frac{1}{2}x+y=-1 \\ x=4 \end{cases}$

Graph both linear equations on the same set of axes.

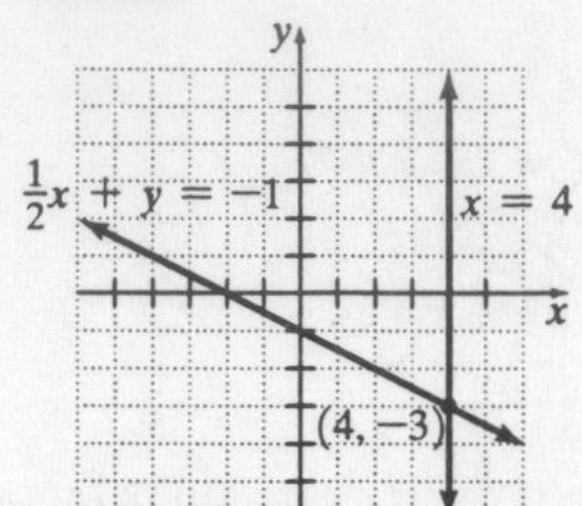

The solution is the intersection point of the two lines, (4, –3). Since this system has only one solution, it is consistent and independent.

25. $\begin{cases} y=x-2 \\ y=2x+3 \end{cases}$

Graph both linear equations on the same set of axes.

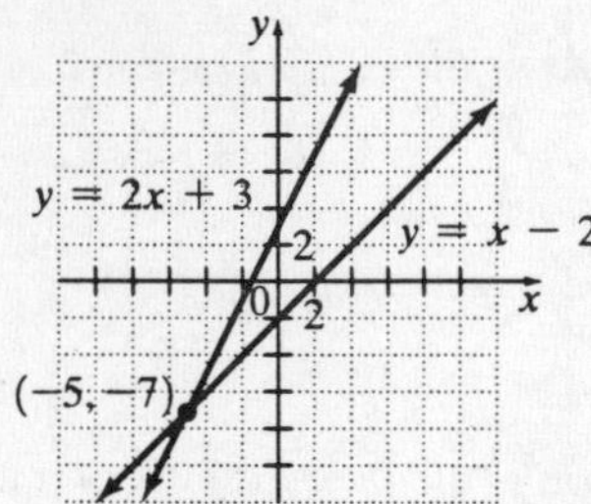

The solution is the intersection point of the two lines, (–5, –7). Since this system has only one solution, it is consistent and independent.

27. $\begin{cases} x+y=7 \\ x-y=3 \end{cases}$

Graph both linear equations on the same set of axes.

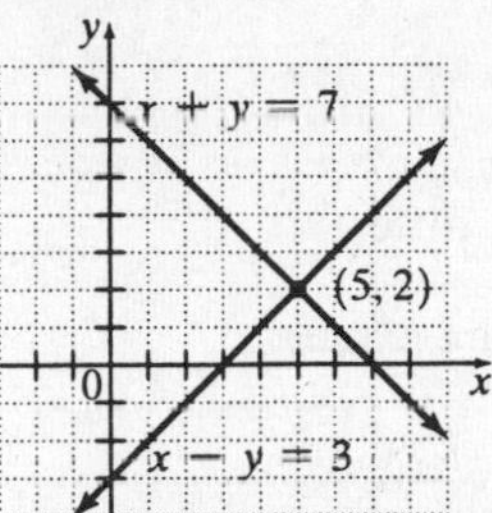

The solution is the intersection point of the two lines, (5, 2). Since this system has only one solution, it is consistent and independent.

29. Answers may vary.

31. $\begin{cases} 4x+y=24 \\ x+2y=2 \end{cases}$

Write the equations in slope-intercept form.

$4x + y = 24$ | $x + 2y = 2$

$y = -4x + 24$ | $2y = -x + 2$

$m = -4$ | $\frac{2y}{2} = \frac{-x}{2} + \frac{2}{2}$

$y = -\frac{1}{2}x + 1$

$m = -\frac{1}{2}$

Slopes are different.

a. Lines intersect at a single point.

b. There is one solution.

33. $\begin{cases} 2x+y=0 \\ 2y=6-4x \end{cases}$

Write the equations in slope-intercept form.

$2x + y = 0$ | $2y = 6 - 4x$

$y = -2x$ | $\frac{2y}{2} = \frac{6}{2} - \frac{4x}{2}$

$m = -2, b = 0$ | $y = 3 - 2x$

$y = -2x + 3$

$m = -2, b = 3$

Slopes are the same.
y-intercepts are different.

a. The lines are parallel.

b. There is no solution.

35. $\begin{cases} 6x-y=4 \\ \frac{1}{2}y=-2+3x \end{cases}$

Write the equations in slope-intercept form.

$6x - y = 4$ | $\frac{1}{2}y = -2 + 3x$

$-y = -6x + 4$ | $2\left(\frac{1}{2}y\right) = 2(-2+3x)$

$\frac{-y}{-1} = \frac{-6x}{-1} + \frac{4}{-1}$ | $y = -4 + 6x$

$y = 6x - 4$ | $y = 6x - 4$

$m = 6, b = -4$ | $m = 6, b = -4$

Slopes are the same.
y-intercepts are the same.

a. Identical lines

b. There are infinitely many solutions.

37. $\begin{cases} x = 5 \\ y = -2 \end{cases}$

$x = 5$ $\quad$ $y = -2$

m is undefined. $\quad$ $m = 0$

Slopes are different.

a. Lines intersect at a single point.

b. There is one solution.

39. $\begin{cases} 3y - 2x = 3 \\ x + 2y = 9 \end{cases}$

Write the equations in slope-intercept form.

$$3y - 2x = 3 \qquad x + 2y = 9$$
$$3y = 2x + 3 \qquad 2y = -x + 9$$
$$\frac{3y}{3} = \frac{2x}{3} + \frac{3}{3} \qquad \frac{2y}{2} = \frac{-y}{2} + \frac{9}{2}$$
$$y = \frac{2}{3}x + 1 \qquad y = -\frac{1}{2}x + \frac{9}{2}$$
$$m = \frac{2}{3} \qquad m = -\frac{1}{2}$$

Slopes are different.

a. The lines intersect at a single point.

b. There is one solution.

41. $\begin{cases} 6y + 4x = 6 \\ 3y - 3 = -2x \end{cases}$

Write the equations in slope-intercept form.

$$6y + 4x = 6 \qquad 3y - 3 = -2x$$
$$6y = -4x + 6 \qquad 3y = -2x + 3$$
$$\frac{6y}{6} = -\frac{4}{6}x + \frac{6}{6} \qquad \frac{3y}{3} = -\frac{2}{3}x + \frac{3}{3}$$
$$y = -\frac{2}{3}x + 1 \qquad y = -\frac{2}{3}x + 1$$
$$m = -\frac{2}{3},\ b = 1 \qquad m = -\frac{2}{3},\ b = 1$$

Slopes are the same.

y-intercepts are the same.

a. Theses are identical lines.

b. There are infinitely many solutions.

43. $\begin{cases} x + y = 4 \\ x + y = 3 \end{cases}$

Write the equations in slope-intercept form.

$$x + y = 4 \qquad x + y = 3$$
$$y = -x + 4 \qquad y = -x + 3$$
$$m = -1,\ b = 4 \qquad m = -1,\ b = 3$$

Slopes are the same.

y-intercepts are different.

a. The lines are parallel.

b. There is no solution.

45. Answers may vary.

47. 1984, 1988; the lines intersect at the approximate points (1984, 6.2) and (1988, 7.3).

49. a. (4, 9); each of the tables has the point (4, 9).

b. Yes

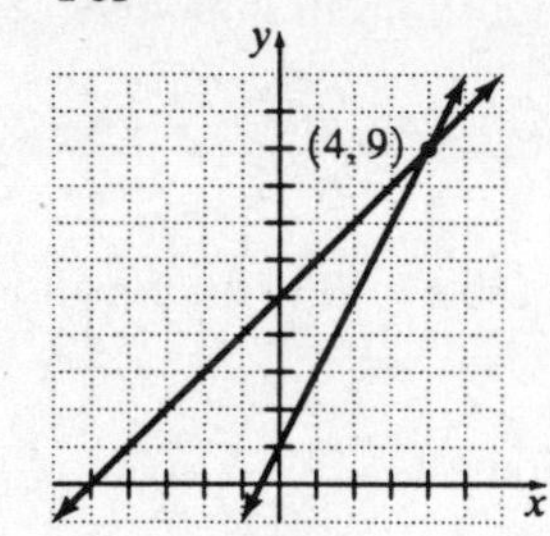

51.
$$-2x + 3(x + 6) = 17$$
$$-2x + 3x + 18 = 17$$
$$x + 18 = 17$$
$$x = -1$$

53.
$$-y + 12\left(\frac{y-1}{4}\right) = 3$$
$$-y + 3(y - 1) = 3$$
$$-y + 3y - 3 = 3$$
$$2y - 3 = 3$$
$$2y = 6$$
$$y = 3$$

55. $3z-(4z-2)=9$
$3z-4z+2=9$
$-z+2=9$
$-z=7$
$z=-7$

Exercise Set 4.2

1. $\begin{cases} x+y=3 \\ x=2y \end{cases}$
Replace x with $2y$ in the first equation.
$2y+y=3$
$3y=3$
$y=1$
To find x, use $y=1$ in the equation $x=2y$.
$x=2(1)$
$x=2$
The solution is (2, 1).

3. $\begin{cases} x+y=6 \\ y=-3x \end{cases}$
Replace y with $-3x$ in the first equation.
$x+(-3x)=6$
$-2x=6$
$x=\frac{6}{-2}=-3$
To find y, use $x=-3$ in the equation $y=-3x$.
$y=-3(-3)$
$y=9$
The solution is (−3, 9).

5. $\begin{cases} 3x+2y=16 \\ x=3y-2 \end{cases}$
Replace x with $3y-2$ in the first equation.
$3(3y-2)+2y=16$
$9y-6+2y=16$
$11y-6=16$
$11y=16+6$
$11y=22$
$y=\frac{22}{11}=2$
To find x, use $y=2$ in the equation $x=3y-2$.
$x=3(2)-2$
$x=6-2$
$x=4$
The solution is (4, 2).

7. $\begin{cases} 3x-4y=10 \\ x=2y \end{cases}$
Replace x with $2y$ in the first equation.
$3(2y)-4y=10$
$6y-4y=10$
$2y=10$
$y=\frac{10}{2}=5$
To find x, use $y=5$ in the equation $x=2y$.
$x=2(5)$
$x=10$
The solution is (10, 5).

9. $\begin{cases} y=3x+1 \\ 4y-8x=12 \end{cases}$
Replace y with $3x+1$ in the second equation.
$4(3x+1)-8x=12$
$12x+4-8x=12$
$4x+4=12$
$4x=12-4$
$4x=8$
$x=\frac{8}{4}=2$
To find y, use $x=2$ in the equation $y=3x+1$.
$y=3(2)+1$
$y=6+1$
$y=7$
The solution is (2, 7).

11. $\begin{cases} x+2y=6 \\ 2x+3y=8 \end{cases}$
Solve the first equation for x.
$x+2y=6$
$x=-2y+6$
Replace x with $-2y+6$ in the second equation.
$2(-2y+6)+3y=8$
$-4y+12+3y=8$
$-y+12=8$
$-y=8-12$
$-y=-4$
$y=4$
To find x, use $y=4$ in the equation $x=-2y+6$.
$x=-2(4)+6$
$x=-8+6$
$x=-2$
The solution is (−2, 4).

13. $\begin{cases} 2x - 5y = 1 \\ 3x + y = -7 \end{cases}$

Solve the second equation for y.

$3x + y = -7$
$y = -3x - 7$

Replace y with $-3x - 7$ in the first equation.

$2x - 5(-3x - 7) = 1$
$2x + 15x + 35 = 1$
$17x + 35 = 1$
$17x = 1 - 35$
$17x = -34$
$x = \frac{-34}{17} = -2$

To find y, use $x = -2$ in the equation $y = -3x - 7$.

$y = -3(-2) - 7$
$y = 6 - 7$
$y = -1$

The solution is $(-2, -1)$.

15. $\begin{cases} 2y = x + 2 \\ 6x - 12y = 0 \end{cases}$

Solve the first equation for x.

$2y = x + 2$
$2y - 2 = x$

Replace x with $2y - 2$ in the second equation.

$6(2y - 2) - 12y = 0$
$12y - 12 - 12y = 0$
$-12 = 0$

This is a contradiction. Therefore, there is no solution.

17. $\begin{cases} \frac{1}{3}x - y = 2 \\ x - 3y = 6 \end{cases}$

Solve the first equation for y.

$\frac{1}{3}x - y = 2$
$\frac{1}{3}x = 2 + y$
$\frac{1}{3}x - 2 = y$

Replace y with $\frac{1}{3}x - 2$ in the second equation.

$x - 3\left(\frac{1}{3}x - 2\right) = 6$
$x - x + 6 = 6$
$6 = 6$

This is an identity.
The equations are dependent and there are infinitely many solutions.

19. $\begin{cases} 4x + y = 11 \\ 2x + 5y = 1 \end{cases}$

Solve the first equation for y.

$4x + y = 11$
$y = -4x + 11$

Replace y with $-4x + 11$ in the second equation.

$2x + 5(-4x + 11) = 1$
$2x - 20x + 55 = 1$
$-18x + 55 = 1$
$-18x = 1 - 55$
$-18x = -54$
$x = \frac{-54}{-18} = 3$

To find y, use $x = 3$ in the equation $y = -4x + 11$.

$y = -4(3) + 11$
$y = -12 + 11$
$y = -1$

The solution is $(3, -1)$.

21. $\begin{cases} 2x - 3y = -9 \\ 3x = y + 4 \end{cases}$

Solve the second equation for y.

$3x = y + 4$

$3x - 4 = y$

Replace y with $3x - 4$ in the first equation.

$2x - 3(3x - 4) = -9$

$2x - 9x + 12 = -9$

$-7x + 12 = -9$

$-7x = -9 - 12$

$-7x = -21$

$x = \frac{-21}{-7} = 3$

To find y, use $x = 3$ in the equation $3x - 4 = y$.

$3(3) - 4 = y$

$9 - 4 = y$

$5 = y$

The solution is (3, 5).

23. $\begin{cases} 6x - 3y = 5 \\ x + 2y = 0 \end{cases}$

Solve the second equation for x.

$x + 2y = 0$

$x = -2y$

Replace x with $-2y$ in the first equation.

$6(-2y) - 3y = 5$

$-12y - 3y = 5$

$-15y = 5$

$y = \frac{5}{-15} = -\frac{1}{3}$

To find x, use $y = -\frac{1}{3}$ in the equation $x = -2y$.

$x = -2\left(-\frac{1}{3}\right)$

$x = \frac{2}{3}$

The solution is $\left(\frac{2}{3}, -\frac{1}{3}\right)$.

25. $\begin{cases} 3x - y = 1 \\ 2x - 3y = 10 \end{cases}$

Solve the first equation for y.

$3x - y = 1$

$-y = -3x + 1$

$y = \frac{-3}{-1}x + \frac{1}{-1}$

$y = 3x - 1$

Replace y with $3x - 1$ in the second equation.

$2x - 3(3x - 1) = 10$

$2x - 9x + 3 = 10$

$-7x + 3 = 10$

$-7x = 10 - 3$

$-7x = 7$

$x = \frac{7}{-7} = -1$

To find y, use $x = -1$ in the equation $y = 3x - 1$.

$y = 3(-1) - 1$

$y = -3 - 1$

$y = -4$

The solution is (–1, –4).

27. $\begin{cases} -x + 2y = 10 \\ -2x + 3y = 18 \end{cases}$

Solve the first equation for x.

$-x + 2y = 10$

$2y = 10 + x$

$2y - 10 = x$

Replace x with $2y - 10$ in the second equation.

$-2(2y - 10) + 3y = 18$

$-4y + 20 + 3y = 18$

$-y + 20 = 18$

$-y = 18 - 20$

$-y = -2$

$y = 2$

To find x, use $y = 2$ in the equation $2y - 10 = x$.

$2(2) - 10 = x$

$4 - 10 = x$

$-6 = x$

The solution is (–6, 2).

29. $\begin{cases} 5x+10y=20 \\ 2x+6y=10 \end{cases}$

Solve the second equation for x.

$$2x+6y=10$$
$$2x=10-6y$$
$$x=\frac{10}{2}-\frac{6}{2}y$$
$$x=5-3y$$

Replace x with $5-3y$ in the first equation.

$$5(5-3y)+10y=20$$
$$25-15y+10y=20$$
$$25-5y=20$$
$$-5y=20-25$$
$$-5y=-5$$
$$y=\frac{-5}{-5}=1$$

To find x, use $y=1$ in the equation $x=5-3y$.

$$x=5-3(1)$$
$$x=5-3$$
$$x=2$$

The solution is (2, 1).

31. $\begin{cases} 3x+6y=9 \\ 4x+8y=16 \end{cases}$

Solve the first equation for x.

$$3x+6y=9$$
$$3x=-6y+9$$
$$x=\frac{-6y}{3}+\frac{9}{3}$$
$$x=-2y+3$$

Replace x with $-2y+3$ in the second equation.

$$4(-2y+3)+8y=16$$
$$-8y+12+8y=16$$
$$12=16$$

This is a contradiction. There is no solution.

33. $\begin{cases} y=2x+9 \\ y=7x+10 \end{cases}$

Replace y with $2x+9$ in the second equation.

$$2x+9=7x+10$$
$$9=7x-2x+10$$
$$9=5x+10$$
$$9-10=5x$$
$$-1=5x$$
$$-\frac{1}{5}=x$$

To find y, use $x=-\frac{1}{5}$ in the equation $y=2x+9$.

$$y=2\left(-\frac{1}{5}\right)+9$$
$$y=-\frac{2}{5}+9$$
$$y=-\frac{2}{5}+\frac{45}{5}$$
$$y=\frac{43}{5}$$

The solution is $\left(-\frac{1}{5}, \frac{43}{5}\right)$.

35. Answers may vary.

37. Simplify the equations.

$$-5y+6y=3x+2(x-5)-3x+5$$
$$-5y+6y=3x+2x-10-3x+5$$
$$y=2x-5$$
$$4(x+y)-x+y=-12$$
$$4x+4y-x+y=-12$$
$$3x+5y=-12$$

Replace y with $2x-5$ in the simplified second equation.

$$3x+5(2x-5)=-12$$
$$3x+10x-25=-12$$
$$13x-25=-12$$
$$13x=13$$
$$x=1$$

To find y, use $x=1$ in the equation $y=2x-5$.

$$y=2(1)-5$$
$$y=2-5$$
$$y=-3$$

The solution is (1, −3).

39. a. $\begin{cases} y = -0.65x + 32.02 \\ y = 0.78x + 1.32 \end{cases}$

Substitute $-0.65x + 32.02$ for y in the second equation. Then solve for x.

$$-0.65x + 32.02 = 0.78x + 1.32$$
$$-1.43x = -30.7$$
$$x \approx 21$$

Substitute 21 for x in the first equation. Then solve for y.

$y \approx -0.65(21) + 32.02$

$y \approx 18$

The solution is approximately (21, 18).

b. Answers may vary.

c.

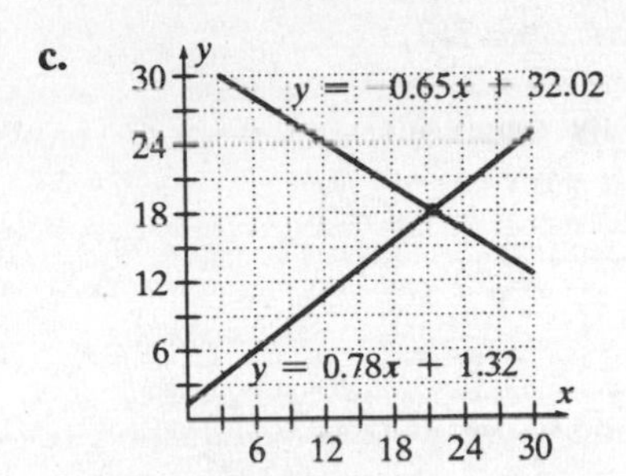

Answers may vary.

41. $\begin{cases} y = 5.1x + 14.56 \\ y = -2x - 3.9 \end{cases}$

Replace y with $5.1x + 14.56$ in the second equation.

$$5.1x + 14.56 = -2x - 3.9$$
$$7.1x = -18.46$$
$$x = -2.6$$

To find y, let $x = -2.6$ in the second equation.

$y = -2x - 3.9$

$y = -2(-2.6) - 3.9$

$y = 5.2 - 3.9$

$y = 1.3$

The solution is (−2.6, 1.3).

43. $\begin{cases} 3x + 2y = 14.05 \\ 5x + y = 18.5 \end{cases}$

Solve the second equation for x.

$$5x + y = 18.5$$
$$5x = 18.5 - y$$
$$x = \frac{18.5}{5} - \frac{y}{5}$$
$$x = 3.7 - 0.2y$$

Replace x with $3.7 - 0.2y$ in the first equation.

$$3x + 2y = 14.05$$
$$3(3.7 - 0.2y) + 2y = 14.05$$
$$11.1 - 0.6y + 2y = 14.05$$
$$1.4y = 2.95$$
$$y \approx 2.11$$

To find x, replace y with 2.11 in the first equation and solve for x.

$$3x + 2y = 14.05$$
$$3x + 2(2.11) \approx 14.05$$
$$3x + 4.22 \approx 14.05$$
$$3x \approx 9.83$$
$$x \approx 3.28$$

The solution is approximately (3.28, 2.11).

45.
$$3x + 2y = 6$$
$$-2(3x + 2y) = -2(6)$$
$$-6x - 4y = -12$$

47.
$$-4x + y = 3$$
$$3(-4x + y) = 3(3)$$
$$-12x + 3y = 9$$

49.
$$\begin{array}{r} 3n + 6m \\ \underline{2n - 6m} \\ 5n \end{array}$$

51.
$$\begin{array}{r} -5a - 7b \\ \underline{5a - 8b} \\ -15b \end{array}$$

Exercise Set 4.3

1. $\begin{cases} 3x + y = 5 \\ 6x - y = 4 \end{cases}$

Add the equations.

$$\begin{array}{l} 3x + y = 5 \\ \underline{6x - y = 4} \\ 9x \quad = 9 \\ x \quad = 1 \end{array}$$

To find y, use $x = 1$ in the equation $3x + y = 5$.

$$\begin{aligned} 3(1) + y &= 5 \\ 3 + y &= 5 \\ y &= 5 - 3 \\ y &= 2 \end{aligned}$$

The solution is (1, 2).

3. $\begin{cases} x - 2y = 8 \\ -x + 5y = -17 \end{cases}$

Add the equations.

$$\begin{array}{l} x - 2y = 8 \\ \underline{-x + 5y = -17} \\ 3y = -9 \\ y = \frac{-9}{3} = -3 \end{array}$$

To find x, use $y = -3$ in the equation $x - 2y = 8$.

$$\begin{aligned} x - 2(-3) &= 8 \\ x + 6 &= 8 \\ x &= 8 - 6 \\ x &= 2 \end{aligned}$$

The solution is (2, –3).

5. $\begin{cases} x + y = 6 \\ x - y = 6 \end{cases}$

Add the equations.

$$\begin{array}{l} x + y = 6 \\ \underline{x - y = 6} \\ 2x \quad = 12 \\ x \quad = 6 \end{array}$$

To find y, use $x = 6$ in the equation $x + y = 6$.

$$\begin{aligned} 6 + y &= 6 \\ y &= 6 - 6 \\ y &= 0 \end{aligned}$$

The solution is (6, 0).

7. $\begin{cases} 3x + y = 4 \\ 9x + 3y = 6 \end{cases}$

Multiply 1st equation by –3.

$$\begin{aligned} -3(3x + y) &= -3(4) \\ -9x - 3y &= -12 \end{aligned}$$

Add the equations.

$$\begin{array}{l} -9x - 3y = -12 \\ \underline{9x + 3y = 6} \\ 0 = -6 \end{array}$$

This is a contradiction. There is no solution.

9. $\begin{cases} 3x - 2y = 7 \\ 5x + 4y = 8 \end{cases}$

Multiply 1st equation by 2.

$$\begin{aligned} 2(3x - 2y) &= 2(7) \\ 6x - 4y &= 14 \end{aligned}$$

Add the equations.

$$\begin{array}{l} 6x - 4y = 14 \\ \underline{5x + 4y = 8} \\ 11x \quad = 22 \\ x \quad = \frac{22}{11} = 2 \end{array}$$

To find y, use $x = 2$ in the equation $3x - 2y = 7$.

$$\begin{aligned} 3(2) - 2y &= 7 \\ 6 - 2y &= 7 \\ -2y &= 7 - 6 \\ -2y &= 1 \\ y &= -\frac{1}{2} \end{aligned}$$

The solution is $\left(2, -\frac{1}{2}\right)$.

11. $\begin{cases} \frac{2}{3}x + 4y = -4 \\ 5x + 6y = 18 \end{cases}$

Multiply 1st equation by 3.

$3\left(\frac{2}{3}x + 4y\right) = 3(-4)$

$2x + 12y = -12$

Since addition now would not eliminate a variable, multiply the 2nd equation by –2.

$-2(5x + 6y) = -2(18)$

$-10x - 12y = -36$

Add the equations.

$$\begin{array}{rl} 2x + 12y &= -12 \\ -10x - 12y &= -36 \\ \hline -8x \quad\quad &= -48 \\ x \quad\quad &= \frac{-48}{-8} = 6 \end{array}$$

To find y, use $x = 6$ in the equation $5x + 6y = 18$.

$5(6) + 6y = 18$

$30 + 6y = 18$

$6y = 18 - 30$

$6y = -12$

$y = -\frac{12}{6} = -2$

The solution is (6, –2).

13. $\begin{cases} 4x - 6y = 8 \\ 6x - 9y = 12 \end{cases}$

Multiply 1st equation by 3.

Multiply 2nd equation by –2.

$3(4x - 6y) = 3(8)$

$-2(6x - 9y) = -2(12)$

Add the resulting equations.

$$\begin{array}{rl} 12x - 18y &= 24 \\ -12x + 18y &= -24 \\ \hline 0 &= 0 \end{array}$$

This is an identity.

The equations are dependent.

There are infinitely many solutions.

15. $\begin{cases} 3x + y = -11 \\ 6x - 2y = -2 \end{cases}$

Multiply 1st equation by 2.

$2(3x + y) = 2(-11)$

$6x + 2y = -22$

Add the two equations.

$$\begin{array}{rl} 6x + 2y &= -22 \\ 6x - 2y &= -2 \\ \hline 12x \quad\quad &= -24 \\ x \quad\quad &= \frac{-24}{12} = -2 \end{array}$$

To find y, use $x = -2$ in the equation $3x + y = -11$.

$3(-2) + y = -11$

$-6 + y = -11$

$y = -11 + 6$

$y = -5$

The solution is (–2, –5).

17. $\begin{cases} 3x + 2y = 11 \\ 5x - 2y = 29 \end{cases}$

Add the two equations.

$$\begin{array}{rl} 3x + 2y &= 11 \\ 5x - 2y &= 29 \\ \hline 8x \quad\quad &= 40 \\ x \quad\quad &= \frac{40}{8} = 5 \end{array}$$

To find y, use $x = 5$ in the equation $3x + 2y = 11$.

$3(5) + 2y = 11$

$15 + 2y = 11$

$2y = 11 - 15$

$2y = -4$

$y = \frac{-4}{2} = -2$

The solution is (5, –2).

19. $\begin{cases} x+5y=18 \\ 3x+2y=-11 \end{cases}$

Multiply 1st equation by –3.

$-3(x+5y)=-3(18)$

$-3x-15y=-54$

Add the two equations.

$$\begin{array}{r} -3x-15y=-54 \\ 3x+\ 2y=-11 \\ \hline -13y=-65 \end{array}$$

$$y=\frac{-65}{-13}=5$$

To find x, use $y=5$ in the equation $x+5y=18$.

$x+5(5)=18$

$x+25=18$

$x=18-25$

$x=-7$

The solution is (–7, 5).

21. $\begin{cases} 2x-5y=4 \\ 3x-2y=4 \end{cases}$

Multiply the 1st equation by –3.

Multiply 2nd equation by 2.

$-3(2x-5y)=-3(4)$

$2(3x-2y)=2(4)$

Add the equations.

$$\begin{array}{r} -6x+15y=-12 \\ 6x-\ 4y=\ \ 8 \\ \hline 11y=-4 \end{array}$$

$$y=-\frac{4}{11}$$

To find x, use $y=-\frac{4}{11}$ in the equation $2x-5y=4$.

$$2x-5\left(-\frac{4}{11}\right)=4$$

$$2x+\frac{20}{11}=4$$

$$2x=4-\frac{20}{11}$$

$$2x=\frac{44}{11}-\frac{20}{11}$$

$$2x=\frac{24}{11}$$

$$\frac{1}{2}(2x)=\frac{1}{2}\left(\frac{24}{11}\right)$$

$$x=\frac{12}{11}$$

The solution is $\left(\frac{12}{11}, -\frac{4}{11}\right)$.

23. $\begin{cases} 2x+3y=0 \\ 4x+6y=3 \end{cases}$

Multiply 1st equation by –2.

$-2(2x+3y)=-2(0)$

$-4x-6y=0$

Add the equations.

$$\begin{array}{r} -4x-6y=0 \\ 4x+6y=3 \\ \hline 0=3 \end{array}$$

This is a contradiction. There is no solution.

25. $\begin{cases} \frac{x}{3}+\frac{y}{6}=1 \\ \frac{x}{2}-\frac{y}{4}=0 \end{cases}$

Multiply 1st equation by 6.
Multiply 2nd equation by 4.

$$6\left(\frac{x}{3}+\frac{y}{6}\right)=6(1)$$
$$4\left(\frac{x}{2}-\frac{y}{4}\right)=4(0)$$

Add the equations.

$$\begin{aligned} 2x+y&=6 \\ 2x-y&=0 \\ \hline 4x\quad&=6 \\ x\quad&=\frac{6}{4}=\frac{3}{2} \end{aligned}$$

To find y, use $x=\frac{3}{2}$ in the equation $2x+y=6$.

$$\begin{aligned} 2\left(\frac{3}{2}\right)+y&=6 \\ 3+y&=6 \\ y&=6-3 \\ y&=3 \end{aligned}$$

The solution is $\left(\frac{3}{2},\ 3\right)$.

27. $\begin{cases} x-\frac{y}{3}=-1 \\ -\frac{x}{2}+\frac{y}{8}=\frac{1}{4} \end{cases}$

Multiply 1st equation by 3.
Multiply 2nd equation by 8.

$$3\left(x-\frac{y}{3}\right)=3(-1)$$
$$8\left(-\frac{x}{2}+\frac{y}{8}\right)=8\left(\frac{1}{4}\right)$$

Add the resulting equations.

$$\begin{aligned} 3x-y&=-3 \\ -4x+y&=\ \ 2 \\ \hline -x\quad&=-1 \\ x\quad&=\frac{-1}{-1}=1 \end{aligned}$$

To find y, use $x=1$ in the equation $-4x+y=2$.

$$\begin{aligned} -4(1)+y&=2 \\ -4+y&=2 \\ y&=2+4 \\ y&=6 \end{aligned}$$

The solution is (1, 6).

29. $\begin{cases} \frac{x}{3}-y=2 \\ -\frac{x}{2}+\frac{3y}{2}=-3 \end{cases}$

Multiply 1st equation by 3.
Multiply 2nd equation by 2.

$$3\left(\frac{x}{3}-y\right)=3(2)$$
$$2\left(-\frac{x}{2}+\frac{3y}{2}\right)=2(-3)$$

Add the equations.

$$\begin{aligned} x-3y&=\ \ 6 \\ -x+3y&=-6 \\ \hline 0&=\ \ 0 \end{aligned}$$

This is an identity.
The equations are dependent.
There are infinitely many solutions.

31. $\begin{cases} 8x=-11y-16 \\ 2x+3y=-4 \end{cases}$

Multiply the 2nd equation by -4.

$$\begin{aligned} -4(2x+3y)&=-4(-4) \\ -8x-12y&=16 \end{aligned}$$

Rearrange the first equation.

$$\begin{aligned} 8x&=-11y-16 \\ 8x+11y&=-16 \end{aligned}$$

Add the equations.

$$\begin{aligned} -8x-12y&=\ \ 16 \\ 8x+11y&=-16 \\ \hline -y&=\ \ 0 \\ y&=\ \ 0 \end{aligned}$$

To find x, use $y=0$ in the equation $2x+3y=-4$.

$$\begin{aligned} 2x+3(0)&=-4 \\ 2x&=-4 \\ x&=\frac{-4}{2}=-2 \end{aligned}$$

The solution is $(-2, 0)$.

33. Answers may vary.

35. $\begin{cases} 2x - 3y = -11 \\ y = 4x - 3 \end{cases}$

Replace y with $4x - 3$ in the 1st equation.

$$\begin{aligned} 2x - 3(4x - 3) &= -11 \\ 2x - 12x + 9 &= -11 \\ -10x + 9 &= -11 \\ -10x &= -11 - 9 \\ -10x &= -20 \\ x &= \frac{-20}{-10} = 2 \end{aligned}$$

To find y, use $x = 2$ in the equation $y = 4x - 3$.

$y = 4(2) - 3$
$y = 8 - 3$
$y = 5$

The solution is (2, 5).

37. $\begin{cases} x + 2y = 1 \\ 3x + 4y = -1 \end{cases}$

Multiply the 1st equation by -3.

$-3(x + 2y) = -3(1)$
$-3x - 6y = -3$

Add the equations.

$$\begin{aligned} -3x - 6y &= -3 \\ 3x + 4y &= -1 \\ \hline -2y &= -4 \\ y &= \frac{-4}{-2} = 2 \end{aligned}$$

To find x, use $y = 2$ in the equation $x + 2y = 1$.

$$\begin{aligned} x + 2(2) &= 1 \\ x + 4 &= 1 \\ x &= 1 - 4 \\ x &= -3 \end{aligned}$$

The solution is (−3, 2).

39. $\begin{cases} 2y = x + 6 \\ 3x - 2y = -6 \end{cases}$

Solve the 1st equation for x.

$2y = x + 6$
$2y - 6 = x$

Replace x with $2y - 6$ in the second equation.

$$\begin{aligned} 3(2y - 6) - 2y &= -6 \\ 6y - 18 - 2y &= -6 \\ 4y - 18 &= -6 \\ 4y &= -6 + 18 \\ 4y &= 12 \\ y &= \frac{12}{4} = 3 \end{aligned}$$

To find x, use $y = 3$ in the equation $2y - 6 = x$.

$$\begin{aligned} 2(3) - 6 &= x \\ 6 - 6 &= x \\ 0 &= x \end{aligned}$$

The solution is (0, 3).

41. $\begin{cases} y = 2x - 3 \\ y = 5x - 18 \end{cases}$

Replace y with $2x - 3$ in the second equation.

$$\begin{aligned} 2x - 3 &= 5x - 18 \\ 2x - 5x - 3 &= -18 \\ -3x - 3 &= -18 \\ -3x &= -18 + 3 \\ -3x &= -15 \\ x &= \frac{-15}{-3} = 5 \end{aligned}$$

To find y, use $x = 5$ in the equation $y = 2x - 3$.

$y = 2(5) - 3$
$y = 10 - 3$
$y = 7$

The solution is (5, 7).

43. $\begin{cases} x+\frac{1}{6}y=\frac{1}{2} \\ 3x+2y=3 \end{cases}$

Multiply 1st equation by 6.

$$6\left(x+\frac{1}{6}y\right)=6\left(\frac{1}{2}\right)$$
$$6x+y=3$$

Since addition now would not eliminate a variable, multiply this equation by –2.

$$-2(6x+y)=-2(3)$$
$$-12x-2y=-6$$

Add the equations.

$$\begin{aligned} -12x-2y&=-6 \\ 3x+2y&=3 \\ \hline -9x\quad&=-3 \\ x\quad&=\frac{-3}{-9}=\frac{1}{3} \end{aligned}$$

To find y, use $x=\frac{1}{3}$ in the equation $3x+2y=3$.

$$3\left(\frac{1}{3}\right)+2y=3$$
$$1+2y=3$$
$$2y=3-1$$
$$2y=2$$
$$y=\frac{2}{2}=1$$

The solution is $\left(\frac{1}{3},\ 1\right)$.

45. $\begin{cases} \frac{x+2}{2}=\frac{y+11}{3} \\ \frac{x}{2}=\frac{2y+16}{6} \end{cases}$

Multiply 1st equation by 6.

$$6\left(\frac{x+2}{2}\right)=6\left(\frac{y+11}{3}\right)$$
$$3(x+2)=2(y+11)$$
$$3x+6=2y+22$$
$$3x-2y=22-6$$
$$3x-2y=16$$

Multiply 2nd equation by 6.

$$6\left(\frac{x}{2}\right)=6\left(\frac{2y+16}{6}\right)$$
$$3x=2y+16$$
$$3x-2y=16$$
$$-1(3x-2y)=-1(16)$$
$$-3x+2y=-16$$

Add the equations.

$$\begin{aligned} 3x-2y&=16 \\ -3x+2y&=-16 \\ \hline 0&=0 \end{aligned}$$

This is an identity.
The equations are dependent.
There are infinitely many solutions.

47. $\begin{cases} 2x+3y=14 \\ 3x-4y=-69.1 \end{cases}$

Multiply the 1st equation by 3 and the 2nd equation by –2.

$$3(2x+3y)=3(14)$$
$$-2(3x-4y)=-2(-69.1)$$

Add the resulting equations.

$$\begin{aligned} 6x+9y&=42 \\ -6x+8y&=138.2 \\ \hline 17y&=180.2 \\ y&=10.6 \end{aligned}$$

To find x, use $y=10.6$ in the equation $2x+3y=14$.

$$2x+3(10.6)=14$$
$$2x+31.8=14$$
$$2x=-17.8$$
$$x=-8.9$$

The solution is (–8.9, 10.6).

49. a. $\begin{cases} 10x - 2y = -986 \\ -21x + y = 295 \end{cases}$

Multiply the 1st equation by $\frac{1}{2}$.

$\frac{1}{2}(10x - 2y) = \frac{1}{2}(-986)$

$5x - y = -493$

Add the equations.

$$\begin{array}{rl} 5x - y = & -493 \\ -21x + y = & 295 \\ \hline -16x \quad = & -198 \end{array}$$

$x = 12.375$

To find y, use $x = 12.375$ in the equation $-21x + y = 295$.

$-21(12.375) + y = 295$

$-259.875 + y = 295$

$y = 554.875$

The solution is approximately (12, 555).

b. Answers may vary.

c. Since they were equal 12 years after 1980, then from 1980 + 13 = 1993 to 1997, there were more UHF stations than VHF stations.

51. a. $x + y = 5$

$3(x + y) = 3(5)$

$3x + 3y = 15$

$b = 15$

b. b can be any real number except 15.

53. Answers may vary.

55. Let x = number.

$2x + 6 = x - 3$

57. Let x = number.

$20 - 3x = 2$

59. Let n = number.

$4(n + 6) = 2n$

Exercise Set 4.4

1. c; length = 9 feet, width = 6 feet
The length is 3 feet longer than the width and the perimeter is
$2(9) + 2(6) = 18 + 12 = 30$ feet.

3. b; notebook = \$3, computer disk = \$4
The cost of 2 computer disks and
3 notebooks $= 2(4) + 3(3)$
$= 8 + 9$
$= \$17$.
The cost of 5 computer disks and
4 notebooks $= 5(4) + 4(3)$
$= 20 + 12$
$= \$32$.

5. a; 80 dimes and 20 quarters
The total number of coins is 80 + 20 = 100.
Their total value is
$80(0.10) + 20(0.25) = 8 + 5 = \$13$.

7. Let x = one number and
y = the other number.

$\begin{cases} x + y = 15 \\ x - y = 7 \end{cases}$

9. Let x = amount invested in larger account and y = amount invested in smaller account.

$\begin{cases} x + y = 6500 \\ x = y + 800 \end{cases}$

11. Let x = one number and
y = the other number.

$$\begin{array}{rl} x + y = & 83 \\ x - y = & 17 \\ \hline 2x \quad = & 100 \\ x \quad = & 50 \end{array}$$

To find y, let $x = 50$ in the first equation.

$50 + y = 83$

$y = 33$

The two numbers are 33 and 50.

13. Let $x =$ first number and
$y =$ second number.

$$\begin{cases} x+2y=8 \\ 2x+y=25 \end{cases}$$

Multiply the 1st equation by –2 and then add the equations.

$$\begin{cases} -2(x+2y)=-2(8) \\ 2x+y=25 \end{cases}$$

$$\begin{cases} -2x-4y=-16 \\ 2x+\ \ y=\ \ 25 \end{cases}$$

$$-3y=9$$
$$y=-3$$

To find x, let $y=-3$ in the first equation.

$$x+2(-3)=8$$
$$x-6=8$$
$$x=14$$

The numbers are 14 and –3.

15. Let $x =$ number of points Cooper scored and $y =$ number of points Swoopes scored.

$$\begin{cases} x=101+y \\ x+y=1271 \end{cases}$$

Substitute $101 + y$ for x in the second equation.

$$101+y+y=1271$$
$$2y=1170$$
$$y=585$$

To find x, let $y = 585$ in the first equation.

$$x = 101 + 585$$
$$x = 686$$

Cooper scored 686 points and Swoopes scored 585 points.

17. Let $x =$ price of adult's ticket and $y =$ price of child's ticket.

$$\begin{cases} 3x+4y=159 \\ 2x+3y=112 \end{cases}$$

Multiply the first equation by –2 and the second equation by 3. Then add.

$$\begin{cases} -2(3x+4y)=-2(159) \\ 3(2x+3y)=3(112) \end{cases}$$

$$\begin{cases} -6x-8y=-318 \\ 6x+9y=\ \ 336 \end{cases}$$

$$y=18$$

To find x, let $y = 18$ in the first equation.

$$3x+4(18)=159$$
$$3x+72=159$$
$$3x=87$$
$$x=29$$

An adult's ticket is \$29 and a child's ticket is \$18.

19. Let $x =$ number of quarters and
$y =$ number of nickels.

$$\begin{cases} x+y=80 \\ 0.25x+0.05y=14.60 \end{cases}$$

Solve the first equation for y.

$$x+y=80$$
$$y=80-x$$

Substitute $80 - x$ for y in the second equation.

$$0.25x+0.05(80-x)=14.60$$
$$0.25x+4-0.05x=14.60$$
$$0.20x=10.60$$
$$x=53$$

To find y, let $x = 53$ in the first equation.

$$53 + y = 80$$
$$y=27$$

There are 53 quarters and 27 nickels.

21. Let x = price of IBM stock and
y = price of GA Financial stock.

$$\begin{cases} 50x + 40y = 6485.90 \\ y = x + 64.25 \end{cases}$$

Substitute $x + 64.25$ for y in the first equation.

$$50x + 40(x + 64.25) = 6485.90$$
$$50x + 40x + 2570 = 6485.90$$
$$90x = 3915.90$$
$$x = 43.51$$

To find y, let $x = 43.51$ in the second equation.

$$y = 43.51 + 64.25$$
$$y = 107.76$$

The stock of IBM was \$43.51 and the stock of GA Financial was \$107.76.

23. Let x = Pratap's rate in still water, in miles per hour, and
y = rate of current in miles per hour.
Then $x + y$ is the rate downstream and $x - y$ is the rate upstream.

$$\begin{cases} 2(x + y) = 18 \\ 4.5(x - y) = 18 \end{cases}$$

$$\begin{cases} 2x + 2y = 18 \\ 4.5x - 4.5y = 18 \end{cases}$$

Multiply the first equation by 2.25 and then add the equations.

$$\begin{cases} 2.25(2x + 2y) = 2.25(18) \\ 4.5x - 4.5y = 18 \end{cases}$$

$$\begin{cases} 4.5x + 4.5y = 40.5 \\ 4.5x - 4.5y = 18 \end{cases}$$
$$9x = 58.5$$
$$x = 6.5$$

To find y, let $x = 6.5$ in the first equation.

$$2(6.5) + 2y = 18$$
$$2y = 5$$
$$y = 2.5$$

Pratap's rate in still water was 6.5 miles per hour and the current's rate was 2.5 miles per hour.

25. Let x = speed of plane in still air, in miles per hour, and
y = speed of wind, in miles per hour.
Then $x - y$ is the speed of the plane against the wind and $x + y$ the speed with the wind.

$$\begin{cases} 780 = 2(x - y) \\ 780 = 1.5(x + y) \end{cases}$$

Divide each side of the first equation by 2 and each side of the second equation by 1.5. Then add the equations.

$$\begin{cases} 390 = x - y \\ 520 = x + y \end{cases}$$
$$910 = 2x$$
$$455 = x$$

To find y, let $x = 455$ in the equation $520 = x + y$.

$$520 = 455 + y$$
$$65 = y$$

The plane's speed is 455 miles per hour in still air and the wind's speed is 65 miles per hour.

27. Let x = amount of 12% solution, and
y = amount of 4% solution.

$$\begin{cases} x + y = 12 \\ 0.12x + 0.04y = 0.09(12) \end{cases}$$

Solve the first equation for y.

$$x + y = 12$$
$$y = 12 - x$$

Substitute $12 - x$ for y in the second equation.

$$0.12x + 0.04(12 - x) = 0.09(12)$$
$$0.12x + 0.48 - 0.04x = 1.08$$
$$0.08x = 0.6$$
$$x = 7.5$$

To find y, let $x = 7.5$ in the first equation.

$$7.5 + y = 12$$
$$y = 4.5$$

She needs $7\frac{1}{2}$ ounces of 12% solution and $4\frac{1}{2}$ ounces of 4% solution.

29. Let x = number of pounds of \$4.95 per pound beans, and
y = number of pounds of \$2.65 per pound beans.
Then $x + y$ is the total weight of the beans and $4.95x + 2.65y$ the total cost.

$$\begin{cases} x+y=200 \\ 4.95x+2.65y=200(3.95) \end{cases}$$

Solve the first equation for y.
$x+y=200$
$y=200-x$
Substitute $200 - x$ for y in the second equation.
$4.95x+2.65(200-x)=200(3.95)$
$4.95x+530-2.65x=790$
$2.30x=260$
$x \approx 113$
To find y, substitute 113 for x in the first equation.
$113 + y \approx 200$
$y \approx 87$
To the nearest pound, Wayne needs 113 pounds of the \$4.95 per pound beans and 87 pounds of the \$2.65 per pound beans.

31. Let x = the measure of the first angle, and
y = the measure of the second angle.

$$\begin{cases} x+y=90 \\ x=2y \end{cases}$$

Substitute $2y$ for x in the first equation.
$2y+y=90$
$3y=90$
$y=30$
To find x, let $y = 30$ in the second equation.
$x=2(30)$
$x=60$
The angles measure 60° and 30°.

33. Let x = the measure of the first angle, and
y = the measure of the second angle.

$$\begin{cases} x+y=90 \\ y=3x+10 \end{cases}$$

Substitute $3x + 10$ for y in the first equation.
$x+3x+10=90$
$4x=80$
$x=20$
To find y, let $x = 20$ in the second equation.
$y = 3(20) + 10$
$y=70$
The angles measure 20° and 70°.

35. Let x = liters of 20% solution, and
y = liters of 70% solution.

$$\begin{cases} x+y=50 \\ 0.20x+0.70y=0.60(50) \end{cases}$$

Multiply 1st equation by −0.20 and add the equations.

$$\begin{array}{r} -0.20x-0.20y=-10 \\ 0.20x+0.70y=30 \\ \hline 0.50y=20 \end{array}$$

$y=40$
To find x, let $y = 40$ in the first equation.
$x+40=50$
$x=10$
Barb needs 10 liters of 20% solution and 40 liters of 70% solution.

37. Let x = number sold at \$9.50, and
y = number sold at \$7.50.
Then $x + y$ is the total number of pieces sold and $9.50x + 7.50y$ is the total revenue.

$$\begin{cases} x + y = 90 \\ 9.5x + 7.5y = 721 \end{cases}$$

Solve the first equation for x.

$x + y = 90$
$x = 90 - y$

Substitute $90 - y$ for x in the second equation.

$9.5(90 - y) + 7.5y = 721$
$855 - 9.5y + 7.5y = 721$
$-2y = -134$
$y = 67$

To find x, let $y = 67$ in the first equation.

$x + 67 = 90$
$x = 23$

They sold 23 at \$9.50 each and 67 at \$7.50 each.

39. Let x = width of the garden, and
y = length of the garden.
Then $2x + y$ is the total amount of fencing.

$$\begin{cases} 2x + y = 33 \\ y = 2x - 3 \end{cases}$$

Substitute $2x - 3$ for y in the first equation.

$2x + 2x - 3 = 33$
$4x = 36$
$x = 9$

To find y, let $x = 9$ in the second equation.

$y = 2(9) - 3$
$y = 15$

The width of the garden is 9 feet and the length is 15 feet.

41. Let x = time bicycling, and
y = time walking.
Then $x + y$ is the total time and $40x + 4y$ the total distance.

$$\begin{cases} x + y = 6 \\ 40x + 4y = 186 \end{cases}$$

Solve the 1st equation for x.

$x = 6 - y$

Replace x with $6 - y$ in the second equation.

$40(6 - y) + 4y = 186$
$240 - 40y + 4y = 186$
$240 - 36y = 186$
$-36y = 186 - 240$
$-36y = -54$
$y = \frac{-54}{-36} = 1.5$

To find x, use $y = 1.5$ in the equation $x = 6 - y$.

$x = 6 - 1.5$
$x = 4.5$

The time bicycling is $4\frac{1}{2}$ hours.

43. Let x = the speed of the eastbound train, and
y = the speed of the westbound train.
Then the total distance the trains travel in $1\frac{1}{4}$ hours is $\frac{5}{4}x + \frac{5}{4}y$.

$$\begin{cases} \frac{5}{4}x + \frac{5}{4}y = 150 \\ y = 2x \end{cases}$$

Substitute $2x$ for y in the first equation.

$\frac{5}{4}x + \frac{5}{4}(2x) = 150$
$\frac{5}{4}x + \frac{10}{4}x = 150$
$\frac{15}{4}x = 150$
$\frac{4}{15}\left(\frac{15}{4}x\right) = \frac{4}{15}(150)$
$x = 40$

To find y, let $x = 40$ in the second equation.

$y = 2(40)$
$y = 80$

The eastbound train is traveling at 40 miles per hour and the westbound train is traveling at 80 miles per hour.

45. b
The resulting mixture must have an acid strength that is between the strengths of the two solutions.

47. $y \ge 4 - 2x$
Graph $y = 4 - 2x$ as a solid line since the inequality symbol is ≥.
Test (0, 0).
$0 \ge 4 - 2(0)$ False
Shade the half-plane that does not contain (0, 0).

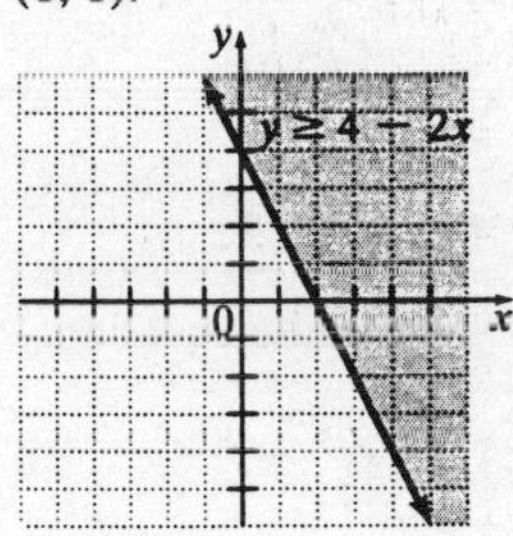

49. $3x + 5y < 15$
Graph $3x + 5y = 15$ as a dashed line since the inequality symbol is <.
Test (0, 0).
$3(0) + 5(0) < 15$ True
Shade the half-plane that contains (0, 0).

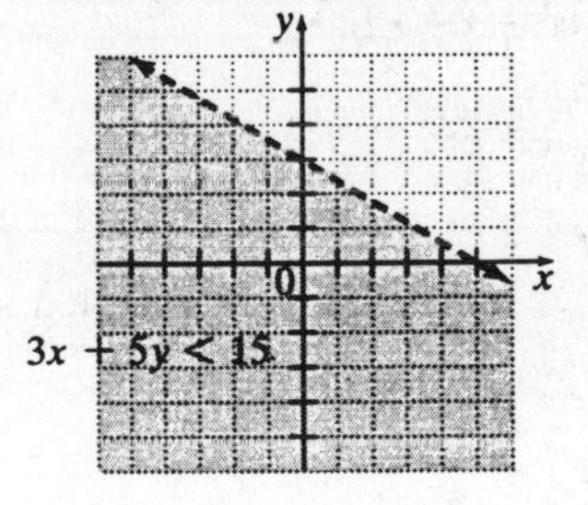

Chapter 4 Review Exercises

1. $\begin{cases} 2x - 3y = 12 \\ 3x + 4y = 1 \end{cases}$

a. (12, 4):
$$2x - 3y = 12$$
$$2(12) - 3(4) \stackrel{?}{=} 12$$
$$24 - 12 \stackrel{?}{=} 12$$
$$12 = 12 \text{ True}$$
$$3x + 4y = 1$$
$$3(12) + 4(4) \stackrel{?}{=} 1$$
$$36 + 16 \stackrel{?}{=} 1$$
$$52 = 1 \text{ False}$$
(12, 4) is not a solution.

b. (3, −2):
$$2x - 3y = 12$$
$$2(3) - 3(-2) \stackrel{?}{=} 12$$
$$6 + 6 \stackrel{?}{=} 12$$
$$12 = 12 \text{ True}$$
$$3x + 4y = 1$$
$$3(3) + 4(-2) \stackrel{?}{=} 1$$
$$9 - 8 \stackrel{?}{=} 1$$
$$1 = 1 \text{ True}$$
(3, −2) is a solution.

c. (−3, 6):
$$2x - 3y = 12$$
$$2(-3) - 3(6) \stackrel{?}{=} 12$$
$$-6 - 18 \stackrel{?}{=} 12$$
$$-24 = 12 \text{ False}$$
$$3x + 4y = 1$$
$$3(-3) + 4(6) \stackrel{?}{=} 1$$
$$-9 + 24 \stackrel{?}{=} 1$$
$$15 = 1 \text{ False}$$
(−3, 6) is not a solution.

2. $\begin{cases} 4x+y=0 \\ -8x-5y=9 \end{cases}$

a. $\left(\frac{3}{4}, -3\right)$:

$$4x+y=0$$
$$4\left(\frac{3}{4}\right)+(-3)\stackrel{?}{=}0$$
$$3+(-3)\stackrel{?}{=}0$$
$$0=0 \text{ True}$$
$$-8x-5y=9$$
$$-8\left(\frac{3}{4}\right)-5(-3)\stackrel{?}{=}9$$
$$-6+15\stackrel{?}{=}9$$
$$9=9 \text{ True}$$

$\left(\frac{3}{4}, -3\right)$ is a solution.

b. $(-2, 8)$:

$$4x+y=0$$
$$4(-2)+8\stackrel{?}{=}0$$
$$-8+8\stackrel{?}{=}0$$
$$0=0 \text{ True}$$
$$-8x-5y=9$$
$$-8(-2)-5(8)\stackrel{?}{=}9$$
$$16-40\stackrel{?}{=}9$$
$$-24=9 \text{ False}$$

$(-2, 8)$ is not a solution.

c. $\left(\frac{1}{2}, -2\right)$:

$$4x+y=0$$
$$4\left(\frac{1}{2}\right)+(-2)\stackrel{?}{=}0$$
$$2+(-2)\stackrel{?}{=}0$$
$$0=0 \text{ True}$$
$$-8x-5y=9$$
$$-8\left(\frac{1}{2}\right)-5(-2)\stackrel{?}{=}9$$
$$-4+10\stackrel{?}{=}9$$
$$6=9 \text{ False}$$

$\left(\frac{1}{2}, -2\right)$ is not a solution.

3. $\begin{cases} 5x-6y=18 \\ 2y-x=-4 \end{cases}$

a. $(-6, -8)$:

$$5x-6y=18$$
$$5(-6)-6(-8)\stackrel{?}{=}18$$
$$-30+48\stackrel{?}{=}18$$
$$18=18 \text{ True}$$
$$2y-x=-4$$
$$2(-8)-(-6)\stackrel{?}{=}-4$$
$$-16+6\stackrel{?}{=}-4$$
$$-10=-4 \text{ False}$$

$(-6, -8)$ is not a solution.

b. $\left(3, \frac{5}{2}\right)$:

$$5x-6y=18$$
$$5(3)-6\left(\frac{5}{2}\right)\stackrel{?}{=}18$$
$$15-15\stackrel{?}{=}18$$
$$0=18 \text{ False}$$
$$2y-x=-4$$
$$2\left(\frac{5}{2}\right)-3\stackrel{?}{=}-4$$
$$5-3\stackrel{?}{=}-4$$
$$2=-4 \text{ False}$$

$\left(3, \frac{5}{2}\right)$ is not a solution.

c. $\left(3, -\frac{1}{2}\right)$:

$$5x-6y=18$$
$$5(3)-6\left(-\frac{1}{2}\right)\stackrel{?}{=}18$$
$$15+3\stackrel{?}{=}18$$
$$18=18 \text{ True}$$
$$2y-x=-4$$
$$2\left(-\frac{1}{2}\right)-3\stackrel{?}{=}-4$$
$$-1-3\stackrel{?}{=}-4$$
$$-4=-4 \text{ True}$$

$\left(3, -\frac{1}{2}\right)$ is a solution.

4. $\begin{cases} 2x+3y=1 \\ 3y-x=4 \end{cases}$

a. (2, 2):

$2x+3y=1$
$2(2)+3(2) \stackrel{?}{=} 1$
$4+6 \stackrel{?}{=} 1$
$10=1$ False

(2, 2) is not a solution.

b. (−1, 1):

$2x+3y=1$
$2(-1)+3(1) \stackrel{?}{=} 1$
$-2+3 \stackrel{?}{=} 1$
$1=1$ True

$3y-x=4$
$3(1)-(-1) \stackrel{?}{=} 4$
$3+1 \stackrel{?}{=} 4$
$4=4$ True

(−1, 1) is a solution.

c. (2, −1):

$2x+3y=1$
$2(2)+3(-1) \stackrel{?}{=} 1$
$4-3 \stackrel{?}{=} 1$
$1=1$ True

$3y-x=4$
$3(-1)-2 \stackrel{?}{=} 4$
$-3-2 \stackrel{?}{=} 4$
$-5=4$ False

(2, −1) is not a solution.

5. $\begin{cases} 2x+y=5 \\ 3y=-x \end{cases}$

Graph both linear equations on the same set of axes.

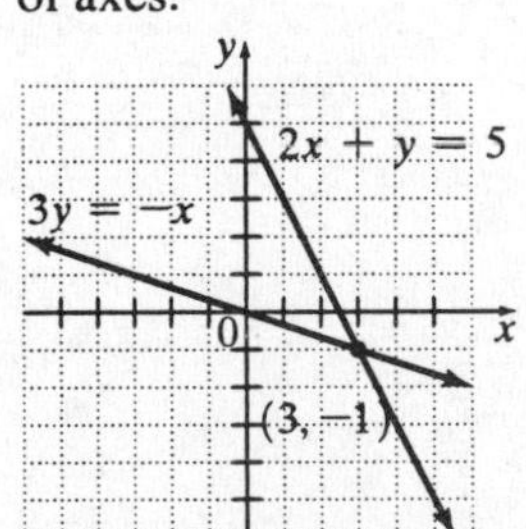

The solution is the intersection point of the two lines, (3, −1).

6. $\begin{cases} 3x+y=-2 \\ 2x-y=-3 \end{cases}$

Graph both linear equations on the same set of axes.

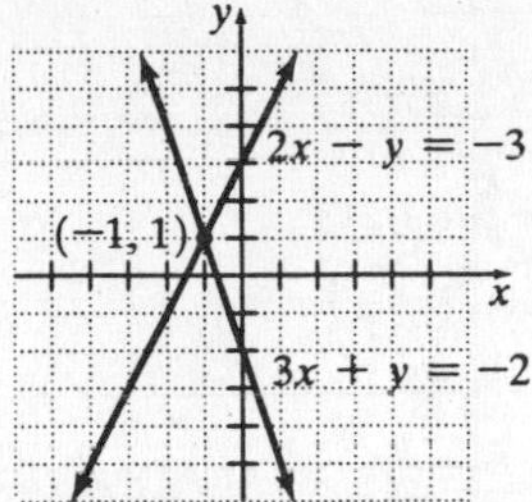

The solution is the intersection point of the two lines, (−1, 1).

7. $\begin{cases} y-2x=4 \\ x+y=-5 \end{cases}$

Graph both linear equations on the same set of axes.

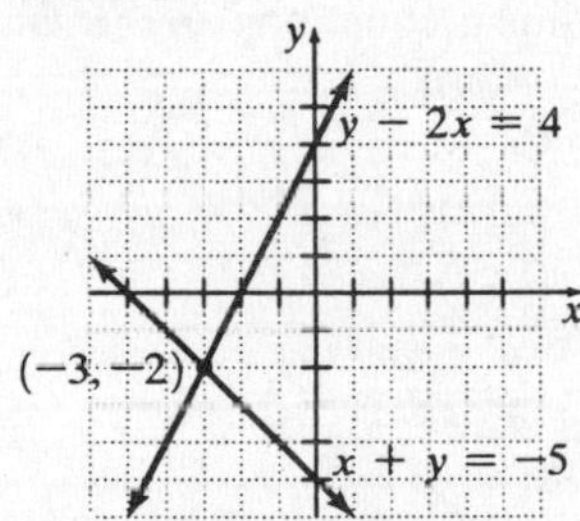

The solution is the intersection point of the two lines (−3, −2).

8. $\begin{cases} y-3x=0 \\ 2y-3=6x \end{cases}$

Graph both linear equations on the same set of axes.

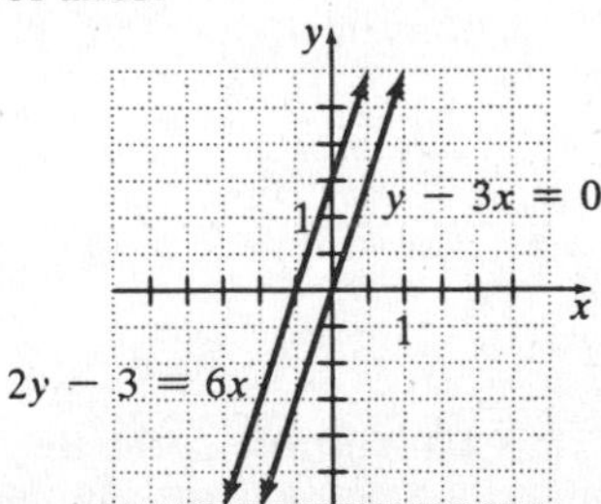

Since the lines are parallel, the system has no solution.

9. $\begin{cases} 3x+y=2 \\ 3x-6=-9y \end{cases}$

Graph both linear equations on the same set of axes.

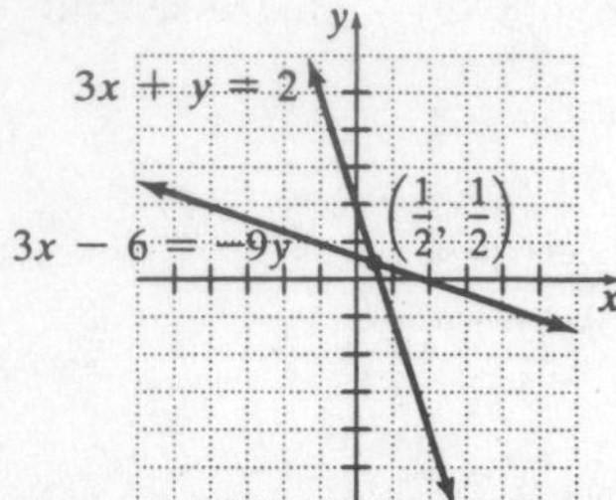

The solution is the intersection point of the two lines, $\left(\frac{1}{2}, \frac{1}{2}\right)$.

10. $\begin{cases} 2y+x=2 \\ x-y=5 \end{cases}$

Graph both linear equations on the same set of axes.

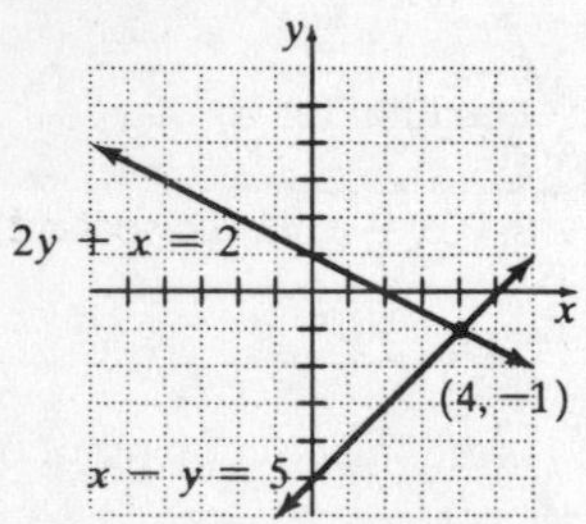

The solution is the intersection point of the two lines (4, –1).

11. $\begin{cases} 2x-y=3 \\ y=3x+1 \end{cases}$

Write the equations in slope-intercept form.

$2x - y = 3$ $\quad$ $y = 3x + 1$

$-y = -2x + 3$ $\quad$ $m = 3$

$\frac{-y}{-1} = \frac{-2}{-1}x + \frac{3}{-1}$

$y = 2x - 3$

$m = 2$

Since the slopes are different, the lines intersect at a single point; one solution.

12. $\begin{cases} 3x+y=4 \\ y=-3x+1 \end{cases}$

Write the equations in slope-intercept form.

$3x + y = 4$ $\quad$ $y = -3x + 1$

$y = -3x + 4$ $\quad$ $m = -3,\ b = 1$

$m = -3, b = 4$

Since the lines have the same slope but different y-intercepts, the lines are parallel; no solution.

13. $\begin{cases} \frac{2}{3}x+\frac{1}{6}y=0 \\ y=-4x \end{cases}$

Write the equations in slope-intercept form.

$\frac{2}{3}x + \frac{1}{6}y = 0$ $\quad$ $y = -4x$

$\frac{1}{6}y = -\frac{2}{3}x$

$6\left(\frac{1}{6}y\right) = 6\left(-\frac{2}{3}x\right)$

$y = -4x$

Since the equations are identical, the lines are the same; infinite number of solutions.

14. $\begin{cases} \frac{1}{4}x+\frac{1}{8}y=0 \\ y=-6x \end{cases}$

Write the equations in slope-intercept form.

$\frac{1}{4}x + \frac{1}{8}y = 0$ $\quad$ $y = -6x$

$2x + y = 0$ $\quad$ $m = -6$

$y = -2x$

$m = -2$

Since the lines have different slopes, they intersect at a single point; one solution.

15. $\begin{cases} y = 2x + 6 \\ 3x - 2y = -11 \end{cases}$

Replace y with $2x + 6$ in the second equation.

$$\begin{aligned} 3x - 2(2x + 6) &= -11 \\ 3x - 4x - 12 &= -11 \\ -x - 12 &= -11 \\ -x &= -11 + 12 \\ -x &= 1 \\ \frac{-x}{-1} &= \frac{1}{-1} \\ x &= -1 \end{aligned}$$

To find y, use $x = -1$ in the equation $y = 2x + 6$.

$y = 2(-1) + 6$
$y = -2 + 6$
$y = 4$

The solution is $(-1, 4)$.

16. $\begin{cases} y = 3x - 7 \\ 2x - 3y = 7 \end{cases}$

Replace y with $3x - 7$ in the second equation.

$$\begin{aligned} 2x - 3(3x - 7) &= 7 \\ 2x - 9x + 21 &= 7 \\ -7x &= -14 \\ \frac{-7x}{-7} &= \frac{-14}{-7} \\ x &= 2 \end{aligned}$$

To find y, use $x = 2$ in the equation $y = 3x - 7$.

$y = 3(2) - 7$
$y = 6 - 7$
$y = -1$

The solution is $(2, -1)$.

17. $\begin{cases} x + 3y = -3 \\ 2x + y = 4 \end{cases}$

Solve the 1st equation for x.

$x = -3y - 3$

Replace x with $-3y - 3$ in the second equation.

$$\begin{aligned} 2(-3y - 3) + y &= 4 \\ -6y - 6 + y &= 4 \\ -5y - 6 &= 4 \\ -5y &= 4 + 6 \\ -5y &= 10 \\ y &= \frac{10}{-5} = -2 \end{aligned}$$

To find x, use $y = -2$ in the equation $x = -3y - 3$.

$x = -3(-2) - 3$
$x = 6 - 3$
$x = 3$

The solution is $(3, -2)$.

18. $\begin{cases} 3x + y = 11 \\ x + 2y = 12 \end{cases}$

Solve the 1st equation for y.

$y = -3x + 11$

Replace y with $-3x + 11$ in the second equation.

$$\begin{aligned} x + 2(-3x + 11) &= 12 \\ x - 6x + 22 &= 12 \\ -5x &= -10 \\ \frac{-5x}{-5} &= \frac{-10}{-5} \\ x &= 2 \end{aligned}$$

To find y, let $x = 2$ in the equation $y = -3x + 11$.

$y = -3(2) + 11$
$y = -6 + 11$
$y = 5$

The solution is $(2, 5)$.

19. $\begin{cases} 4y = 2x - 3 \\ x - 2y = 4 \end{cases}$

Solve the 2nd equation for x.

$x = 2y + 4$

Replace x with $2y + 4$ in the 1st equation.

$$4y = 2(2y+4) - 3$$
$$4y = 4y + 8 - 3$$
$$4y = 4y + 5$$
$$4y - 4y = 5$$
$$0 = 5$$

This is a contradiction.
There is no solution.
The system is inconsistent.

20. $\begin{cases} 2x = 3y - 18 \\ x + 4y = 2 \end{cases}$

Solve the 2nd equation for x.

$x = 2 - 4y$

Replace x with $2 - 4y$ in the 1st equation.

$$2(2-4y) = 3y - 18$$
$$4 - 8y = 3y - 18$$
$$22 = 11y$$
$$\frac{22}{11} = \frac{11y}{11}$$
$$2 = y$$

To find x, use $y = 2$ in the equation $x = 2 - 4y$.

$x = 2 - 4(2)$

$x = 2 - 8$

$x = -6$

The solution is $(-6, 2)$.

21. $\begin{cases} 2(3x - y) = 7x - 5 \\ 3(x - y) = 4x - 6 \end{cases}$

Simplify the equations.

$$2(3x - y) = 7x - 5$$
$$6x - 2y = 7x - 5$$
$$6x - 7x - 2y = -5$$
$$-x - 2y = -5$$
$$3(x - y) = 4x - 6$$
$$3x - 3y = 4x - 6$$
$$3x - 4x - 3y = -6$$
$$-x - 3y = -6$$

Solve the 1st equation for x.

$$-x - 2y = -5$$
$$-x = 2y - 5$$
$$\frac{-x}{-1} = \frac{2y}{-1} - \frac{5}{-1}$$
$$x = -2y + 5$$

Replace x with $-2y + 5$ in the 2nd equation.

$$-(-2y+5) - 3y = -6$$
$$2y - 5 - 3y = -6$$
$$-y - 5 = -6$$
$$-y = -6 + 5$$
$$-y = -1$$
$$\frac{-y}{-1} = \frac{-1}{-1}$$
$$y = 1$$

To find x, use $y = 1$ in the equation $x = -2y + 5$.

$x = -2(1) + 5$

$x = -2 + 5$

$x = 3$

The solution is $(3, 1)$.

22. $\begin{cases} 4(x-3y)=3x-1 \\ 3(4y-3x)=1-8x \end{cases}$

Simplify the 1st equation and solve for x.

$4x - 12y = 3x - 1$
$x = 12y - 1$

Replace x with $12y - 1$ in the 2nd equation.

$$3[4y-3(12y-1)]=1-8(12y-1)$$
$$3[4y-36y+3]=1-96y+8$$
$$12y-108y+9=1-96y+8$$
$$-96y+9=-96y+9$$
$$9=9$$

There is an infinite number of solutions. The system is dependent.

23. $\begin{cases} \frac{3}{4}x+\frac{2}{3}y=2 \\ 3x+y=18 \end{cases}$

Solve the 2nd equation for y.

$y = -3x + 18$

Replace y with $-3x + 18$ in the 1st equation.

$$\frac{3}{4}x+\frac{2}{3}(-3x+18)=2$$
$$\frac{3}{4}x-2x+12=2$$
$$\frac{3}{4}x-\frac{8}{4}x+12=2$$
$$-\frac{5}{4}x+12=2$$
$$-\frac{5}{4}x=2-12$$
$$-\frac{5}{4}x=-10$$
$$-\frac{4}{5}\left(-\frac{5}{4}x\right)=-\frac{4}{5}(-10)$$
$$x=8$$

To find y, use $x = 8$ in the equation $y = -3x + 18$.

$y = -3(8) + 18$
$y = -24 + 18$
$y = -6$

The solution is $(8, -6)$.

24. $\begin{cases} \frac{2}{5}x+\frac{3}{4}y=1 \\ x+3y=-2 \end{cases}$

Solve the 2nd equation for x.

$x = -3y - 2$

Replace x with $-3y - 2$ in the 1st equation.

$$\frac{2}{5}(-3y-2)+\frac{3}{4}y=1$$
$$20\left[\frac{2}{5}(-3y-2)+\frac{3}{4}y=1\right]$$
$$8(-3y-2)+15y=20$$
$$-24y-16+15y=20$$
$$-9y=36$$
$$\frac{-9y}{-9}=\frac{36}{-9}$$
$$y=-4$$

To find x, use $y = -4$ in the equation $x = -3y - 2$.

$x = -3(-4) - 2$
$x = 12 - 2$
$x = 10$

The solution is $(10, -4)$.

25. $\begin{cases} 2x+3y=-6 \\ x-3y=-12 \end{cases}$

Add the equations.

$$\begin{aligned} 2x+3y &= -6 \\ x-3y &= -12 \\ \hline 3x \quad &= -18 \\ x \quad &= \frac{-18}{3}=-6 \end{aligned}$$

To find y, use $x = -6$ in the equation $x - 3y = -12$.

$$-6-3y=-12$$
$$-3y=-12+6$$
$$-3y=-6$$
$$y=\frac{-6}{-3}=2$$

The solution is $(-6, 2)$.

26. $\begin{cases} 4x + y = 15 \\ -4x + 3y = -19 \end{cases}$

Add the equations.

$$\begin{aligned} 4x + y &= 15 \\ -4x + 3y &= -19 \\ \hline 4y &= -4 \\ y &= -1 \end{aligned}$$

To find x, use $y = -1$ in the equation $4x + y = 15$.

$$\begin{aligned} 4x - 1 &= 15 \\ 4x &= 16 \\ x &= 4 \end{aligned}$$

The solution is $(4, -1)$.

27. $\begin{cases} 2x - 3y = -15 \\ x + 4y = 31 \end{cases}$

Multiply the 2nd equation by -2.

$$\begin{aligned} -2(x + 4y) &= -2(31) \\ -2x - 8y &= -62 \end{aligned}$$

Add the equations.

$$\begin{aligned} -2x - 8y &= -62 \\ 2x - 3y &= -15 \\ \hline -11y &= -77 \\ y &= \frac{-77}{-11} = 7 \end{aligned}$$

To find x, use $y = 7$ in the equation $x + 4y = 31$.

$$\begin{aligned} x + 4(7) &= 31 \\ x + 28 &= 31 \\ x &= 31 - 28 \\ x &= 3 \end{aligned}$$

The solution is $(3, 7)$.

28. $\begin{cases} x - 5y = -22 \\ 4x + 3y = 4 \end{cases}$

Multiply the 1st equation by -4 and add to the 2nd equation.

$$\begin{aligned} -4x + 20y &= 88 \\ 4x + 3y &= 4 \\ \hline 23y &= 92 \\ y &= 4 \end{aligned}$$

To find x, use $y = 4$ in the equation $x - 5y = -22$.

$$\begin{aligned} x - 5(4) &= -22 \\ x - 20 &= -22 \\ x &= -2 \end{aligned}$$

The solution is $(-2, 4)$.

29. $\begin{cases} 2x = 6y - 1 \\ \frac{1}{3}x - y = -\frac{1}{6} \end{cases}$

Multiply the 2nd equation by 6.

$$6\left(\frac{1}{3}x - y\right) = 6\left(-\frac{1}{6}\right)$$

$$2x - 6y = -1$$

Multiply the 1st equation by -1.

$$\begin{aligned} -1(2x) &= -(6y - 1) \\ -2x &= -6y + 1 \\ -2x + 6y &= 1 \end{aligned}$$

Add the equations.

$$\begin{aligned} 2x - 6y &= -1 \\ -2x + 6y &= 1 \\ \hline 0 &= 0 \end{aligned}$$

This is an identity.
The system is dependent.
There are infinitely many solutions.

30. $\begin{cases} 8x = 3y - 2 \\ \frac{4}{7}x - y = -\frac{5}{2} \end{cases}$

Multiply the 2nd equation by -14 and add to the 1st equation.

$$\begin{aligned} 8x - 3y &= -2 \\ -8x + 14y &= 35 \\ \hline 11y &= 33 \\ y &= 3 \end{aligned}$$

To find x, use $y = 3$ in the equation $8x = 3y - 2$.

$$\begin{aligned} 8x &= 3(3) - 2 \\ 8x &= 9 - 2 \\ 8x &= 7 \\ \frac{8x}{8} &= \frac{7}{8} \\ x &= \frac{7}{8} \end{aligned}$$

The solution is $\left(\frac{7}{8}, 3\right)$.

31. $\begin{cases} 5x = 6y + 25 \\ -2y = 7x - 9 \end{cases}$

Rearrange the equations.

$5x - 6y = 25$
$-7x - 2y = -9$

Multiply the 2nd equation by –3.

$-3(-7x - 2y) = -3(-9)$
$21x + 6y = 27$

Add the equations.

$$\begin{array}{r} 21x + 6y = 27 \\ 5x - 6y = 25 \\ \hline 26x \quad = 52 \\ x \quad = \frac{52}{26} = 2 \end{array}$$

To find y, use $x = 2$ in the equation $5x - 6y = 25$.

$5(2) - 6y = 25$
$10 - 6y = 25$
$-6y = 25 - 10$
$-6y = 15$
$y = \frac{15}{-6} = -\frac{5}{2}$

The solution is $\left(2, -\frac{5}{2}\right)$.

32. $\begin{cases} -4x = 8 + 6y \\ -3y = 2x - 3 \end{cases}$

Rearrange the equations.

$-4x - 6y = 8$
$-2x - 3y = -3$

Multiply the 2nd equation by –2.

$-2(-2x - 3y) = -2(-3)$
$4x + 6y = 6$

Add the equations.

$$\begin{array}{r} -4x - 6y = 8 \\ 4x + 6y = 6 \\ \hline 0 = 14 \end{array}$$

This is a contradiction.
There is no solution.
The system is inconsistent.

33. $\begin{cases} 3(x - 4) = -2y \\ 2x = 3(y - 19) \end{cases}$

Simplify each equation.

$3(x - 4) = -2y$
$3x - 12 = -2y$
$3x + 2y - 12 = 0$
$3x + 2y = 12$
$2x = 3(y - 19)$
$2x = 3y - 57$
$2x - 3y = -57$

Multiply the 1st equation by –2.

$-2(3x + 2y) = -2(12)$
$-6x - 4y = -24$

Multiply the 2nd equation by 3.

$3(2x - 3y) = 3(-57)$
$6x - 9y = -171$

Add the equations.

$$\begin{array}{r} -6x - 4y = -24 \\ 6x - 9y = -171 \\ \hline -13y = -195 \\ y = \frac{-195}{-13} = 15 \end{array}$$

To find x, use $y = 15$ in the equation $2x - 3y = -57$.

$2x - 3(15) = -57$
$2x - 45 = -57$
$2x = -57 + 45$
$2x = -12$
$x = \frac{-12}{2} = -6$

The solution is (–6, 15).

34. $\begin{cases} 4(x+5) = -3y \\ 3x - 2(y+18) = 0 \end{cases}$

Simplify each equation.

$4(x+5) = -3y$
$4x + 20 = -3y$
$4x + 3y = -20$

$3x - 2(y+18) = 0$
$3x - 2y - 36 = 0$
$3x - 2y = 36$

Multiply the 1st equation by 2.

$2(4x+3y) = 2(-20)$
$8x + 6y = -40$

Multiply the 2nd equation by 3.

$3(3x - 2y) = 3(36)$
$9x - 6y = 108$

Add the equations.

$$\begin{array}{r} 8x + 6y = -40 \\ 9x - 6y = 108 \\ \hline 17x \quad\;\; = 68 \end{array}$$

$x = 4$

To find y, use $x = 4$ in the equation $4x + 3y = -20$.

$4(4) + 3y = -20$
$16 + 3y = -20$
$3y = -36$
$y = -12$

The solution is (4, –12).

35. $\begin{cases} \dfrac{2x+9}{3} = \dfrac{y+1}{2} \\ \dfrac{x}{3} = \dfrac{y-7}{6} \end{cases}$

Simplify the equations.

$6\left(\dfrac{2x+9}{3}\right) = 6\left(\dfrac{y+1}{2}\right)$
$2(2x+9) = 3(y+1)$
$4x + 18 = 3y + 3$
$4x - 3y + 18 = 3$
$4x - 3y = 3 - 18$
$4x - 3y = -15$

$6\left(\dfrac{x}{3}\right) = 6\left(\dfrac{y-7}{6}\right)$
$2x = y - 7$
$2x - y = -7$

Multiply the 2nd equation by –3.

$-3(2x - y) = -3(-7)$
$-6x + 3y = 21$

Add the equations.

$$\begin{array}{r} -6x + 3y = \;\; 21 \\ 4x - 3y = -15 \\ \hline -2x \quad\;\; = \;\; 6 \end{array}$$

$x = \dfrac{6}{-2} = -3$

To find y, use $x = -3$ in the equation $2x - 3y = -7$.

$2(-3) - y = -7$
$-6 - y = -7$
$-y = -7 + 6$
$-y = -1$
$y = \dfrac{-1}{-1} = 1$

The solution is (–3, 1).

36. $\begin{cases} \dfrac{2-5x}{4} = \dfrac{2y-4}{2} \\ \dfrac{x+5}{3} = \dfrac{y}{5} \end{cases}$

Simplify each equation.

$4\left(\dfrac{2-5x}{4}\right) = 4\left(\dfrac{2y-4}{2}\right)$
$2 - 5x = 4y - 8$
$5x + 4y = 10$

$15\left(\dfrac{x+5}{3}\right) = 15\left(\dfrac{y}{5}\right)$
$5x + 25 = 3y$
$-5x + 3y = 25$

Add the equations.

$$\begin{array}{r} 5x + 4y = 10 \\ -5x + 3y = 25 \\ \hline 7y = 35 \end{array}$$

$y = 5$

To find x, use $y = 5$ in the equation $5x + 4y = 10$.

$5x + 4(5) = 10$
$5x + 20 = 10$
$5x = -10$
$x = -2$

The solution is (–2, 5).

37. Let x = smaller number, and
y = larger number.

$$\begin{cases} x+y=16 \\ 3y-x=72 \end{cases}$$

Solve the 1st equation for x.
$x = 16 - y$
Replace x with $16 - y$ in the 2nd equation.

$$3y-(16-y)=72$$
$$3y-16+y=72$$
$$4y-16=72$$
$$4y=72+16$$
$$4y=88$$
$$y=\frac{88}{4}=22$$

To find x, use $y = 22$ in the equation
$x = 16 - y$.
$x = 16 - 22$
$x = -6$
The numbers are 22 and -6.

38. Let x = number of orchestra seats, and
y = number of balcony seats.

$$\begin{cases} x+y=360 \\ 45x+35y=15{,}150 \end{cases}$$

Multiply the 1st equation by -45.
$-45(x+y) = -45(360)$
$-45x - 45y = -16{,}200$
Add the equations.

$$\begin{array}{r} -45x-45y=-16{,}200 \\ 45x+35y=\ \ 15{,}150 \\ \hline -10y=-1050 \end{array}$$

$y = 105$
To find x, let $y = 105$ in the equation
$x + y = 360$.
$x + 105 = 360$
$x = 255$
255 people can be seated in the orchestra section.

39. Let x = speed of boat in still water, and
y = speed of current.

$$\begin{cases} 19(x-y)=340 \\ 14(x+y)=340 \end{cases}$$

Simplify the equations.
$19x - 19y = 340$
$14x + 14y = 340$
Multiply the 1st equation by 14.
$14(19x - 19y) = 14(340)$
$266x - 266y = 4760$
Multiply the second equation by 19.
$19(14x + 14y) = 19(340)$
$266x + 266y = 6460$
Add the equations.

$$\begin{array}{l} 266x-266y=\ \ 4760 \\ 266x+266y=\ \ 6460 \\ \hline 532x\qquad\quad =11220 \end{array}$$

$$x=\frac{11220}{532}$$
$$x \approx 21.1$$

To find y, use $x \approx 21.1$ in the equation
$14x + 14y = 340$.
$14(21.1) + 14y = 340$
$295.4 + 14y = 340$
$14y = 340 - 295.4$
$14y = 44.6$

$$y=\frac{44.6}{14}$$
$$y \approx 3.2$$

Boat's speed $\approx$ 21.1 miles per hour
Current's speed $\approx$ 3.2 miles per hour

40. Let x = amount invested at 6%, and
y = amount invested at 10%.

$$\begin{cases} x+y=9000 \\ 0.06x+0.10y=652.80 \end{cases}$$

Multiply the first equation by -6 and the second equation by 100.
$-6(x + y) = -6(9000)$
$100(0.06x + 0.10y) = 100(652.80)$
Add the resulting equations.

$$\begin{array}{r} -6x-\ \ 6y=-54{,}000 \\ 6x+10y=\ \ 65{,}280 \\ \hline 4y=11{,}280 \end{array}$$

$y = 2820$
To find x, let $y = 2820$ in the equation
$x + y = 9000$.
$x + 2820 = 9000$
$x = 6180$
\$6180 was invested at 6% and \$2820 was invested at 10%.

41. Let x = width of the frame, and
y = length of the frame.

$$\begin{cases} 2x + 2y = 6 \\ 1.6x = y \end{cases}$$

Replace y with $1.6x$ in the 1st equation.

$$\begin{aligned} 2x + 2(1.6x) &= 6 \\ 2x + 3.2x &= 6 \\ 5.2x &= 6 \\ x &= \frac{6}{5.2} \approx 1.15 \end{aligned}$$

To find y, use $x \approx 1.15$ in the equation $y = 1.6x$.

$y \approx 1.6(1.15)$
$y \approx 1.84$
width ≈ 1.15 feet
length ≈ 1.84 feet

42. Let x = amount of 6% solution, and
y = amount of 14% solution.

$$\begin{cases} x + y = 50 \\ 0.06x + 0.14y = 0.12(50) \end{cases}$$

Multiply the first equation by −6 and the second equation by 100.

$-6(x + y) = -6(50)$
$100(0.06x + 0.14y) = 100(0.12)(50)$

Add the resulting equations.

$$\begin{aligned} -6x - 6y &= -300 \\ 6x + 14y &= 600 \\ \hline 8y &= 300 \\ y &= 37.5 \end{aligned}$$

To find x, use $y = 37.5$ in the equation $x + y = 50$.

$$\begin{aligned} x + 37.5 &= 50 \\ x &= 12.5 \end{aligned}$$

12.5 cc of 6% solution;
37.5 cc of 14% solution

43. Let x = cost of an egg, and
y = cost of a strip of bacon.

$$\begin{cases} 3x + 4y = 3.80 \\ 2x + 3y = 2.75 \end{cases}$$

Multiply the 1st equation by −2.

$-2(3x + 4y) = -2(3.80)$
$-6x - 8y = -7.60$

Multiply the 2nd equation by 3.

$3(2x + 3y) = 3(2.75)$
$6x + 9y = 8.25$

Add the equations.

$$\begin{aligned} -6x - 8y &= -7.60 \\ 6x + 9y &= 8.25 \\ \hline y &= 0.65 \end{aligned}$$

To find x, use $y = 0.65$ in the equation $2x + 3y = 2.75$.

$$\begin{aligned} 2x + 3(0.65) &= 2.75 \\ 2x + 1.95 &= 2.75 \\ 2x &= 2.75 - 1.95 \\ 2x &= 0.80 \\ x &= \frac{0.80}{2} = 0.40 \end{aligned}$$

One egg is \$0.40 and a strip of bacon is \$0.65.

44. Let x = time spent jogging, and
y = time spent walking.

$$\begin{cases} 7.5x + 4y = 15 \\ x + y = 3 \end{cases}$$

Multiply the second equation by −4.

$-4(x + y) = -4(3)$
$-4x - 4y = -12$

Add the equations.

$$\begin{aligned} 7.5x + 4y &= 15 \\ -4x - 4y &= -12 \\ \hline 3.5x &= 3 \\ 35x &= 30 \\ x &= \frac{30}{35} \approx 0.86 \end{aligned}$$

He spent about 0.86 hours jogging.

Chapter 4 Test

1. False; the system can have one solution, infinitely many solutions, or no solution.

2. False; if the ordered pair is not a solution of one of the equations, it is not a solution of the system.

3. True; the system is inconsistent.

4. False; if $3x = 0$, then $\frac{3x}{3} = \frac{0}{3}$ or $x = 0$.

5. $\begin{cases} 2x - 3y = 5 \\ 6x + y = 1 \end{cases}$; $(1, -1)$

$$\begin{aligned} 2x - 3y &= 5 \\ 2(1) - 3(-1) &\stackrel{?}{=} 5 \\ 2 + 3 &\stackrel{?}{=} 5 \\ 5 &= 5 \text{ True} \end{aligned}$$

$$\begin{aligned} 6x + y &= 1 \\ 6(1) - 1 &\stackrel{?}{=} 1 \\ 6 - 1 &\stackrel{?}{=} 1 \\ 5 &= 1 \text{ False} \end{aligned}$$

$(1, -1)$ is not a solution.

6. $\begin{cases} 4x - 3y = 24 \\ 4x + 5y = -8 \end{cases}$; $(3, -4)$

$$\begin{aligned} 4x - 3y &= 24 \\ 4(3) - 3(-4) &\stackrel{?}{=} 24 \\ 12 + 12 &\stackrel{?}{=} 24 \\ 24 &= 24 \text{ True} \end{aligned}$$

$$\begin{aligned} 4x + 5y &= -8 \\ 4(3) + 5(-4) &\stackrel{?}{=} -8 \\ 12 - 20 &\stackrel{?}{=} -8 \\ -8 &= -8 \text{ True} \end{aligned}$$

$(3, -4)$ is a solution.

7. $\begin{cases} y - x = 6 \\ y + 2x = -6 \end{cases}$

Graph both equations on the same set of axes.

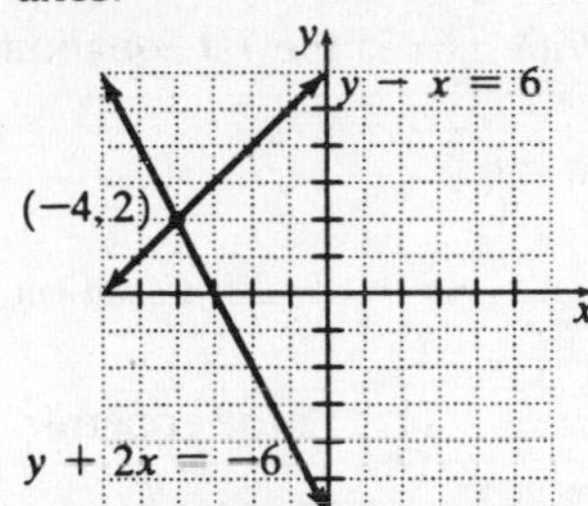

The solution is the intersection point of the two lines, $(-4, 2)$.

8. $\begin{cases} 3x - 2y = -14 \\ x + 3y = -1 \end{cases}$

Solve the 2nd equation for x.

$x = -3y - 1$

Replace x with $-3y - 1$ in the 1st equation.

$$\begin{aligned} 3(-3y - 1) - 2y &= -14 \\ -9y - 3 - 2y &= -14 \\ -11y - 3 &= -14 \\ -11y &= -14 + 3 \\ -11y &= -11 \\ y &= \frac{-11}{-11} = 1 \end{aligned}$$

To find x, use $y = 1$ in the equation $x = -3y - 1$.

$x = -3(1) - 1$

$x = -3 - 1$

$x = -4$

The solution is $(-4, 1)$.

9. $\begin{cases} \frac{1}{2}x + 2y = -\frac{15}{4} \\ 4x = -y \end{cases}$

Solve the 2nd equation for y.

$4x = -y$

$\frac{4x}{-1} = y$

$-4x = y$

Replace y with $-4x$ in the 1st equation.

$\frac{1}{2}x + 2(-4x) = -\frac{15}{4}$

$\frac{1}{2}x - 8x = -\frac{15}{4}$

$-\frac{15}{2}x = -\frac{15}{4}$

$-\frac{2}{15}\left(-\frac{15}{2}x\right) = -\frac{2}{15}\left(-\frac{15}{4}\right)$

$x = \frac{2}{4} = \frac{1}{2}$

To find y, use $x = \frac{1}{2}$ in the equation

$y = -4x.$

$y = -4\left(\frac{1}{2}\right)$

$y = -2$

The solution is $\left(\frac{1}{2}, -2\right)$.

10. $\begin{cases} 3x + 5y = 2 \\ 2x - 3y = 14 \end{cases}$

Multiply the 1st equation by 3.

$3(3x + 5y) = 3(2)$

$9x + 15y = 6$

Multiply the 2nd equation by 5.

$5(2x - 3y) = 5(14)$

$10x - 15y = 70$

Add the equations.

$$\begin{array}{rl} 9x + 15y &= 6 \\ 10x - 15y &= 70 \\ \hline 19x &= 76 \\ x &= \frac{76}{19} = 4 \end{array}$$

To find y, use $x = 4$ in the equation

$2x - 3y = 14.$

$2(4) - 3y = 14$

$8 - 3y = 14$

$-3y = 14 - 8$

$-3y = 6$

$y = \frac{6}{-3} = -2$

The solution is $(4, -2)$.

11. $\begin{cases} 5x - 6y = 7 \\ 7x - 4y = 12 \end{cases}$

Multiply the 1st equation by -7.

$-7(5x - 6y) = -7(7)$

$-35x + 42y = -49$

Multiply the 2nd equation by 5.

$5(7x - 4y) = 5(12)$

$35x - 20y = 60$

Add the equations.

$$\begin{array}{rl} -35x + 42y &= -49 \\ 35x - 20y &= 60 \\ \hline 22y &= 11 \\ y &= \frac{11}{22} = \frac{1}{2} \end{array}$$

To find x, use $y = \frac{1}{2}$ in the equation

$5x - 6y = 7.$

$5x - 6\left(\frac{1}{2}\right) = 7$

$5x - 3 = 7$

$5x = 7 + 3$

$5x = 10$

$x = \frac{10}{5} = 2$

The solution is $\left(2, \frac{1}{2}\right)$.

12. $\begin{cases} 3x + y = 7 \\ 4x + 3y = 1 \end{cases}$

Multiply the 1st equation by –3.

$-3(3x + y) = -3(7)$

$-9x - 3y = -21$

Add the equations.

$$\begin{array}{r} -9x - 3y = -21 \\ 4x + 3y = \quad 1 \\ \hline -5x \qquad = -20 \end{array}$$

$x = \frac{-20}{-5} = 4$

To find y, use $x = 4$ in the equation

$3x + y = 7$

$3(4) + y = 7$

$12 + y = 7$

$y = 7 - 12$

$y = -5$

The solution is (4, –5).

13. $\begin{cases} 3(2x + y) = 4x + 20 \\ x - 2y = 3 \end{cases}$

Simplify the 1st equation.

$6x + 3y = 4x + 20$

$6x - 4x + 3y = 20$

$2x + 3y = 20$

Multiply the 2nd equation by –2.

$-2(x - 2y) = -2(3)$

$-2x + 4y = -6$

Add the equations.

$$\begin{array}{r} 2x + 3y = \ 20 \\ -2x + 4y = -6 \\ \hline 7y = 14 \end{array}$$

$y = \frac{14}{7} = 2$

To find x, use $y = 2$ in the equation $x - 2y = 3$.

$x - 2(2) = 3$

$x - 4 = 3$

$x = 3 + 4$

$x = 7$

The solution is (7, 2).

14. $\begin{cases} \frac{x-3}{2} = \frac{2-y}{4} \\ \frac{7-2x}{3} = \frac{y}{2} \end{cases}$

Simplify each equation.

$4\left(\frac{x-3}{2}\right) = 4\left(\frac{2-y}{4}\right)$

$2(x - 3) = 2 - y$

$2x - 6 = 2 - y$

$2x + y - 6 = 2$

$2x + y = 8$

$6\left(\frac{7-2x}{3}\right) = 6\left(\frac{y}{2}\right)$

$2(7 - 2x) = 3y$

$14 - 4x = 3y$

$14 - 4x - 3y = 0$

$-4x - 3y = -14$

Multiply the 1st equation by 3.

$3(2x + y) = 3(8)$

$6x + 3y = 24$

Add the equations.

$$\begin{array}{r} -4x - 3y = -14 \\ 6x + 3y = \ \ 24 \\ \hline 2x \qquad = 10 \end{array}$$

$x = \frac{10}{2} = 5$

To find y, use $x = 5$ in the equation $2x + y = 8$.

$2(5) + y = 8$

$10 + y = 8$

$y = 8 - 10$

$y = -2$

The solution is (5, –2).

15. Let x = number of $1 bills, and
y = number of $5 bills.

$$\begin{cases} x+y=62 \\ 1x+5y=230 \end{cases}$$

Multiply the 1st equation by −1.

$-1(x+y)=-1(62)$

$-x-y=-62$

Add the equations.

$$\begin{array}{r} x+5y=230 \\ -x-\ y=-62 \\ \hline 4y=168 \end{array}$$

$$y=\frac{168}{4}=42$$

To find x, use $y=42$ in the equation $x+y=62$.

$x+42=62$

$x=62-42$

$x=20$

There are 20 $1 bills and 42 $5 bills.

16. Let x = money at 5%, and
y = money at 9%.

$$\begin{cases} x+y=4000 \\ 0.05x+0.09y=311 \end{cases}$$

Solve the 1st equation for x.

$x=4000-y$

Replace x with $4000-y$ in the second equation.

$0.05(4000-y)+0.09y=311$

$200-0.05y+0.09y=311$

$200+0.04y=311$

$0.04y=311-200$

$0.04y=111$

$$y=\frac{111}{0.04}=2775$$

To find x, use $y=2775$ in the equation $x=4000-y$.

$x=4000-2775$

$x=1225$

$1225 was invested at 5%, and $2775 was invested at 9%.

17. Let x = number of farms in Missouri (in thousands),
and y = number of farms in Texas (in thousands).

$$\begin{cases} x+y=336 \\ y=116+x \end{cases}$$

Substitute $116+x$ for y in the 1st equation.

$x+116+x=336$

$2x=220$

$x=110$

To find y, let $x=110$ in the equation $y=116+x$.

$y=116+110$

$y=226$

Missouri has 110 thousand farms and Texas has 226 thousand farms.

Chapter 5

Section 5.1

Mental Math

1. 3^2
   base: 3; exponent: 2

2. 5^4
   base: 5; exponent: 4

3. $(-3)^6$
   base: –3; exponent: 6

4. -3^7
   base: 3; exponent: 7

5. -4^2
   base: 4; exponent: 2

6. $(-4)^3$
   base: –4; exponent: 3

7. $5\cdot 3^4$
   base: 5, exponent: 1
   base: 3; exponent: 4

8. $9\cdot 7^6$
   base: 9; exponent: 1
   base: 7; exponent: 6

9. $5x^2$
   base: 5; exponent: 1
   base: x; exponent: 2

10. $(5x)^2$
    base: $5x$; exponent: 2

Exercise Set 5.1

1. $7^2 = 7\cdot 7 = 49$

3. $(-5)^1 = -5$

5. $-2^4 = -(2\cdot 2\cdot 2\cdot 2) = -16$

7. $(-2)^4 = (-2)(-2)(-2)(-2) = 16$

9. $\left(\frac{1}{3}\right)^3 = \left(\frac{1}{3}\right)\left(\frac{1}{3}\right)\left(\frac{1}{3}\right) = \frac{1}{27}$

11. $7\cdot 2^4 = 7\cdot 2\cdot 2\cdot 2\cdot 2 = 112$

13. Answers may vary.

15. $x^2 = (-2)^2 = (-2)(-2) = 4$

17. $5x^3 = 5(3)^3 = 5\cdot 3\cdot 3\cdot 3 = 135$

19. $2xy^2 = 2(3)(5)^2 = 2(3)(5)(5) = 150$

21. $\frac{2z^4}{5} = \frac{2(-2)^4}{5} = \frac{2(-2)(-2)(-2)(-2)}{5} = \frac{32}{5}$

23. $V = x^3 = 7^3 = 7\cdot 7\cdot 7 = 343$
    The volume of the cube is 343 cubic meters.

25. We use the volume formula.

27. $x^2\cdot x^5 = x^{2+5} = x^7$

29. $(-3)^3\cdot(-3)^9 = (-3)^{3+9} = (-3)^{12}$

31. $\left(5y^4\right)(3y) = 5(3)y^{4+1} = 15y^5$

33. $(4z^{10})(-6z^7)(z^3) = (4)(-6)z^{10+7+3}$
    $= -24z^{20}$

35. $(pq)^7 = p^7q^7$

37. $\left(\frac{m}{n}\right)^9 = \frac{m^9}{n^9}$

39. $(x^2y^3)^5 = x^{2\cdot 5}y^{3\cdot 5} = x^{10}y^{15}$

41. $\left(\frac{-2xz}{y^5}\right)^2 = \frac{(-2)^2x^2z^2}{y^{5\cdot 2}} = \frac{4x^2z^2}{y^{10}}$

43. $\frac{x^3}{x} = x^{3-1} = x^2$

45. $\frac{(-2)^5}{(-2)^3} = (-2)^{5-3} = (-2)^2 = 4$

47. $\frac{p^7 q^{20}}{pq^{15}} = p^{7-1} q^{20-15} = p^6 q^5$

49. $\frac{7x^2y^6}{14x^2y^3} = \frac{7}{14}x^{2-2}y^{6-3}$
$= \frac{1}{2}x^0y^3$
$= \frac{1}{2}(1)y^3$
$= \frac{y^3}{2}$

51. $(2x)^0 = 1$

53. $-2x^0 = -2(1) = -2$

55. $5^0 + y^0 = 1 + 1 = 2$

57. $\left(\frac{-3a^2}{b^3}\right)^3 = \frac{(-3)^3 a^{2\cdot 3}}{b^{3\cdot 3}} = -\frac{27a^6}{b^9}$

59. $\frac{(x^5)^7 \cdot x^8}{x^4} = \frac{x^{5\cdot 7}x^8}{x^4}$
$= \frac{x^{35}x^8}{x^4}$
$= x^{35+8-4}$
$= x^{39}$

61. $\frac{(z^3)^6}{(5z)^4} = \frac{z^{3\cdot 6}}{5^4 z^4} = \frac{z^{18}}{625z^4} = \frac{z^{18-4}}{625} = \frac{z^{14}}{625}$

63. $\frac{(6mn)^5}{mn^2} = \frac{6^5 m^5 n^5}{mn^2}$
$= 7776m^{5-1}n^{5-2}$
$= 7776m^4n^3$

65. $-5^2 = -1 \cdot 5^2 = -(5)(5) = -25$

67. $\left(\frac{1}{4}\right)^3 = \left(\frac{1}{4}\right)\left(\frac{1}{4}\right)\left(\frac{1}{4}\right) = \frac{1}{64}$

69. $(9xy)^2 = 9^2x^2y^2 = 81x^2y^2$

71. $(6b)^0 = 1$

73. $2^3 + 2^5 = 2\cdot 2\cdot 2 + 2\cdot 2\cdot 2\cdot 2\cdot 2$
$= 8 + 32$
$= 40$

75. $b^4b^2 = b^{4+2} = b^6$

77. $a^2a^3a^4 = a^{2+3+4} = a^9$

79. $(2x^3)(-8x^4) = (2)(-8)x^{3+4} = -16x^7$

81. $(4a)^3 = 4^3a^3 = 64a^3$

83. $(-6xyz^3)^2 = (-6)^2x^2y^2z^{3\cdot 2} = 36x^2y^2z^6$

85. $\left(\frac{3y^5}{6x^4}\right)^3 = \left(\frac{y^5}{2x^4}\right)^3$ Reduce fraction.
$= \frac{y^{5\cdot 3}}{2^3x^{4\cdot 3}}$
$= \frac{y^{15}}{8x^{12}}$

87. $\frac{x^5}{x^4} = x^{5-4} = x^1 = x$

89. $\frac{2x^3y^2z}{xyz} = 2x^{3-1}y^{2-1}z^{1-1}$
$= 2x^2y^1z^0$
$= 2x^2y(1)$
$= 2x^2y$

91. $\frac{(3x^2y^5)^5}{x^3y} = \frac{3^5x^{2\cdot5}y^{5\cdot5}}{x^3y}$
$= \frac{243x^{10}y^{25}}{x^3y}$
$= 243x^{10-3}y^{25-1}$
$= 243x^7y^{24}$

93. Answers may vary.

95. Area $= lw$
$= (5x^3)(4x^2)$
$= (5)(4)x^{3+2}$
$= 20x^5$
The area of the rectangle is $20x^5$ square feet.

97. Area$= \pi r^2 = \pi(5y)^2 = \pi \cdot 5^2y^2 = 25\pi y^2$
The area of the circle is $25\pi y^2$ square centimeters.

99. Volume $= s^3 = (3y^4)^3 = 3^3y^{4\cdot3} = 27y^{12}$
The volume of the safe is $27y^{12}$ cubic feet.

101. $3x - 5x + 7 = (3 - 5)x + 7 = -2x + 7$

103. $y - 10 + y = (1 + 1)y - 10 = 2y - 10$

105. $7x + 2 - 8x - 6 = (7 - 8)x + 2 - 6 = -x - 4$

107. $2(x - 5) + 3(5 - x) = 2x - 10 + 15 - 3x$
$= (2 - 3)x - 10 + 15$
$= -x + 5$

109. $x^{5a}x^{4a} = x^{5a+4a} = x^{9a}$

111. $(a^b)^5 = a^{b\cdot5} = a^{5b}$

113. $\frac{x^{9a}}{x^{4a}} = x^{9a-4a} = x^{5a}$

115. $(x^ay^bz^c)^{5a} = x^{a\cdot5a}y^{b\cdot5a}z^{c\cdot5a}$
$= x^{5a^2}y^{5ab}z^{5ac}$

Section 5.2

Calculator Explorations

1. 5.31×10^3 is entered as 5.31 EE 3.

3. 6.6×10^{-9} is entered as 6.6 EE −9.

5. $3{,}000{,}000 \times 5{,}000{,}000 =$ EE 13

7. $(3.26 \times 10^6)(2.5 \times 10^{13}) = 8.15 \times 10^{19}$

Mental Math

1. $5x^{-2} = \frac{5}{x^2}$

2. $3x^{-3} = \frac{3}{x^3}$

3. $\frac{1}{y^{-6}} = y^6$

4. $\frac{1}{x^{-3}} = x^3$

5. $\frac{4}{y^{-3}} = 4y^3$

6. $\frac{16}{y^{-7}} = 16y^7$

Exercise Set 5.2

1. $4^{-3} = \frac{1}{4^3} = \frac{1}{64}$

3. $7x^{-3} = 7 \cdot \frac{1}{x^3} = \frac{7}{x^3}$

5. $\left(-\frac{1}{4}\right)^{-3} = \frac{(-1)^{-3}}{4^{-3}} = \frac{4^3}{(-1)^3} = \frac{64}{-1} = -64$

7. $3^{-1} + 2^{-1} = \frac{1}{3} + \frac{1}{2} = \frac{2}{6} + \frac{3}{6} = \frac{5}{6}$

9. $\frac{1}{p^{-3}} = p^3$

11. $\frac{p^{-5}}{q^{-4}} = \frac{q^4}{p^5}$

13. $\frac{x^{-2}}{x} = x^{-2-1} = x^{-3} = \frac{1}{x^3}$

15. $\frac{z^{-4}}{z^{-7}} = z^{-4-(-7)} = z^3$

17. $2^0 + 3^{-1} = 1 + \frac{1}{3} = \frac{3}{3} + \frac{1}{3} = \frac{4}{3}$

19. $(-3)^{-2} = \frac{1}{(-3)^2} = \frac{1}{9}$

21. $\frac{-1}{p^{-4}} = -1\left(p^4\right) = -p^4$

23. $-2^0 - 3^0 = -1(2^0) - 3^0 = -1(1) - 1 = -2$

25. $\frac{x^2x^5}{x^3} = x^{2+5-3} = x^4$

27. $\frac{p^2p}{p^{-1}} = p^{2+1-(-1)} = p^{2+1+1} = p^4$

29. $\frac{(m^5)^4m}{m^{10}} = m^{5(4)+1-10} = m^{20+1-10} = m^{11}$

31. $\frac{r}{r^{-3}r^{-2}} = r^{1-(-3)-(-2)} = r^{1+3+2} = r^6$

33. $(x^5y^3)^{-3} = x^{5(-3)}y^{3(-3)} = x^{-15}y^{-9} = \frac{1}{x^{15}y^9}$

35. $\frac{(x^2)^3}{x^{10}} = \frac{x^{2\cdot 3}}{x^{10}} = \frac{x^6}{x^{10}} = x^{6-10} = x^{-4} = \frac{1}{x^4}$

37. $\frac{(a^5)^2}{(a^3)^4} = \frac{a^{5\cdot 2}}{a^{3\cdot 4}} = \frac{a^{10}}{a^{12}} = a^{10-12} = a^{-2} = \frac{1}{a^2}$

39. $\frac{8k^4}{2k} = \frac{8}{2}k^{4-1} = 4k^3$

41. $\frac{-6m^4}{-2m^3} = \frac{-6}{-2}m^{4-3} = 3m$

43. $\frac{-24a^6b}{6ab^2} = \frac{-24}{6}a^{6-1}b^{1-2}$
$= -4a^5b^{-1}$
$= -\frac{4a^5}{b}$

45. $\frac{6x^2y^3}{-7xy^5} = -\frac{6}{7}x^{2-1}y^{3-5}$
$= -\frac{6}{7}x^1y^{-2}$
$= -\frac{6x}{7y^2}$

47. $\left(a^{-5}b^2\right)^{-6} = a^{-5(-6)}b^{2(-6)} = a^{30}b^{-12} = \frac{a^{30}}{b^{12}}$

49. $\left(\frac{x^{-2}y^4}{x^3y^7}\right)^2 = \frac{x^{-2(2)}y^{4(2)}}{x^{3(2)}y^{7(2)}}$
$= \frac{x^{-4}y^8}{x^6y^{14}}$
$= x^{-4-6}y^{8-14}$
$= x^{-10}y^{-6}$
$= \frac{1}{x^{10}y^6}$

51. $\frac{4^2z^{-3}}{4^3z^{-5}} = 4^{2-3}z^{-3-(-5)} = 4^{-1}z^2 = \frac{z^2}{4}$

53. $\frac{2^{-3}x^{-4}}{2^2x} = 2^{-3-2}x^{-4-1}$
$= 2^{-5}x^{-5}$
$= \frac{1}{2^5x^5}$
$= \frac{1}{32x^5}$

55. $\dfrac{7ab^{-4}}{7^{-1}a^{-3}b^2} = 7^{1-(-1)}a^{1-(-3)}b^{-4-2}$
$= 7^2a^4b^{-6}$
$= \dfrac{49a^4}{b^6}$

57. $\left(\dfrac{a^{-5}b}{ab^3}\right)^{-4} = \dfrac{a^{-5(-4)}b^{-4}}{a^{-4}b^{3(-4)}}$
$= \dfrac{a^{20}b^{-4}}{a^{-4}b^{-12}}$
$= a^{20-(-4)}b^{-4-(-12)}$
$= a^{24}b^8$

59. $\dfrac{\left(xy^3\right)^5}{(xy)^{-4}} = \dfrac{x^5y^{3(5)}}{x^{-4}y^{-4}}$
$= \dfrac{x^5y^{15}}{x^{-4}y^{-4}}$
$= x^{5-(-4)}y^{15-(-4)}$
$= x^9y^{19}$

61. $\dfrac{\left(-2xy^{-3}\right)^{-3}}{\left(xy^{-1}\right)^{-1}} = \dfrac{(-2)^{-3}x^{-3}y^{-3(-3)}}{x^{-1}y^{-1(-1)}}$
$= \dfrac{(-2)^{-3}x^{-3}y^9}{x^{-1}y^1}$
$= (-2)^{-3}x^{-3-(-1)}y^{9-1}$
$= (-2)^{-3}x^{-2}y^8$
$= \dfrac{y^8}{(-2)^3x^2}$
$= \dfrac{y^8}{-8x^2}$
$= -\dfrac{y^8}{8x^2}$

63. Volume $= s^3 = \left(\dfrac{3x^{-2}}{z}\right)^3 = \dfrac{3^3x^{-2(3)}}{z^3}$
$= \dfrac{27x^{-6}}{z^3} = \dfrac{27}{x^6z^3}$

The volume of the cube is $\dfrac{27}{x^6z^3}$ cubic inches.

65. $78,000 = 7.8 \times 10^4$

67. $0.00000167 = 1.67 \times 10^{-6}$

69. $0.00635 = 6.35 \times 10^{-3}$

71. $1,160,000 = 1.16 \times 10^6$

73. $20,000,000 = 2.0 \times 10^7$

75. $93,000,000 = 9.3 \times 10^7$

77. $120,000,000 = 1.2 \times 10^8$

79. $8.673 \times 10^{-10} = 0.0000000008673$

81. $3.3 \times 10^{-2} = 0.033$

83. $2.032 \times 10^4 = 20,320$

85. $6.25 \times 10^{18} = 6,250,000,000,000,000,000$

87. $9.460 \times 10^{12} = 9,460,000,000,000$

89. $\left(1.2 \times 10^{-3}\right)\left(3 \times 10^{-2}\right) = 1.2 \cdot 3 \cdot 10^{-3} \cdot 10^{-2}$
$= 3.6 \times 10^{-5}$
$= 0.000036$

91. $\left(4 \times 10^{-10}\right)\left(7 \times 10^{-9}\right)$
$= 4 \cdot 7 \cdot 10^{-10} \cdot 10^{-9}$
$= 28 \times 10^{-19}$
$= 0.0000000000000000028$

93. $\frac{8\times10^{-1}}{16\times10^{5}}=\frac{8}{16}\times10^{-1-5}$
$=0.5\times10^{-6}$
$=0.0000005$

95. $\frac{1.4\times10^{-2}}{7\times10^{-8}}=\frac{1.4}{7}\times10^{-2-(-8)}$
$=0.2\times10^{6}$
$=200,000$

97. To find the amount that flows past in an hour, we multiply the amount that flows past in one second by the number of seconds per hour.

$(3600)(4.2\times10^{6})=(3.6\times10^{3})(4.2\times10^{6})$
$=3.6\cdot4.2\cdot10^{3}\cdot10^{6}$
$=15.12\times10^{9}$
$=1.512\times10^{10}$

In one hour, about 1.512×10^{10} cubic feet of water flow past the mouth of the Amazon River.

99. Answers may vary.

101. $a^{-2}=\left(\frac{1}{10}\right)^{-2}$
$=\frac{1^{-2}}{10^{-2}}$
$=\frac{10^{2}}{1^{2}}$
$=\frac{100}{1}$
$=100$

103. $(2.63\times10^{12})(-1.5\times10^{-10})$
$=2.63\cdot(-1.5)\cdot10^{12}\cdot10^{-10}$
$=-3.945\times10^{2}$
$=-394.5$

105. Since $d=r\cdot t$, $t=\frac{d}{r}$.

$t=\frac{238,857}{1.86\times10^{5}}$
$=\frac{2.38857\times10^{5}}{1.86\times10^{5}}$
$=\frac{2.38857}{1.86}\times10^{5-5}\approx1.3$

It takes the light reflected by the moon about 1.3 seconds to reach the Earth.

107. $-5y+4y-18-y=-5y+4y-y-18$
$=-2y-18$

109. $-3x-(4x-2)=-3x-4x+2$
$=-7x+2$

111. $3(z-4)-2(3z+1)=3z-12-6z-2$
$=3z-6z-12-2$
$=-3z-14$

113. $a^{-4m}\cdot a^{5m}=a^{-4m+5m}=a^{m}$

115. $(3y^{2z})^{3}=3^{3}y^{2z(3)}=27y^{6z}$

117. $\frac{y^{4a}}{y^{-a}}=y^{4a-(-a)}=y^{4a+a}=y^{5a}$

119. $\left(z^{3a+2}\right)^{-2}=\frac{1}{\left(z^{3a+2}\right)^{2}}=\frac{1}{z^{(3a+2)(2)}}=\frac{1}{z^{6a+4}}$

Section 5.3

Graphing Calculator Explorations

1. $(2x^{2}+7x+6)+(x^{3}-6x^{2}-14)$
$=x^{3}+2x^{2}-6x^{2}+7x+6-14$
$=x^{3}-4x^{2}+7x-8$

10
−10 10
−10

3. $(1.8x^2 - 6.8x - 1.7) - (3.9x^2 - 3.6x)$
$= 1.8x^2 - 6.8x - 1.7 - 3.9x^2 + 3.6x$
$= 1.8x^2 - 3.9x^2 - 6.8x + 3.6x - 1.7$
$= -2.1x^2 - 3.2x - 1.7$

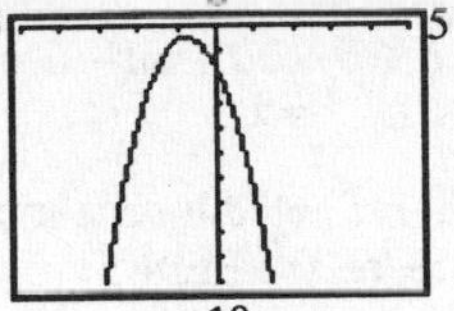

0
−5 5
−10

5. $(1.29x - 5.68) + (7.69x^2 - 2.55x + 10.98)$
$= 7.69x^2 + 1.29x - 2.55x - 5.68 + 10.98$
$= 7.69x^2 - 1.26x + 5.3$

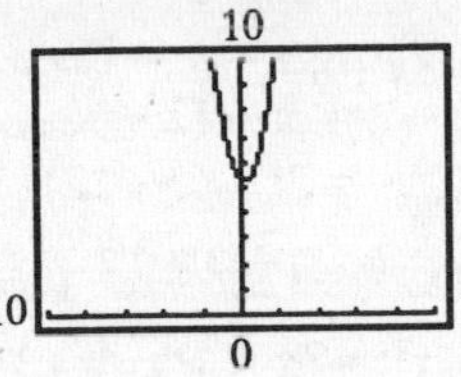

10
−10 10
0

Exercise Set 5.3

1. 4 has degree 0.

3. $5x^2$ has degree 2.

5. $-3xy^2$ has degree 1 + 2 = 3.

7. $6x + 3$ has degree 1 and is a binomial.

9. $3x^2 - 2x + 5$ has degree 2 and is a trinomial.

11. $-xyz$ has degree 1 + 1 + 1 = 3 and is a monomial.

13. $x^2y - 4xy^2 + 5x + y$ has degree 2 + 1 = 3 and is none of these.

15. Answers may vary.

17. $P(x) = x^2 + x + 1$
$P(7) = 7^2 + 7 + 1 = 49 + 7 + 1 = 57$

19. $Q(x) = 5x^2 - 1$
$Q(-10) = 5(-10)^2 - 1$
$= 5(100) - 1$
$= 500 - 1$
$= 499$

21. $P(x) = x^2 + x + 1$
$P(0) = 0^2 + 0 + 1 = 1$

23. To find the height of the object at 2 seconds, we find $P(2)$.
$P(t) = -16t^2 + 1053$
$P(2) = -16(2)^2 + 1053$
$= -16(4) + 1053$
$= -64 + 1053$
$= 989$
When $t = 2$ seconds, the height of the object is 989 feet.

25. To find the height of the object at 6 seconds, we find $P(6)$.
$P(t) = -16t^2 + 1053$
$P(6) = -16(6)^2 + 1053$
$= -16(36) + 1053$
$= -576 + 1053$
$= 477$
When $t = 6$ seconds, the height of the object is 477 feet.

27. $5y + y = (5 + 1)y = 6y$

29. $4x + 7x - 3 = (4 + 7)x - 3 = 11x - 3$

31. $4xy + 2x - 3xy - 1 = (4 - 3)xy + 2x - 1$
$= xy + 2x - 1$

33. $(9y^2 - 8) + (9y^2 - 9) = (9y^2 + 9y^2) - 8 - 9$
$= 18y^2 - 17$

35. $(x^2 + xy - y^2) + (2x^2 - 4xy + 7y^2)$
$= x^2 + 2x^2 + xy - 4xy - y^2 + 7y^2$
$= 3x^2 - 3xy + 6y^2$

37.
$$\begin{array}{r} x^2-6x+3 \\ +\quad (2x+5) \\ \hline x^2-4x+8 \end{array}$$

39. $(9y^2-7y+5)-(8y^2-7y+2)$
$=9y^2-7y+5-8y^2+7y-2$
$=9y^2-8y^2-7y+7y+5-2$
$=y^2+3$

41. $(4x^2+2x)-(6x^2-3x)$
$=4x^2+2x-6x^2+3x$
$=4x^2-6x^2+2x+3x$
$=-2x^2+5x$

43.
$$\begin{array}{r} 3x^2-4x+8 \\ -\quad (5x^2 \qquad -7) \\ \hline \end{array}$$
is equivalent to
$$\begin{array}{r} 3x^2-4x\ +8 \\ +\ (-5x^2 \qquad +7) \\ \hline -2x^2-4x+15 \end{array}$$

45. $(5x-11)+(-x-2)=5x-x-11-2$
$=4x-13$

47. $(7x^2+x+1)-(6x^2+x-1)$
$=7x^2+x+1-6x^2-x+1$
$=7x^2-6x^2+x-x+1+1$
$=x^2+2$

49. $(7x^3-4x+8)+(5x^3+4x+8x)$
$=7x^3-4x+8+5x^3+12x$
$=7x^3+5x^3-4x+12x+8$
$=12x^3+8x+8$

51. $(9x^3-2x^2+4x-7)-(2x^3-6x^2-4x+3)$
$=9x^3-2x^2+4x-7-2x^3+6x^2+4x-3$
$=9x^3-2x^3-2x^2+6x^2+4x+4x-7-3$
$=7x^3+4x^2+8x-10$

53. $(y^2+4yx+7)+(-19y^2+7yx+7)$
$=y^2-19y^2+4yx+7yx+7+7$
$=-18y^2+11yx+14$

55. $(3x^3-b+2a-6)+(-4x^3+b+6a-6)$
$=3x^3-4x^3-b+b+2a+6a-6-6$
$=-x^3+8a-12$

57. $(4x^2-6x+2)-(-x^2+3x+5)$
$=4x^2-6x+2+x^2-3x-5$
$=4x^2+x^2-6x-3x+2-5$
$=5x^2-9x-3$

59. $(-3x+8)+(-3x^2+3x-5)$
$=-3x^2-3x+3x+8-5$
$=-3x^2+3$

61. $(-3+4x^2+7xy^2)+(2x^3-x^2+xy^2)$
$=2x^3+4x^2-x^2+7xy^2+xy^2-3$
$=2x^3+3x^2+8xy^2-3$

63.
$$\begin{array}{r} 6y^2-6y+4 \\ -(-y^2-6y+7) \\ \hline \end{array}$$
is equivalent to
$$\begin{array}{r} 6y^2-6y+4 \\ +\ (\ y^2+6y-7) \\ \hline 7y^2 \qquad -3 \end{array}$$

65.
$$\begin{array}{r} 3x^2+15x+8 \\ +(2x^2+\ 7x+8) \\ \hline 5x^2+22x+16 \end{array}$$

67. $(5q^4-2q^2-3q)+(-6q^4+3q^2+5)$
$=5q^4-6q^4-2q^2+3q^2-3q+5$
$=-q^4+q^2-3q+5$

69. $(7x^2+4x+9)+(8x^2+7x-8)-(3x+7)$
$=7x^2+4x+9+8x^2+7x-8-3x-7$
$=7x^2+8x^2+4x+7x-3x+9-8-7$
$=15x^2+8x-6$

71. $(4x^4-7x^2+3)+(2-3x^4)$
$=4x^4-3x^4-7x^2+3+2$
$=x^4-7x^2+5$

73. $(8x^{2y}-7x^y+3)+(-4x^{2y}+9x^y-14)$
$=8x^{2y}-4x^{2y}-7x^y+9x^y+3-14$
$=4x^{2y}+2x^y-11$

75. The perimeter is the sum of the length of the sides. Perimeter
$=(-x^2+3x)+(2x^2+5)+(4x-1)$
$=-x^2+2x^2+3x+4x+5-1$
$=x^2+7x+4$
The perimeter of the triangle is (x^2+7x+4) feet.

77. To find the length of the remaining piece, subtract (y^2-10) meters from $(4y^2+4y+1)$ meters.
$(4y^2+4y+1)-(y^2-10)$
$=4y^2+4y+1-y^2+10$
$=4y^2-y^2+4y+1+10$
$=3y^2+4y+11$
The length of the remaining piece is $(3y^2+4y+11)$ meters.

79. $P(t)=-32t+500$
$P(3)=-32(3)+500$
$=-96+500$
$=404$
After 3 seconds, the accrued velocity is 404 feet per second.

81. Find $P(1)$, $P(2)$, and $P(3)$.
$P(t)=-16t^2+50t+350$
$P(1)=-16(1)^2+50(1)+350=384$
$P(2)=-16(2)^2+50(2)+350=386$
$P(3)=-16(3)^2+50(3)+350=356$
The height of the projectile when $t=1$ second, $t=2$ seconds, and $t=3$ seconds is 384 feet, 386 feet, and 356 feet, respectively.

83. $P(t)=-16t^2+300t$

a. Find $P(1)$.
$P(1)=-16(1)^2+300(1)=284$
The height of the projectile is 284 feet when $t=1$ second.

b. Find $P(2)$.
$P(2)=-16(2)^2+300(2)=536$
The height of the projectile is 536 feet when $t=2$ seconds.

c. Find $P(3)$.
$P(3)=-16(3)^2+300(3)=756$
The height of the projectile is 756 feet when $t=3$ seconds.

d. Find $P(4)$.
$P(4)=-16(4)^2+300(4)=944$
The height of the projectile is 944 feet when $t=4$ seconds.

85. When the object hits the ground, its height is 0. Find the whole number value of t for which $P(t)$ is closest to 0.
$P(18)=-16(18)^2+300(18)=216$
$P(19)=-16(19)^2+300(19)=-76$
The object hits the ground between 18 and 19 seconds after it is fired, but closer to 19 seconds.

87. $f(x) = 0.43x^2 + 164.6x + 949.3$

a. Find $f(5)$ since 1985 is 5 years after 1980.

$f(5) = 0.43(5)^2 + 164.6(5) + 949.3$
$= 1783.05$

The per capita spending on health care in 1985 was $1783.05.

b. Find $f(15)$ since 1995 is 15 years after 1980.

$f(15) = 0.43(15)^2 + 164.6(15) + 949.3$
$= 3515.05$

The per capita spending on health care in 1995 was $3515.05.

c. Find $f(30)$ since 2010 is 30 years after 1980.

$f(30) = 0.43(30)^2 + 164.6(30) + 949.3$
$= 6274.30$

We predict that $6274.30 will be spent per capita on health care in 2010.

d. No, the amount of money spent is not rising at a steady rate. The first 15 years it increased $3515.05 – $949.3 = $2565.75. The next 15 years it increased $6274.30 – $3515.05 = $2759.25, which is more than the increase for the first 15 years.

89. $f(x) = 3x^2 - 2$ has degree 2 and a positive leading coefficient; A.

91. $g(x) = -2x^3 - 3x^2 + 3x - 2$ has degree 3 and a negative leading coefficient; D.

93. To find the area of each figure, we multiply the length of each rectangle or square by its width. The areas are

$(2x)(2x) = 4x^2$, $7x$, x^2 and $5x$.

To write a polynomial that describes the total area, find the sum of these areas.

$4x^2 + 7x + x^2 + 5x$

Now simplify.

$5x^2 + 12x$

The total area of the figures is $(5x^2 + 12x)$ square units.

95. $5(3x - 2) = 5(3x) - 5(2) = 15x - 10$

97 $-2(x^2 - 5x + 6)$
$= (-2)(x^2) + (-2)(-5x) + (-2)(6)$
$= -2x^2 + 10x - 12$

99. $P(x) = 2x - 3$

a. $P(a) = 2a - 3$

b. $P(-x) = 2(-x) - 3 = -2x - 3$

c. $P(x + h) = 2(x + h) - 3 = 2x + 2h - 3$

101. $P(x) = 4x$

a. $P(a) = 4a$

b. $P(-x) = 4(-x) = -4x$

c. $P(x + h) = 4(x + h) = 4x + 4h$

103. $P(x) = 4x - 1$

a. $P(a) = 4a - 1$

b. $P(-x) = 4(-x) - 1 = -4x - 1$

c. $P(x + h) = 4(x + h) - 1 = 4x + 4h - 1$

Section 5.4

Mental Math

1. $5x(2y) = 5 \cdot 2xy = 10xy$

2. $7a(4b) = 7 \cdot 4ab = 28ab$

3. $x^2 \cdot x^5 = x^{2+5} = x^7$

4. $z \cdot z^4 = z^{1+4} = z^5$

5. $6x(3x^2) = 6 \cdot 3 \cdot x^{1+2} = 18x^3$

6. $5a^2(3a^2) = 5 \cdot 3 \cdot a^{2+2} = 15a^4$

Exercise Set 5.4

1. $2a(2a - 4) = 2a(2a) + 2a(-4) = 4a^2 - 8a$

3. $7x(x^2 + 2x - 1) = 7x(x^2) + 7x(2x) + 7x(-1)$
$= 7x^3 + 14x^2 - 7x$

5. $3x^2(2x^2 - x) = 3x^2(2x^2) + 3x^2(-x)$
$= 6x^4 - 3x^3$

7. The area of the lighter rectangle is $x \cdot x = x^2$, and the area of the darker rectangle is $x \cdot 3 = 3x$. Thus, the total area is $(x^2 + 3x)$ square units.

9. $(a + 7)(a - 2) = a(a) + a(-2) + 7(a) + 7(-2)$
$= a^2 - 2a + 7a - 14$
$= a^2 + 5a - 14$

11. $(2y - 4)^2 = (2y - 4)(2y - 4)$
$= 2y(2y) + 2y(-4) + (-4)(2y) + (-4)(-4)$
$= 4y^2 - 8y - 8y + 16$
$= 4y^2 - 16y + 16$

13. $(5x - 9y)(6x - 5y) = 5x(6x) + 5x(-5y) + (-9y)(6x) + (-9y)(-5y)$
$= 30x^2 - 25xy - 54xy + 45y^2$
$= 30x^2 - 79xy + 45y^2$

15. $(2x^2-5)^2 = (2x^2-5)(2x^2-5)$
$= 2x^2(2x^2)+2x^2(-5)+(-5)(2x^2)+(-5)(-5)$
$= 4x^4-10x^2-10x^2+25$
$= 4x^4-20x^2+25$

17. The area of the square in the upper left corner is $x \cdot x = x^2$, the area of the rectangle in the upper right corner is $x \cdot 3 = 3x$, the area of the rectangle in the lower left corner is $2 \cdot x = 2x$, and the area of the rectangle in the lower right corner is $2 \cdot 3 = 6$. Thus, the total area is $(x^2+3x+2x+6) = (x^2+5x+6)$ square units.

19. $(x-2)(x^2-3x+7) = x(x^2)+x(-3x)+x(7)+(-2)(x^2)+(-2)(-3x)+(-2)(7)$
$= x^3-3x^2+7x-2x^2+6x-14$
$= x^3-5x^2+13x-14$

21. $(x+5)(x^3-3x+4) = x(x^3)+x(-3x)+x(4)+5(x^3)+5(-3x)+5(4)$
$= x^4-3x^2+4x+5x^3-15x+20$
$= x^4+5x^3-3x^2-11x+20$

23. $(2a-3)(5a^2-6a+4) = 2a(5a^2)+2a(-6a)+2a(4)+(-3)(5a^2)+(-3)(-6a)+(-3)(4)$
$= 10a^3-12a^2+8a-15a^2+18a-12$
$= 10a^3-27a^2+26a-12$

25. $(x+2)^3 = (x+2)(x+2)(x+2)$
$= (x^2+2x+2x+4)(x+2)$
$= (x^2+4x+4)(x+2)$
$= (x^2+4x+4)x+(x^2+4x+4)2$
$= x^3+4x^2+4x+2x^2+8x+8$
$= x^3+6x^2+12x+8$

27. $(2y-3)^3 = (2y-3)(2y-3)(2y-3)$
$= (4y^2-6y-6y+9)(2y-3)$
$= (4y^2-12y+9)(2y-3)$
$= (4y^2-12y+9)2y+(4y^2-12y+9)(-3)$
$= 8y^3-24y^2+18y-12y^2+36y-27$
$= 8y^3-36y^2+54y-27$

29.
$$\begin{array}{r} 2x^2 + 4x - 1 \\ \times \qquad\qquad x + 3 \\ \hline 6x^2 + 12x - 3 \\ 2x^3 + 4x^2 - x \quad \\ \hline 2x^3 + 10x^2 + 11x - 3 \end{array}$$

31.
$$\begin{array}{r} x^3 \qquad + 5x - 7 \\ \times \qquad x^2 \qquad\quad - 9 \\ \hline -9x^3 \qquad - 45x + 63 \\ x^5 + 5x^3 - 7x^2 \qquad\qquad\quad \\ \hline x^5 - 4x^3 - 7x^2 - 45x + 63 \end{array}$$

33. **a.** $(2+3)^2 = 5^2 = 25$
$2^2 + 3^2 = 4 + 9 = 13$

b. $(8+10)^2 = (18)^2 = 324$
$8^2 + 10^2 = 64 + 100 = 164$

c. No; answers may vary.

35. $2a(a+4) = 2a(a) + 2a(4) = 2a^2 + 8a$

37. $3x(2x^2 - 3x + 4) = 3x(2x^2) + 3x(-3x) + 3x(4)$
$= 6x^3 - 9x^2 + 12x$

39. $(5x+9y)(3x+2y) = 5x(3x) + 5x(2y) + 9y(3x) + 9y(2y)$
$= 15x^2 + 10xy + 27xy + 18y^2$
$= 15x^2 + 37xy + 18y^2$

41. $(x+2)(x^2+5x+6) = x(x^2) + x(5x) + x(6) + 2(x^2) + 2(5x) + 2(6)$
$= x^3 + 5x^2 + 6x + 2x^2 + 10x + 12$
$= x^3 + 7x^2 + 16x + 12$

43. $(7x+4)^2 = (7x+4)(7x+4)$
$= 7x(7x) + 7x(4) + 4(7x) + 4(4)$
$= 49x^2 + 28x + 28x + 16$
$= 49x^2 + 56x + 16$

45. $-2a^2(3a^2 - 2a + 3) = (-2a^2)(3a^2) + (-2a^2)(-2a) + (-2a^2)(3)$
$= -6a^4 + 4a^3 - 6a^2$

47. $(x+3)(x^2+7x+12) = x(x^2)+x(7x)+x(12)+3(x^2)+3(7x)+3(12)$
$= x^3+7x^2+12x+3x^2+21x+36$
$= x^3+10x^2+33x+36$

49. $(a+1)^3 = (a+1)(a+1)(a+1)$
$= (a^2+a+a+1)(a+1)$
$= (a^2+2a+1)(a+1)$
$= (a^2+2a+1)a+(a^2+2a+1)1$
$= a^3+2a^2+a+a^2+2a+1$
$= a^3+3a^2+3a+1$

51. $(x+y)(x+y) = x(x)+x(y)+y(x)+y(y)$
$= x^2+xy+xy+y^2$
$= x^2+2xy+y^2$

53. $(x-7)(x-6) = x(x)+x(-6)+(-7)(x)+(-7)(-6)$
$= x^2-6x-7x+42$
$= x^2-13x+42$

55. $3a(a^2+2) = 3a(a^2)+3a(2) = 3a^3+6a$

57. $-4y(y^2+3y-11) = (-4y)(y^2)+(-4y)(3y)+(-4y)(-11)$
$= -4y^3-12y^2+44y$

59. $(5x+1)(5x-1) = 5x(5x)+5x(-1)+1(5x)+1(-1)$
$= 25x^2-5x+5x-1$
$= 25x^2-1$

61. $(5x+4)(x^2-x+4) = 5x(x^2)+5x(-x)+5x(4)+4(x^2)+4(-x)+4(4)$
$= 5x^3-5x^2+20x+4x^2-4x+16$
$= 5x^3-x^2+16x+16$

63. $(2x-5)^3 = (2x-5)(2x-5)(2x-5)$
$= (4x^2-10x-10x+25)(2x-5)$
$= (4x^2-20x+25)(2x-5)$
$= (4x^2-20x+25)2x+(4x^2-20x+25)(-5)$
$= 8x^3-40x^2+50x-20x^2+100x-125$
$= 8x^3-60x^2+150x-125$

65. $(4x+5)(8x^2+2x-4) = 4x(8x^2)+4x(2x)+4x(-4)+5(8x^2)+5(2x)+5(-4)$
$= 32x^3+8x^2-16x+40x^2+10x-20$
$= 32x^3+48x^2-6x-20$

67. $(7xy-y)^2 = (7xy-y)(7xy-y)$
$= 7xy(7xy)+7xy(-y)+(-y)(7xy)+(-y)(-y)$
$= 49x^2y^2-7xy^2-7xy^2+y^2$
$= 49x^2y^2-14xy^2+y^2$

69.

$$\begin{array}{r} 5y^2 - y + 3 \\ \times \quad y^2 - 3y - 2 \\ \hline -10y^2 + 2y - 6 \\ -15y^3 + 3y^2 - 9y \\ 5y^4 - y^3 + 3y^2 \\ \hline 5y^4 - 16y^3 - 4y^2 - 7y - 6 \end{array}$$

71.

$$\begin{array}{r} 3x^2 + 2x - 4 \\ \times \quad 2x^2 - 4x + 3 \\ \hline 9x^2 + 6x - 12 \\ -12x^3 - 8x^2 + 16x \\ 6x^4 + 4x^3 - 8x^2 \\ \hline 6x^4 - 8x^3 - 7x^2 + 22x - 12 \end{array}$$

73. $A = lw$
$= (2x+5)(2x-5)$
$= 2x(2x)+2x(-5)+5(2x)+5(-5)$
$= 4x^2-10x+10x-25$
$= 4x^2-25$
The area of the rectangle is $(4x^2-25)$ square yards.

75. $A = \frac{1}{2}bh$
$= \frac{1}{2}(3x-2)(4x)$
$= 2x(3x-2)$
$= 6x^2-4x$
The area of the shaded region is $(6x^2-4x)$ square inches.

77. To write a polynomial that describes the area of the shaded region, subtract the area of the small square, $2 \cdot 2 = 4$, from the area of the l arge square $(x+3)(x+3)$.

$$\begin{aligned}(x+3)(x+3)-4 &= (x^2+3x+3x+9)-4\\ &= x^2+6x+9-4\\ &= x^2+6x+5\end{aligned}$$

The area of the shaded region is (x^2+6x+5) square units.

79. a. $$\begin{aligned}(a+b)(a-b) &= a\cdot a + a(-b) + b\cdot a + b(-b)\\ &= a^2 - ab + ab - b^2\\ &= a^2 - b^2\end{aligned}$$

b. $$\begin{aligned}(2x+3y)(2x-3y) &= 2x(2x)+2x(-3y)+3y(2x)+3y(-3y)\\ &= 4x^2-6xy+6xy-9y^2\\ &= 4x^2-9y^2\end{aligned}$$

c. $$\begin{aligned}(4x+7)(4x-7) &= 4x(4x)+4x(-7)+7(4x)+7(-7)\\ &= 16x^2-28x+28x-49\\ &= 16x^2-49\end{aligned}$$

d. Answers may vary.

81. $(4p)^2 = 4^2p^2 = 16p^2$

83. $(-7m^2)^2 = (-7)^2m^{2\cdot 2} = 49m^4$

85. The point on the line corresponding to 7 years is at a height of 3500. Thus, the value of the machine in 7 years is \$3500.

87. At the end of the first year, the value of the machine is \$6500, and at he end of the second year, the value is \$6000. Thus, the loss in value during the second year is \$6500 – \$6000 = \$500.

89. There is a loss in value each year.

Section 5.5

Graphing Calculator Explorations

1. $(x+4)(x-4) = x^2 - 4^2 = x^2 - 16$

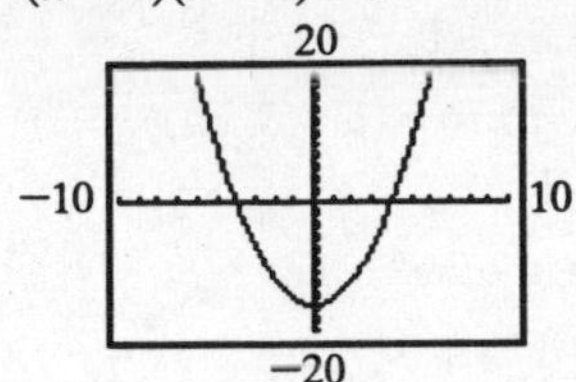

3. $(3x-7)^2 = (3x)^2 - 2(3x)(7) + 7^2$
$= 9x^2 - 42x + 49$

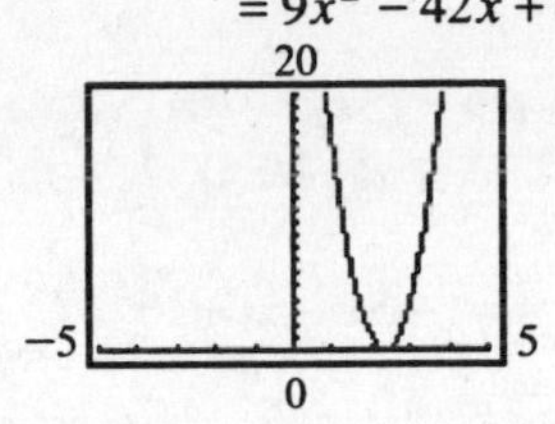

5. $(5x+1)(x^2-3x-2)$
$= 5x(x^2-3x-2) + 1(x^2-3x-2)$
$= 5x^3 - 15x^2 - 10x + x^2 - 3x - 2$
$= 5x^3 - 14x^2 - 13x - 2$

50
−5
5
−50

Mental Math

1. False
$(x+4)^2 = -x^2 + 2(x)(4) + 4^2$
$= x^2 + 8x + 16$

2. True

3. False
$(x+4)(x-4) = x^2 - 4^2 = x^2 - 16$

4. False
The product of $(x-1)(x^3+3x-1)$ is a polynomial of degree 4.

Exercise Set 5.5

1. $(x+3)(x+4)$
F O I L
$= x(x) + x(4) + 3(x) + 3(4)$
$= x^2 + 4x + 3x + 12$
$= x^2 + 7x + 12$

3. $(x-5)(x+10)$
F O I L
$= x(x) + x(10) - 5(x) - 5(10)$
$= x^2 + 10x - 5x - 50$
$= x^2 + 5x - 50$

5. $(5x-6)(x+2)$
F O I L
$= 5x(x) + 5x(2) - 6(x) - 6(2)$
$= 5x^2 + 10x - 6x - 12$
$= 5x^2 + 4x - 12$

7. $(y-6)(4y-1)$
F O I L
$= y(4y) + y(-1) - 6(4y) - 6(-1)$
$= 4y^2 - y - 24y + 6$
$= 4y^2 - 25y + 6$

9. $(2x+5)(3x-1)$
F O I L
$= 2x(3x) + 2x(-1) + 5(3x) + 5(-1)$
$= 6x^2 - 2x + 15x - 5$
$= 6x^2 + 13x - 5$

11. $(x-2)^2 = x^2 - 2(x)(2) + (2)^2$
$= x^2 - 4x + 4$

13. $(2x-1)^2 = (2x)^2 - 2(2x)(1) + (1)^2$
$= 4x^2 - 4x + 1$

15. $(3a-5)^2 = (3a)^2 - 2(3a)(5) + (5)^2$
$= 9a^2 - 30a + 25$

17. $(5x+9)^2 = (5x)^2 + 2(5x)(9) + (9)^2$
$= 25x^2 + 90x + 81$

19. Answers may vary.

21. $(a-7)(a+7)$
$= (a)^2 - (7)^2$
$= a^2 - 49$

23. $(3x-1)(3x+1)$
$= (3x)^2 - (1)^2$
$= 9x^2 - 1$

25. $\left(3x - \frac{1}{2}\right)\left(3x + \frac{1}{2}\right) = (3x)^2 - \left(\frac{1}{2}\right)^2$
$= 9x^2 - \frac{1}{4}$

27. $(9x+y)(9x-y) = (9x)^2 - (y)^2 = 81x^2 - y^2$

29. $A = s^2$
$= (2x+1)^2$
$= (2x)^2 + 2(2x)(1) + (1)2$
$= 4x^2 + 4x + 1$
The area of the rug is
$(4x^2 + 4x + 1)$ square feet.

31. $(a+5)(a+4) = a^2 + 4a + 5a + 20$
$= a^2 + 9a + 20$

33. $(a+7)^2 = (a)^2 + 2(a)(7) + (7)^2$
$= a^2 + 14a + 49$

35. $(4a+1)(3a-1)$
$= 12a^2 - 4a + 3a - 1$
$= 12a^2 - a - 1$

37. $(x+2)(x-2) = (x)^2 - (2)^2$
$= x^2 - 4$

39. $(3a+1)^2 = (3a)^2 + 2(3a)(1) + (1)^2$
$= 9a^2 + 6a + 1$

41. $(x+y)(4x-y)$
$= 4x^2 - xy + 4xy - y^2$
$= 4x^2 + 3xy - y^2$

43. $(x+3)(x^2 - 6x + 1)$
$= x(x^2) + x(-6x) + x(1) + 3(x^2)$
$+3(-6x) + 3(1)$
$= x^3 - 6x^2 + x + 3x^2 - 18x + 3$
$= x^3 - 3x^2 - 17x + 3$

45. $(2a-3)^2 = (2a)^2 - 2(2a)(3) + (3)^2$
$= 4a^2 - 12a + 9$

47. $(5x-6z)(5x+6z) = (5x)^2 - (6z)^2$
$= 25x^2 - 36z^2$

49. $(x-3)(x-5) = x^2 - 5x - 3x + 15$
$= x^2 - 8x + 15$

51. $\left(x - \frac{1}{3}\right)\left(x + \frac{1}{3}\right) = (x)^2 - \left(\frac{1}{3}\right)^2 = x^2 - \frac{1}{9}$

53. $(a+11)(a-3) = a^2 - 3a + 11a - 33$
$= a^2 + 8a - 33$

55. $(x-2)^2 = (x)^2 - 2(x)(2) + (2)^2$
$= x^2 - 4x + 4$

57. $(3b+7)(2b-5) = 6b^2 - 15b + 14b - 35$
$= 6b^2 - b - 35$

59. $(7p-8)(7p+8) = (7p)^2 - (8)^2$
$= 49p^2 - 64$

61. $\left(\frac{1}{3}a^2 - 7\right)\left(\frac{1}{3}a^2 + 7\right) = \left(\frac{1}{3}a^2\right)^2 - (7)^2$
$= \frac{1}{9}a^4 - 49$

63. $5x^2(3x^2 - x + 2)$
$= 5x^2(3x^2) + 5x^2(-x) + 5x^2(2)$
$= 15x^4 - 5x^3 + 10x^2$

65. $(2r-3s)(2r+3s) = (2r)^2 - (3s)^2$
$= 4r^2 - 9s^2$

67. $(3x-7y)^2 = (3x)^2 - 2(3x)(7y) + (7y)^2$
$= 9x^2 - 42xy + 49y^2$

69. $(4x+5)(4x-5) = (4x)^2 - (5)^2$
$= 16x^2 - 25$

71. $(x+4)(x+4) = x^2 + 4x + 4x + 16$
$= x^2 + 8x + 16$

73. $\left(a - \frac{1}{2}y\right)\left(a + \frac{1}{2}y\right) = (a)^2 - \left(\frac{1}{2}y\right)^2$
$= a^2 - \frac{1}{4}y^2$

75. $\left(\frac{1}{5}x - y\right)\left(\frac{1}{5}x + y\right) = \left(\frac{1}{5}x\right)^2 - (y)^2$
$= \frac{1}{25}x^2 - y^2$

77. $(a+1)(3a^2 - a + 1)$
$= a(3a^2) + a(-a) + a(1) + 1(3a^2) + 1(-a) + 1(1)$
$= 3a^3 - a^2 + a + 3a^2 - a + 1$
$= 3a^3 + 2a^2 + 1$

79. The area of the shaded region is the area of the small square, x^2, subtracted from the area of the rectangle, $(2x-3)(2x+3)$.
$(2x-3)(2x+3) - x^2$
$= (2x)^2 - (3)^2 - x^2$
$= 4x^2 - 9 - x^2$
$= 3x^2 - 9$
The area of the shaded region is $(3x^2 - 9)$ square units.

81. The area of the shaded region is the area of the small square, $(x+1)^2$, subtracted from the area of the large square, $(5x-3)^2$.
$(5x-3)^2 - (x+1)^2$
$= [5(x)^2 - 2(5x)(3) + (3)^2]$
$-[(x)^2 + 2(x)(1) + (1)^2]$
$= (25x^2 - 30x + 9) - (x^2 + 2x + 1)$
$= 24x^2 - 32x + 8$
The area of the shaded region is $(24x^2 - 32x + 8)$ square meters.

83. $[3 + (4b+1)]^2$
$= 3^2 + 2(3)(4b+1) + (4b+1)^2$
$= 9 + 6(4b+1) + (4b)^2 + 2(4b)1 + 1^2$
$= 9 + 24b + 6 + 16b^2 + 8b + 1$
$= 16b^2 + 32b + 16$

85. $[(2s-3) - 1][(2s-3) + 1]$
$= (2s-3)^2 - 1^2$
$= (2s)^2 - 2(2s)3 + 3^2 - 1$
$= 4s^2 - 12s + 9 - 1$
$= 4s^2 - 12s + 8$

87. $[(x+4) - y]^2$
$= (x+4)^2 - 2(x+4)y + y^2$
$= x^2 + 2(x)(4) + 4^2 - 2y(x+4) + y^2$
$= x^2 + 8x + 16 - 2xy - 8y + y^2$

89. $[(x+y) - 3][(x+y) + 3]$
$= (x+y)^2 - (3)^2$
$= (x)^2 + 2(x)(y) + (y)^2 - 9$
$= x^2 + 2xy + y^2 - 9$

91. $[(a-3) + b][(a-3) - b]$
$= (a-3)^2 - (b)^2$
$= (a)^2 - 2(a)(3) + (3)^2 - b^2$
$= a^2 - 6a + 9 - b^2$

93. The line passes through the points (–1, 1) and (2, 2). Thus, its slope is
$m = \frac{y_2 - y_1}{x_2 - x_1} = \frac{2-1}{2-(-1)} = \frac{1}{2+1} = \frac{1}{3}.$

95. The line passes through the points (–1, –2) and (1, 0). Thus, its slope is
$m = \frac{y_2 - y_1}{x_2 - x_1} = \frac{0-(-2)}{1-(-1)} = \frac{2}{1+1} = \frac{2}{2} = 1.$

Section 5.6

Mental Math

1. $\frac{a^6}{a^4} = a^{6-4} = a^2$

2. $\frac{y^2}{y} = y^{2-1} = y$

3. $\frac{a^3}{a} = a^{3-1} = a^2$

4. $\frac{p^8}{p^3} = p^{8-3} = p^5$

5. $\frac{k^5}{k^2} = k^{5-2} = k^3$

6. $\frac{k^7}{k^5} = k^{7-5} = k^2$

7. $\frac{p^8}{p^3} = p^{8-3} = p^5$

8. $\frac{k^5}{k^2} = k^{5-2} = k^3$

9. $\frac{k^7}{k^5} = k^{7-5} = k^2$

Exercise Set 5.6

1. $\frac{4a^2+8a}{2a} = \frac{4a^2}{2a} + \frac{8a}{2a} = 2a+4$

3. $\frac{12a^5b^2+16a^4b}{4a^4b} = \frac{12a^5b^2}{4a^4b} + \frac{16a^4b}{4a^4b}$
$= 3ab+4$

5. $\frac{4x^2y^2+6xy^2-4y^2}{2x^2y}$
$= \frac{4x^2y^2}{2x^2y} + \frac{6xy^2}{2x^2y} - \frac{4y^2}{2x^2y}$
$= 2y + \frac{3y}{x} - \frac{2y}{x^2}$

7. $\frac{4x^2+8x+4}{4} = \frac{4x^2}{4} + \frac{8x}{4} + \frac{4}{4}$
$= x^2+2x+1$

9. Since the three pieces have the same length, we divide the total length by 3.
$\frac{3x^4+6x^2-18}{3} = \frac{3x^4}{3} + \frac{6x^2}{3} - \frac{18}{3}$
$= x^4+2x^2-6$
Each piece is (x^4+2x-6) meters long.

11.
$$\begin{array}{r} x+1 \\ x+2\overline{)x^2+3x+2} \\ \underline{x^2+2x} \\ x+2 \\ \underline{x+2} \\ 0 \end{array}$$
Answer: $x + 1$

13.
$$\begin{array}{r} 2x-8 \\ x+1\overline{)2x^2-6x-8} \\ \underline{2x^2+2x} \\ -8x-8 \\ \underline{-8x-8} \\ 0 \end{array}$$
Answer: $2x - 8$

15.
$$\begin{array}{r} x-\frac{1}{2} \\ 2x+4\overline{\big)2x^2+3x-2} \\ \underline{2x^2+4x} \\ -x-2 \\ \underline{-x-2} \\ 0 \end{array}$$

Answer: $x-\frac{1}{2}$

17.
$$\begin{array}{r} 2x^2-\frac{1}{2}x+5 \\ 2x+4\overline{\big)4x^3+7x^2+8x+20} \\ \underline{4x^3+8x^2} \\ -x^2+8x \\ \underline{-x^2-2x} \\ 10x+20 \\ \underline{10x+20} \\ 0 \end{array}$$

Answer: $2x^2-\frac{1}{2}x+5$

19. Recall that $A = lw$, so

$$w=\frac{A}{l}=\frac{15x^2-29x-14}{5x+2}$$

$$\begin{array}{r} 3x-7 \\ 5x+2\overline{\big)15x^2-29x-14} \\ \underline{15x^2+6x} \\ -35x-14 \\ \underline{-35x-14} \\ 0 \end{array}$$

The width is $(3x-7)$ inches

21. $$\frac{25a^2b^{12}}{10a^5b^7}=\frac{5b^5}{2a^3}$$

23. $$\frac{x^6y^6-x^3y^3}{x^3y^3}=\frac{x^6y^6}{x^3y^3}-\frac{x^3y^3}{x^3y^3}=x^3y^3-1$$

25.
$$\begin{array}{r} a+3 \\ a+1\overline{\big)a^2+4a+3} \\ \underline{a^2+a} \\ 3a+3 \\ \underline{3a+3} \\ 0 \end{array}$$

Answer: $a + 3$

27.
$$\begin{array}{r} 2x+5 \\ x-2\overline{\big)2x^2+x-10} \\ \underline{2x^2-4x} \\ 5x-10 \\ \underline{5x-10} \\ 0 \end{array}$$

Answer: $2x + 5$

29. $$\frac{-16y^3+24y^4}{-4y^2}=\frac{-16y^3}{-4y^2}+\frac{24y^4}{-4y^2}=4y-6y^2$$

31.
$$\begin{array}{r} 2x+23 \\ x-5\overline{\big)2x^2+13x+15} \\ \underline{2x^2-10x} \\ 23x+15 \\ \underline{23x-115} \\ 130 \end{array}$$

Answer: $2x+23+\frac{130}{x-5}$

33. $$\frac{20x^2y^3+6xy^4-12x^3y^5}{2xy^3}$$
$$=\frac{20x^2y^3}{2xy^3}+\frac{6xy^4}{2xy^3}-\frac{12x^3y^5}{2xy^3}$$
$$=10x+3y-6x^2y^2$$

35.
$$\begin{array}{r} 2x+4 \\ 3x+2\overline{)6x^2+16x+8} \\ \underline{6x^2+\ \ 4x} \\ 12x+8 \\ \underline{12x+8} \\ 0 \end{array}$$
Answer: $2x+4$

37.
$$\begin{array}{r} y+\ 5 \\ 2y-3\overline{)2y^2+7y-15} \\ \underline{2y^2-3y} \\ 10y-15 \\ \underline{10y-15} \\ 0 \end{array}$$
Answer: $y+5$

39.
$$\begin{array}{r} 2x+3 \\ 2x-3\overline{)4x^2+0x-9} \\ \underline{4x^2-6x} \\ 6x-9 \\ \underline{6x-9} \\ 0 \end{array}$$
Answer: $2x+3$

41.
$$\begin{array}{r} 2x^2-8x\ +38 \\ x+4\overline{)2x^3+0x^2+6x\quad -4} \\ \underline{2x^3+8x^2} \\ -8x^2+\ 6x \\ \underline{-8x^2-32x} \\ 38x-\quad 4 \\ \underline{38x\ +152} \\ -156 \end{array}$$
Answer: $2x^2-8x+38-\dfrac{156}{x+4}$

43.
$$\begin{array}{r} 3x+3 \\ x-1\overline{)3x^2+0x-4} \\ \underline{3x^2-3x} \\ 3x-4 \\ \underline{3x-3} \\ -1 \end{array}$$
Answer: $3x+3-\dfrac{1}{x-1}$

45.
$$\begin{array}{r} -2x^3+\ 3x^2-\ x+4 \\ -x+5\overline{)2x^4-13x^3+16x^2-9x+20} \\ \underline{2x^4-10x^3} \\ -\ 3x^3+16x^2 \\ \underline{-\ 3x^3+15x^2} \\ x^2-9x \\ \underline{x^2-5x} \\ -4x+20 \\ \underline{-4x+20} \\ 0 \end{array}$$
Answer: $-2x^3+3x^2-x+4$

47.
$$\begin{array}{r} 3x^3\qquad\quad +\ 5x+4 \\ x^2+0x-2\overline{)3x^5+0x^4\ -x^3+4x^2-12x-8} \\ \underline{3x^5+0x^4-6x^3} \\ 5x^3+4x^2-12x \\ \underline{5x^3+0x^2-10x} \\ 4x^2-\ 2x-8 \\ \underline{4x^2+\ 0x-8} \\ -2x \end{array}$$
Answer: $3x^3+5x+4-\dfrac{2x}{x^2-2}$

49. $\dfrac{3x^3-5}{3x^2}=\dfrac{3x^3}{3x^2}-\dfrac{5}{3x^2}=x-\dfrac{5}{3x^2}$

51. The perimeter of a square is given by $P = 4s$, where s is the length of a side. Thus, $s = \frac{P}{4}$.

$$s = \frac{12x^3 + 4x - 16}{4}$$
$$= \frac{12x^3}{4} + \frac{4x}{4} - \frac{16}{4}$$
$$= 3x^3 + x - 4$$

Each side of the square is $(3x^3 + x - 4)$ feet long.

53. The area of a parallelogram is given by $A = bh$, so $h = \frac{A}{b} = \frac{10x^2 + 31x + 15}{5x + 3}$.

$$\begin{array}{r|l} & 2x + 5 \\ \hline 5x+3 & 10x^2 + 31x + 15 \\ & \underline{10x^2 + 6x} \\ & \quad 25x + 15 \\ & \quad \underline{25x + 15} \\ & \qquad 0 \end{array}$$

Its height is $(2x + 5)$ meters.

55. $P(x) = 3x^3 + 2x^2 - 4x + 3$

$$P(1) = 3(1)^3 + 2(1)^2 - 4(1) + 3$$
$$= 3 + 2 - 4 + 3$$
$$= 4$$

$$\begin{array}{r|l} & 3x^2 + 5x + 1 \\ \hline x-1 & 3x^3 + 2x^2 - 4x + 3 \\ & \underline{3x^3 - 3x^2} \\ & \quad 5x^2 - 4x \\ & \quad \underline{5x^2 - 5x} \\ & \qquad x + 3 \\ & \qquad \underline{x - 1} \\ & \qquad\quad 4 \end{array}$$

Remainder = 4

57. $P(x) = 5x^4 - 2x^2 + 3x - 6$

$$P(-3) = 5(-3)^4 - 2(-3)^2 + 3(-3) - 6$$
$$= 5(81) - 2(9) - 9 - 6$$
$$= 405 - 18 - 9 - 6$$
$$= 372$$

$$\begin{array}{r|l} & 5x^3 - 15x^2 + 43x - 126 \\ \hline x+3 & 5x^4 + 0x^3 - 2x^2 + 3x - 6 \\ & \underline{5x^4 + 15x^3} \\ & \quad -15x^3 - 2x^2 \\ & \quad \underline{-15x^3 - 45x^2} \\ & \qquad 43x^2 + 3x \\ & \qquad \underline{43x^2 + 129x} \\ & \qquad\quad -126x - 6 \\ & \qquad\quad \underline{-126x - 378} \\ & \qquad\qquad 372 \end{array}$$

Remainder = 372

59. Answers may vary.

61.

$$\begin{array}{r|l} & 3x^2 + 10x + 8 \\ \hline x-2 & 4x^2 - 12x - 12 + 3x^3 \\ & \underline{-6x^2 \qquad\qquad + 3x^3} \\ & 10x^2 - 12x - 12 \\ & \underline{10x^2 - 20x} \\ & \quad 8x - 12 \\ & \quad \underline{8x - 16} \\ & \qquad 4 \end{array}$$

$$\begin{array}{r|l} & 3x^2 + 10x + 8 \\ \hline x-2 & 3x^3 + 4x^2 - 12x - 12 \\ & \underline{3x^3 - 6x^2} \\ & \quad 10x^2 - 12x \\ & \quad \underline{10x^2 - 20x} \\ & \qquad 8x - 12 \\ & \qquad \underline{8x - 16} \\ & \qquad\quad 4 \end{array}$$

Answer: $3x^2 + 10x + 8 + \frac{4}{x - 2}$

63. $(-5)^2 = (-5)(-5) = 25 = (5)(5) = 5^2$
$5^2 = 5 \cdot 5 = 25$
Thus, the answer is $(-5)^2 = 5^2$.

65. $3^4 = 3 \cdot 3 \cdot 3 \cdot 3 = 81$
$(-3)^4 = (-3)(-3)(-3)(-3) = 81$,
Thus, $3^4 = (-3)^4$.

67. $2x + 1 \geq 9$
$2x \geq 8$
$x \geq 4$

69. $2x - 6 \leq x + 11$
$x - 6 \leq 11$
$x \leq 17$

71.
$$\begin{array}{r} 2x^2 + \frac{1}{2}x - 5 \\ x+2\overline{)2x^3 + \frac{9}{2}x^2 - 4x - 10} \\ \underline{2x^3 + 4x^2} \\ \frac{1}{2}x^2 - 4x \\ \underline{\frac{1}{2}x^2 + x} \\ -5x - 10 \\ \underline{-5x - 10} \\ 0 \end{array}$$

Answer: $2x^2 + \frac{1}{2}x - 5$

73.
$$\begin{array}{r} 2x^3 + \frac{9}{2}x^2 + 10x + 21 \\ x-2\overline{)2x^4 + \frac{1}{2}x^3 + x^2 + x + 0} \\ \underline{2x^4 - 4x^3} \\ \frac{9}{2}x^3 + x^2 \\ \underline{\frac{9}{2}x^3 - 9x^2} \\ 10x^2 + x \\ \underline{10x^2 - 20x} \\ 21x + 0 \\ \underline{21x - 42} \\ 42 \end{array}$$

Answer: $2x^3 + \frac{9}{2}x^2 + 10x + 21 + \frac{42}{x-2}$

75.
$$\begin{array}{r} 3x^4 \qquad\qquad -2x \\ 3x+2\overline{)9x^5 + 6x^4 + 0x^3 - 6x^2 - 4x + 0} \\ \underline{9x^5 + 6x^4} \qquad\qquad\qquad \\ -6x^2 - 4x \\ \underline{-6x^2 - 4x} \\ 0 \end{array}$$

Answer: $3x^4 - 2x$

Exercise Set 5.7

1.
$$\begin{array}{c|ccc} 5 & 1 & 3 & -40 \\ & & 5 & 40 \\ \hline & 1 & 8 & 0 \end{array}$$

Answer: $x + 8$

3.
$$\begin{array}{c|ccc} -6 & 1 & 5 & -6 \\ & & -6 & 6 \\ \hline & 1 & -1 & 0 \end{array}$$

Answer: $x - 1$

5.
$$\begin{array}{c|cccc} 2 & 1 & -7 & -13 & 5 \\ & & 2 & -10 & -46 \\ \hline & 1 & -5 & -23 & -41 \end{array}$$

Answer: $x^2 - 5x - 23 - \frac{41}{x-2}$

7.
$$\begin{array}{c|ccc} 2 & 4 & 0 & -9 \\ & & 8 & 16 \\ \hline & 4 & 8 & 7 \end{array}$$

Answer: $4x + 8 + \frac{7}{x-2}$

9. a. $P(2) = 3(2)^2 - 4(2) - 1$
$= 12 - 8 - 1 = 3$

b.
$$\begin{array}{c|ccc} 2 & 3 & -4 & -1 \\ & & 6 & 4 \\ \hline & 3 & 2 & 3 \end{array}$$

$P(2) = 3$

11. a. $P(-2) = 4(-2)^4 + 7(-2)^2 + 9(-2) - 1$
$= 64 + 28 - 18 - 1$
$= 73$

b.
$$\begin{array}{r|rrrrr} -2 & 4 & 0 & 7 & 9 & -1 \\ & & -8 & 16 & -46 & 74 \\ \hline & 4 & -8 & 23 & -37 & 73 \end{array}$$

$P(-2) = 73$

13. a. $P(-1) = (-1)^5 + 3(-1)^4 + 3(-1) - 7$
$= -1 + 3 - 3 - 7$
$= -8$

b.
$$\begin{array}{r|rrrrrr} -1 & 1 & 3 & 0 & 0 & 3 & -7 \\ & & -1 & -2 & 2 & -2 & -1 \\ \hline & 1 & 2 & -2 & 2 & 1 & -8 \end{array}$$

$P(-1) = -8$

15.
$$\begin{array}{r|rrrr} 3 & 1 & -3 & 0 & 2 \\ & & 3 & 0 & 0 \\ \hline & 1 & 0 & 0 & 2 \end{array}$$

Answer: $x^2 + \frac{2}{x-3}$

17.
$$\begin{array}{r|rrr} -1 & 6 & 13 & 8 \\ & & -6 & -7 \\ \hline & 6 & 7 & 1 \end{array}$$

Answer: $6x + 7 + \frac{1}{x+1}$

19.
$$\begin{array}{r|rrrrr} 5 & 2 & -13 & 16 & -9 & 20 \\ & & 10 & -15 & 5 & -20 \\ \hline & 2 & -3 & 1 & -4 & 0 \end{array}$$

Answer: $2x^3 - 3x^2 + x - 4$

21.
$$\begin{array}{r|rrr} -3 & 3 & 0 & -15 \\ & & -9 & 27 \\ \hline & 3 & -9 & 12 \end{array}$$

Answer: $3x - 9 + \frac{12}{x+3}$

23.
$$\begin{array}{r|rrrr} \frac{1}{2} & 3 & -6 & 4 & 5 \\ & & \frac{3}{2} & -\frac{9}{4} & \frac{7}{8} \\ \hline & 3 & -\frac{9}{2} & \frac{7}{4} & \frac{47}{8} \end{array}$$

Answer: $3x^2 - \frac{9}{2}x + \frac{7}{4} + \frac{47}{8\left(x - \frac{1}{2}\right)}$

25.
$$\begin{array}{r|rrrr} \frac{1}{3} & 3 & 2 & -4 & 1 \\ & & 1 & 1 & -1 \\ \hline & 3 & 3 & -3 & 0 \end{array}$$

Answer: $3x^2 + 3x - 3$

27.
$$\begin{array}{r|rrrr} -1 & 3 & 7 & -4 & 12 \\ & & -3 & -4 & 8 \\ \hline & 3 & 4 & -8 & 20 \end{array}$$

Answer: $3x^2 + 4x - 8 + \frac{20}{x+1}$

29.
$$\begin{array}{r|rrrr} 1 & 1 & 0 & 0 & -1 \\ & & 1 & 1 & 1 \\ \hline & 1 & 1 & 1 & 0 \end{array}$$

Answer: $x^2 + x + 1$

31.
$$\begin{array}{r|rrr} -6 & 1 & 0 & -36 \\ & & -6 & 36 \\ \hline & 1 & -6 & 0 \end{array}$$

Answer: $x - 6$

33.
$$\begin{array}{r|rrrr} 1 & 1 & 3 & -7 & 4 \\ & & 1 & 4 & -3 \\ \hline & 1 & 4 & -3 & 1 \end{array}$$

Thus, $P(1) = 1$.

35. $\begin{array}{r|rrrr} -3 & 3 & -7 & -2 & 5 \\ & & -9 & 48 & -138 \\ \hline & 3 & -16 & 46 & -133 \end{array}$

Thus, $P(-3) = -133$.

37. $\begin{array}{r|rrrrr} -1 & 4 & 0 & 1 & 0 & -2 \\ & & -4 & 4 & -5 & 5 \\ \hline & 4 & -4 & 5 & -5 & 3 \end{array}$

Thus, $P(-1) = 3$.

39. $\begin{array}{r|rrrrr} \frac{1}{3} & 2 & 0 & -3 & 0 & -2 \\ & & \frac{2}{3} & \frac{2}{9} & -\frac{25}{27} & -\frac{25}{81} \\ \hline & 2 & \frac{2}{3} & -\frac{25}{9} & -\frac{25}{27} & -\frac{187}{81} \end{array}$

Thus, $P\left(\frac{1}{3}\right) = -\frac{187}{81}$.

41. $\begin{array}{r|rrrrrr} \frac{1}{2} & 1 & 1 & -1 & 0 & 0 & 3 \\ & & \frac{1}{2} & \frac{3}{4} & -\frac{1}{8} & -\frac{1}{16} & -\frac{1}{32} \\ \hline & 1 & \frac{3}{2} & -\frac{1}{4} & -\frac{1}{8} & -\frac{1}{16} & \frac{95}{32} \end{array}$

Thus, $P\left(\frac{1}{2}\right) = \frac{95}{32}$.

43. Answers may vary.

45. $\begin{array}{r|rrrr} -3 & 1 & 3 & 4 & 12 \\ & & -3 & 0 & -12 \\ \hline & 1 & 0 & 4 & 0 \end{array}$

Remainder = 0

47. $P(c)$ is equal to the remainder when $P(x)$ is divided by $x - c$. Therefore, $P(c) = 0$.

49. Multiply $(x^2 - x + 10)$ by $(x + 3)$ and add the remainder, -2.

$(x+3)(x^2 - x + 10) - 2$
$= x^3 - x^2 + 10x + 3x^2 - 3x + 30 - 2$
$= x^3 + 2x^2 + 7x + 28$

51. The volume of a rectangular box is given by $V = lwh$, so

$w = \frac{V}{lh}$.

$w = \frac{x^4 + 6x^3 - 7x^2}{x^2(x+7)}$
$= \frac{x^4 + 6x^3 - 7x^2}{x^3 + 7x^2}$

$$\begin{array}{r} x - 1 \\ x^3 + 7x^2 \overline{\big)\, x^4 + 6x^3 - 7x^2 + 0x + 0} \\ \underline{x^4 + 7x^3} \qquad\qquad\qquad \\ -x^3 - 7x^2 \qquad\quad \\ \underline{-x^3 - 7x^2} \qquad\quad \\ 0 \qquad\quad \end{array}$$

The width is $(x - 1)$ meters.

53. $-4a(3a^2 - 4) = (-4a)(3a^2) + (-4a)(-4)$
$= -12a^3 + 16a$

55 $4y(y^2 - 8y - 4)$
$= 4y(y^2) + -4y(-8y) + 4y(-4)$
$= 4y^3 - 32y^2 - 16y$

57. $-9xy(4xyz + 7xy^2z + 2)$
$= (-9xy)(4xyz) + (-9xy)(7xy^2z) + (-9xy)(2)$
$= -36x^2y^2z - 63x^2y^3z - 18xy$

59. $-7sr(6s^2r + 9sr^2 + 9rs + 8)$
$= (-7sr)(6s^2r) + (-7sr)(9sr^2) + (-7sr)(9rs) + (-7sr)(8)$
$= -42s^3r^2 - 63s^2r^3 - 63s^2r^2 - 56sr$

61. \$103 million

63. The Eagles (1994)

Chapter 5 Review Exercises

1. 3^2
base = 3; exponent = 2

2. $(-5)^4$
base = –5; exponent = 4

3. -5^4
base = 5; exponent = 4

4. $8^3 = 8 \cdot 8 \cdot 8 = 512$

5. $(-6)^2 = (-6)(-6) = 36$

6. $-6^2 = -(6 \cdot 6) = -36$

7. $-4^3 - 4^0 = -(64) - (1) = -64 - 1 = -65$

8. $(3b)^0 = 1$

9. $\frac{8b}{8b} = 1$

10. $5b^3b^5a^6 = 5a^6b^{3+5} = 5a^6b^8$

11. $2^3 \cdot x^0 = 8 \cdot 1 = 8$

12. $\left[(-3)^2\right]^3 = (9)^3 = 9 \cdot 9 \cdot 9 = 729$

13. $(2x^3)(-5x^2) = (2)(-5)x^{3+2} = -10x^5$

14.
$$\left(\frac{mn}{q}\right)^2\left(\frac{mn}{q}\right) = \left(\frac{mn}{q}\right)^{2+1} = \left(\frac{mn}{q}\right)^3 = \frac{m^3n^3}{q^3}$$

15. $\left(\frac{3ab^2}{6ab}\right)^4 = \left(\frac{b}{2}\right)^4 = \frac{b^4}{2^4} = \frac{b^4}{16}$

16. $\frac{x^9}{x^4} = x^{9-4} = x^5$

17. $\frac{2x^7y^8}{8xy^2} = \frac{2x^{7-1}y^{8-2}}{8} = \frac{x^6y^6}{4}$

18. $\frac{12xy^6}{3x^4y^{10}} = \frac{12x^{1-4}y^{6-10}}{3} = 4x^{-3}y^{-4} = \frac{4}{x^3y^4}$

19.
$$5a^7(2a^4)^3 = 5a^7(2^3a^{4\cdot 3}) = 5a^7(8a^{12}) = 5(8)a^{7+12} = 40a^{19}$$

20.
$$(2x)^2(9x) = 2^2x^2 = (4x^2)(9x) = 4(9)x^{2+1} = 36x^3$$

21.
$$\frac{(-4)^2(3^3)}{(4^5)(3^2)} = \frac{(-1\cdot 4)^2(3^3)}{(4^5)(3^2)} = \frac{(-1)^2(4^2)(3^3)}{(4^5)(3^2)} = \frac{(-1)^2(3)}{4^3} = \frac{3}{64}$$

22. $\frac{(-7)^2(3^5)}{(-7)^3(3^4)} = \frac{3}{-7} = -\frac{3}{7}$

23. $\frac{(2x)^0(-4)^2}{16x} = \frac{1\cdot 16}{16x} = \frac{1}{x}$

24. $\frac{(8xy)(3xy)}{18x^2y^2} = \frac{24x^2y^2}{18x^2y^2} = \frac{4}{3}$

25. $m^0 + p^0 + 3q^0 = 1 + 1 + 3(1)$
$= 1 + 1 + 3$
$= 5$

26. $(-5a)^0 + 7^0 + 8^0 = 1 + 1 + 1 = 3$

27. $(3xy^2 + 8x + 9)^0 = 1$

28. $8x^0 + 9^0 = 8(1) + 1 = 9$

29. $6(a^2b^3)^3 = 6(a^{2\cdot 3}b^{3\cdot 3}) = 6a^6b^9$

30. $\frac{(x^3z)^a}{x^2z^2} = \frac{x^{3a}z^a}{x^2z^2} = x^{3a-2}z^{a-2}$

31. $7^{-2} = \frac{1}{7^2} = \frac{1}{49}$

32. $-7^{-2} = -\frac{1}{7^2} = -\frac{1}{49}$

33. $2x^{-4} = 2\left(\frac{1}{x^4}\right) = \frac{2}{x^4}$

34. $(2x)^{-4} = \frac{1}{(2x)^4} = \frac{1}{16x^4}$

35. $\left(\frac{1}{5}\right)^{-3} = \frac{1^{-3}}{5^{-3}} = \frac{5^3}{1^3} = \frac{125}{1} = 125$

36. $\left(\frac{-2}{3}\right)^{-2} = \frac{(-2)^{-2}}{3^{-2}} = \frac{3^2}{(-2)^2} = \frac{9}{4}$

37. $2^0 + 2^{-4} = 1 + \frac{1}{2^4} = 1 + \frac{1}{16} = \frac{16}{16} + \frac{1}{16} = \frac{17}{16}$

38. $6^{-1} - 7^{-1} = \frac{1}{6} - \frac{1}{7} = \frac{7}{42} - \frac{6}{42} = \frac{1}{42}$

39. $\frac{1}{(2q)^{-3}} = 1 \cdot (2q)^3 = 1 \cdot 2^3q^3 = 8q^3$

40. $\frac{-1}{(qr)^{-3}} = -1 \cdot (qr)^3 = -1 \cdot q^3r^3 = -q^3r^3$

41. $\frac{r^{-3}}{s^{-4}} = \frac{s^4}{r^3}$

42. $\frac{rs^{-3}}{r^{-4}} = r^{1-(-4)}s^{-3} = r^5s^{-3} = \frac{r^5}{s^3}$

43. $\frac{-6}{8x^{-3}r^4} = \frac{-6x^3}{8r^4} = -\frac{3x^3}{4r^4}$

44. $\frac{-4s}{16s^{-3}} = -\frac{s^{1-(-3)}}{4} = -\frac{s^4}{4}$

45. $(2x^{-5})^{-3} = 2^{-3}x^{-5(-3)} = \frac{x^{15}}{2^3} = \frac{x^{15}}{8}$

46. $(3y^{-6})^{-1} = 3^{-1}y^{-6(-1)} = \frac{y^6}{3}$

47. $(3a^{-1}b^{-1}c^{-2})^{-2}$
$= 3^{-2}a^{-1(-2)}b^{-1(-2)}c^{-2(-2)}$
$= \frac{a^2b^2c^4}{3^2}$
$= \frac{a^2b^2c^4}{9}$

48. $(4x^{-2}y^{-3}z)^{-3} = 4^{-3}x^{-2(-3)}y^{-3(-3)}z^{-3}$
$= \frac{x^6y^9}{4^3z^3}$
$= \frac{x^6y^9}{64z^3}$

49. $\frac{5^{-2}x^8}{5^{-3}x^{11}} = 5^{-2-(-3)}x^{8-11}$
$= 5^{-2+3}x^{8-11}$
$= 5^1x^{-3}$
$= \frac{5}{x^3}$

50. $\frac{7^5y^{-2}}{7^7y^{-10}} = 7^{5-7} \cdot y^{-2-(-10)}$
$= 7^{-2} \cdot y^{-2+10}$
$= \frac{y^8}{7^2}$
$= \frac{y^8}{49}$

51. $\left(\frac{bc^{-2}}{bc^{-3}}\right)^4 = (b^{1-1}c^{-2-(-3)})^4$
$= (b^0c^{-2+3})^4$
$= (1 \cdot c^1)^4$
$= 1^4c^4$
$= c^4$

52. $\left(\frac{x^{-3}y^{-4}}{x^{-2}y^{-5}}\right)^{-3} = \frac{x^9y^{12}}{x^6y^{15}}$
$= x^{9-6} \cdot y^{12-15}$
$= x^3y^{-3}$
$= \frac{x^3}{y^3}$

53. $\frac{x^{-4}y^{-6}}{x^2y^7} = x^{-4-2}y^{-6-7} = x^{-6}y^{-13} = \frac{1}{x^6y^{13}}$

54. $\frac{a^5b^{-5}}{a^{-5}b^5} = a^{5-(-5)}b^{-5-5} = a^{10}b^{-10} = \frac{a^{10}}{b^{10}}$

55. $-2^0 + 2^{-4} = -1 \cdot 2^0 + \frac{1}{2^4}$
$= -1 \cdot 1 + \frac{1}{16}$
$= -1 + \frac{1}{16}$
$= -\frac{16}{16} + \frac{1}{16}$
$= -\frac{15}{16}$

56. $-3^{-2} - 3^{-3} = -\frac{1}{3^2} - \frac{1}{3^3}$
$= -\frac{1}{9} - \frac{1}{27}$
$= -\frac{3}{27} - \frac{1}{27}$
$= -\frac{4}{27}$

57. $a^{6m}a^{5m} = a^{6m+5m} = a^{11m}$

58. $\frac{\left(x^{5+h}\right)^3}{x^5} = \frac{x^{3(5+h)}}{x^5}$
$= \frac{x^{15+3h}}{x^5}$
$= x^{15+3h-5}$
$= x^{10+3h}$

59. $(3xy^{2z})^3 = 3^3x^3y^{2z(3)} = 27x^3y^{6z}$

60. $a^{m+2}a^{m+3} = a^{m+2+m+3} = a^{2m+5}$

61. $0.00027 = 2.7 \times 10^{-4}$

62. $0.8868 = 8.868 \times 10^{-1}$

63. $80{,}800{,}000 = 8.08 \times 10^7$

64. $-868{,}000 = -8.68 \times 10^5$

65. $32{,}667{,}000 = 3.2667 \times 10^7$

66. $4000 = 4.0 \times 10^3$

67. $8.67 \times 10^5 = 867{,}000$

68. $3.86 \times 10^{-3} = 0.00386$

69. $8.6 \times 10^{-4} = 0.00086$

70. $8.936 \times 10^5 = 893{,}600$

71. $1 \times 10^{20} = 100{,}000{,}000{,}000{,}000{,}000{,}000$

72. 3×10^{-25}
$= 0.0000000000000000000000003$

73. $(8\times10^4)(2\times10^{-7}) = 8\cdot2\cdot10^4\cdot10^{-7}$
$= 16\times10^{4+(-7)}$
$= 16\times10^{-3}$
$= 0.016$

74. $\dfrac{8\times10^4}{2\times10^{-7}} = 4\times10^{4-(-7)}$
$= 4\times10^{11}$
$= 400,000,000,000$

75. **a.** $3a^2b - 2a^2 + ab - b^2 - 6$

Term	Numerical Coefficient	Degree of Term
$3a^2b$	3	$2 + 1 = 3$
$-2a^2$	-2	2
ab	1	$1 + 1 = 2$
$-b^2$	-1	2
-6	-6	0

b. The degree of the polynomial is 3, since the highest degree of the terms is 3.

76. **a.** $x^2y^2 + 5x^2 - 7y^2 + 11xy - 1$

Term	Numerical Coefficient	Degree of Term
x^2y^2	1	$2 + 2 = 4$
$5x^2$	5	2
$-7y^2$	-7	$1 + 1 = 2$
$11xy$	11	2
-1	-1	0

b. The degree of the polynomial is 4, since the highest degree of the terms is 4.

77. $4x + 8x - 6x^2 - 6x^2y$
$= (4+8)x - 6x^2 - 6x^2y$
$= 12x - 6x^2 - 6x^2y$

78. $-8xy^3 + 4xy^3 - 3x^3y$
$= (-8+4)xy^3 - 3x^3y$
$= -4xy^3 - 3x^3y$

79. $(3x + 7y) + (4x^2 - 3x + 7) + (y - 1)$
$= 4x^2 + 3x - 3x + 7y + y + 7 - 1$
$= 4x^2 + 8y + 6$

80. $(4x^2 - 6xy + 9y^2) - (8x^2 - 6xy - y^2)$
$= 4x^2 - 6xy + 9y^2 - 8x^2 + 6xy + y^2$
$= 4x^2 - 8x^2 - 6xy + 6xy + 9y^2 + y^2$
$= -4x^2 + 10y^2$

81. $(3x^2 - 4b + 28) + (9x^2 - 30) - (4x^2 - 6b + 20)$
$= 3x^2 - 4b + 28 + 9x^2 - 30 - 4x^2 + 6b - 20$
$= 3x^2 + 9x^2 - 4x^2 - 4b + 6b + 28 - 30 - 20$
$= 8x^2 + 2b - 22$

82. $(9xy + 4x^2 + 18) + (7xy - 4x^3 - 9x)$
$= -4x^3 + 4x^2 + 9xy + 7xy - 9x + 18$
$= -4x^3 + 4x^2 + 16xy - 9x + 18$

83. $(3x^2y - 7xy - 4) + (9x^2y + x) - (x - 7)$
$= 3x^2y - 7xy - 4 + 9x^2y + x - x + 7$
$= 3x^2y + 9x^2y - 7xy + x - x - 4 + 7$
$= 12x^2y - 7xy + 3$

84.
$$\begin{array}{r} x^2 - 5x + 7 \\ -\quad (x+4) \\ \hline \end{array}$$
is equivalent to
$$\begin{array}{r} x^2 - 5x + 7 \\ +\quad (-x-4) \\ \hline x^2 - 6x + 3 \end{array}$$

85.
$$\begin{array}{r} x^3 \quad + 2xy^2 - y \\ + \quad (x - 4xy^2 \quad - 7) \\ \hline x^3 + x - 2xy^2 - y - 7 \end{array}$$

86. $P(6) = 9(6)^2 - 7(6) + 8 = 290$

87. $P(-2) = 9(-2)^2 - 7(-2) + 8 = 58$

88. $P(-3) = 9(-3)^2 - 7(-3) + 8 = 110$

89. Recall that the perimeter of a rectangle is the sum of twice the length and twice the width.
$2(x^2y + 5) + 2(2x^2y - 6x + 1)$
$= 2x^2y + 10 + 4x^2y - 12x + 2$
$= 2x^2y + 4x^2y - 12x + 10 + 2$
$= 6x^2y - 12x + 12$
The perimeter of the rectangle is $(6x^2y - 12x + 12)$ centimeters.

90. $9x(x^2y) = 9x^{1+2}y = 9x^3y$

91. $-7(8xz^2) = (-7)(8)xz^2 = -56xz^2$

92. $(6xa^2)(xya^3) = 6x^{1+1}a^{2+3}y = 6x^2a^5y$

93. $(4xy)(-3xa^2y^3) = (4)(-3)x^{1+1}a^2y^{1+3}$
$= -12x^2a^2y^4$

94. $6(x + 5) = 6x + 6(5) = 6x + 30$

95. $9(x - 7) = 9x + 9(-7) = 9x - 63$

96. $4(2a + 7) = 4(2a) + 4(7) = 8a + 28$

97. $9(6a - 3) = 9(6a) + 9(-3) = 54a - 27$

98. $-7x(x^2 + 5) = (-7x)(x^2) + (-7x)(5)$
$= -7x^3 - 35x$

99. $-8y(4y^2 - 6) = (-8y)(4y^2) + (-8y)(-6)$
$= -32y^3 + 48y$

100. $-2(x^3 - 9x^2 + x)$
$= (-2)x^3 + (-2)(-9x^2) + (-2)x$
$= -2x^3 + 18x^2 - 2x$

101. $-3a(a^2b + ab + b^2)$
$= (-3a)(a^2b) + (-3a)(ab) + (-3a)(b^2)$
$= -3a^3b - 3a^2b - 3ab^2$

102. $(3a^3 - 4a + 1)(-2a)$
$= (3a^3)(-2a) + (-4a)(-2a) + (1)(-2a)$
$= -6a^4 + 8a^2 - 2a$

103. $(6b^3 - 4b + 2)(7b)$
$= 6b^3(7b) + (-4b)(7b) + (2)(7b)$
$= 42b^4 - 28b^2 + 14b$

104. $(2x + 2)(x - 7)$
$= 2x(x) + 2x(-7) + 2(x) + 2(-7)$
$= 2x^2 - 14x + 2x - 14$
$= 2x^2 - 12x - 14$

105. $(2x - 5)(3x + 2)$
$= 2x(3x) + 2x(2) + (-5)(3x) + (-5)(2)$
$= 6x^2 + 4x - 15x - 10$
$= 6x^2 - 11x - 10$

106. $(4a - 1)(a + 7)$
$= 4a(a) + 4a(7) + (-1)(a) + (-1)(7)$
$= 4a^2 + 28a - a - 7$
$= 4a^2 + 27a - 7$

107. $(6a - 1)(7a + 3)$
$= 6a(7a) + 6a(3) + (-1)(7a) + (-1)(3)$
$= 42a^2 + 18a - 7a - 3$
$= 42a^2 + 11a - 3$

108. $(x + 7)(x^3 + 4x - 5)$
$= x(x^3) + x(4x) + x(-5) + 7(x^3)$
$\quad + 7(4x) + 7(-5)$
$= x^4 + 4x^2 - 5x + 7x^3 + 28x - 35$
$= x^4 + 7x^3 + 4x^2 + 23x - 35$

109. $(x+2)(x^5+x+1)$
$= x(x^5)+x(x)+x(1)+2(x^5)+2(x)+2(1)$
$= x^6+x^2+x+2x^5+2x+2$
$= x^6+2x^5+x^2+3x+2$

110.
$$\begin{array}{r} x^2+2x+4 \\ \times \quad x^2+2x-4 \\ \hline -4x^2-8x-16 \\ 2x^3+4x^2+8x \quad\quad \\ x^4+2x^3+4x^2 \quad\quad\quad\quad \\ \hline x^4+4x^3+4x^2 \quad\quad -16 \end{array}$$

111. $(x^3+4x+4)(x^3+4x-4)$
$= x^3(x^3)+x^3(4x)+x^3(-4)+4x(x^3)$
$+4x(4x)+4x(-4)+4(x^3)+4(4x)+4(-4)$
$= x^6+4x^4-4x^3+4x^4+16x^2$
$-16x+4x^3+16x-16$
$= x^6+8x^4+16x^2-16$

112. $2x(3x^2-7x+1)$
$= 2x(3x^2)+2x(-7x)+2x(1)$
$= 6x^3-14x^2+2x$

113. $3y(5y^2-y+2) = 3y(5y^2)+3y(-y)+3y(2)$
$= 15y^3-3y^2+6y$

114. $(6x-1)(4x+3)$
F O I L
$= (6x)(4x)+)(6x)(3)+(-1)(4x)+(-1)(3)$
$= 24x^2+18x-4x-3$
$= 24x^2+14x-3$

115. $(4a-1)(3a+7)$
F O I L
$= 4a(3a)+4a(7)+(-1)(3a)+(-1)(7)$
$= 12a^2+28a-3a-7$
$= 12a^2+25a-7$

116. $(x+7)^2 = x^2+2(x)(7)+7^2$
$= x^2+14x+49$

117. $(x-5)^2 = x^2-2(x)(5)+5^2$
$= x^2-10x+25$

118. $(3x-7)^2 = (3x)^2-2(3x)(7)+(7)^2$
$= 9x^2-42x+49$

119. $(4x+2)^2 = (4x)^2+2(4x)(2)+2^2$
$= 16x^2+16x+4$

120. $(y+1)(y^2-6y-5)$
$= y(y^2)+y(-6y)+y(-5)+1(y^2)$
$+1(-6y)+1(-5)$
$= y^3-6y^2-5y+y^2-6y-5$
$= y^3-5y^2-11y-5$

121. $(x-2)(x^2-x-2)$
$= x(x^2)+x(-x)+x(-2)+(-2)(x^2)$
$+(-2)(-x)+(-2)(-2)$
$= x^3-x^2-2x-2x^2+2x+4$
$= x^3-3x^2+4$

122. $(5x-9)^2 = (5x)^2-2(5x)(9)+(9)^2$
$= 25x^2-90x+81$

123. $(5x+1)(5x-1) = (5x)^2-(1)^2 = 25x^2-1$

124. $(7x+4)(7x-4) = (7x)^2-(4)^2$
$= 49x^2-16$

125. $(a+2b)(a-2b) = a^2-(2b)^2 = a^2-4b^2$

126. $(2x-6)(2x+6) = (2x)^2-(6)^2 = 4x^2-36$

127. $(4a^2-2b)(4a^2+2b) = (4a^2)^2-(2b)^2$
$= 16a^4-4b^2$

128. $[4+(3a-b)][4-(3a-b)]$
$= 4^2-(3a-b)^2$
$= 16-[(3a)^2-2(3a)(b)+b^2]$
$= 16-(9a^2-6ab+b^2)$
$= 16-9a^2+6ab-b^2$

129. $\dfrac{3x^5yb^9}{9xy^7} = \dfrac{x^{5-1}y^{1-7}b^9}{3} = \dfrac{x^4y^{-6}b^9}{3}$

130. $\dfrac{-9xb^4z^3}{-4axb^2} = \dfrac{(-1)(9)x^{1-1}b^{4-2}z^3}{(-1)(4)a} = \dfrac{9b^2z^3}{4a}$

131. $\dfrac{4xy+2x^2-9}{4xy} = \dfrac{4xy}{4xy} + \dfrac{2x^2}{4xy} - \dfrac{9}{4xy}$

$= 1 + \dfrac{x^{2-1}}{2y} - \dfrac{9}{4xy}$

$= 1 + \dfrac{x}{2y} - \dfrac{9}{4xy}$

132. $\dfrac{12xb^2+16xb^4}{4xb^3}$

$= \dfrac{12xb^2}{4xb^3} + \dfrac{16xb^4}{4xb^3}$

$= 3x^{1-1}b^{2-3} + 4x^{1-1}b^{4-3}$

$= 3x^0b^{-1} + 4x^0b^1$

$= \dfrac{3}{b} + 4b$

133.
$$
\begin{array}{r|l}
 & 3x^3 + 9x^2 + 2x + 6 \\
\hline
x-3 & 3x^4 + 0x^3 - 25x^2 + 0x - 20 \\
 & \underline{3x^4 - 9x^3} \\
 & 9x^3 - 25x^2 \\
 & \underline{9x^3 - 27x^2} \\
 & 2x^2 + 0x \\
 & \underline{2x^2 - 6x} \\
 & 6x - 20 \\
 & \underline{6x - 18} \\
 & -2
\end{array}
$$

Answer: $3x^3 + 9x^2 + 2x + 6 - \dfrac{2}{x-3}$

134.
$$
\begin{array}{r|l}
 & 2x^3 - 4x^2 + 7x - 9 \\
\hline
x+2 & 2x^4 + 0x^3 - x^2 + 5x - 12 \\
 & \underline{2x^4 + 4x^3} \\
 & -4x^3 - x^2 \\
 & \underline{-4x^3 - 8x^2} \\
 & 7x^2 + 5x \\
 & \underline{7x^2 + 14x} \\
 & -9x - 12 \\
 & \underline{-9x - 18} \\
 & 6
\end{array}
$$

Answer: $2x^3 - 4x^2 + 7x - 9 + \dfrac{6}{x+2}$

135.
$$
\begin{array}{r|l}
 & 3x^2 + 2x - 1 \\
\hline
x^2+x+2 & 3x^4 + 5x^3 + 7x^2 + 3x - 2 \\
 & \underline{3x^4 + 3x^3 + 6x^2} \\
 & 2x^3 + x^2 + 3x \\
 & \underline{2x^3 + 2x^2 + 4x} \\
 & -x^2 - x - 2 \\
 & \underline{-x^2 - x - 2} \\
 & 0
\end{array}
$$

Answer: $3x^2 + 2x - 1$

136.
$$
\begin{array}{r|l}
 & 3x^2 + 6 \\
\hline
3x^2-2x-5 & 9x^4 - 6x^3 + 3x^2 - 12x - 30 \\
 & \underline{9x^4 - 6x^3 - 15x^2} \\
 & 18x^2 - 12x - 30 \\
 & \underline{18x^2 - 12x - 30} \\
 & 0
\end{array}
$$

Answer: $3x^2 + 6$

137.
$$
\begin{array}{r|rrrr}
2 & 3 & 0 & 12 & -4 \\
 & & 6 & 12 & 48 \\
\hline
 & 3 & 6 & 24 & 44
\end{array}
$$

Answer: $3x^2 + 6x + 24 + \dfrac{44}{x-2}$

138.
$$\begin{array}{r|rrrr} -\frac{3}{2} & 3 & 2 & -4 & -1 \\ & & -\frac{9}{2} & \frac{15}{4} & \frac{3}{8} \\ \hline & 3 & -\frac{5}{2} & -\frac{1}{4} & -\frac{5}{8} \end{array}$$

Answer: $3x^2 - \frac{5}{2}x - \frac{1}{4} - \frac{5}{8\left(x+\frac{3}{2}\right)}$

139.
$$\begin{array}{r|rrrrrr} -1 & 1 & 0 & 0 & 0 & 0 & -1 \\ & & -1 & 1 & -1 & 1 & -1 \\ \hline & 1 & -1 & 1 & -1 & 1 & -2 \end{array}$$

Answer: $x^4 - x^3 + x^2 - x + 1 - \frac{2}{x+1}$

140.
$$\begin{array}{r|rrrr} 3 & 1 & 0 & 0 & -81 \\ & & 3 & 9 & 27 \\ \hline & 1 & 3 & 9 & -54 \end{array}$$

Answer: $x^2 + 3x + 9 - \frac{54}{x-3}$

141.
$$\begin{array}{r|rrrrr} 4 & 3 & 1 & -1 & 0 & -2 \\ & & 12 & 52 & 204 & 816 \\ \hline & 3 & 13 & 51 & 204 & 814 \end{array}$$

Answer: $3x^3 + 13x^2 + 51x + 204 + \frac{814}{x-4}$

142.
$$\begin{array}{r|rrrrr} -2 & 3 & 0 & -2 & 0 & 10 \\ & & -6 & 12 & -20 & 40 \\ \hline & 3 & -6 & 10 & -20 & 50 \end{array}$$

Answer: $3x^3 - 6x^2 + 10x - 20 + \frac{50}{x+2}$

143.
$$\begin{array}{r|rrrrrr} 4 & 3 & 0 & 0 & 0 & -9 & 7 \\ & & 12 & 48 & 192 & 768 & 3036 \\ \hline & 3 & 12 & 48 & 192 & 759 & 3043 \end{array}$$

Thus, $P(4) = 3043$.

144. $l = \frac{A}{w} = \frac{8^4x^3 - 6x^3 - 6x + 13}{x+5}$.

$$\begin{array}{r|rrrrrr} -5 & 3 & 0 & 0 & 0 & -9 & 7 \\ & & -15 & 75 & -375 & 1875 & -9330 \\ \hline & 3 & -15 & 75 & -375 & 1866 & -9323 \end{array}$$

Answer: $P(-5) = -9323$

145. The area of a rectangle is given by $A = lw$. Thus, its length is

$l = \frac{A}{w} = \frac{x^4 - x^3 - 6x^2 - 6x + 18}{x-3}$.

$$\begin{array}{r|rrrrr} 3 & 1 & -1 & -6 & -6 & 18 \\ & & 3 & 6 & 0 & -18 \\ \hline & 1 & 2 & 0 & -6 & 0 \end{array}$$

The area of the rectangle is $(x^3 + 2x^2 - 6)$ miles.

Chapter 5 Test

1. $2^5 = 2 \cdot 2 \cdot 2 \cdot 2 \cdot 2 = 32$

2. $(-3)^4 = (-3)(-3)(-3)(-3) = 81$

3. $-3^4 = -(3 \cdot 3 \cdot 3 \cdot 3) = -81$

4. $4^{-3} = \frac{1}{4^3} = \frac{1}{64}$

5. $(3x^2)(-5x^9) = (3)(-5)(x^{2+9}) = -15x^{11}$

6. $\frac{y^7}{y^2} = y^{7-2} = y^5$

7. $\frac{r^{-8}}{r^{-3}} = r^{-8-(-3)} = r^{-5} = \frac{1}{r^5}$

8. $\left(\frac{x^2y^3}{x^3y^{-4}}\right)^2 = \frac{x^{2\cdot2}y^{3\cdot2}}{x^{3\cdot2}y^{-4\cdot2}}$
$= \frac{x^4y^6}{x^6y^{-8}}$
$= x^{4-6}y^{6-(-8)}$
$= x^{-2}y^{14}$
$= \frac{y^{14}}{x^2}$

9. $\frac{6^2x^{-4}y^{-1}}{6^3x^{-3}y^7} = 6^{2-3}x^{-4-(-3)}y^{-1-7}$
$= 6^{-1}x^{-1}y^{-8}$
$= \frac{1}{6xy^8}$

10. $563{,}000 = 5.63\times10^5$

11. $0.0000863 = 8.63\times10^{-5}$

12. $1.5\times10^{-3} = 0.0015$

13. $6.23\times10^4 = 62{,}300$

14. $(1.2\times10^5)(3\times10^{-7}) = (1.2)(3)\times10^{5-7}$
$= 3.6\times10^{-2}$
$= 0.036$

15. a. $4xy^2 + 7xyz + x^3y - 2$

Term	Numerical Coefficient	Degree of Term
$4xy^2$	4	$1+2=3$
$7xyz$	7	$1+1+1=3$
x^3y	1	$3+1=4$
-2	-2	0

b. The degree of the polynomial is 4, since the highest degree of the terms is 4.

16. $5x^2 + 4xy - 7x^2 + 11 + 8xy$
$= (5-7)x^2 + (4+8)xy + 11$
$= -2x^2 + 12xy + 11$

17. $(8x^3 + 7x^2 + 4x - 7) + (8x^3 - 7x - 6)$
$= 8x^3 + 8x^3 + 7x^2 + 4x - 7x - 7 - 6$
$= 16x^3 + 7x^2 - 3x - 13$

18.
$$\begin{array}{r} 5x^3 + x^2 + 5x - 2 \\ -(8x^3 - 4x^2 + x - 7) \\ \hline \end{array}$$
is equivalent to
$$\begin{array}{r} 5x^3 + x^2 + 5x - 2 \\ +(-8x^3 + 4x^2 - x + 7 \\ \hline -3x^3 + 5x^2 + 4x + 5 \end{array}$$

19. $(8x^2 + 7x + 5) + (x^3 - 8) - (4x + 2)$
$= 8x^2 + 7x + 5 + x^3 - 8 - 4x - 2$
$= x^3 + 8x^2 + 7x - 4x + 5 - 8 - 2$
$= x^3 + 8x^2 + 3x - 5$

20. $(3x + 7)(x^2 + 5x + 2)$
$= 3x(x^2) + 3x(5x) + 3x(2) + 7(x^2)$
$+7(5x) + 7(2)$
$= 3x^3 + 15x^2 + 6x + 7x^2 + 35x + 14$
$= 3x^3 + 15x^2 + 7x^2 + 6x + 35x + 14$
$= 3x^3 + 22x^2 + 41x + 14$

21. $3x^2(2x^2 - 3x + 7)$
$= 3x^2(2x^2) + 3x^2(-3x) + 3x^2(7)$
$= 6x^4 - 9x^3 + 21x^2$

22. $(x + 7)(3x - 5)$
F O I L
$= x(3x) + x(-5) + 7(3x) + 7(-5)$
$= 3x^2 - 5x + 21x - 35$
$= 3x^2 + 16x - 35$

23. $(3x - 7)(3x + 7) = (3x)^2 - (7)^2 = 9x^2 - 49$

24. $(4x-2)^2 = (4x)^2 - 2(4x)(2) + (2)^2$
$= 16x^2 - 16x + 4$

25. $(8x+3)^2 = (8x)^2 + 2(8x)(3) + (3)^2$
$= 64x^2 + 48x + 9$

26. $(x^2-9b)(x^2+9b) = (x^2)^2 - (9b)^2$
$= x^4 - 81b^2$

27.

t	0 sec	1 sec	3 sec	5 sec
$-16t^2+1001$	1001 ft	985 ft	857 ft	601 ft

$t = 0: -16(0)^2 + 1001 = 1001$
$t = 1: -16(1)^2 + 1001 = 985$
$t = 3: -16(3)^2 + 1001 = 857$
$t = 5: -16(5)^2 + 1000 = 601$

28. $\dfrac{8xy^2}{4x^3y^3z} = \dfrac{8x^{1-3}y^{2-3}}{4z} = \dfrac{2x^{-2}y^{-1}}{z} = \dfrac{2}{x^2yz}$

29. $\dfrac{4x^2+2xy-7x}{8xy} = \dfrac{4x^2}{8xy} + \dfrac{2xy}{8xy} - \dfrac{7x}{8xy}$
$= \dfrac{x}{2y} + \dfrac{1}{4} - \dfrac{7}{8y}$

30.
$$\begin{array}{r} x+2 \\ x+5\overline{)x^2+7x+10} \\ \underline{x^2+5x} \\ 2x+10 \\ \underline{2x+10} \\ 0 \end{array}$$

Answer: $x + 2$

31.
$$\begin{array}{r} 9x^2-6x+4 \\ 3x+2\overline{)27x^3+0x^2+0x-8} \\ \underline{27x^3+18x^2} \\ -18x^2+0x \\ \underline{-18x^2-12x} \\ 12x-8 \\ \underline{12x+8} \\ -16 \end{array}$$

Answer: $9x^2 - 6x + 4 - \dfrac{16}{3x+2}$

32. $f(x) = 62x^2 - 149x + 922$
We find $f(7)$ since 2000 is 7 years after 1993.
$f(7) = 62(7)^2 - 149(7) + 922$
$= 3038 - 1043 + 922$
$= 2917$
The model predicts that there were 2917 thousand cases in 2000.

33.
$$\begin{array}{r|rrrrr} -3 & 4 & -3 & 2 & -1 & -1 \\ & & -12 & 45 & -141 & 426 \\ \hline & 4 & -15 & 47 & -142 & 425 \end{array}$$

Answer: $4x^3 - 15x^2 + 47x - 142 + \dfrac{425}{x+3}$

34.
$$\begin{array}{r|rrrrr} -2 & 4 & 0 & 7 & -2 & -5 \\ & & -8 & 16 & -46 & 96 \\ \hline & 4 & -8 & 23 & -48 & 91 \end{array}$$

$P(-2) = 91$

35. The area of the shaded region is the area of the smaller square $(2y)(2y) = 4y^2$ subtracted from the area of the larger square $x \cdot x = x^2$. Thus, its area is $x^2 - 4y^2$ square units.

36. $h(t) = -16t^2 + 96t + 880$

a. $h(1) = -16(1)^2 + 96(1) + 880$
$= -16 + 96 + 880$
$= 960$
After 1 second, the height of the pebble is 960 feet.

b. $h(5.1) = -16(5.1)^2 + 96(5.1) + 880$
$= -416.16 + 489.6 + 880$
$= 953.44$
After 5.1 seconds, the height of the pebble is 953.44 feet.

Chapter 6

Section 6.1

Mental Math

1. $14 = 2 \cdot 7$

2. $15 = 3 \cdot 5$

3. $10 = 2 \cdot 5$

4. $70 = 2 \cdot 5 \cdot 7$

5. $6 = 2 \cdot 3$
   $15 = 3 \cdot 5$
   $\text{GCF} = 3$

6. $20 = 2 \cdot 2 \cdot 5$
   $15 = 3 \cdot 5$
   $\text{GCF} = 5$

7. $3 = 3$
   $18 = 2 \cdot 3 \cdot 3$
   $\text{GCF} = 3$

8. $14 = 2 \cdot 7$
   $35 = 5 \cdot 7$
   $\text{GCF} = 7$

Exercise Set 6.1

1. $32 = 2 \cdot 2 \cdot 2 \cdot 2 \cdot 2$
   $36 = 2 \cdot 2 \cdot 3 \cdot 3$
   $\text{GCF} = 2 \cdot 2 = 4$

3. $12 = 2 \cdot 2 \cdot 3 = 2^2 \cdot 3$
   $18 = 2 \cdot 3 \cdot 3 = 2 \cdot 3^2$
   $36 = 2 \cdot 2 \cdot 3 \cdot 3 = 2^2 \cdot 3^2$
   $\text{GCF} = 2 \cdot 3 = 6$

5. $y^2,\ y^4,\ y^7$
   $\text{GCF} = y^2$

7. $x^{10}y^2,\ xy^2,\ x^3y^3$
   $\text{GCF} = xy^2$

9. $8x = 2 \cdot 2 \cdot 2 \cdot x$
   $4 = 2 \cdot 2$
   $\text{GCF} = 2 \cdot 2 = 4$

11. $12y^4 = 2 \cdot 2 \cdot 3 \cdot y^4$
    $20y^3 = 2 \cdot 2 \cdot 5 \cdot y^3$
    $\text{GCF} = 2 \cdot 2 \cdot y^3 = 4y^3$

13. $12x^3 = 2 \cdot 2 \cdot 3 \cdot x^3$
    $6x^4 = 2 \cdot 3 \cdot x^4$
    $3x^5 = 3 \cdot x^5$
    $\text{GCF} = 3x^3$

15. $18x^2y = 2 \cdot 3 \cdot 3 \cdot x^2y$
    $9x^3y^3 = 3 \cdot 3 \cdot x^3y^3$
    $36x^3y = 2 \cdot 2 \cdot 3 \cdot 3 \cdot x^3y$
    $\text{GCF} = 3 \cdot 3 \cdot x^2y = 9x^2y$

17. $30x - 15$; $\text{GCF} = 15$
    $15 \cdot 2x - 15 \cdot 1 = 15(2x - 1)$

19. $24cd^3 - 18c^2d$; $\text{GCF} = 6cd$
    $6cd \cdot 4d^2 - 6cd \cdot 3c = 6cd(4d^2 - 3c)$

21. $-24a^4x + 18a^3x$; $\text{GCF} = -6a^3x$
    $-6a^3x(4a) - 6a^3x(-3) = -6a^3x(4a - 3)$

23. $12x^3 + 16x^2 - 8x$; $\text{GCF} = 4x$
    $4x(3x^2) + 4x(4x) + 4x(-2)$
    $= 4x(3x^2 + 4x - 2)$

25. $5x^3y - 15x^2y + 10xy$; $\text{GCF} = 5xy$
    $5xy(x^2) + 5xy(-3x) + 5xy(2)$
    $= 5xy(x^2 - 3x + 2)$

27. Answers may vary.

29. $y(x + 2) + 3(x + 2)$; $\text{GCF} = (x + 2)$
    $(x + 2)(y + 3)$

31. $x(y-3)-4(y-3)$; GCF = $(y-3)$
$(y-3)(x-4)$

33. $2x(x+y)-(x+y)$; GCF = $(x+y)$
$2x(x+y)+(x+y)(-1)=(x+y)(2x-1)$

35. $5x+15+xy+3y=5(x)+5(3)+x(y)+3(y)$
$=5(x+3)+y(x+3)$
$=(x+3)(5+y)$

37. $2y-8+xy-4x=2(y-4)+x(y-4)$
$=(y-4)(2+x)$

39. $3xy-6x+8y-16=3x(y-2)+8(y-2)$
$=(y-2)(3x+8)$

41. $y^3+3y^2+y+3=y^2(y+3)+1(y+3)$
$=(y+3)(y^2+1)$

43. $12x(x^2)-2x=12x^3-2x=2x(6x^2-1)$

45. $(20x)(10)+\pi(5)^2=200x+25\pi=25(8x+\pi)$

47. $3x-6$; GCF = 3
$3\cdot x-3\cdot 2=3(x-2)$

49. $32xy-18x^2$; GCF = $2x$
$2x(16y)-2x(9x)=2x(16y-9x)$

51. $4x-8y+4$; GCF = 4
$4\cdot x-4\cdot 2y+4\cdot 1=4(x-2y+1)$

53. $8(x+2)-y(x+2)$; GCF = $(x+2)$
$(x+2)(8-y)$

55. $-40x^8y^6-16x^9y^5$; GCF = $-8x^8y^5$
$-8x^8y^5\cdot 5y-8x^8y^5\cdot 2x=-8x^8y^5(5y+2x)$

57. $-3x+12$; GCF = -3
$-3\cdot x-3(-4)=-3(x-4)$

59. $18x^3y^3-12x^3y^2+6x^5y^2$; GCF = $6x^3y^2$
$6x^3y^2\cdot 3y+6x^3y^2(-2)+6x^3y^2\cdot x^2$
$=6x^3y^2(3y-2+x^2)$

61. $y^2(x-2)+(x-2)$; GCF = $(x-2)$
$y^2(x-2)+1(x-2)=(x-2)(y^2+1)$

63. $5xy+15x+6y+18=5x(y+3)+6(y+3)$
$=(y+3)(5x+6)$

65. $4x^2-8xy-3x+6y=4x(x-2y)-3(x-2y)$
$=(x-2y)(4x-3)$

67. $126x^3yz+210y^4z^3$; GCF = $42yz$
$42yz\cdot 3x^3+42yz\cdot 5y^3z^2$
$=42yz(3x^3+5y^3z^2)$

69. $3y-5x+15-xy$
$=3y+15-5x-xy$
$=3(y+5)-x(5+y)$
$=3(5+y)-x(5+y)$
$=(3-x)(5+y)$

71. $12x^2y-42x^2-4y+14$; GCF = 2
$2[6x^2y-21x^2-2y+7]$
$=2[3x^2(2y-7)-1(2y-7)]$
$=2[(2y-7)(3x^2-1)]$
$=2(2y-7)(3x^2-1)$

73. Answers may vary.

75. $(a+6)(b-2)$ is factored.

77. $5(2y+z)-b(2y+z)$ is not factored, because it factors further into $(2y+z)(5-b)$.

79. $\dfrac{4n^4-24n}{4n}=\dfrac{4n^4}{4n}-\dfrac{24n}{4n}=(n^3-6)$ units

81. a. $f(x)=60x^2-85x+780$
$f(2)=60(2)^2-85(2)+780$
$=240-170+780$
$=850$
850 million CDs were sold in 1998.

b. $x = 2001 - 1996 = 5$
$f(x) = 60x^2 - 85x + 780$
$f(5) = 60(5)^2 - 85(5) + 780$
$= 1500 - 425 + 780$
$= 1855$
1855 million CDs will be sold in 2001.

c. $60x^2 - 85x + 780$
$= 5 \cdot 12x^2 + 5(-17x) + 5(156)$
$= 5(12x^2 - 17x + 156)$

83. $(x+2)(x+5) = x^2 + 5x + 2x + 10$
$= x^2 + 7x + 10$

85. $(a-7)(a-8) = a^2 - 8a - 7a + 56$
$= a^2 - 15a + 56$

87. The two numbers are 2 and 6.
$2 \cdot 6 = 12;\ 2 + 6 = 8$

89. The two numbers are −1 and −8.
$-1 \cdot (-8) = 8;\ -1 + (-8) = -9$

91. The two numbers are −2 and 5.
$-2 \cdot 5 = -10;\ -2 + 5 = 3$

93. The two numbers are −8 and 3.
$-8 \cdot 3 = -24;\ -8 + 3 = -5$

Section 6.2

Mental Math

1. $x^2 + 9x + 20 = (x+4)(x+5)$

2. $x^2 + 12x + 35 = (x+5)(x+7)$

3. $x^2 - 7x + 12 = (x-4)(x-3)$

4. $x^2 - 13x + 22 = (x-2)(x-11)$

5. $x^2 + 4x + 4 = (x+2)(x+2)$

6. $x^2 + 10x + 24 = (x+6)(x+4)$

Exercise Set 6.2

In exercises 1–61, to factor a trinomial $x^2 + bx + c$, find two numbers whose product is c and whose sum is b.

1. $x^2 + 7x + 6$
The product is 6, and the sum is 7: 6 and 1.
$(x+6)(x+1)$

3. $x^2 + 9x + 8$
The product is 8, and the sum is 9: 8 and 1.
$(x+8)(x+1)$

5. $x^2 - 8x + 15$
The product is 15, and the sum is −8: −5 and −3.
$(x-5)(x-3)$

7. $x^2 - 10x + 9$
The product is 9, and the sum is −10: −9 and −1.
$(x-9)(x-1)$

9. $x^2 - 15x + 5$
The product is 5, and the sum is −15: none exist. The trinomial is prime.

11. $x^2 - 3x - 18$
The product is −18, and the sum is −3: −6 and 3.
$(x-6)(x+3)$

13. $x^2 + 5x + 2$
The product is 2, and the sum is 5: none exist. The trinomial is prime.

15. $x^2 + 8xy + 15y^2$
The product is $15y^2$, and the sum is $8y$: $3y$ and $5y$.
$(x+3y)(x+5y)$

17. $x^2 - 2xy + y^2$
The product is y^2, and the sum is $-2y$: $-y$ and $-y$.
$(x-y)(x-y)$

19. $x^2 - 3xy - 4y^2$
The product is $-4y^2$, and the sum is $-3y$: $-4y$ and y.
$(x - 4y)(x + y)$

21. $2z^2 + 20z + 32$; GCF = 2
$2(z^2 + 10z + 16) = 2(z + 8)(z + 2)$

23. $2x^3 - 18x^2 + 40x$; GCF = $2x$
$2x(x^2 - 9x + 20) = 2x(x - 5)(x - 4)$

25. $7x^2 + 14xy - 21y^2$; GCF = 7
$7(x^2 + 2xy - 3y^2) = 7(x + 3y)(x - y)$

27. product; sum

29. $x^2 + 15x + 36$
The product is 36, and the sum is 15: 12 and 3.
$(x + 12)(x + 3)$

31. $x^2 - x - 2$
The product is −2, and the sum is −1: −2 and 1.
$(x - 2)(x + 1)$

33. $r^2 - 16r + 48$
The product is 48, and the sum is −16: −12 and −4.
$(r - 12)(r - 4)$

35. $x^2 - 4x - 21$
The product is −21, and the sum is −4: −7 and 3.
$(x - 7)(x + 3)$

37. $x^2 + 7xy + 10y^2$
The product is $10y^2$, and the sum is $7y$: $2y$ and $5y$.
$(x + 2y)(x + 5y)$

39. $r^2 - 3r + 6$
The product is 6, and the sum is −3: none exist. The trinomial is prime.

41. $2t^2 + 24t + 64$; GCF = 2
$2(t^2 + 12t + 32) = 2(t + 4)(t + 8)$

43. $x^3 - 2x^2 - 24x$; GCF = x
$x(x^2 - 2x - 24) = x(x - 6)(x + 4)$

45. $x^2 - 16x + 63$
The product is 63, and the sum is −16: −9 and −7.
$(x - 9)(x - 7)$

47. $x^2 + xy - 2y^2$
The product is $-2y^2$, and the sum is y: $2y$ and $-y$.
$(x + 2y)(x - y)$

49. $3x^2 - 60x + 108$; GCF = 3
$3(x^2 - 20x + 36) = 3(x - 18)(x - 2)$

51. $x^2 - 18x - 144$
The product is −144, and the sum is −18: −24 and 6.
$(x - 24)(x + 6)$

53. $6x^3 + 54x^2 + 120x$; GCF = $6x$
$6x(x^2 + 9x + 20) = 6x(x + 5)(x + 4)$

55. $2t^5 - 14t^4 + 24t^3$; GCF = $2t^3$
$2t^3(t^2 - 7t + 12) = 2t^3(t - 4)(t - 3)$

57. $5x^3y - 25x^2y^2 - 120xy^3$; GCF = $5xy$
$5xy(x^2 - 5xy - 24y^2)$
$= 5xy(x - 8y)(x + 3y)$

59. $4x^2y + 4xy - 12y$; GCF = $4y$
$4y(x^2 + x - 3)$, which is not factorable any further.

61. $2a^2b - 20ab^2 + 42b^3$; GCF = $2b$
$2b(a^2 - 10ab + 21b^2) = 2b(a - 7b)(a - 3b)$

63. $x^2 + bx + 15$
Find pairs of numbers whose products are 15 and whose sums are positive.
1, 15 3, 5
b can be the sum of these pairs of numbers. Thus, 8 and 16 are the possible values of b.

65. $m^2 + bm - 27$
Find pairs of numbers whose products are −27 and whose sums are positive.
−1, 27 −3, 9
b can be the sum of either of these pairs of numbers. Thus, 6 and 26 are the possible values of b.

67. $x^2 + 6x + c$
Find pairs of numbers whose sums are 6 and whose products are positive.
1, 5 2, 4 3, 3
c can be the product of any of these pairs of numbers. Thus, 5, 8, and 9 are the possible values of c.

69. $y^2 - 4y + c$
Find pairs of numbers whose sums are −4 and whose products are positive.
−1, −3 −2, −2
c can be the product of either of these pairs of numbers. Thus, 3 and 4 are the possible values of c.

71. Answers may vary.

73. $(2x+1)(x+5) = 2x^2 + 10x + x + 5$
$= 2x^2 + 11x + 5$

75. $(5y-4)(3y-1) = 15y^2 - 5y - 12y + 4$
$= 15y^2 - 17y + 4$

77. $(a+3)(9a-4) = 9a^2 - 4a + 27a - 12$
$= 9a^2 + 23a - 12$

79. $y = -3x$
Find 3 points.

x	y
−1	3
0	0
1	−3

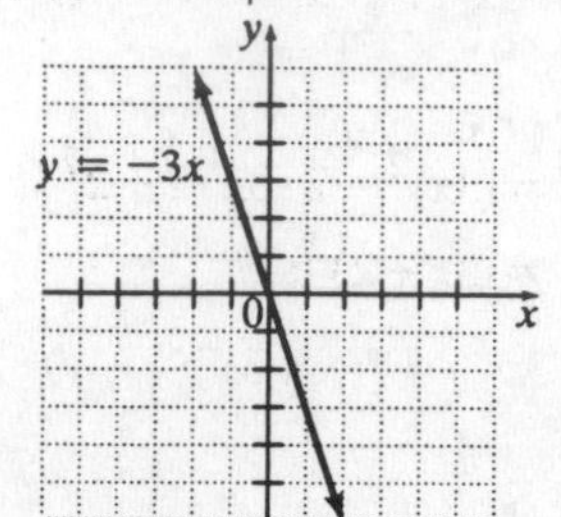

81. $y = 2x - 7$
Find 3 points.

x	y
0	−7
1	−5
2	−3

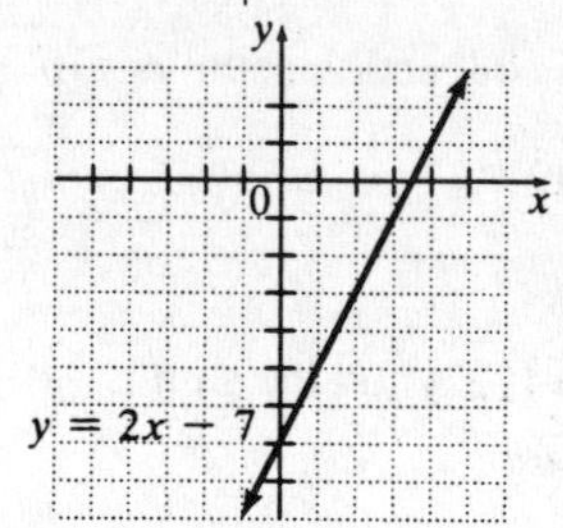

83. $2x^2y + 30xy + 100y$; GCF = $2y$
$2y(x^2 + 15x + 50) = 2y(x+5)(x+10)$

85. $-12x^2y^3 - 24xy^3 - 36y^3$; GCF = $-12y^3$
$-12y^3(x^2 + 2x + 3)$, which can not be factored further.

87. $y^2(x+1)-2y(x+1)-15(x+1);$
GCF = $(x+1)$
$(x+1)(y^2-2y-15)=(x+1)(y-5)(y+3)$

Section 6.3

Mental Math

1. $x^2+14x+49=x^2+2\cdot x\cdot 7+7^2=(x+7)^2;$
Yes

2. $9x^2-12x+4=(3x)^2-2\cdot 3x\cdot 2+2^2$
$=(3x-2)^2;$
Yes

3. No

4. No

5. $9y^2+6y+1=(3y)^2+2\cdot 3y\cdot 1+1^2$
$=(3y+1)^2;$
Yes

6. $y^2-16y+64=y^2-2\cdot y\cdot 8+8=(y-8)^2;$
Yes

Exercise Set 6.3

1. $2x^2+13x+15$
$2x^2=2x\cdot x$
$15=3\cdot 5$
$2x\cdot 5+x\cdot 3=13x$
$(2x+3)(x+5)$

3. $2x^2-9x-5$
$2x^2=2x\cdot x$
$-5=-5\cdot 1$
$2x(-5)+x\cdot 1=-9x$
$(2x+1)(x-5)$

5. $2y^2-y-6$
$2y^2=2y\cdot y$
$-6=-2\cdot 3$
$2y(-2)+y\cdot 3=-y$
$(2y+3)(y-2)$

7. $16a^2-24a+9=(4a-3)(4a-3)$
$=(4a-3)^2$

9. $36r^2-5r-24$
$36r^2=9r\cdot 4r$
$-24=-8\cdot 3$
$9r\cdot 3+4r(-8)=-5r$
$(9r-8)(4r+3)$

11. $10x^2+17x+3=(5x+1)(2x+3)$

13. $21x^2-48x-45;$ GCF = 3
$3(7x^2-16x-15)=3(7x+5)(x-3)$

15. $12x^2-14x-6;$ GCF = 2
$2(6x^2-7x-3)=2(2x-3)(3x+1)$

17. $4x^3-9x^2-9x;$ GCF = x
$x(4x^2-9x-9)=x(4x+3)(x-3)$

19. $x^2+22x+121=(x)^2+2(x)(11)+(11)^2$
$=(x+11)^2$

21. $x^2-16x+64=(x)^2-2(x)(8)+(8)^2$
$=(x-8)^2$

23. $16y^2-40y+25=(4y)^2-2(4y)(5)+(5)^2$
$=(4y-5)^2$

25. $x^2y^2-10xy+25=(xy)^2-2(xy)(5)+(5)^2$
$=(xy-5)^2$

27. Answers may vary.

29. $2x^2 - 7x - 99$
$2x^2 = 2x \cdot x$
$-99 = -9 \cdot 11$
$2x(-9) + x \cdot 11 = -7x$
$(2x+11)(x-9)$

31. $4x^2 - 8x - 21$
$4x^2 = 2x \cdot 2x$
$-21 = -7 \cdot 3$
$2x \cdot 3 + 2x(-7) = -8x$
$(2x-7)(2x+3)$

33. $30x^2 - 53x + 21$
$30x^2 = 6x \cdot 5x$
$21 = -7(-3)$
$6x(-3) + 5x(-7) = -53x$
$(6x-7)(5x-3)$

35. $24x^2 - 58x + 9$
$24x^2 = 4x \cdot 6x$
$9 = -9(-1)$
$4x(-1) + 6x(-9) = -58x$
$(4x-9)(6x-1)$

37. $9x^2 - 24xy + 16y^2 = (3x)^2 - 2(3x)(4y) + (4y)^2$
$= (3x-4y)^2$

39. $x^2 - 14xy + 49y^2 = (x)^2 - 2(x)(7y) + (7y)^2$
$= (x-7y)^2$

41. $2x^2 + 7x + 5$
$2x^2 = 2x \cdot x$
$5 = 5 \cdot 1$
$2x \cdot 1 + x \cdot 5 = 7x$
$(2x+5)(x+1)$

43. $3x^2 - 5x + 1$
not factorable, prime

45. $-2y^2 + y + 10 = 10 + y - 2y^2$
$= (5-2y)(2+y)$

47. $16x^2 + 24xy + 9y^2 = (4x)^2 + 2(4x)(3y) + (3y)^2$
$= (4x+3y)^2$

49. $8x^2y + 34xy - 84y$; GCF $= 2y$
$2y(4x^2 + 17x - 42) = 2y(4x-7)(x+6)$

51. $3x^2 + x - 2$
$3x^2 = 3x \cdot x$
$-2 = -2 \cdot 1$
$3x \cdot 1 + x(-2) = x$
$(3x-2)(x+1)$

53. $x^2y^2 + 4xy + 4 = (xy)^2 + 2(xy)(2) + (2)^2$
$= (xy+2)^2$

55. $49y^2 + 42xy + 9x^2 = (7y)^2 + 2(7y)(3x) + (3x)^2$
$= (7y+3x)^2$

57. $3x^2 - 42x + 63$; GCF $= 3$
$3(x^2 - 14x + 21)$, which can not be factored further.

59. $42a^2 - 43a + 6$
$42a^2 = 7a \cdot 6a$
$6 = -6(-1)$
$7a(-1) + 6a(-6) = -43a$
$(7a-6)(6a-1)$

61. $18x^2 - 9x - 14$
$18x^2 = 6x \cdot 3x$
$-14 = -7 \cdot 2$
$6x \cdot 2 + 3x(-7) = -9x$
$(6x-7)(3x+2)$

63. $25p^2 - 70pq + 49q^2 = (5p)^2 - 2(5p)(7q) + (7q)^2$
$= (5p-7q)^2$

65. $15x^2 - 16x - 15$
$15x^2 = 5x \cdot 3x$
$-15 = 3(-5)$
$5x(-5) + 3x \cdot 3 = -16x$
$(5x+3)(3x-5)$

67. $-27t + 7t^2 - 4 = 7t^2 - 27t - 4$
$= (7t + 1)(t - 4)$

69. $a^2 + ab + ab + b^2 = a^2 + 2ab + b^2$

71. $3x^2 + bx - 5$
Find factorizations of 3 and –5.
$3 = 1 \cdot 3 \qquad -5 = -1 \cdot 5$
Possible values of b are the combinations of a factor of 3 times a factor of –5 plus the other factor of 3 times the other factor of –5.
$1 \cdot -1 + 3 \cdot 5 = -1 + 15 = 14$
$1 \cdot 5 + 3 \cdot -1 = 5 + (-3) = 2$
Thus, 2 and 14 are possible values of b.

73. $2z^2 + bz - 7$
Find factorizations of 2 and –7.
$2 = 1 \cdot 2 \qquad -7 = -1 \cdot 7$
Possible values of b are the combinations of a factor of 2 times a factor of –7 plus the other factor of 2 times the other factor of –7.
$1 \cdot -1 + 2 \cdot 7 = -1 + 14 = 13$
$1 \cdot 7 + 2 \cdot -1 = 7 + (-2) = 5$
Thus, 5 and 13 are possible values of b.

75. $5x^2 + 7x + c$
Find pairs of numbers whose sums are 7, one of which is a multiple of 5.
5, 2
Write a product of two binomials where the product of the first terms is $5x^2$, the product of the outside term is $5x$, and the product of the inside terms is $2x$.
$(5x + 2)(x + 1)$
The product of the last terms is 2. Thus, 2 is a possible value of c.

77. $3x^2 - 8x + c$
Find pairs of negative numbers whose sums are –8, one of which is a multiple of 3.
–3, –5   –6, –2
For each pair, write a product of two binomials where the product of the first terms is $3x^2$. For the first pair, the product of the outside terms should be $-3x$ and the product of the inside terms $-5x$.
$(3x - 5)(x - 1)$
For the second pair, the product of the outside terms should be $-6x$ and the product of the inside terms $-2x$.
$(3x - 2)(x - 2)$
The product of the last terms in the two cases are 5 and 4. Thus, 4 and 5 are possible values of c.

79. $(x-2)(x+2) = x^2 - 2x + 2x - 4 = x^2 - 4$

81. $(a+3)(a^2 - 3a + 9)$
$= a^3 - 3a^2 + 9a + 3a^2 - 9a + 27$
$= a^3 + 27$

83. $(y-5)(y^2 + 5y + 25)$
$= y^3 + 5y^2 + 25y - 5y^2 - 25y - 125$
$= y^3 - 125$

85. $75,000 and above because the graph for this income is the tallest.

87. Answers may vary.

89. $-12x^3y^2 + 3x^2y^2 + 15xy^2$; GCF = $-3xy^2$
$-3xy^2(4x^2 - x - 5) = -3xy^2(4x - 5)(x + 1)$

91. $-30p^3q + 88p^2q^2 + 6pq^3$; GCF $= -2pq$
$-2pq(15p^2 - 44pq - 3q^2)$
$= -2pq(15p + q)(p - 3q)$

93. $4x^2(y-1)^2 + 10x(y-1)^2 + 25(y-1)^2$
GCF $= (y-1)^2$
$(y-1)^2(4x^2 + 10x + 25)$, which can not be factored further.

Section 6.4

Calculator Explorations

	$x^2 - 2x + 1$	$x^2 - 2x - 1$	$(x-1)^2$
$x = 5$	16	14	16
$x = -3$	16	14	16
$x = 2.7$	2.89	0.89	2.89
$x = -12.1$	171.61	169.61	171.61
$x = 0$	1	-1	1

Mental Math

1. $1 = 1^2$

2. $25 = 5^2$

3. $81 = 9^2$

4. $64 = 8^2$

5. $9 = 3^2$

6. $100 = 10^2$

7. $1 = 1^3$

8. $64 = 4^3$

9. $8 = 2^3$

10. $27 = 3^3$

Exercise Set 6.4

1. $x^2 - 4 = x^2 - 2^2 = (x+2)(x-2)$

3. $y^2 - 49 = y^2 - 7^2 = (y+7)(y-7)$

5. $25y^2 - 9 = (5y)^2 - 3^2 = (5y-3)(5y+3)$

7. $121 - 100x^2 = 11^2 - (10x)^2$
$= (11 - 10x)(11 + 10x)$

9. $12x^2 - 27$; GCF $= 3$
$3(4x^2 - 9) = 3[(2x)^2 - 3^2]$
$= 3(2x-3)(2x+3)$

11. $169a^2 - 49b^2 = (13a)^2 - (7b)^2$
$= (13a - 7b)(13a + 7b)$

13. $x^2y^2 - 1 = (xy)^2 - 1^2$
$= (xy-1)(xy+1)$

15. $x^4 - 9 = (x^2)^2 - 3^2 = (x^2+3)(x^2-3)$

17. $49a^4 - 16 = (7a^2)^2 - 4^2$
$= (7a^2 + 4)(7a^2 - 4)$

19. $x^4 - y^{10} = (x^2)^2 - (y^5)^2$
$= (x^2 + y^5)(x^2 - y^5)$

21. $x + 6$ because $(x-6)(x+6) = x^2 - 36$

23. $a^3 + 27 = a^3 + 3^3 = (a+3)(a^2 - 3a + 9)$

25. $8a^3 + 1 = (2a)^3 + 1^3$
$= (2a+1)(4a^2 - 2a + 1)$

27. $5k^3 + 40$; GCF $= 5$
$5(k^3 + 8) = 5(k^3 + 2^3)$
$= 5(k+2)(k^2 - 2k + 4)$

29. $x^3y^3 - 64 = (xy)^3 - 4^3$
$= (xy-4)(x^2y^2 + 4xy + 16)$

31. $x^3+125=x^3+5^3=(x+5)(x^2-5x+25)$

33. $24x^4-81xy^3$; GCF $=3x$

$3x(8x^3-27y^3)$
$=3x[(2x)^3-(3y)^3]$
$=3x(2x-3y)(4x^2+6xy+9y^2)$

35. $(2x+y)$ because

$(2x+y)(4x^2-2xy+y^2)=(2x)^3+y^3$
$=8x^3+y^3$

37. $x^2-4=x^2-2^2=(x-2)(x+2)$

39. $81-p^2=9^2-p^2=(9-p)(9+p)$

41. $4r^2-1=(2r)^2-1^2=(2r-1)(2r+1)$

43. $9x^2-16^2=(3x)^2-4^2=(3x-4)(3x+4)$

45. $16r^2+1$

not factorable, prime

47. $27-t^3=3^3-t^3=(3-t)(9+3t+t^2)$

49. $8r^3-64$; GCF $=8$

$8(r^3-8)=8(r^3-2^3)$
$=8(r-2)(r^2+2r+4)$

51. $t^3-343=t^3-7^3=(t-7)(t^2+7t+49)$

53. $x^2-169y^2=x^2-(13y)^2$
$=(x-13y)(x+13y)$

55. $x^2y^2-z^2=(xy)^2-z^2=(xy-z)(xy+z)$

57. $x^3y^3+1=(xy)^3+1^3$
$=(xy+1)(x^2y^2-xy+1)$

59. $s^3-64t^3=s^3-(4t)^3$
$=(s-4t)(s^2+4st+16t^2)$

61. $18r^2-8$; GCF $=2$

$2(9r^2-4)=2[(3r)^2-2^2]$
$=2(3r-2)(3r+2)$

63. $9xy^2-4x$; GCF $=x$

$x(9y^2-4)=x[(3y)^2-2^2]$
$=x(3y+2)(3y-2)$

65. $25y^4-100y^2$; GCF $=25y^2$

$25y^2(y^2-4)=25y^2(y^2-2^2)$
$=25y^2(y+2)(y-2)$

67. x^3y-4xy^3; GCF $=xy$

$xy(x^2-4y^2)=xy[x^2-(2y)^2]$
$=xy(x-2y)(x+2y)$

69. $8s^6t^3+100s^3t^6$; GCF $=4s^3t^3$

$4s^3t^3(2s^3+25t^3)$

71. $27x^2y^3-xy^2$; GCF $=xy^2$

$xy^2(27xy-1)$

73. $f(t)=841-16t^2$

a. Let $t=2$.

$f(2)=841-16(2)^2$
$=841-16(4)$
$=841-64$
$=777$

After 2 seconds, the height of the object is 777 feet.

b. Let $t=5$.

$f(5)=841-16(5)^2$
$=841-16(25)$
$=841-400$
$=441$

After 5 seconds, the height of the object is 441 feet.

c. When the object hits the ground, its height is zero feet. Thus, to find the time, t, when the object's height is zero feet above the ground, we set the expression $841-16t^2$ equal to 0 and solve for t.

$$841-16t^2=0$$
$$841-16t^2+16t^2=0+16t^2$$
$$841=16t^2$$
$$\frac{841}{16}=\frac{16t^2}{16}$$
$$52.5625=t^2$$
$$\sqrt{52.5625}=\sqrt{t^2}$$
$$7.25=t$$

Thus, the object will hit the ground after approximately 7 seconds.

d. $841-16t^2=29^2-(4t)^2$
$=(29+4t)(29-4t)$

75. Answers may vary.

77. $\dfrac{8x^4+4x^3-2x+6}{2x}=\dfrac{8x^4}{2x}+\dfrac{4x^3}{2x}-\dfrac{2x}{2x}+\dfrac{6}{2x}$
$=4x^3+2x^2-1+\dfrac{3}{x}$

79.
$$\begin{array}{r} 2x+1 \\ x-2\overline{)2x^2-3x-2} \\ \underline{2x^2-4x} \\ x-2 \\ \underline{x-2} \\ 0 \end{array}$$

$2x+1$

81.
$$\begin{array}{r} 3x+4 \\ x+3\overline{)3x^2+13x+10} \\ \underline{3x^2+\ 9x} \\ 4x+10 \\ \underline{4x+12} \\ -2 \end{array}$$

$3x+4-\dfrac{2}{x+3}$

83. $a^2-(2+b)^2=[a+(2+b)][a-(2+b)]$
$=(a+2+b)(a-2-b)$

85. $(x^2-4)^2-(x-2)^2$
$=[(x^2-4)+(x-2)][(x^2-4)-(x-2)]$
$=(x^2-4+x-2)(x^2-4-x+2)$
$=(x^2+x-6)(x^2-x-2)$
$=(x+3)(x-2)(x-2)(x+1)$
$=(x-2)^2(x+1)(x+3)$

Exercise Set 6.5

1. $a^2+2ab+b^2=(a+b)^2$

3. a^2+a-12
The product is –12, and the sum is 1: –3 and 4.
$(a-3)(a+4)$

5. a^2-a-6
The product is –6, and the sum is –1: –3 and 2.
$(a+2)(a-3)$

7. $x^2+2x+1=(x)^2+2(x)(1)+(1)^2$
$=(x+1)^2$

9. x^2+4x+3
The product is 3, and the sum is 4: 3 and 1.
$(x+1)(x+3)$

11. $x^2+7x+12$
The product is 12, and the sum is 7: 3 and 4.
$(x+3)(x+4)$

13. x^2+3x-4
The product is –4, and the sum is 3: 4 and –1.
$(x+4)(x-1)$

15. $x^2+2x-15$
The product is –15, and the sum is 2: 5 and –3.
$(x+5)(x-3)$

17. x^2-x-30
The product is –30, and the sum is –1: –6 and 5.
$(x-6)(x+5)$

19. $2x^2 - 98$; GCF = 2
$2(x^2 - 49) = 2(x^2 - 7^2) = 2(x - 7)(x + 7)$

21. $x^2 + 3x + xy + 3y = x(x + 3) + y(x + 3)$
$= (x + 3)(x + y)$

23. $x^2 + 6x - 16$
The product is -16, and the sum is 6: 8 and -2.
$(x + 8)(x - 2)$

25. $4x^3 + 20x^2 - 56x$; GCF = $4x$
$4x(x^2 + 5x - 14) = 4x(x + 7)(x - 2)$

27. $12x^2 + 34x + 24$; GCF = 2
$2(6x^2 + 17x + 12) = 2(3x + 4)(2x + 3)$

29. $4a^2 - b^2 = (2a)^2 - b^2 = (2a - b)(2a + b)$

31. $20 - 3x - 2x^2$
$20 = 5 \cdot 4$
$-2x^2 = -2x \cdot x$
$5 \cdot x + 4(-2x) = -3x$
$(5 - 2x)(4 + x)$

33. $a^2 + a - 3$
not factorable, prime

35. $4x^2 - x - 5$
$4x^2 = 4x \cdot x$
$-5 = -5 \cdot 1$
$4x \cdot 1 + x(-5) = -x$
$(4x - 5)(x + 1)$

37. $4t^2 + 36$; GCF = 4
$4(t^2 + 9)$

39. $ax + 2x + a + 2 = x(a + 2) + 1(a + 2)$
$= (x + 1)(a + 2)$

41. $12a^3 - 24a^2 + 4a$; GCF = $4a$
$4a(3a^2 - 6a + 1)$

43. $x^2 - 14x - 48$
not factorable, prime

45. $25p^2 - 70pq + 49q^2 = (5p)^2 - 2(5p)(7q) + (7q)^2$
$= (5p - 7q)^2$

47. $125 - 8y^3 = 5^3 - (2y)^3$
$= (5 - 2y)(25 + 10y + 4y^2)$

49. $-x^2 - x + 30 = 30 - x - x^2 = (5 - x)(6 + x)$

51. $14 + 5x - x^2$
$14 = 7 \cdot 2$
$-x^2 = -x \cdot x$
$7 \cdot x + 2(-x) = 5x$
$(7 - x)(2 + x)$

53. $3x^4y + 6x^3y - 72x^2y$; GCF = $3x^2y$
$3x^2y(x^2 + 2x - 24) = 3x^2y(x + 6)(x - 4)$

55. $5x^3y^2 - 40x^2y^3 + 35xy^4$; GCF = $5xy^2$
$5xy^2(x^2 - 8xy + 7y^2)$
$= 5xy^2(x - 7y)(x - y)$

57. $12x^3y + 243xy$; GCF = $3xy$
$3xy(4x^2 + 81)$

59. $(x - y)^2 - z^2 = [(x - y) - z][(x - y) + z]$
$= (x - y - z)(x - y + z)$

61. $3rs - s + 12r - 4 = s(3r - 1) + 4(3r - 1)$
$= (s + 4)(3r - 1)$

63. $4x^2 - 8xy - 3x + 6y = 4x(x - 2y) - 3(x - 2y)$
$= (4x - 3)(x - 2y)$

65. $6x^2 + 18xy + 12y^2$; GCF = 6
$6(x^2 + 3xy + 2y^2) = 6(x + 2y)(x + y)$

67. $xy^2 - 4x + 3y^2 - 12 = x(y^2 - 4) + 3(y^2 - 4)$
$= (y^2 - 4)(x + 3)$
$= (y^2 - 2^2)(x + 3)$
$= (y + 2)(y - 2)(x + 3)$

69. $5(x + y) + x(x + y) = (5 + x)(x + y)$

71. $14t^2 - 9t + 1$
$14t^2 = 7t \cdot 2t$
$1 = -1(-1)$
$7t(-1) + 2t(-1) = -9t$
$(7t - 1)(2t - 1)$

73. $3x^2 + 2x - 5$
$3x^2 = 3x \cdot x$
$-5 = 5(-1)$
$3x(-1) + x \cdot 5 = 2x$
$(3x + 5)(x - 1)$

75. $x^2 + 9xy - 36y^2$
$x^2 = x \cdot x$
$-36y^2 = 12y(-3y)$
$x(-3y) + x \cdot 12y = 9xy$
$(x + 12y)(x - 3y)$

77. $1 - 8ab - 20a^2b^2 = (1 - 10ab)(1 + 2ab)$

79. $x^4 - 10x^2 + 9 = (x^2 - 1)(x^2 - 9)$
$= (x^2 - 1^2)(x^2 - 3^2)$
$= (x + 1)(x - 1)(x + 3)(x - 3)$

81. $x^4 - 14x^2 - 32 = (x^2 - 16)(x^2 + 2)$
$= (x^2 - 4^2)(x^2 + 2)$
$= (x - 4)(x + 4)(x^2 + 2)$

83. $x^2 - 23x + 120$
The product is 120, and the sum is –23: –15 and –8.
$(x - 15)(x - 8)$

85. $6x^3 - 28x^2 + 16x$; GCF = $2x$
$2x(3x^2 - 14x + 8) = 2x(3x - 2)(x - 4)$

87. $27x^3 - 125y^3$
$= (3x)^3 - (5y)^3$
$= (3x - 5y)(9x^2 + 15xy + 25y^2)$

89. $x^3y^3 + 8z^3 = (xy)^3 + (2z)^3$
$= (xy + 2z)(x^2y^2 - 2xyz + 4z^2)$

91. $2xy - 72x^3y$; GCF = $2xy$
$2xy(1 - 36x^2) = 2xy(1 - 6x)(1 + 6x)$

93. $x^3 + 6x^2 - 4x - 24 = x^2(x + 6) - 4(x + 6)$
$= (x + 6)(x^2 - 4)$
$= (x + 6)(x + 2)(x - 2)$

95. $6a^3 + 10a^2$; GCF = $2a^2$
$2a^2(3a + 5)$

97. $a^2(a + 2) + 2(a + 2) = (a^2 + 2)(a + 2)$

99. $x^3 - 28 + 7x^2 - 4x = x^3 + 7x^2 - 4x - 28$
$= x^2(x + 7) - 4(x + 7)$
$= (x + 7)(x^2 - 4)$
$= (x + 7)(x + 2)(x - 2)$

101. Answers may vary.

103.
$$x - 6 = 0$$
$$x - 6 + 6 = 0 + 6$$
$$x = 6$$

105.
$$2m + 4 = 0$$
$$2m + 4 - 4 = 0 - 4$$
$$2m = -4$$
$$\frac{2m}{2} = \frac{-4}{2}$$
$$m = -2$$

107. $5z - 1 = 0$
$5z - 1 + 1 = 0 + 1$
$5z = 1$
$\frac{5z}{5} = \frac{1}{5}$
$z = \frac{1}{5}$

109. $V = lwh$
$960 = (12)(x)(10)$
$960 = 120x$
$\frac{960}{120} = \frac{120x}{120}$
$8 = x$
8 inches

111. x-intercepts: $(-2, 0)$;
y-intercepts: $(4, 0)$, $(0, 2)$, $(0, -2)$

113. x-intercepts: $(2, 0)$, $(4, 0)$;
y-intercept: $(0, 4)$

Section 6.6

Mental Math

1. $(a - 3)(a - 7) = 0$
$a - 3 = 0$ or $a - 7 = 0$
$a = 3$ or $a = 7$

2. $(a - 5)(a - 2) = 0$
$a - 5 = 0$ or $a - 2 = 0$
$a = 5$ or $a = 2$

3. $(x + 8)(x + 6) = 0$
$x + 8 = 0$ or $x + 6 = 0$
$x = -8$ or $x = -6$

4. $(x + 2)(x + 3) = 0$
$x + 2 = 0$ or $x + 3 = 0$
$x = -2$ or $x = -3$

5. $(x + 1)(x - 3) = 0$
$x + 1 = 0$ or $x - 3 = 0$
$x = -1$ or $x = 3$

6. $(x - 1)(x + 2) = 0$
$x - 1 = 0$ or $x + 2 = 0$
$x = 1$ or $x = -2$

Exercise Set 6.6

1. $(x - 2)(x + 1) = 0$
$x - 2 = 0$ or $x + 1 = 0$
$x = 2$ or $x = -1$

3. $x(x + 6) = 0$
$x = 0$ or $x + 6 = 0$
$x = 0$ or $x = -6$

5. $(2x + 3)(4x - 5) = 0$
$2x + 3 = 0$ or $4x - 5 = 0$
$2x = -3$ or $4x = 5$
$x = -\frac{3}{2}$ or $x = \frac{5}{4}$

7. $(2x - 7)(7x + 2) = 0$
$2x - 7 = 0$ or $7x + 2 = 0$
$2x = 7$ or $7x = -2$
$x = \frac{7}{2}$ or $x = -\frac{2}{7}$

9. $x = 6$ or $x = -1$
$x - 6 = 0$ or $x + 1 = 0$
When $x = -1$, $x + 1 = 0$.
$(x - 6)(x + 1) = 0$

11. $x^2 - 13x + 36 = 0$
$(x - 4)(x - 9) = 0$
$x - 4 = 0$ or $x - 9 = 0$
$x = 4$ or $x = 9$

13. $x^2 + 2x - 8 = 0$
$(x + 4)(x - 2) = 0$
$x + 4 = 0$ or $x - 2 = 0$
$x = -4$ or $x = 2$

15. $x^2 - 4x = 32$
$x^2 - 4x - 32 = 0$
$(x - 8)(x + 4) = 0$
$x - 8 = 0$ or $x + 4 = 0$
$x = 8$ or $x = -4$

17.
$$x(3x-1)=14$$
$$3x^2-x=14$$
$$3x^2-x-14=0$$
$$(3x-7)(x+2)=0$$
$$3x-7=0 \quad \text{or} \quad x+2=0$$
$$3x=7 \quad \text{or} \quad x=-2$$
$$x=\frac{7}{3} \quad \text{or} \quad x=-2$$

19. $3x^2+19x-72=0$
$$(3x-8)(x+9)=0$$
$$3x-8=0 \quad \text{or} \quad x+9=0$$
$$3x=8 \quad \text{or} \quad x=-9$$
$$x=\frac{8}{3} \quad \text{or} \quad x=-9$$

21. Two solutions, 5 and 7.
$$x=5 \quad \text{or} \quad x=7$$
$$x-5=0 \quad \text{or} \quad x-7=0$$
$$(x-5)(x-7)=0$$
$$x^2-7x-5x+35=0$$
$$x^2-12x+35=0$$

23. $x^3-12x^2+32x=0$
$$x(x^2-12x+32)=0$$
$$x(x-8)(x-4)=0$$
$$x=0 \quad \text{or} \quad x-8=0 \quad \text{or} \quad x-4=0$$
$$x=0 \quad \text{or} \quad x=8 \quad \text{or} \quad x=4$$

25. $(4x-3)(16x^2-24x+9)=0$
$$(4x-3)(4x-3)(4x-3)=0$$
Since all factors are the same:
$$4x-3=0$$
$$4x=3$$
$$x=\frac{3}{4}$$

27.
$$4x^3-x=0$$
$$x(4x^2-1)=0$$
$$x(2x+1)(2x-1)=0$$
$$x=0 \quad \text{or} \quad 2x+1=0 \quad \text{or} \quad 2x-1=0$$
$$x=0 \quad \text{or} \quad 2x=-1 \quad \text{or} \quad 2x=1$$
$$x=0 \quad \text{or} \quad x=-\frac{1}{2} \quad \text{or} \quad x=\frac{1}{2}$$

29. $32x^3-4x^2-6x=0$
$$2x(16x^2-2x-3)=0$$
$$2x(2x-1)(8x+3)=0$$
$$2x=0 \quad \text{or} \quad 2x-1=0 \quad \text{or} \quad 8x+3=0$$
$$x=\frac{0}{2} \quad \text{or} \quad 2x=1 \quad \text{or} \quad 8x=-3$$
$$x=0 \quad \text{or} \quad x=\frac{1}{2} \quad \text{or} \quad x=-\frac{3}{8}$$

31. $x(x+7)=0$
$$x=0 \quad \text{or} \quad x+7=0$$
$$x=0 \quad \text{or} \quad x=-7$$

33. $(x+5)(x-4)=0$
$$x+5=0 \quad \text{or} \quad x-4=0$$
$$x=-5 \quad \text{or} \quad x=4$$

35.
$$x^2-x=30$$
$$x^2-x-30=0$$
$$(x-6)(x+5)=0$$
$$x-6=0 \quad \text{or} \quad x+5=0$$
$$x=6 \quad \text{or} \quad x=-5$$

37.
$$6y^2-22y-40=0$$
$$2(3y^2-11y-20)=0$$
$$2(3y+4)(y-5)=0$$
$$3y+4=0 \quad \text{or} \quad y-5=0$$
$$3y=-4 \quad \text{or} \quad y=5$$
$$y=-\frac{4}{3} \quad \text{or} \quad y=5$$

39. $(2x+3)(2x^2-5x-3)=0$
$$(2x+3)(2x+1)(x-3)=0$$
$$2x+3=0 \quad \text{or} \quad 2x+1=0 \quad \text{or} \quad x-3=0$$
$$2x=-3 \quad \text{or} \quad 2x=-1 \quad \text{or} \quad x=3$$
$$x=-\frac{3}{2} \quad \text{or} \quad x=-\frac{1}{2} \quad \text{or} \quad x=3$$

41.
$$x^2-15=-2x$$
$$x^2+2x-15=0$$
$$(x+5)(x-3)=0$$
$$x+5=0 \quad \text{or} \quad x-3=0$$
$$x=-5 \quad \text{or} \quad x=3$$

43. $x^2 - 16x = 0$
$x(x - 16) = 0$
$x = 0$ or $x - 16 = 0$
$x = 0$ or $x = 16$

45. $-18y^2 - 33y + 216 = 0$
$-3(6y^2 + 11y - 72) = 0$
$-3(3y - 8)(2y + 9) = 0$
$3y - 8 = 0$ or $2y + 9 = 0$
$3y = 8$ or $2y = -9$
$y = \frac{8}{3}$ or $y = -\frac{9}{2}$

47. $12x^2 - 59x + 55 = 0$
$(4x - 5)(3x - 11) = 0$
$4x - 5 = 0$ or $3x - 11 = 0$
$4x = 5$ or $3x = 11$
$x = \frac{5}{4}$ or $x = \frac{11}{3}$

49. $18x^2 + 9x - 2 = 0$
$(3x + 2)(6x - 1) = 0$
$3x + 2 = 0$ or $6x - 1 = 0$
$3x = -2$ or $6x = 1$
$x = -\frac{2}{3}$ or $x = \frac{1}{6}$

51. $x(6x + 7) = 5$
$6x^2 + 7x = 5$
$6x^2 + 7x - 5 = 0$
$(3x + 5)(2x - 1) = 0$
$3x + 5 = 0$ or $2x - 1 = 0$
$3x = -5$ or $2x = 1$
$x = -\frac{5}{3}$ or $x = \frac{1}{2}$

53. $4(x - 7) = 6$
$4x - 28 = 6$
$4x = 6 + 28$
$4x = 34$
$x = \frac{34}{4} = \frac{17}{2}$

55. $5x^2 - 6x - 8 = 0$
$(5x + 4)(x - 2) = 0$
$5x + 4 = 0$ or $x - 2 = 0$
$5x = -4$ or $x = 2$
$x = -\frac{4}{5}$ or $x = 2$

57. $(y - 2)(y + 3) = 6$
$y^2 + 3y - 2y - 6 = 6$
$y^2 + y - 6 = 6$
$y^2 + y - 6 - 6 = 0$
$y^2 + y - 12 = 0$
$(y + 4)(y - 3) = 0$
$y + 4 = 0$ or $y - 3 = 0$
$y = -4$ or $y = 3$

59. $4y^2 - 1 = 0$
$(2y + 1)(2y - 1) = 0$
$2y + 1 = 0$ or $2y - 1 = 0$
$2y = -1$ or $2y = 1$
$y = -\frac{1}{2}$ or $y = \frac{1}{2}$

61. $t^2 + 13t + 22 = 0$
$(t + 11)(t + 2) = 0$
$t + 11 = 0$ or $t + 2 = 0$
$t = -11$ or $t = -2$

63. $5t - 3 = 12$
$5t = 12 + 3$
$5t = 15$
$t = \frac{15}{5}$
$t = 3$

65. $x^2 + 6x - 17 = -26$
$x^2 + 6x - 17 + 26 = 0$
$x^2 + 6x + 9 = 0$
$(x + 3)(x + 3) = 0$
$x + 3 = 0$
$x = -3$

67. $12x^2 + 7x - 12 = 0$
$(3x+4)(4x-3) = 0$
$3x + 4 = 0$ or $4x - 3 = 0$
$3x = -4$ or $4x = 3$
$x = -\frac{4}{3}$ or $x = \frac{3}{4}$

69. $10t^3 - 25t - 15t^2 = 0$
$10t^3 - 15t^2 - 25t = 0$
$5t(2t^2 - 3t - 5) = 0$
$5t(2t-5)(t+1) = 0$
$5t = 0$ or $2t - 5 = 0$ or $t + 1 = 0$
$t = \frac{0}{5}$ or $2t = 5$ or $t = -1$
$t = 0$ or $t = \frac{5}{2}$ or $t = -1$

71. Let $y = 0$ and solve for x.
$y = (3x+4)(x-1)$
$0 = (3x+4)(x-1)$
$3x + 4 = 0$ or $x - 1 = 0$
$3x = -4$ or $x = 1$
$x = -\frac{4}{3}$
The x-intercepts are $\left(-\frac{4}{3}, 0\right)$ and $(1, 0)$.

73. Let $y = 0$ and solve for x.
$y = x^2 - 3x - 10$
$0 = x^2 - 3x - 10$
$0 = (x-5)(x+2)$
$x - 5 = 0$ or $x + 2 = 0$
$x = 5$ or $x = -2$
The x-intercepts are $(5, 0)$ and $(-2, 0)$.

75. Let $y = 0$ and solve for x.
$y = 2x^2 + 11x - 6$
$0 = 2x^2 + 11x - 6$
$0 = (2x-1)(x+6)$
$2x - 1 = 0$ or $x + 6 = 0$
$2x = 1$ or $x = -6$
$x = \frac{1}{2}$
The x-intercepts are $\left(\frac{1}{2}, 0\right)$ and $(-6, 0)$.

77. E; x-intercepts are $(-2, 0)$, $(1, 0)$.

79. B; x-intercepts are $(0, 0)$, $(-3, 0)$.

81. C; $y = 2x^2 - 8 = 2(x-2)(x+2)$
x-intercepts are $(2, 0)$ and $(-2, 0)$

83. **a.** $y = -16x^2 + 20x + 300$

x	0	1	2	3	4	5	6
y	300	304	276	216	124	0	−156

b. At 5 seconds, when $y = 0$.

c. 304 feet

d.

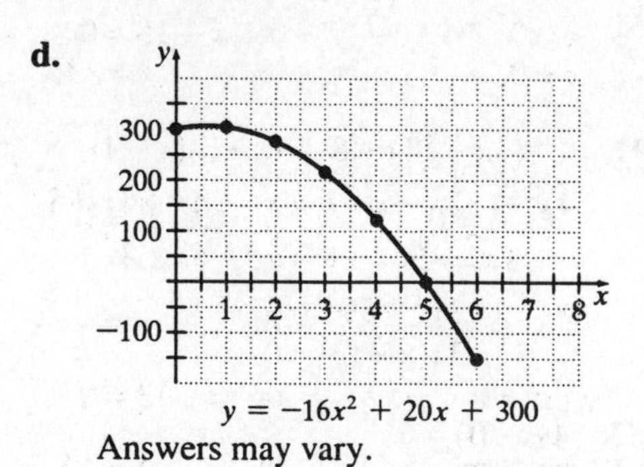

Answers may vary.

85. $\frac{3}{5}+\frac{4}{9}=\frac{3}{5}\cdot\frac{9}{9}+\frac{4}{9}\cdot\frac{5}{5}=\frac{27}{45}+\frac{20}{45}=\frac{47}{45}$

87. $\frac{7}{10}-\frac{5}{12}=\frac{7}{10}\cdot\frac{6}{6}-\frac{5}{12}\cdot\frac{5}{5}=\frac{42}{60}-\frac{25}{60}=\frac{17}{60}$

89. $\frac{7}{8}\div\frac{7}{15}=\frac{7}{8}\cdot\frac{15}{7}=\frac{15}{8}$

91. $\frac{4}{5}\cdot\frac{7}{8}=\frac{4\cdot 7}{5\cdot 8}=\frac{28}{40}=\frac{7}{10}$

93.
$$(x-3)(3x+4)=(x+2)(x-6)$$
$$3x^2+4x-9x-12=x^2-6x+2x-12$$
$$3x^2-5x-12=x^2-4x-12$$
$$2x^2-x=0$$
$$x(2x-1)=0$$
$x=0$ or $2x-1=0$
$x=0$ or $2x=1$
$x=0$ or $x=\frac{1}{2}$

95.
$$(2x-3)(x+8)=(x-6)(x+4)$$
$$2x^2+16x-3x-24=x^2+4x-6x-24$$
$$2x^2+13x-24=x^2-2x-24$$
$$x^2+15x=0$$
$$x(x+15)=0$$
$x=0$ or $x+15=0$
$x=0$ or $x=-15$

97.
$$(4x-1)(x-8)=(x+2)(x+4)$$
$$4x^2-32x-x+8=x^2+4x+2x+8$$
$$4x^2-33x+8=x^2+6x+8$$
$$3x^2-39x=0$$
$$3x(x-13)=0$$
$3x=0$ or $x-13=0$
$\frac{3x}{3}=\frac{0}{3}$ or $x=13$
$x=0$ or $x=13$

Section 6.7

Graphing Calculator Explorations

1.

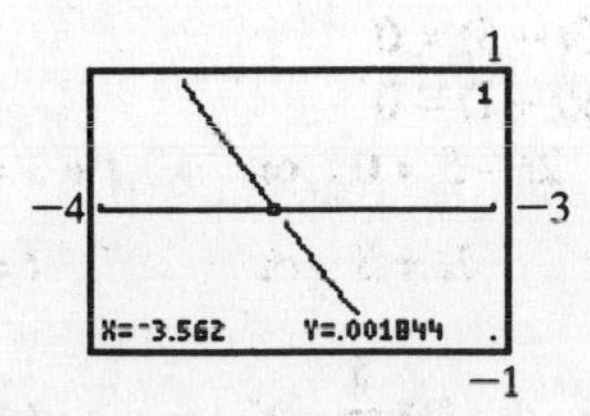

The intercepts are 0.562, –3.562.

3.

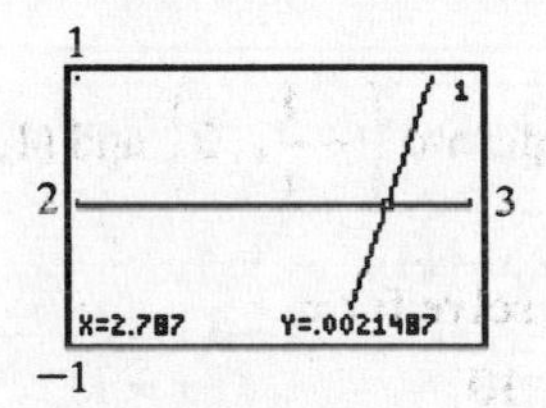

The intercepts are –0.874, 2.787.

5.

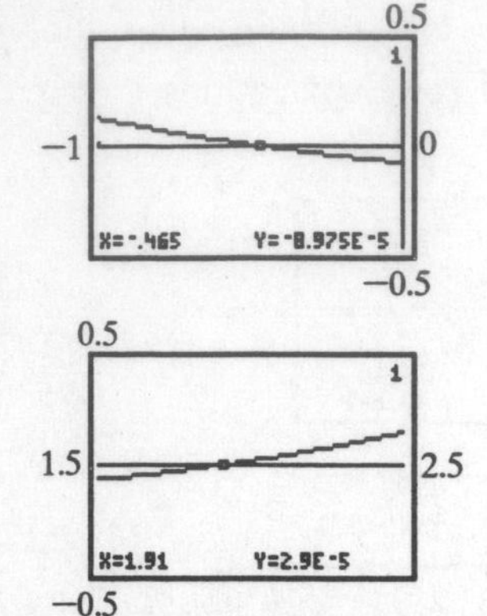

The intercepts are −0.465, 1.910.

Exercise Set 6.7

1. Let x = the width. Then $x + 4$ = the length.

3. Let x = the first odd integer. Then $x + 2$ = the next consecutive odd integer.

5. Let x = the base. Then $4x + 1$ = the height.

7. Let x = the length of one side.
$$A = s^2$$
$$121 = x^2$$
$$0 = x^2 - 121$$
$$0 = x^2 - 11^2$$
$$0 = (x + 11)(x - 11)$$
$$x + 11 = 0 \quad \text{or} \quad x - 11 = 0$$
$$x = -11 \qquad\qquad x = 11$$
Since the length cannot be negative, the sides are 11 units long.

9. The perimeter is the sum of the lengths of the sides.
$$120 = (x + 5) + (x^2 - 3x) + (3x - 8) + (x + 3)$$
$$120 = x + 5 + x^2 - 3x + 3x - 8 + x + 3$$
$$120 = x^2 + 2x$$
$$0 = x^2 + 2x - 120$$
$$0 = (x + 12)(x - 10)$$
$$x + 12 = 0 \quad \text{or} \quad x - 10 = 0$$
$$x = -12 \qquad\qquad x = 10$$
Since the dimensions cannot be negative, the lengths of the sides are: $10 + 5 = 15$ cm, $10^2 - 3(10) = 70$ cm, $3(10) - 8 = 22$ cm, and $10 + 3 = 13$ cm.

11. $x + 5$ = the base, and $x - 5$ = the height.
$$A = bh$$
$$96 = (x + 5)(x - 5)$$
$$96 = x^2 + 5x - 5x - 25$$
$$96 = x^2 - 25$$
$$0 = x^2 - 121$$
$$0 = (x + 11)(x - 11)$$
$$x + 11 = 0 \quad \text{or} \quad x - 11 = 0$$
$$x = -11 \qquad\qquad x = 11$$
Since the dimensions cannot be negative, $x = 11$. The base is $11 + 5 = 16$ miles, and the height is $11 - 5 = 6$ miles.

13. Find t when $h = 0$.
$$h = -16t^2 + 64t + 80$$
$$0 = -16t^2 + 64t + 80$$
$$0 = -16(t^2 - 4t - 5)$$
$$0 = -16(t - 5)(t + 1)$$
$$t - 5 = 0 \quad \text{or} \quad t + 1 = 0$$
$$t = 5 \qquad\qquad t = -1$$
Since the time t cannot be negative, the object hits the ground after 5 seconds.

15. Let x = the width. Then $2x - 7$ = the length.

$A = lw$
$30 = (2x-7)(x)$
$30 = 2x^2 - 7x$
$0 = 2x^2 - 7x - 30$
$0 = (2x+5)(x-6)$
$2x + 5 = 0$ or $x - 6 = 0$
$2x = -5$ $\quad x = 6$
$x = -\frac{5}{2}$

Since the dimensions cannot be negative, the width is 6 centimeters and the length is $2(6) - 7 = 5$ centimeters.

17. Let $n = 12$.

$D = \frac{1}{2}n(n-3)$

$D = \frac{1}{2} \cdot 12(12-3) = 6(9) = 54$

A polygon with 12 sides has 54 diagonals.

19. Let $D = 35$ and solve for n.

$D = \frac{1}{2}n(n-3)$

$35 = \frac{1}{2}n(n-3)$

$35 = \frac{1}{2}n^2 - \frac{3}{2}n$

$0 = \frac{1}{2}n^2 - \frac{3}{2}n - 35$

$0 = \frac{1}{2}(n^2 - 3n - 70)$

$0 = \frac{1}{2}(n-10)(n+7)$

$n - 10 = 0$ or $n + 7 = 0$
$n = 10$ $\quad n = -7$

The polygon has 10 sides.

21. Let x = the unknown number.

$x + x^2 = 132$
$x^2 + x - 132 = 0$
$(x+12)(x-11) = 0$
$x + 12 = 0$ or $x - 11 = 0$
$x = -12$ $\quad x = 11$

There are two numbers: -12 and 11.

23. Let x = the rate (in mph) of the slower boat. Then $x + 7$ = the rate (in mph) of the faster boat. After one hour, the slower boat has traveled x miles and the faster boat has traveled $x + 7$ miles. By the Pythagorean theorem,

$x^2 + (x+7)^2 = 17^2$
$x^2 + x^2 + 14x + 49 = 289$
$2x^2 + 14x + 49 = 289$
$2x^2 + 14x - 240 = 0$
$2(x^2 + 7x - 120) = 0$
$2(x+15)(x-8) = 0$
$x + 15 = 0$ or $x - 8 = 0$
$x = -15$ $\quad x = 8$

Since the rate cannot be negative, the slower boat travels at 8 mph. The faster boat travels at $8 + 7 = 15$ mph.

25. Let x = the first number. Then $20 - x$ = the other number.

$x^2 + (20-x)^2 = 218$
$x^2 + 400 - 40x + x^2 = 218$
$2x^2 - 40x + 400 = 218$
$2x^2 - 40x + 182 = 0$
$2(x^2 - 20x + 91) = 0$
$2(x-13)(x-7) = 0$
$x - 13 = 0$ or $x - 7 = 0$
$x = 13$ $\quad x = 7$

The numbers are 13 and 7.

27. Let x = the length of a side of the original square. Then $x + 3$ = the length of a side of the larger square.

$64 = (x+3)^2$
$64 = x^2 + 6x + 9$
$0 = x^2 + 6x - 55$
$0 = (x+11)(x-5)$
$x + 11 = 0$ or $x - 5 = 0$
$x = -11$ $\quad x = 5$

Since the length cannot be negative, the sides of the original square are 5 inches long.

29. Let x = the length of the shorter leg. Then $x+4$ = the length of the longer leg and $x+8$ = the length of the hypotenuse.
By the Pythagorean theorem

$$x^2+(x+4)^2=(x+8)^2$$
$$x^2+x^2+8x+16=x^2+16x+64$$
$$2x^2+8x+16=x^2+16x+64$$
$$x^2-8x-48=0$$
$$(x-12)(x+4)=0$$
$$x-12=0 \quad \text{or} \quad x+4=0$$
$$x=12 \qquad x=-4$$

Since the length cannot be negative, the sides of the triangle are 12 mm, 12 + 4 = 16 mm, and 12 + 8 = 20 mm.

31. Let x = the height of the triangle. Then $2x$ = the base.

$$A=\frac{1}{2}bh$$
$$100=\frac{1}{2}(2x)(x)$$
$$100=x^2$$
$$0=x^2-100$$
$$0=(x+10)(x-10)$$
$$x+10=0 \quad \text{or} \quad x-10=0$$
$$x=-10 \qquad x=10$$

Since the altitude cannot be negative, the height of the triangle is 10 kilometers.

33. Let x = the length of the shorter leg. Then $x+12$ = the length of the longer leg and $2x-12$ = the length of the hypotenuse.
By the Pythagorean theorem,

$$x^2+(x+12)^2=(2x-12)^2$$
$$x^2+x^2+24x+144=4x^2-48x+144$$
$$2x^2+24x+144=4x^2-48x+144$$
$$0=2x^2-72x$$
$$0=2x(x-36)$$
$$2x=0 \quad \text{or} \quad x-36=0$$
$$x=0 \qquad x=36$$

Since the length cannot be zero feet, the shorter leg is 36 feet long.

35. Find t when $h=0$.

$$h=-16t^2+625$$
$$0=-16t^2+625$$
$$0=-(16t^2-625)$$
$$0=-(4t+25)(4t-25)$$
$$4t+25=0 \quad \text{or} \quad 4t-25=0$$
$$4t=-25 \qquad 4t=25$$
$$t=-\frac{25}{4} \qquad t=\frac{25}{4}$$
$$t=-6.25 \qquad t=6.25$$

Since the time cannot be negative, the object will reach the ground after 6.25 seconds.

37. The sunglasses will hit the ground when the height $h(t)$ equals 0.

$$-16t^2+1600=0$$
$$-16(t^2-100)=0$$
$$-16(t-10)(t+10)=0$$
$$t-10=0 \quad \text{or} \quad t+10=0$$
$$t=10 \quad \text{or} \quad t=-10$$

The sunglasses hit the ground 10 seconds after being dropped.

39. Let the width of the floor = w. Then the length is $w+6$. So the area is

$$91=w(w+6)$$
$$91=w^2+6w$$
$$w^2+6w-91=0$$
$$(w+13)(w-7)=0$$
$$w+13=0 \quad \text{or} \quad w-7=0$$
$$w=-13 \quad \text{or} \quad w=7.$$

Since the width can't be negative, the width is 7ft and the length is 13 ft.

41.

$$0.5x^2=50$$
$$0.5x^2-50=0$$
$$x^2-100=0$$
$$(x+10)(x-10)=0$$
$$x+10=0 \quad \text{or} \quad x-10=0$$
$$x=-10 \quad \text{or} \quad x=10$$

Disregard the negative solution. A 10-inch square tier is needed, provided each person has one serving.

43. $A = P(1+r)^2$
$144 = 100(1+r)^2$
$144 = 100(1+2r+r^2)$
$144 = 100 + 200r + 100r^2$
$0 = 100r^2 + 200r - 44$
$0 = 4(25r^2 + 50r - 11)$
$0 = 4(5r-1)(5r+11)$
$5r - 1 = 0$ or $5r + 11 = 0$
$5r = 1$ or $5r = -11$
$r = \frac{1}{5}$ or $r = -\frac{11}{5}$
r cannot be negative.
$r = \frac{1}{5} = 0.20 = 20\%$

45. length $= x$
width $= x - 7$
$x(x-7) = 120$
$x^2 - 7x - 120 = 0$
$(x-15)(x+8) = 0$
$x - 15 = 0$ or $x + 8 = 0$
$x = 15$ or $x = -8$
length = 15 miles
width = 8 miles

47. $C = x^2 - 15x + 50$
$9500 = x^2 - 15x + 50$
$0 = x^2 - 15x - 9450$
$0 = (x-105)(x+90)$
$x - 105 = 0$ or $x + 90 = 0$
$x = 105$ or $x = -90$
105 units were manufactured.

49. x = length of boom
$2x + 28$ = height of mainsail
$\frac{1}{2}bh = A$
$\frac{1}{2}x(2x+28) = 0.60(3000)$
$\frac{1}{2}x(2x) + \frac{1}{2}x(28) = 1800$
$x^2 + 14x = 1800$
$x^2 + 14x - 1800 = 1800 - 1800$
$x^2 + 14x - 1800 = 0$
$(x+50)(x-36) = 0$
$x + 50 = 0$ or $x - 36 = 0$
$x = -50$ or $x = 36$
$x = 36$
$2x + 28 = 100$
boom length: 36 ft
height of mainsail: 100 ft

51. $g(x) = 0$ for $x = -1, 6$.
D

53. $F(x) = 0$ for $-1, 2, -5$.
A

55. $H(0) = -4$
C

57. Answers may vary.
Ex: $f(x) = (x-6)(x-7) = x^2 - 13x + 42$

59. Answers may vary.
Ex: $f(x) = (x-4)(x+3) = x^2 - x - 12$

61. $(-4, 0)$, $(0, 0)$, $(3, 0)$
Function; no vertical line intersects the graph in more than one point.

63. (–5, 0), (5, 0), (0, –4)
Function; no vertical line intersects the graph in more than one point.

65. Answers may vary.

Section 6.8

Mental Math

1. $f(x) = 2x^2 + 7x + 10$ opens upward because 2 is positive.

2. $f(x) = -3x^2 - 5x$ opens downward because –3 is negative.

3. $f(x) = -x^2 + 5$ opens downward because –1 is negative.

4. $f(x) = x^2 + 3x + 7$ opens upward because 1 is positive.

Exercise Set 6.8

1. a. Domain: $(-\infty, \infty)$
Range: $(-\infty, 5]$

b. x-intercepts: (–2, 0), (6, 0)
y-intercept: (0, 5)

c. (0, 5)

d. There is no such point.

e. –2, 6

f. $\{x|-2 < x < 6\}$

g. –2, 6

3. a. Domain: $(-\infty, \infty)$
Range: $[-4, \infty)$

b. x-intercepts: (–3, 0), (1, 0)
y-intercept: (0, –3)

c. There is no such point.

d. (–1, –4)

e. –3, 1

f. $\{x|x < -3 \text{ or } x > 1\}$

g. –3, 1

5. a. Domain: $(-\infty, \infty)$
Range: $(-\infty, \infty)$

b. x-intercepts: (–2, 0), (0, 0), (2, 0)
y-intercept: (0, 0)

c. There is no such point.

d. There is no such point.

e. –2, 0, 2

f. $\{x|-2 < x < 0 \text{ or } x > 2\}$

g. –2, 0, 2

7. $f(x) = 2x^2$

x	$f(x)$
–1	2
0	0
1	2

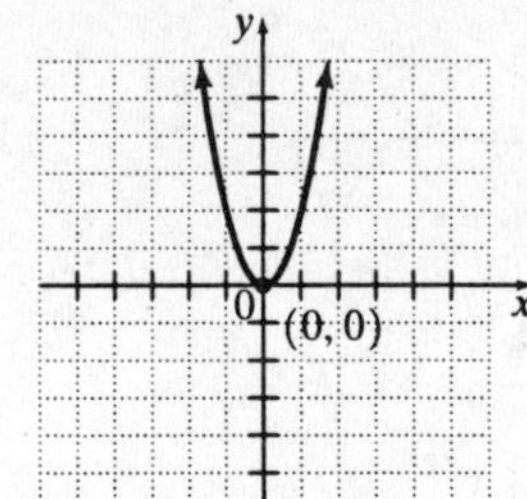

9. $f(x) = x^2 + 1$

x	$f(x)$
−2	5
0	1
2	5

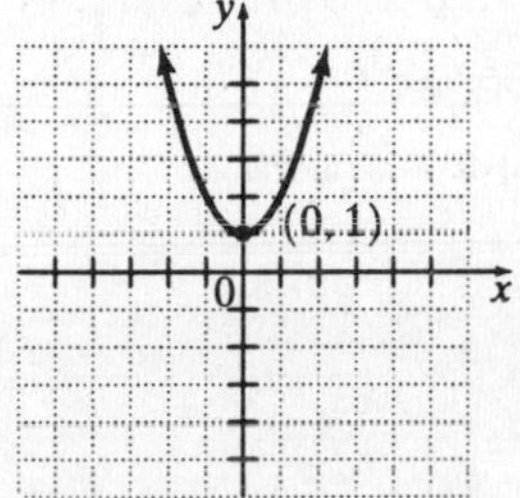

11. $f(x) = -x^2$

x	$f(x)$
−2	−4
0	0
2	−4

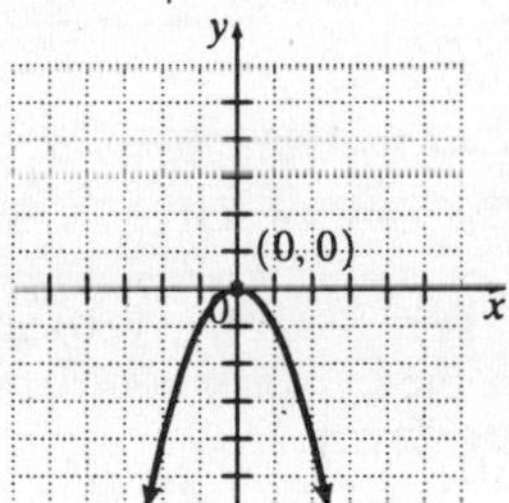

13. $f(x) = x^2 + 8x + 7$
The x-coordinate of the vertex is given by
$x = -\dfrac{b}{2a} = -\dfrac{8}{2(1)} = -4$. The y-coordinate is then $f(-4) = (-4)^2 + 8(-4) + 7 = -9$. The vertex is $(-4, -9)$

15. $f(x) = 3x^2 + 6x + 4$
The x-coordinate of the vertex is given by
$x = -\dfrac{b}{2a} = -\dfrac{6}{2(3)} = -1$. The y-coordinate is then $f(-1) = 3(-1)^2 + 6(-1) + 4 = 1$. The vertex is $(-1, 1)$.

17. $f(x) = -x^2 + 10x + 5$
The x-coordinate of the vertex is given by
$x = -\dfrac{b}{2a} = -\dfrac{10}{2(-1)} = 5$. The y-coordinate is then $f(5) = -5^2 + 10(5) + 5 = 30$. The vertex is $(5, 30)$.

19. 2, since each branch will cross the x-axis.

21. 0, since neither branch will cross the x-axis.

23. One x-intercept and one y-intercept, since the parabola will not cross the axes anywhere else.

25. $f(x) = x^2 + 6x + 5$
$x = \dfrac{-b}{2a} = \dfrac{-6}{2(1)} = -3$
$f(-3) = 9 - 18 + 5 = -4$
The vertex is $(-3, -4)$.
If $x^2 + 6x + 5 = 0$, then
$(x + 1)(x + 5) = 0$
$x + 1 = 0$ or $x + 5 = 0$
$x = -1$ or $x = -5$.
The x-intercepts are $(-1, 0)$ and $(-5, 0)$.
If $x = 0$, then $y = f(0) = 5$.
The y-intercept is $(0, 5)$.

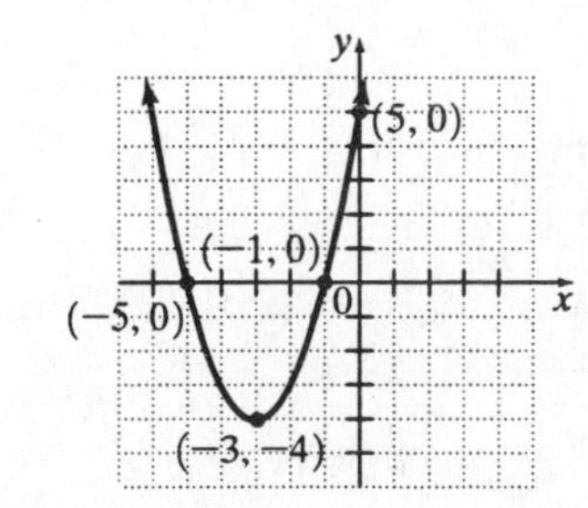

27. $f(x) = x^2 - 12x + 35$

$x = \frac{-b}{2a} = \frac{12}{2(1)} = 6$

$f(6) = 36 - 72 + 35 = -1$

The vertex is $(6, -1)$.

If $x^2 - 12x + 35 = 0$, then

$(x-5)(x-7) = 0$

$x - 5 = 0$ or $x - 7 = 0$

$x = 5$ or $x = 7$.

The x-intercepts are $(5, 0)$ and $(7, 0)$.

If $x = 0$, then $y = f(0) = 35$.

The y-intercept is $(0, 35)$.

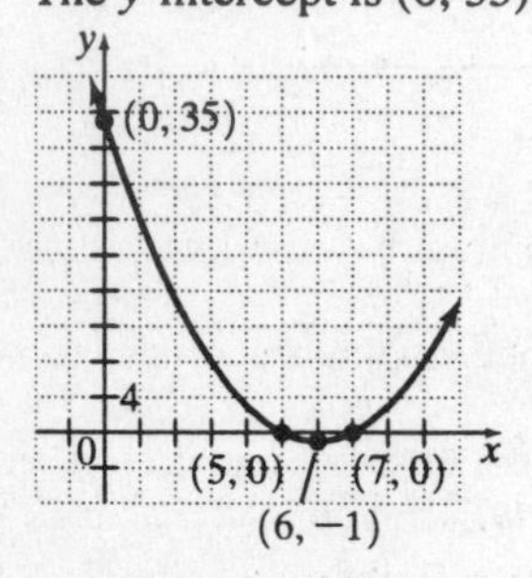

29. $f(x) = -3x^2 + 6x$

$x = \frac{-b}{2a} = \frac{-6}{2(-3)} = 1$

$f(1) = -3 + 6 = 3$

The vertex is $(1, 3)$.

If $-3x^2 + 6x = 0$, then

$3x(-x+2) = 0$

$3x = 0$ or $-x + 2 = 0$

$x = 0$ or $x = 2$.

The x-intercepts are $(0, 0)$ and $(2, 0)$.

The y-intercept is $(0, 0)$.

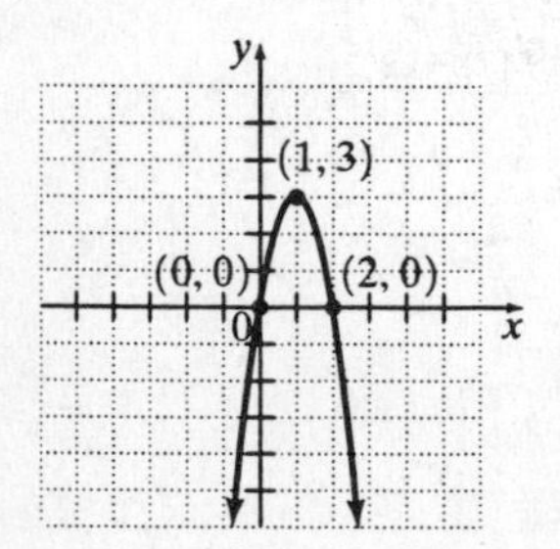

31. $f(x) = 2x^3 - 5x^2 - 3x$

If $2x^3 - 5x^2 - 3x = 0$, then

$x(2x^2 - 5x - 3) = 0$

$x(2x+1)(x-3) = 0$

$x = 0$ or $2x + 1 = 0$ or $x - 3 = 0$

$2x = -1$ or $x = 3$.

$x = -\frac{1}{2}$

The x-intercepts are $\left(-\frac{1}{2}, 0\right)$, $(0, 0)$, and $(3, 0)$.

The y-intercept is $(0, 0)$.

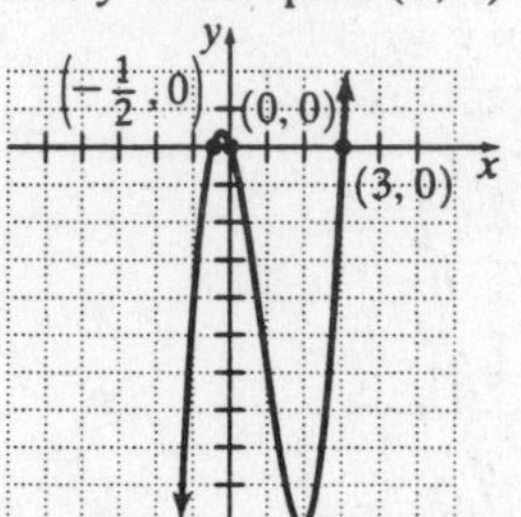

33. $f(x) = x^3 + x^2 - 4x - 4$

If $x^3 + x^2 - 4x - 4 = 0$, then

$x^2(x+1) - 4(x+1) = 0$

$(x+1)(x^2 - 4) = 0$

$(x+1)(x+2)(x-2) = 0$

$x + 1 = 0$ or $x + 2 = 0$ or $x - 2 = 0$

$x = -1$ or $x = -2$ or $x = 2$.

The x-intercepts are $(-2, 0)$, $(-1, 0)$ and $(2. 0)$. If $x = 0$, then $y = f(0) = -4$. The y-intercept is $(0, -4)$.

35. No; there are functions whose graph crosses the x-axis infinitely many times.

37. $f(x) = x^2 + 2x - 3$

$x = \frac{-b}{2a} = \frac{-2}{2(1)} = -1$

$f(-1) = 1 - 2 - 3 = -4$

The vertex is $(-1, -4)$.

If $x^2 + 2x - 3 = 0$, then

$(x-1)(x+3) = 0$

$x - 1 = 0$ or $x + 3 = 0$

$x = 1$ or $x = -3$.

The x-intercepts are $(-3, 0)$ and $(1, 0)$.

If $x = 0$, then $y = f(0) = -3$.

The y-intercept is $(0, -3)$.

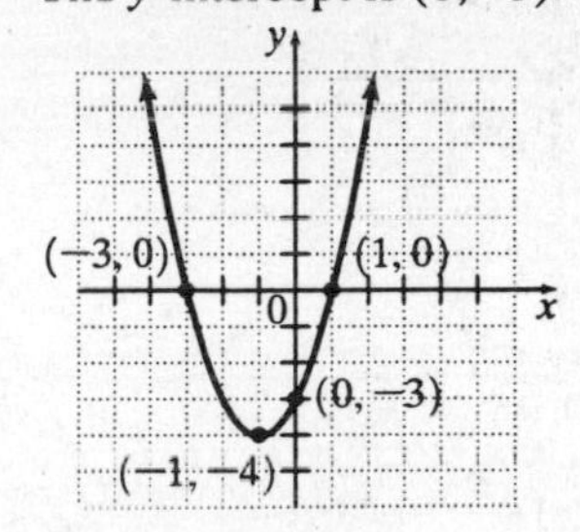

39. $f(x) = x^3 - 4x^2 + 3x$

If $x^3 - 4x^2 + 3x = 0$, then

$x(x^2 - 4x + 3) = 0$

$x(x-1)(x-3) = 0$

$x = 0$ or $x - 1 = 0$ or $x - 3 = 0$

$x = 1$ or $x = 3$.

The x-intercepts are $(0, 0)$, $(1, 0)$, and $(3, 0)$.

The y-intercept is $(0, 0)$.

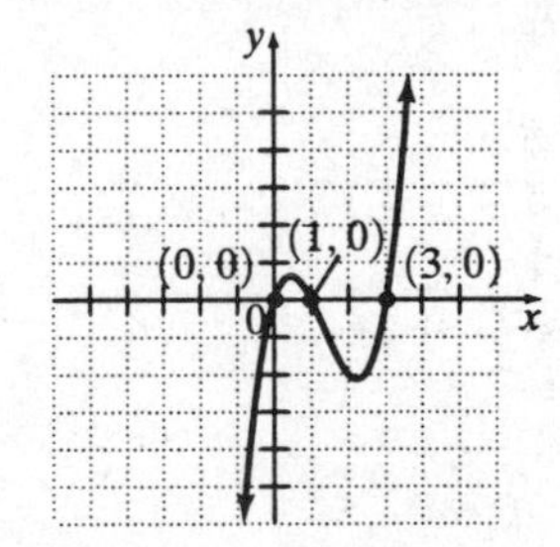

41. $f(x) = x^2 + 4$

$x = \frac{-b}{2a} = 0$

$f(0) = 4$

The vertex is $(0, 4)$.

There are no x-intercepts.

The y-intercept is $(0, 4)$.

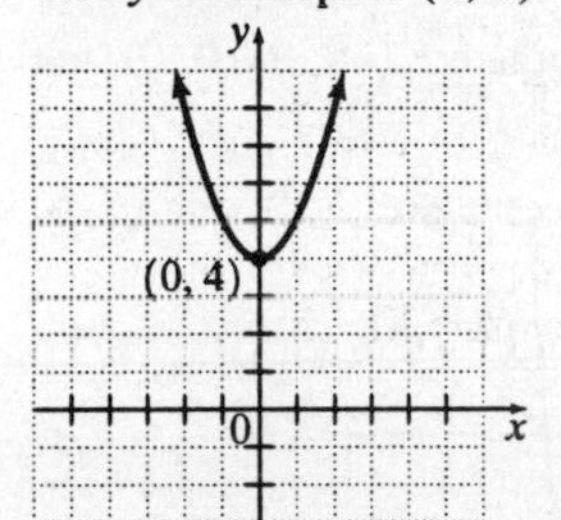

43. $f(x) = 3x^2 - 12x$

$x = \frac{-b}{2a} = \frac{12}{2(3)} = 2$

$f(2) = 12 - 24 = -12$

The vertex is $(2, -12)$.

If $3x^2 - 12x = 0$, then

$3x(x-4) = 0$

$3x = 0$ or $x - 4 = 0$

$x = 0$ or $x = 4$.

The x-intercepts are $(0, 0)$ and $(4, 0)$.

The y-intercept is $(0, 0)$.

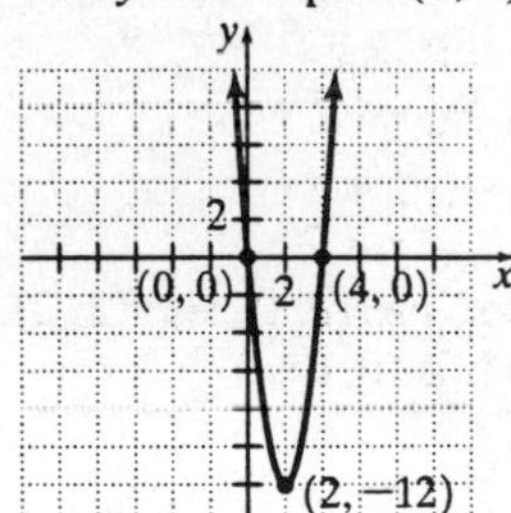

45. $f(x) = x^3 + x^2 - 12x$

If $x^3 + x^2 - 12x = 0$, then

$$x(x^2 + x - 12) = 0$$
$$x(x-3)(x+4) = 0$$
$$x = 0 \quad \text{or} \quad x - 3 = 0 \quad \text{or} \quad x + 4 = 0$$
$$x = 3 \quad \text{or} \quad x = -4.$$

The x-intercepts are (–4, 0) (0, 0), and (3, 0).
The y-intercept is (0, 0).

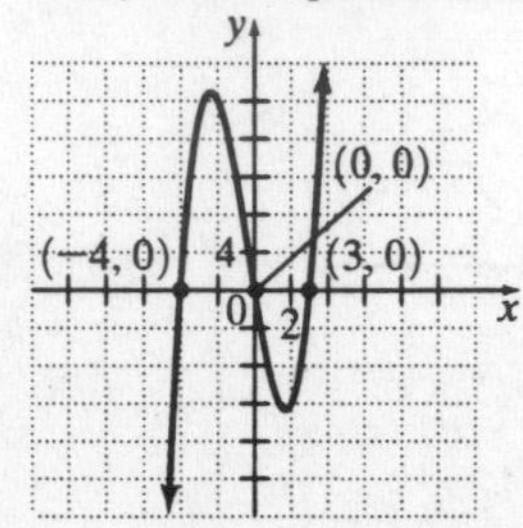

47. $f(x) = x^3 + x^2 - 9x - 9$

If $x^3 + x^2 - 9x - 9 = 0$, then

$$x^2(x+1) - 9(x+1) = 0$$
$$(x+1)(x^2 - 9) = 0$$
$$(x+1)(x+3)(x-3) = 0$$
$$x + 1 = 0 \quad \text{or} \quad x + 3 = 0 \quad \text{or} \quad x - 3 = 0$$
$$x = -1 \quad \text{or} \quad x = -3 \quad \text{or} \quad x = 3.$$

The x-intercepts are (–3, 0), (–1, 0), and (3, 0). If $x = 0$, then $y = f(0) = -9$.
The y-intercept is (0, –9).

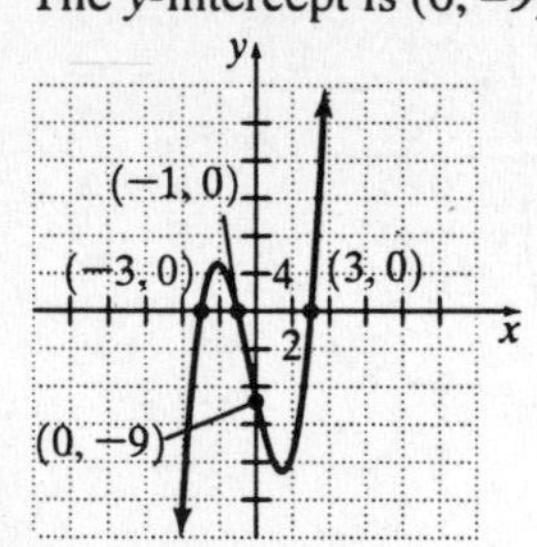

49. $f(x) = x^2 - 2x + 1$

$$x = \frac{-b}{2a} = \frac{2}{2(1)} = 1$$

$f(1) = 1 - 2 + 1 = 0$
The vertex is (1, 0).

If $x^2 - 2x + 1 = 0$, then

$$(x-1)(x-1) = 0$$
$$x - 1 = 0$$
$$x = 1$$

The x-intercept is (1, 0). If $x = 0$, then $y = f(0) = 1$. The y-intercept is (0, 1).

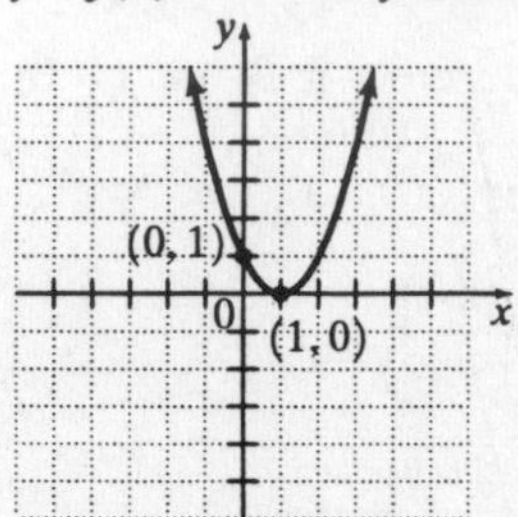

51. $f(x) = x^2 + 6x$

$$x = \frac{-b}{2a} = \frac{-6}{2(1)} = -3$$

$f(-3) = 9 - 18 = -9$
The vertex is (–3, –9).

If $x^2 + 6x = 0$, then

$$x(x+6) = 0$$
$$x = 0 \quad \text{or} \quad x + 6 = 0$$
$$x = -6.$$

The x-intercepts are (–6, 0) and (0, 0).
The y-intercept is (0, 0).

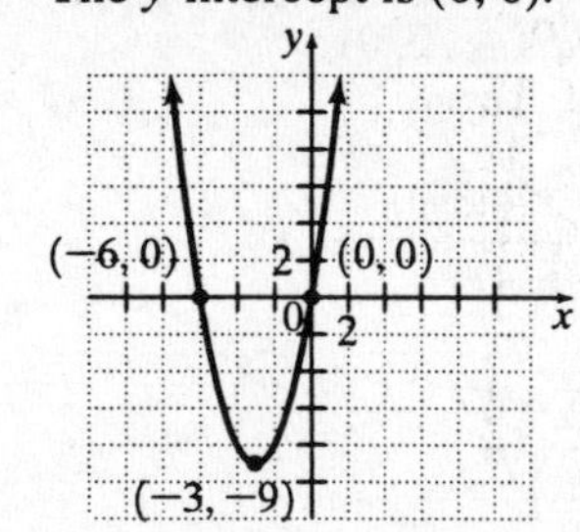

53. $f(x) = (x+2)(x-2) = x^2 - 4$

$x = \frac{-b}{2a} = \frac{0}{2(1)} = 0$

$f(0) = (0)^2 - 4 = -4$

The vertex is (0, –4).

If $(x+2)(x-2) = 0$, then

$x + 2 = 0$ or $x - 2 = 0$

$x = -2$ or $x = 2$.

The x-intercepts are (–2, 0) and (2, 0).

If $x = 0$, then $y = f(0) = -4$.

The y-intercept is (0, –4).

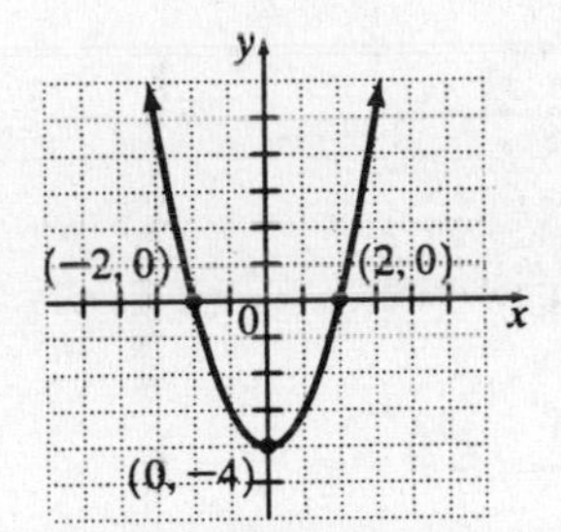

55. $f(x) = -x^3 + 25x$

If $-x^3 + 25x = 0$, then

$x(-x^2 + 25) = 0$

$x(5 - x)(5 + x) = 0$

$x = 0$ or $5 - x = 0$ or $5 + x = 0$

$x = 5$ or $x = -5$.

The x-intercepts are (–5,0), (0, 0), and (5, 0).

The y-intercept is (0, 0).

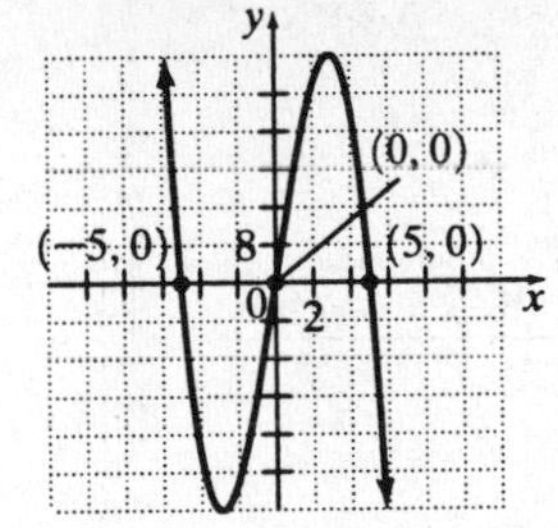

57. $f(x) = x^2 + 2x + 4$

$x = \frac{-b}{2a} = \frac{-2}{2(1)} = -1$

$f(-1) = 1 - 2 + 4 = 3$

The vertex is (–1, 3).

There are no x-intercepts.

If $x = 0$, then $y = f(0) = 4$.

The y-intercept is (0, 4).

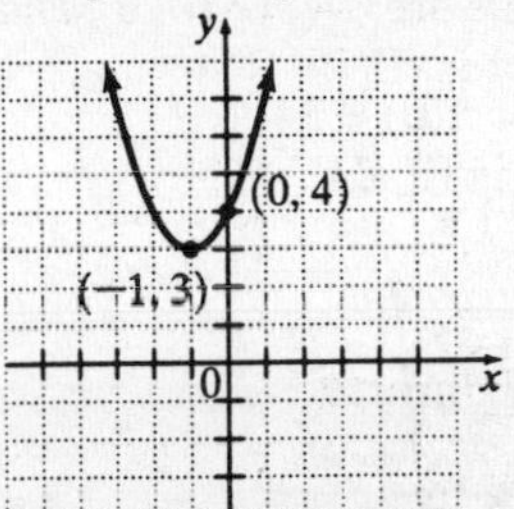

59. $f(x) = 3x(x-3)(x+5)$

If $3x(x-3)(x+5) = 0$, then

$3x = 0$ or $x - 3 = 0$ or $x + 5 = 0$

$x = 0$ or $x = 3$ or $x = -5$.

The x-intercepts are (–5, 0), (0, 0), and (3, 0).

The y-intercept is (0, 0).

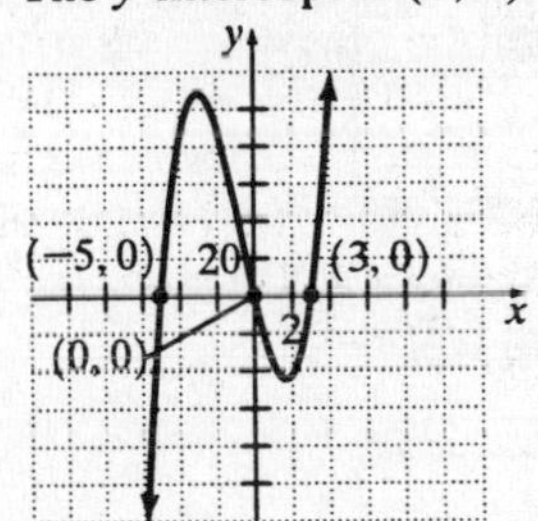

61. $h(x) = (x-4)(x-2)(2x+1)(x+3)$

$x-4=0$ or $x-2=0$

$x=4$ or $x=2$

or $2x+1=0$ or $x+3=0$

$x=-\frac{1}{2}$ or $x=-3$

The x-intercepts are (4, 0), (2, 0), $\left(-\frac{1}{2}, 0\right)$ and (−3, 0).

If $x = 0$, then $y = h(0) = 24$.

The y-intercept is (0, 24).

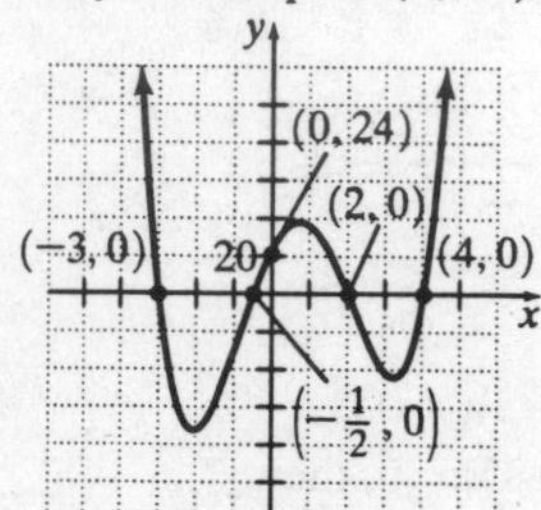

63. $f(x) = x^2 + 2x - 3$

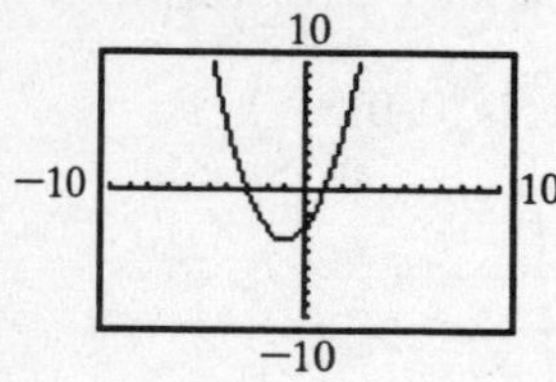

65. $f(x) = -x^3 + 25x$

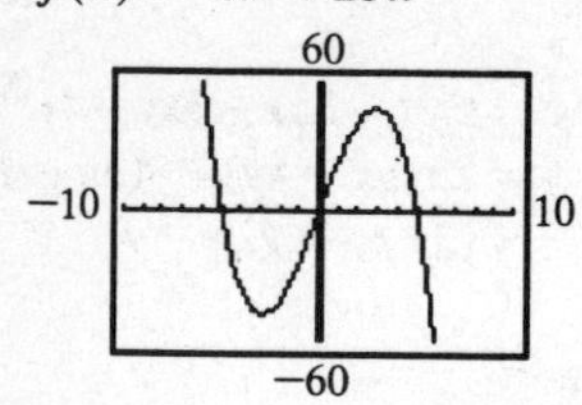

67. $G(x) = x^4 - 6.2x^2 - 6.2$

$x = -2.7, x = 2.7$

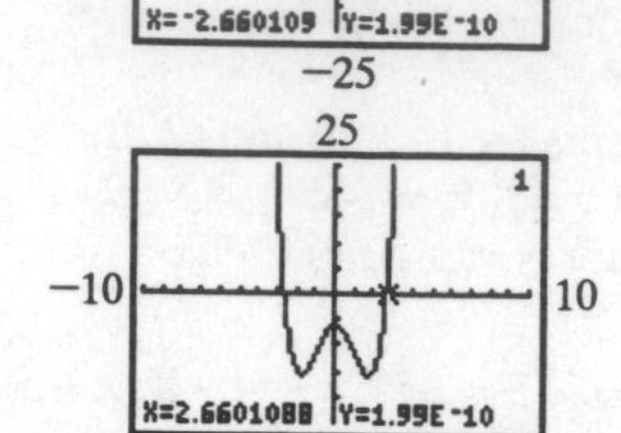

69. $-\frac{45}{100} = -\frac{9(5)}{20(5)} = -\frac{9}{20}$

71. $\frac{a^{14}b^2}{ab^4} = \frac{a^{14-1}}{b^{4-2}} = \frac{a^{13}}{b^2}$

73. $\frac{20x^{-3}y^5}{25y^{-2}x} = \frac{4(5)y^{5+2}}{5(5)x^{1+3}} = \frac{4y^7}{5x^4}$

Chapter 6 Review Exercises

1. $6x^2 - 15x = 3x(2x-5)$

2. $2x^3y - 6x^2y^2 - 8xy^3 = 2xy(x^2 - 3xy - 4y^2)$
$= 2xy(x-4y)(x+y)$

3. $20x^2 + 12x$; GCF = $4x$
$4x(5x+3)$

4. $6x^2y^2 - 3xy^3$; GCF = $3xy^2$
$3xy^2(2x-y)$

5. $-8x^3y+6x^2y^2$; GCF $= -2x^2y$
$-2x^2y(4x-3y)$

6. $3x(2x+3)-5(2x+3)$; GCF $= (2x+3)$
$(2x+3)(3x-5)$

7. $5x(x+1)-(x+1)$; GCF $= (x+1)$
$5x(x+1)-1(x+1)=(x+1)(5x-1)$

8. $3x^2-3x+2x-2=3x(x-1)+2(x-1)$
$=(x-1)(3x+2)$

9. $6x^2+10x-3x-5=2x(3x+5)-1(3x+5)$
$=(3x+5)(2x-1)$

10. $3a^2+9ab+3b^2+ab$
$=3a(a+3b)+b(a+3b)$
$=(a+3b)(3a+b)$

11. x^2+6x+8
The product is 8, and the sum is 6: 4 and 2.
$(x+4)(x+2)$

12. $x^2-11x+24$
The product is 24, and the sum is –11: –8 and –3.
$(x-8)(x-3)$

13. x^2+x+2
The product is 2, and the sum is 1: none exist. The trinomial is prime.

14. x^2-5x-6
The product is –6, and the sum is –5: –6 and 1.
$(x-6)(x+1)$

15. x^2+2x-8
The product is –8, and the sum is 2: 4 and –2.
$(x+4)(x-2)$

16. $x^2+4xy-12y^2$
The product is $-12y^2$, and the sum is $4y$: $6y$ and $-2y$.
$(x+6y)(x-2y)$

17. $x^2+8xy+15y^2$
The product is $15y^2$, and the sum is $8y$: $3y$ and $5y$.
$(x+3y)(x+5y)$

18. $3x^2y+6xy^2+3y^3=3y(x^2+2xy+y^2)$
$=3y(x+y)^2$

19. $72-18x-2x^2$; GCF $= 2$
$2(36-9x-x^2)=2(12+x)(3-x)$

20. $32+12x-4x^2=4(8+3x-x^2)$

21. $2x^2+11x-6$
$2x^2=2x\cdot x$
$-6=-1\cdot 6$
$2x\cdot 6+x(-1)=11x$
$(2x-1)(x+6)$

22. $4x^2-7x+4$
not factorable, prime

23. $4x^2+4x-3$
$4x^2=2x\cdot 2x$
$-3=3(-1)$
$2x(-1)+2x\cdot 3=4x$
$(2x+3)(2x-1)$

24. $6x^2+5xy-4y^2=6x^2+8xy-3xy-4y^2$
$=2x(3x+4y)-y(3x+4y)$
$=(3x+4y)(2x-y)$

25. $6x^2-25xy+4y^2$
$6x^2=6x\cdot x$
$4y^2=-y(-4y)$
$6x(-4y)+x(-y)=-25xy$
$(6x-y)(x-4y)$

26. $18x^2-60x+50=2(9x^2-30x+25)$
$=2(3x-5)(3x-5)$
$=2(3x-5)^2$

27. $2x^2 - 23xy - 39y^2$
$2x^2 = 2x \cdot x$
$-39y^2 = 3y(-13y)$
$2x(-13y) + x \cdot 3y = -23xy$
$(2x + 3y)(x - 13y)$

28. $4x^2 - 28xy + 49y^2 = (2x - 7y)(2x - 7y)$
$= (2x - 7y)^2$

29. $18x^2 - 9xy - 20y^2$
$18x^2 = 6x \cdot 3x$
$-20y^2 = 5y(-4y)$
$6x(-4y) + 3x \cdot 5y = -9xy$
$(6x + 5y)(3x - 4y)$

30. $36x^3y + 24x^2y^2 - 45xy^3$
$= 3xy(12x^2 + 8xy - 15y^2)$
$= 3xy(12x^2 + 18xy - 10xy - 15y^2)$
$= 3xy[6x(2x + 3y) - 5y(2x + 3y)]$
$= 3xy(2x + 3y)(6x - 5y)$

31. $4x^2 - 9 = (2x)^2 - 3^2 = (2x + 3)(2x - 3)$

32. $9t^2 - 25s^2 = (3t)^2 - (5s)^2 = (3t - 5s)(3t + 5s)$

33. $16x^2 + y^2$
not factorable, prime

34. $x^3 - 8y^3 = x^3 - (2y)^3$
$= (x - 2y)(x^2 + 2xy + 4y^2)$

35. $8x^3 + 27 = (2x)^3 + 3^3$
$= (2x + 3)(4x^2 - 6x + 9)$

36. $2x^3 + 8x = 2x(x^2 + 4)$

37. $54 - 2x^3y^3$; GCF = 2
$2(27 - x^3y^3) = 2[3^3 - (xy)^3]$
$= 2(3 - xy)(9 + 3xy + x^2y^2)$

38. $9x^2 - 4y^2$
$= (3x)^2 - (2y)^2$
$= (3x - 2y)(3x + 2y)$

39. $16x^4 - 1 = (4x^2)^2 - 1^2$
$= (4x^2 + 1)(4x^2 - 1)$
$= (4x^2 + 1)[(2x)^2 - 1^2]$
$= (4x^2 + 1)(2x + 1)(2x - 1)$

40. $x^4 + 16$
not factorable, prime

41. $2x^2 + 5x - 12$
$2x^2 = 2x \cdot x$
$-12 = -3 \cdot 4$
$2x \cdot 4 + x(-3) = 5x$
$(2x - 3)(x + 4)$

42. $3x^2 - 12 = 3(x^2 - 4) = 3(x - 2)(x + 2)$

43. $x(x - 1) + 3(x - 1)$; GCF = $(x - 1)$
$(x - 1)(x + 3)$

44. $x^2 + xy - 3x - 3y = x(x + y) - 3(x + y)$
$= (x + y)(x - 3)$

45. $4x^2y - 6xy^2$; GCF = $2xy$
$2xy(2x - 3y)$

46. $8x^2 - 15x - x^3 = -x(-8x + 15 + x^2)$
$= -x(x^2 - 8x + 15)$
$= -x(x - 5)(x - 3)$

47. $125x^3 + 27 = (5x)^3 + 3^3$
$= (5x + 3)(25x^2 - 15x + 9)$

48. $24x^2 - 3x - 18$; GCF = 3
$3(8x^2 - x - 6)$

49. $(x + 7)^2 - y^2 = [(x + 7) + y][(x + 7) - y]$
$= (x + 7 + y)(x + 7 - y)$

50. $x^2(x+3)-4(x+3)=(x+3)(x^2-4)$
$=(x+3)(x-2)(x+2)$

51. $(x+6)(x-2)=0$
$x+6=0$ or $x-2=0$
$x=-6$ or $x=2$

52. $3x(x+1)(7x-2)=0$
$3x=0$ or $x+1=0$ or $7x-2=0$
$x=0$ or $x=-1$ or $7x=2$
$x=\frac{2}{7}$

53. $4(5x+1)(x+3)=0$
$5x+1=0$ or $x+3=0$
$5x=-1$ or $x=-3$
$x=-\frac{1}{5}$ or $x=-3$

54. $x^2+8x+7=0$
$(x+7)(x+1)=0$
$x+7=0$ or $x+1=0$
$x=-7$ or $x=-1$

55. $x^2-2x-24=0$
$(x-6)(x+4)=0$
$x-6=0$ or $x+4=0$
$x=6$ or $x=-4$

56. $x^2+10x=-25$
$x^2+10x+25=0$
$(x+5)(x+5)=0$
$x+5=0$
$x=-5$

57. $x(x-10)=-16$
$x^2-10x=-16$
$x^2-10x+16=0$
$(x-8)(x-2)=0$
$x-8=0$ or $x-2=0$
$x=8$ or $x=2$

58. $(3x-1)(9x^2+3x+1)=0$
$3x-1=0$ or $9x^2+3x+1=0$
$3x=1$ does not factor
$x=\frac{1}{3}$

59. $56x^2-5x-6=0$
$(7x+2)(8x-3)=0$
$7x+2=0$ or $8x-3=0$
$7x=-2$ or $8x=3$
$x=-\frac{2}{7}$ or $x=\frac{3}{8}$

60. $20x^2-7x-6=0$
$(4x-3)(5x+2)=0$
$4x-3=0$ or $5x+2=0$
$4x=3$ or $5x=-2$
$x=\frac{3}{4}$ or $x=-\frac{2}{5}$

61. $5(3x+2)=4$
$15x+10=4$
$15x=4-10$
$15x=-6$
$x=-\frac{6}{15}=-\frac{2}{5}$

62. $6x^2-3x+8=0$
does not factor

63. $12-5t=-3$
$-5t=-3-12$
$-5t=-15$
$t=\frac{-15}{-5}$
$t=3$

64. $5x^3+20x^2+20x=0$
$5x(x^2+4x+4)=0$
$5x(x+2)(x+2)=0$
$x+2=0$ or $5x=0$
$x=-2$ or $x=0$

65. $4t^3 - 5t^2 - 21t = 0$

$t(4t^2 - 5t - 21) = 0$

$t(4t + 7)(t - 3) = 0$

$t = 0$ or $4t + 7 = 0$ or $t - 3 = 0$

$t = 0$ or $4t = -7$ or $t = 3$

$t = 0$ or $t = -\frac{7}{4}$ or $t = 3$

66. Let x = width.
Then $2x - 15$ = length.

$A = lw$

$500 = (2x - 15)(x)$

$500 = 2x^2 - 15x$

$0 = 2x^2 - 15x - 500$

$0 = (2x + 25)(x - 20)$

$2x + 25 = 0$ or $x - 20 = 0$

$x = -\frac{25}{2}$ or $x = 20$

Since the width cannot be negative, the width is 20 inches and the length is 25 inches.

67. base = $4x$
height = x

$A = \frac{1}{2}bh$

$162 = \frac{1}{2}(4x)(x)$

$162 = 2x^2$

$81 = x^2$

$x^2 - 81 = 0$

$(x + 9)(x - 9) = 0$

$x - 9 = 0$ or $x + 9 = 0$

$x = 9$ or $x = -9$

Since the length cannot be negative, the base is 4(9) = 36 yards.

68. 1st positive integer = x
2nd positive integer = $x + 1$

$x(x + 1) = 380$

$x^2 + x - 380 = 0$

$(x + 20)(x - 19) = 0$

$x + 20 = 0$ or $x - 19 = 0$

$x = -20$ or $x = 19$

Since –20 is not positive, the 1st integer is 19 and the 2nd integer is 20.

69. a. $h = -16t^2 + 440t$

$2800 = -16t^2 + 440t$

$0 = -16t^2 + 440t - 2800$

$0 = -8(2t^2 - 55t + 350)$

$0 = -8(2t - 35)(t - 10)$

$2t - 35 = 0$ or $t - 10 = 0$

$t = \frac{35}{2} = 17\frac{1}{2}$ or $t = 10$

10 seconds and 17.5 seconds
The rocket reaches a height of 2800 ft on its way up and on its way back down.

b. $h = -16t^2 + 440t$

$0 = -16t^2 + 440t$

$0 = -16t(t - 27.5)$

$t - 27.5 = 0$

$t = 27.5$ seconds

70. short leg = $x - 8$
long leg = x
hypotenuse = $x + 8$
By the Pythagorean theorem,

$(x - 8)^2 + x^2 = (x + 8)^2$

$x^2 + 2(x)(-8) + 8^2 + x^2 = x^2 + 2(x)(8) + 8^2$

$x^2 - 16x + 64 + x^2 = x^2 + 16x + 64$

$2x^2 - 16x + 64 = x^2 + 16x + 64$

$x^2 - 32x = 0$

$x(x - 32) = 0$

$x = 0$ or $x - 32 = 0$

$x = 0$ or $x = 32$

The long leg is 32 centimeters.

71. Domain: $(-\infty, \infty)$
Range: $(-\infty, 4]$

72. x-intercepts: $(-4, 0)$, $(2, 0)$
y-intercept: $(0, 3)$

73. $(-1, 4)$

74. $\{x|-4 < x < 2\}$

75. $f(x) = x^2 + 6x + 9$
$f(x) = (x+3)^2$
Solve $(x+3)^2 = 0$.
$x + 3 = 0$
$x = -3$
The x-intercept is $(-3, 0)$. If $x = 0$, then $y = f(0) = 9$. The y-intercept is $(0, 9)$.
$x = \dfrac{-b}{2a} = \dfrac{-6}{2(1)} = -3$ and $y = f(-3) = 0$.
Thus, the vertex is $(-3, 0)$.

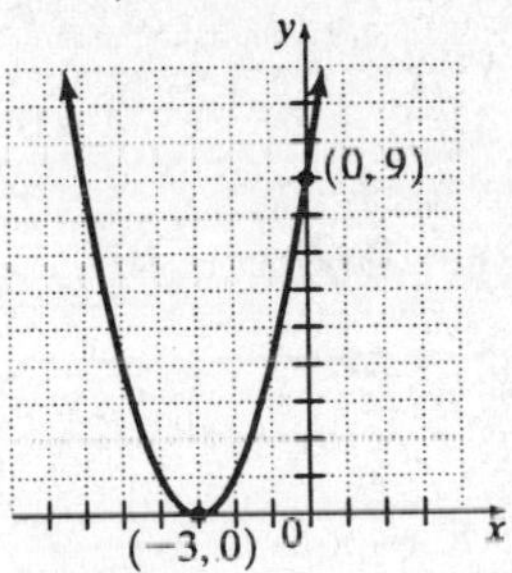

76. $f(x) = x^2 - 5x + 4$
$x = \dfrac{-b}{2a} = \dfrac{5}{2(1)} = \dfrac{5}{2}$
$f\left(\frac{5}{2}\right) = \left(\frac{5}{2}\right)^2 - 5\left(\frac{5}{2}\right) + 4 = -\frac{9}{4}$
The vertex is $\left(\frac{5}{2}, -\frac{9}{4}\right)$.
If $x^2 - 5x + 4 = 0$, then
$(x-1)(x-4) = 0$
$x - 1 = 0$ or $x - 4 = 0$
$x = 1$ or $x = 4$.
The x-intercepts are $(1, 0)$ and $(4, 0)$. If $x = 0$, then $y = f(0) = 4$. The y-intercept is $(0, 4)$.

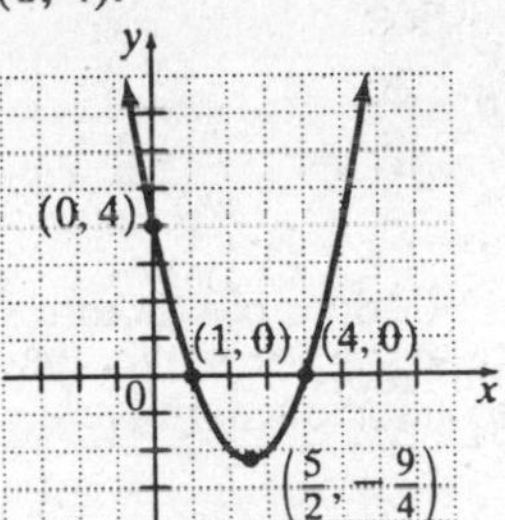

77. $f(x) = (x-1)(x^2 - 2x - 3)$
$= (x-1)(x-3)(x+1)$
Solve $(x-1)(x-3)(x+1) = 0$.
$x - 1 = 0$ or $x - 3 = 0$ or $x + 1 = 0$
$x = 1$ or $x = 3$ or $x = -1$
The x-intercepts are $(1, 0)$, $(3, 0)$, and $(-1, 0)$. If $x = 0$, then $y = f(0) = 3$. The y-intercept is $(0, 3)$.

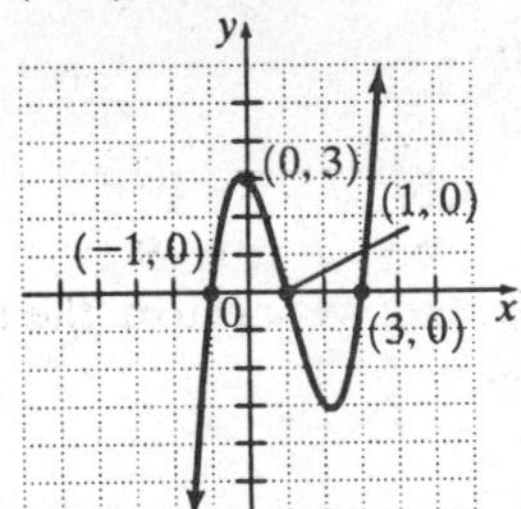

78. $f(x) = (x+3)(x^2 - 4x + 3)$
$f(x) = (x+3)(x-1)(x-3)$
If $(x+3)(x-1)(x-3) = 0$
$x+3=0$ or $x-1=0$ or $x-3=0$
$x=-3$ or $x=1$ or $x=3$
The x-intercepts are (–3, 0), (1, 0), and (3, 0).
If $x = 0$, then $y = f(0) = (3)(3) = 9$.
The y-intercept is (0, 9).

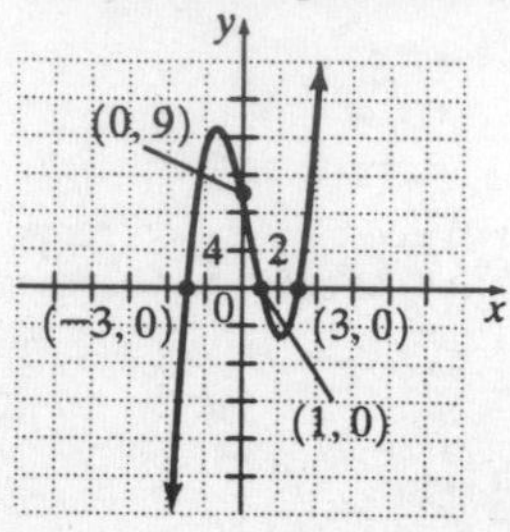

79. $f(x) = 2x^2 - 4x + 5$ does not factor.
There are no x-intercepts.
If $x = 0$, $y = f(0) = 5$.
The y-intercept is (0, 5).
$x = \frac{-b}{2a} = \frac{-(-4)}{2(2)} = 1$
$y = f(1) = 2(1)^2 - 4(1) + 5 = 3$
The vertex is (1, 3).

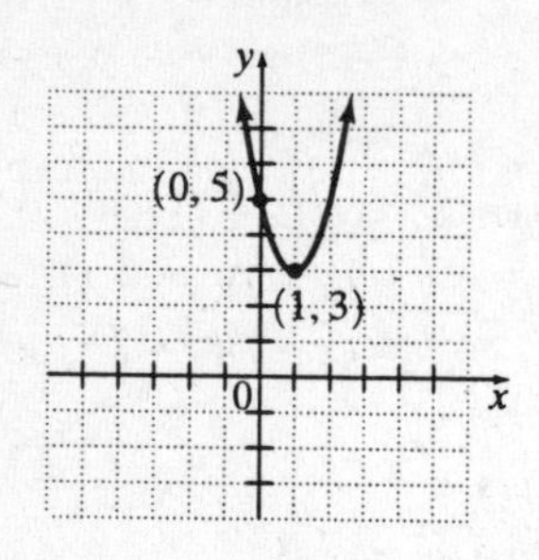

80. $f(x) = x^2 - 2x + 3$ does not factor.
There are no x-intercepts.
$x = \frac{-b}{2a} = \frac{2}{2(1)} = 1$
$y = f(1) = 1 - 2 + 3 = 2$
The vertex is (1, 2).
If $x = 0$, then $y = f(0) = 3$.
The y-intercept is (0, 3).

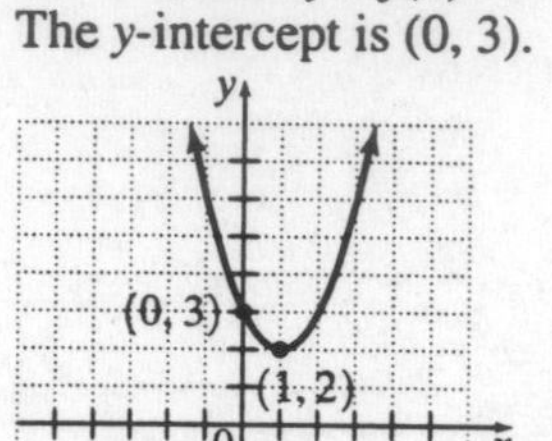

81. $f(x) = x^3 - 16x$
$= x(x^2 - 16)$
$= x(x+4)(x-4)$
Solve $x(x+4)(x-4) = 0$.
$x = 0$ or $x+4=0$ or $x-4=0$
$x=-4$ or $x=4$
The x-intercepts are (0, 0), (–4, 0) and (4, 0).
The y-intercept is (0, 0).

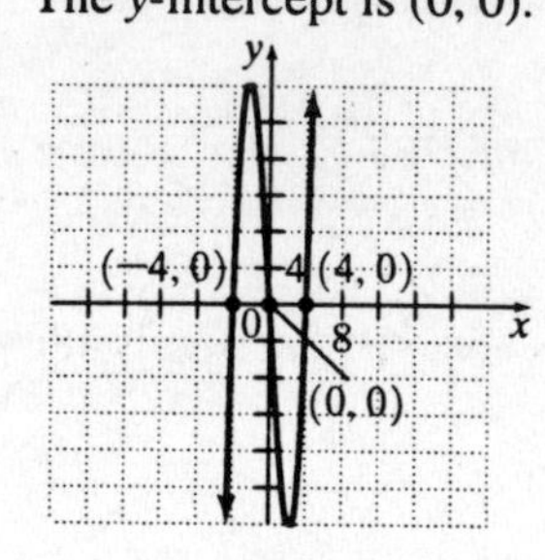

82. $f(x) = x^3 + 5x^2 + 6x$

$f(x) = x(x^2 + 5x + 6) = x(x+3)(x+2)$

If $x(x+3)(x+2) = 0$, then

$x = 0$ or $x + 3 = 0$ or $x + 2 = 0$

$x = 0$ or $x = -3$ or $x = -2.$

The x-intercepts are $(-3, 0)$, $(-2, 0)$, and $(0, 0)$.

The y-intercept is $(0, 0)$.

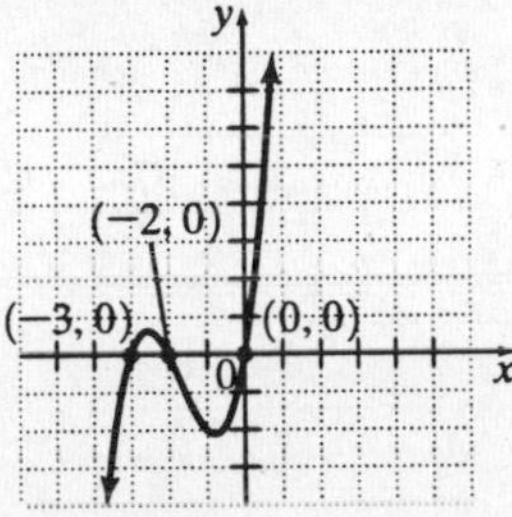

Chapter 6 Test

1. $9x^3 + 39x^2 + 12x$; GCF $= 3x$

$3x(3x^2 + 13x + 4) = 3x(3x+1)(x+4)$

2. $x^2 + x - 10$

The product is -10 and the sum is 1: none exist. The trinomial is prime.

3. $x^2 + 4$

not factorable, prime

4. $y^2 - 8y - 48$

The product is -48, and the sum is -8: -12 and 4.

$(y-12)(y+4)$

5. $3a^2 + 3ab - 7a - 7b = 3a(a+b) - 7(a+b)$

$= (a+b)(3a-7)$

6. $3x^2 - 5x + 2$

$3x^2 = 3x \cdot x$

$2 = -2(-1)$

$3x(-1) + x(-2): -5x$

$(3x-2)(x-1)$

7. $x^2 + 20x + 90$

The product is 90, and the sum is 20: none exist. The trinomial is prime.

8. $x^2 + 14xy + 24y^2$

The product is $24y^2$, and the sum is $14y$: $12y$ and $2y$.

$(x+12y)(x+2y)$

9. $26x^6 - x^4$; GCF $= x^4$

$x^4(26x^2 - 1)$

10. $50x^3 + 10x^2 - 35x$; GCF $= 5x$

$5x(10x^2 + 2x - 7)$

11. $180 - 5x^2$; GCF $= 5$

$5(36 - x^2) = 5(6-x)(6+x)$

12. $64x^3 - 1 = (4x)^3 - 1^3$

$= (4x-1)(16x^2 + 4x + 1)$

13. $6t^2 - t - 5$

$6t^2 = 6t \cdot t$

$-5 = 5(-1)$

$6t(-1) + t \cdot 5 = -t$

$(6t+5)(t-1)$

14. $xy^2 - 7y^2 - 4x + 28 = y^2(x-7) - 4(x-7)$

$= (x-7)(y^2 - 4)$

$= (x-7)(y+2)(y-2)$

15. $x - x^5$; GCF $= x$

$x(1 - x^4) = x[(1^2)^2 - (x^2)^2]$

$= x(1^2 - x^2)(1^2 + x^2)$

$= x(1-x)(1+x)(1+x^2)$

16. $-xy^3 - x^3y$; GCF $= -xy$

$-xy(y^2 + x^2)$

17. $x^2 + 5x = 14$
$x^2 + 5x - 14 = 0$
$(x+7)(x-2) = 0$
$x + 7 = 0$ or $x - 2 = 0$
$x = -7$ or $x = 2$

18. $(x+3)^2 = 16$
$x^2 + 2(x)(3) + 3^2 = 16$
$x^2 + 6x + 9 = 16$
$x^2 + 6x + 9 - 16 = 0$
$x^2 + 6x - 7 = 0$
$(x+7)(x-1) = 0$
$x + 7 = 0$ or $x - 1 = 0$
$x = -7$ or $x = 1$

19. $3x(2x-3)(3x+4) = 0$
$3x = 0$ or $2x - 3 = 0$ or $3x + 4 = 0$
$x = \frac{0}{3}$ or $2x = 3$ or $3x = -4$
$x = 0$ or $x = \frac{3}{2}$ or $x = -\frac{4}{3}$

20. $5t^3 - 45t = 0$
$5t(t^2 - 9) = 0$
$5t(t+3)(t-3) = 0$
$5t = 0$ or $t + 3 = 0$ or $t - 3 = 0$
$t = 0$ or $t = -3$ or $t = 3$

21. $3x^2 = -12x$
$3x^2 + 12x = 0$
$3x(x+4) = 0$
$3x = 0$ or $x + 4 = 0$
$x = 0$ or $x = -4$

22. $t^2 - 2t - 15 = 0$
$(t-5)(t+3) = 0$
$t - 5 = 0$ or $t + 3 = 0$
$t = 5$ or $t = -3$

23. $7x^2 = 168 + 35x$
$7x^2 - 35x - 168 = 0$
$7(x^2 - 5x - 24) = 0$
$7(x-8)(x+3) = 0$
$x - 8 = 0$ or $x + 3 = 0$
$x = 8$ or $x = -3$

24. $6x^2 = 15x$
$6x^2 - 15x = 0$
$3x(2x-5) = 0$
$3x = 0$ or $2x - 5 = 0$
$x = 0$ or $2x = 5$
$x = \frac{5}{2}$

25. width $= x$
length $= x + 5$
$A = lw$
$66 = (x+5)(x)$
$66 = x^2 + 5x$
$0 = x^2 + 5x - 66$
$0 = (x+11)(x-6)$
$x + 11 = 0$ or $x - 6 = 0$
$x = -11$ or $x = 6$
The dimensions are 6 ft by 11 ft.

26. altitude $= x$
base $= x + 9$
$A = \frac{1}{2}bh$
$68 = \frac{1}{2}(x+9)(x)$
$2(68) = 2\left[\frac{1}{2}(x+9)(x)\right]$
$136 = (x+9)(x)$
$136 = x^2 + 9x$
$0 = x^2 + 9x - 136$
$0 = (x+17)(x-8)$
$x + 17 = 0$ or $x - 8 = 0$
$x = -17$ or $x = 8$
The base is 8 + 9 = 17 feet.

27. one number = x
other number = $17 - x$

$$x^2 + (17 - x)^2 = 145$$
$$x^2 + 289 + 2(17)(-x) + x^2 = 145$$
$$x^2 + 289 - 34x + x^2 = 145$$
$$2x^2 - 34x + 289 = 145$$
$$2x^2 - 34x + 289 - 145 = 0$$
$$2x^2 - 34x + 144 = 0$$
$$2(x^2 - 17x + 72) = 0$$
$$2(x - 8)(x - 9) = 0$$

$x - 8 = 0$ or $x - 9 = 0$
$x = 8$ or $x = 9$

The numbers are 8 and 9.

28. $x^2 - (2y)^2 = (x - 2y)(x + 2y)$

29. $h(t) = -16t^2 + 96t + 880$

a. $h(1) = -16(1)^2 + 96(1) + 880$
$= -16 + 96 + 880$
$= 960$
960 feet

b. $h(5.1) = -16(5.1)^2 + 96(5.1) + 880$
$= 416.16 + 489.6 + 880$
$= 953.44$
953.44 feet

c. $0 = -16t^2 + 96t + 880$
$0 = (-16t - 80)(t - 11)$
$-16t - 80 = 0$ or $t - 11 = 0$
$t = -5$ or $t = 11$
Disregard the negative.
11 seconds

30. $f(x) = x^2 - 4x - 5$

$$x = \frac{-b}{2a} = \frac{4}{2} = 2$$

$f(2) = 2^2 - 4(2) - 5 = -9$
The vertex is (2, −9).
$f(x) = x^2 - 4x - 5 = (x + 1)(x - 5)$
If $f(x) = 0$, $x + 1 = 0$ or $x - 5 = 0$
$x = -1$ or $x = 0.$
The x-intercepts are (−1, 0), (5, 0).
If $x = 0$, $y = f(0) = -5$.
The y-intercept is (0, −5).

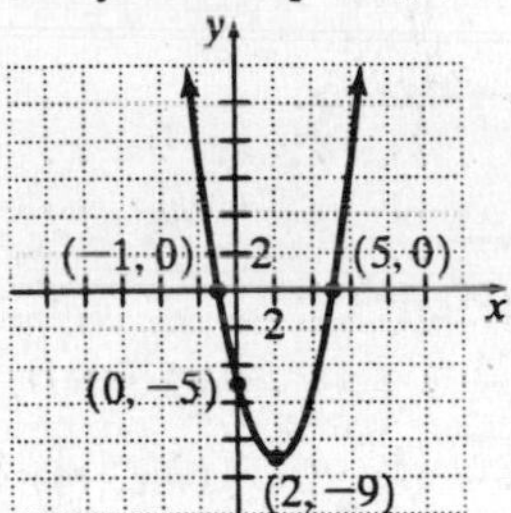

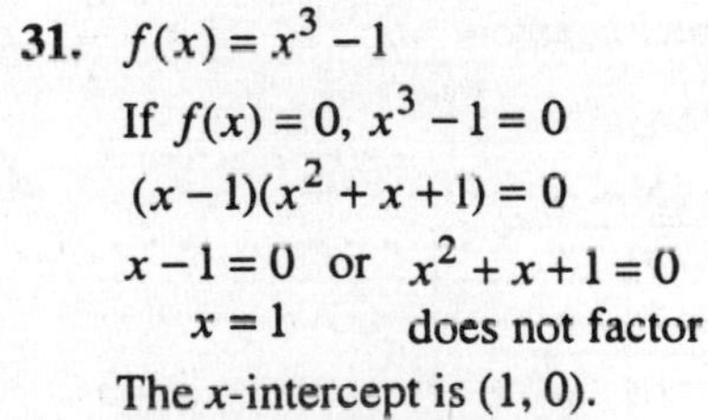
31. $f(x) = x^3 - 1$
If $f(x) = 0$, $x^3 - 1 = 0$
$(x - 1)(x^2 + x + 1) = 0$
$x - 1 = 0$ or $x^2 + x + 1 = 0$
$x = 1$ does not factor
The x-intercept is (1, 0).
If $x = 0$, $y = f(0) = -1$.
The y-intercept is (0, −1).

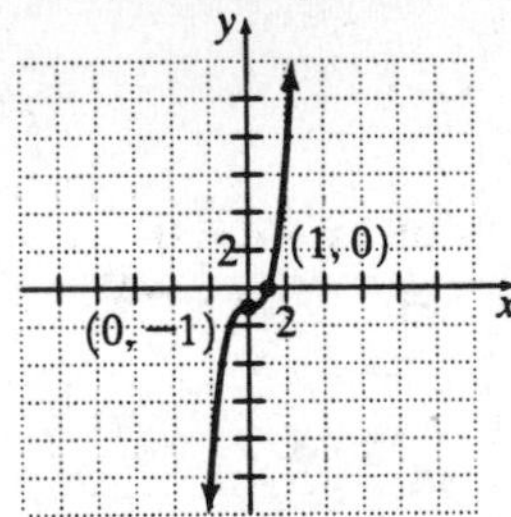

Chapter 7

Section 7.1

Graphing Calculator Explorations

1. $f(x) = \dfrac{x+1}{x^2-4} = \dfrac{x+1}{(x+2)(x-2)}$

Domain:
$\{x|x \text{ is a real number and } x \neq -2,\ x \neq 2\}$

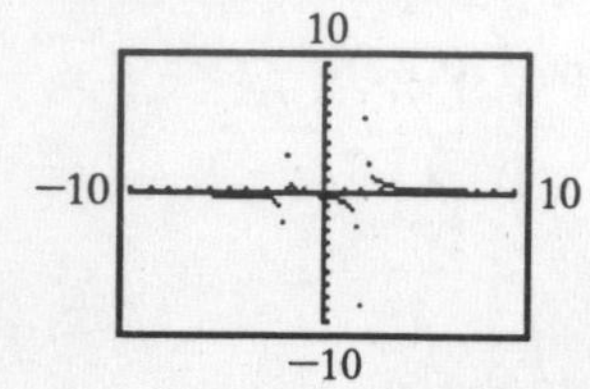

3. $h(x) = \dfrac{x^2}{2x^2+7x-4} = \dfrac{x^2}{(2x-1)(x+4)}$

Domain:
$\{x|x \text{ is a real number}$

$\text{and } x \neq -4 \text{ and } x \neq \frac{1}{2}\}$

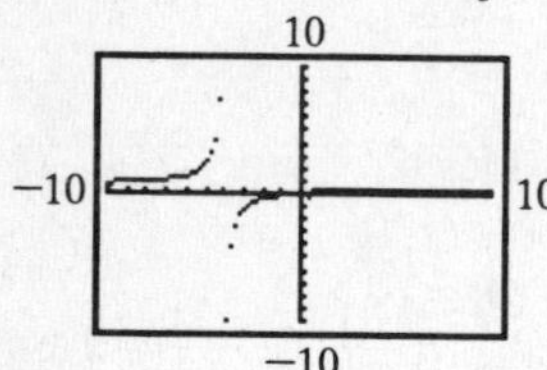

Exercise Set 7.1

1. $f(x) = \dfrac{x+8}{2x-1}$

$f(2) = \dfrac{2+8}{2(2)-1} = \dfrac{10}{4-1} = \dfrac{10}{3}$

$f(0) = \dfrac{0+8}{2(0)-1} = \dfrac{8}{-1} = -8$

$f(-1) = \dfrac{-1+8}{2(-1)-1} = \dfrac{7}{-3} = -\dfrac{7}{3}$

3. $g(x) = \dfrac{x^2+8}{x^3-25x}$

$g(3) = \dfrac{3^2+8}{3^3-25(3)}$
$= \dfrac{9+8}{27-75}$
$= \dfrac{17}{-48}$
$= -\dfrac{17}{48}$

$g(-2) = \dfrac{(-2)^2+8}{(-2)^3-25(-2)}$
$= \dfrac{4+8}{-8+50}$
$= \dfrac{12}{42}$
$= \dfrac{2}{7}$

$g(1) = \dfrac{1^2+8}{1^3-25(1)} = \dfrac{1+8}{1-25} = \dfrac{9}{-24} = -\dfrac{3}{8}$

5. $f(x) = \dfrac{5x-7}{4}$

Domain: $\{x|x \text{ is a real number}\}$

7. $s(t) = \dfrac{t^2+1}{2t}$

Undefined values when
$2t = 0$
$t = \dfrac{0}{2} = 0$

Domain: $\{t|t \text{ is a real number and } t \neq 0\}$

9. $f(x) = \dfrac{3x}{7-x}$

Undefined values when
$7 - x = 0$
$7 = x$

Domain: $\{x|x \text{ is a real number and } x \neq 7\}$

11. $R(x) = \dfrac{3+2x}{x^3+x^2-2x}$

Undefined values when

$$x^3+x^2-2x=0$$
$$x(x^2+x-2)=0$$
$$x(x+2)(x-1)=0$$
$$x=0 \quad \text{or} \quad x+2=0 \quad \text{or} \quad x-1=0$$
$$x=-2 \quad \text{or} \quad x=1$$

Domain: $\{x|x \text{ is a real number and } x \neq 0,\ x \neq -2,\ x \neq 1\}$

13. $C(x) = \dfrac{x+3}{x^2-4}$

Undefined values when

$$x^2-4=0$$
$$(x+2)(x-2)=0$$
$$x+2=0 \quad \text{or} \quad x-2=0$$
$$x=-2 \quad \text{or} \quad x=2$$

Domain: $\{x|x \text{ is a real number } x \neq 2,\ x \neq -2\}$

15. Answers may vary.

17. The GCF of the numerator and denominator is $2x$.

$$\frac{10x^3}{18x} = \frac{2x(5x^2)}{2x(9)} = \frac{5x^2}{9}$$

19. The GCF of the numerator and denominator is $9x^2y^3$.

$$\frac{9x^6y^3}{18x^2y^5} = \frac{9x^2y^3(x^4)}{9x^2y^3(2y^2)} = \frac{x^4}{2y^2}$$

21. $\dfrac{x+5}{5+x} = \dfrac{x+5}{x+5} = 1$

23.
$$\frac{x-1}{1-x^2} = \frac{x-1}{(1+x)(1-x)}$$
$$= \frac{x-1}{(1+x)(-1)(x-1)}$$
$$= \frac{1}{(1+x)(-1)}$$
$$= -\frac{1}{1+x}$$

25. $\dfrac{4x-8}{3x-6} = \dfrac{4(x-2)}{3(x-2)} = \dfrac{4}{3}$

27. $\dfrac{2x-14}{7-x} = \dfrac{2(x-7)}{-(x-7)} = -2$

29. $\dfrac{x^2-2x-3}{x^2-6x+9} = \dfrac{(x-3)(x+1)}{(x-3)^2} = \dfrac{x+1}{x-3}$

31.
$$\frac{2x^2+12x+18}{x^2-9} = \frac{2(x^2+6x+9)}{x^2-3^2}$$
$$= \frac{2(x+3)^2}{(x+3)(x-3)}$$
$$= \frac{2(x+3)}{x-3}$$

33. $\dfrac{3x+6}{x^2+2x} = \dfrac{3(x+2)}{x(x+2)} = \dfrac{3}{x}$

35.
$$\frac{2x^2-x-3}{2x^3-3x^2+2x-3} = \frac{(2x-3)(x+1)}{(2x-3)(x^2+1)}$$
$$= \frac{x+1}{x^2+1}$$

37. $\dfrac{8q^2}{16q^3-16q^2} = \dfrac{8q^2}{16q^2(q-1)} = \dfrac{1}{2(q-1)}$

39. $\dfrac{x^2+6x-40}{10+x} = \dfrac{(x+10)(x-4)}{x+10} = x-4$

41.
$$\frac{x^3-125}{5-x} = \frac{x^3-5^3}{-(x-5)}$$
$$= \frac{(x-5)(x^2+5x+25)}{-(x-5)}$$
$$= -(x^2+5x+25)$$
$$= -x^2-5x-25$$

43.
$$\frac{8x^3-27}{4x-6} = \frac{(2x)^3-3^3}{2(2x-3)}$$
$$= \frac{(2x-3)(4x^2+6x+9)}{2(2x-3)}$$
$$= \frac{4x^2+6x+9}{2}$$

45. $C = \frac{DA}{A+12}$; $A = 8, D = 1000$

$C = \frac{1000 \cdot 8}{8+12} = \frac{8000}{20} = 400$

The child should receive a dose of 400 milligrams.

47. $B = \frac{705w}{h^2}$; $w = 148, h = 66$

$B = \frac{705 \cdot 148}{(66)^2} = \frac{104,340}{4356} \approx 24$

No, a 148-pound person who is 5 feet 6 inches tall should not lose weight.

49. a. $R(x) = \frac{150x^2}{x^2+3}$

$R(1) = \frac{150(1)^2}{1^2+3}$

$= \frac{150}{4}$

$= \$37.5$ million

b. $R(x) = \frac{150x^2}{x^2+3}$

$R(2) = \frac{150(2)^2}{2^2+3}$

$= \frac{600}{7}$

$\approx \$85.7$ million

c. $R(2) - R(1) = 85.7 - 37.5 = \$48.2$ million

51. a, c

Answers may vary.

53. $f(x) = \frac{1}{x}$

x	$\frac{1}{4}$	$\frac{1}{2}$	1	2	4
y	4	2	1	$\frac{1}{2}$	$\frac{1}{4}$

x	-4	-2	-1	$-\frac{1}{2}$	$-\frac{1}{4}$
y	$-\frac{1}{4}$	$-\frac{1}{2}$	-1	-2	-4

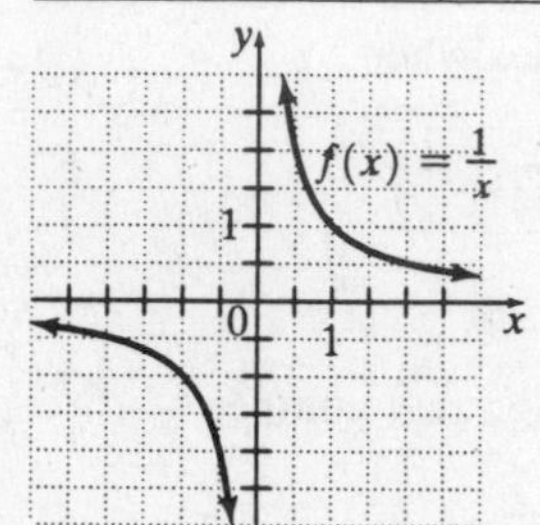

55. a. $R(1) = \frac{1000(1)^2}{1^2+4} = \frac{1000}{5} = 200$

$200 million

b. $R(2) = \frac{1000(2)^2}{2^2+4} = \frac{4000}{8} = 500$

$500 million

c. $R(2) - R(1) = 300$

$300 million

57. $y = \frac{x^2-16}{x-4} = \frac{(x+4)(x-4)}{x-4} = x+4,\ x \neq 4$

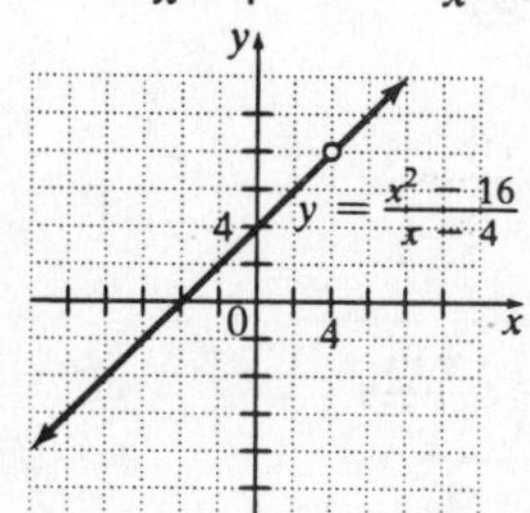

59. $y = \dfrac{x^2 - 6x + 8}{x - 2}$
$= \dfrac{(x-2)(x-4)}{x-2}$
$= x - 4,\ x \neq 2$

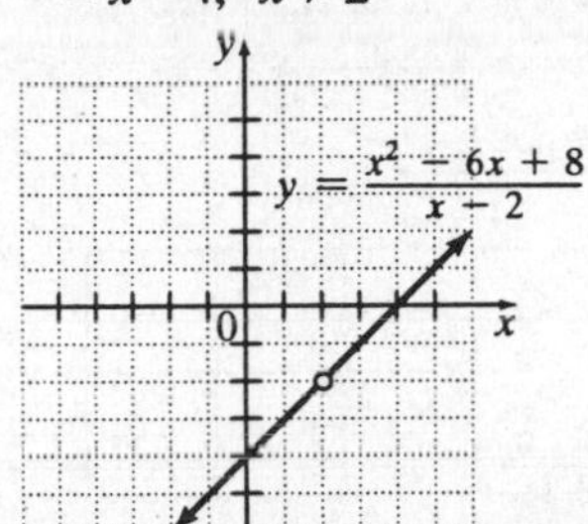

61. $\dfrac{5}{27} \cdot \dfrac{2}{5} = \dfrac{5 \cdot 2}{3 \cdot 3 \cdot 3 \cdot 5} = \dfrac{2}{27}$

63. $\dfrac{7}{8} \div \dfrac{1}{2} = \dfrac{7}{8} \cdot \dfrac{2}{1} = \dfrac{7 \cdot 2}{2 \cdot 2 \cdot 2} = \dfrac{7}{4}$

65. $\dfrac{4}{3} \cdot \dfrac{1}{7} \cdot \dfrac{10}{13} = \dfrac{2 \cdot 2 \cdot 1 \cdot 2 \cdot 5}{3 \cdot 7 \cdot 13} = \dfrac{40}{273}$

67. $\dfrac{8}{15} \div \dfrac{5}{8} = \dfrac{8}{15} \cdot \dfrac{8}{5} = \dfrac{2 \cdot 2 \cdot 2 \cdot 2 \cdot 2 \cdot 2}{3 \cdot 5 \cdot 5} = \dfrac{64}{75}$

Section 7.2

Mental Math

1. $\dfrac{2}{y} \cdot \dfrac{x}{3} = \dfrac{2 \cdot x}{y \cdot 3} = \dfrac{2x}{3y}$

2. $\dfrac{3x}{4} \cdot \dfrac{1}{y} = \dfrac{3x \cdot 1}{4 \cdot y} = \dfrac{3x}{4y}$

3. $\dfrac{5}{7} \cdot \dfrac{y^2}{x^2} = \dfrac{5 \cdot y^2}{7 \cdot x^2} = \dfrac{5y^2}{7x^2}$

4. $\dfrac{x^5}{11} \cdot \dfrac{4}{z^3} = \dfrac{x^5 \cdot 4}{11 \cdot z^3} = \dfrac{4x^5}{11z^3}$

5. $\dfrac{9}{x} \cdot \dfrac{x}{5} = \dfrac{9 \cdot x}{x \cdot 5} = \dfrac{9}{5}$

6. $\dfrac{y}{7} \cdot \dfrac{3}{y} = \dfrac{y \cdot 3}{7 \cdot y} = \dfrac{3}{7}$

Exercise Set 7.2

1. $\dfrac{3x}{y^2} \cdot \dfrac{7y}{4x} = \dfrac{21xy}{4x \cdot y \cdot y} = \dfrac{21}{4y}$

3. $\dfrac{8x}{2} \cdot \dfrac{x^5}{4x^2} = \dfrac{8x^6}{8x^2} = \dfrac{8 \cdot x^2 \cdot x^4}{8 \cdot x^2} = x^4$

5. $-\dfrac{5a^2b}{30a^2b^2} \cdot b^3 = -\dfrac{5a^2b^4}{30a^2b^2}$
$= -\dfrac{5 \cdot a^2 \cdot b^2 \cdot b^2}{5 \cdot 6 \cdot a^2 \cdot b^2}$
$= -\dfrac{b^2}{6}$

7. $\dfrac{x}{2x - 14} \cdot \dfrac{x^2 - 7x}{5} = \dfrac{x}{2(x-7)} \cdot \dfrac{x(x-7)}{5}$
$= \dfrac{x \cdot x \cdot (x-7)}{2 \cdot 5 \cdot (x-7)}$
$= \dfrac{x^2}{10}$

9. $\dfrac{6x+6}{5} \cdot \dfrac{10}{36x+36} = \dfrac{6(x+1)}{5} \cdot \dfrac{10}{36(x+1)}$
$= \dfrac{6 \cdot 2 \cdot 5 \cdot (x+1)}{5 \cdot 6 \cdot 2 \cdot 3 \cdot (x+1)}$
$= \dfrac{1}{3}$

11. $\dfrac{m^2 - n^2}{m+n} \cdot \dfrac{m}{m^2 - mn}$
$= \dfrac{(m+n)(m-n)}{m+n} \cdot \dfrac{m}{m(m-n)}$
$= \dfrac{m \cdot (m+n) \cdot (m-n)}{m \cdot (m+n) \cdot (m-n)}$
$= 1$

13. $\frac{x^2-25}{x^2-3x-10}\cdot\frac{x+2}{x}$
$=\frac{(x+5)(x-5)}{(x+2)(x-5)}\cdot\frac{x+2}{x}$
$=\frac{(x+5)\cdot(x-5)\cdot(x+2)}{(x+2)\cdot(x-5)\cdot x}$
$=\frac{x+5}{x}$

15. $A=l\cdot w$
$A=\frac{2x}{x^2-25}\cdot\frac{x+5}{9x^3}$
$A=\frac{2x}{(x+5)(x-5)}\cdot\frac{x+5}{9x^3}$
$A=\frac{2x\cdot(x+5)}{9\cdot x\cdot x^2\cdot(x+5)(x-5)}$
$A=\frac{2}{9x^2(x-5)}$ square feet

17. $\frac{5x^7}{2x^5}\div\frac{10x}{4x^3}=\frac{5x^7}{2x^5}\cdot\frac{4x^3}{10x}$
$=\frac{20x^{10}}{20x^6}$
$=\frac{20x^6\cdot x^4}{20x^6}$
$=x^4$

19. $\frac{8x^2}{y^3}\div\frac{4x^2y^3}{6}=\frac{8x^2}{y^3}\cdot\frac{6}{4x^2y^3}$
$=\frac{48x^2}{4x^2y^6}$
$=\frac{12\cdot 4x^2}{4x^2\cdot y^6}$
$=\frac{12}{y^6}$

21. $\frac{(x-6)(x+4)}{4x}\div\frac{2x-12}{8x^2}$
$=\frac{(x-6)(x+4)}{4x}\cdot\frac{8x^2}{2x-12}$
$=\frac{(x-6)(x+4)}{4x}\cdot\frac{8x^2}{2(x-6)}$
$=\frac{8x\cdot x(x-6)(x+4)}{8x\cdot(x-6)}$
$=x(x+4)$

23. $\frac{3x^2}{x^2-1}\div\frac{x^5}{(x+1)^2}$
$=\frac{3x^2}{(x+1)(x-1)}\cdot\frac{(x+1)(x+1)}{x^5}$
$=\frac{3x^2\cdot(x+1)\cdot(x+1)}{x^2\cdot x^3\cdot(x+1)(x-1)}$
$=\frac{3(x+1)}{x^3(x-1)}$

25. $\frac{m^2-n^2}{m+n}\div\frac{m}{m^2+nm}$
$=\frac{m^2-n^2}{m+n}\cdot\frac{m^2+nm}{m}$
$=\frac{(m+n)(m-n)}{m+n}\cdot\frac{m(m+n)}{m}$
$=\frac{m\cdot(m+n)\cdot(m-n)\cdot(m+n)}{m(m+n)}$
$=m^2-n^2$

27. $\frac{x+2}{7-x}\div\frac{x^2-5x+6}{x^2-9x+14}$
$=\frac{x+2}{7-x}\cdot\frac{x^2-9x+14}{x^2-5x+6}$
$=\frac{x+2}{-1(x-7)}\cdot\frac{(x-7)(x-2)}{(x-3)(x-2)}$
$=\frac{(x+2)\cdot(x-7)\cdot(x-2)}{-1\cdot(x-7)\cdot(x-3)\cdot(x-2)}$
$=-\frac{x+2}{x-3}$

29. $\dfrac{x^2+7x+10}{1-x} \div \dfrac{x^2+2x-15}{x-1}$

$= \dfrac{x^2+7x+10}{1-x} \cdot \dfrac{x-1}{x^2+2x-15}$

$= \dfrac{(x+5)(x+2)}{-1(x-1)} \cdot \dfrac{x-1}{(x+5)(x-3)}$

$= \dfrac{(x+5)\cdot(x+2)\cdot(x-1)}{-1\cdot(x-1)\cdot(x+5)(x-3)}$

$= -\dfrac{x+2}{x-3}$

31. Answers may vary.

33. $\dfrac{5a^2b}{30a^2b^2} \cdot \dfrac{1}{b^3} = \dfrac{5a^2b}{30a^2b^5} = \dfrac{1}{6b^4}$

35. $\dfrac{12x^3y}{8xy^7} \div \dfrac{7x^5y}{6x} = \dfrac{12x^3y}{8xy^7} \cdot \dfrac{6x}{7x^5y}$

$= \dfrac{72x^4y}{56x^6y^8}$

$= \dfrac{9}{7x^2y^7}$

37. $\dfrac{5x-10}{12} \div \dfrac{4x-8}{8} = \dfrac{5x-10}{12} \cdot \dfrac{8}{4x-8}$

$= \dfrac{5(x-2)}{12} \cdot \dfrac{8}{4(x-2)}$

$= \dfrac{5\cdot 4\cdot 2\cdot (x-2)}{2\cdot 6\cdot 4\cdot (x-2)}$

$= \dfrac{5}{6}$

39. $\dfrac{x^2+5x}{8} \cdot \dfrac{9}{3x+15} = \dfrac{x(x+5)}{8} \cdot \dfrac{9}{3(x+5)}$

$= \dfrac{3\cdot 3\cdot x\cdot (x+5)}{3\cdot 8\cdot (x+5)}$

$= \dfrac{3x}{8}$

41. $\dfrac{7}{6p^2+q} \div \dfrac{14}{18p^2+3q}$

$= \dfrac{7}{6p^2+q} \cdot \dfrac{18p^2+3q}{14}$

$= \dfrac{7}{6p^2+q} \cdot \dfrac{3(6p^2+q)}{14}$

$= \dfrac{7\cdot 3\cdot (6p^2+q)}{2\cdot 7\cdot (6p^2+q)}$

$= \dfrac{3}{2}$

43. $\dfrac{3x+4y}{x^2+4xy+4y^2} \cdot \dfrac{x+2y}{2}$

$= \dfrac{3x+4y}{(x+2y)(x+2y)} \cdot \dfrac{x+2y}{2}$

$= \dfrac{(3x+4y)\cdot(x+2y)}{2\cdot(x+2y)\cdot(x+2y)}$

$= \dfrac{3x+4y}{2(x+2y)}$

45. $\dfrac{x^2-9}{x^2+8} \div \dfrac{3-x}{2x^2+16}$

$= \dfrac{x^2-9}{x^2+8} \cdot \dfrac{2x^2+16}{3-x}$

$= \dfrac{(x+3)(x-3)}{x^2+8} \cdot \dfrac{2(x^2+8)}{-1(x-3)}$

$= \dfrac{2\cdot(x+3)\cdot(x-3)\cdot(x^2+8)}{-1\cdot(x^2+8)\cdot(x-3)}$

$= -2(x+3)$

47. $\dfrac{(x+2)^2}{x-2} \div \dfrac{x^2-4}{2x-4}$

$= \dfrac{(x+2)^2}{(x-2)} \cdot \dfrac{2x-4}{x^2-4}$

$= \dfrac{(x+2)(x+2)}{(x-2)} \cdot \dfrac{2(x-2)}{(x+2)(x-2)}$

$= \dfrac{2\cdot(x+2)(x+2)\cdot(x-2)}{(x-2)\cdot(x+2)(x-2)}$

$= \dfrac{2(x+2)}{x-2}$

49. $\frac{a^2+7a+12}{a^2+5a+6}\cdot\frac{a^2+8a+15}{a^2+5a+4}$
$=\frac{(a+3)(a+4)}{(a+3)(a+2)}\cdot\frac{(a+3)(a+5)}{(a+1)(a+4)}$
$=\frac{(a+3)\cdot(a+4)\cdot(a+3)\cdot(a+5)}{(a+3)\cdot(a+2)\cdot(a+1)\cdot(a+4)}$
$=\frac{(a+3)(a+5)}{(a+2)(a+1)}$

51. $\frac{1}{-x-4}\div\frac{x^2-7x}{x^2-3x-28}$
$=\frac{1}{-x-4}\cdot\frac{x^2-3x-28}{x^2-7x}$
$=\frac{1}{-1(x+4)}\cdot\frac{(x-7)(x+4)}{x(x-7)}$
$=\frac{(x-7)\cdot(x+4)}{-1\cdot x\cdot(x+4)\cdot(x-7)}$
$=-\frac{1}{x}$

53. $\frac{x^2-5x-24}{2x^2-2x-24}\cdot\frac{4x^2+4x-24}{x^2-10x+16}$
$=\frac{(x-8)(x+3)}{2(x^2-x-12)}\cdot\frac{4(x^2+x-6)}{(x-8)(x-2)}$
$=\frac{(x-8)(x+3)}{2(x-4)(x+3)}\cdot\frac{4(x+3)(x-2)}{(x-8)(x-2)}$
$=\frac{2\cdot2\cdot(x-8)\cdot(x+3)\cdot(x+3)\cdot(x-2)}{2\cdot(x-4)\cdot(x+3)\cdot(x-8)\cdot(x-2)}$
$=\frac{2(x+3)}{x-4}$

55. $(x-5)\div\frac{5-x}{x^2+2}=(x-5)\cdot\frac{x^2+2}{5-x}$
$=(x-5)\cdot\frac{x^2+2}{-1(x-5)}$
$=\frac{(x-5)\cdot(x^2+2)}{-1\cdot(x-5)}$
$=-(x^2+2)$

57. $\frac{x^2-y^2}{x^2-2xy+y^2}\cdot\frac{y-x}{x+y}$
$=\frac{(x+y)(x-y)}{(x-y)(x-y)}\cdot\frac{-1(x-y)}{x+y}$
$=\frac{-1\cdot(x+y)\cdot(x-y)\cdot(x-y)}{(x-y)\cdot(x-y)\cdot(x+y)}$
$=-1$

59. $\frac{x^2-9}{2x}\div\frac{x+3}{8x^4}=\frac{x^2-9}{2x}\cdot\frac{8x^4}{x+3}$
$=\frac{(x+3)(x-3)}{2x}\cdot\frac{8x^4}{x+3}$
$=\frac{2x\cdot4x^3\cdot(x+3)(x-3)}{2x\cdot(x+3)}$
$=4x^3(x-3)$

61. 10 square feet $=\frac{10\text{ sq. ft}}{1}\cdot\frac{144\text{ sq. in.}}{1\text{ sq. ft}}$
$=1440$ square inches

63. 3,707,745 square feet
$=\frac{3,707,745\text{ sq. ft}}{1}\cdot\frac{1\text{ sq. yd}}{9\text{ sq. ft}}$
$\approx 411,972$ square yards

65. 50 miles per hour
$=\frac{50\text{ mi}}{1\text{ hr}}\cdot\frac{5280\text{ ft}}{1\text{ mi}}\cdot\frac{1\text{ hr}}{3600\text{ sec}}$
≈ 73 feet per second

67. 5023 feet per second
$=\frac{5023\text{ ft}}{1\text{ sec}}\cdot\frac{1\text{ mi}}{5280\text{ ft}}\cdot\frac{3600\text{ sec}}{1\text{ hr}}$
≈ 3424.8 miles per hour

69. $\frac{a^2+ac+ba+bc}{a-b} \div \frac{a+c}{a+b}$
$= \frac{a(a+c)+b(a+c)}{a-b} \cdot \frac{a+b}{a+c}$
$= \frac{(a+c)(a+b)}{a-b} \cdot \frac{a+b}{a+c}$
$= \frac{(a+c)\cdot(a+b)\cdot(a+b)}{(a-b)\cdot(a+c)}$
$= \frac{(a+b)^2}{a-b}$

71. $\frac{3x^2+8x+5}{x^2+8x+7} \cdot \frac{x+7}{x^2+4}$
$= \frac{(3x+5)(x+1)}{(x+7)(x+1)} \cdot \frac{x+7}{x^2+4}$
$= \frac{(3x+5)\cdot(x+1)\cdot(x+7)}{(x+7)\cdot(x+1)\cdot(x^2+4)}$
$= \frac{3x+5}{x^2+4}$

73. $\frac{x^3+8}{x^2-2x+4} \cdot \frac{4}{x^2-4}$
$= \frac{(x+2)(x^2-2x+4)}{x^2-2x+4} \cdot \frac{4}{(x+2)(x-2)}$
$= \frac{4\cdot(x+2)\cdot(x^2-2x+4)}{(x+2)(x-2)(x^2-2x+4)}$
$= \frac{4}{x-2}$

75. $\frac{a^2-ab}{6a^2+6ab} \div \frac{a^3-b^3}{a^2-b^2}$
$= \frac{a^2-ab}{6a^2+6ab} \cdot \frac{a^2-b^2}{a^3-b^3}$
$= \frac{a(a-b)}{6a(a+b)} \cdot \frac{(a-b)(a+b)}{(a-b)(a^2+ab+b^2)}$
$= \frac{a\cdot(a-b)\cdot(a-b)\cdot(a+b)}{6\cdot a\cdot(a+b)\cdot(a-b)\cdot(a^2+ab+b^2)}$
$= \frac{a-b}{6(a^2+ab+b^2)}$

77. $\frac{1}{5}+\frac{4}{5}=\frac{5}{5}=1$

79. $\frac{9}{9}-\frac{19}{9}=-\frac{10}{9}$

81. $\frac{6}{5}+\left(\frac{1}{5}-\frac{8}{5}\right)=\frac{6}{5}+\left(-\frac{7}{5}\right)=-\frac{1}{5}$

83.

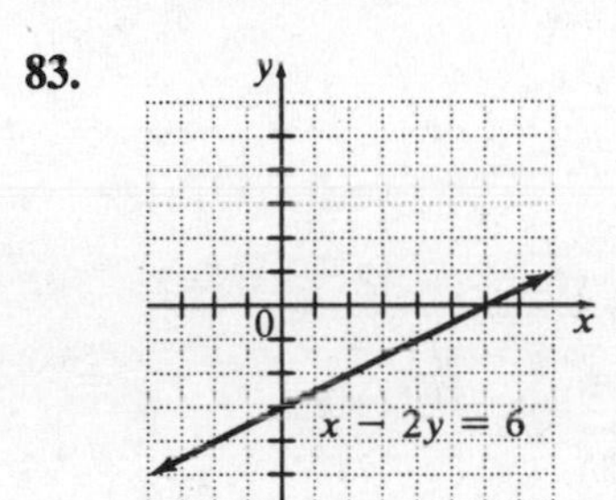

85. $\left(\frac{x^2-y^2}{x^2+y^2} \div \frac{x^2-y^2}{3x}\right) \cdot \frac{x^2+y^2}{6}$
$= \frac{x^2-y^2}{x^2+y^2} \cdot \frac{3x}{x^2-y^2} \cdot \frac{x^2+y^2}{6}$
$= \frac{3x}{6}$
$= \frac{x}{2}$

87. $\left(\frac{2a+b}{b^2} \cdot \frac{3a^2-2ab}{ab+2b^2}\right) \div \frac{a^2-3ab+2b^2}{5ab-10b^2}$
$= \frac{2a+b}{b^2} \cdot \frac{3a^2-2ab}{ab+2b^2} \cdot \frac{5ab-10b^2}{a^2-3ab+2b^2}$
$= \frac{2a+b}{b^2} \cdot \frac{a(3a-2b)}{b(a+2b)} \cdot \frac{5b(a-2b)}{(a-2b)(a-b)}$
$= \frac{5\cdot a\cdot b\cdot(2a+b)\cdot(3a-2b)\cdot(a-2b)}{b\cdot b^2\cdot(a+2b)\cdot(a-2b)\cdot(a-b)}$
$= \frac{5a(2a+b)(3a-2b)}{b^2(a+2b)(a-b)}$

Section 7.3

Mental Math

1. $\frac{2}{3}+\frac{1}{3}=\frac{3}{3}=1$

2. $\frac{5}{11}+\frac{1}{11}=\frac{6}{11}$

3. $\frac{3x}{9}+\frac{4x}{9}=\frac{7x}{9}$

4. $\frac{3y}{8}+\frac{2y}{8}=\frac{5y}{8}$

5. $\frac{8}{9}-\frac{7}{9}=\frac{1}{9}$

6. $-\frac{4}{12}-\frac{3}{12}=-\frac{7}{12}$

7. $\frac{7}{5}-\frac{10y}{5}=\frac{7-10y}{5}$

8. $\frac{12x}{7}-\frac{4x}{7}=\frac{8x}{7}$

Exercise Set 7.3

1. $\frac{a}{13}+\frac{9}{13}=\frac{a+9}{13}$

3. $\frac{9}{3+y}+\frac{y+1}{3+y}=\frac{9+y+1}{3+y}=\frac{10+y}{3+y}$

5. $\frac{4m}{3n}+\frac{5m}{3n}=\frac{4m+5m}{3n}=\frac{9m}{3n}=\frac{3m}{n}$

7. $\frac{2x+1}{x-3}+\frac{3x+6}{x-3}=\frac{2x+1+3x+6}{x-3}=\frac{5x+7}{x-3}$

9. $\frac{7}{8}-\frac{3}{8}=\frac{4}{8}=\frac{1}{2}$

11. $\frac{4m}{m-6}-\frac{24}{m-6}=\frac{4m-24}{m-6}=\frac{4(m-6)}{m-6}=4$

13. $$\begin{aligned}\frac{2x^2}{x-5}-\frac{25+x^2}{x-5}&=\frac{2x^2-(25+x^2)}{x-5}\\&=\frac{2x^2-25-x^2}{x-5}\\&=\frac{x^2-25}{x-5}\\&=\frac{(x+5)(x-5)}{x-5}\\&=x+5\end{aligned}$$

15. $$\begin{aligned}&\frac{-3x^2-4}{x-4}-\frac{12-4x^2}{x-4}\\&=\frac{-3x^2-4-(12-4x^2)}{x-4}\\&=\frac{-3x^2-4-12+4x^2}{x-4}\\&=\frac{x^2-16}{x-4}\\&=\frac{(x+4)(x-4)}{x-4}\\&=x+4\end{aligned}$$

17. $$\begin{aligned}\frac{2x+3}{x+1}-\frac{x+2}{x+1}&=\frac{2x+3-(x+2)}{x+1}\\&=\frac{2x+3-x-2}{x+1}\\&=\frac{x+1}{x+1}\\&=1\end{aligned}$$

19. $\frac{3}{x^3}+\frac{9}{x^3}=\frac{3+9}{x^3}=\frac{12}{x^3}$

21. $\frac{5}{x+4}-\frac{10}{x+4}=\frac{5-10}{x+4}=-\frac{5}{x+4}$

23. $\frac{x}{x+y}-\frac{2}{x+y}=\frac{x-2}{x+y}$

25. $\frac{8x}{2x+5}+\frac{20}{2x+5}=\frac{8x+20}{2x+5}=\frac{4(2x+5)}{2x+5}=4$

27. $\frac{5x+4}{x-1}-\frac{2x+7}{x-1}=\frac{5x+4-(2x+7)}{x-1}$
$=\frac{5x+4-2x-7}{x-1}$
$=\frac{3x-3}{x-1}$
$=\frac{3(x-1)}{x-1}$
$=3$

29. $\frac{a}{a^2+2a-15}-\frac{3}{a^2+2a-15}=\frac{a-3}{a^2+2a-15}$
$=\frac{a-3}{(a+5)(a-3)}$
$=\frac{1}{a+5}$

31. $\frac{2x+3}{x^2-x-30}-\frac{x-2}{x^2-x-30}=\frac{2x+3-(x-2)}{x^2-x-30}$
$=\frac{2x+3-x+2}{x^2-x-30}$
$=\frac{x+5}{x^2-x-30}$
$=\frac{x+5}{(x-6)(x+5)}$
$=\frac{1}{x-6}$

33. $P=4s$
$P=4\left(\frac{5}{x-2}\right)=\frac{20}{x-2}$ meters

35. Answers may vary.

37. $3=3$
$33=3\cdot 11$
$\text{LCD}=3\cdot 11=33$

39. $2x=2\cdot x$
$4x^3=2^2\cdot x^3$
$\text{LCD}=2^2\cdot x^3=4x^3$

41. $8x=2^3\cdot x$
$2x+4=2(x+2)$
$\text{LCD}=2^3\cdot x\cdot(x+2)=8x(x+2)$

43. $3x+3=3(x+1)$
$2x^2+4x+2=2(x+1)^2$
$\text{LCD}=3\cdot 2\cdot(x+1)^2=6(x+1)^2$

45. $x-8=x-8$
$8-x=-1(x-8)$
$\text{LCD}=x-8$ or $8-x$

47. $8x^2(x-1)^2=2^3\cdot x^2\cdot(x-1)^2$
$10x^3(x-1)=2\cdot 5\cdot x^3\cdot(x-1)$
$\text{LCD}=2^3\cdot 5\cdot x^3\cdot(x-1)^2=40x^3(x-1)^2$

49. $2x+1=(2x+1)$
$2x-1=(2x-1)$
$\text{LCD}=(2x+1)(2x-1)$

51. $2x^2+7x-4=(2x-1)(x+4)$
$2x^2+5x-3=(2x-1)(x+3)$
$\text{LCD}=(2x-1)(x+4)(x+3)$

53. Answers may vary.

55. $\frac{3}{2x}\cdot\frac{2x}{2x}=\frac{6x}{4x^2}$

57. $\frac{6}{3a}\cdot\frac{4b^2}{4b^2}=\frac{24b^2}{12ab^2}$

59. $\frac{9}{x+3}\cdot\frac{2}{2}=\frac{18}{2(x+3)}$

61. $\frac{9a+2}{5a+10}\cdot\frac{b}{b}=\frac{9ab+2b}{5b(a+2)}$

63. $\frac{x}{x^2+6x+8}=\frac{x}{(x+4)(x+2)}$
$=\frac{x(x+1)}{(x+4)(x+2)(x+1)}$
$=\frac{x^2+x}{(x+4)(x+2)(x+1)}$

65. $\frac{9y-1}{15x^2-30}=\frac{(9y-1)(2)}{(15x^2-30)(2)}=\frac{18y-2}{30x^2-60}$

67. $\frac{5}{2x^2-9x-5} = \frac{5}{(2x+1)(x-5)} \cdot \frac{3x(x-7)}{3x(x-7)}$
$= \frac{15x(x-7)}{3x(2x+1)(x-7)(x-5)}$

69. $\frac{5}{2-x} = \frac{5(-1)}{(2-x)(-1)} = -\frac{5}{x-2}$

71. $-\frac{7+x}{2-x} = \frac{7+x}{(-1)(2-x)} = \frac{7+x}{x-2}$

73. Since $88 = 2^3 \cdot 11$ and $4332 = 2^2 \cdot 3 \cdot 19^2$, the least common multiple is $2^3 \cdot 3 \cdot 11 \cdot 19^2 = 95{,}304$. It will take 95,304 Earth days for Jupiter and Mercury to align again.

75. Answers may vary.

77. $2x(x+5) = 0$
$2x = 0$ or $x + 5 = 0$
$x = 0$ or $x = -5$

79. $x^2 - 6x + 5 = 0$
$(x-5)(x-1) = 0$
$x - 5 = 0$ or $x - 1 = 0$
$x = 5$ or $x = 1$

81. $\frac{9}{10} - \frac{3}{5} = \frac{9}{10} - \frac{3}{5} \cdot \frac{2}{2} = \frac{9}{10} - \frac{6}{10} = \frac{3}{10}$

83. $\frac{11}{15} + \frac{5}{9} = \frac{11}{15} \cdot \frac{3}{3} + \frac{5}{9} \cdot \frac{5}{5} = \frac{33}{45} + \frac{25}{45} = \frac{58}{45}$

Section 7.4

Mental Math

1. D
2. C
3. A
4. B

Exercise Set 7.4

1. $\frac{4}{2x} + \frac{9}{3x}$; LCD $= 6x$
$\frac{4}{2x} \cdot \frac{3}{3} + \frac{9}{3x} \cdot \frac{2}{2} = \frac{12}{6x} + \frac{18}{6x} = \frac{30}{6x} = \frac{5}{x}$

3. $\frac{15a}{b} + \frac{6b}{5}$; LCD $= 5b$
$\frac{15a}{b} \cdot \frac{5}{5} + \frac{6b}{5} \cdot \frac{b}{b} = \frac{75a}{5b} + \frac{6b^2}{5b} = \frac{75a + 6b^2}{5b}$

5. $\frac{3}{x} + \frac{5}{2x^2}$; LCD $= 2x^2$
$\frac{3}{x} \cdot \frac{2x}{2x} + \frac{5}{2x^2} = \frac{6x}{2x^2} + \frac{5}{2x^2} = \frac{6x+5}{2x^2}$

7. $\frac{6}{x+1} + \frac{9}{2x+2} = \frac{6}{x+1} + \frac{9}{2(x+1)}$;
LCD $= 2(x+1)$
$\frac{6}{(x+1)} \cdot \frac{2}{2} + \frac{9}{2(x+1)} = \frac{12}{2(x+1)} + \frac{9}{2(x+1)}$
$= \frac{21}{2(x+1)}$

9. $\frac{15}{2x-4} + \frac{x}{x^2-4} = \frac{15}{2(x-2)} + \frac{x}{(x+2)(x-2)}$
LCD $= 2(x-2)(x+2)$
$\frac{15}{2(x-2)} \cdot \frac{(x+2)}{(x+2)} + \frac{x}{(x+2)(x-2)} \cdot \frac{2}{2}$
$= \frac{15(x+2)}{2(x-2)(x+2)} + \frac{2x}{2(x-2)(x+2)}$
$= \frac{15x + 30 + 2x}{2(x-2)(x+2)}$
$= \frac{17x + 30}{2(x-2)(x+2)}$

11. $\dfrac{3}{4x}+\dfrac{8}{x-2}$; LCD $= 4x(x-2)$

$\dfrac{3}{4x}\cdot\dfrac{(x-2)}{(x-2)}+\dfrac{8}{(x-2)}\cdot\dfrac{4x}{4x}$

$=\dfrac{3(x-2)}{4x(x-2)}+\dfrac{32x}{4x(x-2)}$

$=\dfrac{3x-6+32x}{4x(x-2)}$

$=\dfrac{35x-6}{4x(x-2)}$

13. $\dfrac{5}{y^2}-\dfrac{y}{2y+1}$; LCD $= y^2(2y+1)$

$\dfrac{5}{y^2}\cdot\dfrac{(2y+1)}{(2y+1)}-\dfrac{y}{(2y+1)}\cdot\dfrac{y^2}{y^2}$

$=\dfrac{5(2y+1)}{y^2(2y+1)}-\dfrac{y^3}{y^2(2y+1)}$

$=\dfrac{10y+5-y^3}{y^2(2y+1)}$

15. Answers may vary.

17. $\dfrac{6}{x-3}+\dfrac{8}{3-x}$

$=\dfrac{6}{x-3}-\dfrac{8}{x-3}$

$=\dfrac{6-8}{x-3}$

$=-\dfrac{2}{x-3}$

19. $\dfrac{-8}{x^2-1}-\dfrac{7}{1-x^2}=\dfrac{-8}{x^2-1}+\dfrac{7}{x^2-1}=-\dfrac{1}{x^2-1}$

21. $\dfrac{x}{x^2-4}-\dfrac{2}{4-x^2}=\dfrac{x}{x^2-4}+\dfrac{2}{x^2-4}$

$=\dfrac{x+2}{x^2-4}$

$=\dfrac{x+2}{(x+2)(x-2)}$

$=\dfrac{1}{x-2}$

23. $\dfrac{5}{x}+2$; LCD $= x$

$\dfrac{5}{x}+2\cdot\dfrac{x}{x}=\dfrac{5}{x}+\dfrac{2x}{x}=\dfrac{5+2x}{x}$

25. $\dfrac{5}{x-2}+6$; LCD $= x-2$

$\dfrac{5}{x-2}+6\cdot\dfrac{(x-2)}{(x-2)}=\dfrac{5}{x-2}+\dfrac{6(x-2)}{x-2}$

$=\dfrac{5+6x-12}{x-2}$

$=\dfrac{6x-7}{x-2}$

27. $\dfrac{y+2}{y+3}-2$; LCD $= y+3$

$\dfrac{y+2}{y+3}-2\cdot\dfrac{(y+3)}{(y+3)}=\dfrac{y+2}{y+3}-\dfrac{2(y+3)}{y+3}$

$=\dfrac{y+2-2y-6}{y+3}$

$=\dfrac{-y-4}{y+3}$

$=\dfrac{-1(y+4)}{y+3}$

$=-\dfrac{y+4}{y+3}$

29. $90°-\left(\dfrac{40}{x}\right)^\circ=\left(90-\dfrac{40}{x}\right)^\circ$

LCD $= x$

$\left(90\cdot\dfrac{x}{x}-\dfrac{40}{x}\right)^\circ=\left(\dfrac{90x}{x}-\dfrac{40}{x}\right)^\circ$

$=\left(\dfrac{90x-40}{x}\right)^\circ$

31. $\frac{5x}{x+2}-\frac{3x-4}{x+2}=\frac{5x-(3x-4)}{x+2}$
$=\frac{5x-3x+4}{x+2}$
$=\frac{2x+4}{x+2}$
$=\frac{2(x+2)}{x+2}$
$=2$

33. $\frac{3x^4}{x}-\frac{4x^2}{x^2}=3x^3-4$

35. $\frac{1}{x+3}-\frac{1}{(x+3)^2}$; LCD $=(x+3)^2$
$\frac{1}{x+3}\cdot\frac{(x+3)}{(x+3)}-\frac{1}{(x+3)^2}$
$=\frac{x+3}{(x+3)^2}-\frac{1}{(x+3)^2}$
$=\frac{x+3-1}{(x+3)^2}$
$=\frac{x+2}{(x+3)^2}$

37. $\frac{4}{5b}+\frac{1}{b-1}$; LCD $=5b(b-1)$
$\frac{4}{5b}\cdot\frac{(b-1)}{(b-1)}+\frac{1}{(b-1)}\cdot\frac{5b}{5b}$
$=\frac{4(b-1)}{5b(b-1)}+\frac{5b}{5b(b-1)}$
$=\frac{4b-4+5b}{5b(b-1)}$
$=\frac{9b-4}{5b(b-1)}$

39. $\frac{2}{m}+1$; LCD $=m$
$\frac{2}{m}+1\cdot\frac{m}{m}=\frac{2}{m}+\frac{m}{m}=\frac{2+m}{m}$

41. $\frac{6}{1-2x}-\frac{4}{2x-1}=\frac{6}{1-2x}+\frac{4}{1-2x}=\frac{10}{1-2x}$

43. $\frac{7}{(x+1)(x-1)}+\frac{8}{(x+1)^2}$;
LCD $=(x+1)^2(x-1)$
$\frac{7}{(x+1)(x-1)}\cdot\frac{(x+1)}{(x+1)}+\frac{8}{(x+1)^2}\cdot\frac{(x-1)}{(x-1)}$
$=\frac{7(x+1)}{(x+1)^2(x-1)}+\frac{8(x-1)}{(x+1)^2(x-1)}$
$=\frac{7x+7+8x-8}{(x+1)^2(x-1)}$
$=\frac{15x-1}{(x+1)^2(x-1)}$

45. $\frac{x}{x^2-1}-\frac{2}{x^2-2x+1}$
$=\frac{x}{(x+1)(x-1)}-\frac{2}{(x-1)^2}$
LCD $=(x+1)(x-1)^2$
$\frac{x}{(x+1)(x-1)}\cdot\frac{(x-1)}{(x-1)}-\frac{2}{(x-1)^2}\cdot\frac{(x+1)}{(x+1)}$
$=\frac{x(x-1)}{(x+1)(x-1)^2}-\frac{2(x+1)}{(x-1)^2(x+1)}$
$=\frac{x^2-x-2x-2}{(x+1)(x-1)^2}$
$=\frac{x^2-3x-2}{(x+1)(x-1)^2}$

47. $\frac{3a}{2a+6}-\frac{a-1}{a+3}=\frac{3a}{2(a+3)}-\frac{a-1}{a+3}$
LCD $=2(a+3)$
$\frac{3a}{2(a+3)}-\frac{(a-1)}{(a+3)}\cdot\frac{2}{2}=\frac{3a}{2(a+3)}-\frac{2(a-1)}{2(a+3)}$
$=\frac{3a-2a+2}{2(a+3)}$
$=\frac{a+2}{2(a+3)}$

49. $\dfrac{5}{2-x}+\dfrac{x}{2x-4}=-\dfrac{5}{x-2}+\dfrac{x}{2(x-2)}$

LCD $= 2(x-2)$

$-\dfrac{5}{(x-2)}\cdot\dfrac{2}{2}+\dfrac{x}{2(x-2)}$

$=\dfrac{-10}{2(x-2)}+\dfrac{x}{2(x-2)}$

$=\dfrac{x-10}{2(x-2)}$

51. $\dfrac{-7}{y^2-3y+2}-\dfrac{2}{y-1}=\dfrac{-7}{(y-2)(y-1)}-\dfrac{2}{y-1}$

LCD $= (y-2)(y-1)$

$\dfrac{-7}{(y-2)(y-1)}-\dfrac{2}{(y-1)}\cdot\dfrac{(y-2)}{(y-2)}$

$=\dfrac{-7}{(y-2)(y-1)}-\dfrac{2(y-2)}{(y-1)(y-2)}$

$=\dfrac{-7-2y+4}{(y-2)(y-1)}$

$=\dfrac{-2y-3}{(y-2)(y-1)}$

53. $\dfrac{13}{x^2-5x+6}-\dfrac{5}{x-3}=\dfrac{13}{(x-3)(x-2)}-\dfrac{5}{x-3}$

LCD $= (x-3)(x-2)$

$\dfrac{13}{(x-3)(x-2)}-\dfrac{5}{(x-3)}\cdot\dfrac{(x-2)}{(x-2)}$

$=\dfrac{13}{(x-3)(x-2)}-\dfrac{5(x-2)}{(x-3)(x-2)}$

$=\dfrac{13-5x+10}{(x-3)(x-2)}$

$=\dfrac{-5x+23}{(x-3)(x-2)}$

55. $\dfrac{8}{(x+2)(x-2)}+\dfrac{4}{(x+2)(x-3)}$

LCD $= (x+2)(x-2)(x-3)$

$\dfrac{8}{(x+2)(x-2)}\cdot\dfrac{(x-3)}{(x-3)}+\dfrac{4}{(x+2)(x-3)}\cdot\dfrac{(x-2)}{(x-2)}$

$=\dfrac{8(x-3)}{(x+2)(x-2)(x-3)}+\dfrac{4(x-2)}{(x+2)(x-3)(x-2)}$

$=\dfrac{8x-24+4x-8}{(x+2)(x-2)(x-3)}$

$=\dfrac{12x-32}{(x+2)(x-2)(x-3)}$

57. $\dfrac{5}{9x^2-4}+\dfrac{2}{3x-2}=\dfrac{5}{(3x+2)(3x-2)}+\dfrac{2}{3x-2}$

LCD $= (3x+2)(3x-2)$

$\dfrac{5}{(3x+2)(3x-2)}+\dfrac{2}{(3x-2)}\cdot\dfrac{(3x+2)}{(3x+2)}$

$=\dfrac{5}{(3x+2)(3x-2)}+\dfrac{2(3x+2)}{(3x-2)(3x+2)}$

$=\dfrac{5+6x+4}{(3x+2)(3x-2)}$

$=\dfrac{6x+9}{(3x+2)(3x-2)}$

59. $\dfrac{x+8}{x^2-5x-6}+\dfrac{x+1}{x^2-4x-5}$

$=\dfrac{x+8}{(x-6)(x+1)}+\dfrac{x+1}{(x-5)(x+1)}$

LCD $= (x-6)(x+1)(x-5)$

$\dfrac{(x+8)}{(x-6)(x+1)}\cdot\dfrac{(x-5)}{(x-5)}+\dfrac{(x+1)}{(x-5)(x+1)}\cdot\dfrac{(x-6)}{(x-6)}$

$=\dfrac{(x+8)(x-5)}{(x-6)(x+1)(x-5)}+\dfrac{(x+1)(x-6)}{(x-5)(x+1)(x-6)}$

$=\dfrac{x^2-5x+8x-40+x^2-6x+x-6}{(x-6)(x+1)(x-5)}$

$=\dfrac{2x^2-2x-46}{(x-6)(x+1)(x-5)}$

61. $\frac{3}{x+4}-\frac{1}{x-4}$
LCD = $(x+4)(x-4)$
$$\frac{3}{(x+4)}\cdot\frac{(x-4)}{(x-4)}-\frac{1}{(x-4)}\cdot\frac{(x+4)}{(x+4)}$$
$$=\frac{3(x-4)}{(x+4)(x-4)}-\frac{1(x+4)}{(x-4)(x+4)}$$
$$=\frac{3x-12-x-4}{(x+4)(x-4)}$$
$$=\frac{2x-16}{(x+4)(x-4)}\text{ inches}$$

63. C
$$\frac{3}{x}+\frac{y}{x}=\frac{3+y}{x}$$

65. B
$$\frac{3}{x}\cdot\frac{y}{x}=\frac{3y}{x^2}$$

67. $\frac{15x}{x+8}\cdot\frac{2x+16}{3x}=\frac{15x}{x+8}\cdot\frac{2(x+8)}{3x}$
$$=\frac{30x}{3x}$$
$$=10$$

69. $\frac{8x+7}{3x+5}-\frac{2x-3}{3x+5}=\frac{8x+7-(2x-3)}{3x+5}$
$$=\frac{8x+7-2x+3}{3x+5}$$
$$=\frac{6x+10}{3x+5}$$
$$=\frac{2(3x+5)}{3x+5}$$
$$=2$$

71. $\frac{5a+10}{18}\div\frac{a^2-4}{10a}=\frac{5(a+2)}{2\cdot 9}\cdot\frac{2\cdot 5a}{(a-2)(a+2)}$
$$=\frac{25a}{9(a-2)}$$

73. $\frac{5}{x^2-3x+2}+\frac{1}{x-2}=\frac{5}{(x-2)(x-1)}+\frac{1}{x-2}$
LCD = $(x-2)(x-1)$
$$\frac{5}{(x-2)(x-1)}+\frac{1}{(x-2)}\cdot\frac{(x-1)}{(x-1)}$$
$$=\frac{5}{(x-2)(x-1)}+\frac{x-1}{(x-2)(x-1)}$$
$$=\frac{5+x-1}{(x-2)(x-1)}$$
$$=\frac{x+4}{(x-2)(x-1)}$$

75. $\left(\frac{2}{3}-\frac{1}{x}\right)\cdot\left(\frac{3}{x}+\frac{1}{2}\right)$
$$=\left(\frac{2x}{3x}-\frac{3}{3x}\right)\cdot\left(\frac{6}{2x}+\frac{x}{2x}\right)$$
$$=\frac{2x-3}{3x}\cdot\frac{x+6}{2x}$$
$$=\frac{2x^2+9x-18}{6x^2}$$

77. $\left(\frac{1}{x}+\frac{2}{3}\right)-\left(\frac{1}{x}-\frac{2}{3}\right)$
$$=\frac{1}{x}+\frac{2}{3}-\frac{1}{x}+\frac{2}{3}$$
$$=\frac{4}{3}$$

79. $\left(\frac{2a}{3}\right)^2\div\left(\frac{a^2}{a+1}-\frac{1}{a+1}\right)$
$$=\frac{4a^2}{9}\div\frac{a^2-1}{a+1}$$
$$=\frac{4a^2}{9}\cdot\frac{a+1}{(a+1)(a-1)}$$
$$=\frac{4a^2}{9}\cdot\frac{1}{a-1}$$
$$=\frac{4a^2}{9(a-1)}$$

81. $\left(\frac{2x}{3}\right)^2 \div \left(\frac{x}{3}\right)^2 = \left(\frac{\frac{2x}{3}}{\frac{x}{3}}\right)^2 = 2^2 = 4$

83.

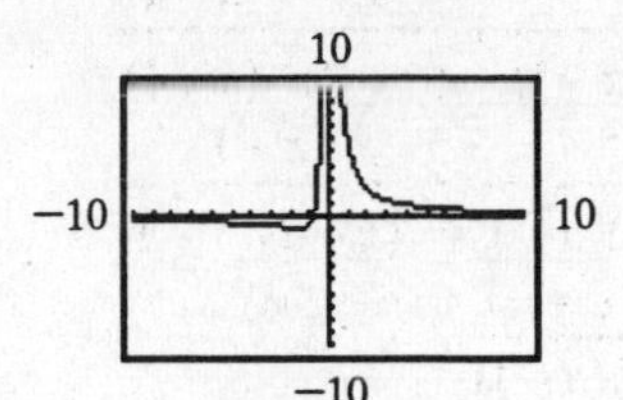

85.

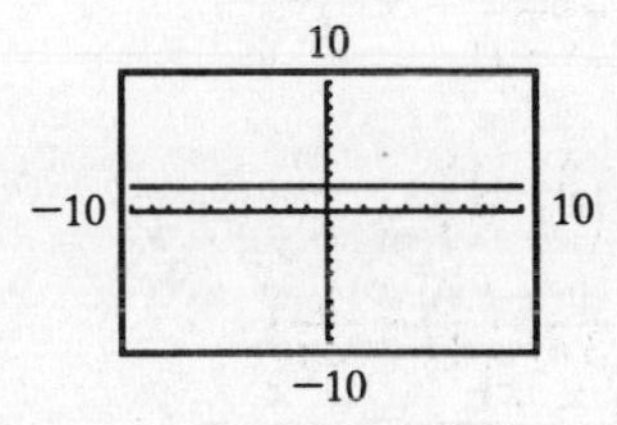

87. Answers may vary.

89. $\frac{\frac{3}{4}+\frac{1}{4}}{\frac{3}{8}+\frac{13}{8}} = \frac{\frac{4}{4}}{\frac{16}{8}} = \frac{4}{4}\cdot\frac{8}{16} = \frac{4\cdot 8}{4\cdot 8\cdot 2} = \frac{1}{2}$

91. $\frac{\frac{2}{5}+\frac{1}{5}}{\frac{7}{10}+\frac{7}{10}} = \frac{\frac{3}{5}}{\frac{14}{10}} = \frac{3}{5}\cdot\frac{10}{14} = \frac{3\cdot 2\cdot 5}{5\cdot 2\cdot 7} = \frac{3}{7}$

93. (1, 2) and (–1, –2)

$m = \frac{-2-2}{-1-1} = \frac{-4}{-2} = 2$

95. (0, 0) and (3, –1)

$m = \frac{-1-0}{3-0} = -\frac{1}{3}$

97.
$$\begin{aligned} x^{-1}+(2x)^{-1} &= \frac{1}{x}+\frac{1}{2x} \\ &= \frac{2}{2x}+\frac{1}{2x} \\ &= \frac{3}{2x} \end{aligned}$$

99.
$$\begin{aligned} 4x^{-2}-3x^{-1} &= \frac{4}{x^2}-\frac{3}{x} \\ &= \frac{4}{x^2}-\frac{3x}{x^2} \\ &= \frac{4-3x}{x^2} \end{aligned}$$

101.
$$\begin{aligned} x^{-3}(2x+1)-5x^{-2} &= \frac{2x+1}{x^3}-\frac{5}{x^2} \\ &= \frac{2x+1}{x^3}-\frac{5x}{x^3} \\ &= \frac{2x+1-5x}{x^3} \\ &= \frac{1-3x}{x^3} \end{aligned}$$

Exercise Set 7.5

1. $\frac{\frac{1}{3}}{\frac{2}{5}} = \frac{1}{3}\cdot\frac{5}{2} = \frac{5}{6}$

3. $\frac{\frac{4}{x}}{\frac{5}{2x}} = \frac{4}{x}\cdot\frac{2x}{5} = \frac{4}{1}\cdot\frac{2}{5} = \frac{8}{5}$

5. $\frac{\frac{10}{3x}}{\frac{5}{6x}} = \frac{10}{3x}\cdot\frac{6x}{5} = \frac{2}{1}\cdot\frac{2}{1} = 4$

7. $\frac{1+\frac{2}{5}}{2+\frac{3}{5}} = \frac{\left(1+\frac{2}{5}\right)5}{\left(2+\frac{3}{5}\right)5} = \frac{5+2}{10+3} = \frac{7}{13}$

9. $\frac{\frac{4}{x-1}}{\frac{x}{x-1}} = \frac{4}{x-1}\cdot\frac{x-1}{x} = \frac{4}{x}$

11. $\frac{1-\frac{2}{x}}{x-\frac{4}{9x}} = \frac{\left(1-\frac{2}{x}\right)9x}{\left(x-\frac{4}{9x}\right)9x} = \frac{9x-18}{9x^2-4}$

13. $\dfrac{\frac{1}{x+1}-1}{\frac{1}{x-1}+1}=\dfrac{\left(\frac{1}{x+1}-1\right)(x+1)(x-1)}{\left(\frac{1}{x-1}+1\right)(x-1)(x+1)}$
$=\dfrac{[1-(x+1)](x-1)}{[1+(x-1)](x+1)}$
$=\dfrac{(1-x-1)(x-1)}{(1+x-1)(x+1)}$
$=\dfrac{-x(x-1)}{x(x+1)}$
$=\dfrac{-(x-1)}{x+1}$
$=\dfrac{1-x}{x+1}$

15. $\dfrac{x^{-1}}{x^{-2}+y^{-2}}=\dfrac{\frac{1}{x}}{\frac{1}{x^2}+\frac{1}{y^2}}$
$=\dfrac{\frac{1}{x}(x^2y^2)}{\left(\frac{1}{x^2}+\frac{1}{y^2}\right)(x^2y^2)}$
$=\dfrac{xy^2}{y^2+x^2}$

17. $\dfrac{2a^{-1}+3b^{-2}}{a^{-1}-b^{-1}}=\dfrac{\frac{2}{a}+\frac{3}{b^2}}{\frac{1}{a}-\frac{1}{b}}$
$=\dfrac{\left(\frac{2}{a}+\frac{3}{b^2}\right)ab^2}{\left(\frac{1}{a}-\frac{1}{b}\right)ab^2}$
$=\dfrac{2b^2+3a}{b^2-ab}$

19. $\dfrac{1}{x-x^{-1}}=\dfrac{1}{x-\frac{1}{x}}$
$=\dfrac{1\cdot x}{\left(x-\frac{1}{x}\right)\cdot x}$
$=\dfrac{x}{x^2-1}$

21. $\dfrac{\frac{x+1}{7}}{\frac{x+2}{7}}=\dfrac{x+1}{7}\cdot\dfrac{7}{x+2}=\dfrac{x+1}{x+2}$

23. $\dfrac{\frac{1}{2}-\frac{1}{3}}{\frac{3}{4}+\frac{2}{5}}=\dfrac{\left(\frac{1}{2}-\frac{1}{3}\right)60}{\left(\frac{3}{4}+\frac{2}{5}\right)60}=\dfrac{30-20}{45+24}=\dfrac{10}{69}$

25. $\dfrac{\frac{x+1}{3}}{\frac{2x-1}{6}}=\dfrac{x+1}{3}\cdot\dfrac{6}{2x-1}$
$=\dfrac{x+1}{1}\cdot\dfrac{2}{2x-1}=\dfrac{2(x+1)}{2x-1}$

27. $\dfrac{\frac{x}{3}}{\frac{2}{x+1}}=\dfrac{x}{3}\cdot\dfrac{x+1}{2}=\dfrac{x(x+1)}{6}$

29. $\dfrac{\frac{2}{x}+3}{\frac{4}{x^2}-9}=\dfrac{\left(\frac{2}{x}+3\right)x^2}{\left(\frac{4}{x^2}-9\right)x^2}=\dfrac{2x+3x^2}{4-9x^2}$
$=\dfrac{x(2+3x)}{(2-3x)(2+3x)}=\dfrac{x}{2-3x}$

31. $\dfrac{1-\frac{x}{y}}{\frac{x^2}{y^2}-1}=\dfrac{\left(1-\frac{x}{y}\right)y^2}{\left(\frac{x^2}{y^2}-1\right)y^2}$
$=\dfrac{y^2-xy}{x^2-y^2}$
$=\dfrac{y(y-x)}{(x+y)(x-y)}$
$=\dfrac{y(-1)}{x+y}$
$=-\dfrac{y}{x+y}$

33. $\dfrac{\frac{-2x}{x-y}}{\frac{y}{x^2}}=\dfrac{-2x}{x-y}\cdot\dfrac{x^2}{y}=-\dfrac{2x^3}{y(x-y)}$

35. $\dfrac{\frac{2}{x}+\frac{1}{x^2}}{\frac{y}{x^2}}=\dfrac{\left(\frac{2}{x}+\frac{1}{x^2}\right)x^2}{\left(\frac{y}{x^2}\right)x^2}=\dfrac{2x+1}{y}$

37. $\dfrac{\frac{x}{9}-\frac{1}{x}}{1+\frac{3}{x}}=\dfrac{\left(\frac{x}{9}-\frac{1}{x}\right)9x}{\left(1+\frac{3}{x}\right)9x}=\dfrac{x^2-9}{9x+27}$

$=\dfrac{(x+3)(x-3)}{9(x+3)}=\dfrac{x-3}{9}$

39. $\dfrac{\frac{x-1}{x^2-4}}{1+\frac{1}{x-2}}=\dfrac{\frac{x-1}{(x+2)(x-2)}(x+2)(x-2)}{\left(1+\frac{1}{x-2}\right)(x-2)(x+2)}$

$=\dfrac{x-1}{(x-2+1)(x+2)}$

$=\dfrac{x-1}{(x-1)(x+2)}$

$=\dfrac{1}{x+2}$

41. $\dfrac{\frac{4}{5-x}+\frac{5}{x-5}}{\frac{2}{x}+\frac{3}{x-5}}=\dfrac{-\frac{4}{x-5}+\frac{5}{x-5}}{\frac{2(x-5)+3x}{x(x-5)}}$

$=\dfrac{\frac{1}{x-5}}{\frac{2x-10+3x}{x(x-5)}}$

$=\dfrac{1}{x-5}\cdot\dfrac{x(x-5)}{5x-10}$

$=\dfrac{x}{5x-10}$

43. $\dfrac{\frac{x+2}{x}-\frac{2}{x-1}}{\frac{x+1}{x}+\frac{x+1}{x-1}}=\dfrac{\frac{(x+2)(x-1)-2x}{x(x-1)}}{\frac{(x+1)(x-1)+(x+1)x}{x(x-1)}}$

$=\dfrac{x^2+x-2-2x}{x^2-1+x^2+x}$

$=\dfrac{x^2-x-2}{2x^2+x-1}$

$=\dfrac{(x-2)(x+1)}{(2x-1)(x+1)}$

$=\dfrac{x-2}{2x-1}$

45. $\dfrac{\frac{x-2}{x+2}+\frac{x+2}{x-2}}{\frac{x-2}{x+2}-\frac{x+2}{x-2}}=\dfrac{\frac{(x-2)(x-2)+(x+2)(x+2)}{(x+2)(x-2)}}{\frac{(x-2)(x-2)-(x+2)(x+2)}{(x+2)(x-2)}}$

$=\dfrac{x^2-4x+4+x^2+4x+4}{x^2-4x+4-x^2-4x-4}$

$=\dfrac{2x^2+8}{-8x}$

$=-\dfrac{x^2+4}{4x}$

47. $\dfrac{\frac{2}{y^2}-\frac{5}{xy}-\frac{3}{x^2}}{\frac{2}{y^2}+\frac{7}{xy}+\frac{3}{x^2}}=\dfrac{\frac{2x^2-5xy-3y^2}{x^2y^2}}{\frac{2x^2+7xy+3y^2}{x^2y^2}}$

$=\dfrac{(2x+y)(x-3y)}{(2x+y)(x+3y)}$

$=\dfrac{x-3y}{x+3y}$

49. $\dfrac{a^{-1}+1}{a^{-1}-1}=\dfrac{\left(\frac{1}{a}+1\right)a}{\left(\frac{1}{a}-1\right)a}=\dfrac{1+a}{1-a}$

51. $\dfrac{3x^{-1}+(2y)^{-1}}{x^{2}}=\dfrac{\left(\frac{3}{x}+\frac{1}{2y}\right)\cdot 2x^2y}{\left(\frac{1}{x^2}\right)\cdot 2x^2y}$

$=\dfrac{6xy+x^2}{2y}$

53. $\dfrac{2a^{-1}+(2a)^{-1}}{a^{-1}+2a^{-2}}=\dfrac{\left(\frac{2}{a}+\frac{1}{2a}\right)\cdot 2a^2}{\left(\frac{1}{a}+\frac{2}{a^2}\right)\cdot 2a^2}$

$=\dfrac{4a+a}{2a+4}$

$=\dfrac{5a}{2a+4}$

55. $\dfrac{5x^{-1}+2y^{-1}}{x^{-2}y^{-2}}=\dfrac{\left(\frac{5}{x}+\frac{2}{y}\right)\cdot x^2y^2}{\left(\frac{1}{x^2}\cdot\frac{1}{y^2}\right)\cdot x^2y^2}$

$=\dfrac{5xy^2+2x^2y}{1}$

$=5xy^2+2x^2y$

57. $\dfrac{5x^{-1}-2y^{-1}}{25x^{-2}-4y^{-2}}$
$=\dfrac{\left(\frac{5}{x}-\frac{2}{y}\right)\cdot x^2y^2}{\left(\frac{25}{x^2}-\frac{4}{y^2}\right)\cdot x^2y^2}$
$=\dfrac{5xy^2-2x^2y}{25y^2-4x^2}$
$=\dfrac{xy(5y-2x)}{(5y-2x)(5y+2x)}$
$=\dfrac{xy}{5y+2x}$

59. $(x^{-1}+y^{-1})^{-1}=\dfrac{1}{\frac{1}{x}+\frac{1}{y}}\cdot\dfrac{xy}{xy}=\dfrac{xy}{y+x}$

61. $\dfrac{x}{1-\frac{1}{1+\frac{1}{x}}}=\dfrac{x}{1-\frac{x}{x+1}}$
$=\dfrac{x}{\frac{x+1-x}{x+1}}$
$=x\cdot\dfrac{x+1}{1}$
$=x^2+x$

63. $\dfrac{a}{1-\frac{s}{770}}=\dfrac{a\cdot 770}{\left(1-\frac{s}{770}\right)\cdot 770}=\dfrac{770a}{770-s}$

65. $\dfrac{\frac{1}{3}+\frac{3}{4}}{2}=\dfrac{\frac{1}{3}\cdot\frac{4}{4}+\frac{3}{4}\cdot\frac{3}{3}}{2}$
$=\dfrac{\frac{4}{12}+\frac{9}{12}}{2}$
$=\dfrac{\frac{13}{12}}{2}$
$=\dfrac{13}{12}\cdot\dfrac{1}{2}$
$=\dfrac{13}{24}$

67. $\dfrac{1}{\frac{1}{R_1}+\frac{1}{R_2}}=\dfrac{1}{\frac{R_2}{R_1R_2}+\frac{R_1}{R_1R_2}}=\dfrac{1}{\frac{R_2+R_1}{R_1R_2}}=\dfrac{R_1R_2}{R_2+R_1}$

69. $t=\dfrac{d}{r}$
$t=\dfrac{\frac{20x}{3}}{\frac{5x}{9}}=\dfrac{20x}{3}\cdot\dfrac{9}{5x}=\dfrac{4\cdot 5x\cdot 3\cdot 3}{3\cdot 5x}=12$ hours

71. $f(x)=\dfrac{1}{x}$

a. $f(a+h)=\dfrac{1}{a+h}$

b. $f(a)=\dfrac{1}{a}$

c. $\dfrac{f(a+h)-f(a)}{h}=\dfrac{\frac{1}{a+h}-\frac{1}{a}}{h}$

d. $\dfrac{f(a+h)-f(a)}{h}$
$=\dfrac{\frac{1}{a+h}\cdot\frac{a}{a}-\frac{1}{a}\cdot\frac{(a+h)}{(a+h)}}{h}$
$=\dfrac{\frac{a}{a(a+h)}-\frac{a+h}{a(a+h)}}{h}$
$=\dfrac{\frac{a-(a+h)}{a(a+h)}}{h}$
$=\dfrac{\frac{a-a-h}{a(a+h)}}{h}$
$=\dfrac{\frac{-h}{a(a+h)}}{h}$
$=\dfrac{-h}{a(a+h)}\cdot\dfrac{1}{h}$
$=\dfrac{-1}{a(a+h)}$

73. $f(x)=\dfrac{3}{x+1}$

a. $f(a+h)=\dfrac{3}{a+h+1}$

b. $f(a)=\dfrac{3}{a+1}$

c. $\dfrac{f(a+h)-f(a)}{h}=\dfrac{\frac{3}{a+h+1}-\frac{3}{a+1}}{h}$

d. $\dfrac{f(a+h)-f(a)}{h}$

$= \dfrac{h(a+h+1)(a+1)\left[\frac{3}{(a+h+1)}-\frac{3}{(a+1)}\right]}{h(a+h+1)(a+1)\cdot[h]}$

$= \dfrac{3h(a+1)-3h(a+h+1)}{h^2(a+h+1)(a+1)}$

$= \dfrac{3ah+3h-3ah-3h^2-3h}{h^2(a+h+1)(a+1)}$

$= \dfrac{-3h^2}{h^2(a+h+1)(a+1)}$

$= \dfrac{-3}{(a+h+1)(a+1)}$

75. $\dfrac{3x^3y^2}{12x}=\dfrac{x^{3-1}y^2}{4}=\dfrac{x^2y^2}{4}$

77. $\dfrac{144x^5y^5}{-16x^2y}=-9x^{5-2}y^{5-1}=-9x^3y^4$

79. $\dfrac{1}{1-(1-x)^{-1}}=\dfrac{1\cdot(1-x)}{\left(1-\frac{1}{1-x}\right)\cdot(1-x)}$

$= \dfrac{1-x}{1-x-1}$

$= \dfrac{1-x}{-x}$

$= \dfrac{x-1}{x}$

81. $\dfrac{(x+2)^{-1}+(x-2)^{-1}}{(x^2-4)^{-1}}$

$= \dfrac{\left(\frac{1}{x+2}+\frac{1}{x-2}\right)\cdot(x+2)(x-2)}{\left(\frac{1}{x^2-4}\right)\cdot(x+2)(x-2)}$

$= \dfrac{x-2+x+2}{1}$

$= 2x$

83. $\dfrac{3(a+1)^{-1}+4a^{-2}}{(a^3+a^2)^{-1}}$

$= \dfrac{\left(\frac{3}{a+1}+\frac{4}{a^2}\right)\cdot a^2(a+1)}{\left(\frac{1}{a^3+a^2}\right)\cdot a^2(a+1)}$

$= \dfrac{3a^2+4(a+1)}{1}$

$= 3a^2+4a+4$

Section 7.6

Calculator Explorations

1. Graph $y_1=\dfrac{x-4}{2}-\dfrac{x-3}{9}$ and $y_2=\dfrac{5}{18}$.
Use INTERSECT to find the point of intersection.

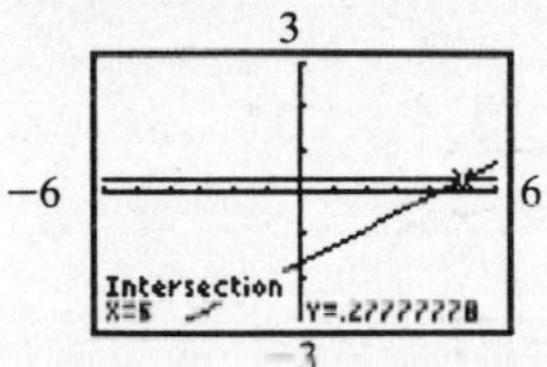

The point of intersection has an x-value of 5, so the solution of the equation is 5.

3. Graph $y_1=3-\dfrac{6}{x}$ and $y_2=x+8$.
Use INTERSECT to find the point of intersection.

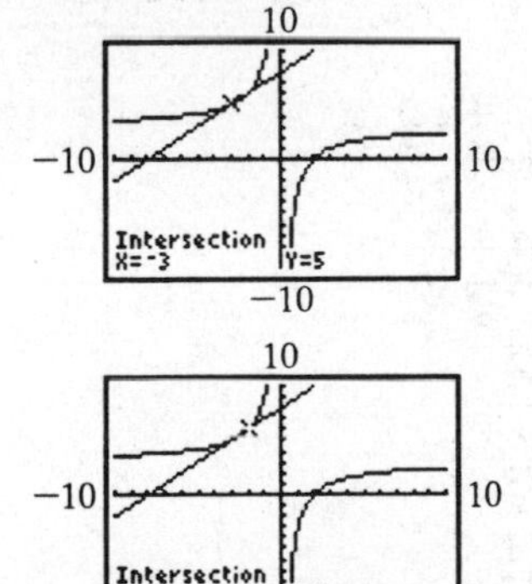

The first graph shows a point of intersection that has an x-value of –3. The second graph shows a point of intersection that has an x-value of –2. Thus, the equation has 2 solutions, $x=-3$ and $x=-2$.

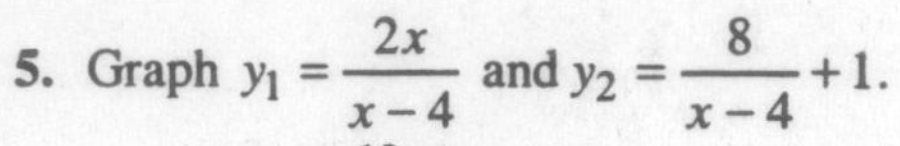

5. Graph $y_1 = \frac{2x}{x-4}$ and $y_2 = \frac{8}{x-4} + 1$.

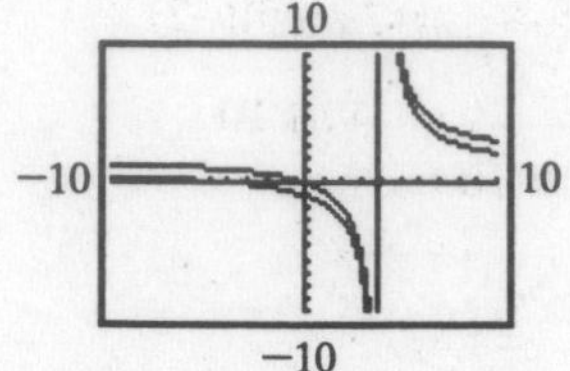

The graphs get closer and closer as they approach $x = 4$ but they never intersect.

Mental Math

1. $$\frac{x}{5} = 2$$
$$5 \cdot \frac{x}{5} = 5 \cdot 2$$
$$x = 10$$

2. $$\frac{x}{8} = 4$$
$$8 \cdot \frac{x}{8} = 8 \cdot 4$$
$$x = 32$$

3. $$\frac{z}{6} = 6$$
$$6 \cdot \frac{z}{6} = 6 \cdot 6$$
$$z = 36$$

4. $$\frac{y}{7} = 8$$
$$7 \cdot \frac{y}{7} = 7 \cdot 8$$
$$y = 56$$

Exercise Set 7.6

1. $$\frac{x}{5} + 3 = 9$$
$$5\left(\frac{x}{5} + 3\right) = 5(9)$$
$$5\left(\frac{x}{5}\right) + 5(3) = 5(9)$$
$$x + 15 = 45$$
$$x = 45 - 15$$
$$x = 30$$

3. $$\frac{x}{2} + \frac{5x}{4} = \frac{x}{12}$$
$$12\left(\frac{x}{2} + \frac{5x}{4}\right) = 12\left(\frac{x}{12}\right)$$
$$12\left(\frac{x}{2}\right) + 12\left(\frac{5x}{4}\right) = 12\left(\frac{x}{12}\right)$$
$$6x + 15x = x$$
$$21x = x$$
$$21x - x = 0$$
$$20x = 0$$
$$x = \frac{0}{20} = 0$$

5. $$2 + \frac{10}{x} = x + 5$$
$$x\left(2 + \frac{10}{x}\right) = x(x+5)$$
$$2x + 10 = x^2 + 5x$$
$$0 = x^2 + 3x - 10$$
$$0 = (x+5)(x-2)$$
$$x + 5 = 0 \quad \text{or} \quad x - 2 = 0$$
$$x = -5 \quad \text{or} \quad x = 2$$

7. $$\frac{a}{5} = \frac{a-3}{2}$$
$$10\left(\frac{a}{5}\right) = 10\left(\frac{a-3}{2}\right)$$
$$2a = 5(a-3)$$
$$2a = 5a - 15$$
$$2a - 5a = -15$$
$$-3a = -15$$
$$a = \frac{-15}{-3} = 5$$

9.

$$\frac{x-3}{5}+\frac{x-2}{2}=\frac{1}{2}$$
$$10\left(\frac{x-3}{5}+\frac{x-2}{2}\right)=10\left(\frac{1}{2}\right)$$
$$10\left(\frac{x-3}{5}\right)+10\left(\frac{x-2}{2}\right)=10\left(\frac{1}{2}\right)$$
$$2(x-3)+5(x-2)=5$$
$$2x-6+5x-10=5$$
$$7x-16=5$$
$$7x=5+16$$
$$7x=21$$
$$x=\frac{21}{7}=3$$

11.

$$\frac{20x}{3}+\frac{32x}{6}=180$$
$$6\left(\frac{20x}{3}+\frac{32x}{6}\right)=6\cdot 180$$
$$40x+32x=1080$$
$$72x=1080$$
$$x=15$$
$$\left(\frac{20x}{3}\right)^{\circ}=\left(\frac{20\cdot 15}{3}\right)^{\circ}=100^{\circ}$$
$$\left(\frac{32x}{6}\right)^{\circ}=\left(\frac{32\cdot 15}{6}\right)^{\circ}=80^{\circ}$$

13.

$$\frac{150}{x}+\frac{450}{x}=90$$
$$x\left(\frac{150}{x}+\frac{450}{x}\right)=x\cdot 90$$
$$150+450=90x$$
$$600=90x$$
$$\frac{600}{90}=x$$
$$\frac{20}{3}=x$$
$$\left(\frac{150}{x}\right)^{\circ}=\left(\frac{150}{\frac{20}{3}}\right)^{\circ}=\left(150\cdot\frac{3}{20}\right)^{\circ}=22.5^{\circ}$$
$$\left(\frac{450}{x}\right)^{\circ}=\left(\frac{450}{\frac{20}{3}}\right)^{\circ}=\left(450\cdot\frac{3}{20}\right)^{\circ}=67.5^{\circ}$$

15.

$$\frac{9}{2a-5}=-2$$
$$(2a-5)\left(\frac{9}{2a-5}\right)=(2a-5)(-2)$$
$$9=-4a+10$$
$$9-10=-4a$$
$$-1=-4a$$
$$\frac{-1}{-4}=a$$
$$\frac{1}{4}=a$$

17.

$$\frac{y}{y+4}+\frac{4}{y+4}=3$$
$$(y+4)\left(\frac{y}{y+4}+\frac{4}{y+4}\right)=(y+4)(3)$$
$$(y+4)\left(\frac{y}{y+4}\right)+(y+4)\left(\frac{4}{y+4}\right)=(y+4)(3)$$
$$y+4=3y+12$$
$$y-3y=12-4$$
$$-2y=8$$
$$y=\frac{8}{-2}=-4$$

-4 is an extraneous solution. If $y=-4$, the denominator would equal zero. The equation has no solution.

19.

$$\begin{aligned}
\frac{2x}{x+2}-2&=\frac{x-8}{x-2}\\
(x+2)(x-2)\left(\frac{2x}{x+2}-2\right)&=(x+2)(x-2)\left(\frac{x-8}{x-2}\right)\\
(x+2)(x-2)\left(\frac{2x}{x+2}\right)+(x+2)(x-2)(-2)&=(x+2)(x-8)\\
(x-2)(2x)+(x+2)(x-2)(-2)&=(x+2)(x-8)\\
2x^2-4x+(x^2-4)(-2)&=x^2-6x-16\\
2x^2-4x-2x^2+8&=x^2-6x-16\\
-4x+8&=x^2-6x-16\\
0&=x^2-6x+4x-16-8\\
0&=x^2-2x-24\\
0&=(x-6)(x+4)\\
x-6=0 \quad &\text{or} \quad x+4=0\\
x=6 \qquad & \qquad x=-4
\end{aligned}$$

21.

$$\begin{aligned}
\frac{4y}{y-4}+5&=\frac{5y}{y-4}\\
(y-4)\left(\frac{4y}{y-4}+5\right)&=(y-4)\left(\frac{5y}{y-4}\right)\\
(y-4)\left(\frac{4y}{y-4}\right)+(y-4)(5)&=(y-4)\left(\frac{5y}{y-4}\right)\\
4y+5y-20&=5y\\
4y+5y-5y&=20\\
4y&=20\\
y&=\frac{20}{4}=5
\end{aligned}$$

23.

$$\begin{aligned}
\frac{7}{x-2}+1&=\frac{x}{x+2}\\
(x-2)(x+2)\left(\frac{7}{x-2}+1\right)&=(x-2)(x+2)\left(\frac{x}{x+2}\right)\\
(x-2)(x+2)\left(\frac{7}{x-2}\right)+(x-2)(x+2)(1)&=(x-2)(x)\\
(x+2)(7)+(x-2)(x+2)(1)&=(x-2)(x)\\
7x+14+x^2-4&=x^2-2x\\
x^2+7x+10&=x^2-2x\\
x^2-x^2+7x+2x&=-10\\
9x&=-10\\
x&=-\frac{10}{9}
\end{aligned}$$

25.

$$\frac{x+1}{x+3}=\frac{2x^2-15x}{x^2+x-6}-\frac{x-3}{x-2}$$
$$\frac{x+1}{x+3}=\frac{2x^2-15x}{(x+3)(x-2)}-\frac{x-3}{x-2}$$
$$(x+3)(x-2)\left(\frac{x+1}{x+3}\right)=(x+3)(x-2)\left(\frac{2x^2-15x}{(x+3)(x-2)}-\frac{x-3}{x-2}\right)$$
$$(x-2)(x+1)=(x+3)(x-2)\left(\frac{2x^2-15x}{(x+3)(x-2)}\right)-(x+3)(x-2)\left(\frac{x-3}{x-2}\right)$$
$$(x-2)(x+1)=2x^2-15x-(x+3)(x-3)$$
$$x^2-x-2=2x^2-15x-(x^2-9)$$
$$x^2-x-2=2x^2-15x-x^2+9$$
$$x^2-x-2=x^2-15x+9$$
$$x^2-x^2-x+15x=9+2$$
$$14x=11$$
$$x=\frac{11}{14}$$

27.

$$\frac{y}{2y+2}+\frac{2y-16}{4y+4}=\frac{2y-3}{y+1}$$
$$\frac{y}{2(y+1)}+\frac{2y-16}{4(y+1)}=\frac{2y-3}{y+1}$$
$$4(y+1)\left(\frac{y}{2(y+1)}+\frac{2y-16}{4(y+1)}\right)=4(y+1)\left(\frac{2y-3}{y+1}\right)$$
$$4(y+1)\left(\frac{y}{2(y+1)}\right)+4(y+1)\left(\frac{2y-16}{4(y+1)}\right)=4(2y-3)$$
$$2y+2y-16=8y-12$$
$$4y-16=8y-12$$
$$4y-8y=-12+16$$
$$-4y=4$$
$$y=\frac{4}{-4}=-1$$

-1 is an extraneous solution. If $y=-1$, the denominator would equal zero. The equation has no solution.

29. Expression;

$$\frac{1}{x}+\frac{2}{3}=\frac{1}{x}\cdot\frac{3}{3}+\frac{2}{3}\cdot\frac{x}{x}=\frac{3}{3x}+\frac{2x}{3x}=\frac{3+2x}{3x}$$

31. Equation;

$$\frac{1}{x}+\frac{2}{3}=\frac{3}{x}$$
$$3x\left(\frac{1}{x}+\frac{2}{3}\right)=3x\left(\frac{3}{x}\right)$$
$$3x\left(\frac{1}{x}\right)+3x\left(\frac{2}{3}\right)=3x\left(\frac{3}{x}\right)$$
$$3+2x=9$$
$$2x=9-3$$
$$2x=6$$
$$x=3$$

33. Expression;

$$\frac{2}{x+1}-\frac{1}{x}=\frac{2}{x+1}\cdot\frac{x}{x}-\frac{1}{x}\cdot\frac{(x+1)}{(x+1)}$$
$$=\frac{2x}{x(x+1)}-\frac{x+1}{x(x+1)}$$
$$=\frac{2x-(x+1)}{x(x+1)}$$
$$=\frac{2x-x-1}{x(x+1)}$$
$$=\frac{x-1}{x(x+1)}$$

35. Equation;

$$\frac{2}{x+1}-\frac{1}{x}=1$$
$$x(x+1)\left(\frac{2}{x+1}-\frac{1}{x}\right)=x(x+1)\cdot 1$$
$$x(x+1)\left(\frac{2}{x+1}\right)+x(x+1)\left(\frac{-1}{x}\right)=x(x+1)$$
$$2x-(x+1)=x(x+1)$$
$$2x-x-1=x^2+x$$
$$x-1=x^2+x$$
$$-1=x^2$$

No solution

37. Answers may vary.

39.

$$\frac{2x}{7}-5x=9$$
$$7\left(\frac{2x}{7}-5x\right)=7(9)$$
$$7\left(\frac{2x}{7}\right)+7(-5x)=7(9)$$
$$2x-35x=63$$
$$-33x=63$$
$$x=-\frac{63}{33}=-\frac{21}{11}$$

41.

$$\frac{2}{y}+\frac{1}{2}=\frac{5}{2y}$$
$$2y\left(\frac{2}{y}+\frac{1}{2}\right)=2y\left(\frac{5}{2y}\right)$$
$$2y\left(\frac{2}{y}\right)+2y\left(\frac{1}{2}\right)=2y\left(\frac{5}{2y}\right)$$
$$4+y=5$$
$$y=5-4$$
$$y=1$$

43.

$$\frac{4x+10}{7}=\frac{8}{2}$$
$$14\left(\frac{4x+10}{7}\right)=14\left(\frac{8}{2}\right)$$
$$2(4x+10)=7(8)$$
$$8x+20=56$$
$$8x=56-20$$
$$8x=36$$
$$x=\frac{36}{8}=\frac{9}{2}$$

45.
$$2+\frac{3}{a-3}=\frac{a}{a-3}$$
$$(a-3)\left(2+\frac{3}{a-3}\right)=(a-3)\left(\frac{a}{a-3}\right)$$
$$(a-3)(2)+(a-3)\left(\frac{3}{a-3}\right)=a$$
$$2a-6+3=a$$
$$2a-3=a$$
$$-3=a-2a$$
$$-3=-a$$
$$\frac{-3}{-1}=a$$
$$3=a$$

3 is an extraneous solution. If $a = 3$, the denominator would equal zero. The equation has no solution.

47.
$$\frac{5}{x}+\frac{2}{3}=\frac{7}{2x}$$
$$6x\left(\frac{5}{x}+\frac{2}{3}\right)=6x\left(\frac{7}{2x}\right)$$
$$6x\left(\frac{5}{x}\right)+6x\left(\frac{2}{3}\right)=21$$
$$30+4x=21$$
$$4x=21-30$$
$$4x=-9$$
$$x=-\frac{9}{4}$$

49.
$$\frac{2a}{a+4}=\frac{3}{a-1}$$
$$(a+4)(a-1)\left(\frac{2a}{a+4}\right)=(a+4)(a-1)\left(\frac{3}{a-1}\right)$$
$$(a-1)(2a)=(a+4)(3)$$
$$2a^2-2a=3a+12$$
$$2a^2-2a-3a-12=0$$
$$2a^2-5a-12=0$$
$$(2a+3)(a-4)=0$$
$$2a+3=0 \quad \text{or} \quad a-4=0$$
$$2a=-3 \qquad a=4$$
$$a=-\frac{3}{2}$$

51.
$$\frac{x+1}{3}-\frac{x-1}{6}=\frac{1}{6}$$
$$6\left(\frac{x+1}{3}-\frac{x-1}{6}\right)=6\left(\frac{1}{6}\right)$$
$$6\left(\frac{x+1}{3}\right)-6\left(\frac{x-1}{6}\right)=1$$
$$2(x+1)-(x-1)=1$$
$$2x+2-x+1=1$$
$$x+3=1$$
$$x=1-3$$
$$x=-2$$

53.

$$\frac{4r-1}{r^2+5r-14}+\frac{2}{r+7}=\frac{1}{r-2}$$
$$(r+7)(r-2)\left(\frac{4r-1}{(r+7)(r-2)}+\frac{2}{r+7}\right)=(r+7)(r-2)\left(\frac{1}{r-2}\right)$$
$$(r+7)(r-2)\left(\frac{4r-1}{(r+7)(r-2)}\right)+(r+7)(r-2)\left(\frac{2}{r+7}\right)=(r+7)(1)$$
$$4r-1+(r-2)(2)=r+7$$
$$4r-1+2r-4=r+7$$
$$6r-5=r+7$$
$$6r-r=7+5$$
$$5r=12$$
$$r=\frac{12}{5}$$

55.

$$\frac{t}{t-4}=\frac{t+4}{6}$$
$$6(t-4)\left(\frac{t}{t-4}\right)=6(t-4)\left(\frac{t+4}{6}\right)$$
$$6t=(t-4)(t+4)$$
$$6t=t^2-16$$
$$0=t^2-6t-16$$
$$0=(t-8)(t+2)$$
$$t-8=0 \quad \text{or} \quad t+2=0$$
$$t=8 \qquad\qquad t=-2$$

57.

$$\frac{x}{2x+6}+\frac{x+1}{3x+9}=\frac{2}{4x+12}$$
$$\frac{x}{2(x+3)}+\frac{x+1}{3(x+3)}=\frac{2}{4(x+3)}$$
$$12(x+3)\left(\frac{x}{2(x+3)}+\frac{x+1}{3(x+3)}\right)=12(x+3)\left(\frac{2}{4(x+3)}\right)$$
$$12(x+3)\left(\frac{x}{2(x+3)}\right)+12(x+3)\left(\frac{x+1}{3(x+3)}\right)=3(2)$$
$$6x+4(x+1)=6$$
$$6x+4x+4=6$$
$$10x+4=6$$
$$10x=6-4$$
$$10x=2$$
$$x=\frac{2}{10}=\frac{1}{5}$$

59.
$$\frac{D}{R} = T$$
$$R\left(\frac{D}{R}\right) = R(T)$$
$$D = RT$$
$$\frac{D}{T} = R$$

61.
$$\frac{3}{x} = \frac{5y}{x+2}$$
$$x(x+2)\left(\frac{3}{x}\right) = x(x+2)\left(\frac{5y}{x+2}\right)$$
$$(x+2)(3) = x(5y)$$
$$3x+6 = 5xy$$
$$\frac{3x+6}{5x} = y$$

63.
$$\frac{3a+2}{3b-2} = -\frac{4}{2a}$$
$$2a(3b-2)\left(\frac{3a+2}{3b-2}\right) = 2a(3b-2)\left(-\frac{4}{2a}\right)$$
$$2a(3a+2) = (3b-2)(-4)$$
$$6a^2+4a = -12b+8$$
$$6a^2+4a-8 = -12b$$
$$-\frac{6a^2+4a-8}{2 \cdot 6} = b$$
$$-\frac{2(3a^2+2a-4)}{2 \cdot 6} = b$$
$$-\frac{3a^2+2a-4}{6} = b$$

65.
$$\frac{A}{BH} = \frac{1}{2}$$
$$2BH\left(\frac{A}{BH}\right) = 2BH\left(\frac{1}{2}\right)$$
$$2A = BH$$
$$\frac{2A}{H} = B$$

67.
$$\frac{C}{\pi r} = 2$$
$$\pi r\left(\frac{C}{\pi r}\right) = \pi r(2)$$
$$C = 2\pi r$$
$$\frac{C}{2\pi} = r$$

69.
$$\frac{1}{a} = \frac{1}{b} + \frac{1}{c}$$
$$abc\left(\frac{1}{a}\right) = abc\left(\frac{1}{b} + \frac{1}{c}\right)$$
$$bc = abc\left(\frac{1}{b}\right) + abc\left(\frac{1}{c}\right)$$
$$bc = ac + ab$$
$$bc = a(c+b)$$
$$\frac{bc}{c+b} = a$$

71.
$$\frac{m^2}{6} - \frac{n}{3} = \frac{p}{2}$$
$$6\left(\frac{m^2}{6} - \frac{n}{3}\right) = 6\left(\frac{p}{2}\right)$$
$$6\left(\frac{m^2}{6}\right) - 6\left(\frac{n}{3}\right) = 3p$$
$$m^2 - 2n = 3p$$
$$m^2 - 3p = 2n$$
$$\frac{m^2-3p}{2} = n$$

73.
$$\frac{5}{a^2+4a+3}+\frac{2}{a^2+a-6}-\frac{3}{a^2-a-2}=0$$
$$\frac{5}{(a+3)(a+1)}+\frac{2}{(a+3)(a-2)}-\frac{3}{(a-2)(a+1)}=0$$
$$(a+3)(a+1)(a-2)\left(\frac{5}{(a+3)(a+1)}+\frac{2}{(a+3)(a-2)}-\frac{3}{(a-2)(a+1)}\right)=(a+3)(a+1)(a-2)(0)$$
$$(a+3)(a+1)(a-2)\left(\frac{5}{(a+3)(a+1)}\right)+(a+3)(a+1)(a-2)\left(\frac{2}{(a+3)(a-2)}\right)$$
$$-(a+3)(a+1)(a-2)\left(\frac{3}{(a-2)(a+1)}\right)=0$$
$$5(a-2)+2(a+1)-3(a+3)=0$$
$$5a-10+2a+2-3a-9=0$$
$$4a-17=0$$
$$4a=17$$
$$a=\frac{17}{4}$$

75. $x^{-2}-19x^{-1}+48=0$
$$\frac{1}{x^2}-\frac{19}{x}+48=0$$
$$1-19x+48x^2=0$$
$$(16x-1)(3x-1)=0$$
$$16x-1=0 \quad\text{or}\quad 3x-1=0$$
$$16x=1 \quad\text{or}\quad 3x=1$$
$$x=\frac{1}{16} \quad\text{or}\quad x=\frac{1}{3}$$

77. $p^{-2}+4p^{-1}-5=0$
$$\frac{1}{p^2}+4\frac{1}{p}-5=0$$
$$1+4p-5p^2=0$$
$$5p^2-4p-1=0$$
$$(5p+1)(p-1)=0$$
$$5p+1=0 \quad\text{or}\quad p-1=0$$
$$5p=-1 \quad\text{or}\quad p=1$$
$$p=-\frac{1}{5} \quad\text{or}\quad p=1$$

79.
$$\frac{1.4}{x-2.6}=\frac{-3.5}{x+7.1}$$
$$1.4(x+7.1)=-3.5(x-2.6)$$
$$1.4x+9.94=-3.5x+9.1$$
$$4.9x=-0.84$$
$$x\approx-0.17$$

81.
$$\frac{10.6}{y}-14.7=\frac{9.92}{3.2}+7.6$$
$$33.92-47.04y=9.92y+24.32y$$
$$33.92=81.28y$$
$$0.42\approx y$$

83. The graph crosses the x-axis at $x=2$. It crosses the y-axis at $y=-2$. The x-intercept is (2, 0) and the y-intercept is (0, –2).

85. The graph crosses the x-axis at $x=-4$, $x=-2$, and $x=3$. It crosses the y-axis at $y=4$. The x-intercepts are (–4, 0), (–2, 0), and (3, 0) and the y-intercept is (0, 4).

Exercise Set 7.7

1. $\frac{2}{15}$

3. $\frac{10}{12} = \frac{5}{6}$

5. 3 gallons = 12 quarts
$\frac{5}{12}$

7. 2 dollars = 40 nickels
$\frac{4}{40} = \frac{1}{10}$

9. 5 meters = 500 centimeters
$\frac{175}{500} = \frac{7}{20}$

11. 3 hours = 180 minutes
$\frac{190}{180} = \frac{19}{18}$

13. Answers may vary.

15. $$\begin{aligned} \frac{2}{3} &= \frac{x}{6} \\ 2 \cdot 6 &= 3 \cdot x \\ 12 &= 3x \\ \frac{12}{3} &= x \\ 4 &= x \end{aligned}$$

17. $$\begin{aligned} \frac{x}{10} &= \frac{5}{9} \\ x \cdot 9 &= 10 \cdot 5 \\ 9x &= 50 \\ x &= \frac{50}{9} \end{aligned}$$

19. $$\begin{aligned} \frac{4x}{6} &= \frac{7}{2} \\ 4x \cdot 2 &= 6 \cdot 7 \\ 8x &= 42 \\ x &= \frac{42}{8} = \frac{21}{4} \end{aligned}$$

21. $$\begin{aligned} \frac{a}{25} &= \frac{12}{10} \\ a \cdot 10 &= 25 \cdot 12 \\ 10a &= 300 \\ a &= \frac{300}{10} = 30 \end{aligned}$$

23. $$\begin{aligned} \frac{x-3}{x} &= \frac{4}{7} \\ 7(x-3) &= 4 \cdot x \\ 7x - 21 &= 4x \\ -21 &= 4x - 7x \\ -21 &= -3x \\ \frac{-21}{-3} &= x \\ 7 &= x \end{aligned}$$

25. $$\begin{aligned} \frac{5x+1}{x} &= \frac{6}{3} \\ 3(5x+1) &= 6 \cdot x \\ 15x + 3 &= 6x \\ 3 &= 6x - 15x \\ 3 &= -9x \\ \frac{3}{-9} &= x \\ -\frac{1}{3} &= x \end{aligned}$$

27. $$\begin{aligned} \frac{x+1}{2x+3} &= \frac{2}{3} \\ 3(x+1) &= 2(2x+3) \\ 3x + 3 &= 4x + 6 \\ 3 - 6 &= 4x - 3x \\ -3 &= x \end{aligned}$$

29. $$\begin{aligned} \frac{9}{5} &= \frac{12}{3x+2} \\ 9(3x+2) &= 5 \cdot 12 \\ 27x + 18 &= 60 \\ 27x &= 60 - 18 \\ 27x &= 42 \\ x &= \frac{42}{27} = \frac{14}{9} \end{aligned}$$

31. $\frac{3}{x+1} = \frac{5}{2x}$
$3 \cdot 2x = 5(x+1)$
$6x = 5x + 5$
$6x - 5x = 5$
$x = 5$

33. $\frac{3}{100} = \frac{x}{4100}$
$3 \cdot 4100 = 100 \cdot x$
$12,300 = 100x$
$123 = x$
The elephant's weight on Pluto is 123 pounds.

35. $\frac{110}{28.4} = \frac{x}{42.6}$
$110 \cdot 42.6 = 28.4 \cdot x$
$4686 = 28.4x$
$\frac{4686}{28.4} = x$
$165 = x$
42.6 grams of this cereal has 165 calories.

37. $\frac{1}{6} = \frac{x}{23,000}$
$1 \cdot 23,000 = 6 \cdot x$
$23,000 = 6x$
$\frac{23,000}{6} = x$
$3833 \approx x$
I would expect 3833 women to earn bigger paychecks.

39. $\frac{8}{2} = \frac{36}{x}$
$8 \cdot x = 2 \cdot 36$
$8x = 72$
$x = \frac{72}{8}$
$x = 9$
9 gallons of water are needed.

41. $\frac{39}{250} = \frac{x}{50,000}$
$39 \cdot 50,000 = 250 \cdot x$
$1,950,000 = 250x$
$\frac{1,950,000}{250} = x$
$7800 = x$
We would expect 7800 people to have no health insurance.

43. $\frac{1280}{14} = \frac{x}{2}$
$1280 \cdot 2 = 14 \cdot x$
$2560 = 14x$
$\frac{2560}{14} = x$
$\frac{1280}{7} = x$
$182\frac{6}{7} = x$
2 ounces of Eagle Brand Milk has $182\frac{6}{7}$ calories.

45. $\frac{\$5.79}{110} \approx \$0.0526$
$\frac{\$13.99}{240} \approx \$0.0583$
The best buy is 110 ounces for $5.79.

47. $\frac{\$0.69}{6} = \$0.115$
$\frac{\$0.90}{8} = \$0.1125$
$\frac{\$1.89}{16} \approx \$0.1181$
The best buy is 8 ounces for $0.90.

49. $\frac{\$8.99}{4} = \$2.2475$
$\frac{\$13.99}{6} \approx \$2.3317$
The best buy is the 4-pack for $8.99.

51. $\frac{\$1.57}{1} = \$1.57$

$\frac{\$2.10}{2} = \$1.05$

$\frac{\$3.99}{4} = \$0.9975$

The best buy is 1 gallon for \$3.99.

53. The capacity in 1982 is approximately 70 megawatts. The capacity in 1984 is approximately 600 megawatts. The increase is approximately
$600 - 70 = 530$ megawatts.

55. The megawatt capacity in 2000 is approximately 2650 megawatts.

$$\frac{1000}{560{,}000} = \frac{2650}{x}$$
$$1000 \cdot x = 560{,}000 \cdot 2650$$
$$1000x = 1{,}484{,}000{,}000$$
$$x = 1{,}484{,}000$$

In 2000, the number of megawatts that can be generated from wind will serve the electricity needs of approximately 1,484,000 people.

57. Yes; answers may vary.

59. Notice that **a.** is a proportion and **b.** is not a proportion. We can immediately use cross products to solve for x in equation **a.**

61. (0, 4), (2, 10)

$$m = \frac{10-4}{2-0} = \frac{6}{2} = 3$$

Since the slope is positive, the line moves upward.

63. (–2, 7), (3, –2)

$$m = \frac{-2-7}{3-(-2)} = -\frac{9}{5}$$

Since the slope is negative, the line moves downward.

65. (0, –4), (2, –4)

$$m = \frac{-4-(-4)}{2-0} = \frac{0}{2} = 0$$

Since the slope is 0, the line is horizontal.

Exercise Set 7.8

1. Let x = the unknown number.

Its reciprocal $= \frac{1}{x}$.

$$3\left(\frac{1}{x}\right) = 9\left(\frac{1}{6}\right)$$
$$\frac{3}{x} = \frac{9}{6}$$
$$3 \cdot 6 = x \cdot 9$$
$$18 = 9x$$
$$\frac{18}{9} = x$$
$$2 = x$$

3. Let x = the unknown number.

$$\frac{2x+3}{x+1} = \frac{3}{2}$$
$$2(2x+3) = 3(x+1)$$
$$4x + 6 = 3x + 3$$
$$4x - 3x = 3 - 6$$
$$x = -3$$

5.

	Time	In one hour
Experienced	4	$\frac{1}{4}$
Apprentice	5	$\frac{1}{5}$
Together	x	$\frac{1}{x}$

$$\frac{1}{4} + \frac{1}{5} = \frac{1}{x}$$
$$20x\left(\frac{1}{4} + \frac{1}{5}\right) = 20x\left(\frac{1}{x}\right)$$
$$20x\left(\frac{1}{4}\right) + 20x\left(\frac{1}{5}\right) = 20$$
$$5x + 4x = 20$$
$$9x = 20$$
$$x = \frac{20}{9} = 2\frac{2}{9}$$

Together it will take them $2\frac{2}{9}$ hours.

7.

	Time	In one minute
Belt	2	$\frac{1}{2}$
Smaller	6	$\frac{1}{6}$
Together	x	$\frac{1}{x}$

$$\frac{1}{2}+\frac{1}{6}=\frac{1}{x}$$
$$6x\left(\frac{1}{2}+\frac{1}{6}\right)=6x\left(\frac{1}{x}\right)$$
$$6x\left(\frac{1}{2}\right)+6x\left(\frac{1}{6}\right)=6$$
$$3x+x=6$$
$$4x=6$$
$$x=\frac{6}{4}=\frac{3}{2}=1\frac{1}{2}$$

If both are used, it takes $1\frac{1}{2}$ minutes.

9.

	distance	= rate	· time
Trip	12	$\frac{12}{x}$	x
Return Trip	18	$\frac{18}{x+1}$	$x+1$

$$\frac{12}{x}=\frac{18}{x+1}$$
$$x(x+1)\left(\frac{12}{x}\right)=x(x+1)\left(\frac{18}{x+1}\right)$$
$$12(x+1)=18x$$
$$12x+12=18x$$
$$12=18x-12x$$
$$12=6x$$
$$\frac{12}{6}=\frac{6x}{6}$$
$$2=x$$

Her jogging speed is

$\frac{12}{x}=\frac{12}{2}=6$ miles per hour.

11.

	distance	= rate	· time
First part	20	r	$\frac{20}{r}$
Cooldown	16	$r-2$	$\frac{16}{r-2}$

$$\frac{20}{r}=\frac{16}{r-2}$$
$$20(r-2)=r\cdot 16$$
$$20r-40=16r$$
$$20r-16r=40$$
$$4r=40$$
$$r=\frac{40}{4}=10$$

The first part is at 10 miles per hour, and the cooldown is at 8 miles per hour.

13. $\frac{4}{12}=\frac{x}{18}$
$$12\cdot x=4\cdot 18$$
$$12x=72$$
$$x=\frac{72}{12}=6$$

15. $\frac{x}{3.75}=\frac{12}{9}$
$$x\cdot 9=3.75\cdot 12$$
$$9x=45$$
$$x=\frac{45}{9}=5$$

17. Let x = the unknown number.
$$\frac{1}{4}=\frac{x}{8}$$
$$1\cdot 8=4\cdot x$$
$$8=4x$$
$$\frac{8}{4}=x$$
$$2=x$$

19.

	Time	In one hour
Marcus	6	$\frac{1}{6}$
Tony	4	$\frac{1}{4}$
Together	x	$\frac{1}{x}$

$$\frac{1}{6}+\frac{1}{4}=\frac{1}{x}$$
$$12x\left(\frac{1}{6}+\frac{1}{4}\right)=12x\left(\frac{1}{x}\right)$$
$$12x\left(\frac{1}{6}\right)+12x\left(\frac{1}{4}\right)=12$$
$$2x+3x=12$$
$$5x=12$$
$$x=\frac{12}{5}=2.4 \text{ hours}$$

2.4 hours · \$45/hour = \$108

21. Let x = speed of car in still air.

	distance	= rate	· time
With the wind	11	$x+3$	$\frac{11}{x+3}$
Into wind	10	$x-3$	$\frac{10}{x-3}$

$$\frac{10}{x-3}=\frac{11}{x+3}$$
$$10(x+3)=11(x-3)$$
$$10x+30=11x-33$$
$$30+33=11x-10x$$
$$63=x$$

The speed of the car in still air is 63 miles per hour.

23.
$$\frac{10}{16}=\frac{y}{34}$$
$$16\cdot y=10\cdot 34$$
$$16y=340$$
$$y=\frac{340}{16}=21.25$$

25.
$$\frac{28\text{ ft}}{20\text{ ft}}=\frac{8\text{ ft}}{y}$$
$$28\text{ ft}\cdot y=20\text{ ft}\cdot 8\text{ ft}$$
$$28\text{ ft}\cdot y=160\text{ sq. ft}$$
$$y=\frac{160\text{ sq. ft}}{28\text{ ft}}$$
$$y=\frac{40}{7}\text{ ft}=5\frac{5}{7}\text{ ft}$$

27.
$$\frac{y}{25\text{ ft}}=\frac{3\text{ ft}}{2\text{ ft}}$$
$$y\cdot 2\text{ ft}=25\text{ ft}\cdot 3\text{ ft}$$
$$y\cdot 2\text{ ft}=75\text{ sq. ft}$$
$$y=\frac{75\text{ sq. ft}}{2\text{ ft}}$$
$$y=37\frac{1}{2}\text{ ft}$$

29. Let x = the unknown number.

$$\frac{2}{x-3}-\frac{4}{x+3}=8\left(\frac{1}{x^2-9}\right)$$
$$\frac{2}{x-3}-\frac{4}{x+3}=\frac{8}{x^2-9}$$
$$(x-3)(x-3)\left(\frac{2}{x-3}-\frac{4}{x+3}\right)=(x+3)(x-3)\left(\frac{8}{(x+3)(x-3)}\right)$$
$$(x-3)(x+3)\left(\frac{2}{x-3}\right)-(x-3)(x+3)\left(\frac{4}{x+3}\right)=8$$
$$2(x+3)-4(x-3)=8$$
$$2x+6-4x+12=8$$
$$-2x+18=8$$
$$-2x=8-18$$
$$-2x=-10$$
$$x=\frac{-10}{-2}=5$$

31. Let x = the speed of plane in still air.

	distance = rate · time		
With the wind	630	$x+35$	$\frac{630}{x+35}$
Against the wind	455	$x-35$	$\frac{455}{x-35}$

$$\frac{630}{x+35}=\frac{455}{x-35}$$
$$630(x-35)=455(x+35)$$
$$630x-22{,}050=455x+15{,}925$$
$$630x-455x=15{,}925+22{,}050$$
$$175x=37{,}975$$
$$x=\frac{37{,}975}{175}=217$$

The speed of the plane in still air is 217 miles per hour.

33. Let x = the speed of the wind.

	distance = rate · time		
With the wind	48	$16+x$	$\frac{48}{16+x}$
Against the wind	16	$16-x$	$\frac{16}{16-x}$

$$\frac{48}{16+x}=\frac{16}{16-x}$$
$$48(16-x)=16(16+x)$$
$$768-48x=256+16x$$
$$768-256=16x+48x$$
$$512=64x$$
$$\frac{512}{64}=x$$
$$8=x$$

The wind is 8 miles per hour.

35.

	Time	In one hour
1st custodian	3	$\frac{1}{3}$
2nd custodian	x	$\frac{1}{x}$
Together	$1\frac{1}{2}=\frac{3}{2}$	$\frac{1}{\frac{3}{2}}=\frac{2}{3}$

$$\frac{1}{3}+\frac{1}{x}=\frac{2}{3}$$
$$3x\left(\frac{1}{3}+\frac{1}{x}\right)=3x\left(\frac{2}{3}\right)$$
$$3x\left(\frac{1}{3}\right)+3x\left(\frac{1}{x}\right)=2x$$
$$x+3=2x$$
$$3=2x-x$$
$$3=x$$

It takes the 2nd custodian 3 hours to do the job alone.

37.

	Time	In one hour
First pipe	20	$\frac{1}{20}$
Second pipe	15	$\frac{1}{15}$
Third pipe	x	$\frac{1}{x}$
Together	6	$\frac{1}{6}$

$$\frac{1}{20}+\frac{1}{15}+\frac{1}{x}=\frac{1}{6}$$
$$60x\left(\frac{1}{20}+\frac{1}{15}+\frac{1}{x}\right)=60x\left(\frac{1}{6}\right)$$
$$60x\left(\frac{1}{20}\right)+60x\left(\frac{1}{15}\right)+60x\left(\frac{1}{x}\right)=60x\left(\frac{1}{6}\right)$$
$$3x+4x+60=10x$$
$$7x+60=10x$$
$$60=10x-7x$$
$$60=3x$$
$$20=x$$

The third pipe does the job in 20 hours.

39.
$$\frac{20 \text{ feet}}{6 \text{ inches}} = \frac{x}{8 \text{ inches}}$$
$$20 \text{ feet} \cdot 8 \text{ inches} = 6 \text{ inches} \cdot x$$
$$160 \text{ feet} \cdot \text{inches} = 6 \text{ inches} \cdot x$$
$$\frac{160 \text{ feet} \cdot \text{inches}}{6 \text{ inches}} = x$$
$$\frac{80}{3} \text{ feet} = x$$
$$26\frac{2}{3} \text{ feet} = x$$

41.

	Time	In one hour
Andrew	2	$\frac{1}{2}$
Timothy	3	$\frac{1}{3}$
Together	x	$\frac{1}{x}$

$$\frac{1}{2} + \frac{1}{3} = \frac{1}{x}$$
$$6x\left(\frac{1}{2} + \frac{1}{3}\right) = 6x\left(\frac{1}{x}\right)$$
$$6x\left(\frac{1}{2}\right) + 6x\left(\frac{1}{3}\right) = 6$$
$$3x + 2x = 6$$
$$5x = 6$$
$$\frac{5x}{5} = \frac{6}{5}$$
$$x = \frac{6}{5} = 1\frac{1}{5}$$

Together it will take them $1\frac{1}{5}$ hours.

43.

	Time	In one hour
First cook	6	$\frac{1}{6}$
Second Cook	7	$\frac{1}{7}$
Third cook	x	$\frac{1}{x}$
Together	2	$\frac{1}{2}$

$$\frac{1}{6} + \frac{1}{7} + \frac{1}{x} = \frac{1}{2}$$
$$42x\left(\frac{1}{6} + \frac{1}{7} + \frac{1}{x}\right) = 42x\left(\frac{1}{2}\right)$$
$$42x\left(\frac{1}{6}\right) + 42x\left(\frac{1}{7}\right) + 42x\left(\frac{1}{x}\right) = 21x$$
$$7x + 6x + 42 = 21x$$
$$13x + 42 = 21x$$
$$42 = 21x - 13x$$
$$42 = 8x$$
$$\frac{42}{8} = x$$
$$\frac{21}{4} = x$$
$$5\frac{1}{4} = x$$

The third cook can prepare the pies in $5\frac{1}{4}$ hours.

45. Let x = the time for faster pump.
Then $3x$ = the time for slower pump.

	Time	In one minute
Faster pump	x	$\frac{1}{x}$
Slower pump	$3x$	$\frac{1}{3x}$
Together	21	$\frac{1}{21}$

$$\frac{1}{x}+\frac{1}{3x}=\frac{1}{21}$$
$$21x\left(\frac{1}{x}+\frac{1}{3x}\right)=21x\left(\frac{1}{21}\right)$$
$$21x\left(\frac{1}{x}\right)+21x\left(\frac{1}{3x}\right)=x$$
$$21+7=x$$
$$28=x$$

The faster pump takes 28 minutes and the slower pump takes 84 minutes.

47. From Exercise 46, we know that the age of his death is 84.
Age when son was born
$$=\frac{1}{6}x+\frac{1}{12}x+\frac{1}{7}x+5$$
$$=\frac{1}{6}(84)+\frac{1}{12}(84)+\frac{1}{7}(84)+5$$
$$=14+7+12+5$$
$$=38 \text{ years}$$

Age of son when he died $=\frac{1}{2}x$
$$=\frac{1}{2}(84)$$
$$=42 \text{ years}$$

49. Answers may vary.

51. Let d = the distance the giraffe runs.

	distance =	rate	time
Hyena	$d+0.5$	40	$\frac{d+0.5}{40}$
Giraffe	d	32	$\frac{d}{32}$

$$\frac{d+0.5}{40}=\frac{d}{32}$$
$$32(d+0.5)=40\cdot d$$
$$32d+16=40d$$
$$16=40d-32d$$
$$16=8d$$
$$\frac{16}{8}=d$$
$$2=d$$
$$\frac{d}{32}=\frac{2}{32}=\frac{1}{16}$$

It will take the hyena $\frac{1}{16}$ hour, or 3.75 minutes, to overtake the giraffe.

53. $-x+3y=6$

If $x=0$, then
$$-0+3y=6$$
$$3y=6$$
$$y=2.$$
If $y=0$, then
$$-x+3(0)=6$$
$$-x=6$$
$$x=-6.$$
If $x=-3$, then
$$-(-3)+3y=6$$
$$3+3y=6$$
$$3y=3$$
$$y=1.$$

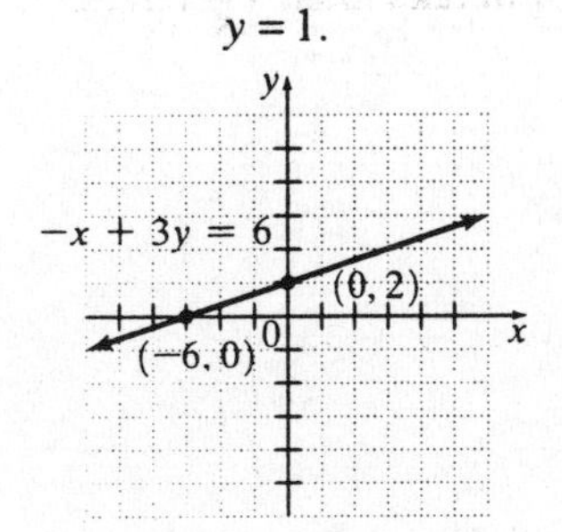

55. $y = 2x$

If $x = 0$, then
$y = 2(0)$
$y = 0$.
If $y = 0$, then
$0 = 2x$
$0 = x$.
If $x = 1$, then
$y = 2(1)$
$y = 2$.

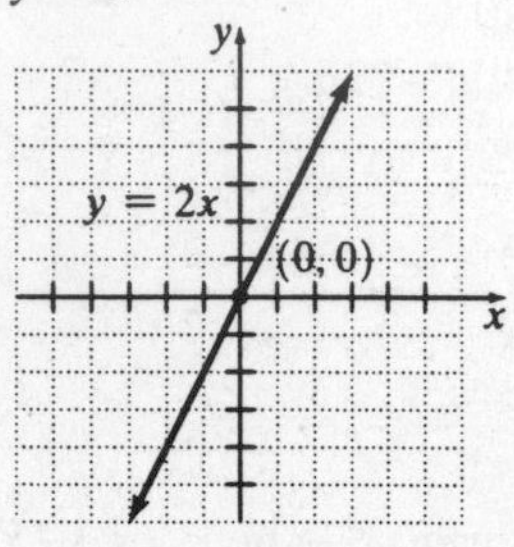

57. $y - x = -5$

If $x = 0$, then
$y - 0 = -5$
$y = -5$.
If $y = 0$, then
$0 - x = -5$
$-x = -5$
$x = 5$.
If $x = 1$, then
$y - 1 = -5$
$y = -4$.

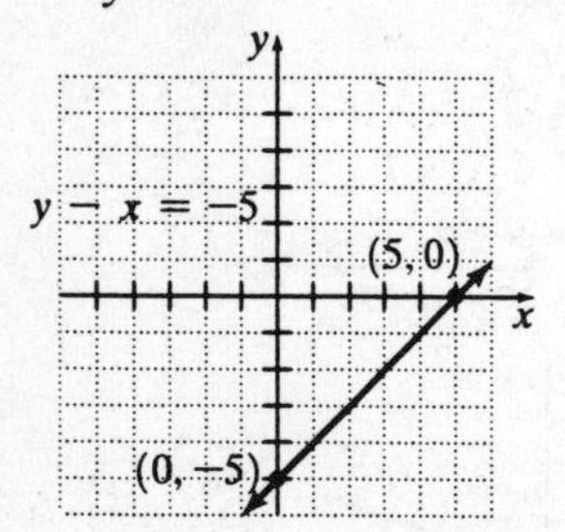

Chapter 7 Review Exercises

1. $F(x) = \dfrac{-3x^2}{x-5}$
Undefined values when
$x - 5 = 0$
$x = 5$
Domain: $\{x \mid x \text{ is a real number and } x \neq 5\}$

2. $h(x) = \dfrac{4x}{3x-12}$
Undefined values when
$3x - 12 = 0$
$3x = 12$
$x = 4$
Domain: $\{x \mid x \text{ is a real number and } x \neq 4\}$

3. $f(x) = \dfrac{x^3+2}{x^2+8x}$
Undefined values when
$x^2 + 8x = 0$
$x(x+8) = 0$
$x = 0$ or $x + 8 = 0$
$x = -8$
Domain :
$\{x \mid x \text{ is a real number and } x \neq 0, x \neq -8\}$

4. $G(x) = \dfrac{20}{3x^2-48}$
Undefined values when
$3x^2 - 48 = 0$
$3(x^2 - 16) = 0$
$3(x+4)(x-4) = 0$
$x + 4 = 0$ or $x - 4 = 0$
$x = -4$ or $x = 4$
Domain: $\{x \mid x \text{ is a real number and } x \neq -4, x \neq 4\}$

5. $\dfrac{15x^4}{45x^2} = \dfrac{15x^{4-2}}{15 \cdot 3} = \dfrac{x^2}{3}$

6. $\dfrac{x+2}{2+x} = \dfrac{x+2}{x+2} = 1$

7. $\dfrac{18m^6p^2}{10m^4p} = \dfrac{2 \cdot 9m^{6-4}p^{2-1}}{2 \cdot 5} = \dfrac{9m^2p}{5}$

8. $\frac{x-12}{12-x} = \frac{-1(12-x)}{12-x} = -1$

9. $\frac{5x-15}{25x-75} = \frac{5(x-3)}{5\cdot 5(x-3)} = \frac{1}{5}$

10. $\frac{22x+8}{11x+4} = \frac{2(11x+4)}{11x+4} = 2$

11. $\frac{2x}{2x^2-2x} = \frac{2x}{2x(x-1)} = \frac{1}{x-1}$

12. $\frac{x+7}{x^2-49} = \frac{x+7}{(x+7)(x-7)} = \frac{1}{x-7}$

13. $\frac{2x^2+4x-30}{x^2+x-20} = \frac{2(x+5)(x-3)}{(x+5)(x-4)}$
$= \frac{2(x-3)}{x-4}$

14. $\frac{xy-3x+2y-6}{x^2+4x+4} = \frac{x(y-3)+2(y-3)}{(x+2)^2}$
$= \frac{(x+2)(y-3)}{(x+2)(x+2)}$
$= \frac{y-3}{x+2}$

15. $C(x) = \frac{35x+4200}{x}$

(a) $C(50) = \frac{35(50)+4200}{50} = \$119$

(b) $C(100) = \frac{35(100)+4200}{100} = \$77$

(c) As the number of bookcases increases, the average cost per bokkcase decreases.

16. $\frac{15x^3y^2}{z}\cdot\frac{z}{5xy^3} = \frac{3\cdot 5\cdot x\cdot x^2\cdot y^2\cdot z}{z\cdot 5\cdot x\cdot y^2\cdot y} = \frac{3x^2}{y}$.

17. $\frac{-y^3}{8}\cdot\frac{9x^2}{y^3} = -\frac{9x^2y^3}{8y^3} = -\frac{9x^2}{8}$

18. $\frac{x^2-9}{x^2-4}\cdot\frac{x-2}{x+3} = \frac{(x-3)(x+3)(x-2)}{(x-2)(x+2)(x+3)}$
$= \frac{x-3}{x+2}$

19. $\frac{2x+5}{x-6}\cdot\frac{2x}{-x+6} = \frac{(2x+5)\cdot 2\cdot x}{(x-6)\cdot(-1)(x-6)}$
$= -\frac{2x(2x+5)}{(x-6)^2}$

20. $\frac{x^2-5x-24}{x^2-x-12} \div \frac{x^2-10x+16}{x^2+x-6}$
$= \frac{(x-8)(x+3)}{(x-4)(x+3)}\cdot\frac{(x+3)(x-2)}{(x-8)(x-2)}$
$= \frac{(x-8)(x+3)(x+3)(x-2)}{(x-4)(x+3)(x-8)(x-2)}$
$= \frac{x+3}{x-4}$

21. $\frac{4x+4y}{xy^2} \div \frac{3x+3y}{x^2y} = \frac{4x+4y}{xy^2}\cdot\frac{x^2y}{3x+3y}$
$= \frac{4(x+y)}{xy^2}\cdot\frac{x^2y}{3(x+y)}$
$= \frac{4(x+y)\cdot x\cdot x\cdot y}{x\cdot y\cdot y\cdot 3(x+y)}$
$= \frac{4x}{3y}$

22. $\frac{x^2+x-42}{x-3}\cdot\frac{(x-3)^2}{x+7}$
$= \frac{(x+7)(x-6)(x-3)(x-3)}{(x-3)(x+7)}$
$= (x-6)(x-3)$

23. $\frac{2a+2b}{3}\cdot\frac{a-b}{a^2-b^2} = \frac{2(a+b)(a-b)}{3(a+b)(a-b)} = \frac{2}{3}$

24. $\dfrac{x^2-9x+14}{x^2-5x+6}\cdot\dfrac{x+2}{x^2-5x-14}$
$=\dfrac{(x-7)(x-2)(x+2)}{(x-3)(x-2)(x-7)(x+2)}$
$=\dfrac{1}{x-3}$

25. $(x-3)\cdot\dfrac{x}{x^2+3x-18}=\dfrac{(x-3)\cdot x}{(x+6)(x-3)}$
$=\dfrac{x}{x+6}$

26. $\dfrac{2x^2-9x+9}{8x-12}\div\dfrac{x^2-3x}{2x}$
$=\dfrac{(2x-3)(x-3)}{2\cdot 2(2x-3)}\cdot\dfrac{2x}{x(x-3)}$
$=\dfrac{(2x-3)(x-3)\cdot 2\cdot x}{2\cdot 2(2x-3)\cdot x(x-3)}$
$=\dfrac{1}{2}$

27. $\dfrac{x^2-y^2}{x^2+xy}\div\dfrac{3x^2-2xy-y^2}{3x^2+6x}$
$=\dfrac{x^2-y^2}{x^2+xy}\cdot\dfrac{3x^2+6x}{3x^2-2xy-y^2}$
$=\dfrac{(x+y)(x-y)}{x(x+y)}\cdot\dfrac{3x(x+2)}{(3x+y)(x-y)}$
$=\dfrac{(x+y)(x-y)\cdot 3\cdot x(x+2)}{x(x+y)(3x+y)(x-y)}$
$=\dfrac{3(x+2)}{3x+y}$

28. $\dfrac{x^2-y^2}{8x^2-16xy+8y^2}\div\dfrac{x+y}{4x-y}$
$=\dfrac{(x-y)(x+y)}{8(x-y)(x-y)}\cdot\dfrac{4x-y}{x+y}$
$=\dfrac{(x-y)(x+y)(4x-y)}{8(x-y)(x-y)(x+y)}$
$=\dfrac{4x-y}{8(x-y)}$

29. $\dfrac{x-y}{4}\div\dfrac{y^2-2y-xy+2x}{16x+24}$
$=\dfrac{x-y}{4}\cdot\dfrac{16x+24}{y^2-2y-xy+2x}$
$=\dfrac{x-y}{4}\cdot\dfrac{8(2x+3)}{y(y-2)-x(y-2)}$
$=\dfrac{x-y}{4}\cdot\dfrac{8(2x+3)}{(y-2)(y-x)}$
$=-\dfrac{y-x}{4}\cdot\dfrac{8(2x+3)}{(y-2)(y-x)}$
$=-\dfrac{2\cdot 4(y-x)(2x+3)}{4(y-2)(y-x)}$
$=-\dfrac{2(2x+3)}{y-2}$

30. $\dfrac{y-3}{4x+3}\div\dfrac{9-y^2}{4x^2-x-3}$
$=\dfrac{y-3}{4x+3}\cdot\dfrac{4x^2-x-3}{9-y^2}$
$=\dfrac{y-3}{4x+3}\cdot\dfrac{(4x+3)(x-1)}{-1(y-3)(y+3)}$
$=\dfrac{(y-3)(4x+3)(x-1)}{-(4x+3)(y-3)(y+3)}$
$=-\dfrac{x-1}{y+3}$

31. $\dfrac{5x-4}{3x-1}+\dfrac{6}{3x-1}=\dfrac{5x-4+6}{3x-1}=\dfrac{5x+2}{3x-1}$

32. $\dfrac{4x-5}{3x^2}-\dfrac{2x+5}{3x^2}=\dfrac{4x-5-(2x+5)}{3x^2}$
$=\dfrac{4x-5-2x-5}{3x^2}$
$=\dfrac{2x-10}{3x^2}$

33. $\frac{9x+7}{6x^2}-\frac{3x+4}{6x^2}=\frac{9x+7-(3x+4)}{6x^2}$
$=\frac{9x+7-3x-4}{6x^2}$
$=\frac{6x+3}{6x^2}$
$=\frac{3(2x+1)}{6x^2}$
$=\frac{2x+1}{2x^2}$

34. $2x=2\cdot x$
$7x=7\cdot x$
$\text{LCD}=2\cdot 7\cdot x=14x$

35. $x^2-5x-24=(x-8)(x+3)$
$x^2+11x+24=(x+8)(x+3)$
$\text{LCD}=(x-8)(x+8)(x+3)$

36. $\frac{x+2}{x^2+11x+18}=\frac{x+2}{(x+2)(x+9)}\cdot\frac{x-5}{x-5}$
$=\frac{(x+2)(x-5)}{(x+2)(x+9)(x-5)}$
$=\frac{x^2-3x-10}{(x+2)(x-5)(x+9)}$

37. $\frac{3x-5}{x^2+4x+4}=\frac{3x-5}{(x+2)^2}\cdot\frac{x+3}{x+3}$
$=\frac{(3x-5)(x+3)}{(x+2)^2(x+3)}$
$=\frac{3x^2+4x-15}{(x+2)^2(x+3)}$

38. $\frac{4}{5x^2}-\frac{6}{y}=\frac{4}{5x^2}\cdot\frac{y}{y}-\frac{6}{y}\cdot\frac{5x^2}{5x^2}$
$=\frac{4y}{5x^2y}-\frac{30x^2}{5x^2y}$
$=\frac{4y-30x^2}{5x^2y}$

39. $\frac{2}{x-3}-\frac{4}{x-1}$
$=\frac{2}{x-3}\cdot\frac{x-1}{x-1}-\frac{4}{x-1}\cdot\frac{x-3}{x-3}$
$=\frac{2(x-1)}{(x-3)(x-1)}-\frac{4(x-3)}{(x-1)(x-3)}$
$=\frac{2(x-1)-4(x-3)}{(x-3)(x-1)}$
$=\frac{2x-2-4x+12}{(x-3)(x-1)}$
$=\frac{-2x+10}{(x-3)(x-1)}$

40. $\frac{x+7}{x+3}-\frac{x-3}{x+7}$
$=\frac{x+7}{x+3}\cdot\frac{x+7}{x+7}-\frac{x-3}{x+7}\cdot\frac{x+3}{x+3}$
$=\frac{(x+7)^2}{(x+3)(x+7)}-\frac{(x-3)(x+3)}{(x+3)(x+7)}$
$=\frac{(x+7)^2-(x-3)(x+3)}{(x+3)(x+7)}$
$=\frac{x^2+14x+49-x^2+9}{(x+3)(x+7)}$
$=\frac{14x+58}{(x+3)(x+7)}$

41. $\frac{4}{x+3}-2=\frac{4}{x+3}-2\cdot\frac{(x+3)}{(x+3)}$
$=\frac{4}{x+3}-\frac{2(x+3)}{x+3}$
$=\frac{4-2(x+3)}{x+3}$
$=\frac{4-2x-6}{x+3}$
$=\frac{-2x-2}{x+3}$

42. $\frac{3}{x^2+2x-8}+\frac{2}{x^2-3x+2}$
$=\frac{3}{(x+4)(x-2)}+\frac{2}{(x-2)(x-1)}$
$=\frac{3}{(x+4)(x-2)}\cdot\frac{x-1}{x-1}+\frac{2}{(x-2)(x-1)}\cdot\frac{x+4}{x+4}$
$=\frac{3(x-1)}{(x+4)(x-2)(x-1)}+\frac{2(x+4)}{(x+4)(x-2)(x-1)}$
$=\frac{3x-3+2x+8}{(x+4)(x-2)(x-1)}$
$=\frac{5x+5}{(x+4)(x-2)(x-1)}$

43. $\frac{2x-5}{6x+9}-\frac{4}{2x^2+3x}$
$=\frac{2x-5}{3(2x+3)}-\frac{4}{x(2x+3)}$
$=\frac{2x-5}{3(2x+3)}\cdot\frac{x}{x}-\frac{4}{x(2x+3)}\cdot\frac{3}{3}$
$=\frac{x(2x-5)}{3x(2x+3)}-\frac{4(3)}{3x(2x+3)}$
$=\frac{2x^2-5x-12}{3x(2x+3)}$
$=\frac{(2x+3)(x-4)}{3x(2x+3)}$
$=\frac{x-4}{3x}$

44. $\frac{x-1}{x^2-2x+1}-\frac{x+1}{x-1}=\frac{x-1}{(x-1)(x-1)}-\frac{x+1}{x-1}$
$=\frac{1}{x-1}-\frac{x+1}{x-1}$
$=\frac{1-(x+1)}{x-1}$
$=\frac{1-x-1}{x-1}$
$=-\frac{x}{x-1}$

45. $\frac{x-1}{x^2+4x+4}+\frac{x-1}{x+2}$
$=\frac{x-1}{(x+2)^2}+\frac{x-1}{x+2}$
$=\frac{x-1}{(x+2)^2}+\frac{(x-1)}{(x+2)}\cdot\frac{(x+2)}{(x+2)}$
$=\frac{x-1}{(x+2)^2}+\frac{(x-1)(x+2)}{(x+2)^2}$
$=\frac{x-1+(x-1)(x+2)}{(x+2)^2}$
$=\frac{x-1+x^2+x-2}{(x+2)^2}$
$=\frac{x^2+2x-3}{(x+2)^2}$

46. $P=2l+2w$
$P=2\left(\frac{x}{8}\right)+2\left(\frac{x+2}{4x}\right)$
$P=\frac{x}{4}+\frac{x+2}{2x}$
$P=\frac{x}{4}\cdot\frac{x}{x}+\frac{x+2}{2x}\cdot\frac{2}{2}$
$P=\frac{x^2}{4x}+\frac{2(x+2)}{4x}$
$P=\frac{x^2+2x+4}{4x}$
$A=l\cdot w$
$A=\left(\frac{x}{8}\right)\left(\frac{x+2}{4x}\right)$
$A=\frac{x(x+2)}{8\cdot 4x}$
$A=\frac{x+2}{32}$

47. $P=\frac{3x}{4x-4}+\frac{2x}{3x-3}+\frac{x}{x-1}$

$P=\frac{3x}{4(x-1)}\cdot\frac{3}{3}+\frac{2x}{3(x-1)}\cdot\frac{4}{4}+\frac{x}{x-1}\cdot\frac{12}{12}$

$P=\frac{9x}{12(x-1)}+\frac{8x}{12(x-1)}+\frac{12x}{12(x-1)}$

$P=\frac{9x+8x+12x}{12(x-1)}$

$P=\frac{29x}{12(x-1)}$

$A=\frac{1}{2}bh$

$A=\frac{1}{2}\left(\frac{x}{x-1}\right)\left(\frac{6y}{5}\right)$

$A=\frac{6xy}{2\cdot 5(x-1)}$

$A=\frac{3xy}{5(x-1)}$

48. $\frac{\frac{5x}{27}}{-\frac{10xy}{21}}=\frac{5x}{27}\cdot\frac{-21}{10xy}=\frac{5x(3)(-7)}{3\cdot 9\cdot 2\cdot 5xy}=-\frac{7}{18y}$

49. $\frac{\frac{8x}{x^2-9}}{\frac{4}{x+3}}=\frac{8x}{x^2-9}\cdot\frac{x+3}{4}$

$=\frac{8x(x+3)}{(x+3)(x-3)4}$

$=\frac{2x}{x-3}$

50. $\frac{\frac{3}{5}+\frac{2}{7}}{\frac{1}{5}+\frac{5}{6}}=\frac{210\left(\frac{3}{5}+\frac{2}{7}\right)}{210\left(\frac{1}{5}+\frac{5}{6}\right)}$

$=\frac{210\left(\frac{3}{5}\right)+210\left(\frac{2}{7}\right)}{210\left(\frac{1}{5}\right)+210\left(\frac{5}{6}\right)}$

$=\frac{126+60}{42+175}$

$=\frac{186}{217}$

$=\frac{6\cdot 31}{7\cdot 31}$

$=\frac{6}{7}$

51. $\frac{\frac{2}{a}+\frac{1}{2a}}{a+\frac{a}{2}}=\frac{2a\left(\frac{2}{a}+\frac{1}{2a}\right)}{2a\left(a+\frac{a}{2}\right)}=\frac{4+1}{2a^2+a^2}=\frac{5}{3a^2}$

52. $\frac{3-y^{-1}}{2-y^{-1}}=\frac{3-\frac{1}{y}}{2-\frac{1}{y}}=\frac{y\left(3-\frac{1}{y}\right)}{y\left(2-\frac{1}{y}\right)}=\frac{3y-1}{2y-1}$

53. $\frac{2+x^{-2}}{x^{-1}+2x^{-2}}=\frac{2+\frac{1}{x^2}}{\frac{1}{x}+\frac{2}{x^2}}=\frac{x^2\left(2+\frac{1}{x^2}\right)}{x^2\left(\frac{1}{x}+\frac{2}{x^2}\right)}=\frac{2x^2+1}{x+2}$

54. $\frac{\frac{6}{x+2}+4}{\frac{8}{x+2}-4}=\frac{(x+2)\left(\frac{6}{x+2}+4\right)}{(x+2)\left(\frac{8}{x+2}-4\right)}$

$=\frac{6+4(x+2)}{8-4(x+2)}$

$=\frac{6+4x+8}{8-4x-8}$

$=\frac{14+4x}{-4x}$

$=-\frac{7+2x}{2x}$

55. a. $f(a+h) = \dfrac{3}{a+h}$

b. $f(a) = \dfrac{3}{a}$

c. $\dfrac{f(a+h) - f(a)}{h} = \dfrac{\frac{3}{a+h} - \frac{3}{a}}{h}$

d.
$$\begin{aligned}\frac{f(a+h) - f(a)}{h} &= \frac{\frac{3}{a+h} - \frac{3}{a}}{h} \\ &= \frac{1}{h}\left(\frac{3a}{a(a+h)} - \frac{3(a+h)}{a(a+h)}\right) \\ &= \frac{1}{h}\left(\frac{3a - 3(a+h)}{a(a+h)}\right) \\ &= \frac{1}{h}\left(\frac{3a - 3a - 3h}{a(a+h)}\right) \\ &= \frac{1}{h}\left(\frac{-3h}{a(a+h)}\right) \\ &= \frac{-3}{a(a+h)}\end{aligned}$$

56.
$$\begin{aligned}\frac{x+4}{9} &= \frac{5}{9} \\ 9x + 36 &= 45 \\ 9x &= 9 \\ \frac{9x}{9} &= \frac{9}{9} \\ x &= 1\end{aligned}$$

57.
$$\begin{aligned}\frac{n}{10} &= 9 - \frac{n}{5} \\ 10\left(\frac{n}{10}\right) &= 10\left(9 - \frac{n}{5}\right) \\ n &= 10(9) - 10\left(\frac{n}{5}\right) \\ n &= 90 - 2n \\ n + 2n &= 90 \\ 3n &= 90 \\ \frac{3n}{3} &= \frac{90}{3} \\ n &= 30\end{aligned}$$

58.
$$\begin{aligned}\frac{5y-3}{7} &= \frac{15y-2}{28} \\ 28(5y-3) &= 7(15y-2) \\ 140y - 84 &= 105y - 14 \\ 35y &= 70 \\ \frac{35y}{35} &= \frac{70}{35} \\ y &= 2\end{aligned}$$

59.
$$\frac{2}{x+1}-\frac{1}{x-2}=-\frac{1}{2}$$
$$2(x+1)(x-2)\left(\frac{2}{x+1}-\frac{1}{x-2}\right)=2(x+1)(x-2)\left(-\frac{1}{2}\right)$$
$$2(x+1)(x-2)\left(\frac{2}{x+1}\right)-2(x+1)(x-2)\left(\frac{1}{x-2}\right)=-(x+1)(x-2)$$
$$4(x-2)-2(x+1)=-(x+1)(x-2)$$
$$4x-8-2x-2=-(x^2-x-2)$$
$$2x-10=-x^2+x+2$$
$$x^2+2x-x-10-2=0$$
$$x^2+x-12=0$$
$$(x+4)(x-3)=0$$
$$x+4=0 \quad \text{or} \quad x-3=0$$
$$x=-4 \qquad\qquad x=3$$

60.
$$\frac{1}{a+3}+\frac{1}{a-3}=-\frac{5}{a^2-9}$$
$$\frac{1}{a+3}+\frac{1}{a-3}=-\frac{5}{(a-3)(a+3)}$$
$$(a-3)(a+3)\left(\frac{1}{a+3}+\frac{1}{a-3}\right)=(a-3)(a+3)\left(-\frac{5}{(a-3)(a+3)}\right)$$
$$a-3+a+3=-5$$
$$2a=-5$$
$$\frac{2a}{2}=-\frac{5}{2}$$
$$a=-\frac{5}{2}$$

61.
$$\frac{y}{2y+2}+\frac{2y-16}{4y+4}=\frac{y-3}{y+1}$$
$$\frac{y}{2(y+1)}+\frac{2(y-8)}{4(y+1)}=\frac{y-3}{y+1}$$
$$\frac{y}{2(y+1)}+\frac{y-8}{2(y+1)}=\frac{y-3}{y+1}$$
$$2(y+1)\left(\frac{y}{2(y+1)}+\frac{y-8}{2(y+1)}\right)=2(y+1)\left(\frac{y-3}{y+1}\right)$$
$$2(y+1)\left(\frac{y}{2(y+1)}\right)+2(y+1)\left(\frac{y-8}{2(y+1)}\right)=2(y-3)$$
$$y+y-8=2y-6$$
$$2y-8=2y-6$$
$$2y-2y=-6+8$$
$$0=2 \quad \text{False}$$

No solution

62.

$$\frac{4}{x+3}+\frac{8}{x^2-9}=0$$
$$\frac{4}{x+3}+\frac{8}{(x-3)(x+3)}=0$$
$$(x-3)(x+3)\left(\frac{4}{x+3}+\frac{8}{(x-3)(x+3)}\right)=(x-3)(x+3)(0)$$
$$(x-3)(x+3)\left(\frac{4}{x+3}\right)+(x-3)(x+3)\left(\frac{8}{(x-3)(x+3)}\right)=0$$
$$4(x-3)+8=0$$
$$4x-12+8=0$$
$$4x-4=0$$
$$4x=4$$
$$\frac{4x}{4}=\frac{4}{4}$$
$$x=1$$

63.

$$\frac{2}{x-3}-\frac{4}{x+3}=\frac{8}{x^2-9}$$
$$\frac{2}{x-3}-\frac{4}{x+3}=\frac{8}{(x+3)(x-3)}$$
$$(x+3)(x-3)\left(\frac{2}{x-3}-\frac{4}{x+3}\right)=(x+3)(x-3)\left(\frac{8}{(x+3)(x-3)}\right)$$
$$(x+3)(x-3)\left(\frac{2}{x-3}\right)-(x+3)(x-3)\left(\frac{4}{x+3}\right)=8$$
$$2(x+3)-4(x-3)=8$$
$$2x+6-4x+12=8$$
$$-2x+18=8$$
$$-2x=8-18$$
$$-2x=-10$$
$$\frac{-2x}{-2}=\frac{-10}{-2}$$
$$x=5$$

64.
$$\frac{x-3}{x+1}-\frac{x-6}{x+5}=0$$
$$(x+1)(x+5)\left(\frac{x-3}{x+1}-\frac{x-6}{x+5}\right)=(x+1)(x+5)(0)$$
$$(x+1)(x+5)\left(\frac{x-3}{x+1}\right)-(x+1)(x+5)\left(\frac{x-6}{x+5}\right)=0$$
$$(x-3)(x+5)-(x-6)(x+1)=0$$
$$x^2+2x-15-x^2+5x+6=0$$
$$7x-9=0$$
$$7x=9$$
$$\frac{7x}{7}=\frac{9}{7}$$
$$x=\frac{9}{7}$$

65.
$$x+5=\frac{6}{x}$$
$$x(x+5)=x\left(\frac{6}{x}\right)$$
$$x^2+5x=6$$
$$x^2+5x-6=0$$
$$(x+6)(x-1)=0$$
$$x+6=0 \quad \text{or} \quad x-1=0$$
$$x=-6 \qquad\qquad x=1$$

66.
$$\frac{4A}{5b}=x^2$$
$$5b\left(\frac{4A}{5b}\right)=5b\cdot x^2$$
$$4A=5bx^2$$
$$\frac{4A}{5x^2}=\frac{5bx^2}{5x^2}$$
$$\frac{4A}{5x^2}=b$$

67.
$$\frac{x}{7}+\frac{y}{8}=10$$
$$56\left(\frac{x}{7}+\frac{y}{8}\right)=56(10)$$
$$56\left(\frac{x}{7}\right)+56\left(\frac{y}{8}\right)=560$$
$$8x+7y=560$$
$$7y=560-8x$$
$$y=\frac{560-8x}{7}$$

68. 1 dollar = 100 cents
$$\frac{20}{100}=\frac{1}{5}$$

69. $\frac{4}{6}=\frac{2}{3}$

70.
$$\frac{x}{2}=\frac{12}{4}$$
$$4x=24$$
$$\frac{4x}{4}=\frac{24}{4}$$
$$x=6$$

71. $\frac{20}{1}=\frac{x}{25}$
$1x = 20\cdot 25$
$x = 500$

72. $\frac{32}{100}=\frac{100}{x}$
$32x = 10,000$
$\frac{32x}{32}=\frac{10,000}{32}$
$x = 312.5$

73. $\frac{20}{2}=\frac{c}{5}$
$2c = 100$
$\frac{2c}{2}=\frac{100}{2}$
$c = 50$

74. $\frac{2}{x-1}=\frac{3}{x+3}$
$2(x+3) = 3(x-1)$
$2x+6 = 3x-3$
$6+3 = 3x-2x$
$9 = x$

75. $\frac{4}{y-3}=\frac{2}{y-3}$
$4(y-3) = 2(y-3)$
$4y-12 = 2y-6$
$4y-2y = -6+12$
$2y = 6$
$\frac{2y}{2}=\frac{6}{2}$
$y = 3$
3 is an extraneous solution.
If $y = 3$, the denominator would be zero.
The equation has no solution.

76. $\frac{y+2}{y}=\frac{5}{3}$
$3(y+2) = 5y$
$3y+6 = 5y$
$6 = 2y$
$\frac{6}{2}=\frac{2y}{2}$
$3 = y$

77. $\frac{x-3}{3x+2}=\frac{2}{6}$
$6(x-3) = 2(3x+2)$
$6x-18 = 6x+4$
$6x-6x = 4+18$
$0 = 22$ False
The equation has no solution.

78. $\frac{\$1.29}{10} = \$0.129$
$\frac{\$2.15}{16} \approx \$0.134$
The best buy is 10 ounces for $1.29.

79. $\frac{\$0.89}{8} \approx \$0.111$
$\frac{\$1.63}{15} \approx \$0.109$
$\frac{\$2.36}{20} = \$0.118$
The best buy is 15 ounces for $1.63.

80. $\frac{300}{20}=\frac{x}{45}$
$20x = 300\cdot 45$
$20x = 13,500$
$\frac{20x}{20}=\frac{13,500}{20}$
$x = 675$
675 parts can be processed in 45 minutes.

81. $\frac{90}{8}=\frac{x}{3}$
$8x=90\cdot 3$
$8x=270$
$\frac{8x}{8}=\frac{270}{8}$
$x=33.75$
He charges $33.75.

82. $\frac{100}{35}=\frac{x}{55}$
$35x=5500$
$\frac{35x}{35}=\frac{5500}{35}$
$x=\frac{1100}{7}=157\frac{1}{7}$
He can address 157 letters in 55 minutes.

83. Let x = the unknown number.
$5\left(\frac{1}{x}\right)=\frac{3}{2}\left(\frac{1}{x}\right)+\frac{7}{6}$
$\frac{5}{x}=\frac{3}{2x}+\frac{7}{6}$
$6x\left(\frac{5}{x}\right)=6x\left(\frac{3}{2x}+\frac{7}{6}\right)$
$30=6x\left(\frac{3}{2x}\right)+6x\left(\frac{7}{6}\right)$
$30=9+7x$
$30-9=7x$
$21=7x$
$\frac{21}{7}=\frac{7x}{7}$
$3=x$

84. Let x = the unknown number.
$\frac{1}{x}=\frac{1}{4-x}$
$x=4-x$
$2x=4$
$\frac{2x}{2}=\frac{4}{2}$
$x=2$

85.

	distance	= rate	· time
Slower car	60	$x-10$	$\frac{60}{x-10}$
Faster car	90	x	$\frac{90}{x}$

$\frac{60}{x-10}=\frac{90}{x}$
$60x=90(x-10)$
$60x=90x-900$
$60x-90x=-900$
$-30x=-900$
$\frac{-30x}{-30}=\frac{-900}{-30}$
$x=30$
The speed of the faster car is 30 mph and the speed of the slower car is 20 mph.

86.

	distance	= rate	· time
Upstream	48	$x-4$	$\frac{48}{x-4}$
Downstream	72	$x+4$	$\frac{72}{x+4}$

$\frac{48}{x-4}=\frac{72}{x+4}$
$48(x+4)=72(x-4)$
$48x+192=72x-288$
$480=24x$
$\frac{480}{24}=\frac{24x}{24}$
$20=x$
The speed of the boat is 20 miles per hour in still water.

87.

	Time	In one hour
Mark	7	$\frac{1}{7}$
Maria	x	$\frac{1}{x}$
Together	5	$\frac{1}{5}$

$$\frac{1}{7}+\frac{1}{x}=\frac{1}{5}$$
$$35x\left(\frac{1}{7}+\frac{1}{x}\right)=35x\left(\frac{1}{5}\right)$$
$$35x\left(\frac{1}{7}\right)+35x\left(\frac{1}{x}\right)=7x$$
$$5x+35=7x$$
$$35=7x-5x$$
$$35=2x$$
$$\frac{35}{2}=\frac{2x}{2}$$
$$17\frac{1}{2}=x$$

It takes Maria $17\frac{1}{2}$ hours to do it.

88.

	Time	In one day
Pipe A	20	$\frac{1}{20}$
Pipe B	15	$\frac{1}{15}$
Together	x	$\frac{1}{x}$

$$\frac{1}{20}+\frac{1}{15}=\frac{1}{x}$$
$$60x\left(\frac{1}{20}+\frac{1}{15}\right)=60x\left(\frac{1}{x}\right)$$
$$60x\left(\frac{1}{20}\right)+60x\left(\frac{1}{15}\right)=60$$
$$3x+4x=60$$
$$7x=60$$
$$\frac{7x}{7}=\frac{60}{7}$$
$$x=8\frac{4}{7}$$

It takes them $8\frac{4}{7}$ days to fill the pond.

89.
$$\frac{2}{3}=\frac{10}{x}$$
$$2x=30$$
$$\frac{2x}{2}=\frac{30}{2}$$
$$x=15$$

90. $\frac{12}{4}=\frac{18}{x}$
$12x=72$
$\frac{12x}{12}=\frac{72}{12}$
$x=6$

91. $\frac{9}{7\frac{1}{5}}=\frac{x}{12}$
$108=7\frac{1}{5}x$
$108=\frac{36}{5}x$
$540=36x$
$\frac{540}{36}=\frac{36x}{36}$
$15=x$

92. $\frac{x}{5}=\frac{30}{2.5}$
$2.5x=150$
$\frac{2.5x}{2.5}=\frac{150}{2.5}$
$x=60$

Chapter 7 Test

1. $f(x)=\frac{x+5}{x^2-6x}$
Undefined values when
$x^2-6x=0$
$x(x-6)=0$
$x=0$ or $x-6=0$
$x=6$
Domain: $\{x|x$ is a real number and $x\neq 0,$ $x\neq 6\}$

2. a. $C(x)=\frac{100x+3000}{x}$
$C(200)=\frac{100(200)+3000}{200}=\$115$

b. $C(x)=\frac{100x+3000}{x}$
$C(1000)=\frac{100(1000)+3000}{1000}=\$103$

3. $\frac{3x-6}{5x-10}=\frac{3(x-2)}{5(x-2)}=\frac{3}{5}$

4. $\frac{x+10}{x^2-100}=\frac{x+10}{(x+10)(x-10)}=\frac{1}{x-10}$

5. $\frac{x+6}{x^2+12x+36}=\frac{x+6}{(x+6)(x+6)}=\frac{1}{x+6}$

6. $\frac{x+3}{x^3+27}=\frac{x+3}{(x+3)(x^2-3x+9)}$
$=\frac{1}{x^2-3x+9}$

7. $\frac{2m^3-2m^2-12m}{m^2-5m+6}=\frac{2m(m^2-m-6)}{(m-3)(m-2)}$
$=\frac{2m(m-3)(m+2)}{(m-3)(m-2)}$
$=\frac{2m(m+2)}{m-2}$

8. $\frac{ay+3a+2y+6}{ay+3a+5y+15}=\frac{a(y+3)+2(y+3)}{a(y+3)+5(y+3)}$
$=\frac{(y+3)(a+2)}{(y+3)(a+5)}$
$=\frac{a+2}{a+5}$

9. $\frac{y-x}{x^2-y^2}=-\frac{x-y}{x^2-y^2}$
$=-\frac{x-y}{(x+y)(x-y)}$
$=-\frac{1}{x+y}$

10. $\dfrac{x^2-13x+42}{x^2+10x+21} \div \dfrac{x^2-4}{x^2+x-6}$
$= \dfrac{x^2-13x+42}{x^2+10x+21} \cdot \dfrac{x^2+x-6}{x^2-4}$
$= \dfrac{(x-6)(x-7)}{(x+7)(x+3)} \cdot \dfrac{(x+3)(x-2)}{(x-2)(x+2)}$
$= \dfrac{(x-6)(x-7)(x+3)(x-2)}{(x+7)(x+3)(x-2)(x+2)}$
$= \dfrac{(x-6)(x-7)}{(x+7)(x+2)}$

11. $\dfrac{3}{x-1} \cdot (5x-5) = \dfrac{3}{x-1} \cdot \dfrac{5(x-1)}{1}$
$= \dfrac{15(x-1)}{x-1}$
$= 15$

12. $\dfrac{y^2-5y+6}{2y+4} \cdot \dfrac{y+2}{2y-6}$
$= \dfrac{(y-3)(y-2)}{2(y+2)} \cdot \dfrac{y+2}{2(y-3)}$
$= \dfrac{(y-3)(y-2)(y+2)}{4(y+2)(y-3)}$
$= \dfrac{y-2}{4}$

13. $\dfrac{5}{2x+5} - \dfrac{6}{2x+5} = \dfrac{5-6}{2x+5}$
$= \dfrac{-1}{2x+5}$
$= -\dfrac{1}{2x+5}$

14. $\dfrac{5a}{a^2-a-6} - \dfrac{2}{a-3}$
$= \dfrac{5a}{(a-3)(a+2)} - \dfrac{2}{a-3}$
$= \dfrac{5a}{(a-3)(a+2)} - \dfrac{2}{a-3} \cdot \dfrac{a+2}{a+2}$
$= \dfrac{5a}{(a-3)(a+2)} - \dfrac{2(a+2)}{(a-3)(a+2)}$
$= \dfrac{5a-2(a+2)}{(a-3)(a+2)}$
$= \dfrac{5a-2a-4}{(a-3)(a+2)}$
$= \dfrac{3a-4}{(a-3)(a+2)}$

15. $\dfrac{6}{x^2-1} + \dfrac{3}{x+1}$
$= \dfrac{6}{(x+1)(x-1)} + \dfrac{3}{x+1}$
$= \dfrac{6}{(x+1)(x-1)} + \dfrac{3}{x+1} \cdot \dfrac{x-1}{x-1}$
$= \dfrac{6}{(x+1)(x-1)} + \dfrac{3(x-1)}{(x+1)(x-1)}$
$= \dfrac{6+3(x-1)}{(x+1)(x-1)}$
$= \dfrac{6+3x-3}{(x+1)(x-1)}$
$= \dfrac{3x+3}{(x+1)(x-1)}$
$= \dfrac{3(x+1)}{(x+1)(x-1)}$
$= \dfrac{3}{x-1}$

16. $\dfrac{x^2-9}{x^2-3x} \div \dfrac{xy+5x+3y+15}{2x+10}$

$= \dfrac{x^2-9}{x^2-3x} \cdot \dfrac{2x+10}{xy+5x+3y+15}$

$= \dfrac{(x+3)(x-3)}{x(x-3)} \cdot \dfrac{2(x+5)}{x(y+5)+3(y+5)}$

$= \dfrac{(x+3)(x-3)}{x(x-3)} \cdot \dfrac{2(x+5)}{(y+5)(x+3)}$

$= \dfrac{2(x+3)(x-3)(x+5)}{x(x-3)(y+5)(x+3)}$

$= \dfrac{2(x+5)}{x(y+5)}$

17. $\dfrac{x+2}{x^2+11x+18} + \dfrac{5}{x^2-3x-10}$

$= \dfrac{x+2}{(x+9)(x+2)} + \dfrac{5}{(x-5)(x+2)}$

$= \dfrac{(x+2)}{(x+9)(x+2)} \cdot \dfrac{x-5}{x-5} + \dfrac{5}{(x-5)(x+2)} \cdot \dfrac{x+9}{x+9}$

$= \dfrac{(x+2)(x-5)}{(x+9)(x+2)(x-5)} + \dfrac{5(x+9)}{(x-5)(x+2)(x+9)}$

$= \dfrac{x^2-3x-10}{(x+9)(x+2)(x-5)} + \dfrac{5x+45}{(x+9)(x+2)(x-5)}$

$= \dfrac{x^2-3x-10+5x+45}{(x+9)(x+2)(x-5)}$

$= \dfrac{x^2+2x+35}{(x+9)(x+2)(x-5)}$

18. $\dfrac{4y}{y^2+6y+5} - \dfrac{3}{y^2+5y+4}$

$= \dfrac{4y}{(y+5)(y+1)} - \dfrac{3}{(y+4)(y+1)}$

$= \dfrac{4y}{(y+5)(y+1)} \cdot \dfrac{(y+4)}{(y+4)} - \dfrac{3}{(y+4)(y+1)} \cdot \dfrac{(y+5)}{(y+5)}$

$= \dfrac{4y(y+4)}{(y+5)(y+1)(y+4)} - \dfrac{3(y+5)}{(y+4)(y+1)(y+5)}$

$= \dfrac{4y(y+4)-3(y+5)}{(y+5)(y+1)(y+4)}$

$= \dfrac{4y^2+16y-3y-15}{(y+5)(y+1)(y+4)}$

$= \dfrac{4y^2+13y-15}{(y+5)(y+1)(y+4)}$

19. $\dfrac{4}{y} - \dfrac{5}{3} = \dfrac{-1}{5}$

$15y\left(\dfrac{4}{y} - \dfrac{5}{3}\right) = 15y\left(-\dfrac{1}{5}\right)$

$15y\left(\dfrac{4}{y}\right) + 15y\left(-\dfrac{5}{3}\right) = -3y$

$60 - 25y = -3y$

$60 = -3y + 25y$

$60 = 22y$

$\dfrac{60}{22} = \dfrac{22y}{22}$

$\dfrac{30}{11} = y$

20. $\dfrac{5}{y+1} = \dfrac{4}{y+2}$

$5(y+2) = 4(y+1)$

$5y + 10 = 4y + 4$

$5y - 4y = 4 - 10$

$y = -6$

21.

$$\frac{a}{a-3} = \frac{3}{a-3} - \frac{3}{2}$$
$$2(a-3)\left(\frac{a}{a-3}\right) = 2(a-3)\left(\frac{3}{a-3} - \frac{3}{2}\right)$$
$$2(a-3)\left(\frac{a}{a-3}\right) = 2(a-3)\left(\frac{3}{a-3}\right) + 2(a-3)\left(-\frac{3}{2}\right)$$
$$2a = 6 - 3(a-3)$$
$$2a = 6 - 3a + 9$$
$$2a = 15 - 3a$$
$$2a + 3a = 15$$
$$5a = 15$$
$$\frac{5a}{5} = \frac{15}{5}$$
$$a = 3$$

3 is an extraneous solution. If $a = 3$, the denominator would be zero. The equation has no solution.

22.

$$\frac{10}{x^2 - 25} = \frac{3}{x+5} + \frac{1}{x-5}$$
$$\frac{10}{(x+5)(x-5)} = \frac{3}{x+5} + \frac{1}{x-5}$$
$$(x+5)(x-5)\left(\frac{10}{(x+5)(x-5)}\right) = (x+5)(x-5)\left(\frac{3}{x+5} + \frac{1}{x-5}\right)$$
$$10 = (x+5)(x-5)\left(\frac{3}{x+5}\right) + (x+5)(x-5)\left(\frac{1}{x-5}\right)$$
$$10 = 3(x-5) + 1(x+5)$$
$$10 = 3x - 15 + x + 5$$
$$10 = 4x - 10$$
$$10 + 10 = 4x$$
$$20 = 4x$$
$$\frac{20}{4} = \frac{4x}{4}$$
$$5 = x$$

5 is an extraneous solution. If $x = 5$, the denominator would be zero. The equation has no solution.

23. $\dfrac{\frac{5x^2}{yz^2}}{\frac{10x}{z^3}} = \dfrac{5x^2}{yz^2} \cdot \dfrac{z^3}{10x} = \dfrac{5 \cdot x \cdot x \cdot z \cdot z^2}{y \cdot z^2 \cdot 2 \cdot 5 \cdot x} = \dfrac{xz}{2y}$

24.
$$\frac{\frac{b}{a}-\frac{a}{b}}{\frac{b}{a}+\frac{b}{a}} = \frac{ab\left(\frac{b}{a}-\frac{a}{b}\right)}{ab\left(\frac{b}{a}+\frac{b}{a}\right)}$$
$$= \frac{ab\left(\frac{b}{a}\right)-ab\left(\frac{a}{b}\right)}{ab\left(\frac{b}{a}\right)+ab\left(\frac{b}{a}\right)}$$
$$= \frac{b^2-a^2}{b^2+b^2}$$
$$= \frac{b^2-a^2}{2b^2}$$

25.
$$\frac{5-y^{-2}}{y^{-1}+2y^{-2}} = \frac{5-\frac{1}{y^2}}{\frac{1}{y}+\frac{2}{y^2}}$$
$$= \frac{y^2\left(5-\frac{1}{y^2}\right)}{y^2\left(\frac{1}{y}+\frac{2}{y^2}\right)}$$
$$= \frac{5y^2-1}{y+2}$$

26.
$$\frac{3}{85} = \frac{x}{510}$$
$$3 \cdot 510 = 85x$$
$$1530 = 85x$$
$$\frac{1530}{85} = \frac{85x}{85}$$
$$18 = x$$

27. Let x = the unknown number.
$$x+5\left(\frac{1}{x}\right) = 6$$
$$x+\frac{5}{x} = 6$$
$$x\left(x+\frac{5}{x}\right) = x(6)$$
$$x(x)+x\left(\frac{5}{x}\right) = 6x$$
$$x^2+5 = 6x$$
$$x^2-6x+5 = 0$$
$$(x-1)(x-5) = 0$$
$$x-1=0 \quad \text{or} \quad x-5=0$$
$$x=1 \qquad\qquad x=5$$

28.

	distance	= rate	· time
downstream	16	$x+2$	$\frac{16}{x+2}$
upstream	14	$x-2$	$\frac{14}{x-2}$

$$\frac{16}{x+2} = \frac{14}{x-2}$$
$$16(x-2) = 14(x+2)$$
$$16x-32 = 14x+28$$
$$16x-14x = 28+32$$
$$2x = 60$$
$$\frac{2x}{2} = \frac{60}{2}$$
$$x = 30$$

The speed of the boat is 30 miles per hour in still water.

29.

	Time	In one hour
First pipe	12	$\frac{1}{12}$
Second pipe	15	$\frac{1}{15}$
Together	x	$\frac{1}{x}$

$$\frac{1}{12}+\frac{1}{15}=\frac{1}{x}$$
$$60x\left(\frac{1}{12}+\frac{1}{15}\right)=60x\left(\frac{1}{x}\right)$$
$$60x\left(\frac{1}{12}\right)+60x\left(\frac{1}{15}\right)=60x\left(\frac{1}{x}\right)$$
$$5x+4x=60$$
$$9x=60$$
$$x=\frac{60}{9}=\frac{20}{3}=6\frac{2}{3}$$

It takes them $6\frac{2}{3}$ hours to fill the tank.

30. $\frac{\$1.19}{6}\approx\$0.198$

$\frac{\$2.15}{10}=\$0.215$

$\frac{\$3.25}{16}\approx\$0.203$

The best buy is 6 ounces for $1.19.

31.
$$\frac{8}{x}=\frac{10}{15}$$
$$8\cdot 15=10x$$
$$120=10x$$
$$\frac{120}{10}=\frac{10x}{10}$$
$$12=x$$

Chapter 8

Exercise Set 8.1

1. $x^2+11x+24=0$
$(x+3)(x+8)=0$
$x+3=0$ or $x+8=0$
$x=-3$ or $x=-8$
The solutions are –3 and –8.

3. $3x-4-5x=x+4+x$
$-2x-4=2x+4$
$-4x=8$
$x=-2$
The solution is –2.

5. $12x^2+5x-2=0$
$(4x-1)(3x+2)=0$
$4x-1=0$ or $3x+2=0$
$4x=1$ or $3x=-2$
$x=\frac{1}{4}$ or $x=-\frac{2}{3}$
The solutions are $\frac{1}{4}$ and $-\frac{2}{3}$.

7. $z^2+9=10z$
$z^2-10z+9=0$
$(z-1)(z-9)=0$
$z-1=0$ or $z-9=0$
$z=1$ or $z=9$
The solutions are 1 and 9.

9. $5(y+4)=4(y+5)$
$5y+20=4y+20$
$y=0$
The solution is 0.

11. $0.6x-10=1.4x-14$
$4=0.8x$
$x=5$
The solution is 5.

13. $x(5x+2)=3$
$5x^2+2x-3=0$
$(5x-3)(x+1)=0$
$5x-3=0$ or $x+1=0$
$5x=3$ or $x=-1$
$x=\frac{3}{5}$
The solutions are $\frac{3}{5}$ and –1.

15. $6x-2(x-3)=4(x+1)+4$
$6x-2x+6=4x+4+4$
$4x+6=4x+8$
$6=8$
This is a false statement. Therefore, no solution exists.

17. $\frac{3}{8}+\frac{b}{3}=\frac{5}{12}$

$24\left(\frac{3}{8}+\frac{b}{3}\right)=24\left(\frac{5}{12}\right)$
$9+8b=10$
$8b=1$
$b=\frac{1}{8}$
The solution is $\frac{1}{8}$.

19. $x^2-6x=x(8+x)$
$x^2-6x=8x+x^2$
$0=14x$
$x=0$
The solution is 0.

21. $\frac{z^2}{6}-\frac{z}{2}-3=0$
$z^2-3z-18=0$
$(z-6)(z+3)=0$
$z-6=0$ or $z+3=0$
$z=6$ or $z=-3$
The solutions are 6 and –3.

23. $z+3(2+4z)=6(z+1)+5z$
$z+6+12z=6z+6+5z$
$13z+6=11z+6$
$2z=0$
$z=0$
The solution is 0.

25. $\frac{x^2}{2}+\frac{x}{20}=\frac{1}{10}$
$10x^2+x=2$
$10x^2+x-2=0$
$(5x-2)(2x+1)=0$
$5x-2=0$ or $2x+1=0$
$5x=2$ or $2x=-1$
$x=\frac{2}{5}$ or $x=-\frac{1}{2}$
The solutions are $\frac{2}{5}$ and $-\frac{1}{2}$.

27. $\frac{4t^2}{5}=\frac{t}{5}+\frac{3}{10}$
$8t^2=2t+3$
$8t^2-2t-3=0$
$(4t-3)(2t+1)=0$
$4t-3=0$ or $2t+1=0$
$4t=3$ or $2t=-1$
$t=\frac{3}{4}$ or $t=-\frac{1}{2}$
The solutions are $\frac{3}{4}$ and $-\frac{1}{2}$.

29. $\frac{3t+1}{8}=\frac{5+2t}{7}+2$
$56\left(\frac{3t+1}{8}\right)=56\left(\frac{5+2t}{7}+2\right)$
$7(3t+1)=8(5+2t)+56(2)$
$21t+7=40+16t+112$
$21t+7=16t+152$
$5t=145$
$t=29$
The solution is 29.

31. $\frac{m-4}{3}-\frac{3m-1}{5}=1$
$15\left(\frac{m-4}{3}-\frac{3m-1}{5}\right)=15(1)$
$5(m-4)-3(3m-1)=15$
$5m-20-9m+3=15$
$-4m-17=15$
$-4m=32$
$m=-8$
The solution is –8.

33. $3x^2=-x$
$3x^2+x=0$
$x(3x+1)=0$
$x=0$ or $3x+1=0$
$3x=-1$
$x=-\frac{1}{3}$
The solutions are $-\frac{1}{3}$ and 0.

35. $x(x-3)=x^2+5x+7$
$x^2-3x=x^2+5x+7$
$-7=8x$
$x=-\frac{7}{8}$
The solution is $-\frac{7}{8}$.

37. $3(t-8)+2t=7+t$
$3t-24+2t=7+t$
$5t-24=7+t$
$4t=31$
$t=\frac{31}{4}$
The solution is $\frac{31}{4}$.

39. $-3(x-4)+x=5(3-x)$
$-3x+12+x=15-5x$
$-2x+12=15-5x$
$3x=3$
$x=1$
The solution is 1.

41. $(x-1)(x+4)=24$
$x^2+3x-4=24$
$x^2+3x-28=0$
$(x+7)(x-4)=0$
$x+7=0$ or $x-4=0$
$x=-7$ or $x=4$
The solutions are –7 and 4.

43. $\frac{x^2}{4}-\frac{5}{2}x+6=0$
$x^2-10x+24=0$
$(x-4)(x-6)=0$
$x-4=0$ or $x-6=0$
$x=4$ or $x=6$
The solutions are 4 and 6.

45. $y^2+\frac{1}{4}=-y$
$4y^2+1=-4y$
$4y^2+4y+1=0$
$(2y+1)^2=0$
$2y+1=0$
$2y=-1$
$y=-\frac{1}{2}$
The solution is $-\frac{1}{2}$.

47. Strategies **a** and **d** are incorrect because the right side of the equation is not zero.

49. $3.2x+4=5.4x-7$
$3.2x+4-4=5.4x-7-4$
$3.2x=5.4x-11$
From this we see that $K=-11$.

51. $\frac{x}{6}+4=\frac{x}{3}$
$6\left(\frac{x}{6}+4\right)=6\left(\frac{x}{3}\right)$
$x+24=2x$
From this we see that $K=24$.

53. $2.569x=-12.48534$
$\frac{2.569x}{2.569}=\frac{-12.48534}{2.569}$
$x=-4.86$
Check: $2.569x=-12.48534$
$2.569(-4.86)\stackrel{?}{=}-12.48534$
$-12.48534=-12.48534$
The solution is –4.86.

55. $2.86z-8.1258=-3.75$
$2.86z=4.3758$
$\frac{2.86z}{2.86}=\frac{4.3758}{2.86}$
$z=1.53$
Check: $2.86z-8.1258=-3.75$
$2.86(1.53)-8.1258\stackrel{?}{=}-3.75$
$4.3758-8.1258\stackrel{?}{=}-3.75$
$-3.75=-3.75$
The solution is 1.53.

57. The quotient of 8 and a number is $\frac{8}{x}$.

59. The product of 8 and a number is $8x$.

61. 2 more than three times a number is $3x+2$.

Exercise Set 8.2

1. $y+y+y+y=4y$
The perimeter is $4y$ units.

3. $z+(z+1)+(z+2)=3z+3$
The sum is $3z+3$.

5. $5x+10(x+3)=5x+10x+30$
$=15x+30$
The total amount is $(15x+30)$ cents.

7. $4x+3(2x+1)=4x+6x+3=10x+3$
$(10x+3)$ units of fencing are needed.

9. $4(x-2)=2+6x$
$4x-8=2+6x$
$-10=2x$
$-5=x$
The number is –5.

11. Let x = one number; then $5x$ = the other number.

$$x + 5x = 270$$
$$6x = 270$$
$$x = 45$$
$$5x = 5(45) = 225$$

The numbers are 45 and 225.

13. $29\% \cdot 2271 = 0.29 \cdot 2271 = 658.59$

Approximately 658.59 million acres are federally owned.

15. $47\% \cdot 110{,}000 = 0.47 \cdot 110{,}000 = 51{,}700$

You would expect 51,700 homes to have computers.

17. $100 - 12 - 39 - 8 - 21 = 20$

20% of credit union loans are for credit cards and other unsecured loans.

19. $39\% \cdot 300 = 0.39 \cdot 300 = 117$

We would expect 117 of the loans to be automobile loans.

21.
$$3x + 17.5 = 199$$
$$3x = 181.5$$
$$x = 60.5$$
$$x + 7.7 = 60.5 + 7.7 = 68.2$$
$$x + 9.8 = 60.5 + 9.8 = 70.3$$

The Dallas/Ft. Worth airport has 60.5 million annual arrivals and departures. The Atlanta airport has 68.2 million annual arrivals and departures. The Chicago airport has 70.3 million annual arrivals and departures.

23. Let x = seats in B737-200; then $x + 104$ = seats in B767-300ER.

$$x + x + 104 = 328$$
$$2x + 104 = 328$$
$$2x = 224$$
$$x = 112$$
$$x + 104 = 112 + 104 = 216$$

The B737-200 aircraft has 112 seats. The B767-300ER aircraft has 216 seats.

25. Let x = price before taxes.

$$x + 0.08x = 464.40$$
$$1.08x = 464.40$$
$$x = 430$$

The price was $430 before taxes.

27. a. Let x = number of telephone company operators in 1996.

$$x - 0.47x = 26{,}000$$
$$0.53x = 26{,}000$$
$$x \approx 49{,}057$$

There were approximately 49,057 operators in 1996.

b. Answers may vary.

29.
$$20 - 75\% \cdot 20 = 20 - 0.75 \cdot 20$$
$$= 20 - 15$$
$$= 5$$

The life span is about 5 years.

31. Let x = number of millions of returns filed electronically in 1998; then $x + 0.24x$ = number in 1999.

$$x + 0.24x = 21.1$$
$$1.24x = 21.1$$
$$x \approx 17$$

There were approximately 17 million returns filed electronically in 1998.

33. Let x = length of a side of the square; then $x + 6$ = length of a side of the triangle.

$$4x = 3(x + 6)$$
$$4x = 3x + 18$$
$$x = 18$$
$$x + 6 = 24$$

The square's sides are 18 centimeters. The triangle's sides are 24 centimeters.

35. Let x = width of room; then $2x + 2$ = length of room.

$$2x + 2(2x + 2) = 40$$
$$2x + 4x + 4 = 40$$
$$6x + 4 = 40$$
$$6x = 36$$
$$x = 6$$
$$2x + 2 = 14$$

The width is 6 centimeters and the length is 14 centimeters.

37. Let x = width; then $5(x + 1)$ = height.
$x+5(x+1)=55.4$
$x+5x+5=55.4$
$6x+5=55.4$
$6x=50.4$
$x=8.4$
$5(x + 1) = 47$
The width is 8.4 meters and the height is 47 meters.

39. Let x = measure of second angle; then $2x$ = measure of first angle, and $3x - 12$ = measure of third angle.
$x+2x+3x-12=180$
$6x-12=180$
$6x=192$
$x=32$
$2x=2\cdot 32=64$
$3x-12=3\cdot 32-12=84$
The angles measure 64°, 32°, and 84°.

41. $x+20+x=180$
$2x+20=180$
$2x=160$
$x=80$
$x + 20 = 80 + 20 = 100$
The angles measure 80° and 100°.

43. $x+5x=90$
$6x=90$
$x=15$
$5x=5\cdot 15=75$
The angles measure 15° and 75°.

45. Let x = measure of the angle; then $180 - x$ = measure of its supplement.
$x=3(180-x)+20$
$x=540-3x+20$
$4x=560$
$x=140$
$180-x=180-140=40$
The angles measure 140° and 40°.

47. Let x = first integer; then $(x + 1)$ = next consecutive integer, and $(x + 2)$ = third consecutive integer.
$x+(x+1)+(x+2)=228$
$3x+3=228$
$3x=225$
$x=75$
$x+1=75+1=76$
$x+2=75+2=77$
The integers are 75, 76 and 77.

49. Let x = 1st integer; then $x + 2$ = 2nd integer, and $x + 4$ = 3rd integer.
$2x+x+4=268{,}222$
$3x+4=268{,}222$
$3x=268{,}218$
$x=89{,}406$
$x+2=89{,}406+2=89{,}408$
$x+4=89{,}406+4=89{,}410$
Fallon's zip code is 89406, Fernley's is 89408, and Gardnerville Ranchos' is 89410.

51. Let x = 1st integer; then $x + 1$ = 2nd integer, and $x + 2$ = 3rd integer.
$x+x+1+x+2=3(x+1)$
$3x+3=3x+3$
Since this is an identity, any three consecutive integers will work.

53. To find the number of skateboards, set $R = C$ and solve for x.
$24x=100+20x$
$4x=100$
$x=25$
To break even, 25 skateboards must be produced.

55. To find the number of books, set $R = C$ and solve for x.
$7.50x=4.50x+2400$
$3x=2400$
$x=800$
To break even, 800 books must be produced and sold.

57. Answers may vary.

59. Let x = number (in millions) of trees' worth of newsprint recycled; then $x + 30$ = number (in millions) of trees' worth of newsprint either recycled or discarded.

$x = 0.27(x+30)$
$x = 0.27x + 8.1$
$0.73x = 8.1$
$x \approx 11$

About 11 million trees' worth of newsprint is recycled.

61. Let n = the one number; then $n + 5$ = the other number.

$n(n+5) = 66$
$n^2 + 5n - 66 = 0$
$(n+11)(n-6) = 0$
$n + 11 = 0$ or $n - 6 = 0$
$n = -11$ or $n = 6$

There are two solutions: –11 and –6 or 6 and 11.

63. Let d = the amount of cable needed. Then from the Pythagorean theorem,

$d^2 = 45^2 + 60^2 = 5625.$
$d^2 - 5625 = 0$
$(d-75)(d+75) = 0$
$d - 75 = 0$ or $d + 75 = 0$
$d = 75$ or $d = -75$

Since the cable length must be positive, disregard –75. The cable should be 75 feet long.

65. $C(x) = x^2 - 15x + 50$

$9500 = x^2 - 15x + 50$
$0 = x^2 - 15x - 9450$
$0 = (x-105)(x+90)$
$x - 105 = 0$ or $x + 90 = 0$
$x = 105$ or $x = -90$

Since a negative number of units cannot be produced, disregard –90. 105 units are manufactured for $9500.

67. Let x = the length of the longer leg of the right triangle and $x - 3$ = the length of the shorter leg of the right triangle. Then from the Pythagorean theorem,

$15^2 = x^2 + (x-3)^2$
$225 = x^2 + x^2 - 6x + 9$
$0 = 2x^2 - 6x - 216$
$0 = x^2 - 3x - 108$
$0 = (x-12)(x+9)$
$x - 12 = 0$ or $x + 9 = 0$
$x = 12$ or $x = -9.$

Discarding –9, we find that the shorter leg of the right triangle is 12 centimeters and the other leg is 9 centimeters.

69. Note that the outer rectangle has dimensions of $2x + 12$ and $2x + 16$. Thus, the area of the border is $(2x + 12)(2x + 16) - 12 \cdot 16$.

$(2x+12)(2x+16) - 12 \cdot 16 = 128$
$4x^2 + 56x = 128$
$4x^2 + 56x - 128 = 0$
$x^2 + 14x - 32 = 0$
$(x+16)(x-2) = 0$
$x + 16 = 0$ or $x - 2 = 0$
$x = -16$ or $x = 2$

Since x must be positive, we see that the border should be 2 inches wide.

71. $\{x \mid 0 \le x \le 5\}$

[0, 5]

0 5

73. $\left\{x \middle| -\frac{1}{2} < x < \frac{3}{2}\right\}$

$\left(-\frac{1}{2}, \frac{3}{2}\right)$

$-\frac{1}{2}$ $\frac{3}{2}$

Exercise Set 8.3

1. $C \cup D = \{2, 3, 4, 5, 6, 7\}$

3. $A \cap D = \{4, 6\}$

5. $A \cup B = \{..., -2, -1, 0, 1, ...\}$

7. $B \cap D = \{5, 7\}$

9. $B \cup C = \{x | x \text{ is an odd integer or } x = 2 \text{ or } x = 4\}$

11. $A \cap C = \{2, 4\}$

13. $x < 5$ and $x > -2$
$-2 < x < 5$
$(-2, 5)$

15. $x + 1 \geq 7$ and $3x - 1 \geq 5$
$x \geq 6$ and $3x \geq 6$
$x \geq 2$
$x \geq 6$
$[6, \infty)$

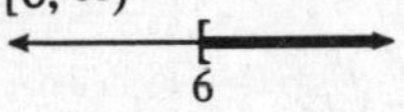

17. $4x + 2 \leq -10$ and $2x \leq 0$
$4x \leq -12$ and $x \leq 0$
$x \leq -3$
$x \leq -3$
$(-\infty, -3]$

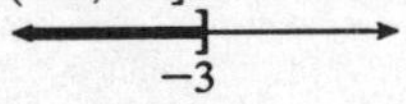

19. $5 < x - 6 < 11$
$11 < x < 17$
$(11, 17)$

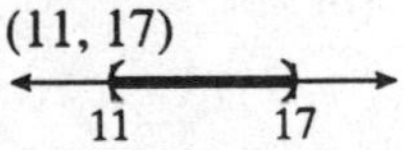

21. $-2 \leq 3x - 5 \leq 7$
$3 \leq 3x \leq 12$
$1 \leq x \leq 4$
$[1, 4]$

23. $1 \leq \frac{2}{3}x + 3 \leq 4$
$-2 \leq \frac{2}{3}x \leq 1$
$-3 \leq x \leq \frac{3}{2}$
$\left[-3, \frac{3}{2}\right]$

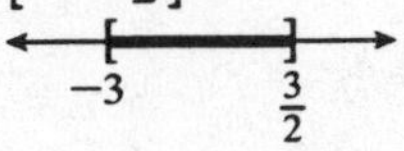

25. $-5 \leq \frac{x+1}{4} \leq -2$
$-20 \leq x + 1 \leq -8$
$-21 \leq x \leq -9$
$[-21, -9]$

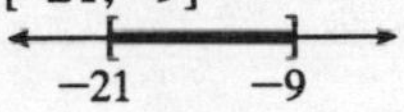

27. $x < -1$ or $x > 0$
$(-\infty, -1) \cup (0, \infty)$

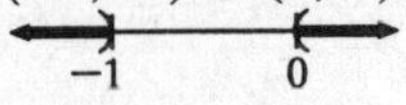

29. $-2x \leq -4$ or $5x - 20 \geq 5$
$x \geq 2$ or $5x \geq 25$
$x \geq 5$
$x \geq 2$
$[2, \infty)$

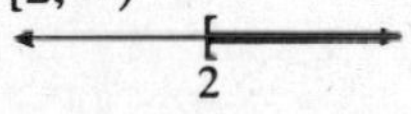

31. $3(x - 1) < 12$ or $x + 7 > 10$
$x - 1 < 4$ or $x > 3$
$x < 5$
All real numbers.
$(-\infty, \infty)$

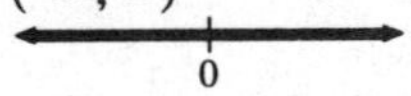

33. Answers may vary.

35. $x < 2$ and $x > -1$
$-1 < x < 2$
$(-1, 2)$

37. $x < 2$ or $x > -1$
All real numbers
$(-\infty, \infty)$

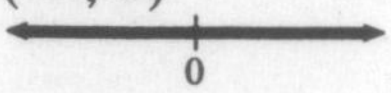

39. $x \geq -5$ and $x \geq -1$
$x \geq -1$
$[-1, \infty)$

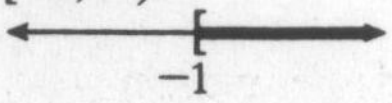

41. $x \geq -5$ or $x \geq -1$
$x \geq -5$
$[-5, \infty)$

43.
$$0 \leq 2x - 3 \leq 9$$
$$3 \leq 2x \leq 12$$
$$\frac{3}{2} \leq x \leq 6$$
$$\left[\frac{3}{2}, 6\right]$$

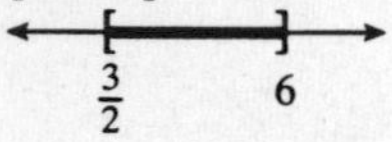

45.
$$\frac{1}{2} < x - \frac{3}{4} < 2$$
$$\frac{5}{4} < x < \frac{11}{4}$$
$$\left(\frac{5}{4}, \frac{11}{4}\right)$$

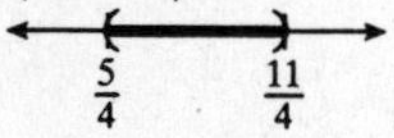

47. $x + 3 \geq 3$ and $x + 3 \leq 2$
$x \geq 0$ and $x \leq -1$
No solutions exist.
$\varnothing$

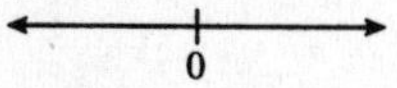

49. $3x \geq 5$ or $-x - 6 < 1$
$x \geq \frac{5}{3}$ or $-x < 7$
$x > -7$

$x > -7$
$(-7, \infty)$

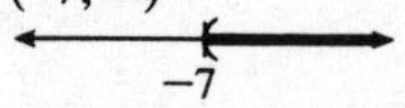

51.
$$0 < \frac{5 - 2x}{3} < 5$$
$$0 < 5 - 2x < 15$$
$$-5 < -2x < 10$$
$$\frac{-5}{-2} > \frac{-2x}{-2} > \frac{10}{-2}$$
$$\frac{5}{2} > x > -5$$
$$-5 < x < \frac{5}{2}$$
$$\left(-5, \frac{5}{2}\right)$$

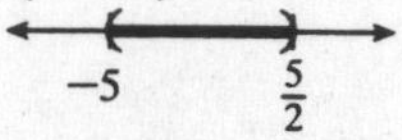

53.
$$-6 < 3(x - 2) \leq 8$$
$$-2 < x - 2 \leq \frac{8}{3}$$
$$0 < x \leq \frac{14}{3}$$
$$\left(0, \frac{14}{3}\right]$$

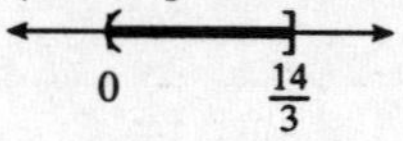

55. $-x + 5 > 6$ and $1 + 2x \leq -5$
$-x > 1$ and $2x \leq -6$
$x < -1$ and $x \leq -3$
$x \leq -3$
$(-\infty, -3]$

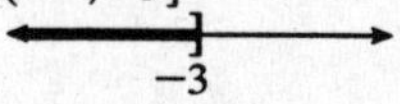

57. $3x + 2 \leq 5$ or $7x > 29$
$3x \leq 3$ or $x > \frac{29}{7}$
$x \leq 1$
$$(-\infty, 1] \cup \left(\frac{29}{7}, \infty\right)$$

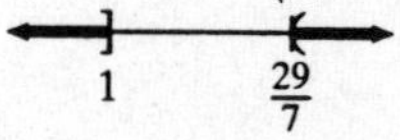

59. $5-x>7$ and $2x+3\geq 13$
$-x>2$ and $2x\geq 10$
$x<-2$ and $x\geq 5$
No solutions exist.
$\varnothing$

0

61. $-\dfrac{1}{2}\leq\dfrac{4x-1}{6}<\dfrac{5}{6}$
$-3\leq 4x-1<5$
$-2\leq 4x<6$
$-\dfrac{1}{2}\leq x<\dfrac{3}{2}$
$\left[-\dfrac{1}{2},\dfrac{3}{2}\right)$

$-\frac{1}{2}$ $\frac{3}{2}$

63. $\dfrac{1}{15}<\dfrac{8-3x}{15}<\dfrac{4}{5}$
$1<8-3x<12$
$-7<-3x<4$
$\dfrac{-7}{-3}>x>\dfrac{4}{-3}$
$\dfrac{7}{3}>x>-\dfrac{4}{3}$
$-\dfrac{4}{3}<x<\dfrac{7}{3}$
$\left(-\dfrac{4}{3},\dfrac{7}{3}\right)$

$-\frac{4}{3}$ $\frac{7}{3}$

65. $0.3<0.2x-0.9<1.5$
$1.2<0.2x<2.4$
$6<x<12$
$(6, 12)$

6 12

67. $-29\leq C\leq 35$
$-29\leq\dfrac{5}{9}(F-32)\leq 35$
$-52.2\leq F-32\leq 63$
$-20.2\leq F\leq 95$
$-20.2°\leq F\leq 95°$

69. $70\leq\dfrac{68+65+75+78+2x}{6}\leq 79$
$420\leq 286+2x\leq 474$
$134\leq 2x\leq 188$
$67\leq x\leq 94$
If Christian scores between 67 and 94 on his final exam, he will receive a C in the course.

71. The years that the consumption of pork was greater than 48 pounds per person were 1992, 1993, 1994, and 1995.
The years that the consumption of chicken was greater than 48 pounds per person were 1993, 1994, 1995, 1996, and 1997.
The years in common are 1993, 1994, and 1995.

73. $|-7|-|19|=7-19=-12$

75. $-(-6)-|-10|=6-10=-4$

77. $|x|=7$
$x=-7, 7$

79. $|x|=0$
$x=0$

81. $2x-3<3x+1<4x-5$
$2x-3<3x+1$ and $3x+1<4x-5$
$-4<x$ and $6<x$
$x>6$
$(6,\infty)$

6

83. $-3(x-2)\leq 3-2x\leq 10-3x$
$-3x+6\leq 3-2x$ and $3-2x\leq 10-3x$
$3\leq x$ and $x\leq 7$
$3\leq x\leq 7$
$[3, 7]$

3 7

85. $5x - 8 < 2(2 + x) < -2(1 + 2x)$

$$\begin{array}{lcl} 5x-8<4+2x & \text{and} & 4+2x<-2-4x \\ 3x<12 & \text{and} & 6<-6x \\ x<4 & \text{and} & -1>x \end{array}$$

$x < -1$

$(-\infty, -1)$

Section 8.4

Mental Math

1. $|-7| = 7$

2. $|-8| = 8$

3. $-|5| = -5$

4. $-|10| = -10$

5. $-|-6| = -6$

6. $-|-3| = -3$

7. $|-3| + |-2| + |-7| = 3 + 2 + 7 = 12$

8. $|-1| + |-6| + |-8| = 1 + 6 + 8 = 15$

Exercise Set 8.4

1. $|x| = 7$

$x = 7$ or $x = -7$

The solutions are 7 and –7.

3. $|3x| = 12.6$

$3x = 12.6$ or $3x = -12.6$

$x = 4.2$ or $x = -4.2$

The solutions are 4.2 and –4.2.

5. $|2x - 5| = 9$

$2x - 5 = 9$ or $2x - 5 = -9$

$2x = 14$ or $2x = -4$

$x = 7$ or $x = -2$

The solutions are 7 and –2.

7. $\left|\frac{x}{2} - 3\right| = 1$

$\frac{x}{2} - 3 = 1$ or $\frac{x}{2} - 3 = -1$

$\frac{x}{2} = 4$ or $\frac{x}{2} = 2$

$x = 8$ or $x = 4$

The solutions are 8 and 4.

9. $|z| + 4 = 9$

$|z| = 5$

$z = 5$ or $z = -5$

The solutions are 5 and –5.

11. $|3x| + 5 = 14$

$|3x| = 9$

$3x = 9$ or $3x = -9$

$x = 3$ or $x = -3$

The solutions are 3 and –3.

13. $|2x| = 0$

$2x = 0$

$x = 0$

The solution is 0.

15. $|4n + 1| + 10 = 4$

$|4n + 1| = -6$

The absolute value of an expression is never negative.

There is no solution.

17. $|5x - 1| = 0$

$5x - 1 = 0$

$5x = 1$

$x = \frac{1}{5}$

The solution is $\frac{1}{5}$.

19. $|x| = 5$

21. $|5x-7|=|3x+11|$

$5x-7=3x+11$ or $5x-7=-(3x+11)$

$2x=18$ or $5x-7=-3x-11$

$x=9$ or $8x=-4$

$x=9$ or $x=-\frac{1}{2}$

The solutions are 9 and $-\frac{1}{2}$.

23. $|z+8|=|z-3|$

$z+8=z-3$ or $z+8=-(z-3)$

$8=-3$ or $z+8=-z+3$

False or $2z=-5$

$z=-\frac{5}{2}$

The solution is $-\frac{5}{2}$.

25. Answers may vary.

27. $|x|=4$

$x=4$ or $x=-4$

The solutions are 4 and –4.

29. $|y|=0$

$y=0$

The solution is 0.

31. $|z|=-2$

The absolute value of an expression can never be negative.
There is no solution.

33. $|7-3x|=7$

$7-3x=7$ or $7-3x=-7$

$-3x=0$ or $-3x=-14$

$x=0$ or $x=\frac{14}{3}$

The solutions are 0 and $\frac{14}{3}$.

35. $|6x|-1=11$

$|6x|=12$

$6x=12$ or $6x=-12$

$x=2$ or $x=-2$

The solutions are 2 and –2.

37. $|4p|=-8$

The absolute value of an expression can never be negative.
There is no solution.

39. $|x-3|+3-7$

$|x-3|=4$

$x-3=4$ or $x-3=-4$

$x=7$ or $x=-1$

The solutions are 7 and –1.

41. $\left|\frac{z}{4}+5\right|=-7$

The absolute value of an expression can never be negative.
There is no solution.

43. $|9v-3|=-8$

The absolute value of an expression can never be negative.
There is no solution.

45. $|8n+1|=0$

$8n+1=0$

$8n=-1$

$n=-\frac{1}{8}$

The solution is $-\frac{1}{8}$.

47. $|1+6c|-7=-3$

$|1+6c|=4$

$1+6c=4$ or $1+6c=-4$

$6c=3$ or $6c=-5$

$c=\frac{1}{2}$ or $c=-\frac{5}{6}$

The solutions are $\frac{1}{2}$ and $-\frac{5}{6}$.

49. $|5x+1|=11$

$5x+1=11$ or $5x+1=-11$

$5x=10$ or $5x=-12$

$x=2$ or $x=-\frac{12}{5}$

The solutions are 2 and $-\frac{12}{5}$.

51. $|4x-2|=|-10|$

$4x-2=10$ or $4x-2=-10$

$4x=12$ or $4x=-8$

$x=3$ or $x=-2$

The solutions are 3 and –2.

53. $|5x+1|=|4x-7|$

$5x+1=4x-7$ or $5x+1=-(4x-7)$

$x=-8$ or $5x+1=-4x+7$

$x=-8$ or $9x=6$

$x=-8$ or $x=\frac{2}{3}$

The solutions are –8 and $\frac{2}{3}$.

55. $|6+2x|=-|-7|$

The absolute value of an expression can never be negative.
There is no solution.

57. $|2x-6|=|10-2x|$

$2x-6=10-2x$ or $2x-6=-(10-2x)$

$4x=16$ or $2x-6=-10+2x$

$x=4$ or $-6=-10$

False

The solution is 4.

59. $\left|\frac{2x-5}{3}\right|=7$

$\frac{2x-5}{3}=7$ or $\frac{2x-5}{3}=-7$

$2x-5=21$ or $2x-5=-21$

$2x=26$ or $2x=-16$

$x=13$ or $x=-8$

The solutions are 13 and –8.

61. $2+|5n|=17$

$|5n|=15$

$5n=15$ or $5n=-15$

$n=3$ or $n=-3$

The solutions are 3 and –3.

63. $\left|\frac{2x-1}{3}\right|=|-5|$

$\left|\frac{2x-1}{3}\right|=5$

$\frac{2x-1}{3}=5$ or $\frac{2x-1}{3}=-5$

$2x-1=15$ or $2x-1=-15$

$2x=16$ or $2x=-14$

$x=8$ or $x=-7$

The solutions are 8 and –7.

65. $|2y-3|=|9-4y|$

$2y-3=9-4y$ or $2y-3=-(9-4y)$

$6y=12$ or $2y-3=-9+4y$

$y=2$ or $6=2y$

$y=2$ or $3=y$

The solutions are 2 and 3.

67. $\left|\frac{3n+2}{8}\right|=|-1|$

$\frac{3n+2}{8}=1$ or $\frac{3n+2}{8}=-1$

$3n+2=8$ or $3n+2=-8$

$3n=6$ or $3n=-10$

$n=2$ or $n=-\frac{10}{3}$

The solutions are 2 and $-\frac{10}{3}$.

69. $|x+4|=|7-x|$

$x+4=7-x$ or $x+4=-(7-x)$

$2x=3$ or $x+4=-7+x$

$x=\frac{3}{2}$ or $4=-7$

False

The solution is $\frac{3}{2}$.

71. $\left|\frac{8c-7}{3}\right|=-|-5|$

$\left|\frac{8c-7}{3}\right|=-5$

The absolute value of an expression can never be negative.
There is no solution.

73. Answers may vary.

75. 13% of Disney's operating income came from consumer products.

77. $32\%(3.4 \text{ billion}) = 0.32(3.4 \text{ billion}) = 1.088 \text{ billion}$
\$1.088 billion is expected from the media networks segment.

79. $|x| \geq -2$
$-2, -1, 0, 1, 2$, for example

81. The solution set is $\varnothing$.

Exercise Set 8.5

1. $|x| \leq 4$
$-4 \leq x \leq 4$
The solution set is $[-4, 4]$.

3. $|x-3| < 2$
$-2 < x-3 < 2$
$1 < x < 5$
The solution set is $(1, 5)$.

5. $|x+3| < 2$
$-2 < x+3 < 2$
$-5 < x < -1$
The solution set is $(-5, -1)$.

7. $|2x+7| \leq 13$
$-13 \leq 2x+7 \leq 13$
$-20 \leq 2x \leq 6$
$-10 \leq x \leq 3$
The solution set is $[-10, 3]$.

9. $|x|+7 \leq 12$
$|x| \leq 5$
$-5 \leq x \leq 5$
The solution set is $[-5, 5]$.

11. $|3x-1| < -5$
An absolute value expression can never be less than a negative number.
The solution set is $\varnothing$.

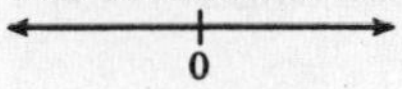

13. $|x-6|-7 \leq -1$
$|x-6| \leq 6$
$-6 \leq x-6 \leq 6$
$0 \leq x \leq 12$
The solution set is $[0, 12]$.

15. $|x| > 3$
$x < -3$ or $x > 3$
The solution set is $(-\infty, -3) \cup (3, \infty)$.

17. $|x+10| \geq 14$
$x+10 \leq -14$ or $x+10 \geq 14$
$x \leq -24$ or $x \geq 4$
The solution set is $(-\infty, -24] \cup [4, \infty)$.

19. $|x|+2 > 6$
$|x| > 4$
$x < -4$ or $x > 4$
The solution set is $(-\infty, -4) \cup (4, \infty)$.

21. $|5x| > -4$
An absolute value expression is always greater than a negative number.
All real numbers
The solution set is $(-\infty, \infty)$.

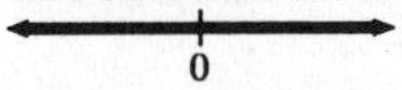

23. $|6x-8|+3>7$
$|6x-8|>4$
$6x-8<-4$ or $6x-8>4$
$6x<4$ or $6x>12$
$x<\frac{2}{3}$ or $x>2$
The solution set is $\left(-\infty, \frac{2}{3}\right)\cup(2, \infty)$.

25. $|x|\le 0$
An absolute value expression can never be less than 0, so $|x|\le 0$ is only true when $|x|=0$.
$x=0$
The solution set is $\{0\}$.

27. $|8x+3|>0$ only excludes $|8x+3|=0$.
$8x+3=0$
$8x=-3$
$x=-\frac{3}{8}$
All real numbers except $-\frac{3}{8}$.
The solution set is $\left(\infty, -\frac{3}{8}\right)\cup\left(-\frac{3}{8}, \infty\right)$.

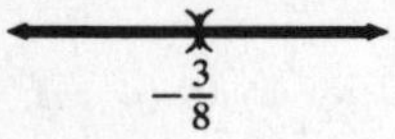

29. $|x|<7$

31. $|x|\le 5$

33. $|x|\le 2$
$-2\le x\le 2$
The solution set is $[-2, 2]$.

35. $|y|>1$
$y<-1$ or $y>1$
The solution set is $(-\infty, -1)\cup(1, \infty)$.

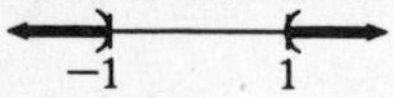

37. $|x-3|<8$
$-8<x-3<8$
$-5<x<11$
The solution set is $(-5, 11)$.

39. $|0.6x-3|>0.6$
$0.6x-3<-0.6$ or $0.6x-3>0.6$
$0.6x<2.4$ or $0.6x>3.6$
$x<4$ or $x>6$
The solution set is $(-\infty, 4)\cup(6, \infty)$.

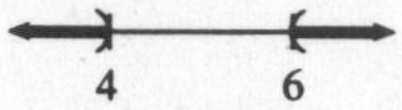

41. $5+|x|\le 2$
$|x|\le -3$
An absolute value expression can never be less than a negative number.
The solution set is $\varnothing$.

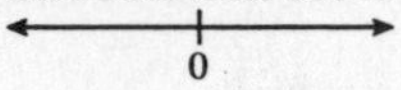

43. $|x|>-4$
An absolute value expression is always greater than a negative number.
All real numbers
The solution is $(-\infty, \infty)$.

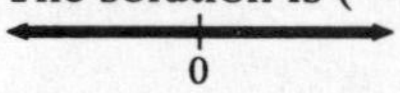

45. $|2x-7|\le 11$
$-11\le 2x-7\le 11$
$-4\le 2x\le 18$
$-2\le x\le 9$
The solution set is $[-2, 9]$.

47. $|x+5|+2\ge 8$
$|x+5|\ge 6$
$x+5\le -6$ or $x+5\ge 6$
$x\le -11$ or $x\ge 1$
The solution set is $(-\infty, -11]\cup[1, \infty)$.

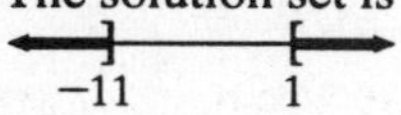

49. $|x| > 0$ excludes only $|x| = 0$, or $x = 0$.
The solution set is $(-\infty, 0) \cup (0, \infty)$.

51.
$$9 + |x| > 7$$
$$|x| > -2$$
An absolute value expression is always greater than a negative number.
All real numbers
The solution set is $(-\infty, \infty)$.

53.
$$6 + |4x - 1| \le 9$$
$$|4x - 1| \le 3$$
$$-3 \le 4x - 1 \le 3$$
$$-2 \le 4x \le 4$$
$$-\frac{1}{2} \le x \le 1$$
The solution set is $\left[-\frac{1}{2}, 1\right]$.

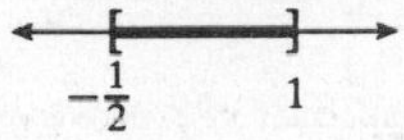

55.
$$\left|\frac{2}{3}x + 1\right| > 1$$
$$\frac{2}{3}x + 1 < -1 \quad \text{or} \quad \frac{2}{3}x + 1 > 1$$
$$\frac{2}{3}x < -2 \quad \text{or} \quad \frac{2}{3}x > 0$$
$$x < -3 \quad \text{or} \quad x > 0$$
The solution set is $(-\infty, -3) \cup (0, \infty)$.

57. $|5x + 3| < -6$
An absolute value expression is never less than a negative number.
The solution set is $\varnothing$.

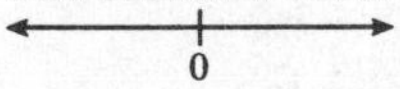

59. $|8x + 3| \ge 0$
An absolute value expression is always greater than or equal to 0.
All real numbers
The solution set is $(-\infty, \infty)$.

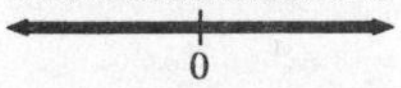

61.
$$|1 + 3x| + 4 < 5$$
$$|1 + 3x| < 1$$
$$-1 < 1 + 3x < 1$$
$$-2 < 3x < 0$$
$$-\frac{2}{3} < x < 0$$
The solution set is $\left(-\frac{2}{3}, 0\right)$.

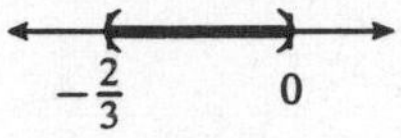

63.
$$|x| - 3 \ge -3$$
$$|x| \ge 0$$
An absolute value expression is always greater than or equal to 0.
All real numbers
The solution set is $(-\infty, \infty)$.

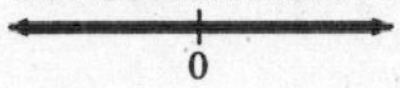

65.
$$|8x| - 10 > -2$$
$$|8x| > 8$$
$$8x < -8 \quad \text{or} \quad 8x > 8$$
$$x < -1 \quad \text{or} \quad x > 1$$
The solution set is $(-\infty, -1) \cup (1, \infty)$.

67.
$$\left|\frac{x+6}{3}\right| > 2$$
$$\frac{x+6}{3} < -2 \quad \text{or} \quad \frac{x+6}{3} > 2$$
$$x + 6 < -6 \quad \text{or} \quad x + 6 > 6$$
$$x < -12 \quad \text{or} \quad x > 0$$
The solution set is $(-\infty, -12) \cup (0, \infty)$.

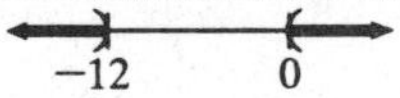

69.
$$|2(3 + x)| > 6$$
$$2|3 + x| > 6$$
$$|3 + x| > 3$$
$$3 + x < -3 \quad \text{or} \quad 3 + x > 3$$
$$x < -6 \quad \text{or} \quad x > 0$$
The solution set is $(-\infty, -6) \cup (0, \infty)$.

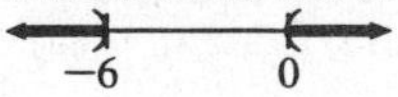

71. $\left|\frac{5(x+2)}{3}\right| < 7$

$\frac{5}{3}|x+2| < 7$

$|x+2| < \frac{21}{5}$

$-\frac{21}{5} < x+2 < \frac{21}{5}$

$-\frac{31}{5} < x < \frac{11}{5}$

The solution set is $\left(-\frac{31}{5}, \frac{11}{5}\right)$.

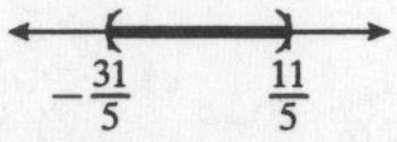

73. $-15 + |2x-7| \le -6$

$|2x-7| \le 9$

$-9 \le 2x-7 \le 9$

$-2 \le 2x \le 16$

$-1 \le x \le 8$

The solution set is [–1, 8].

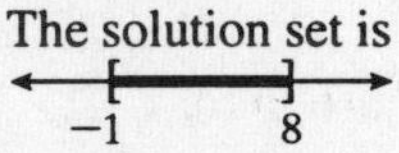

75. $\left|2x+\frac{3}{4}\right| - 7 \le -2$

$\left|2x+\frac{3}{4}\right| \le 5$

$-5 \le 2x + \frac{3}{4} \le 5$

$-\frac{23}{4} \le 2x \le \frac{17}{4}$

$-\frac{23}{8} \le x \le \frac{17}{8}$

The solution set is $\left[-\frac{23}{8}, \frac{17}{8}\right]$.

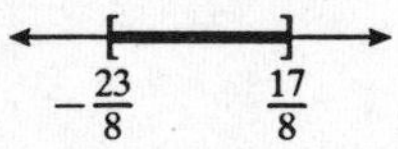

77. $|2x-3| < 7$

$-7 < 2x-3 < 7$

$-4 < 2x < 10$

$-2 < x < 5$

The solution set is (–2, 5).

79. $|2x-3| = 7$

$2x-3=7$ or $2x-3=-7$

$2x=10$ or $2x=-4$

$x=5$ or $x=-2$

The solutions are 5 and –2.

81. $|x-5| \ge 12$

$x-5 \le -12$ or $x-5 \ge 12$

$x \le -7$ or $x \ge 17$

The solution set is $(-\infty, -7] \cup [17, \infty)$.

83. $|9+4x| = 0$

$9+4x=0$

$4x=-9$

$x=-\frac{9}{4}$

The solution is $-\frac{9}{4}$.

85. $|2x+1| + 4 < 7$

$|2x+1| < 3$

$-3 < 2x+1 < 3$

$-4 < 2x < 2$

$-2 < x < 1$

The solution set is (–2, 1).

87. $|3x-5| + 4 = 5$

$|3x-5| = 1$

$3x-5=-1$ or $3x-5=1$

$3x=4$ or $3x=6$

$x=\frac{4}{3}$ or $x=2$

The solutions are 2 and $\frac{4}{3}$.

89. $|x+11|=-1$
An absolute value expression can never be negative.
There are no solutions.

91. $\left|\frac{2x-1}{3}\right|=6$

$$\frac{2x-1}{3}=-6 \quad \text{or} \quad \frac{2x-1}{3}=6$$
$$2x-1=-18 \quad \text{or} \quad 2x-1=18$$
$$2x=-17 \quad \text{or} \quad 2x=19$$
$$x=-\frac{17}{2} \quad \text{or} \quad x=\frac{19}{2}$$

The solutions are $\frac{19}{2}$ and $-\frac{17}{2}$.

93. $\left|\frac{3x-5}{6}\right|>5$

$$\frac{3x-5}{6}<-5 \quad \text{or} \quad \frac{3x-5}{6}>5$$
$$3x-5<-30 \quad \text{or} \quad 3x-5>30$$
$$3x<-25 \quad \text{or} \quad 3x>35$$
$$x<-\frac{25}{3} \quad \text{or} \quad x>\frac{35}{3}$$

The solution set is $\left(-\infty, -\frac{25}{3}\right)\cup\left(\frac{35}{3}, \infty\right)$.

95. Answers may vary.

97. $|3.5-x|<0.05$

$$-0.05<3.5-x<0.05$$
$$-3.55<-x<-3.45$$
$$3.55>x>3.45$$
$$3.45<x<3.55$$

99.
$$3x-4y=12$$
$$3(2)-4y=12$$
$$6-4y=12$$
$$-4y=6$$
$$y=-\frac{6}{4}$$
$$y=-\frac{3}{2}=-1.5$$

101.
$$3x-4y=12$$
$$3x-4(-3)=12$$
$$3x+12=12$$
$$3x=0$$
$$x=0$$

Exercise Set 8.6

1.
$$\begin{cases} x+y=3 & (1) \\ 2y=10 & (2) \\ 3x+2y-3z=1 & (3) \end{cases}$$

Solve the second equation for y.
$y=5$
Replace y with 5 in the first equation.
$x+5=3$
$x=-2$
Replace x with -2 and y with 5 in the third equation.

$$3(-2)+2(5)-3z=1$$
$$-6+10-3z=1$$
$$4-3z=1$$
$$-3z=-3$$
$$z=1$$

The solution is $(-2, 5, 1)$.

3. $\begin{cases} 2x+2y+\ z=1 & (1) \\ -x+\ y+2z=3 & (2) \\ x+2y+4z=0 & (3) \end{cases}$

Add equations (2) and (3).
$3y + 6z = 3$ or $y + 2z = 1$
Add twice equation (2) to equation (1).

$$\begin{array}{r} -2x+2y+4z=6 \\ 2x+2y+z=1 \\ \hline 4y+5z=7 \end{array}$$

Solve the new system:

$\begin{cases} y+2z=1 \\ 4y+5z=7 \end{cases}$

Multiply the first equation by –4.

$\begin{cases} -4y-8z=-4 \\ 4y+5z=7 \end{cases}$

Add the equations.

$$\begin{array}{r} -4y-8z=-4 \\ 4y+5z=7 \\ \hline -3z=3 \end{array}$$

$z=-1$
Replace z with –1 in the equation
$y + 2z = 1$.

$$\begin{aligned} y+2(-1)&=1 \\ y-2&=1 \\ y&=3 \end{aligned}$$

Replace y with 3 and z with –1 in equation (3).

$$\begin{aligned} x+2(3)+4(-1)&=0 \\ x+6-4&=0 \\ x+2&=0 \\ x&=-2 \end{aligned}$$

The solution is (–2, 3, –1).

5. $\begin{cases} x-2y+\ z=-5 & (1) \\ -3x+6y-3z=15 & (2) \\ 2x-4y+2z=-10 & (3) \end{cases}$

Multiply equation (2) by $-\frac{1}{3}$ and equation (3) by $\frac{1}{2}$.

$\begin{cases} x-2y+z=-5 \\ x-2y+z=-5 \\ x-2y+z=-5 \end{cases}$

All three equations are identical. There are infinitely many solutions.
The solution is $\{(x, y, z) \mid x-2y+z=-5\}$.

7. $\begin{cases} 4x-y+2z=5 & (1) \\ 2y+\ z=4 & (2) \\ 4x+y+3z=10 & (3) \end{cases}$

Multiply equation (1) by –1 and add to equation (3).

$$\begin{array}{r} -4x+y-2z=-5 \\ 4x+y+3z=10 \\ \hline 2y+\ z=\ 5 \quad (4) \end{array}$$

Multiply equation (4) by –1 and add to equation (2).

$$\begin{array}{r} -2y-z=-5 \\ 2y+z=\ 4 \\ \hline 0=-1 \quad \text{False} \end{array}$$

Inconsistent system
There is no solution.

9. Answers may vary.

11. $\begin{cases} x + \quad 5z = 0 & (1) \\ 5x + y \quad = 0 & (2) \\ \quad y - 3z = 0 & (3) \end{cases}$

Multiply equation (3) by –1 and add to equation (2).

$$\begin{array}{r} -y + 3z = 0 \\ \underline{5x + y \qquad = 0} \\ 5x + \quad 3z = 0 \quad (4) \end{array}$$

Multiply equation (1) by –5 and add to equation (4).

$$\begin{array}{r} -5x - 25z = 0 \\ \underline{5x + \ 3z = 0} \\ -22z = 0 \end{array}$$

$z = 0$

Replace z with 0 in equation (4).

$5x + 3(0) = 0$

$5x = 0$

$x = 0$

Replace x with 0 in equation (2).

$5(0) + y = 0$

$y = 0$

The solution is (0, 0, 0).

13. $\begin{cases} 6x - \quad 5z = 17 & (1) \\ 5x - y + 3z = -1 & (2) \\ 2x + y \quad = -41 & (3) \end{cases}$

Add equations (2) and (3).

$7x + 3z = -42 \quad (4)$

Multiply equation (4) by 5, multiply equation (1) by 3, and add.

$$\begin{array}{r} 35x + 15z = -210 \\ \underline{18x - 15z = \quad 51} \\ 53x \qquad = -159 \end{array}$$

$x = -3$

Replace x with –3 in equation (1).

$6(-3) - 5z = 17$

$-18 - 5z = 17$

$-5z = 35$

$z = -7$

Replace x with –3 in equation (3).

$2(-3) + y = -41$

$-6 + y = -41$

$y = -35$

The solution is (–3, –35, –7).

15. $\begin{cases} x + \ y + \ z = 8 & (1) \\ 2x - \ y - \ z = 10 & (2) \\ x - 2y - 3z = 22 & (3) \end{cases}$

Add equations (1) and (2).

$3x = 18$ or $x = 6$

Add twice equation (1) to equation (3).

$$\begin{array}{r} 2x + 2y + 2z = 16 \\ \underline{x - 2y - 3z = 22} \\ 3x - \qquad z = 38 \end{array}$$

Replace x with 6 in this equation.

$3(6) - z = 38$

$18 - z = 38$

$-z = 20$

$z = -20$

Replace x with 6 and z with –20 in equation (1).

$6 + y + (-20) = 8$

$y - 14 = 8$

$y = 22$

The solution is (6, 22, –20).

17. $\begin{cases} x + 2y - \ z = 5 & (1) \\ 6x + \ y + \ z = 7 & (2) \\ 2x + 4y - 2z = 5 & (3) \end{cases}$

Add equations (1) and (2).

$7x + 3y = 12 \quad (4)$

Add twice equation (2) to equation (3).

$$\begin{array}{r} 12x + 2y + 2z = 14 \\ \underline{2x + 4y - 2z = \ 5} \\ 14x + 6y \qquad = 19 \quad (5) \end{array}$$

Multiply equation (4) by –2 and add to equation (5).

$$\begin{array}{r} -14x - 6y = -24 \\ \underline{14x + 6y = \ 19} \\ 0 = \ -5 \end{array}$$ False

Inconsistent system

There is no solution.

19. $\begin{cases} 2x-3y+z=2 & (1) \\ x-5y+5z=3 & (2) \\ 3x+y-3z=5 & (3) \end{cases}$

Add –2 times equation (2) to equation (1).

$$\begin{array}{r} 2x-3y+z=2 \\ -2x+10y-10z=-6 \\ \hline 7y-9z=-4 \end{array}$$

Add –3 times equation (2) to equation (3).

$$\begin{array}{r} -3x+15y-15z=-9 \\ 3x+y-3z=5 \\ \hline 16y-18z=-4 \end{array}$$

We now have the system:

$\begin{cases} 7y-9z=-4 & (4) \\ 16y-18z=-4 & (5) \end{cases}$

Multiply equation (4) by –2 and add to equation (5).

$$\begin{array}{r} -14y+18z=8 \\ 16y-18z=-4 \\ \hline 2y=4 \end{array}$$

$y=2$

Replace y with 2 in equation (4).

$7(2)-9z=-4$

$-9z=-18$

$z=2$

Replace y with 2 and z with 2 in equation (1).

$2x-3(2)+2=2$

$x=3$

The solution is (3, 2, 2).

21. $\begin{cases} -2x-4y+6z=-8 & (1) \\ x+2y-3z=4 & (2) \\ 4x+8y-12z=16 & (3) \end{cases}$

Add 2 times equation (2) to equation (1).

$$\begin{array}{r} 2x+4y-6z=8 \\ -2x-4y+6z=-8 \\ \hline 0=0 \end{array}$$

Add –4 times equation (2) to equation (3).

$$\begin{array}{r} -4x-8y+12z=-16 \\ 4x+8y-12z=16 \\ \hline 0=0 \end{array}$$

The system is dependent.

The solution is $\{(x, y, z) \mid x+2y-3z=4\}$.

23. $\begin{cases} 2x+2y-3z=1 & (1) \\ y+2z=-14 & (2) \\ 3x-2y=-1 & (3) \end{cases}$

Add equations (1) and (3).

$5x-3z=0$ (4)

Add twice equation (2) to equation (3).

$$\begin{array}{r} 2y+4z=-28 \\ 3x-2y=-1 \\ \hline 3x+4z=-29 \end{array} \quad (5)$$

Multiply equation (4) by 4, multiply equation (5) by 3 and add.

$$\begin{array}{r} 20x-12z=0 \\ 9x+12z=-87 \\ \hline 29x=-87 \end{array}$$

$x=-3$

Replace x with –3 in equation (4).

$5(-3)-3z=0$

$3z=-15$

$z=-5$

Replace z with –5 in equation (2).

$y+2(-5)=-14$

$y-10=-14$

$y=-4$

The solution is (–3, –4, –5).

25. $\begin{cases} \frac{3}{4}x - \frac{1}{3}y + \frac{1}{2}z = 9 & (1) \\ \frac{1}{6}x + \frac{1}{3}y - \frac{1}{2}z = 2 & (2) \\ \frac{1}{2}x - y + \frac{1}{2}z = 2 & (3) \end{cases}$

Multiply equation (1) by 12, multiply equation (2) by 6, and multiply equation (3) by 2.

$\begin{cases} 9x - 4y + 6z = 108 & (4) \\ x + 2y - 3z = 12 & (5) \\ x - 2y + z = 4 & (6) \end{cases}$

Add twice equation (5) to equation (4).

$$\begin{aligned} 2x + 4y - 6z &= 24 \\ 9x - 4y + 6z &= 108 \\ \hline 11x \quad\quad &= 132 \end{aligned}$$

$x = 12$

Add equations (5) and (6).

$2x - 2z = 16$ or $x - z = 8$

Replace x with 12 in this equation.

$12 - z = 8$

$z = 4$

Replace x with 12 and z with 4 in equation (6).

$12 - 2y + 4 = 4$

$12 - 2y = 0$

$12 = 2y$

$y = 6$

The solution is (12, 6, 4).

27. $\begin{cases} x + y + z = 1 & (1) \\ 2x - y + z = 0 & (2) \\ -x + 2y + 2z = -1 & (3) \end{cases}$

Multiply equation (3) by 2 and add to equation (2).

$$\begin{aligned} 2x - y + z &= 0 \\ -2x + 4y + 4z &= -2 \\ \hline 3y + 5z &= -2 \quad (4) \end{aligned}$$

Multiply equation (1) by –2 and add to equation (2).

$$\begin{aligned} -2x - 2y - 2z &= -2 \\ 2x - y + z &= 0 \\ \hline -3y - z &= -2 \quad (5) \end{aligned}$$

Add equations (4) and (5).

$$\begin{aligned} 3y + 5z &= -2 \\ -3y - z &= -2 \\ \hline 4z &= -4 \end{aligned}$$

$z = -1$

Replace z with –1 in equation (4).

$3y + 5(-1) = -2$

$3y - 5 = -2$

$3y = 3$

$y = 1$

Replace y with 1 and z with –1 in equation (1).

$x + 1 + (-1) = 1$

$x = 1$

The solution is (1, 1, –1).

$\frac{1}{24} = \frac{x}{8} + \frac{y}{4} + \frac{z}{3}$

$\frac{1}{24} = \frac{1}{8} + \frac{1}{4} - \frac{1}{3}$

$\frac{1}{24} = \frac{3}{24} + \frac{6}{24} - \frac{8}{24}$

$\frac{1}{24} = \frac{1}{24}$ True

29. $\begin{cases} 4A+6B+4C=30 & (1) \\ 6A+B+C=16 & (2) \\ 3A+2B+12C=24 & (3) \end{cases}$

Multiply equation (2) by –6 and add to equation (1).

$$\begin{array}{l} 4A+6B+4C=30 \\ \underline{-36A-6B-6C=-96} \\ -32A \quad -2C=-66 \quad (4) \end{array}$$

Multiply equation (3) by –3 and add to equation (1).

$$\begin{array}{l} 4A+6B+4C=30 \\ \underline{-9A-6B-36C=-72} \\ -5A \quad -32C=-42 \quad (5) \end{array}$$

Multiply equation (4) by –16 and add to equation (5).

$$\begin{array}{l} 512A+32C=1056 \\ \underline{-5A-32C=-42} \\ 507A \quad =1014 \\ A=2 \end{array}$$

Replace A with 2 in equation (5).

$$\begin{array}{l} -5(2)-32C=-42 \\ -10-32C=-42 \\ -32C=-32 \\ C=1 \end{array}$$

Replace A with 2 and C with 1 in equation (2).

$$\begin{array}{l} 6(2)+B+1=16 \\ 12+B+1=16 \\ B+13=16 \\ B=3 \end{array}$$

2 units of Mix A
3 units of Mix B
1 unit of Mix C

31. Let x = number of free throws,
y = number of two-point field goals,
and z = number of three-point field goals.

$\begin{cases} x+2y+3z=686 \\ y=3z-20 \\ x=y+50 \end{cases}$

$\begin{cases} x+2y+3z=686 & (1) \\ y-3z=-20 & (2) \\ x-y=50 & (3) \end{cases}$

Add equations (1) and (2).

$$\begin{array}{l} x+2y+3z=686 \\ \underline{y-3z=-20} \\ x+3y \quad =666 \quad (4) \end{array}$$

Multiply equation (3) by –1 and add to equation (4).

$$\begin{array}{l} -x+y=-50 \\ \underline{x+3y=666} \\ 4y=616 \\ y=154 \end{array}$$

Replace y with 154 in equation (3).

$$\begin{array}{l} x-154=50 \\ x=204 \end{array}$$

Replace y with 154 in equation (2).

$$\begin{array}{l} 154-3z=-20 \\ -3z=-174 \\ z=58 \end{array}$$

She scored 204 free throws, 154 two-point goals, and 58 three-point field goals.

33. Let x = the measure of the smallest angle,
y = the measure of the second smallest,
and z = the measure of the largest angle.

$$\begin{cases} x+y+z=180 \\ z=x+90 \\ y=x+30 \end{cases}$$

$$\begin{cases} x+y+z=180 & (1) \\ -x+z=90 & (2) \\ -x+y=30 & (3) \end{cases}$$

Multiply equation (2) by −1 and add to equation (1).

$$\begin{array}{l} x+y+z=180 \\ x\quad -z=-90 \\ \hline 2x+y\quad = 90 \end{array}$$

Multiply equation (3) by −1 and add to equation (4).

$$\begin{array}{l} 2x+y=\ 90 \\ x-y=-30 \\ \hline 3x\quad =\ 60 \\ x\quad =\ 20 \end{array}$$

Replace x with 20 in equation (3).

$$-20+y=30$$
$$y=50$$

Replace x with 20 in equation (2).

$$-20+z=90$$
$$z=110$$

The measures of the angles are 20°, 50° and 110°.

35. $y=ax^2+bx+c$

(1, 6): $6=a(1)^2+b(1)+c$

(−1, −2): $-2=a(-1)^2+b(-1)+c$

(0, −1): $-1=a(0)^2+b(0)+c$

$$\begin{cases} a+b+c=\ 6 & (1) \\ a-b+c=-2 & (2) \\ c=-1 & (3) \end{cases}$$

Add equation (1) to equation (2).

$$\begin{array}{l} a+b+c=\ 6 \\ a-b+c=-2 \\ \hline 2a\quad +2c=\ 4 \quad (4) \end{array}$$

Replace c with −1 in equation (4).

$$2a+2(-1)=4$$
$$2a-2=4$$
$$2a=6$$
$$a=3$$

Replace a with 3 and c with −1 in equation (1).

$$3+b+(-1)=6$$
$$b+2=6$$
$$b=4$$

Thus the coefficients are a = 3, b = 4 and c = −1.

37. $$\begin{cases} x+y+z=180 \\ (2x+5)+y=180 \\ (2x-5)+z=180 \end{cases}$$

$$\begin{cases} x+y+z=180 & (1) \\ 2x+y=175 & (2) \\ 2x+z=185 & (3) \end{cases}$$

Multiply equation (1) by −1 and add to equation (2).

$$\begin{array}{l} -x-y-z=-180 \\ 2x+y\quad =\ 175 \\ \hline x\quad -z=\ -5 \quad (4) \end{array}$$

Add equation (3) and equation (4).

$$\begin{array}{l} 2x+z=185 \\ x-z=-5 \\ \hline 3x\quad =180 \\ x\quad =\ 60 \end{array}$$

Replace x with 60 in equation (2).

$$2(60)+y=175$$
$$120+y=175$$
$$y=\ 55$$

Replace x with 60 in equation (3).

$$2(60)+z=185$$
$$120+z=185$$
$$z=65$$

$x=60$, $y=55$, $z=65$

39. $y = ax^2 + bx + c$

$(5,945)$: $945 = a(5)^2 + b(5) + c$

$(6,925)$: $925 = a(6)^2 + b(6) + c$

$(9,1019)$: $1019 = a(9)^2 + b(9) + c$

$$\begin{cases} 25a + 5b + c = 945 & (1) \\ 36a + 6b + c = 925 & (2) \\ 81a + 9b + c = 1019 & (3) \end{cases}$$

Multiply equation (1) by –1 and add to equation (2).

$$\begin{array}{r} -25a - 5b - c = -945 \\ 36a + 6b + c = 925 \\ \hline 11a + b = -20 \quad (4) \end{array}$$

Multiply equation (2) by –1 and add to equation (3)

$$\begin{array}{r} -36a - 6b - c = -925 \\ 81a + 9b + c = 1019 \\ \hline 45a + 3b = 94 \quad (5) \end{array}$$

Multiply equation (4) by –3 and add to equation (5).

$$\begin{array}{r} -33a - 3b = 60 \\ 45a + 3b = 94 \\ \hline 12a = 154 \end{array}$$

$$a = \frac{154}{12} = \frac{77}{6} = 12\frac{5}{6}$$

Replace a with $\frac{77}{6}$ in equation (4)

$$11\left(\frac{77}{6}\right) + b = -20$$

$$\frac{847}{6} + b = -20$$

$$b = -\frac{120}{6} - \frac{847}{6}$$

$$b = -\frac{967}{6} = -161\frac{1}{6}$$

Replace a with $\frac{77}{6}$ and b with $-\frac{967}{6}$ in equation (2).

$$36\left(\frac{77}{6}\right) + 6\left(\frac{-967}{6}\right) + c = 925$$

$$6(77) + (-967) + c = 925$$

$$462 - 967 + c = 925$$

$$-505 + c = 925$$

$$c = 1430$$

Thus, the coefficients are $a = 12\frac{5}{6}$, $b = -161\frac{1}{6}$, and $c = 1430$.

The model is $y = 12\frac{5}{6}x^2 - 161\frac{1}{6}x + 1430$.

Using $x = 15$, then

$y = 12\frac{5}{6}(15)^2 - 161\frac{1}{6}(15) + 1430 = 1900$.

In 2005, 1900 students will take the ACT.

41. $3x - y + z = 2$ (1)

$-x + 2y + 3z = 6$ (2)

Multiply equation (1) by 2 and add to equation (2).

$$\begin{array}{r} 6x - 2y + 2z = 4 \\ -x + 2y + 3z = 6 \\ \hline 5x \quad + 5z = 10 \end{array}$$

43. $x + 2y - z = 0$ (1)

$3x + y - z = 2$ (2)

Multiply equation (1) by –3 and add to equation (2).

$$\begin{array}{r} -3x - 6y + 3z = 0 \\ 3x + y - z = 2 \\ \hline -5y + 2z = 2 \end{array}$$

45. $\begin{cases} x+y \quad\quad -w = 0 & (1) \\ y+2z+w = 3 & (2) \\ x \quad -z \quad = 1 & (3) \\ 2x-y \quad -w = -1 & (4) \end{cases}$

Add equations (2) and (4).

$$\begin{array}{l} y+2z+w=\ 3 \\ \underline{2x-y \quad\ -w=-1} \\ 2x \quad +2z \quad = \ 2 \quad (5) \end{array}$$

Multiply equation (3) by 2 and add to equation (5).

$$\begin{array}{l} 2x-2z=2 \\ \underline{2x+2z=2} \\ 4x \quad\ =4 \\ x \quad\ =1 \end{array}$$

Replace x with 1 in equation (3).

$1-z=1$

$z=0$

Replace x with 1 in equation (1).

$1+y-w=\ 0$

$y-w=-1 \quad (6)$

Replace x with 1 in equation (4).

$2-y-w=-1$

$-y-w=-3 \quad (7)$

Add equations (6) and (7).

$$\begin{array}{l} y-w=-1 \\ \underline{-y-w=-3} \\ -2w=-4 \\ w=\ 2 \end{array}$$

Replace w with 2 in equation (6).

$y-2=-1$

$y=1$

The solution is (1, 1, 0, 2).

47. $\begin{cases} x+y+z+w=5 & (1) \\ 2x+y+z+w=6 & (2) \\ x+y+z \quad =2 & (3) \\ x+y \quad\quad =0 & (4) \end{cases}$

Multiply equation (1) by −1 and add to equation (2).

$$\begin{array}{l} -x-y-z-w=-5 \\ \underline{2x+y+z+w=6} \\ x \quad\quad\quad =1 \end{array}$$

Replace x with 1 in equation (4).

$1+y=0$

$y=-1$

Replace x with 1 and y with −1 in equation (3).

$1+(-1)+z=2$

$z=2$

Replace x with 1, y with −1 and z with 2 in equation (1).

$1+(-1)+2+w=5$

$2+w=5$

$w=3$

The solution is (1, −1, 2, 3).

Exercise Set 8.7

1. $\begin{cases} x+\ y=1 \\ x-2y=4 \end{cases}$

$$\left[\begin{array}{cc|c} 1 & 1 & 1 \\ 1 & -2 & 4 \end{array}\right]$$

Multiply row 1 by −1 and add to row 2.

$$\left[\begin{array}{cc|c} 1 & 1 & 1 \\ 0 & -3 & 3 \end{array}\right]$$

Divide row 2 by −3.

$$\left[\begin{array}{cc|c} 1 & 1 & 1 \\ 0 & 1 & -1 \end{array}\right]$$

This corresponds to $\begin{cases} x+y=\ 1 \\ \quad\ y=-1 \end{cases}$.

$x+(-1)=1$

$x-1=1$

$x=2$

The solution is (2, −1).

3. $\begin{cases} x+3y=2 \\ x+2y=0 \end{cases}$

$\left[\begin{array}{cc|c} 1 & 3 & 2 \\ 1 & 2 & 0 \end{array}\right]$

Multiply row 1 by –1 and add to row 2.

$\left[\begin{array}{cc|c} 1 & 3 & 2 \\ 0 & -1 & -2 \end{array}\right]$

Multiply row 2 by –1.

$\left[\begin{array}{cc|c} 1 & 3 & 2 \\ 0 & 1 & 2 \end{array}\right]$

This corresponds to $\begin{cases} x+3y=2 \\ y=2 \end{cases}$

$x+3(2)=2$
$x+6=2$
$x=-4$

The solution is (–4, 2).

5. $\begin{cases} x-2y=4 \\ 2x-4y=4 \end{cases}$

$\left[\begin{array}{cc|c} 1 & -2 & 4 \\ 2 & -4 & 4 \end{array}\right]$

Multiply row 1 by –2 and add to row 2.

$\left[\begin{array}{cc|c} 1 & -2 & 4 \\ 0 & 0 & -4 \end{array}\right]$

This corresponds to $\begin{cases} x-2y=4 \\ 0=-4 \end{cases}$.

This is an inconsistent system.
There is no solution.

7. $\begin{cases} 3x-3y=9 \\ 2x-2y=6 \end{cases}$

$\left[\begin{array}{cc|c} 3 & -3 & 9 \\ 2 & -2 & 6 \end{array}\right]$

Divide row 1 by 3.

$\left[\begin{array}{cc|c} 1 & -1 & 3 \\ 2 & -2 & 6 \end{array}\right]$

Multiply row 1 by –2 and add to row 2.

$\left[\begin{array}{cc|c} 1 & -1 & 3 \\ 0 & 0 & 0 \end{array}\right]$

This corresponds to $\begin{cases} x-y=3 \\ 0=0 \end{cases}$.

This is a dependent system.
The solution is $\{(x, y) \mid x-y=3\}$.

9. $\begin{cases} x+y=3 \\ 2y=10 \\ 3x+2y-4z=12 \end{cases}$

$\left[\begin{array}{ccc|c} 1 & 1 & 0 & 3 \\ 0 & 2 & 0 & 10 \\ 3 & 2 & -4 & 12 \end{array}\right]$

Multiply row 1 by –3 and add to row 3.

$\left[\begin{array}{ccc|c} 1 & 1 & 0 & 3 \\ 0 & 2 & 0 & 10 \\ 0 & -1 & -4 & 3 \end{array}\right]$

Divide row 2 by 2.

$\left[\begin{array}{ccc|c} 1 & 1 & 0 & 3 \\ 0 & 1 & 0 & 5 \\ 0 & -1 & -4 & 3 \end{array}\right]$

Add row 2 to row 3.

$\left[\begin{array}{ccc|c} 1 & 1 & 0 & 3 \\ 0 & 1 & 0 & 5 \\ 0 & 0 & -4 & 8 \end{array}\right]$

Divide row 3 by –4.

$\left[\begin{array}{ccc|c} 1 & 1 & 0 & 3 \\ 0 & 1 & 0 & 5 \\ 0 & 0 & 1 & -2 \end{array}\right]$

This corresponds to

$\begin{cases} x+y=3 \\ y=5 \\ z=-2 \end{cases}$.

$x+5=3$
$x=-2$

The solution is (–2, 5, –2).

11. $\begin{cases} 2y - z = -7 \\ x + 4y + z = -4 \\ 5x - y + 2z = 13 \end{cases}$

$\left[\begin{array}{ccc|c} 0 & 2 & -1 & -7 \\ 1 & 4 & 1 & -4 \\ 5 & -1 & 2 & 13 \end{array}\right]$

Interchange rows 1 and 2.

$\left[\begin{array}{ccc|c} 1 & 4 & 1 & -4 \\ 0 & 2 & -1 & -7 \\ 5 & -1 & 2 & 13 \end{array}\right]$

Multiply row 1 by –5 and add to row 3.

$\left[\begin{array}{ccc|c} 1 & 4 & 1 & -4 \\ 0 & 2 & -1 & -7 \\ 0 & -21 & -3 & 33 \end{array}\right]$

Divide row 2 by 2.

$\left[\begin{array}{ccc|c} 1 & 4 & 1 & -4 \\ 0 & 1 & -\frac{1}{2} & -\frac{7}{2} \\ 0 & -21 & -3 & 33 \end{array}\right]$

Multiply row 2 by 21 and add to row 3.

$\left[\begin{array}{ccc|c} 1 & 4 & 1 & -4 \\ 0 & 1 & -\frac{1}{2} & -\frac{7}{2} \\ 0 & 0 & -\frac{27}{2} & -\frac{81}{2} \end{array}\right]$

Multiply row 3 by $-\frac{2}{27}$.

$\left[\begin{array}{ccc|c} 1 & 4 & 1 & -4 \\ 0 & 1 & -\frac{1}{2} & -\frac{7}{2} \\ 0 & 0 & 1 & 3 \end{array}\right]$

This corresponds to

$\begin{cases} x + 4y + z = -4 \\ y - \frac{1}{2}z = -\frac{7}{2} \\ z = 3 \end{cases}.$

$$y - \frac{1}{2}(3) = -\frac{7}{2}$$
$$y - \frac{3}{2} = -\frac{7}{2}$$
$$y = -2$$
$$x + 4(-2) + 3 = -4$$
$$x - 8 + 3 = -4$$
$$x = 1$$

The solution is (1, –2, 3).

13. $\begin{cases} x - 4 = 0 \\ x + y = 1 \end{cases}$ or $\begin{cases} x = 4 \\ x + y = 1 \end{cases}$

$\left[\begin{array}{cc|c} 1 & 0 & 4 \\ 1 & 1 & 1 \end{array}\right]$

Multiply row 1 by –1 and add to row 2.

$\left[\begin{array}{cc|c} 1 & 0 & 4 \\ 0 & 1 & -3 \end{array}\right]$

This corresponds to $\begin{cases} x = 4 \\ y = -3 \end{cases}.$

The solution is (4, –3).

15. $\begin{cases} x + y + z = 2 \\ 2x - z = 5 \\ 3y + z = 2 \end{cases}$

$\left[\begin{array}{ccc|c} 1 & 1 & 1 & 2 \\ 2 & 0 & -1 & 5 \\ 0 & 3 & 1 & 2 \end{array}\right]$

Multiply row 1 by –2 and add to row 2.

$\left[\begin{array}{ccc|c} 1 & 1 & 1 & 2 \\ 0 & -2 & -3 & 1 \\ 0 & 3 & 1 & 2 \end{array}\right]$

Divide row 2 by –2.

$\left[\begin{array}{ccc|c} 1 & 1 & 1 & 2 \\ 0 & 1 & \frac{3}{2} & -\frac{1}{2} \\ 0 & 3 & 1 & 2 \end{array}\right]$

Multiply row 2 by –3 and add to row 3.

$\left[\begin{array}{ccc|c} 1 & 1 & 1 & 2 \\ 0 & 1 & \frac{3}{2} & -\frac{1}{2} \\ 0 & 0 & -\frac{7}{2} & \frac{7}{2} \end{array}\right]$

Multiply row 3 by $-\frac{2}{7}$.

$\left[\begin{array}{ccc|c} 1 & 1 & 1 & 2 \\ 0 & 1 & \frac{3}{2} & -\frac{1}{2} \\ 0 & 0 & 1 & -1 \end{array}\right]$

This corresponds to

$\begin{cases} x + y + z = 2 \\ y + \frac{3}{2}z = -\frac{1}{2} \\ z = -1 \end{cases}.$

$$y+\frac{3}{2}(-1)=-\frac{1}{2}$$
$$y-\frac{3}{2}=-\frac{1}{2}$$
$$y=1$$

$$x+1+(-1)=2$$
$$x=2$$

The solution is (2, 1, –1).

17. $\begin{cases} 5x-2y=27 \\ -3x+5y=18 \end{cases}$

$$\left[\begin{array}{cc|c} 5 & -2 & 27 \\ -3 & 5 & 18 \end{array}\right]$$

Divide row 1 by 5.

$$\left[\begin{array}{cc|c} 1 & -\frac{2}{5} & \frac{27}{5} \\ -3 & 5 & 18 \end{array}\right]$$

Multiply row 1 by 3 and add to row 2.

$$\left[\begin{array}{cc|c} 1 & -\frac{2}{5} & \frac{27}{5} \\ 0 & \frac{19}{5} & \frac{171}{5} \end{array}\right]$$

Multiply row 2 by $\frac{5}{19}$.

$$\left[\begin{array}{cc|c} 1 & -\frac{2}{5} & \frac{27}{5} \\ 0 & 1 & 9 \end{array}\right]$$

Multiply row 2 by $\frac{2}{5}$ and add to row 1.

$$\left[\begin{array}{cc|c} 1 & 0 & 9 \\ 0 & 1 & 9 \end{array}\right]$$

This corresponds to $\begin{cases} x=9 \\ y=9 \end{cases}$.

The solution is (9, 9).

19. $\begin{cases} 4x-7y=7 \\ 12x-21y=24 \end{cases}$

$$\left[\begin{array}{cc|c} 4 & -7 & 7 \\ 12 & -21 & 24 \end{array}\right]$$

Divide row 1 by 4.

$$\left[\begin{array}{cc|c} 1 & -\frac{7}{4} & \frac{7}{4} \\ 12 & -21 & 24 \end{array}\right]$$

Multiply row 1 by –12 and add to row 2.

$$\left[\begin{array}{cc|c} 1 & -\frac{7}{4} & \frac{7}{4} \\ 0 & 0 & 3 \end{array}\right]$$

This corresponds to $\begin{cases} x-\frac{7}{4}y=\frac{7}{4} \\ 0=3 \end{cases}$.

This is an inconsistent system.
There is no solution.

21. $\begin{cases} 4x-y+2z=5 \\ 2y+z=4 \\ 4x+y+3z=10 \end{cases}$

$$\left[\begin{array}{ccc|c} 4 & -1 & 2 & 5 \\ 0 & 2 & 1 & 4 \\ 4 & 1 & 3 & 10 \end{array}\right]$$

Divide row 1 by 4.

$$\left[\begin{array}{ccc|c} 1 & -\frac{1}{4} & \frac{1}{2} & \frac{5}{4} \\ 0 & 2 & 1 & 4 \\ 4 & 1 & 3 & 10 \end{array}\right]$$

Multiply row 1 by –4 and add to row 3.

$$\left[\begin{array}{ccc|c} 1 & -\frac{1}{4} & \frac{1}{2} & \frac{5}{4} \\ 0 & 2 & 1 & 4 \\ 0 & 2 & 1 & 5 \end{array}\right]$$

Divide row 2 by 2.

$$\left[\begin{array}{ccc|c} 1 & -\frac{1}{4} & \frac{1}{2} & \frac{5}{4} \\ 0 & 1 & \frac{1}{2} & 2 \\ 0 & 2 & 1 & 5 \end{array}\right]$$

Multiply row 2 by –2 and add to row 3.

$$\left[\begin{array}{ccc|c} 1 & -\frac{1}{4} & \frac{1}{2} & \frac{5}{4} \\ 0 & 1 & \frac{1}{2} & 2 \\ 0 & 0 & 0 & 1 \end{array}\right]$$

This corresponds to $\begin{cases} x-\frac{1}{4}y+\frac{1}{2}z=\frac{5}{4} \\ y+\frac{1}{2}z=2 \\ 0=1 \end{cases}$.

This is an inconsistent system.
There is no solution.

23. $\begin{cases} 4x + y + z = 3 \\ -x + y - 2z = -11 \\ x + 2y + 2z = -1 \end{cases}$

$$\left[\begin{array}{ccc|c} 4 & 1 & 1 & 3 \\ -1 & 1 & -2 & -11 \\ 1 & 2 & 2 & -1 \end{array}\right]$$

Interchange rows 1 and 3.

$$\left[\begin{array}{ccc|c} 1 & 2 & 2 & -1 \\ -1 & 1 & -2 & -11 \\ 4 & 1 & 1 & 3 \end{array}\right]$$

Multiply row 1 by 1 and add to row 2.
Multiply row 1 by –4 and add to row 3.

$$\left[\begin{array}{ccc|c} 1 & 2 & 2 & -1 \\ 0 & 3 & 0 & -12 \\ 0 & -7 & -7 & 7 \end{array}\right]$$

Divide row 2 by 3.

$$\left[\begin{array}{ccc|c} 1 & 2 & 2 & -1 \\ 0 & 1 & 0 & -4 \\ 0 & -7 & -7 & 7 \end{array}\right]$$

Multiply row 2 by 7 and add to row 3.

$$\left[\begin{array}{ccc|c} 1 & 2 & 2 & -1 \\ 0 & 1 & 0 & -4 \\ 0 & 0 & -7 & -21 \end{array}\right]$$

Divide row 3 by –7.

$$\left[\begin{array}{ccc|c} 1 & 2 & 2 & -1 \\ 0 & 1 & 0 & -4 \\ 0 & 0 & 1 & 3 \end{array}\right]$$

This corresponds to

$$\begin{cases} x + 2y + 2z = -1 \\ y = -4. \\ z = 3 \end{cases}$$

$$x + 2(-4) + 2(3) = -1$$
$$x - 8 + 6 = -1$$
$$x = 1$$

The solution is (1, –4, 3).

25. The matrix should be $\left[\begin{array}{cc|c} 2 & -3 & 8 \\ 1 & 5 & -3 \end{array}\right]$.

Answers may vary.

27. Function; each vertical line intersects the graph in exactly one point.

29. Not a function; the vertical line $x = 0$ intersects the graph in more than one point.

31. $(-1)(-5) - (6)(3) = 5 - 18 = -13$

33. $(4)(-10) - (2)(-2) = -40 + 4 = -36$

35. $(-3)(-3) - (-1)(-9) = 9 - 9 = 0$

Exercise Set 8.8

1. $\begin{vmatrix} 3 & 5 \\ -1 & 7 \end{vmatrix} = 3(7) - 5(-1)$
$= 21 + 5$
$= 26$

3. $\begin{vmatrix} 9 & -2 \\ 4 & -3 \end{vmatrix} = 9(-3) - 4(-2)$
$= -27 + 8$
$= -19$

5. $\begin{vmatrix} -2 & 9 \\ 4 & -18 \end{vmatrix} = -2(-18) - 9(4)$
$= 36 - 36$
$= 0$

7. $\begin{cases} 2y - 4 = 0 \\ x + 2y = 5 \end{cases}$ or $\begin{cases} 2y = 4 \\ x + 2y = 5 \end{cases}$

$D = \begin{vmatrix} 0 & 2 \\ 1 & 2 \end{vmatrix} = 0(2) - 2(1) = 0 - 2 = -2$

$D_x = \begin{vmatrix} 4 & 2 \\ 5 & 2 \end{vmatrix} = 4(2) - 2(5) = 8 - 10 = -2$

$D_y = \begin{vmatrix} 0 & 4 \\ 1 & 5 \end{vmatrix} = 0(5) - 4(1) = 0 - 4 = -4$

$x = \frac{-2}{-2} = 1$ and $y = \frac{-4}{-2} = 2$

The solution is (1, 2).

9. $\begin{cases} 3x+y=1 \\ 2y=2-6x \end{cases}$ or $\begin{cases} 3x+\ y=1 \\ 6x+2y=2 \end{cases}$

$D=\begin{vmatrix} 3 & 1 \\ 6 & 2 \end{vmatrix}=3(2)-1(6)=6-6=0$

Thus, this system cannot be solved using Cramer's rule. Since equation 2 is 2 times equation 1, the system is dependent.
The solution is $\{(x,\ y)|\ 3x+y=1\}$.

11. $\begin{cases} 5x-2y=27 \\ -3x+5y=18 \end{cases}$

$D=\begin{vmatrix} 5 & -2 \\ -3 & 5 \end{vmatrix}$
$=5(5)-(-2)(-3)$
$=25-6$
$=19$

$D_x=\begin{vmatrix} 27 & -2 \\ 18 & 5 \end{vmatrix}$
$=27(5)-(-2)18$
$=135+36$
$=171$

$D_y=\begin{vmatrix} 5 & 27 \\ -3 & 18 \end{vmatrix}$
$=5(18)-27(-3)$
$=90+81$
$=171$

$x=\frac{171}{19}=9$ and $y=\frac{171}{19}=9$
The solution is (9, 9).

13. $\begin{vmatrix} 2 & 1 & 0 \\ 0 & 5 & -3 \\ 4 & 0 & 2 \end{vmatrix}$
$=2\begin{vmatrix} 5 & -3 \\ 0 & 2 \end{vmatrix}-1\begin{vmatrix} 0 & -3 \\ 4 & 2 \end{vmatrix}+0\begin{vmatrix} 0 & 5 \\ 4 & 0 \end{vmatrix}$
$=2[5(2)-(-3)(0)]-[0(2)-4(-3)]+0$
$=2(10)-12$
$=8$

15. $\begin{vmatrix} 4 & -6 & 0 \\ -2 & 3 & 0 \\ 4 & -6 & 1 \end{vmatrix}$
$=0\begin{vmatrix} -2 & 3 \\ 4 & -6 \end{vmatrix}-0\begin{vmatrix} 4 & -6 \\ 4 & -6 \end{vmatrix}+1\begin{vmatrix} 4 & -6 \\ -2 & 3 \end{vmatrix}$
$=0-0+[4(3)-(-6)(-2)]$
$=0$

17. $\begin{vmatrix} 3 & 6 & -3 \\ -1 & -2 & 3 \\ 4 & -1 & 6 \end{vmatrix}$
$=3\begin{vmatrix} -2 & 3 \\ -1 & 6 \end{vmatrix}-6\begin{vmatrix} -1 & 3 \\ 4 & 6 \end{vmatrix}+(-3)\begin{vmatrix} -1 & -2 \\ 4 & -1 \end{vmatrix}$
$=3[-2(6)-3(-1)]-6[-1(6)-3(4)]$
$\quad -3[(-1)(-1)-(-2)4]$
$=3(-9)-6(-18)-3(9)$
$=-27+108-27$
$=54$

19. $\begin{cases} 3x \quad + z = -1 \\ -x - 3y + z = 7 \\ \quad 3y + z = 5 \end{cases}$

$$D = \begin{vmatrix} 3 & 0 & 1 \\ -1 & -3 & 1 \\ 0 & 3 & 1 \end{vmatrix}$$

$$= 3\begin{vmatrix} -3 & 1 \\ 3 & 1 \end{vmatrix} - 0\begin{vmatrix} -1 & 1 \\ 0 & 1 \end{vmatrix} + 1\begin{vmatrix} -1 & -3 \\ 0 & 3 \end{vmatrix}$$

$$= 3[(-3)(1) - 1(3)] - 0 + [(-1)(3) - (-3)0]$$

$$= 3(-6) - 3$$

$$= -21$$

$$D_x = \begin{vmatrix} -1 & 0 & 1 \\ 7 & -3 & 1 \\ 5 & 3 & 1 \end{vmatrix}$$

$$= -1\begin{vmatrix} -3 & 1 \\ 3 & 1 \end{vmatrix} - 0\begin{vmatrix} 7 & 1 \\ 5 & 1 \end{vmatrix} + 1\begin{vmatrix} 7 & -3 \\ 5 & 3 \end{vmatrix}$$

$$= -[(-3)(1) - 1(3)] - 0 + [7(3) - (-3)5]$$

$$= 6 + 36$$

$$= 42$$

$$D_y = \begin{vmatrix} 3 & -1 & 1 \\ -1 & 7 & 1 \\ 0 & 5 & 1 \end{vmatrix}$$

$$= 3\begin{vmatrix} 7 & 1 \\ 5 & 1 \end{vmatrix} - (-1)\begin{vmatrix} -1 & 1 \\ 0 & 1 \end{vmatrix} + 1\begin{vmatrix} -1 & 7 \\ 0 & 5 \end{vmatrix}$$

$$= 3[7(1) - 1(5)] + 1[(-1)1 - 1(0)] + [(-1)(5) - 7(0)]$$

$$= 3(2) + (-1) + (-5)$$

$$= 0$$

$$D_z = \begin{vmatrix} 3 & 0 & -1 \\ -1 & -3 & 7 \\ 0 & 3 & 5 \end{vmatrix}$$

$$= 3\begin{vmatrix} -3 & 7 \\ 3 & 5 \end{vmatrix} - 0\begin{vmatrix} -1 & 7 \\ 0 & 5 \end{vmatrix} + (-1)\begin{vmatrix} -1 & -3 \\ 0 & 3 \end{vmatrix}$$

$$= 3[(-3)(5) - 7(3)] - 0 - [(-1)(3) - (-3)0]$$

$$= 3(-36) - (-3)$$

$$= -105$$

$$x = \frac{42}{-21} = -2,\ y = \frac{0}{-21} = 0,\ z = \frac{-105}{-21} = 5$$

The solution is (–2, 0, 5).

21. $\begin{cases} x + y + z = 8 \\ 2x - y - z = 10 \\ x - 2y + 3z = 22 \end{cases}$

$$D = \begin{vmatrix} 1 & 1 & 1 \\ 2 & -1 & -1 \\ 1 & -2 & 3 \end{vmatrix}$$

$$= 1\begin{vmatrix} -1 & -1 \\ -2 & 3 \end{vmatrix} - 1\begin{vmatrix} 2 & -1 \\ 1 & 3 \end{vmatrix} + 1\begin{vmatrix} 2 & -1 \\ 1 & -2 \end{vmatrix}$$

$$= (-3 - 2) - [6 - (-1)] + [-4 - (-1)]$$

$$= -5 - 7 - 3$$

$$= -15$$

$$D_x = \begin{vmatrix} 8 & 1 & 1 \\ 10 & -1 & -1 \\ 22 & -2 & 3 \end{vmatrix}$$

$$= 8\begin{vmatrix} -1 & -1 \\ -2 & 3 \end{vmatrix} - 1\begin{vmatrix} 10 & -1 \\ 22 & 3 \end{vmatrix} + 1\begin{vmatrix} 10 & -1 \\ 22 & -2 \end{vmatrix}$$

$$= 8(-3 - 2) - [30 - (-22)] + [-20 - (-22)]$$

$$= 8(-5) - 52 + 2$$

$$= -40 - 52 + 2$$

$$= -90$$

$$D_y = \begin{vmatrix} 1 & 8 & 1 \\ 2 & 10 & -1 \\ 1 & 22 & 3 \end{vmatrix}$$

$$= 1\begin{vmatrix} 10 & -1 \\ 22 & 3 \end{vmatrix} - 8\begin{vmatrix} 2 & -1 \\ 1 & 3 \end{vmatrix} + 1\begin{vmatrix} 2 & 10 \\ 1 & 22 \end{vmatrix}$$

$$= [30 - (-22)] - 8[6 - (-1)] + [44 - 10]$$

$$= 52 - 8(7) + 34$$

$$= 52 - 56 + 34$$

$$= 30$$

$$D_z = \begin{vmatrix} 1 & 1 & 8 \\ 2 & -1 & 10 \\ 1 & -2 & 22 \end{vmatrix}$$

$$= 1\begin{vmatrix} -1 & 10 \\ -2 & 22 \end{vmatrix} - 1\begin{vmatrix} 2 & 10 \\ 1 & 22 \end{vmatrix} + 8\begin{vmatrix} 2 & -1 \\ 1 & -2 \end{vmatrix}$$

$$= [-22 - (-20)] - (44 - 10) + 8[-4 - (-1)]$$

$$= -2 - 34 + 8(-3)$$

$$= -36 - 24$$

$$= -60$$

$$x = \frac{-90}{-15} = 6,\ y = \frac{30}{-15} = -2,\ z = \frac{-60}{-15} = 4$$

The solution is (6, –2, 4).

23. $\begin{vmatrix} 10 & -1 \\ -4 & 2 \end{vmatrix} = 10(2) - (-1)(-4)$
$= 20 - 4$
$= 16$

25. $\begin{vmatrix} 1 & 0 & 4 \\ 1 & -1 & 2 \\ 3 & 2 & 1 \end{vmatrix} = 1\begin{vmatrix} -1 & 2 \\ 2 & 1 \end{vmatrix} - 0\begin{vmatrix} 1 & 2 \\ 3 & 1 \end{vmatrix} + 4\begin{vmatrix} 1 & -1 \\ 3 & 2 \end{vmatrix}$
$= 1[-1-4] - 0 + 4[2-(-3)]$
$= -5 + 4(5)$
$= -5 + 20$
$= 15$

27. $\begin{vmatrix} \frac{3}{4} & \frac{5}{2} \\ -\frac{1}{6} & \frac{7}{3} \end{vmatrix} = \frac{3}{4}\left(\frac{7}{3}\right) - \frac{5}{2}\left(-\frac{1}{6}\right)$
$= \frac{21}{12} + \frac{5}{12}$
$= \frac{26}{12}$
$= \frac{13}{6}$

29. $\begin{vmatrix} 4 & -2 & 2 \\ 6 & -1 & 3 \\ 2 & 1 & 1 \end{vmatrix}$
$= 4\begin{vmatrix} -1 & 3 \\ 1 & 1 \end{vmatrix} - (-2)\begin{vmatrix} 6 & 3 \\ 2 & 1 \end{vmatrix} + 2\begin{vmatrix} 6 & -1 \\ 2 & 1 \end{vmatrix}$
$= 4[-1-3] + 2[6-6] + 2[6-(-2)]$
$= 4(-4) + 2 \cdot 0 + 2(8)$
$= -16 + 0 + 16$
$= 0$

31. $\begin{vmatrix} -2 & 5 & 4 \\ 5 & -1 & 3 \\ 4 & 1 & 2 \end{vmatrix}$
$= -2\begin{vmatrix} -1 & 3 \\ 1 & 2 \end{vmatrix} - 5\begin{vmatrix} 5 & 3 \\ 4 & 2 \end{vmatrix} + 4\begin{vmatrix} 5 & -1 \\ 4 & 1 \end{vmatrix}$
$= -2[-2-3] - 5[10-12] + 4[5-(-4)]$
$= -2(-5) - 5(-2) + 4(9)$
$= 10 + 10 + 36$
$= 56$

33. If the elements of a single row of a determinant are all zero, the value of the determinant will be zero. To see this, consider expanding on that row containing all zeros.

35. $\begin{vmatrix} 1 & x \\ 2 & 7 \end{vmatrix} = -3$
$(1)(7) - 2x = -3$
$7 - 2x = -3$
$2x = 10$
$x = 5$

37. $\begin{cases} 2x - 5y = 4 \\ x + 2y = -7 \end{cases}$

$D = \begin{vmatrix} 2 & -5 \\ 1 & 2 \end{vmatrix} = 2(2) - (-5)(1) = 4 + 5 = 9$

$D_x = \begin{vmatrix} 4 & -5 \\ -7 & 2 \end{vmatrix}$
$= 4(2) - (-5)(-7)$
$= 8 - 35$
$= -27$

$D_y = \begin{vmatrix} 2 & 4 \\ 1 & -7 \end{vmatrix} = 2(-7) - 4(1) = -14 - 4 = -18$

$x = \frac{-27}{9} = -3$, and $y = \frac{-18}{9} = -2$

The solution is (–3, –2).

39. $\begin{cases} 4x + 2y = 5 \\ 2x + y = -1 \end{cases}$

$D = \begin{vmatrix} 4 & 2 \\ 2 & 1 \end{vmatrix} = 4(1) - 2(2) = 4 - 4 = 0$

Thus, Cramer's rule cannot be used to solve the system. Multiply equation 2 by 2 yielding the new system:

$\begin{cases} 4x + 2y = 5 \\ 4x + 2y = -2 \end{cases}$

Therefore, the system is inconsistent. There is no solution.

41. $\begin{cases} 2x + 2y + z = 1 \\ -x + y + 2z = 3 \\ x + 2y + 4z = 0 \end{cases}$

$$D = \begin{vmatrix} 2 & 2 & 1 \\ -1 & 1 & 2 \\ 1 & 2 & 4 \end{vmatrix}$$

$$= 2\begin{vmatrix} 1 & 2 \\ 2 & 4 \end{vmatrix} - 2\begin{vmatrix} -1 & 2 \\ 1 & 4 \end{vmatrix} + 1\begin{vmatrix} -1 & 1 \\ 1 & 2 \end{vmatrix}$$

$= 2(4-4) - 2(-4-2) + (-2-1)$

$= 2(0) - 2(-6) + (-3)$

$= 0 + 12 - 3$

$= 9$

$$D_x = \begin{vmatrix} 1 & 2 & 1 \\ 3 & 1 & 2 \\ 0 & 2 & 4 \end{vmatrix}$$

$$= 1\begin{vmatrix} 1 & 2 \\ 2 & 4 \end{vmatrix} - 3\begin{vmatrix} 2 & 1 \\ 2 & 4 \end{vmatrix} + 0\begin{vmatrix} 2 & 1 \\ 1 & 2 \end{vmatrix}$$

$= (4-4) - 3(8-2) + 0$

$= 0 - 3(6)$

$= -18$

$$D_y = \begin{vmatrix} 2 & 1 & 1 \\ -1 & 3 & 2 \\ 1 & 0 & 4 \end{vmatrix}$$

$$= 1\begin{vmatrix} 1 & 1 \\ 3 & 2 \end{vmatrix} - 0\begin{vmatrix} 2 & 1 \\ -1 & 2 \end{vmatrix} + 4\begin{vmatrix} 2 & 1 \\ -1 & 3 \end{vmatrix}$$

$= (2-3) - 0 + 4[6-(-1)]$

$= -1 + 4(7)$

$= -1 + 28$

$= 27$

$$D_z = \begin{vmatrix} 2 & 2 & 1 \\ -1 & 1 & 3 \\ 1 & 2 & 0 \end{vmatrix}$$

$$= 1\begin{vmatrix} 2 & 1 \\ 1 & 3 \end{vmatrix} - 2\begin{vmatrix} 2 & 1 \\ -1 & 3 \end{vmatrix} + 0\begin{vmatrix} 2 & 2 \\ -1 & 1 \end{vmatrix}$$

$= (6-1) - 2[6-(-1)] + 0$

$= 5 - 2(7)$

$= 5 - 14$

$= -9$

$x = -\frac{18}{9} = -2,\ y = \frac{27}{9} = 3,$

$z = \frac{-9}{9} = -1$

The solution is (–2, 3, –1).

43. $\begin{cases} \frac{2}{3}x - \frac{3}{4}y = -1 \\ -\frac{1}{6}x + \frac{3}{4}y = \frac{5}{2} \end{cases}$

$$D = \begin{vmatrix} \frac{2}{3} & -\frac{3}{4} \\ -\frac{1}{6} & \frac{3}{4} \end{vmatrix}$$

$= \frac{2}{3} \cdot \frac{3}{4} - \left(-\frac{3}{4}\right)\left(-\frac{1}{6}\right)$

$= \frac{1}{2} - \frac{1}{8}$

$= \frac{3}{8}$

$$D_x = \begin{vmatrix} -1 & -\frac{3}{4} \\ \frac{5}{2} & \frac{3}{4} \end{vmatrix}$$

$= (-1)\frac{3}{4} - \left(-\frac{3}{4}\right)\frac{5}{2}$

$= -\frac{3}{4} + \frac{15}{8}$

$= \frac{9}{8}$

$$D_y = \begin{vmatrix} \frac{2}{3} & -1 \\ -\frac{1}{6} & \frac{5}{2} \end{vmatrix}$$

$= \frac{2}{3} \cdot \frac{5}{2} - (-1)\left(-\frac{1}{6}\right)$

$= \frac{5}{3} - \frac{1}{6}$

$= \frac{3}{2}$

$x = \frac{\frac{9}{8}}{\frac{3}{8}} = 3$ and $y = \frac{\frac{3}{2}}{\frac{3}{8}} = 4$

The solution is (3, 4).

45. $\begin{cases} 0.7x - 0.2y = -1.6 \\ 0.2x - y = -1.4 \end{cases}$

$$D = \begin{vmatrix} 0.7 & -0.2 \\ 0.2 & -1 \end{vmatrix} = 0.7(-1) - (-0.2)(0.2) = -0.7 + 0.04 = -0.66$$

$$D_x = \begin{vmatrix} -1.6 & -0.2 \\ -1.4 & -1 \end{vmatrix} = (-1.6)(-1) - (-0.2)(-1.4) = 1.6 - 0.28 = 1.32$$

$$D_y = \begin{vmatrix} 0.7 & -1.6 \\ 0.2 & -1.4 \end{vmatrix} = (0.7)(-1.4) - (-1.6)(0.2) = -0.98 + 0.32 = -0.66$$

$x = \dfrac{1.32}{-0.66} = -2$ and $y = \dfrac{-0.66}{-0.66} = 1$

The solution is (–2, 1).

47. $\begin{cases} -2x + 4y - 2z = 6 \\ x - 2y + z = -3 \\ 3x - 6y + 3z = -9 \end{cases}$

$$D = \begin{vmatrix} -2 & 4 & -2 \\ 1 & -2 & 1 \\ 3 & -6 & 3 \end{vmatrix}$$

$$= -2\begin{vmatrix} -2 & 1 \\ -6 & 3 \end{vmatrix} - 4\begin{vmatrix} 1 & 1 \\ 3 & 3 \end{vmatrix} + (-2)\begin{vmatrix} 1 & -2 \\ 3 & -6 \end{vmatrix}$$

$$= -2[-6 - (-6)] - 4[3 - 3] - 2[-6 - (-6)] = -2(0) - 4(0) - 2(0) = 0$$

Therefore, Cramer's rule will not provide the solution. Note that equation 1 is –2 times equation 2 and that equation 3 is 3 times equation 2. Thus, the system is dependent. The solution is $\{(x, y, z) | x - 2y + z = -3\}$.

49. $\begin{cases} x - 2y + z = -5 \\ 3y + 2z = 4 \\ 3x - y = -2 \end{cases}$

$$D = \begin{vmatrix} 1 & -2 & 1 \\ 0 & 3 & 2 \\ 3 & -1 & 0 \end{vmatrix}$$

$$= 1\begin{vmatrix} 3 & 2 \\ -1 & 0 \end{vmatrix} - 0\begin{vmatrix} -2 & 1 \\ -1 & 0 \end{vmatrix} + 3\begin{vmatrix} -2 & 1 \\ 3 & 2 \end{vmatrix}$$

$$= [0 - (-2)] - 0 + 3[-4 - 3] = 2 + 3(-7) = 2 - 21 = -19$$

$$D_x = \begin{vmatrix} -5 & -2 & 1 \\ 4 & 3 & 2 \\ -2 & -1 & 0 \end{vmatrix}$$

$$= 1\begin{vmatrix} 4 & 3 \\ -2 & -1 \end{vmatrix} - 2\begin{vmatrix} -5 & -2 \\ -2 & -1 \end{vmatrix} + 0\begin{vmatrix} -5 & -2 \\ 4 & 3 \end{vmatrix}$$

$$= [-4 - (-6)] - 2[5 - 4] + 0 = 2 - 2(1) = 0$$

$$D_y = \begin{vmatrix} 1 & -5 & 1 \\ 0 & 4 & 2 \\ 3 & -2 & 0 \end{vmatrix}$$

$$= 1\begin{vmatrix} 4 & 2 \\ -2 & 0 \end{vmatrix} - 0\begin{vmatrix} -5 & 1 \\ -2 & 0 \end{vmatrix} + 3\begin{vmatrix} -5 & 1 \\ 4 & 2 \end{vmatrix}$$

$$= [0 - (-4)] - 0 + 3[-10 - 4] = 4 + 3(-14) = 4 - 42 = -38$$

$$D_z = \begin{vmatrix} 1 & -2 & -5 \\ 0 & 3 & 4 \\ 3 & -1 & -2 \end{vmatrix}$$

$$= 1\begin{vmatrix} 3 & 4 \\ -1 & -2 \end{vmatrix} - 0\begin{vmatrix} -2 & -5 \\ -1 & -2 \end{vmatrix} + 3\begin{vmatrix} -2 & -5 \\ 3 & 4 \end{vmatrix}$$

$$= [-6 - (-4)] - 0 + 3[-8 - (-15)] = -2 + 3(7) = -2 + 21 = 19$$

$x = \dfrac{0}{-19} = 0$, $y = \dfrac{-38}{-19} = 2$,

$z = \dfrac{19}{-19} = -1$

The solution is (0, 2, –1).

51. The array of signs for use with a 4×4 matrix is

$$\begin{matrix} + & - & + & - \\ - & + & - & + \\ + & - & + & - \\ - & + & - & + \end{matrix}$$

53. $5x - 6 + x - 12 = 6x - 18$

55. $2(3x-6)+3(x-1) = 6x - 12 + 3x - 3$
$= 9x - 15$

57. $f(x) = 5x - 6$ or $y = 5x - 6$

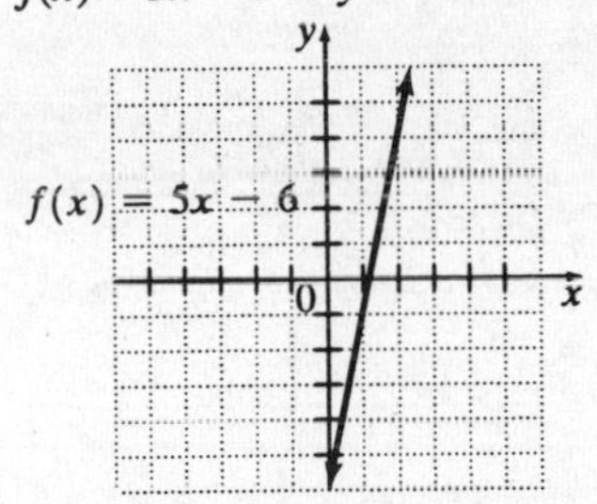

59. $h(x) = 3$ or $y = 3$

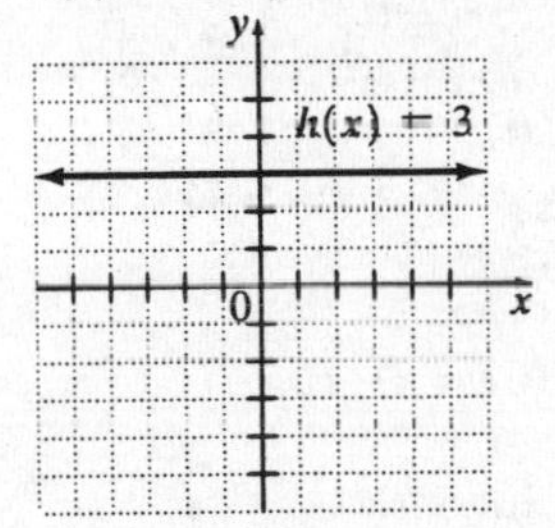

61.
$$\begin{vmatrix}5&0&0&0\\0&4&2&-1\\1&3&-2&0\\0&-3&1&2\end{vmatrix} = 5\begin{vmatrix}4&2&-1\\3&-2&0\\-3&1&2\end{vmatrix} - 0\begin{vmatrix}0&2&-1\\1&-2&0\\0&1&2\end{vmatrix} + 0\begin{vmatrix}0&4&-1\\1&3&0\\0&-3&2\end{vmatrix} - 0\begin{vmatrix}0&4&2\\1&3&-2\\0&-3&1\end{vmatrix}$$
$$= 5\left[(-1)\begin{vmatrix}3&-2\\-3&1\end{vmatrix} - 0\cdot\begin{vmatrix}4&2\\-3&1\end{vmatrix} + 2\begin{vmatrix}4&2\\3&-2\end{vmatrix}\right]$$
$= 5[-(3-6) - 0 + 2(-8-6)]$
$= 5[3 + 2(-14)]$
$= 5[3 - 28]$
$= 5[-25]$
$= -125$

63.
$$\begin{vmatrix}4&0&2&5\\0&3&-1&1\\0&0&2&0\\0&0&0&1\end{vmatrix} = 4\begin{vmatrix}3&-1&1\\0&2&0\\0&0&1\end{vmatrix} - 0\begin{vmatrix}0&2&5\\0&2&0\\0&0&1\end{vmatrix} + 0\begin{vmatrix}0&2&5\\3&-1&1\\0&0&1\end{vmatrix} - 0\begin{vmatrix}0&2&5\\3&-1&1\\0&2&0\end{vmatrix}$$
$$= 4\left[3\begin{vmatrix}2&0\\0&1\end{vmatrix} - 0\begin{vmatrix}-1&1\\0&1\end{vmatrix} + 0\begin{vmatrix}-1&1\\2&0\end{vmatrix}\right]$$
$= 4[3(2-0) - 0 + 0]$
$= 12(2)$
$= 24$

Chapter 8 Review

1. $4(x-5)=2x-14$
$4x-20=2x-14$
$2x=6$
$x=3$
The solution is 3.

2. $x+7=-2(x+8)$
$x+7=-2x-16$
$3x=-23$
$x=-\frac{23}{3}$
The solution is $-\frac{23}{3}$.

3. $3(2y-1)=-8(6+y)$
$6y-3=-48-8y$
$14y=-45$
$y=-\frac{45}{14}$
The solution is $-\frac{45}{14}$.

4. $-(z+12)=5(2z-1)$
$-z-12=10z-5$
$-7=11z$
$-\frac{7}{11}=z$
The solution is $-\frac{7}{11}$.

5. $w^2-5w=36$
$w^2-5w-36=0$
$(w-9)(w+4)=0$
$w-9=0$ or $w+4=0$
$w=9$ or $w=-4$
The solutions are –4 and 9.

6. $x^2+32=12x$
$x^2-12x+32=0$
$(x-8)(x-4)=0$
$x-8=0$ or $x-4=0$
$x=8$ or $x=4$
The solutions are 4 and 8.

7. $0.3(x-2)=1.2$
$10[0.3(x-2)]=10(1.2)$
$3(x-2)=12$
$3x-6=12$
$3x=18$
$x=6$
The solution is 6.

8. $1.5=0.2(c-0.3)$
$1.5=0.2c-0.06$
$100(1.5)=100(0.2c-0.06)$
$150=20c-6$
$156=20c$
$7.8=c$
The solution is 7.8.

9. $-4(2-3h)=2(3h-4)+6h$
$-8+12h=6h-8+6h$
$-8+12h=12h-8$
$-8=-8$
All real numbers are solutions.

10. $6(m-1)+3(2-m)=0$
$6m-6+6-3m=0$
$3m=0$
$m=0$
The solution is 0.

11. $6-3(2g+4)-4g=5(1-2g)$
$6-6g-12-4g=5-10g$
$-6-10g=5-10g$
$-6=5$
There are no solutions.

12. $20-5(p+1)+3p=-(2p-15)$
$20-5p-5+3p=-2p+15$
$15-2p=-2p+15$
$15=15$
All real numbers are solutions.

13. $x^2-2x-15=0$
$(x-5)(x+3)=0$
$x-5=0$ or $x+3=0$
$x=5$ or $x=-3$
The solutions are –3 and 5.

14.
$$x^2+6x-7=0$$
$$(x+7)(x-1)=0$$
$$x+7=0 \text{ or } x-1=0$$
$$x=-7 \text{ or } x=1$$
The solutions are –7 and 1.

15.
$$12x^2+2x-2=0$$
$$6x^2+x-1=0$$
$$(3x-1)(2x+1)=0$$
$$3x-1=0 \text{ or } 2x+1=0$$
$$3x=1 \text{ or } 2x=-1$$
$$x=\frac{1}{3} \text{ or } x=-\frac{1}{2}$$
The solutions are $-\frac{1}{2}$ and $\frac{1}{3}$.

16.
$$8x^2+13x+5=0$$
$$(8x+5)(x+1)=0$$
$$8x+5=0 \text{ or } x+1=0$$
$$8x=-5 \text{ or } x=-1$$
$$x=-\frac{5}{8} \text{ or } x=-1$$
The solutions are –1 and $-\frac{5}{8}$.

17.
$$\frac{y}{4}-\frac{y}{2}=-8$$
$$4\left(\frac{y}{4}-\frac{y}{2}\right)=4(-8)$$
$$y-2y=-32$$
$$-y=-32$$
$$y=32$$
The solution is 32.

18.
$$\frac{2x}{3}-\frac{8}{3}=x$$
$$2x-8=3x$$
$$-8=x$$
The solution is –8.

19.
$$\frac{b-2}{3}=\frac{b+2}{5}$$
$$15\left(\frac{b-2}{3}\right)=15\left(\frac{b+2}{5}\right)$$
$$5b-10=3b+6$$
$$2b=16$$
$$b=8$$
The solution is 8.

20.
$$\frac{2t-1}{3}=\frac{3t+2}{15}$$
$$15\left(\frac{2t-1}{3}\right)=15\left(\frac{3t+2}{15}\right)$$
$$5(2t-1)=3t+2$$
$$10t-5=3t+2$$
$$7t=7$$
$$t=1$$
The solution is 1.

21.
$$x^2+(x+1)^2=61$$
$$x^2+x^2+2x+1-61=0$$
$$2x^2+2x-60=0$$
$$x^2+x-30=0$$
$$(x+6)(x-5)=0$$
$$x+6=0 \text{ or } x-5=0$$
$$x=-6 \text{ or } x=5$$
The solutions are –6 and 5.

22.
$$y^2+(y+2)^2=34$$
$$y^2+y^2+4y+4-34=0$$
$$2y^2+4y-30=0$$
$$y^2+2y-15=0$$
$$(y+5)(y-3)=0$$
$$y+5=0 \text{ or } y-3=0$$
$$y=-5 \text{ or } y=3$$
The solutions are 3 and –5.

23. $\frac{x-2}{5}+\frac{x+2}{2}=\frac{x+4}{3}$

$30\left(\frac{x-2}{5}+\frac{x+2}{2}\right)=30\left(\frac{x+4}{3}\right)$

$6(x-2)+15(x+2)=10(x+4)$

$6x-12+15x+30=10x+40$

$21x+18=10x+40$

$11x=22$

$x=2$

The solution is 2.

24. $\frac{2z-3}{4}-\frac{4-z}{2}=\frac{z+1}{3}$

$12\left(\frac{2z-3}{4}-\frac{4-z}{2}\right)=12\left(\frac{z+1}{3}\right)$

$3(2z-3)-6(4-z)=4(z+1)$

$6z-9-24+6z=4z+4$

$12z-33=4z+4$

$8z=37$

$z=\frac{37}{8}$

The solution is $\frac{37}{8}$.

25. Let x = the number.

$2(x-3)=3x+1$

$2x-6=3x+1$

$-7=x$

The number is –7.

26. Let x = smaller number.
Then $x + 5$ = larger number.

$x+x+5=285$

$2x=280$

$x=140$

$x+5=145$

The numbers are 140 and 145.

27. Let x = 1998 earnings for a high school graduate.

$x+0.3047x=29{,}872$

$1.3047x=29{,}872$

$x\approx 22{,}896$

Rounded to the nearest dollar, the average annual salary for a high school graduate in 1998 was $22,896.

28. Let n = the first integer. Then
$n + 1$ = the second integer,
$n + 2$ = the third integer, and
$n + 4$ = the fourth integer.

$(n+1)+(n+2)+(n+3)-2n=16$

$n+6=16$

$n=10$

Therefore, the integers are 10, 11, 12, and 13.

29. Let x = smaller odd integer.
Then $x + 2$ = larger odd integer.

$5x=3(x+2)+54$

$5x=3x+6+54$

$2x=60$

$x=30$

Since this is not odd, no such consecutive odd integers exist.

30. Let w = the width of the playing field. Then $2w - 5$ = the length of the playing field.

$2w+2(2w-5)=230$

$2w+4w-10=230$

$6w=240$

$w=40$

$2w-5=2(40)-5$

$=80-5$

$=75$

Therefore, the field is 75 meters long and 40 meters wide.

31. Let n = the number of miles driven.

$2(29.95)+0.15(n-200)=83.6$

$59.9+0.15n-30=83.6$

$29.9+0.15n=83.6$

$0.15n=53.7$

$n=358$

He drove 358 miles.

32. Solve $R = C$.

$16.50x=4.50x+3000$

$12x=3000$

$x=250$

Thus, 250 calculators must be produced and sold in order to break even.

33. Let x be the number.

$$x + 2x^2 = 105$$
$$2x^2 + x - 105 = 0$$
$$(2x+15)(x-7) = 0$$
$$2x + 15 = 0 \quad \text{or} \quad x - 7 = 0$$
$$x = -\frac{15}{2} \quad \text{or} \quad x = 7$$

The solutions are $-\frac{15}{2}$ and 7.

34. Let x be the width. Then $5x - 2$ is the length.

$$x(5x-2) = 16$$
$$5x^2 - 2x - 16 = 0$$
$$(5x+8)(x-2) = 0$$
$$5x + 8 = 0 \quad \text{or} \quad x - 2 = 0$$
$$x = -\frac{8}{5} \quad \text{or} \quad x = 2$$

Disregard the negative.
$5x - 2 = 5(2) - 2 = 10 - 2 = 8$
The width is 2 meters, and the length is 8 meters.

35. $h(t) = -16t^2 + 400$

$$0 = -16t^2 + 400$$
$$0 = -16(t^2 - 25)$$
$$0 = t^2 - 25$$
$$0 = (t-5)(t+5)$$
$$t - 5 = 0 \text{ or } t + 5 = 0$$
$$t = 5 \text{ or } t = -5$$

Discard the negative. The dummy will hit the ground 5 seconds after being dropped.

36. $1 \le 4x - 7 \le 3$

$$8 \le 4x \le 10$$
$$2 \le x \le \frac{5}{2}$$

The solution set is $\left[2, \frac{5}{2}\right]$.

37. $-2 \le 8 + 5x < -1$

$$-10 \le 5x < -9$$
$$-2 \le x < -\frac{9}{5}$$

The solution set is $\left[-2, -\frac{9}{5}\right)$.

38. $-3 < 4(2x-1) < 12$

$$-\frac{3}{4} < 2x - 1 < 3$$
$$\frac{1}{4} < 2x < 4$$
$$\frac{1}{8} < x < 2$$

The solution set is $\left(\frac{1}{8}, 2\right)$.

39. $-6 < x - (3 - 4x) < -3$

$$-6 < x - 3 + 4x < -3$$
$$-6 < 5x - 3 < -3$$
$$-3 < 5x < 0$$
$$-\frac{3}{5} < x < 0$$

The solution set is $\left(-\frac{3}{5}, 0\right)$.

40. $\frac{1}{6} < \frac{4x-3}{3} \le \frac{4}{5}$

$$30\left(\frac{1}{6}\right) < 30\left(\frac{4x-3}{3}\right) \le 30\left(\frac{4}{5}\right)$$
$$5 < 10(4x-3) \le 24$$
$$5 < 40x - 30 \le 24$$
$$35 < 40x \le 54$$
$$\frac{7}{8} < x \le \frac{27}{20}$$

The solution set is $\left(\frac{7}{8}, \frac{27}{20}\right]$.

41. $0 \le \frac{2(3x+4)}{5} \le 3$

$$5(0) \le 5\left[\frac{2(3x+4)}{5}\right] \le 5(3)$$
$$0 \le 6x + 8 \le 15$$
$$-8 \le 6x \le 7$$
$$-\frac{4}{3} \le x \le \frac{7}{6}$$

The solution set is $\left[-\frac{4}{3}, \frac{7}{6}\right]$.

42. $x \le 2$ and $x > -5$

$$-5 < x \le 2$$

The solution set is $(-5, 2]$.

43. $x \le 2$ or $x > -5$

The solution set is $(-\infty, \infty)$.

44. $3x - 5 > 6$ or $-x < -5$

$3x > 11$ or $x > 5$

$x > \frac{11}{3}$ or $x > 5$

The solution set is $\left(\frac{11}{3}, \infty\right)$.

45. $-2x \le 6$ and $-2x + 3 < -7$

$x \ge -3$ and $-2x < -10$

$x \ge -3$ and $x > 5$

The solution set is $(5, \infty)$.

46. $|x - 7| = 9$

$x - 7 = 9$ or $x - 7 = -9$

$x = 16$ or $x = -2$

The solutions are 16 and –2.

47. $|8 - x| = 3$

$8 - x = 3$ or $8 - x = -3$

$5 = x$ or $11 = x$

The solutions are 5 and 11.

48. $|2x + 9| = 9$

$2x + 9 = 9$ or $2x + 9 = -9$

$2x = 0$ or $2x = -18$

$x = 0$ or $x = -9$

The solutions are 0 and –9.

49. $|-3x + 4| = 7$

$-3x + 4 = 7$ or $-3x + 4 = -7$

$-3x = 3$ or $-3x = -11$

$x = -1$ or $x = \frac{11}{3}$

The solutions are -1 and $\frac{11}{3}$.

50. $|3x - 2| + 6 = 10$

$|3x - 2| = 4$

$3x - 2 = 4$ or $3x - 2 = -4$

$3x = 6$ or $3x = -2$

$x = 2$ or $x = -\frac{2}{3}$

The solutions are 2 and $-\frac{2}{3}$.

51. $5 + |6x + 1| = 5$

$|6x + 1| = 0$

$6x + 1 = 0$

$6x = -1$

$x = -\frac{1}{6}$

The solution is $-\frac{1}{6}$.

52. $-5 = |4x - 3|$

An absolute value expression can never be negative.

There is no solution.

53. $|5 - 6x| + 8 = 3$

$|5 - 6x| = -5$

An absolute value expression can never be negative.

There is no solution.

54. $|7x| - 26 = -5$

$|7x| = 21$

$7x = 21$ or $7x = -21$

$x = 3$ or $x = -3$

The solutions are 3 and –3.

55. $-8 = |x - 3| - 10$

$2 = |x - 3|$

$x - 3 = 2$ or $x - 3 = -2$

$x = 5$ or $x = 1$

The solutions are 1 and 5.

56. $\left|\frac{3x - 7}{4}\right| = 2$

$\frac{3x - 7}{4} = 2$ or $\frac{3x - 7}{4} = -2$

$3x - 7 = 8$ or $3x - 7 = -8$

$3x = 15$ or $3x = -1$

$x = 5$ or $x = -\frac{1}{3}$

The solutions are 5 and $-\frac{1}{3}$.

57. $\left|\frac{9-2x}{5}\right| = -3$

An absolute value expression can never be negative.
There is no solution.

58. $|6x+1| = |15+4x|$

$6x+1 = 15+4x$ or $6x+1 = -(15+4x)$
$2x = 14$ or $6x+1 = -15-4x$
$x = 7$ or $10x = -16$
$x = 7$ or $x = -\frac{16}{10} = -\frac{8}{5}$

The solutions are 7 and $-\frac{8}{5}$.

59. $|x-3| = |7+2x|$

$x-3 = 7+2x$ or $x-3 = -(7+2x)$
$-10 = x$ or $x-3 = -7-2x$
$x = -10$ or $3x = -4$
$x = -10$ or $x = -\frac{4}{3}$

The solutions are -10 and $-\frac{4}{3}$.

60. $|5x-1| < 9$

$-9 < 5x-1 < 9$
$-8 < 5x < 10$
$-\frac{8}{5} < x < 2$

The solution set is $\left(-\frac{8}{5}, 2\right)$.

$-\frac{8}{5}$ 2

61. $|6+4x| \ge 10$

$6+4x \le -10$ or $6+4x \ge 10$
$4x \le -16$ or $4x \ge 4$
$x \le -4$ or $x \ge 1$

The solution set is $(-\infty, -4] \cup [1, \infty)$.

-4 1

62. $|3x| - 8 > 1$

$|3x| > 9$
$3x < -9$ or $3x > 9$
$x < -3$ or $x > 3$

The solution set is $(-\infty, -3) \cup (3, \infty)$.

-3 3

63. $9 + |5x| < 24$

$|5x| < 15$
$-15 < 5x < 15$
$-3 < x < 3$

The solution set is $(-3, 3)$.

-3 3

64. $|6x-5| \le -1$

The solution set is $\varnothing$.

65. $|6x-5| \ge -1$

An absolute value expression is always greater than any negative number. The solution set
is $(-\infty, \infty)$.

0

66. $\left|3x + \frac{2}{5}\right| \ge 4$

$3x + \frac{2}{5} \le -4$ or $3x + \frac{2}{5} \ge 4$
$3x \le -\frac{22}{5}$ or $3x \ge \frac{18}{5}$
$x \le -\frac{22}{15}$ or $x \ge \frac{6}{5}$

The solution set is $\left(-\infty, -\frac{22}{15}\right] \cup \left[\frac{6}{5}, \infty\right)$.

$-\frac{22}{5}$ $\frac{6}{5}$

67. $\left|\frac{4x-3}{5}\right| < 1$

$-1 < \frac{4x-3}{5} < 1$

$-5 < 4x - 3 < 5$

$-2 < 4x < 8$

$-\frac{1}{2} < x < 2$

The solution set is $\left(-\frac{1}{2}, 2\right)$.

68. $\left|\frac{x}{3}+6\right| - 8 > -5$

$\left|\frac{x}{3}+6\right| > 3$

$\frac{x}{3}+6 < -3$ or $\frac{x}{3}+6 > 3$

$\frac{x}{3} < -9$ or $\frac{x}{3} > -3$

$x < -27$ or $x > -9$

The solution set is $(-\infty, -27) \cup (-9, \infty)$.

69. $\left|\frac{4(x-1)}{7}\right| + 10 < 2$

$\left|\frac{4(x-1)}{7}\right| < -8$

The solution set is $\varnothing$.

70. $\begin{cases} x + z = 4 & (1) \\ 2x - y = 4 & (2) \\ x + y - z = 0 & (3) \end{cases}$

Add equations (2) and (3).

$$\begin{array}{l} 2x - y = 4 \\ \underline{x + y - z = 0} \\ 3x - z = 4 \quad (4) \end{array}$$

Add equations (1) and (4).

$$\begin{array}{l} x + z = 4 \\ \underline{3x - z = 4} \\ 4x = 8 \\ x = 2 \end{array}$$

Replace x with 2 in equation (1).

$2 + z = 4$

$z = 2$

Replace x with 2 in equation (2).

$2(2) - y = 4$

$4 - y = 4$

$0 = y$

The solution is (2, 0, 2).

71. $\begin{cases} 2x + 5y = 4 & (1) \\ x - 5y + z = -1 & (2) \\ 4x - z = 11 & (3) \end{cases}$

Add equations (2) and (3).

$$\begin{array}{l} x - 5y + z = -1 \\ \underline{4x - z = 11} \\ 5x - 5y = 10 \quad (4) \end{array}$$

Add equations (1) and (4).

$$\begin{array}{l} 2x + 5y = 4 \\ \underline{5x - 5y = 10} \\ 7x = 14 \\ x = 2 \end{array}$$

Replace x with 2 in equation (1).

$2(2) + 5y = 4$

$4 + 5y = 4$

$5y = 0$

$y = 0$

Replace x with 2 in equation (3).

$4(2) - z = 11$

$8 - z = 11$

$-z = 3$

$z = -3$

The solution is (2, 0, −3).

72. $\begin{cases} \quad 4y+2z=5 & (1) \\ 2x+8y \quad =5 & (2) \\ 6x \quad +4z=1 & (3) \end{cases}$

Multiply equation (1) by –2 and add to equation (2).

$$\begin{array}{r} -8y-4z=-10 \\ 2x+8y \quad = \quad 5 \\ \hline 2x \quad -4z=-5 \quad (4) \end{array}$$

Add equations (3) and (4).

$$\begin{array}{l} 6x+4z=1 \\ 2x-4z=-5 \\ \hline 8x \quad =-4 \\ x \quad =-\frac{1}{2} \end{array}$$

Replace x with $-\frac{1}{2}$ in equation (2).

$$2\left(-\frac{1}{2}\right)+8y=5$$
$$-1+8y=5$$
$$8y=6$$
$$y=\frac{3}{4}$$

Replace x with $-\frac{1}{2}$ in equation (3).

$$6\left(-\frac{1}{2}\right)+4z=1$$
$$-3+4z=1$$
$$4z=4$$
$$z=1$$

The solution is $\left(-\frac{1}{2}, \frac{3}{4}, 1\right)$.

73. $\begin{cases} 5x+7y \quad = 9 & (1) \\ \quad 14y-z=28 & (2) \\ 4x \quad +2z=-4 & (3) \end{cases}$

Multiply equation (1) by –2 and add to equation (2).

$$\begin{array}{r} -10x-14y \quad =-18 \\ 14y-z= \quad 28 \\ \hline -10x \quad -z= \quad 10 \quad (4) \end{array}$$

Multiply equation (4) by 2 and add to equation (3).

$$\begin{array}{r} -20x-2z=20 \\ 4x+2z=-4 \\ \hline -16x \quad =16 \\ x \quad =-1 \end{array}$$

Replace x with –1 in equation (1).

$$5(-1)+7y=9$$
$$-5+7y=9$$
$$7y=14$$
$$y=2$$

Replace x with –1 in equation (3).

$$4(-1)+2z=-4$$
$$-4+2z=-4$$
$$2z=0$$
$$z=0$$

The solution is (–1, 2, 0).

74. $\begin{cases} 3x \quad -2y+2z= \ 5 & (1) \\ -x \quad +6y \ +z= \ 4 & (2) \\ 3x+14y+7z=20 & (3) \end{cases}$

Multiply equation (2) by 3 and add to equation (1).

$$\begin{array}{r} 3x-2y+2z=5 \\ -3x+18y+3z=12 \\ \hline 16y+5z=17 \quad (4) \end{array}$$

Multiply equation (3) by –1 and add to equation (1).

$$\begin{array}{r} 3x- \ 2y+2z= \quad 5 \\ -3x-14y-7z=-20 \\ \hline -16y-5z=-15 \quad (5) \end{array}$$

Add equations (4) and (5).

$$\begin{array}{r} 16y+5z= \quad 17 \\ -16y-5z=-15 \\ \hline 0= \quad 2 \quad \text{False} \end{array}$$

The system is inconsistent.
There is no solution.

75. $\begin{cases} x + 2y + 3z = 11 & (1) \\ y + 2z = 3 & (2) \\ 2x + 2z = 10 & (3) \end{cases}$

Multiply equation (2) by –2 and add to equation (1).

$$\begin{array}{l} x + 2y + 3z = 11 \\ \quad -2y - 4z = -6 \\ \hline x - \quad z = 5 \quad (4) \end{array}$$

Multiply equation (4) by 2 and add to equation (3).

$$\begin{array}{l} 2x + 2z = 10 \\ 2x - 2z = 10 \\ \hline 4x \quad = 20 \end{array}$$

$x = 5$

Replace x with 5 in equation (3).

$2(5) + 2z = 10$
$10 + 2z = 10$
$2z = 0$
$z = 0$

Replace z with 0 in equation (2).

$y + 2(0) = 3$
$y + 0 = 3$
$y = 3$

The solution is (5, 3, 0).

76. $\begin{cases} 7x - 3y + 2z = 0 & (1) \\ 4x - 4y - z = 2 & (2) \\ 5x + 2y + 3z = 1 & (3) \end{cases}$

Multiply equation (2) by 2 and add to equation (1).

$$\begin{array}{l} 7x - 3y + 2z = 0 \\ 8x - 8y - 2z = 4 \\ \hline 15x - 11y \quad = 4 \quad (4) \end{array}$$

Multiply equation (2) by 3 and add to equation (3).

$$\begin{array}{l} 12x - 12y - 3z = 6 \\ 5x + 2y + 3z = 1 \\ \hline 17x - 10y \quad = 7 \quad (5) \end{array}$$

Multiply equation (4) by –10, multiply equation (5) by 11, and add.

$$\begin{array}{l} -150x + 110y = -40 \\ 187x - 110y = 77 \\ \hline 37x \quad = 37 \end{array}$$

$x = 1$

Replace x with 1 in equation (4).

$15(1) - 11y = 4$
$15 - 11y = 4$
$-11y = -11$
$y = 1$

Replace x with 1 and y with 1 in equation (1).

$7(1) - 3(1) + 2z = 0$
$4 + 2z = 0$
$2z = -4$
$z = -2$

The solution is (1, 1, –2).

77. $\begin{cases} x - 3y - 5z = -5 & (1) \\ 4x - 2y + 3z = 13 & (2) \\ 5x + 3y + 4z = 22 & (3) \end{cases}$

Multiply equation (1) by –4 and add to equation (2).

$$\begin{array}{l} -4x + 12y + 20z = 20 \\ 4x - 2y + 3z = 13 \\ \hline 10y + 23z = 33 \quad (4) \end{array}$$

Multiply equation (1) by –5 and add to equation (3).

$$\begin{array}{l} -5x + 15y + 25z = 25 \\ 5x + 3y + 4z = 22 \\ \hline 18y + 29z = 47 \end{array}$$

Multiply equation (4) by 9, multiply equation (5) by –5 and add.

$$\begin{array}{l} 90y + 207z = 297 \\ -90y - 145z = -235 \\ \hline 62z = 62 \end{array}$$

$z = 1$

Replace z with 1 in equation (4).

$10y + 23(1) = 33$
$10y + 23 = 33$
$10y = 10$
$y = 1$

Replace y with 1 and z with 1 in equation (1).

$x - 3(1) - 5(1) = -5$
$x - 8 = -5$
$x = 3$

The solution is (3, 1, 1).

78. Let x = the first number,
y = the second number, and
z = the third number.

$$\begin{cases} x+y+z=98 \\ x+y=z+2 \\ y=4x \end{cases}$$

$$\begin{cases} x+y+z=98 & (1) \\ x+y-z=2 & (2) \\ -4x+y=0 & (3) \end{cases}$$

Add equations (1) and (2).

$$\begin{array}{l} x+y+z=98 \\ \underline{x+y-z=2} \\ 2x+2y=100 \quad (4) \end{array}$$

Multiply equation (4) by 2 and add to equation (3).

$$\begin{array}{r} -4x+y=0 \\ \underline{4x+4y=200} \\ 5y=200 \\ y=40 \end{array}$$

Replace y with 40 in equation (4).

$$2x+2(40)=100$$
$$2x+80=100$$
$$2x=20$$
$$x=10$$

Replace x with 10 and y with 40 in equation (1).

$$10+40+z=98$$
$$50+z=98$$
$$z=48$$

The numbers are 10, 40, and 48.

79. Let x = pounds of chocolates used,
y = pounds of nuts used, and
z = pounds of raisins used.

$$\begin{cases} z=2y \\ x+y+z=45 \\ 3.00x+2.70y+2.25z=2.80(45) \end{cases}$$

$$\begin{cases} -2y+z=0 & (1) \\ x+y+z=45 & (2) \\ 3x+2.7y+2.25z=126 & (3) \end{cases}$$

Multiply equation (2) by –3 and add to equation (3).

$$\begin{array}{r} -3x-3y-3z=-135 \\ \underline{3x+2.7y+2.25z=126} \\ -0.3y-0.75z=-9 \quad (4) \end{array}$$

Multiply equation (1) by –3, multiply equation (4) by 20, and add.

$$\begin{array}{r} 6y-3z=0 \\ \underline{-6y-15z=-180} \\ -18z=-180 \\ z=10 \end{array}$$

Replace z with 10 in equation (1).

$$-2y+10=0$$
$$10=2y$$
$$5=y$$

Replace y with 5 and z with 10 in equation (1).

$$x+5+10=45$$
$$x+15=45$$
$$x=30$$

She should use 30 pounds of creme-filled chocolates, 5 pounds of chocolate-covered nuts, and 10 pounds of chocolate-covered raisins.

80. Let x = the number of pennies,
y = the number of nickels, and
z = the number of dimes.

$$\begin{cases} x+y+z=53 \\ 0.01x+0.05y+0.10z=2.77 \\ y=z+4 \end{cases}$$

$$\begin{cases} x+y+z=53 & (1) \\ x+5y+10z=277 & (2) \\ y-z=4 & (3) \end{cases}$$

Multiply equation (1) by −1 and add to equation (2).

$$\begin{array}{r} -x-y-z=-53 \\ x+5y+10z=277 \\ \hline 4y+9z=224 \end{array} \quad (4)$$

Multiply equation (3) by 9 and add to equation (4).

$$\begin{array}{r} 9y-9z=36 \\ 4y+9z=224 \\ \hline 13y=260 \\ y=20 \end{array}$$

Replace y with 20 in equation (4).

$$20-z=4$$
$$-z=-16$$
$$z=16$$

Replace y with 20 and z with 16 in equation (1).

$$x+20+16=53$$
$$x+36=53$$
$$x=17$$

He has 17 pennies, 20 nickels, and 16 dimes in his jar.

81. Let x and y be the lengths of the two equal sides and z the length of the third side.

$$\begin{cases} x+y+z=73 \\ x=y \\ z=x+7 \end{cases}$$

$$\begin{cases} x+y+z=73 & (1) \\ x-y=0 & (2) \\ -x+z=7 & (3) \end{cases}$$

Add equation (1) to equation (2).

$$\begin{array}{r} x+y+z=73 \\ x-y=0 \\ \hline 2x+z=73 \end{array} \quad (4)$$

Multiply equation (3) by −1 and add to equation (4).

$$\begin{array}{r} x-z=-7 \\ 2x+z=73 \\ \hline 3x=66 \\ x=22 \end{array}$$

Replace x with 22 in equation (2).

$$22-y=0$$
$$22=y$$

Replace x with 22 in equation (3).

$$-22+z=7$$
$$z=29$$

Two sides of the triangle have length 22 cm, the third side has length 29 cm.

82. Let x = the first number,
y = the second number, and
z = the third number.

$$\begin{cases} x+y+z=295 \\ x=y+5 \\ x=2z \end{cases}$$

$$\begin{cases} x+y+z=295 & (1) \\ x-y=5 & (2) \\ x-2z=0 & (3) \end{cases}$$

Add equations (1) and (2).

$$\begin{array}{r} x+y+z=295 \\ x-y=5 \\ \hline 2x+z=300 \end{array} \quad (4)$$

Multiply equation (4) by 2 and add to equation (3).

$$\begin{array}{r} x-2z=0 \\ 4x+2z=600 \\ \hline 5x=600 \\ x=120 \end{array}$$

Replace x with 120 in equation (2).

$$120-y=5$$
$$115=y$$

Replace x with 120 in equation (3).

$$120-2z=0$$
$$120=2z$$
$$60=z$$

The first number is 120, the second is 115, and the third is 60.

83. $\begin{cases} 3x+10y=1 \\ x+2y=-1 \end{cases}$

$\left[\begin{array}{cc|c} 3 & 10 & 1 \\ 1 & 2 & -1 \end{array}\right]$

Interchange row 1 and row 2.

$\left[\begin{array}{cc|c} 1 & 2 & -1 \\ 3 & 10 & 1 \end{array}\right]$

Multiply row 1 by –3 and add to row 2.

$\left[\begin{array}{cc|c} 1 & 2 & -1 \\ 0 & 4 & 4 \end{array}\right]$

Divide row 2 by 4.

$\left[\begin{array}{cc|c} 1 & 2 & -1 \\ 0 & 1 & 1 \end{array}\right]$

This corresponds to

$\begin{cases} x+2y=-1 \\ y=1. \end{cases}$

$x+2(1)=-1$
$x=-3$

The solution is (–3, 1).

84. $\begin{cases} 3x-6y=12 \\ 2y=x-4 \end{cases}$

$\left[\begin{array}{cc|c} 3 & -6 & 12 \\ -1 & 2 & -4 \end{array}\right]$

Divide row 1 by 3.

$\left[\begin{array}{cc|c} 1 & -2 & 4 \\ -1 & 2 & -4 \end{array}\right]$

Add row 1 to row 2.

$\left[\begin{array}{cc|c} 1 & -2 & 4 \\ 0 & 0 & 0 \end{array}\right]$

This corresponds to

$\begin{cases} x-2y=4 \\ 0=0 \end{cases}.$

This is a dependent system.
The solution is $\{(x, y) \mid x-2y=4\}$.

85. $\begin{cases} 3x-2y=-8 \\ 6x+5y=11 \end{cases}$

$\left[\begin{array}{cc|c} 3 & -2 & -8 \\ 6 & 5 & 11 \end{array}\right]$

Divide row 1 by 3.

$\left[\begin{array}{cc|c} 1 & -\frac{2}{3} & -\frac{8}{3} \\ 6 & 5 & 11 \end{array}\right]$

Multiply row 1 by –6 and add to row 2.

$\left[\begin{array}{cc|c} 1 & -\frac{2}{3} & -\frac{8}{3} \\ 0 & 9 & 27 \end{array}\right]$

Divide row 2 by 9.

$\left[\begin{array}{cc|c} 1 & -\frac{2}{3} & -\frac{8}{3} \\ 0 & 1 & 3 \end{array}\right]$

This corresponds to

$\begin{cases} x-\frac{2}{3}y=-\frac{8}{3} \\ y=3 \end{cases}.$

$x-\frac{2}{3}(3)=-\frac{8}{3}$
$x-2=-\frac{8}{3}$
$x=-\frac{2}{3}$

The solution is $\left(-\frac{2}{3}, 3\right)$.

86. $\begin{cases} 6x-6y=-5 \\ 10x-2y=1 \end{cases}$

$\left[\begin{array}{cc|c} 6 & -6 & -5 \\ 10 & -2 & 1 \end{array}\right]$

Divide row 1 by 6.

$\left[\begin{array}{cc|c} 1 & -1 & -\frac{5}{6} \\ 10 & -2 & 1 \end{array}\right]$

Multiply row 1 by –10 and add to row 2.

$\left[\begin{array}{cc|c} 1 & -1 & -\frac{5}{6} \\ 0 & 8 & \frac{28}{3} \end{array}\right]$

Divide row 2 by 8.

$$\begin{bmatrix} 1 & -1 & | & -\frac{5}{6} \\ 0 & 1 & | & \frac{7}{6} \end{bmatrix}$$

Add row 1 to row 2.

$$\begin{bmatrix} 1 & 0 & | & \frac{1}{3} \\ 0 & 1 & | & \frac{7}{6} \end{bmatrix}$$

This corresponds to

$$\begin{cases} x = \frac{1}{3} \\ y = \frac{7}{6} \end{cases}$$

The solution is $\left(\frac{1}{3}, \frac{7}{6}\right)$.

87. $\begin{cases} 3x - 6y = 0 \\ 2x + 4y = 5 \end{cases}$

$$\begin{bmatrix} 3 & -6 & | & 0 \\ 2 & 4 & | & 5 \end{bmatrix}$$

Divide row 1 by 3.

$$\begin{bmatrix} 1 & -2 & | & 0 \\ 2 & 4 & | & 5 \end{bmatrix}$$

Multiply row 1 by –2 and add to row 2.

$$\begin{bmatrix} 1 & -2 & | & 0 \\ 0 & 8 & | & 5 \end{bmatrix}$$

Divide row 2 by 8.

$$\begin{bmatrix} 1 & -2 & | & 0 \\ 0 & 1 & | & \frac{5}{8} \end{bmatrix}$$

This corresponds to

$$\begin{cases} x - 2y = 0 \\ y = \frac{5}{8}. \end{cases}$$

$$x - 2\left(\frac{5}{8}\right) = 0$$
$$x - \frac{5}{4} = 0$$
$$x = \frac{5}{4}$$

The solution is $\left(\frac{5}{4}, \frac{5}{8}\right)$.

88. $\begin{cases} 5x - 3y = 10 \\ -2x + y = -1 \end{cases}$

$$\begin{bmatrix} 5 & -3 & | & 10 \\ -2 & 1 & | & -1 \end{bmatrix}$$

Divide row 1 by 5.

$$\begin{bmatrix} 1 & -\frac{3}{5} & | & 2 \\ -2 & 1 & | & -1 \end{bmatrix}$$

Multiply row 1 by 2 and add to row 2.

$$\begin{bmatrix} 1 & -\frac{3}{5} & | & 2 \\ 0 & -\frac{1}{5} & | & 3 \end{bmatrix}$$

Multiply row 2 by –5.

$$\begin{bmatrix} 1 & -\frac{3}{5} & | & 2 \\ 0 & 1 & | & -15 \end{bmatrix}$$

This corresponds to

$$\begin{cases} x - \frac{3}{5}y = 2 \\ y = -15 \end{cases}.$$

$$x - \frac{3}{5}(-15) = 2$$
$$x + 9 = 2$$
$$x = -7$$

The solution is (–7, –15).

89. $\begin{cases} 0.2x - 0.3y = -0.7 \\ 0.5x + 0.3y = 1.4 \end{cases}$

$$\begin{bmatrix} 0.2 & -0.3 & | & -0.7 \\ 0.5 & 0.3 & | & 1.4 \end{bmatrix}$$

Multiply both rows by 10 to clear decimals.

$$\begin{bmatrix} 2 & -3 & | & -7 \\ 5 & 3 & | & 14 \end{bmatrix}$$

Divide row 1 by 2.

$$\begin{bmatrix} 1 & -\frac{3}{2} & | & -\frac{7}{2} \\ 5 & 3 & | & 14 \end{bmatrix}$$

Multiply row 1 by –5 and add to row 2.

$$\begin{bmatrix} 1 & -\frac{3}{2} & | & -\frac{7}{2} \\ 0 & \frac{21}{2} & | & \frac{63}{2} \end{bmatrix}$$

Multiply row 2 by $\frac{2}{21}$.

$$\left[\begin{array}{cc|c} 1 & -\frac{3}{2} & -\frac{7}{2} \\ 0 & 1 & 3 \end{array}\right]$$

This corresponds to

$$\begin{cases} x - \frac{3}{2}y = -\frac{7}{2} \\ y = 3 \end{cases}.$$

$$x - \frac{3}{2}(3) = -\frac{7}{2}$$
$$x - \frac{9}{2} = -\frac{7}{2}$$
$$x = 1$$

The solution is (1, 3).

90. $\begin{cases} 3x + 2y = 8 \\ 3x - y = 5 \end{cases}$

$$\left[\begin{array}{cc|c} 3 & 2 & 8 \\ 3 & -1 & 5 \end{array}\right]$$

Divide row 1 by 3.

$$\left[\begin{array}{cc|c} 1 & \frac{2}{3} & \frac{8}{3} \\ 3 & -1 & 5 \end{array}\right]$$

Multiply row 1 by –3 and add to row 2.

$$\left[\begin{array}{cc|c} 1 & \frac{2}{3} & \frac{8}{3} \\ 0 & -3 & -3 \end{array}\right]$$

Divide row 2 by –3.

$$\left[\begin{array}{cc|c} 1 & \frac{2}{3} & \frac{8}{3} \\ 0 & 1 & 1 \end{array}\right]$$

This corresponds to

$$\begin{cases} x + \frac{2}{3}y = \frac{8}{3} \\ y = 1 \end{cases}.$$

$$x + \frac{2}{3}(1) = \frac{8}{3}$$
$$x = 2$$

The solution is (2, 1).

91. $\begin{cases} x + z = 4 \\ 2x - y = 0 \\ x + y - z = 0 \end{cases}$

$$\left[\begin{array}{ccc|c} 1 & 0 & 1 & 4 \\ 2 & -1 & 0 & 0 \\ 1 & 1 & -1 & 0 \end{array}\right]$$

Multiply row 1 by –2 and add to row 2.
Multiply row 1 by –1 and add to row 3.

$$\left[\begin{array}{ccc|c} 1 & 0 & 1 & 4 \\ 0 & -1 & -2 & -8 \\ 0 & 1 & -2 & -4 \end{array}\right]$$

Multiply row 2 by –1.

$$\left[\begin{array}{ccc|c} 1 & 0 & 1 & 4 \\ 0 & 1 & 2 & 8 \\ 0 & 1 & -2 & -4 \end{array}\right]$$

Multiply row 2 by –1 and add to row 3.

$$\left[\begin{array}{ccc|c} 1 & 0 & 1 & 4 \\ 0 & 1 & 2 & 8 \\ 0 & 0 & -4 & -12 \end{array}\right]$$

Divide row 3 by –4.

$$\left[\begin{array}{ccc|c} 1 & 0 & 1 & 4 \\ 0 & 1 & 2 & 8 \\ 0 & 0 & 1 & 3 \end{array}\right]$$

This corresponds to

$$\begin{cases} x + z = 4 \\ y + 2z = 8 \\ z = 3 \end{cases}$$

$$y + 2(3) = 8$$
$$y + 6 = 8$$
$$y = 2$$
$$x + 3 = 4$$
$$x = 1$$

The solution is (1, 2, 3).

92. $\begin{cases} 2x+5y = 4 \\ x-5y+z=-1 \\ 4x -z=11 \end{cases}$

$\left[\begin{array}{ccc|c} 2 & 5 & 0 & 4 \\ 1 & -5 & 1 & -1 \\ 4 & 0 & -1 & 11 \end{array}\right]$

Interchange row 1 and row 2.

$\left[\begin{array}{ccc|c} 1 & -5 & 1 & -1 \\ 2 & 5 & 0 & 4 \\ 4 & 0 & -1 & 11 \end{array}\right]$

Multiply row 1 by –2 and add to row 2.
Multiply row 1 by –4 and add to row 3.

$\left[\begin{array}{ccc|c} 1 & -5 & 1 & -1 \\ 0 & 15 & -2 & 6 \\ 0 & 20 & -5 & 15 \end{array}\right]$

Divide row 2 by 15.

$\left[\begin{array}{ccc|c} 1 & -5 & 1 & -1 \\ 0 & 1 & -\frac{2}{15} & \frac{2}{5} \\ 0 & 20 & -5 & 15 \end{array}\right]$

Multiply row 2 by –20 and add to row 3.

$\left[\begin{array}{ccc|c} 1 & -5 & 1 & -1 \\ 0 & 1 & -\frac{2}{15} & \frac{2}{5} \\ 0 & 0 & -\frac{7}{3} & 7 \end{array}\right]$

Multiply row 3 by $-\frac{3}{7}$.

$\left[\begin{array}{ccc|c} 1 & -5 & 1 & -1 \\ 0 & 1 & -\frac{2}{15} & \frac{2}{5} \\ 0 & 0 & 1 & -3 \end{array}\right]$

This corresponds to

$\begin{cases} x-5y +z=-1 \\ y-\frac{2}{15}z=\frac{2}{5} \\ z=-3 \end{cases}.$

$y-\frac{2}{15}(-3)=\frac{2}{5}$

$y+\frac{2}{5}=\frac{2}{5}$

$y=0$

$x-5(0)+(-3)=-1$

$x-3=-1$

$x=2$

The solution is (2, 0, –3).

93. $\begin{cases} 3x-y = 11 \\ x+ 2z=13 \\ y-z=-7 \end{cases}$

$\left[\begin{array}{ccc|c} 3 & -1 & 0 & 11 \\ 1 & 0 & 2 & 13 \\ 0 & 1 & -1 & -7 \end{array}\right]$

Interchange row 1 and row 2.

$\left[\begin{array}{ccc|c} 1 & 0 & 2 & 13 \\ 3 & -1 & 0 & 11 \\ 0 & 1 & -1 & -7 \end{array}\right]$

Interchange row 2 and row 3.

$\left[\begin{array}{ccc|c} 1 & 0 & 2 & 13 \\ 0 & 1 & -1 & -7 \\ 3 & -1 & 0 & 11 \end{array}\right]$

Multiply row 1 by –3 and add to row 3.

$\left[\begin{array}{ccc|c} 1 & 0 & 2 & 13 \\ 0 & 1 & -1 & -7 \\ 0 & -1 & -6 & -28 \end{array}\right]$

Add row 2 to row 3.

$\left[\begin{array}{ccc|c} 1 & 0 & 2 & 13 \\ 0 & 1 & -1 & -7 \\ 0 & 0 & -7 & -35 \end{array}\right]$

Divide row 3 by –7.

$\left[\begin{array}{ccc|c} 1 & 0 & 2 & 13 \\ 0 & 1 & -1 & -7 \\ 0 & 0 & 1 & 5 \end{array}\right]$

This corresponds to

$\begin{cases} x+2z=13 \\ y-z=-7 \\ z=5 \end{cases}$

$y-5=-7$

$y=-2$

$x+2(5)=13$

$x+10=13$

$x=3$

The solution is (3, –2, 5).

94. $\begin{cases} 5x + 7y + 3z = 9 \\ 14y - z = 28 \\ 4x + 2z = -4 \end{cases}$

$$\left[\begin{array}{ccc|c} 5 & 7 & 3 & 9 \\ 0 & 14 & -1 & 28 \\ 4 & 0 & 2 & -4 \end{array}\right]$$

Divide row 1 by 5.

$$\left[\begin{array}{ccc|c} 1 & \frac{7}{5} & \frac{3}{5} & \frac{9}{5} \\ 0 & 14 & -1 & 28 \\ 4 & 0 & 2 & -4 \end{array}\right]$$

Multiply row 1 by −4 and add to row 3.

$$\left[\begin{array}{ccc|c} 1 & \frac{7}{5} & \frac{3}{5} & \frac{9}{5} \\ 0 & 14 & -1 & 28 \\ 0 & -\frac{28}{5} & -\frac{2}{5} & -\frac{56}{5} \end{array}\right]$$

Divide row 2 by 14.

$$\left[\begin{array}{ccc|c} 1 & \frac{7}{5} & \frac{3}{5} & \frac{9}{5} \\ 0 & 1 & -\frac{1}{14} & 2 \\ 0 & -\frac{28}{5} & -\frac{2}{5} & -\frac{56}{5} \end{array}\right]$$

Multiply row 2 by $\frac{28}{5}$ and add to row 3.

$$\left[\begin{array}{ccc|c} 1 & \frac{7}{5} & \frac{3}{5} & \frac{9}{5} \\ 0 & 1 & -\frac{1}{14} & 2 \\ 0 & 0 & -\frac{4}{5} & 0 \end{array}\right]$$

Multiply row 3 by $-\frac{5}{4}$.

$$\left[\begin{array}{ccc|c} 1 & \frac{7}{5} & \frac{3}{5} & \frac{9}{5} \\ 0 & 1 & -\frac{1}{14} & 2 \\ 0 & 0 & 1 & 0 \end{array}\right]$$

This corresponds to

$$\begin{cases} x + \frac{7}{5}y + \frac{3}{5}z = \frac{9}{5} \\ y - \frac{1}{14}z = 2 \\ z = 0 \end{cases}.$$

$$y - \frac{1}{14}(0) = 2$$
$$y = 2$$

$$x + \frac{7}{5}(2) + \frac{3}{5}(0) = \frac{9}{5}$$
$$x + \frac{14}{5} = \frac{9}{5}$$
$$x = -1$$

The solution is (−1, 2, 0).

95. $\begin{cases} 7x - 3y + 2z = 0 \\ 4x - 4y - z = 2 \\ 5x + 2y + 3z = 1 \end{cases}$

$$\left[\begin{array}{ccc|c} 7 & -3 & 2 & 0 \\ 4 & -4 & -1 & 2 \\ 5 & 2 & 3 & 1 \end{array}\right]$$

Interchange row 1 and row 2.

$$\left[\begin{array}{ccc|c} 4 & -4 & -1 & 2 \\ 7 & -3 & 2 & 0 \\ 5 & 2 & 3 & 1 \end{array}\right]$$

Divide row 1 by 4.

$$\left[\begin{array}{ccc|c} 1 & -1 & -\frac{1}{4} & \frac{1}{2} \\ 7 & -3 & 2 & 0 \\ 5 & 2 & 3 & 1 \end{array}\right]$$

Multiply row 1 by −7 and add to row 2.
Multiply row 1 by −5 and add to row 3.

$$\left[\begin{array}{ccc|c} 1 & -1 & -\frac{1}{4} & \frac{1}{2} \\ 0 & 4 & \frac{15}{4} & -\frac{7}{2} \\ 0 & 7 & \frac{17}{4} & -\frac{3}{2} \end{array}\right]$$

Divide row 2 by 4.

$$\left[\begin{array}{ccc|c} 1 & -1 & -\frac{1}{4} & \frac{1}{2} \\ 0 & 1 & \frac{15}{16} & -\frac{7}{8} \\ 0 & 7 & \frac{17}{4} & -\frac{3}{2} \end{array}\right]$$

Multiply row 2 by −7 and add to row 3.

$$\left[\begin{array}{ccc|c} 1 & -1 & -\frac{1}{4} & \frac{1}{2} \\ 0 & 1 & \frac{15}{16} & -\frac{7}{8} \\ 0 & 0 & -\frac{37}{16} & \frac{37}{8} \end{array}\right]$$

Multiply row 3 by $-\frac{16}{37}$.

$$\left[\begin{array}{ccc|c} 1 & -1 & -\frac{1}{4} & \frac{1}{2} \\ 0 & 1 & \frac{15}{16} & -\frac{7}{8} \\ 0 & 0 & 1 & -2 \end{array}\right]$$

This corresponds to

$$\begin{cases} x - y - \frac{1}{4}z = \frac{1}{2} \\ y + \frac{15}{16}z = -\frac{7}{8} \\ z = -2 \end{cases}$$

$$y + \frac{15}{16}(-2) = -\frac{7}{8}$$
$$y - \frac{15}{8} = -\frac{7}{8}$$
$$y = 1$$
$$x - 1 - \frac{1}{4}(-2) = \frac{1}{2}$$
$$x - 1 + \frac{1}{2} = \frac{1}{2}$$
$$x = 1$$

The solution is (1, 1, –2).

96. $\begin{cases} x + 2y + 3z = 14 \\ y + 2z = 3 \\ 2x - 2z = 10 \end{cases}$

$$\left[\begin{array}{ccc|c} 1 & 2 & 3 & 14 \\ 0 & 1 & 2 & 3 \\ 2 & 0 & -2 & 10 \end{array}\right]$$

Multiply row 1 by –2 and add to row 3.

$$\left[\begin{array}{ccc|c} 1 & 2 & 3 & 14 \\ 0 & 1 & 2 & 3 \\ 0 & -4 & -8 & -18 \end{array}\right]$$

Multiply row 2 by 4 and add to row 3.

$$\left[\begin{array}{ccc|c} 1 & 2 & 3 & 14 \\ 0 & 1 & 2 & 3 \\ 0 & 0 & 0 & -6 \end{array}\right]$$

This corresponds to

$$\begin{cases} x + 2y + 3z = 14 \\ y + 2z = 3 \\ 0 = -6 \end{cases}.$$

The system is inconsistent.
There is no solution.

97. $\begin{vmatrix} -1 & 3 \\ 5 & 2 \end{vmatrix} = (-1)(2) - 3(5) = -2 - 15 = -17$

98. $\begin{vmatrix} 3 & -1 \\ 2 & 5 \end{vmatrix} = 3(5) - (-1)2 = 15 + 2 = 17$

99. $\begin{vmatrix} 2 & -1 & -3 \\ 1 & 2 & 0 \\ 3 & -2 & 2 \end{vmatrix}$

$$= 2\begin{vmatrix} 2 & 0 \\ -2 & 2 \end{vmatrix} - (-1)\begin{vmatrix} 1 & 0 \\ 3 & 2 \end{vmatrix} + (-3)\begin{vmatrix} 1 & 2 \\ 3 & -2 \end{vmatrix}$$
$$= 2(4 - 0) + (2 - 0) - 3(-2 - 6)$$
$$= 2(4) + (2) - 3(-8)$$
$$= 34$$

100. $\begin{vmatrix} -2 & 3 & 1 \\ 4 & 4 & 0 \\ 1 & -2 & 3 \end{vmatrix}$

$$= 1\begin{vmatrix} 4 & 4 \\ 1 & -2 \end{vmatrix} - 0\begin{vmatrix} -2 & 3 \\ 1 & -2 \end{vmatrix} + 3\begin{vmatrix} -2 & 3 \\ 4 & 4 \end{vmatrix}$$
$$= (-8 - 4) - 0 + 3(-8 - 12)$$
$$= -12 + 3(-20)$$
$$= -12 - 60$$
$$= -72$$

101. $\begin{cases} 3x - 2y = -8 \\ 6x + 5y = 11 \end{cases}$

$$D = \begin{vmatrix} 3 & -2 \\ 6 & 5 \end{vmatrix} = 15 - (-12) = 27$$

$$D_x = \begin{vmatrix} -8 & -2 \\ 11 & 5 \end{vmatrix} = -40 - (-22) = -18$$

$$D_y = \begin{vmatrix} 3 & -8 \\ 6 & 11 \end{vmatrix} = 33 - (-48) = 81$$

$$x = \frac{D_x}{D} = \frac{-18}{27} = -\frac{2}{3}$$

$$y = \frac{D_y}{D} = \frac{81}{27} = 3$$

The solution is $\left(-\frac{2}{3}, 3\right)$.

102. $\begin{cases} 6x-6y=-5 \\ 10x-2y=\ \ 1 \end{cases}$

$D=\begin{vmatrix} 6 & -6 \\ 10 & -2 \end{vmatrix}=-12-(-60)=48$

$D_x=\begin{vmatrix} -5 & -6 \\ 1 & -2 \end{vmatrix}=10-(-6)=16$

$D_y=\begin{vmatrix} 6 & -5 \\ 10 & 1 \end{vmatrix}=6-(-50)=56$

$x=\frac{D_x}{D}=\frac{16}{48}=\frac{1}{3}$ and $y=\frac{D_y}{D}=\frac{56}{48}=\frac{7}{6}$

The solution is $\left(\frac{1}{3}, \frac{7}{6}\right)$.

103. $\begin{cases} 3x+10y=\ \ 1 \\ x+\ 2y=-1 \end{cases}$

$D=\begin{vmatrix} 3 & 10 \\ 1 & 2 \end{vmatrix}=6-10=-4$

$D_x=\begin{vmatrix} 1 & 10 \\ -1 & 2 \end{vmatrix}=2-(-10)=12$

$D_y=\begin{vmatrix} 3 & 1 \\ 1 & -1 \end{vmatrix}=-3-1=-4$

$x=\frac{D_x}{D}=\frac{12}{-4}=-3$

$y=\frac{D_y}{D}=\frac{-4}{-4}=1$

The solution is (–3, 1).

104. $\begin{cases} y=\frac{1}{2}x+\frac{2}{3} \\ 4x+6y=4 \end{cases}$ or $\begin{cases} -\frac{1}{2}x+y=\frac{2}{3} \\ 4x+6y=4 \end{cases}$

$D=\begin{vmatrix} -\frac{1}{2} & 1 \\ 4 & 6 \end{vmatrix}=-3-4=-7$

$D_x=\begin{vmatrix} \frac{2}{3} & 1 \\ 4 & 6 \end{vmatrix}=4-4=0$

$D_y=\begin{vmatrix} -\frac{1}{2} & \frac{2}{3} \\ 4 & 4 \end{vmatrix}=-2-\frac{8}{3}=-\frac{14}{3}$

$x=\frac{D_x}{D}=\frac{0}{-7}=0$ and $y=\frac{D_y}{D}=\frac{-\frac{14}{3}}{-7}=\frac{2}{3}$

The solution is $\left(0, \frac{2}{3}\right)$.

105. $\begin{cases} 2x-4y=22 & (1) \\ 5x-10y=16 & (2) \end{cases}$

$D=\begin{vmatrix} 2 & -4 \\ 5 & -10 \end{vmatrix}=-20-(-20)=0$

This cannot be solved by Cramer's rule. Multiply equation (1) by –5, multiply equation (2) by 2, and add.

$$\begin{array}{r} -10x+20y=-110 \\ 10x-20y=\ \ \ 32 \\ \hline 0=\ -78 \end{array}$$

There is no solution.

106. $\begin{cases} 3x-6y=12 \\ 2y=x-4 \end{cases}$ or $\begin{cases} 3x-6y=12 & (1) \\ -x+2y=-4 & (2) \end{cases}$

$D=\begin{vmatrix} 3 & -6 \\ -1 & 2 \end{vmatrix}=6-6=0$

Cramer's Rule cannot be used to solve this system. Since equation (1) is –3 times equation (2), the system is dependent. The solution is $\{(x,\ y) \mid x-2y=4\}$.

107. $\begin{cases} x \quad\;\; +z=4 \\ 2x-y \quad\;\, =0 \\ x+y-z=0 \end{cases}$

$$D=\begin{vmatrix} 1 & 0 & 1 \\ 2 & -1 & 0 \\ 1 & 1 & -1 \end{vmatrix}$$
$$=1\begin{vmatrix} -1 & 0 \\ 1 & -1 \end{vmatrix}-0\begin{vmatrix} 2 & 0 \\ 1 & -1 \end{vmatrix}+1\begin{vmatrix} 2 & -1 \\ 1 & 1 \end{vmatrix}$$
$$=(1-0)-0+[2-(-1)]$$
$$=1+3$$
$$=4$$
$$D_x=\begin{vmatrix} 4 & 0 & 1 \\ 0 & -1 & 0 \\ 0 & 1 & -1 \end{vmatrix}$$
$$=4\begin{vmatrix} -1 & 0 \\ 1 & -1 \end{vmatrix}-0\begin{vmatrix} 0 & 1 \\ 1 & -1 \end{vmatrix}+0\begin{vmatrix} 0 & 1 \\ -1 & 0 \end{vmatrix}$$
$$=4(1-0)-0+0=4$$
$$D_y=\begin{vmatrix} 1 & 4 & 1 \\ 2 & 0 & 0 \\ 1 & 0 & -1 \end{vmatrix}$$
$$=-4\begin{vmatrix} 2 & 0 \\ 1 & -1 \end{vmatrix}+0\begin{vmatrix} 1 & 1 \\ 1 & -1 \end{vmatrix}-0\begin{vmatrix} 1 & 1 \\ 2 & 0 \end{vmatrix}$$
$$=(-4)(-2-0)$$
$$=8$$
$$D_z=\begin{vmatrix} 1 & 0 & 4 \\ 2 & -1 & 0 \\ 1 & 1 & 0 \end{vmatrix}$$
$$=4\begin{vmatrix} 2 & -1 \\ 1 & 1 \end{vmatrix}-0\begin{vmatrix} 1 & 0 \\ 1 & 1 \end{vmatrix}+0\begin{vmatrix} 1 & 0 \\ 2 & -1 \end{vmatrix}$$
$$=4[2-(-1)]$$
$$=12$$
$$x=\frac{D_x}{D}=\frac{4}{4}=1$$
$$y=\frac{D_y}{D}=\frac{8}{4}=2$$
$$z=\frac{D_z}{D}=\frac{12}{4}=3$$

The solution is (1, 2, 3).

108. $\begin{cases} 2x+5y \quad\;\; = 4 \\ x-5y+z=-1 \\ 4x \quad\;\; -z=11 \end{cases}$

$$D=\begin{vmatrix} 2 & 5 & 0 \\ 1 & -5 & 1 \\ 4 & 0 & -1 \end{vmatrix}$$
$$=0\begin{vmatrix} 1 & -5 \\ 4 & 0 \end{vmatrix}-1\begin{vmatrix} 2 & 5 \\ 4 & 0 \end{vmatrix}+(-1)\begin{vmatrix} 2 & 5 \\ 1 & -5 \end{vmatrix}$$
$$=0-(0-20)-(-10-5)$$
$$=20+15$$
$$=35$$
$$D_x=\begin{vmatrix} 4 & 5 & 0 \\ -1 & -5 & 1 \\ 11 & 0 & -1 \end{vmatrix}$$
$$=0\begin{vmatrix} -1 & -5 \\ 11 & 0 \end{vmatrix}-1\begin{vmatrix} 4 & 5 \\ 11 & 0 \end{vmatrix}+(-1)\begin{vmatrix} 4 & 5 \\ -1 & -5 \end{vmatrix}$$
$$=0-(0-55)-[-20-(-5)]$$
$$=55+15$$
$$=70$$
$$D_y=\begin{vmatrix} 2 & 4 & 0 \\ 1 & -1 & 1 \\ 4 & 11 & -1 \end{vmatrix}$$
$$=0\begin{vmatrix} 1 & -1 \\ 4 & 11 \end{vmatrix}-1\begin{vmatrix} 2 & 4 \\ 4 & 11 \end{vmatrix}+(-1)\begin{vmatrix} 2 & 4 \\ 1 & -1 \end{vmatrix}$$
$$=0-(22-16)-(-2-4)$$
$$=-6+6$$
$$=0$$
$$D_z=\begin{vmatrix} 2 & 5 & 4 \\ 1 & -5 & -1 \\ 4 & 0 & 11 \end{vmatrix}$$
$$=4\begin{vmatrix} 1 & -5 \\ 4 & 0 \end{vmatrix}-(-1)\begin{vmatrix} 2 & 5 \\ 4 & 0 \end{vmatrix}+11\begin{vmatrix} 2 & 5 \\ 1 & -5 \end{vmatrix}$$
$$=4[0-(-20)]+1(0-20)+11(-10-5)$$
$$=4(20)-20+11(-15)$$
$$=80-20-165$$
$$=-105$$
$$x=\frac{D_x}{D}=\frac{70}{35}=2,\ y=\frac{D_y}{D}=\frac{0}{35}=0,$$
$$z=\frac{D_z}{D}=\frac{-105}{35}=-3$$

The solution is (2, 0, –3).

109. $\begin{cases} x+3y-z=5 \\ 2x-y-2z=3 \\ x+2y+3z=4 \end{cases}$

$$D=\begin{vmatrix} 1 & 3 & -1 \\ 2 & -1 & -2 \\ 1 & 2 & 3 \end{vmatrix}$$
$$=1\begin{vmatrix} -1 & -2 \\ 2 & 3 \end{vmatrix}-3\begin{vmatrix} 2 & -2 \\ 1 & 3 \end{vmatrix}+(-1)\begin{vmatrix} 2 & -1 \\ 1 & 2 \end{vmatrix}$$
$$=[-3-(-4)]-3[6-(-2)]-[4-(-1)]$$
$$=1-3(8)-5$$
$$=1-24-5$$
$$=-28$$

$$D_x=\begin{vmatrix} 5 & 3 & -1 \\ 3 & -1 & -2 \\ 4 & 2 & 3 \end{vmatrix}$$
$$=5\begin{vmatrix} -1 & -2 \\ 2 & 3 \end{vmatrix}-3\begin{vmatrix} 3 & -2 \\ 4 & 3 \end{vmatrix}+(-1)\begin{vmatrix} 3 & -1 \\ 4 & 2 \end{vmatrix}$$
$$=5[-3-(-4)]-3[9-(-8)]-[6-(-4)]$$
$$=5(1)-3(17)-10$$
$$=5-51-10$$
$$=-56$$

$$D_y=\begin{vmatrix} 1 & 5 & -1 \\ 2 & 3 & -2 \\ 1 & 4 & 3 \end{vmatrix}$$
$$=1\begin{vmatrix} 3 & -2 \\ 4 & 3 \end{vmatrix}-5\begin{vmatrix} 2 & -2 \\ 1 & 3 \end{vmatrix}+(-1)\begin{vmatrix} 2 & 3 \\ 1 & 4 \end{vmatrix}$$
$$=[9-(-8)]-5[6-(-2)]-(8-3)$$
$$=17-5(8)-5$$
$$=17-40-5$$
$$=-28$$

$$D_z=\begin{vmatrix} 1 & 3 & 5 \\ 2 & -1 & 3 \\ 1 & 2 & 4 \end{vmatrix}$$
$$=1\begin{vmatrix} -1 & 3 \\ 2 & 4 \end{vmatrix}-3\begin{vmatrix} 2 & 3 \\ 1 & 4 \end{vmatrix}+5\begin{vmatrix} 2 & -1 \\ 1 & 2 \end{vmatrix}$$
$$=(-4-6)-3(8-3)+5[4-(-1)]$$
$$=-10-3(5)+5(5)$$
$$=-10-15+25$$
$$=0$$

$$x=\frac{D_x}{D}=\frac{-56}{-28}=2$$
$$y=\frac{D_y}{D}=\frac{-28}{-28}=1$$
$$z=\frac{D_z}{D}=\frac{0}{-28}=0$$

The solution is (2, 1, 0).

110. $\begin{cases} 2x \quad\; -z=1 \\ 3x-y+2z=3 \\ x+y+3z=-2 \end{cases}$

$$D=\begin{vmatrix} 2 & 0 & -1 \\ 3 & -1 & 2 \\ 1 & 1 & 3 \end{vmatrix}$$
$$=2\begin{vmatrix} -1 & 2 \\ 1 & 3 \end{vmatrix}-0\begin{vmatrix} 3 & 2 \\ 1 & 3 \end{vmatrix}+(-1)\begin{vmatrix} 3 & -1 \\ 1 & 1 \end{vmatrix}$$
$$=2(-3-2)-0-[3-(-1)]$$
$$=2(-5)-4$$
$$=-10-4$$
$$=-14$$

$$D_x=\begin{vmatrix} 1 & 0 & -1 \\ 3 & -1 & 2 \\ -2 & 1 & 3 \end{vmatrix}$$
$$=1\begin{vmatrix} -1 & 2 \\ 1 & 3 \end{vmatrix}-0\begin{vmatrix} 3 & 2 \\ -2 & 3 \end{vmatrix}+(-1)\begin{vmatrix} 3 & -1 \\ -2 & 1 \end{vmatrix}$$
$$=(-3-2)-0-(3-2)$$
$$=-5-1$$
$$=-6$$

$$D_y=\begin{vmatrix} 2 & 1 & -1 \\ 3 & 3 & 2 \\ 1 & -2 & 3 \end{vmatrix}$$
$$=2\begin{vmatrix} 3 & 2 \\ -2 & 3 \end{vmatrix}-1\begin{vmatrix} 3 & 2 \\ 1 & 3 \end{vmatrix}+(-1)\begin{vmatrix} 3 & 3 \\ 1 & -2 \end{vmatrix}$$
$$=2[9-(-4)]-(9-2)-(-6-3)$$
$$=2(13)-7+9$$
$$=26+2$$
$$=28$$

$$D_z = \begin{vmatrix} 2 & 0 & 1 \\ 3 & -1 & 3 \\ 1 & 1 & -2 \end{vmatrix}$$
$$= 2\begin{vmatrix} -1 & 3 \\ 1 & -2 \end{vmatrix} - 0\begin{vmatrix} 3 & 3 \\ 1 & -2 \end{vmatrix} + 1\begin{vmatrix} 3 & -1 \\ 1 & 1 \end{vmatrix}$$
$$= 2(2-3) - 0 + [3-(-1)]$$
$$= 2(-1) + 4$$
$$= -2 + 4$$
$$= 2$$
$$x = \frac{D_x}{D} = \frac{-6}{-14} = \frac{3}{7},\ y = \frac{D_y}{D} = \frac{28}{-14} = -2,$$
$$z = \frac{D_z}{D} = \frac{2}{-14} = -\frac{1}{7}$$
The solution is $\left(\frac{3}{7}, -2, -\frac{1}{7}\right)$.

111. $\begin{cases} x + 2y + 3z = 14 & (1) \\ \quad y + 2z = 3 & (2) \\ 2x \quad - 2z = 10 & (3) \end{cases}$

$$D = \begin{vmatrix} 1 & 2 & 3 \\ 0 & 1 & 2 \\ 2 & 0 & -2 \end{vmatrix}$$
$$= 1\begin{vmatrix} 1 & 2 \\ 0 & -2 \end{vmatrix} - 0\begin{vmatrix} 2 & 3 \\ 0 & -2 \end{vmatrix} + 2\begin{vmatrix} 2 & 3 \\ 1 & 2 \end{vmatrix}$$
$$= (-2-0) - 0 + 2(4-3)$$
$$= -2 + 2$$
$$= 0$$
This cannot be solved by Cramer's rule.
Solve equation (2) for y.
$y = -2z + 3$
Solve equation (3) for x.
$2x = 2z + 10$
$x = z + 5$
Replace x with $z + 5$ and y with $-2z + 3$ in equation (1).
$z + 5 + 2(-2z + 3) + 3z = 14$
$z + 5 - 4z + 6 + 3z = 14$
$11 = 14$ False
The system is inconsistent.
There is no solution.

112. $\begin{cases} 5x + 7y \quad = 9 \\ \quad 14y - z = 28 \\ 4x \quad + 2z = -4 \end{cases}$

$$D = \begin{vmatrix} 5 & 7 & 0 \\ 0 & 14 & -1 \\ 4 & 0 & 2 \end{vmatrix}$$
$$= 5\begin{vmatrix} 14 & -1 \\ 0 & 2 \end{vmatrix} - 7\begin{vmatrix} 0 & -1 \\ 4 & 2 \end{vmatrix} + 0\begin{vmatrix} 0 & 14 \\ 4 & 0 \end{vmatrix}$$
$$= 5(28-0) - 7[0-(-4)] + 0$$
$$= 140 - 28$$
$$= 112$$
$$D_x = \begin{vmatrix} 9 & 7 & 0 \\ 28 & 14 & -1 \\ -4 & 0 & 2 \end{vmatrix}$$
$$= 9\begin{vmatrix} 14 & -1 \\ 0 & 2 \end{vmatrix} - 7\begin{vmatrix} 28 & -1 \\ -4 & 2 \end{vmatrix} + 0\begin{vmatrix} 28 & 14 \\ -4 & 0 \end{vmatrix}$$
$$= 9(28-0) - 7(56-4) + 0$$
$$= 252 - 7(52)$$
$$= 252 - 364$$
$$= -112$$
$$D_y = \begin{vmatrix} 5 & 9 & 0 \\ 0 & 28 & -1 \\ 4 & -4 & 2 \end{vmatrix}$$
$$= 5\begin{vmatrix} 28 & -1 \\ -4 & 2 \end{vmatrix} - 9\begin{vmatrix} 0 & -1 \\ 4 & 2 \end{vmatrix} + 0\begin{vmatrix} 0 & 28 \\ 4 & -4 \end{vmatrix}$$
$$= 5(56-4) - 9[0-(-4)] + 0$$
$$= 5(52) - 9(4)$$
$$= 260 - 36$$
$$= 224$$
$$D_z = \begin{vmatrix} 5 & 7 & 9 \\ 0 & 14 & 28 \\ 4 & 0 & -4 \end{vmatrix}$$
$$= 5\begin{vmatrix} 14 & 28 \\ 0 & -4 \end{vmatrix} - 0\begin{vmatrix} 7 & 9 \\ 0 & -4 \end{vmatrix} + 4\begin{vmatrix} 7 & 9 \\ 14 & 28 \end{vmatrix}$$
$$= 5(-56-0) - 0 + 4(196-126)$$
$$= -280 + 4(70)$$
$$= -280 + 280$$
$$= 0$$

$$x = \frac{D_x}{D} = \frac{-112}{112} = -1,\ y = \frac{D_y}{D} = \frac{224}{112} = 2,$$
$$z = \frac{D_z}{D} = \frac{0}{112} = 0$$
The solution is (–1, 2, 0).

Chapter 8 Test

1. $8x + 14 = 5x + 44$
$3x = 30$
$x = 10$
The solution is 10.

2. $3(x+2) = 11 - 2(2-x)$
$3x + 6 = 11 - 4 + 2x$
$3x + 6 = 7 + 2x$
$x = 1$
The solution is 1.

3. $\frac{x^2}{5} - \frac{2x}{5} = 3$
$x^2 - 2x = 15$
$x^2 - 2x - 15 = 0$
$(x-5)(x+3) = 0$
$x - 5 = 0$ or $x + 3 = 0$
$x = 5$ or $x = -3$
The solutions are –3 and 5.

4. $\frac{x^2}{7} + \frac{6x}{7} = 1$
$x^2 + 6x = 7$
$x^2 + 6x - 7 = 0$
$(x+7)(x-1) = 0$
$x + 7 = 0$ or $x - 1 = 0$
$x = -7$ or $x = 1$
The solutions are –7 and 1.

5. $\frac{z}{2} + \frac{z}{3} = 10$
$6\left(\frac{z}{2} + \frac{z}{3}\right) = 6(10)$
$3z + 2z = 60$
$5z = 60$
$z = 12$
The solution is 12.

6. $\frac{7w}{4} + 5 = \frac{3w}{10} + 1$
$20\left(\frac{7w}{4} + 5\right) = 20\left(\frac{3w}{10} + 1\right)$
$35w + 100 = 6w + 20$
$29w = -80$
$w = -\frac{80}{29}$
The solution is $-\frac{80}{29}$.

7. $|6x - 5| = 1$
$6x - 5 = -1$ or $6x - 5 = 1$
$6x = 4$ or $6x = 6$
$x = \frac{2}{3}$ or $x = 1$
The solutions are 1 and $\frac{2}{3}$.

8. $|8 - 2t| = -6$
An absolute value expression can never be negative.
There is no solution.

9. $-3 < 2(x-3) \le 4$
$-3 < 2x - 6 \le 4$
$3 < 2x \le 10$
$\frac{3}{2} < x \le 5$
The solution set is $\left(\frac{3}{2}, 5\right]$.

10. $|3x + 1| > 5$
$3x + 1 < -5$ or $3x + 1 > 5$
$3x < -6$ or $3x > 4$
$x < -2$ or $x > \frac{4}{3}$
The solution set is $(-\infty, -2) \cup \left(\frac{4}{3}, \infty\right)$.

11. $x \ge 5$ and $x \ge 4$
$x \ge 5$
The solution set is $[5, \infty)$.

12. $x \ge 5$ or $x \ge 4$
$x \ge 4$
The solution set is $[4, \infty)$.

13. $-x > 1$ and $3x + 3 \geq x - 3$

$x < -1$ and $2x \geq -6$

$x < -1$ and $x \geq -3$

$-3 \leq x < -1$

The solution set is $[-3, -1)$.

14. $6x + 1 > 5x + 4$ or $1 - x > -4$

$x > 3$ or $5 > x$

$x > 3$ or $x < 5$

The solution set is $(-\infty, \infty)$.

15. $$\begin{vmatrix} 4 & -7 \\ 2 & 5 \end{vmatrix} = 4(5) - (-7)(2) = 20 + 14 = 34$$

16. $$\begin{vmatrix} 4 & 0 & 2 \\ 1 & -3 & 5 \\ 0 & -1 & 2 \end{vmatrix}$$

$$= 4\begin{vmatrix} -3 & 5 \\ -1 & 2 \end{vmatrix} - 1\begin{vmatrix} 0 & 2 \\ -1 & 2 \end{vmatrix} + 0\begin{vmatrix} 0 & 2 \\ -3 & 5 \end{vmatrix}$$

$= 4[-6 - (-5)] - [0 - (-2)] + 0$

$= 4(-1) - 2$

$= -4 - 2$

$= -6$

17. $$\begin{cases} 2x - 3y = 4 & (1) \\ 3y + 2z = 2 & (2) \\ x - z = -5 & (3) \end{cases}$$

Add equations (1) and (2).

$$\begin{array}{r} 2x - 3y = 4 \\ \underline{3y + 2z = 2} \\ 2x + 2z = 6 \quad (4) \end{array}$$

Multiply equation (3) by 2 and add to equation (4).

$$\begin{array}{r} 2x - 2z = -10 \\ \underline{2x + 2z = 6} \\ 4x = -4 \\ x = -1 \end{array}$$

Replace x with -1 in equation (3).

$-1 - z = -5$

$-z = -4$

$z = 4$

Replace x with -1 in equation (1).

$2(-1) - 3y = 4$

$-2 - 3y = 4$

$-3y = 6$

$y = -2$

The solution is $(-1, -2, 4)$.

18. $$\begin{cases} 3x - 2y - z = -1 & (1) \\ 2x - 2y = 4 & (2) \\ 2x - 2z = -12 & (3) \end{cases}$$

Multiply equation (2) by -1 and add to equation (1).

$$\begin{array}{r} 3x - 2y - z = -1 \\ \underline{-2x + 2y = -4} \\ x - z = -5 \quad (4) \end{array}$$

Multiply equation (4) by -2 and add to equation (3).

$$\begin{array}{r} 2x - 2z = -12 \\ \underline{-2x + 2z = 10} \\ 0 = -2 \end{array} \quad \text{False}$$

The system is inconsistent.
There is no solution.

19. $$\begin{cases} \frac{x}{2} + \frac{y}{4} = -\frac{3}{4} \\ x + \frac{3}{4}y = -4 \end{cases}$$

Clear fractions by multiplying both equations by 4.

$$\begin{cases} 2x + y = -3 & (1) \\ 4x + 3y = -16 & (2) \end{cases}$$

Multiply equation (1) by -2 and add to equation (2).

$$\begin{array}{r} -4x - 2y = 6 \\ \underline{4x + 3y = -16} \\ y = -10 \end{array}$$

Replace y with -10 in equation (1).

$2x + (-10) = -3$

$2x = 7$

$x = \frac{7}{2}$

The solution is $\left(\frac{7}{2}, -10\right)$.

20. $\begin{cases} 3x - y = 7 \\ 2x + 5y = -1 \end{cases}$

$$D = \begin{vmatrix} 3 & -1 \\ 2 & 5 \end{vmatrix} = 3(5) - (-1)(2) = 15 + 2 = 17$$

$$D_x = \begin{vmatrix} 7 & -1 \\ -1 & 5 \end{vmatrix} = 7(5) - (-1)(-1) = 35 - 1 = 34$$

$$D_y = \begin{vmatrix} 3 & 7 \\ 2 & -1 \end{vmatrix} = 3(-1) - 7(2) = -3 - 14 = -17$$

$$x = \frac{D_x}{D} = \frac{34}{17} = 2 \text{ and } y = \frac{D_y}{D} = \frac{-17}{17} = -1$$

The solution is (2, –1).

21. $\begin{cases} 4x - 3y = -6 \\ -2x + y = 0 \end{cases}$

$$D = \begin{vmatrix} 4 & -3 \\ -2 & 1 \end{vmatrix} = 4(1) - (-3)(-2) = 4 - 6 = -2$$

$$D_x = \begin{vmatrix} -6 & -3 \\ 0 & 1 \end{vmatrix} = (-6)(1) - (-3)(0) = -6 - 0 = -6$$

$$D_y = \begin{vmatrix} 4 & -6 \\ -2 & 0 \end{vmatrix} = 4(0) - (-6)(-2) = 0 - 12 = -12$$

$$x = \frac{D_x}{D} = \frac{-6}{-2} = 3 \text{ and } y = \frac{D_y}{D} = \frac{-12}{-2} = 6$$

The solution is (3, 6).

22. $\begin{cases} x + y + z = 4 \\ 2x + 5y = 1 \\ x - y - 2z = 0 \end{cases}$

$$D = \begin{vmatrix} 1 & 1 & 1 \\ 2 & 5 & 0 \\ 1 & -1 & -2 \end{vmatrix} = 1\begin{vmatrix} 2 & 5 \\ 1 & -1 \end{vmatrix} - 0\begin{vmatrix} 1 & 1 \\ 1 & -1 \end{vmatrix} + (-2)\begin{vmatrix} 1 & 1 \\ 2 & 5 \end{vmatrix} = (-2-5) - 0 - 2(5-2) = -7 - 2(3) = -7 - 6 = -13$$

$$D_x = \begin{vmatrix} 4 & 1 & 1 \\ 1 & 5 & 0 \\ 0 & -1 & -2 \end{vmatrix} = 1\begin{vmatrix} 1 & 5 \\ 0 & -1 \end{vmatrix} - 0\begin{vmatrix} 4 & 1 \\ 0 & -1 \end{vmatrix} + (-2)\begin{vmatrix} 4 & 1 \\ 1 & 5 \end{vmatrix} = (-1-0) - 0 - 2(20-1) = -1 - 2(19) = -1 - 38 = -39$$

$$D_y = \begin{vmatrix} 1 & 4 & 1 \\ 2 & 1 & 0 \\ 1 & 0 & -2 \end{vmatrix} = 1\begin{vmatrix} 2 & 1 \\ 1 & 0 \end{vmatrix} - 0\begin{vmatrix} 1 & 4 \\ 1 & 0 \end{vmatrix} + (-2)\begin{vmatrix} 1 & 4 \\ 2 & 1 \end{vmatrix} = (0-1) - 0 - 2(1-8) = -1 - 2(-7) = -1 + 14 = 13$$

$$D_z = \begin{vmatrix} 1 & 1 & 4 \\ 2 & 5 & 1 \\ 1 & -1 & 0 \end{vmatrix} = 1\begin{vmatrix} 1 & 4 \\ 5 & 1 \end{vmatrix} - (-1)\begin{vmatrix} 1 & 4 \\ 2 & 1 \end{vmatrix} + 0\begin{vmatrix} 1 & 1 \\ 2 & 5 \end{vmatrix} = (1-20) + (1-8) + 0 = -19 - 7 = -26$$

$$x = \frac{D_x}{D} = \frac{-39}{-13} = 3,\ y = \frac{D_y}{D} = \frac{13}{-13} = -1,$$

$$z = \frac{D_z}{D} = \frac{-26}{-13} = 2$$

The solution is (3, –1, 2).

23. $\begin{cases} 3x+2y+3z=3 \\ x \quad\quad -z=9 \\ \quad 4y+z=-4 \end{cases}$

$$D=\begin{vmatrix} 3 & 2 & 3 \\ 1 & 0 & -1 \\ 0 & 4 & 1 \end{vmatrix}$$
$$=-1\begin{vmatrix} 2 & 3 \\ 4 & 1 \end{vmatrix}+0\begin{vmatrix} 3 & 3 \\ 0 & 1 \end{vmatrix}-(-1)\begin{vmatrix} 3 & 2 \\ 0 & 4 \end{vmatrix}$$
$$=-(2-12)+0+(12-0)$$
$$=-(-10)+12$$
$$=10+12$$
$$=22$$

$$D_x=\begin{vmatrix} 3 & 2 & 3 \\ 9 & 0 & -1 \\ -4 & 4 & 1 \end{vmatrix}$$
$$=-9\begin{vmatrix} 2 & 3 \\ 4 & 1 \end{vmatrix}+0\begin{vmatrix} 3 & 3 \\ -4 & 1 \end{vmatrix}-(-1)\begin{vmatrix} 3 & 2 \\ -4 & 4 \end{vmatrix}$$
$$=-9(2-12)+0+[12-(-8)]$$
$$=-9(-10)+20$$
$$=90+20$$
$$=110$$

$$D_y=\begin{vmatrix} 3 & 3 & 3 \\ 1 & 9 & -1 \\ 0 & -4 & 1 \end{vmatrix}$$
$$=3\begin{vmatrix} 9 & -1 \\ -4 & 1 \end{vmatrix}-1\begin{vmatrix} 3 & 3 \\ -4 & 1 \end{vmatrix}+0\begin{vmatrix} 3 & 3 \\ 9 & -1 \end{vmatrix}$$
$$=3(9-4)-[3-(-12)]+0$$
$$=3(5)-(15)$$
$$=15-15$$
$$=0$$

$$D_z=\begin{vmatrix} 3 & 2 & 3 \\ 1 & 0 & 9 \\ 0 & 4 & -4 \end{vmatrix}$$
$$=3\begin{vmatrix} 0 & 9 \\ 4 & -4 \end{vmatrix}-1\begin{vmatrix} 2 & 3 \\ 4 & -4 \end{vmatrix}+0\begin{vmatrix} 2 & 3 \\ 0 & 9 \end{vmatrix}$$
$$=3(0-36)-(-8-12)+0$$
$$=3(-36)-(-20)$$
$$=-108+20$$
$$=-88$$

$$x=\frac{D_x}{D}=\frac{110}{22}=5,\ y=\frac{D_y}{D}=\frac{0}{22}=0,$$
$$z=\frac{D_z}{D}=\frac{-88}{22}=-4$$

The solution is (5, 0, –4).

24. $\begin{cases} x-y=-2 \\ 3x-3y=-6 \end{cases}$

$$\left[\begin{array}{cc|c} 1 & -1 & -2 \\ 3 & -3 & -6 \end{array}\right]$$

Multiply row 1 by –3 and add to row 2.

$$\left[\begin{array}{cc|c} 1 & -1 & -2 \\ 0 & 0 & 0 \end{array}\right]$$

This corresponds to

$\begin{cases} x-y=-2 \\ 0=0 \end{cases}$

This is a dependent system.

The solution is $\{(x, y) | x-y=-2\}$.

25. $\begin{cases} x+2y=-1 \\ 2x+5y=-5 \end{cases}$

$$\left[\begin{array}{cc|c} 1 & 2 & -1 \\ 2 & 5 & -5 \end{array}\right]$$

Multiply row 1 by –2 and add to row 2.

$$\left[\begin{array}{cc|c} 1 & 2 & -1 \\ 0 & 1 & -3 \end{array}\right]$$

Multiply row 2 by –2 and add to row 1.

$$\left[\begin{array}{cc|c} 1 & 0 & 5 \\ 0 & 1 & -3 \end{array}\right]$$

The solution is (5, –3).

26. $\begin{cases} x - y - z = 0 \\ 3x - y - 5z = -2 \\ 2x + 3y = -5 \end{cases}$

$$\left[\begin{array}{ccc|c} 1 & -1 & -1 & 0 \\ 3 & -1 & -5 & -2 \\ 2 & 3 & 0 & -5 \end{array}\right]$$

Multiply row 1 by –3 and add to row 2.
Multiply row 1 by –2 and add to row 3.

$$\left[\begin{array}{ccc|c} 1 & -1 & -1 & 0 \\ 0 & 2 & -2 & -2 \\ 0 & 5 & 2 & -5 \end{array}\right]$$

Divide row 2 by 2.

$$\left[\begin{array}{ccc|c} 1 & -1 & -1 & 0 \\ 0 & 1 & -1 & -1 \\ 0 & 5 & 2 & -5 \end{array}\right]$$

Multiply row 2 by –5 and add to row 3.

$$\left[\begin{array}{ccc|c} 1 & -1 & -1 & 0 \\ 0 & 1 & -1 & -1 \\ 0 & 0 & 7 & 0 \end{array}\right]$$

Divide row 3 by 7.

$$\left[\begin{array}{ccc|c} 1 & -1 & -1 & 0 \\ 0 & 1 & -1 & -1 \\ 0 & 0 & 1 & 0 \end{array}\right]$$

This corresponds to

$$\begin{cases} x - y - z = 0 \\ y - z = -1. \\ z = 0 \end{cases}$$

$y - 0 = -1$
$y = -1$
$x - (-1) - 0 = 0$
$x + 1 = 0$
$x = -1$

The solution is (–1, –1, 0).

27. $\begin{cases} 2x - y + 3z = 4 \\ 3x - 3z = -2 \\ -5x + y = 0 \end{cases}$

$$\left[\begin{array}{ccc|c} 2 & -1 & 3 & 4 \\ 3 & 0 & -3 & -2 \\ -5 & 1 & 0 & 0 \end{array}\right]$$

Divide row 1 by 2.

$$\left[\begin{array}{ccc|c} 1 & -\frac{1}{2} & \frac{3}{2} & 2 \\ 3 & 0 & -3 & -2 \\ -5 & 1 & 0 & 0 \end{array}\right]$$

Multiply row 1 by –3 and add to row 2.
Multiply row 1 by 5 and add to row 3.

$$\left[\begin{array}{ccc|c} 1 & -\frac{1}{2} & \frac{3}{2} & 2 \\ 0 & \frac{3}{2} & -\frac{15}{2} & -8 \\ 0 & -\frac{3}{2} & \frac{15}{2} & 10 \end{array}\right]$$

Multiply row 2 by $\frac{2}{3}$.

$$\left[\begin{array}{ccc|c} 1 & -\frac{1}{2} & \frac{3}{2} & 2 \\ 0 & 1 & -5 & -\frac{16}{3} \\ 0 & -\frac{3}{2} & \frac{15}{2} & 10 \end{array}\right]$$

Multiply row 2 by $\frac{3}{2}$ and add to row 3.

$$\left[\begin{array}{ccc|c} 1 & -\frac{1}{2} & \frac{3}{2} & 2 \\ 0 & 1 & -5 & -\frac{16}{3} \\ 0 & 0 & 0 & 2 \end{array}\right]$$

This corresponds to

$$\begin{cases} x - \frac{1}{2}y + \frac{3}{2}z = 2 \\ y - 5z = -\frac{16}{3} \\ 0 = 2 \end{cases}$$

This is an inconsistent system.
There is no solution.

28. Let x = number of double occupancy rooms, and
y = the number of single occupancy rooms.

$$\begin{cases} x + \quad y = 80 \\ 90x + 80y = 6930 \end{cases}$$

Multiply the first equation by –80 and add to the second equation.

$$\begin{aligned} -80x - 80y &= -6400 \\ 90x + 80y &= \ 6930 \\ \hline 10x \qquad &= \ 530 \\ x \qquad &= \ 53 \end{aligned}$$

Replace x with 53 in the first equation.

$53 + y = 80$
$y = 27$

53 double-occupancy and
27 single-occupancy rooms are occupied.

29. $R(x) = 4x$ and $C(x) = 1.5x + 2000$
Break even occurs when $R(x) = C(x)$.

$4x = 1.5x + 2000$
$2.5x = 2000$
$x = 800$

The company must sell 800 packages to break even.

30. Let x = employees in 1996.

$x + 1.18x = 461,000$
$2.18x = 461,000$
$x \approx 211,468$

The number of people employed in these occupations in 1996 was 211,468.

31. Recall that $C = 2\pi r$. Here $C = 78.5$.

$78.5 = 2\pi r$

$$r = \frac{78.5}{2\pi} = \frac{39.25}{\pi}$$

Also, recall that $A = \pi r^2$.

$$A = \pi\left(\frac{39.25}{\pi}\right)^2 \approx \frac{39.25^2}{3.14} \approx 490.63$$

Dividing this by 60 yields approximately 8.18. Therefore, about 8 hunting dogs could safely be kept in the pen.

32. Solve $R > C$.

$7.4x > 3910 + 2.8x$
$4.6x > 3910$
$x > 850$

Therefore, more than 850 sunglasses must be produced and sold in order for them to yield a profit.

33. Let x = one of the numbers. Then the other number is $5 - x$.

$$x^2 + (5 - x)^2 = 73$$
$$x^2 + 25 - 10x + x^2 = 73$$
$$2x^2 - 10x - 48 = 0$$
$$x^2 - 5x - 24 = 0$$
$$(x - 8)(x + 3) = 0$$
$$x - 8 = 0 \quad \text{or} \quad x + 3 = 0$$
$$x = 8 \quad \text{or} \quad x = -3$$

If $x = 8$, $5 - x = 5 - 8 = -3$.
If $x = -3$, $5 - x = 5 - (-3) = 8$.
The numbers are –3 and 8.

Chapter 9

Exercise Set 9.1

1. $\sqrt{100} = 10$ because $10^2 = 100$.

3. $\sqrt{\frac{1}{4}} = \frac{1}{2}$ because $\left(\frac{1}{2}\right)^2 = \frac{1}{4}$.

5. $\sqrt{0.0001} = 0.01$ because $(0.01)^2 = 0.0001$.

7. $-\sqrt{36} = -6$ because $(6)^2 = 36$.

9. $\sqrt{x^{10}} = x^5$ because $(x^5)^2 = x^{10}$.

11. $\sqrt{16y^6} = 4y^3$ because $(4y^3)^2 = 16y^6$.

13. $\sqrt{7} \approx 2.646$
Since $4 < 7 < 9$, then
$\sqrt{4} < \sqrt{7} < \sqrt{9}$, or $2 < \sqrt{7} < 3$. The approximation is between 2 and 3 and thus is reasonable.

15. $\sqrt{38} \approx 6.164$
Since $36 < 38 < 49$, then
$\sqrt{36} < \sqrt{38} < \sqrt{49}$, or $6 < \sqrt{38} < 7$. The approximation is between 6 and 7 and thus is reasonable.

17. $\sqrt{200} \approx 14.142$
Since $196 < 200 < 225$, then
$\sqrt{196} < \sqrt{200} < \sqrt{225}$, or
$14 < \sqrt{200} < 15$. The approximation is between 14 and 15 and thus is reasonable.

19. $\sqrt[3]{64} = 4$ because $(4)^3 = 64$.

21. $\sqrt[3]{\frac{1}{8}} = \frac{1}{2}$ because $\left(\frac{1}{2}\right)^3 = \frac{1}{8}$.

23. $\sqrt[3]{-1} = -1$ because $(-1)^3 = -1$.

25. $\sqrt[3]{x^{12}} = x^4$ because $(x^4)^3 = x^{12}$.

27. $\sqrt[3]{-27x^9} = -3x^3$ because $(-3x^3)^3 = -27x^9$.

29. $-\sqrt[4]{16} = -2$ because $(2)^4 = 16$.

31. $\sqrt[4]{-16}$ is not a real number. There is no real number that, when raised to the fourth power, is -16.

33. $\sqrt[5]{-32} = -2$ because $(-2)^5 = -32$.

35. $\sqrt[5]{x^{20}} = x^4$ because $(x^4)^5 = x^{20}$.

37. $\sqrt[6]{64x^{12}} = 2x^2$ because $(2x^2)^6 = 64x^{12}$.

39. $\sqrt{81x^4} = 9x^2$ because $(9x^2)^2 = 81x^4$.

41. $\sqrt[4]{256x^8} = 4x^2$ because $(4x^2)^4 = 256x^8$.

43. $\sqrt{(-8)^2} = |-8| = 8$

45. $\sqrt[3]{(-8)^3} = -8$

47. $\sqrt{4x^2} = \sqrt{(2x)^2} = |2x| = 2|x|$

49. $\sqrt[3]{x^3} = x$

51. $\sqrt[4]{(x-5)^4} = |x-5|$

53. $\sqrt{x^2 + 4x + 4} = \sqrt{(x+2)^2} = |x+2|$

55. $-\sqrt{121} = -\sqrt{11^2} = -11$

57. $\sqrt[3]{8x^3} = \sqrt[3]{(2x)^3} = 2x$

59. $\sqrt{y^{12}} = \sqrt{(y^6)^2} = y^6$

61. $\sqrt{25a^2b^{20}} = \sqrt{(5ab^{10})^2} = 5ab^{10}$

63. $\sqrt[3]{-27x^{12}y^9} = \sqrt[3]{(-3x^4y^3)^3} = -3x^4y^3$

65. $\sqrt[4]{a^{16}b^4} = \sqrt[4]{(a^4b)^4} = a^4b$

67. $\sqrt[5]{-32x^{10}y^5} = \sqrt[5]{(-2x^2y)^5} = -2x^2y$

69. $\sqrt{\frac{25}{49}} = \sqrt{\left(\frac{5}{7}\right)^2} = \frac{5}{7}$

71. $\sqrt{\frac{x^2}{4y^2}} = \sqrt{\left(\frac{x}{2y}\right)^2} = \frac{x}{2y}$

73. $-\sqrt[3]{\frac{z^{21}}{27x^3}} = -\sqrt[3]{\left(\frac{z^7}{3x}\right)^3} = -\frac{z^7}{3x}$

75. $\sqrt[4]{\frac{x^4}{16}} = \sqrt[4]{\left(\frac{x}{2}\right)^4} = \frac{x}{2}$

77. $f(x) = \sqrt{2x+3}$
$f(0) = \sqrt{2(0)+3} = \sqrt{0+3} = \sqrt{3}$

79. $g(x) = \sqrt[3]{x-8}$
$g(7) = \sqrt[3]{7-8} = \sqrt[3]{-1} = -1$

81. $g(x) = \sqrt[3]{x-8}$
$g(-19) = \sqrt[3]{-19-8} = \sqrt[3]{-27} = -3$

83. $f(x) = \sqrt{2x+3}$
$f(2) = \sqrt{2(2)+3} = \sqrt{4+7} = \sqrt{7}$

85. $f(x) = \sqrt{x}+2$
Domain: $[0, \infty)$

x	$f(x) = \sqrt{x}+2$
0	$\sqrt{0}+2 = 0+2 = 2$
1	$\sqrt{1}+2 = 1+2 = 3$
3	$\sqrt{3}+2 \approx 1.7+2 \approx 3.7$
4	$\sqrt{4}+2 = 2+2 = 4$

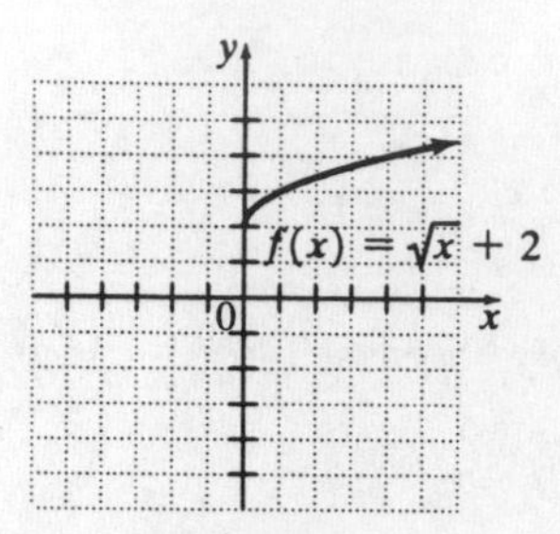

87. $f(x) = \sqrt{x-3}$
Domain: $[3, \infty)$

x	$f(x) = \sqrt{x-3}$
3	$\sqrt{3-3} = \sqrt{0} = 0$
4	$\sqrt{4-3} = \sqrt{1} = 1$
7	$\sqrt{7-3} = \sqrt{4} = 2$
12	$\sqrt{12-3} = \sqrt{9} = 3$

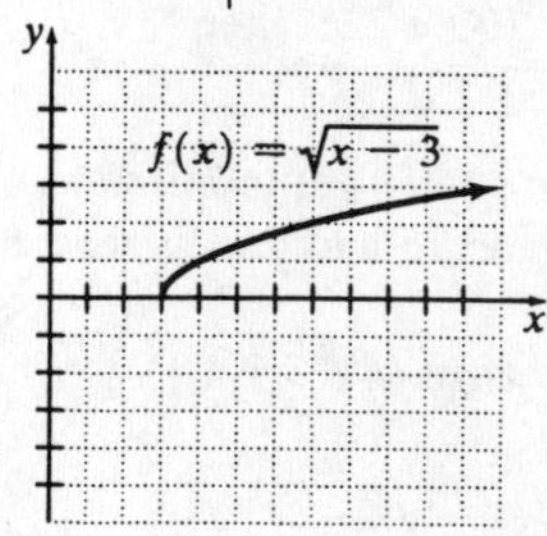

89. $f(x) = \sqrt[3]{x} + 1$
Domain: $(-\infty, \infty)$

x	$f(x) = \sqrt[3]{x} + 1$
−8	$\sqrt[3]{-8} + 1 = -2 + 1 = -1$
−1	$\sqrt[3]{-1} + 1 = -1 + 1 = 0$
0	$\sqrt[3]{0} + 1 = 0 + 1 = 1$
1	$\sqrt[3]{1} + 1 = 1 + 1 = 2$
8	$\sqrt[3]{8} + 1 = 2 + 1 = 3$

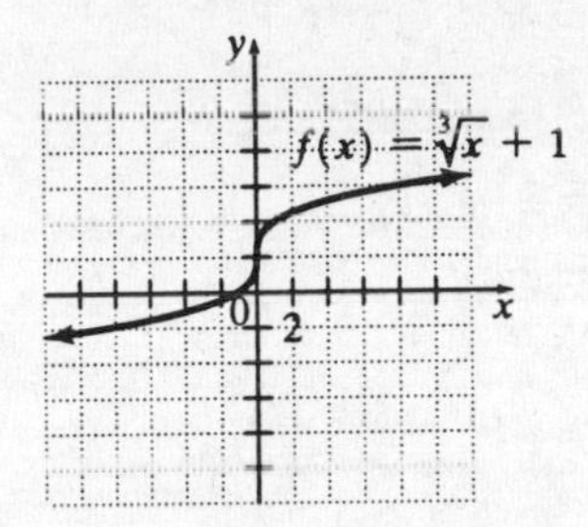

91. $g(x) = \sqrt[3]{x - 1}$
Domain: $(-\infty, \infty)$

x	$g(x) = \sqrt[3]{x - 1}$
1	$\sqrt[3]{1 - 1} = \sqrt[3]{0} = 0$
2	$\sqrt[3]{2 - 1} = \sqrt[3]{1} = 1$
0	$\sqrt[3]{0 - 1} = \sqrt[3]{-1} = -1$
9	$\sqrt[3]{9 - 1} = \sqrt[3]{8} = 2$
−7	$\sqrt[3]{-7 - 1} = \sqrt[3]{-8} = -2$

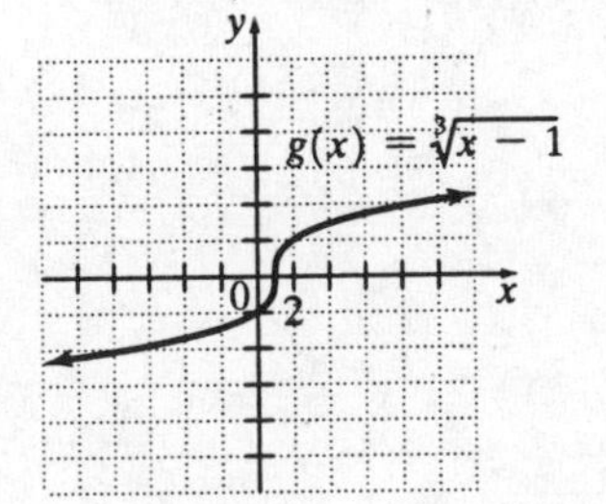

93. Answers may vary.

95.

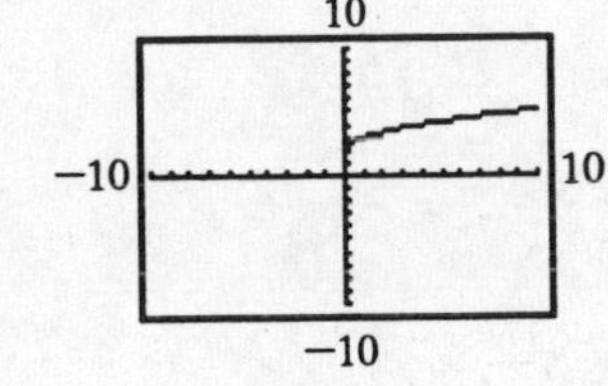

Domain: $[0, \infty)$

97.

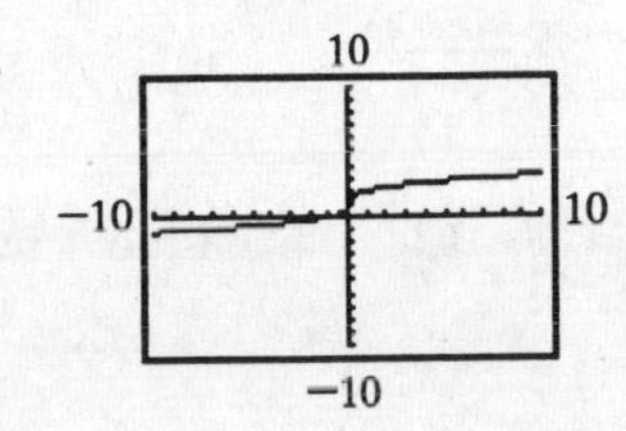

Domain: $(-\infty, \infty)$

99. $(-2x^3y^2)^5 = (-2)^5 x^{3\cdot 5} y^{2\cdot 5} = -32x^{15}y^{10}$

101. $(-3x^2y^3z^5)(20x^5y^7)$
$= (-3)(20)x^{2+5}y^{3+7}z^5$
$= -60x^7y^{10}z^5$

103. $\dfrac{7x^{-1}y}{14(x^5y^2)^{-2}} = \dfrac{7x^{-1}y}{14x^{-10}y^{-4}}$
$= \dfrac{x^{-1+10}y^{1+4}}{2}$
$= \dfrac{x^9y^5}{2}$

Exercise Set 9.2

1. $49^{1/2} = \sqrt{49} = 7$

3. $27^{1/3} = \sqrt[3]{27} = 3$

5. $\left(\frac{1}{16}\right)^{1/4} = \sqrt[4]{\frac{1}{16}} = \frac{1}{2}$

7. $169^{1/2} = \sqrt{169} = 13$

9. $2m^{1/3} = 2\sqrt[3]{m}$

11. $(9x^4)^{1/2} = \sqrt{9x^4} = 3x^2$

13. $(-27)^{1/3} = \sqrt[3]{-27} = -3$

15. $-16^{1/4} = -\sqrt[4]{16} = -2$

17. $16^{3/4} = (\sqrt[4]{16})^3 = 2^3 = 8$

19. $(-64)^{2/3} = \left(\sqrt[3]{-64}\right)^2 = (-4)^2 = 16$

21. $(-16)^{3/4} = \left(\sqrt[4]{-16}\right)^3$, which is not a real number.

23. $(2x)^{3/5} = \left(\sqrt[5]{2x}\right)^3 = \sqrt[5]{(2x)^3}$

25. $(7x+2)^{2/3} = \left(\sqrt[3]{7x+2}\right)^2 = \sqrt[3]{(7x+2)^2}$

27. $\left(\frac{16}{9}\right)^{3/2} = \left(\sqrt{\frac{16}{9}}\right)^3 = \left(\frac{4}{3}\right)^3 = \frac{64}{27}$

29. $8^{-4/3} = \frac{1}{8^{4/3}} = \frac{1}{(\sqrt[3]{8})^4} = \frac{1}{2^4} = \frac{1}{16}$

31. $(-64)^{-2/3} = \frac{1}{(-64)^{2/3}}$

$= \frac{1}{(\sqrt[3]{-64})^2}$

$= \frac{1}{(-4)^2}$

$= \frac{1}{16}$

33. $(-4)^{-3/2} = \frac{1}{(-4)^{3/2}} = \frac{1}{(\sqrt{-4})^3}$ is not a real number.

35. $x^{-1/4} = \frac{1}{x^{1/4}}$

37. $\frac{1}{a^{-2/3}} = a^{2/3}$

39. $\frac{5}{7x^{-3/4}} = \frac{5x^{3/4}}{7}$

41. Answers may vary.

43. $a^{2/3}a^{5/3} = a^{2/3+5/3} = a^{7/3}$

45. $x^{-2/5} \cdot x^{7/5} = x^{-2/5+7/5} = x^{5/5} = x$

47. $3^{1/4} \cdot 3^{3/8} = 3^{1/4+3/8} = 3^{2/8+3/8} = 3^{5/8}$

49. $\frac{y^{1/3}}{y^{1/6}} = y^{1/3-1/6} = y^{2/6-1/6} = y^{1/6}$

51. $(4u^2)^{3/2} = 4^{3/2} \cdot u^{2 \cdot 3/2}$

$= \left(\sqrt{4}\right)^3 \cdot u^3$

$= 2^3u^3$

$= 8u^3$

53. $\frac{b^{1/2}b^{3/4}}{-b^{1/4}} = -b^{1/2+3/4-1/4}$

$= -b^{1/2+2/4}$

$= -b^{1/2+1/2}$

$= -b^1$

$= -b$

55. $\frac{(3x^{1/4})^3}{x^{1/12}} = \frac{3^3x^{3/4}}{x^{1/12}}$

$= 27x^{3/4-1/12}$

$= 27x^{9/12-1/12}$

$= 27x^{8/12}$

$= 27x^{2/3}$

57. $y^{1/2}(y^{1/2} - y^{2/3}) = y^{1/2}y^{1/2} - y^{1/2}y^{2/3}$

$= y^{1/2+1/2} - y^{1/2+2/3}$

$= y^1 - y^{3/6+4/6}$

$= y - y^{7/6}$

59. $x^{2/3}(2x-2) = 2x^1x^{2/3} - 2x^{2/3}$

$= 2x^{3/3+2/3} - 2x^{2/3}$

$= 2x^{5/3} - 2x^{2/3}$

61. $(2x^{1/3}+3)(2x^{1/3}-3)=(2x^{1/3})^2-3^2$
$=2^2(x^{1/3})^2-9$
$=4x^{2/3}-9$

63. $x^{8/3}+x^{10/3}=x^{8/3}+x^{8/3}x^{2/3}$
$=x^{8/3}(1+x^{2/3})$

65. $x^{2/5}-3x^{1/5}=x^{1/5}x^{1/5}-3x^{1/5}$
$=x^{1/5}(x^{1/5}-3)$

67. $5x^{-1/3}+x^{2/3}=5x^{-1/3}+x^{3/3}x^{-1/3}$
$=x^{-1/3}(5+x^1)$
$=x^{-1/3}(5+x)$

69. $\sqrt[6]{x^3}=x^{3/6}=x^{1/2}=\sqrt{x}$

71. $\sqrt[6]{4}=4^{1/6}=(2^2)^{1/6}=2^{2/6}=2^{1/3}=\sqrt[3]{2}$

73. $\sqrt[4]{16x^2}=16^{1/4}x^{2/4}=2x^{1/2}=2\sqrt{x}$

75. $\sqrt[8]{x^4y^4}=x^{4/8}y^{4/8}=x^{1/2}y^{1/2}=\sqrt{xy}$

77. $\sqrt[3]{y}\sqrt[5]{y^2}=y^{1/3}\cdot y^{2/5}$
$=y^{5/15}\cdot y^{6/15}$
$=y^{5/15+6/15}$
$=y^{11/15}$
$=\sqrt[15]{y^{11}}$

79. $\dfrac{\sqrt[3]{b^2}}{\sqrt[4]{b}}=\dfrac{b^{2/3}}{b^{1/4}}$
$=b^{\frac{2}{3}-\frac{1}{4}}$
$=b^{\frac{8}{12}-\frac{3}{12}}$
$=b^{5/12}$
$=\sqrt[12]{b^5}$

81. $\dfrac{\sqrt[3]{a^2}}{\sqrt[6]{a}}=\dfrac{a^{2/3}}{a^{1/6}}$
$=a^{2/3-1/6}$
$=a^{4/6-1/6}$
$=a^{3/6}$
$=a^{1/2}$
$=\sqrt{a}$

83. $\sqrt{3}\cdot\sqrt[3]{4}=3^{1/2}\cdot 4^{1/3}$
$=3^{3/6}\cdot 4^{2/6}$
$=(3^3\cdot 4^2)^{1/6}$
$=(432)^{1/6}$
$=\sqrt[6]{432}$

85. $\sqrt[5]{7}\cdot\sqrt[3]{y}=7^{1/5}\cdot y^{1/3}$
$=7^{3/15}\cdot y^{5/15}$
$=(7^3\cdot y^5)^{1/15}$
$=(343y^5)^{1/15}$
$=\sqrt[15]{343y^5}$

87. $\dfrac{\sqrt{t}}{\sqrt{u}}=\dfrac{t^{1/2}}{u^{1/2}}$

89. $\text{BMR}=70(50)^{3/4}\approx 1316$ calories

91. $f(x)=6550x^{43/50}$

a. We find $f(9)$ since 2004 is 9 years after 1990.

$f(9)=6550(9^{43/50})$
$\approx 43,340$

HP's net revenue for 1999 was about $43,340 million.

b. We find $f(14)$ since 2004 is 14 years after 1990.

$f(14)=6550(14^{43/50})$
$\approx 63,374$

We predict that HP's net revenue will be about $63,374 million in 2004.

93. $x^{3/8}$

$x^{3/8} \cdot x^{1/8} = x^{3/8+1/8} = x^{4/8} = x^{1/2}$

95. $y^{1/4}$

$\frac{y^{1/4}}{y^{-3/4}} = y^{1/4+3/4} = y^{4/4} = y$

97. $20^{1/5} \approx 1.8206$

99. $76^{5/7} \approx 22.0515$

101. $20 = (2 \cdot 2) \cdot 5 = 4 \cdot 5$

103. $45 = (3 \cdot 3) \cdot 5 = 9 \cdot 5$

105. $56 = (2 \cdot 2 \cdot 2) \cdot 7 = 8 \cdot 7$

107. $80 = (2 \cdot 2 \cdot 2) \cdot 2 \cdot 5 = 8 \cdot 10$

Exercise Set 9.3

1. $\sqrt{7} \cdot \sqrt{2} = \sqrt{7 \cdot 2} = \sqrt{14}$

3. $\sqrt[4]{8} \cdot \sqrt[4]{2} = \sqrt[4]{8 \cdot 2} = \sqrt[4]{16} = 2$

5. $\sqrt[3]{4} \cdot \sqrt[3]{9} = \sqrt[3]{4 \cdot 9} = \sqrt[3]{36}$

7. $\sqrt{2} \cdot \sqrt{3x} = \sqrt{2 \cdot 3x} = \sqrt{6x}$

9. $\sqrt{\frac{7}{x}} \cdot \sqrt{\frac{2}{y}} = \sqrt{\frac{7 \cdot 2}{x \cdot y}} = \sqrt{\frac{14}{xy}}$

11. $\sqrt[4]{4x^3} \cdot \sqrt[4]{5} = \sqrt[4]{4x^3 \cdot 5} = \sqrt[4]{20x^3}$

13. $\sqrt{\frac{6}{49}} = \frac{\sqrt{6}}{\sqrt{49}} = \frac{\sqrt{6}}{7}$

15. $\sqrt{\frac{2}{49}} = \frac{\sqrt{2}}{\sqrt{49}} = \frac{\sqrt{2}}{7}$

17. $\sqrt[4]{\frac{x^3}{16}} = \frac{\sqrt[4]{x^3}}{\sqrt[4]{16}} = \frac{\sqrt[4]{x^3}}{2}$

19. $\sqrt[3]{\frac{4}{27}} = \frac{\sqrt[3]{4}}{\sqrt[3]{27}} = \frac{\sqrt[3]{4}}{3}$

21. $\sqrt[4]{\frac{8}{x^8}} = \frac{\sqrt[4]{8}}{\sqrt[4]{x^8}} = \frac{\sqrt[4]{8}}{x^2}$

23. $\sqrt[3]{\frac{2x}{81y^{12}}} = \frac{\sqrt[3]{2x}}{\sqrt[3]{81y^{12}}}$

$= \frac{\sqrt[3]{2x}}{\sqrt[3]{27y^{12}} \cdot \sqrt[3]{3}}$

$= \frac{\sqrt[3]{2x}}{3y^4 \sqrt[3]{3}}$

25. $\sqrt{\frac{x^2y}{100}} = \frac{\sqrt{x^2y}}{\sqrt{100}} = \frac{\sqrt{x^2} \cdot \sqrt{y}}{\sqrt{100}} = \frac{x\sqrt{y}}{10}$

27. $\sqrt{\frac{5x^2}{4y^2}} = \frac{\sqrt{5x^2}}{\sqrt{4y^2}} = \frac{\sqrt{5}\sqrt{x^2}}{\sqrt{4}\sqrt{y^2}} = \frac{\sqrt{5}x}{2y}$

29. $-\sqrt[3]{\frac{z^7}{27x^3}} = -\frac{\sqrt[3]{z^7}}{\sqrt[3]{27x^3}}$

$= -\frac{\sqrt[3]{z^6 z}}{\sqrt[3]{27} \sqrt[3]{x^3}}$

$= -\frac{\sqrt[3]{z^6} \sqrt[3]{z}}{3x}$

$= -\frac{z^2 \sqrt[3]{z}}{3x}$

31. $\sqrt{32} = \sqrt{16 \cdot 2} = \sqrt{16} \cdot \sqrt{2} = 4\sqrt{2}$

33. $\sqrt[3]{192} = \sqrt[3]{64 \cdot 3} = \sqrt[3]{64} \cdot \sqrt[3]{3} = 4\sqrt[3]{3}$

35. $5\sqrt{75} = 5\sqrt{25 \cdot 3}$

$= 5\sqrt{25} \cdot \sqrt{3}$

$= 5 \cdot 5\sqrt{3}$

$= 25\sqrt{3}$

37. $\sqrt{24} = \sqrt{4 \cdot 6} = \sqrt{4} \cdot \sqrt{6} = 2\sqrt{6}$

39. $\sqrt{100x^5} = \sqrt{100x^4 \cdot x}$
$= \sqrt{100} \cdot \sqrt{x^4} \cdot \sqrt{x}$
$= 10x^2\sqrt{x}$

41. $\sqrt[3]{16y^7} = \sqrt[3]{(8y^6)(2y)}$
$= \sqrt[3]{8} \cdot \sqrt[3]{y^6} \cdot \sqrt[3]{2y}$
$= 2y^2\sqrt[3]{2y}$

43. $\sqrt[4]{a^8b^7} = \sqrt[4]{a^8b^4b^3}$
$= \sqrt[4]{a^8} \cdot \sqrt[4]{b^4} \cdot \sqrt[4]{b^3}$
$= a^2b\sqrt[4]{b^3}$

45. $\sqrt{y^5} = \sqrt{y^4y} = \sqrt{y^4} \cdot \sqrt{y} = y^2\sqrt{y}$

47. $\sqrt{25a^2b^3} = \sqrt{25} \cdot \sqrt{a^2} \cdot \sqrt{b^3}$
$= 5a\sqrt{b^2b}$
$= 5a\sqrt{b^2} \cdot \sqrt{b}$
$= 5ab\sqrt{b}$

49. $\sqrt[5]{-32x^{10}y} = \sqrt[5]{-32} \cdot \sqrt[5]{x^{10}} \cdot \sqrt[5]{y}$
$= -2x^2\sqrt[5]{y}$

51. $\sqrt[3]{50x^{14}} = \sqrt[3]{x^{12}(50x^2)}$
$= \sqrt[3]{x^{12}} \cdot \sqrt[3]{50x^2}$
$= x^4\sqrt[3]{50x^2}$

53. $-\sqrt{32a^8b^7} = -\sqrt{16a^8b^6(2b)}$
$= -\sqrt{16} \cdot \sqrt{a^8} \cdot \sqrt{b^6} \cdot \sqrt{2b}$
$= -4a^4b^3\sqrt{2b}$

55. $\sqrt{9x^7y^9} = \sqrt{9x^6y^8 \cdot xy}$
$= \sqrt{9x^6y^8} \cdot \sqrt{xy}$
$= 3x^3y^4\sqrt{xy}$

57. $\sqrt[3]{125r^9s^{12}} = 5r^3s^4$

59. $\frac{\sqrt{14}}{\sqrt{7}} = \sqrt{\frac{14}{7}} = \sqrt{2}$

61. $\frac{\sqrt[3]{24}}{\sqrt[3]{3}} = \sqrt[3]{\frac{24}{3}} = \sqrt[3]{8} = 2$

63. $\frac{5\sqrt[4]{48}}{\sqrt[4]{3}} = 5\sqrt[4]{\frac{48}{3}} = 5\sqrt[4]{16} = 5 \cdot 2 = 10$

65. $\frac{\sqrt{x^5y^3}}{\sqrt{xy}} = \sqrt{\frac{x^5y^3}{xy}}$
$= \sqrt{x^4y^2}$
$= x^2y$

67. $\frac{8\sqrt[3]{54m^7}}{\sqrt[3]{2m}} = 8\sqrt[3]{\frac{54m^7}{2m}}$
$= 8\sqrt[3]{27m^6}$
$= 8 \cdot 3m^2$
$= 24m^2$

69. $\frac{3\sqrt{100x^2}}{2\sqrt{2x^{-1}}} = \frac{3}{2}\sqrt{\frac{100x^2}{2x^{-1}}}$
$= \frac{3}{2}\sqrt{50x^3}$
$= \frac{3}{2}\sqrt{25x^2 \cdot 2x}$
$= \frac{3}{2} \cdot 5x\sqrt{2x}$
$= \frac{15x\sqrt{2x}}{2}$ or $\frac{15x}{2}\sqrt{2x}$

71. $\frac{\sqrt[4]{96a^{10}b^3}}{\sqrt[4]{3a^2b^3}} = \sqrt[4]{\frac{96a^{10}b^3}{3a^2b^3}}$
$= \sqrt[4]{32a^8}$
$= \sqrt[4]{16a^8 \cdot 2}$
$= 2a^2\sqrt[4]{2}$

73. $A = \pi r\sqrt{r^2 + h^2}$

a. Let $h = 3$ and $r = 4$.

$A = \pi(4)\sqrt{4^2 + 3^2}$
$= 4\pi\sqrt{16+9}$
$= 4\pi\sqrt{25}$
$= 4 \cdot 5\pi$
$= 20\pi$

The surface area of the cone is 20π square centimeters.

b. Let $h = 7.2$ and $r = 6.8$.

$A = \pi(6.8)\sqrt{(6.8)^2 + (7.2)^2}$
$= 6.8\pi\sqrt{46.24 + 51.84}$
$= 6.8\pi\sqrt{98.08}$
≈ 211.57

The surface area of the cone is about 211.57 square centimeters.

75. $F(x) = 0.6\sqrt{49 - x^2}$

a. Find $F(3)$.

$F(3) = 0.6\sqrt{49 - 3^2}$
$= 0.6\sqrt{49 - 9}$
$= 0.6\sqrt{40}$
≈ 3.8

The demand for an older release will be about 3.8 times per week if the rental price is $3.

b. Find $F(5)$.

$F(5) = 0.6\sqrt{49 - 5^2}$
$= 0.6\sqrt{49 - 25}$
$= 0.6\sqrt{24}$
≈ 2.9

The demand for an older release will be about 2.9 times per week if the rental price is $5.

c. Answers may vary.

77. $(6x)(8x) = 6 \cdot 8 \cdot x \cdot x = 48x^2$

79. $(2x+3)+(x-5) = 2x+3+x-5$
$= (2x+x)+(3-5)$
$= 3x-2$

81. $(9y^2)(-8y^2) = 9(-8)y^{2+2} = -72y^4$

83. $-3 + x + 5 = x + (-3 + 5) = x + 2$

85. $(2x+1)^2 = (2x)^2 + 2(2x)(1) + 1^2$
$= 4x^2 + 4x + 1$

Section 9.4

Mental Math

1. $2\sqrt{3} + 4\sqrt{3} = (2+4)\sqrt{3} = 6\sqrt{3}$

2. $5\sqrt{7} + 3\sqrt{7} = (5+3)\sqrt{7} = 8\sqrt{7}$

3. $8\sqrt{x} - 5\sqrt{x} = (8-5)\sqrt{x} = 3\sqrt{x}$

4. $3\sqrt{y} + 10\sqrt{y} = (3+10)\sqrt{y} = 13\sqrt{y}$

5. $7\sqrt[3]{x} + 5\sqrt[3]{x} = (7+5)\sqrt[3]{x} = 12\sqrt[3]{x}$

6. $8\sqrt[3]{z} - 2\sqrt[3]{z} = (8-2)\sqrt[3]{z} = 6\sqrt[3]{z}$

Exercise Set 9.4

1. $\sqrt{8} - \sqrt{32} = \sqrt{4 \cdot 2} - \sqrt{16 \cdot 2}$
$= \sqrt{4}\sqrt{2} - \sqrt{16}\sqrt{2}$
$= 2\sqrt{2} - 4\sqrt{2}$
$= -2\sqrt{2}$

3. $2\sqrt{2x^3} + 4x\sqrt{8x}$
$= 2\sqrt{x^2 \cdot 2x} + 4x\sqrt{4 \cdot 2x}$
$= 2\sqrt{x^2}\sqrt{2x} + 4x\sqrt{4}\sqrt{2x}$
$= 2x\sqrt{2x} + 4x \cdot 2\sqrt{2x}$
$= 2x\sqrt{2x} + 8x\sqrt{2x}$
$= 10x\sqrt{2x}$

5. $2\sqrt{50}-3\sqrt{125}+\sqrt{98}$
$=2\sqrt{25\cdot 2}-3\sqrt{25\cdot 5}+\sqrt{49\cdot 2}$
$=2\sqrt{25}\sqrt{2}-3\sqrt{25}\sqrt{5}+\sqrt{49}\sqrt{2}$
$=2\cdot 5\sqrt{2}-3\cdot 5\sqrt{5}+7\sqrt{2}$
$=10\sqrt{2}-15\sqrt{5}+7\sqrt{2}$
$=17\sqrt{2}-15\sqrt{5}$

7. $\sqrt[3]{16x}-\sqrt[3]{54x}=\sqrt[3]{8\cdot 2x}-\sqrt[3]{27\cdot 2x}$
$=\sqrt[3]{8}\sqrt[3]{2x}-\sqrt[3]{27}\sqrt[3]{2x}$
$=2\sqrt[3]{2x}-3\sqrt[3]{2x}$
$=-\sqrt[3]{2x}$

9. $\sqrt{9b^3}-\sqrt{25b^3}+\sqrt{49b^3}$
$=\sqrt{9b^2\cdot b}-\sqrt{25b^2\cdot b}+\sqrt{49b^2\cdot b}$
$=\sqrt{9b^2}\sqrt{b}-\sqrt{25b^2}\sqrt{b}+\sqrt{49b^2}\sqrt{b}$
$=3b\sqrt{b}-5b\sqrt{b}+7b\sqrt{b}$
$=5b\sqrt{b}$

11. $\frac{5\sqrt{2}}{3}+\frac{2\sqrt{2}}{5}=\frac{5(5\sqrt{2})+3(2\sqrt{2})}{3\cdot 5}$
$=\frac{25\sqrt{2}+6\sqrt{2}}{15}$
$=\frac{31\sqrt{2}}{15}$

13. $\sqrt[3]{\frac{11}{8}}-\frac{\sqrt[3]{11}}{6}=\frac{\sqrt[3]{11}}{\sqrt[3]{8}}-\frac{\sqrt[3]{11}}{6}$
$=\frac{\sqrt[3]{11}}{2}-\frac{\sqrt[3]{11}}{6}$
$=\frac{3\sqrt[3]{11}-\sqrt[3]{11}}{6}$
$=\frac{2\sqrt[3]{11}}{6}$
$=\frac{\sqrt[3]{11}}{3}$

15. $\frac{\sqrt{20x}}{9}+\sqrt{\frac{5x}{9}}=\frac{\sqrt{4\cdot 5x}}{9}+\frac{\sqrt{5x}}{\sqrt{9}}$
$=\frac{\sqrt{4}\sqrt{5x}}{9}+\frac{\sqrt{5x}}{3}$
$=\frac{2\sqrt{5x}+3\sqrt{5x}}{9}$
$=\frac{5\sqrt{5x}}{9}$

17. $7\sqrt{9}-7+\sqrt{3}=7\cdot 3-7+\sqrt{3}$
$=21-7+\sqrt{3}$
$=14+\sqrt{3}$

19. $2+3\sqrt{y^2}-6\sqrt{y^2}+5=7-3\sqrt{y^2}$
$=7-3y$

21. $3\sqrt{108}-2\sqrt{18}-3\sqrt{48}$
$=3\sqrt{36}\sqrt{3}-2\sqrt{9}\sqrt{2}-3\sqrt{16}\sqrt{3}$
$=3\cdot 6\sqrt{3}-2\cdot 3\sqrt{2}-3\cdot 4\sqrt{3}$
$=18\sqrt{3}-6\sqrt{2}-12\sqrt{3}$
$=6\sqrt{3}-6\sqrt{2}$

23. $-5\sqrt[3]{625}+\sqrt[3]{40}=-5\sqrt[3]{125}\sqrt[3]{5}+\sqrt[3]{8}\sqrt[3]{5}$
$=-5\cdot 5\sqrt[3]{5}+2\sqrt[3]{5}$
$=-25\sqrt[3]{5}+2\sqrt[3]{5}$
$=-23\sqrt[3]{5}$

25. $\sqrt{9b^3}-\sqrt{25b^3}+\sqrt{16b^3}$
$=\sqrt{9b^2}\sqrt{b}-\sqrt{25b^2}\sqrt{b}+\sqrt{16b^2}\sqrt{b}$
$=3b\sqrt{b}-5b\sqrt{b}+4b\sqrt{b}$
$=(3-5+4)b\sqrt{b}$
$=2b\sqrt{b}$

27. $5y\sqrt{8y}+2\sqrt{50y^3}$
$=5y\sqrt{4}\sqrt{2y}+2\sqrt{25y^2}\sqrt{2y}$
$=5y\cdot 2\sqrt{2y}+2(5y)\sqrt{2y}$
$=10y\sqrt{2y}+10y\sqrt{2y}$
$=20y\sqrt{2y}$

29. $\sqrt[3]{54xy^3} - 5\sqrt[3]{2xy^3} + y\sqrt[3]{128x}$
$= \sqrt[3]{27y^3}\sqrt[3]{2x} - 5\sqrt[3]{y^3}\sqrt[3]{2x} + y\sqrt[3]{64}\sqrt[3]{2x}$
$= 3y\sqrt[3]{2x} - 5y\sqrt[3]{2x} + y \cdot 4\sqrt[3]{2x}$
$= -2y\sqrt[3]{2x} + 4y\sqrt[3]{2x}$
$= 2y\sqrt[3]{2x}$

31. $6\sqrt[3]{11} + 8\sqrt{11} - 12\sqrt{11}$
$= 6\sqrt[3]{11} + (8-12)\sqrt{11}$
$= 6\sqrt[3]{11} - 4\sqrt{11}$

33. $-2\sqrt[4]{x^7} + 3\sqrt[4]{16x^7}$
$= -2\sqrt[4]{x^4}\sqrt[4]{x^3} + 3\sqrt[4]{16x^4}\sqrt[4]{x^3}$
$= -2x\sqrt[4]{x^3} + 3(2x)\sqrt[4]{x^3}$
$= -2x\sqrt[4]{x^3} + 6x\sqrt[4]{x^3}$
$= 4x\sqrt[4]{x^3}$

35. $\frac{4\sqrt{3}}{3} - \frac{\sqrt{12}}{3} = \frac{4\sqrt{3}}{3} - \frac{\sqrt{4}\sqrt{3}}{3}$
$= \frac{4\sqrt{3} - 2\sqrt{3}}{3}$
$= \frac{2\sqrt{3}}{3}$

37. $\frac{\sqrt[3]{8x^4}}{7} + \frac{3x\sqrt[3]{x}}{7} = \frac{\sqrt[3]{8x^3}\sqrt[3]{x} + 3x\sqrt[3]{x}}{7}$
$= \frac{2x\sqrt[3]{x} + 3x\sqrt[3]{x}}{7}$
$= \frac{5x\sqrt[3]{x}}{7}$

39. $\sqrt{\frac{28}{x^2}} + \sqrt{\frac{7}{4x^2}} = \frac{\sqrt{28}}{\sqrt{x^2}} + \frac{\sqrt{7}}{\sqrt{4x^2}}$
$= \frac{\sqrt{4}\sqrt{7}}{x} + \frac{\sqrt{7}}{2x}$
$= \frac{2\sqrt{7}}{x} + \frac{\sqrt{7}}{2x}$
$= \frac{4\sqrt{7} + \sqrt{7}}{2x}$
$= \frac{5\sqrt{7}}{2x}$

41. $\sqrt[3]{\frac{16}{27}} - \frac{\sqrt[3]{54}}{6} = \frac{\sqrt[3]{16}}{\sqrt[3]{27}} - \frac{\sqrt[3]{27}\sqrt[3]{2}}{6}$
$= \frac{\sqrt[3]{8}\sqrt[3]{2}}{3} - \frac{3\sqrt[3]{2}}{6}$
$= \frac{2(2)\sqrt[3]{2}}{6} - \frac{3\sqrt[3]{2}}{6}$
$= \frac{4\sqrt[3]{2} - 3\sqrt[3]{2}}{6}$
$= \frac{\sqrt[3]{2}}{6}$

43. $-\frac{\sqrt[3]{2x^4}}{9} + \sqrt[3]{\frac{250x^4}{27}}$
$= \frac{-\sqrt[3]{x^3}\sqrt[3]{2x}}{9} + \frac{\sqrt[3]{250x^4}}{\sqrt[3]{27}}$
$= \frac{-x\sqrt[3]{2x}}{9} + \frac{\sqrt[3]{125x^3}\sqrt[3]{2x}}{3}$
$= \frac{-x\sqrt[3]{2x}}{9} + \frac{5x\sqrt[3]{2x}}{3}$
$= \frac{-x\sqrt[3]{2x} + 15x\sqrt[3]{2x}}{9}$
$= \frac{14x\sqrt[3]{2x}}{9}$

45. $P = 2\sqrt{12} + \sqrt{12} + 2\sqrt{27} + 3\sqrt{3}$
$= 2\sqrt{4}\sqrt{3} + \sqrt{4}\sqrt{3} + 2\sqrt{9}\sqrt{3} + 3\sqrt{3}$
$= 2 \cdot 2\sqrt{3} + 2\sqrt{3} + 2 \cdot 3\sqrt{3} + 3\sqrt{3}$
$= (4+2+6+3)\sqrt{3}$
$= 15\sqrt{3}$

The perimeter of the trapezoid is $15\sqrt{3}$ inches.

47. $\sqrt{7}\left(\sqrt{5}+\sqrt{3}\right) = \sqrt{7}\sqrt{5} + \sqrt{7}\sqrt{3}$
$= \sqrt{35} + \sqrt{21}$

49. $\left(\sqrt{5}-\sqrt{2}\right)^2 = \sqrt{5}^2 - 2\sqrt{5}\sqrt{2} + \sqrt{2}^2$
$= 5 - 2\sqrt{10} + 2$
$= 7 - 2\sqrt{10}$

51. $\sqrt{3x}\left(\sqrt{3}-\sqrt{x}\right) = \sqrt{3x}\sqrt{3} - \sqrt{3x}\sqrt{x}$
$= \sqrt{3}\sqrt{x}\sqrt{3} - \sqrt{3}\sqrt{x}\sqrt{x}$
$= 3\sqrt{x} - x\sqrt{3}$

53. $(2\sqrt{x}-5)(3\sqrt{x}+1) = (2\sqrt{x})(3\sqrt{x}) + (2\sqrt{x})1 - 5(3\sqrt{x}) - 5 \cdot 1$
$= 6x + 2\sqrt{x} - 15\sqrt{x} - 5$
$= 6x - 13\sqrt{x} - 5$

55. $(\sqrt[3]{a}-4)(\sqrt[3]{a}+5) = (\sqrt[3]{a})^2 + 5\sqrt[3]{a} - 4\sqrt[3]{a} - 4 \cdot 5$
$= \sqrt[3]{a^2} + \sqrt[3]{a} - 20$

57. $6(\sqrt{2}-2) = 6\sqrt{2} - 6 \cdot 2 = 6\sqrt{2} - 12$

59. $\sqrt{2}\left(\sqrt{2}+x\sqrt{6}\right) = \left(\sqrt{2}\right)^2 + \sqrt{2}\left(x\sqrt{6}\right) = 2 + \sqrt{2}\left(x\sqrt{2}\sqrt{3}\right) = 2 + 2x\sqrt{3}$

61. $\left(2\sqrt{7}+3\sqrt{5}\right)\left(\sqrt{7}-2\sqrt{5}\right) = 2\left(\sqrt{7}\right)^2 - \left(2\sqrt{7}\right)\left(2\sqrt{5}\right) + \left(3\sqrt{5}\right)\sqrt{7} - 3 \cdot 2\left(\sqrt{5}\right)^2$
$= 2 \cdot 7 - 4\sqrt{35} + 3\sqrt{35} - 6 \cdot 5$
$= 14 - \sqrt{35} - 30$
$= -16 - \sqrt{35}$

63. $(\sqrt{x}-y)(\sqrt{x}+y) = \left(\sqrt{x}\right)^2 - y^2 = x - y^2$

65. $\left(\sqrt{3}+x\right)^2 = \left(\sqrt{3}\right)^2 + 2\sqrt{3}x + x^2 = 3 + 2\sqrt{3}x + x^2$

67. $\left(\sqrt{5x}-3\sqrt{2}\right)\left(\sqrt{5x}-3\sqrt{3}\right)=\left(\sqrt{5x}\right)^2-3\sqrt{3}\sqrt{5x}-3\sqrt{2}\sqrt{5x}+\left(3\sqrt{2}\right)\left(3\sqrt{3}\right)$
$=5x-3\sqrt{15x}-3\sqrt{10x}+9\sqrt{6}$

69. $\left(\sqrt[3]{4}+2\right)\left(\sqrt[3]{2}-1\right)=\sqrt[3]{4}\sqrt[3]{2}-\sqrt[3]{4}\cdot 1+2\sqrt[3]{2}-2\cdot 1$
$=\sqrt[3]{8}-\sqrt[3]{4}+2\sqrt[3]{2}-2$
$=2-\sqrt[3]{4}+2\sqrt[3]{2}-2$
$=2\sqrt[3]{2}-\sqrt[3]{4}$

71. $\left(\sqrt[3]{x}+1\right)\left(\sqrt[3]{x}-4\sqrt{x}+7\right)=\sqrt[3]{x}\left(\sqrt[3]{x}-4\sqrt{x}+7\right)+1\left(\sqrt[3]{x}-4\sqrt{x}+7\right)$
$=\left(\sqrt[3]{x}\right)^2-\sqrt[3]{x}(4\sqrt{x})+\sqrt[3]{x}(7)+\sqrt[3]{x}-4\sqrt{x}+7$
$=\left(\sqrt[3]{x}\right)^2-x^{1/3}(4x^{1/2})+8\sqrt[3]{x}-4\sqrt{x}+7$
$=\sqrt[3]{x^2}-4x^{5/6}+8\sqrt[3]{x}-4\sqrt{x}+7$
$=-4\sqrt[6]{x^5}+\sqrt[3]{x^2}+8\sqrt[3]{x}-4\sqrt{x}+7$

73. a. $P=2(3\sqrt{20})+2\sqrt{125}$
$=6\sqrt{4}\sqrt{5}+2\sqrt{25}\sqrt{5}$
$=6(2)\sqrt{5}+2(5)\sqrt{5}$
$=12\sqrt{5}+10\sqrt{5}$
$=22\sqrt{5}$

$22\sqrt{5}$ feet of baseboard should be ordered.

b. $A=3\sqrt{20}\sqrt{125}$
$=3\sqrt{4}\sqrt{5}\sqrt{5}\sqrt{25}$
$=3\cdot 2\cdot 5\cdot 5$
$=150$

The area of the room is 150 square feet.

75. Answers may vary.

77. $\dfrac{2x-14}{2}=\dfrac{2(x-7)}{2}=x-7$

79. $\dfrac{7x-7y}{x^2-y^2}=\dfrac{7(x-y)}{(x-y)(x+y)}=\dfrac{7}{x+y}$

81. $\dfrac{6a^2b-9ab}{3ab}=\dfrac{3ab(2a-3)}{3ab}=2a-3$

83. $\dfrac{-4+2\sqrt{3}}{6}=\dfrac{2(-2+\sqrt{3})}{2\cdot 3}=\dfrac{-2+\sqrt{3}}{3}$

Section 9.5

Mental Math

1. The conjugate of $\sqrt{2}+x$ is $\sqrt{2}-x$.

2. The conjugate of $\sqrt{3}+y$ is $\sqrt{3}-y$.

3. The conjugate of $5-\sqrt{a}$ is $5+\sqrt{a}$.

4. The conjugate of $6-\sqrt{b}$ is $6+\sqrt{b}$.

5. The conjugate of $7\sqrt{5}+8\sqrt{x}$ is $7\sqrt{5}-8\sqrt{x}$.

6. The conjugate of $9\sqrt{2}-6\sqrt{y}$ is $9\sqrt{2}+6\sqrt{y}$.

Exercise Set 9.5

1. $\frac{\sqrt{2}}{\sqrt{7}}=\frac{\sqrt{2}\cdot\sqrt{7}}{\sqrt{7}\cdot\sqrt{7}}=\frac{\sqrt{14}}{7}$

3. $\sqrt{\frac{1}{5}}=\frac{\sqrt{1}}{\sqrt{5}}=\frac{1\cdot\sqrt{5}}{\sqrt{5}\cdot\sqrt{5}}=\frac{\sqrt{5}}{5}$

5. $\sqrt[3]{\frac{3}{4}}=\sqrt[3]{\frac{3}{4}\cdot\frac{16}{16}}$
$=\sqrt[3]{\frac{48}{64}}$
$=\frac{\sqrt[3]{8\cdot 6}}{4}$
$=\frac{2\sqrt[3]{6}}{4}$
$=\frac{\sqrt[3]{6}}{2}$

7. $\frac{4}{\sqrt[3]{3}}=\frac{4}{\sqrt[3]{3}}\cdot\frac{\sqrt[3]{9}}{\sqrt[3]{9}}=\frac{4\sqrt[3]{9}}{\sqrt[3]{27}}=\frac{4\sqrt[3]{9}}{3}$

9. $\frac{3}{\sqrt{8x}}=\frac{3}{\sqrt{8x}}\cdot\frac{\sqrt{2x}}{\sqrt{2x}}=\frac{3\sqrt{2x}}{\sqrt{16x^2}}=\frac{3\sqrt{2x}}{4x}$

11. $\frac{3}{\sqrt[3]{4x^2}}=\frac{3}{\sqrt[3]{4x^2}}\cdot\frac{\sqrt[3]{2x}}{\sqrt[3]{2x}}=\frac{3\sqrt[3]{2x}}{\sqrt[3]{8x^3}}=\frac{3\sqrt[3]{2x}}{2x}$

13. $\sqrt{\frac{4}{x}}=\sqrt{\frac{4}{x}\cdot\frac{x}{x}}=\frac{\sqrt{4x}}{\sqrt{x^2}}=\frac{2\sqrt{x}}{x}$

15. $\frac{9}{\sqrt{3a}}=\frac{9}{\sqrt{3a}}\cdot\frac{\sqrt{3a}}{\sqrt{3a}}=\frac{9\sqrt{3a}}{3a}=\frac{3\sqrt{3a}}{a}$

17. $\frac{3}{\sqrt[3]{2}}=\frac{3}{\sqrt[3]{2}}\cdot\frac{\sqrt[3]{4}}{\sqrt[3]{4}}=\frac{3\sqrt[3]{4}}{\sqrt[3]{8}}=\frac{3\sqrt[3]{4}}{2}$

19. $\frac{2\sqrt{3}}{\sqrt{7}}=\frac{2\sqrt{3}}{\sqrt{7}}\cdot\frac{\sqrt{7}}{\sqrt{7}}=\frac{2\sqrt{21}}{7}$

21. $\sqrt{\frac{2x}{5y}}=\frac{\sqrt{2x}}{\sqrt{5y}}=\frac{\sqrt{2x}\cdot\sqrt{5y}}{\sqrt{5y}\cdot\sqrt{5y}}=\frac{\sqrt{10xy}}{5y}$

23. $\sqrt[4]{\frac{81}{8}}=\sqrt[4]{\frac{81\cdot 2}{8\cdot 2}}=\sqrt[4]{\frac{3^4\cdot 2}{2^4}}=\frac{3\sqrt[4]{2}}{2}$

25. $\sqrt[4]{\frac{16}{9x^7}}=\frac{\sqrt[4]{16}}{\sqrt[4]{9x^7}}$
$=\frac{2}{x\sqrt[4]{9x^3}}$
$=\frac{2\cdot\sqrt[4]{9x}}{x\sqrt[4]{9x^3}\cdot\sqrt[4]{9x}}$
$=\frac{2\sqrt[4]{9x}}{3x^2}$

27. $\frac{5a}{\sqrt[5]{8a^9b^{11}}}=\frac{5a}{ab^2\sqrt[5]{8a^4b}}$
$=\frac{5a\sqrt[5]{4ab^4}}{ab^2\sqrt[5]{8a^4b}\cdot\sqrt[5]{4ab^4}}$
$=\frac{5a\sqrt[5]{4ab^4}}{2a^2b^3}$
$=\frac{5\sqrt[5]{4ab^4}}{2ab^3}$

29. $\sqrt{\frac{5}{3}} = \frac{\sqrt{5}}{\sqrt{3}} \cdot \frac{\sqrt{5}}{\sqrt{5}} = \frac{\sqrt{25}}{\sqrt{15}} = \frac{5}{\sqrt{15}}$

31. $\sqrt{\frac{18}{5}} = \frac{\sqrt{18}}{\sqrt{5}}$
$= \frac{\sqrt{9}\sqrt{2}}{\sqrt{5}}$
$= \frac{3\sqrt{2}}{\sqrt{5}}$
$= \frac{3\sqrt{2}}{\sqrt{5}} \cdot \frac{\sqrt{2}}{\sqrt{2}}$
$= \frac{3\sqrt{4}}{\sqrt{10}}$
$= \frac{3 \cdot 2}{\sqrt{10}}$
$= \frac{6}{\sqrt{10}}$

33. $\frac{\sqrt{4x}}{7} = \frac{2\sqrt{x}}{7} = \frac{2\sqrt{x}}{7} \cdot \frac{\sqrt{x}}{\sqrt{x}} = \frac{2x}{7\sqrt{x}}$

35. $\frac{\sqrt[3]{5y^2}}{\sqrt[3]{4x}} = \frac{\sqrt[3]{5y^2}}{\sqrt[3]{4x}} \cdot \frac{\sqrt[3]{5^2 y}}{\sqrt[3]{5^2 y}}$
$= \frac{\sqrt[3]{5^3 y^3}}{\sqrt[3]{4(5^2)xy}}$
$= \frac{5y}{\sqrt[3]{100xy}}$

37. $\sqrt{\frac{2}{5}} = \frac{\sqrt{2}}{\sqrt{5}} \cdot \frac{\sqrt{2}}{\sqrt{2}} = \frac{\sqrt{4}}{\sqrt{10}} = \frac{2}{\sqrt{10}}$

39. $\frac{\sqrt{2x}}{11} = \frac{\sqrt{2x}}{11} \cdot \frac{\sqrt{2x}}{\sqrt{2x}} = \frac{\sqrt{4x^2}}{11\sqrt{2x}} = \frac{2x}{11\sqrt{2x}}$

41. $\sqrt[3]{\frac{7}{8}} = \frac{\sqrt[3]{7}}{\sqrt[3]{8}} = \frac{\sqrt[3]{7}}{2} \cdot \frac{\sqrt[3]{7^2}}{\sqrt[3]{7^2}} = \frac{\sqrt[3]{7^3}}{2\sqrt[3]{7^2}} = \frac{7}{2\sqrt[3]{49}}$

43. $\frac{\sqrt[3]{3x^5}}{10} = \frac{\sqrt[3]{3x^5}}{10} \cdot \frac{\sqrt[3]{3^2 x}}{\sqrt[3]{3^2 x}}$
$= \frac{\sqrt[3]{3^3 x^6}}{10\sqrt[3]{3^2 x}}$
$= \frac{3x^2}{10\sqrt[3]{9x}}$

45. $\sqrt{\frac{18x^4y^6}{3z}} = \sqrt{\frac{6x^4y^6}{z}}$
$= \frac{x^2y^3\sqrt{6}}{\sqrt{z}}$
$= \frac{x^2y^3\sqrt{6}}{\sqrt{z}} \cdot \frac{\sqrt{6}}{\sqrt{6}}$
$= \frac{x^2y^3\sqrt{36}}{\sqrt{6z}}$
$= \frac{6x^2y^3}{\sqrt{6z}}$

47. Answers may vary.

49. $\frac{6}{2-\sqrt{7}} = \frac{6}{2-\sqrt{7}} \cdot \frac{2+\sqrt{7}}{2+\sqrt{7}}$
$= \frac{6(2+\sqrt{7})}{2^2 - (\sqrt{7})^2}$
$= \frac{6(2+\sqrt{7})}{4-7}$
$= \frac{6(2+\sqrt{7})}{-3}$
$= -2(2+\sqrt{7})$

51. $\frac{-7}{\sqrt{x}-3}=\frac{-7}{\sqrt{x}-3}\cdot\frac{\sqrt{x}+3}{\sqrt{x}+3}$
$=\frac{-7(\sqrt{x}+3)}{(\sqrt{x})^2-3^2}$
$=\frac{-7(\sqrt{x}+3)}{x-9}$
$=\frac{7(3+\sqrt{x})}{9-x}$

53. $\frac{\sqrt{2}-\sqrt{3}}{\sqrt{2}+\sqrt{3}}=\frac{\sqrt{2}-\sqrt{3}}{\sqrt{2}+\sqrt{3}}\cdot\frac{\sqrt{2}-\sqrt{3}}{\sqrt{2}-\sqrt{3}}$
$=\frac{(\sqrt{2})^2-2\sqrt{2}\sqrt{3}+(\sqrt{3})^2}{(\sqrt{2})^2-(\sqrt{3})^2}$
$=\frac{2-2\sqrt{6}+3}{2-3}$
$=\frac{5-2\sqrt{6}}{-1}$
$=-5+2\sqrt{6}$

55. $\frac{\sqrt{a}+1}{2\sqrt{a}-\sqrt{b}}$
$=\frac{\sqrt{a}+1}{2\sqrt{a}-\sqrt{b}}\cdot\frac{2\sqrt{a}+\sqrt{b}}{2\sqrt{a}+\sqrt{b}}$
$=\frac{\sqrt{a}(2\sqrt{a})+\sqrt{a}\sqrt{b}+1(2\sqrt{a})+1\sqrt{b}}{(2\sqrt{a})^2-(\sqrt{b})^2}$
$=\frac{2a+2\sqrt{a}+\sqrt{ab}+\sqrt{b}}{4a-b}$

57. $\frac{8}{1+\sqrt{10}}=\frac{8}{1+\sqrt{10}}\cdot\frac{1-\sqrt{10}}{1-\sqrt{10}}$
$=\frac{8(1-\sqrt{10})}{1^2-(\sqrt{10})^2}$
$=\frac{8(1-\sqrt{10})}{1-10}$
$=\frac{8(1-\sqrt{10})}{-9}$
$=-\frac{8(1-\sqrt{10})}{9}$

59. $\frac{\sqrt{x}}{\sqrt{x}+\sqrt{y}}=\frac{\sqrt{x}}{\sqrt{x}+\sqrt{y}}\cdot\frac{\sqrt{x}-\sqrt{y}}{\sqrt{x}-\sqrt{y}}$
$=\frac{\sqrt{x}\sqrt{x}-\sqrt{x}\sqrt{y}}{(\sqrt{x})^2-(\sqrt{y})^2}$
$=\frac{x-\sqrt{xy}}{x-y}$

61. $\frac{2\sqrt{3}+\sqrt{6}}{4\sqrt{3}-\sqrt{6}}$
$=\frac{2\sqrt{3}+\sqrt{6}}{4\sqrt{3}-\sqrt{6}}\cdot\frac{4\sqrt{3}+\sqrt{6}}{4\sqrt{3}+\sqrt{6}}$
$=\frac{8(\sqrt{3})^2+2\sqrt{3}\sqrt{6}+4\sqrt{3}\sqrt{6}+(\sqrt{6})^2}{(4\sqrt{3})^2-(\sqrt{6})^2}$
$=\frac{8\cdot 3+2\sqrt{18}+4\sqrt{18}+6}{16\cdot 3-6}$
$=\frac{24+6\sqrt{18}+6}{48-6}$
$=\frac{30+6\sqrt{18}}{42}$
$=\frac{30+18\sqrt{2}}{42}$
$=\frac{6(5+3\sqrt{2})}{6\cdot 7}$
$=\frac{5+3\sqrt{2}}{7}$

63. $$\frac{2-\sqrt{11}}{6}\cdot\frac{2+\sqrt{11}}{2+\sqrt{11}}=\frac{(2-\sqrt{11})(2+\sqrt{11})}{6(2+\sqrt{11})}=\frac{4-\sqrt{121}}{12+6\sqrt{11}}=\frac{4-11}{12+6\sqrt{11}}=\frac{-7}{12+6\sqrt{11}}$$

65. $$\frac{2-\sqrt{7}}{-5}\cdot\frac{2+\sqrt{7}}{2+\sqrt{7}}=\frac{(2-\sqrt{7})(2+\sqrt{7})}{-5(2+\sqrt{7})}=\frac{4-\sqrt{49}}{-10-5\sqrt{7}}=\frac{4-7}{-10-5\sqrt{7}}=\frac{-3}{-10-5\sqrt{7}}=\frac{-1(3)}{-1(10+5\sqrt{7})}=\frac{3}{10+5\sqrt{7}}$$

67. $$\frac{\sqrt{x}+3}{\sqrt{x}}\cdot\frac{\sqrt{x}-3}{\sqrt{x}-3}=\frac{(\sqrt{x}+3)(\sqrt{x}-3)}{\sqrt{x}(\sqrt{x}-3)}=\frac{\left(\sqrt{x}\right)^2-9}{\left(\sqrt{x}\right)^2-3\sqrt{x}}=\frac{x-9}{x-3\sqrt{x}}$$

69. $$\frac{\sqrt{2}-1}{\sqrt{2}+1}\cdot\frac{\sqrt{2}+1}{\sqrt{2}+1}=\frac{(\sqrt{2}-1)(\sqrt{2}+1)}{(\sqrt{2}+1)(\sqrt{2}+1)}=\frac{2-1}{2+2\sqrt{2}+1}=\frac{1}{3+2\sqrt{2}}$$

71. $$\frac{\sqrt{x}+1}{\sqrt{x}-1}\cdot\frac{\sqrt{x}-1}{\sqrt{x}-1}=\frac{(\sqrt{x}+1)(\sqrt{x}-1)}{(\sqrt{x}-1)(\sqrt{x}-1)}=\frac{\left(\sqrt{x}\right)^2-1}{\left(\sqrt{x}\right)^2-2\sqrt{x}+1}=\frac{x-1}{x-2\sqrt{x}+1}$$

73. $$r=\sqrt{\frac{A}{4\pi}}=\frac{\sqrt{A}}{\sqrt{4\pi}}=\frac{\sqrt{A}}{2\sqrt{\pi}}=\frac{\sqrt{A}\sqrt{\pi}}{2\sqrt{\pi}\sqrt{\pi}}=\frac{\sqrt{A\pi}}{2\pi}$$

75. Answers may vary.

77. $2x-7=3(x-4)$
$2x-7=3x-12$
$12-7=3x-2x$
$5=x$
The solution is 5.

79. $(x-6)(2x+1)=0$
$x-6=0$ or $2x+1=0$
$x=6$ or $x=-\frac{1}{2}$
The solutions are $-\frac{1}{2}$ and 6.

81. $x^2-8x=-12$
$x^2-8x+12=0$
$(x-6)(x-2)=0$
$x-6=0$ or $x-2=0$
$x=6$ or $x=2$
The solutions are 2 and 6.

Section 9.6

Graphing Calculator Explorations

1.

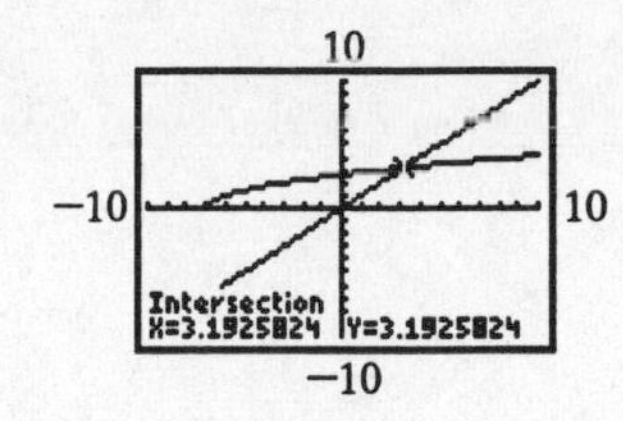

The solution is 3.19.

3.

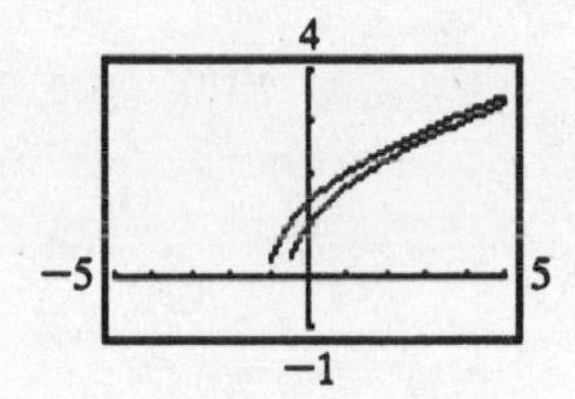

There is no solution.

5.

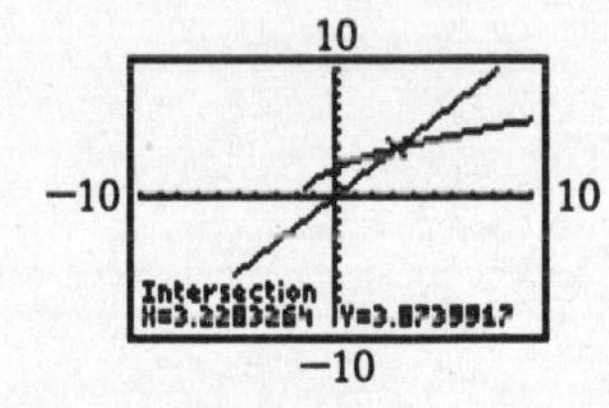

The solution is 3.23.

Exercise Set 9.6

1. $\sqrt{2x} = 4$

$$\left(\sqrt{2x}\right)^2 = 4^2$$
$$2x = 16$$
$$x = 8$$

The solution is 8.

3. $\sqrt{x-3} = 2$

$$\left(\sqrt{x-3}\right)^2 = 2^2$$
$$x - 3 = 4$$
$$x = 7$$

The solution is 7.

5. $\sqrt{2x} = -4$

$$\left(\sqrt{2x}\right)^2 = (-4)^2$$
$$2x = 16$$
$$x = 8$$

We discard 8 as extraneous, so there is no solution.

7. $\sqrt{4x-3} - 5 = 0$

$$\sqrt{4x-3} = 5$$
$$\left(\sqrt{4x-3}\right)^2 = 5^2$$
$$4x - 3 = 25$$
$$4x = 28$$
$$x = 7$$

The solution is 7.

9. $\sqrt{2x-3} - 2 = 1$

$$\sqrt{2x-3} = 3$$
$$\left(\sqrt{2x-3}\right)^2 = 3^2$$
$$2x - 3 = 9$$
$$2x = 12$$
$$x = 6$$

The solution is 6.

11. $\sqrt[3]{6x} = -3$

$$\left(\sqrt[3]{6x}\right)^3 = (-3)^3$$
$$6x = -27$$
$$x = \frac{-27}{6} = -\frac{9}{2}$$

The solution is $-\frac{9}{2}$.

13. $\sqrt[3]{x-2} - 3 = 0$

$$\sqrt[3]{x-2} = 3$$
$$\left(\sqrt[3]{x-2}\right)^3 = 3^3$$
$$x - 2 = 27$$
$$x = 29$$

The solution is 29.

15.
$$\sqrt{13-x}=x-1$$
$$\left(\sqrt{13-x}\right)^2=(x-1)^2$$
$$13-x=x^2-2(x)(1)+1^2$$
$$13-x=x^2-2x+1$$
$$0=x^2-x-12$$
$$0=(x-4)(x+3)$$
$$x-4=0 \quad \text{or} \quad x+3=0$$
$$x=4 \quad \text{or} \quad x=-3$$
We discard –3 as extraneous, leaving 4 as the only solution.

17.
$$x-\sqrt{4-3x}=-8$$
$$x+8=\sqrt{4-3x}$$
$$(x+8)^2=\left(\sqrt{4-3x}\right)^2$$
$$x^2+2(x)(8)+8^2=4-3x$$
$$x^2+16x+64=4-3x$$
$$x^2+19x+60=0$$
$$(x+4)(x+15)=0$$
$$x+4=0 \quad \text{or} \quad x+15=0$$
$$x=-4 \quad \text{or} \quad x=-15$$
We discard –15 as extraneous, leaving –4 as the only solution.

19.
$$\sqrt{y+5}=2-\sqrt{y-4}$$
$$\left(\sqrt{y+5}\right)^2=\left(2-\sqrt{y-4}\right)^2$$
$$y+5=2^2-2(2)\sqrt{y-4}+\left(\sqrt{y-4}\right)^2$$
$$y+5=4-4\sqrt{y-4}+y-4$$
$$5=-4\sqrt{y-4}$$
$$5^2=\left(-4\sqrt{y-4}\right)^2$$
$$25=(-4)^2\left(\sqrt{y-4}\right)^2$$
$$25=16(y-4)$$
$$\frac{25}{16}=y-4$$
$$y=4+\frac{25}{16}=\frac{64}{16}+\frac{25}{16}=\frac{89}{16}=5\frac{9}{16}$$
which we discard as extraneous. There is no solution.

21. $\sqrt{x-3}+\sqrt{x+2}=5$

$$\sqrt{x-3}=5-\sqrt{x+2}$$
$$\left(\sqrt{x-3}\right)^2=\left(5-\sqrt{x+2}\right)^2$$
$$x-3=5^2-2(5)\sqrt{x+2}+\left(\sqrt{x+2}\right)^2$$
$$x-3=25-10\sqrt{x+2}+x+2$$
$$-3=27-10\sqrt{x+2}$$
$$-30=-10\sqrt{x+2}$$
$$3=\sqrt{x+2}$$
$$3^2=\left(\sqrt{x+2}\right)^2$$
$$9=x+2$$
$$7=x$$

The solution is 7.

23. $\sqrt{3x-2}=5$

$$\left(\sqrt{3x-2}\right)^2=5^2$$
$$3x-2=25$$
$$3x=27$$
$$x=9$$

The solution is 9.

25. $-\sqrt{2x}+4=-6$

$$10=\sqrt{2x}$$
$$10^2=\left(\sqrt{2x}\right)^2$$
$$100=2x$$
$$x=50$$

The solution is 50.

27. $\sqrt{3x+1}+2=0$

$$\sqrt{3x+1}=-2$$
$$\left(\sqrt{3x+1}\right)^2=(-2)^2$$
$$3x+1=4$$
$$3x=3$$
$$x=1$$

We discard 1 as extraneous, so there is no solution.

29. $\sqrt[4]{4x+1}-2=0$

$$\sqrt[4]{4x+1}=2$$
$$\left(\sqrt[4]{4x+1}\right)^4=2^4$$
$$4x+1=16$$
$$4x=15$$
$$x=\frac{15}{4}$$

The solution is $\frac{15}{4}$.

31. $\sqrt{4x-3}=7$

$$\left(\sqrt{4x-3}\right)^2=7^2$$
$$4x-3=49$$
$$4x=52$$
$$x=13$$

The solution is 13.

33. $\sqrt[3]{6x-3}-3=0$

$$\sqrt[3]{6x-3}=3$$
$$\left(\sqrt[3]{6x-3}\right)^3=3^3$$
$$6x-3=27$$
$$6x=30$$
$$x=5$$

The solution is 5.

35. $\sqrt[3]{2x-3}-2=-5$

$$\sqrt[3]{2x-3}=-3$$
$$\left(\sqrt[3]{2x-3}\right)^3=(-3)^3$$
$$2x-3=-27$$
$$2x=-24$$
$$x=-12$$

The solution is –12.

37. $\sqrt{x+4}=\sqrt{2x-5}$

$$\left(\sqrt{x+4}\right)^2=\left(\sqrt{2x-5}\right)^2$$
$$x+4=2x-5$$
$$9=x$$

The solution is 9.

39. $x-\sqrt{1-x}=-5$

$$x+5=\sqrt{1-x}$$
$$(x+5)^2=\left(\sqrt{1-x}\right)^2$$
$$x^2+2(5)(x)+5^2=1-x$$
$$x^2+10x+25=1-x$$
$$x^2+11x+24=0$$
$$(x+3)(x+8)=0$$
$$x+3=0 \quad \text{or} \quad x+8=0$$
$$x=-3 \quad \text{or} \quad x=-8$$

We discard –8 as extraneous, leaving –3 as the only solution.

41. $\sqrt[3]{-6x-1}=\sqrt[3]{-2x-5}$

$$\left(\sqrt[3]{-6x-1}\right)^3=\left(\sqrt[3]{-2x-5}\right)^3$$
$$-6x-1=-2x-5$$
$$4=4x$$
$$x=1$$

The solution is 1.

43.

$$\sqrt{5x-1}-\sqrt{x}+2=3$$
$$\sqrt{5x-1}=\sqrt{x}+1$$
$$\left(\sqrt{5x-1}\right)^2=\left(\sqrt{x}+1\right)^2$$
$$5x-1=\left(\sqrt{x}\right)^2+2\left(\sqrt{x}\right)(1)+1^2$$
$$5x-1=x+2\sqrt{x}+1$$
$$4x-2=2\sqrt{x}$$
$$2x-1=\sqrt{x}$$
$$(2x-1)^2=\left(\sqrt{x}\right)^2$$
$$(2x)^2-2(2x)(1)+1^2=x$$
$$4x^2-4x+1=x$$
$$4x^2-5x+1=0$$
$$(4x-1)(x-1)=0$$
$$4x-1=0 \quad \text{or} \quad x-1=0$$
$$4x=1 \quad \text{or} \quad x=1$$
$$x=\frac{1}{4}$$

We discard $\frac{1}{4}$ as extraneous, leaving 1 as the only solution.

45.

$$\sqrt{2x-1}=\sqrt{1-2x}$$
$$\left(\sqrt{2x-1}\right)^2=\left(\sqrt{1-2x}\right)^2$$
$$2x-1=1-2x$$
$$4x=2$$
$$x=\frac{2}{4}=\frac{1}{2}$$

The solution is $\frac{1}{2}$.

47.

$$\sqrt{3x+4}-1=\sqrt{2x+1}$$
$$\left(\sqrt{3x+4}-1\right)^2=\left(\sqrt{2x+1}\right)^2$$
$$\left(\sqrt{3x+4}\right)^2-2\left(\sqrt{3x+4}\right)(1)+1^2=2x+1$$
$$3x+4-2\sqrt{3x+4}+1=2x+1$$
$$x+4=2\sqrt{3x+4}$$
$$(x+4)^2=\left(2\sqrt{3x+4}\right)^2$$
$$x^2+2(x)(4)+4^2=4(3x+4)$$
$$x^2+8x+16=12x+16$$
$$x^2-4x=0$$
$$x(x-4)=0$$
$$x=0 \quad \text{or} \quad x=4$$

The solutions are 0 and 4.

49. $\sqrt{y+3}-\sqrt{y-3}=1$

$$\sqrt{y+3}=1+\sqrt{y-3}$$
$$\left(\sqrt{y+3}\right)^2=\left(1+\sqrt{y-3}\right)^2$$
$$y+3=1^2+2(1)\sqrt{y-3}+\left(\sqrt{y-3}\right)^2$$
$$y+3=1+2\sqrt{y-3}+y-3$$
$$5=2\sqrt{y-3}$$
$$5^2=\left(2\sqrt{y-3}\right)^2$$
$$25=4(y-3)$$
$$\frac{25}{4}=y-3$$
$$\frac{25}{4}+\frac{12}{4}=y$$
$$\frac{37}{4}=y$$

The solution is $\frac{37}{4}$.

51. Answers may vary.

53. Let c = the length of the hypotenuse of the right triangle. By the Pythagorean theorem,

$c^2 = 6^2 + 3^2 = 36 + 9 = 45.$

Thus, $c = \sqrt{45} = \sqrt{9}\sqrt{5} = 3\sqrt{5}.$

The length of the unknown side is $3\sqrt{5}$ feet.

55. Let b = the length of the unknown leg of the right triangle. By the Pythagorean theorem,

$$7^2 = 3^2 + b^2$$
$$49 = 9 + b^2$$
$$b^2 = 40.$$

Thus, $b = \sqrt{40} = \sqrt{4}\sqrt{10} = 2\sqrt{10}.$

The length of the unknown side is $2\sqrt{10}$ meters.

57. Let b = the length of the unknown leg of the right triangle. By the Pythagorean theorem,

$$\left(11\sqrt{5}\right)^2 = 9^2 + b^2$$
$$605 = 81 + b^2$$
$$b^2 = 524.$$

Thus, $b = \sqrt{524} = \sqrt{4}\sqrt{131} = 2\sqrt{131} \approx 22.9.$

The length of the unknown side is $2\sqrt{131}$ or about 22.9 meters.

59. Let c = the length of the hypotenuse of the right triangle. By the Pythagorean theorem,

$c^2 = 7^2 + (7.2)^2 = 100.84.$

So, $c = \sqrt{100.84} \approx 10.0$

The length of the unknown side is $\sqrt{100.84}$ or about 10.0 millimeters.

61. Let c = the length of the hypotenuse of the right triangle as shown in the figure in the text. By the Pythagorean theorem,

$c^2 = 8^2 + 15^2 = 64 + 225 = 289.$

So, $c = \sqrt{289} = 17$ feet.

Thus, 17 feet of cable are needed.

63. Let c = the length of the hypotenuse of the right triangle as shown in the figure in the text. By the Pythagorean theorem,

$c^2 = 12^2 + 5^2 = 144 + 25 = 169.$

So, $x = \sqrt{169} = 13$ feet.

Thus, a 13-foot ladder is needed.

65.

$$r = \sqrt{\frac{A}{4\pi}}$$
$$1080 = \sqrt{\frac{A}{4\pi}}$$
$$(1080)^2 = \left(\sqrt{\frac{A}{4\pi}}\right)^2$$
$$1{,}166{,}400 = \frac{A}{4\pi}$$
$$14{,}657{,}415 \approx A$$

The surface area is 14,657,415 square miles.

67.

$$v = \sqrt{2gh}$$
$$80 = \sqrt{2(32)h}$$
$$(80)^2 = \left(\sqrt{64h}\right)^2$$
$$6400 = 64 \cdot h$$
$$100 = h$$

The object fell 100 feet.

69.

$$\sqrt{\sqrt{x+3} + \sqrt{x}} = \sqrt{3}$$
$$\left(\sqrt{\sqrt{x+3} + \sqrt{x}}\right)^2 = \left(\sqrt{3}\right)^2$$
$$\sqrt{x+3} + \sqrt{x} = 3$$
$$\sqrt{x+3} = 3 - \sqrt{x}$$
$$\left(\sqrt{x+3}\right)^2 = \left(3 - \sqrt{x}\right)^2$$
$$x + 3 = 3^2 - 2(3)\sqrt{x} + \left(\sqrt{x}\right)^2$$
$$x + 3 = 9 - 6\sqrt{x} + x$$
$$6\sqrt{x} = 6$$
$$\sqrt{x} = 1$$
$$x = 1$$

The solution is 1.

71. $C(x) = 80\sqrt[3]{x} + 500$
We replace $C(x)$ with 1620 and solve for x.

$$80\sqrt[3]{x} + 500 = 1620$$
$$80\sqrt[3]{x} = 1120$$
$$\sqrt[3]{x} = 14$$
$$\left(\sqrt[3]{x}\right)^3 = 14^3$$
$$x = 2744$$

Thus, the company needs to make fewer than 2744 deliveries per day, or 2743 deliveries.

73. Answers may vary.

75. Not a function; the vertical line $x = 1$ intersects the graph in more than one point.

77. Not a function; the graph itself is a vertical line that intersects the graph in more than one point.

79. Not a function; the vertical line $x = 0$ intersects the graph in more than one point.

81.
$$\frac{\frac{1}{y}+\frac{4}{5}}{\frac{-3}{20}} = \frac{20y\left(\frac{1}{y}+\frac{4}{5}\right)}{20y\left(\frac{-3}{20}\right)}$$
$$= \frac{20y\left(\frac{1}{y}\right)+20y\left(\frac{4}{5}\right)}{20y\left(\frac{-3}{20}\right)}$$
$$= \frac{20+16y}{-3y}$$
$$= -\frac{20+16y}{3y}$$

83.
$$\frac{\frac{1}{y}+\frac{1}{x}}{\frac{1}{y}-\frac{1}{x}} = \frac{xy\left(\frac{1}{y}+\frac{1}{x}\right)}{xy\left(\frac{1}{y}-\frac{1}{x}\right)}$$
$$= \frac{xy\left(\frac{1}{y}\right)+xy\left(\frac{1}{x}\right)}{xy\left(\frac{1}{y}\right)-xy\left(\frac{1}{x}\right)}$$
$$= \frac{x+y}{x-y}$$

85. $\sqrt{(x^2-x)+7} = 2(x^2-x)-1$

Let $t = x^2 - x$.

$$\sqrt{t+7} = 2t-1$$
$$(\sqrt{t+7})^2 = (2t-1)^2$$
$$t+7 = 4t^2-4t+1$$
$$0 = 4t^2-5t-6$$
$$0 = (4t+3)(t-2)$$

$4t+3=0$ or $t-2=0$
$4t=-3$ or $t=2$
$t=-\frac{3}{4}$ or $t=2$

$t=2$: Replace t with x^2-x.

$$x^2-x=2$$
$$x^2-x-2=0$$
$$(x-2)(x+1)=0$$

$x-2=0$ or $x+1=0$
$x=2$ or $x=-1$

$t=-\frac{3}{4}$: Substitute $-\frac{3}{4}$ for t in the equation $\sqrt{t+7}=2t-1$, and you will find $-\frac{3}{4}$ to be an extraneous solution.

$$\sqrt{-\frac{3}{4}+7} = 2\left(-\frac{3}{4}\right)-1$$
$$\sqrt{-\frac{3}{4}+\frac{28}{4}} = -\frac{3}{2}-\frac{2}{2}$$
$$\sqrt{\frac{25}{4}} = -\frac{5}{2}$$
$$\frac{5}{2} = -\frac{5}{2} \quad \text{False}$$

Thus, $t=-\frac{3}{4}$ does not lead to any solutions of the original equation. The solutions are -1 and 2.

87. $x^2+6x=4\sqrt{x^2+6x}$

Let $t=x^2+6x$.

$$t=4\sqrt{t}$$
$$(t)^2=(4\sqrt{t})^2$$
$$t^2=16t$$
$$t^2-16t=0$$
$$t(t-16)=0$$
$$t=0 \quad \text{or} \quad t-16=0$$
$$t=16$$

$t=0$: Replace t with x^2+6x.

$$x^2+6x=0$$
$$x(x+6)=0$$
$$x=0 \text{ or } x+6=0$$
$$x=-6$$

$t=16$: Replace t with x^2+6x.

$$x^2+6x=16$$
$$x^2+6x-16=0$$
$$(x+8)(x-2)=0$$
$$x+8=0 \quad \text{or} \quad x-2=0$$
$$x=-8 \quad \text{or} \quad x=2$$

The solutions are −8, −6, 0, and 2.

Section 9.7

Mental Math

1. $\sqrt{-81}=\sqrt{-1\cdot 81}=\sqrt{-1}\cdot\sqrt{81}=9i$

2. $\sqrt{-49}=\sqrt{-1\cdot 49}=\sqrt{-1}\cdot\sqrt{49}=7i$

3. $\sqrt{-7}=\sqrt{-1\cdot 7}=\sqrt{-1}\cdot\sqrt{7}=i\sqrt{7}$

4. $\sqrt{-3}=\sqrt{-1\cdot 3}=\sqrt{-1}\cdot\sqrt{3}=i\sqrt{3}$

5. $-\sqrt{16}=-4$

6. $-\sqrt{4}=-2$

7. $\sqrt{-64}=\sqrt{-1\cdot 64}=\sqrt{-1}\cdot\sqrt{64}=8i$

8. $\sqrt{-100}=\sqrt{-1\cdot 100}=\sqrt{-1}\cdot\sqrt{100}=10i$

Exercise Set 9.7

1. $\sqrt{-24}=\sqrt{4}\sqrt{6}\sqrt{-1}=2i\sqrt{6}$

3. $-\sqrt{-36}=-\sqrt{36}\sqrt{-1}=-6i$

5. $8\sqrt{-63}=8\sqrt{9}\sqrt{7}\sqrt{-1}$
$$=8\cdot 3i\sqrt{7}$$
$$=24i\sqrt{7}$$

7. $-\sqrt{54}=-\sqrt{9}\sqrt{6}=-3\sqrt{6}$

9. $\sqrt{-2}\cdot\sqrt{-7}=(i\sqrt{2})(i\sqrt{7})$
$$=i^2\sqrt{14}$$
$$=(-1)\sqrt{14}$$
$$=-\sqrt{14}$$

11. $\sqrt{-5}\cdot\sqrt{-10}=(i\sqrt{5})(i\sqrt{10})$
$$=i^2\sqrt{5}\sqrt{5}\sqrt{2}$$
$$=(-1)5\sqrt{2}$$
$$=-5\sqrt{2}$$

13. $\sqrt{16}\cdot\sqrt{-1}=4i$

15. $\dfrac{\sqrt{-9}}{\sqrt{3}}=\dfrac{\sqrt{9}\cdot\sqrt{-1}}{\sqrt{3}}=i\sqrt{\dfrac{9}{3}}=i\sqrt{3}$

17. $\dfrac{\sqrt{-80}}{\sqrt{-10}}=\dfrac{i\sqrt{80}}{i\sqrt{10}}$
$$=\sqrt{\frac{80}{10}}$$
$$=\sqrt{8}$$
$$=\sqrt{4}\sqrt{2}$$
$$=2\sqrt{2}$$

19. $(4-7i)+(2+3i)=(4+2)+(-7+3)i$
$$=6-4i$$

21. $(6+5i)-(8-i)=(6-8)+[5-(-1)]i$
$$=-2+6i$$

23. $6-(8+4i)=6-8-4i=-2-4i$

25. $6i(2-3i) = 6i(2) - 6i(3i)$
$= 12i - 18i^2$
$= 12i - 18(-1)$
$= 18 + 12i$

27. $(\sqrt{3}+2i)(\sqrt{3}-2i) = (\sqrt{3})^2 - (2i)^2$
$= 3 - 4i^2$
$= 3 - 4(-1)$
$= 3 + 4$
$= 7$

29. $(4-2i)^2 = 4^2 - 2(4)(2i) + (2i)^2$
$= 16 - 16i + 4i^2$
$= 16 - 16i + 4(-1)$
$= 16 - 4 - 16i$
$= 12 - 16i$

31. $\frac{4}{i} = \frac{4}{i} \cdot \frac{-i}{-i} = \frac{-4i}{-i^2} = \frac{-4i}{-(-1)} = \frac{-4i}{1} = -4i$

33. $\frac{7}{4+3i} = \frac{7}{4+3i} \cdot \frac{4-3i}{4-3i}$
$= \frac{7(4-3i)}{4^2-(3i)^2}$
$= \frac{28-21i}{4^2-9i^2}$
$= \frac{28-21i}{16+9}$
$= \frac{28-21i}{25}$
$= \frac{28}{25} - \frac{21}{25}i$

35. $\frac{3+5i}{1+i} = \frac{3+5i}{1+i} \cdot \frac{1-i}{1-i}$
$= \frac{3-3i+5i-5i^2}{1^2-i^2}$
$= \frac{3+2i-5(-1)}{1+1}$
$= \frac{3+5+2i}{2}$
$= \frac{8+2i}{2}$
$= 4 + i$

37. $\frac{5-i}{3-2i} = \frac{5-i}{3-2i} \cdot \frac{3+2i}{3+2i}$
$= \frac{15+10i-3i-2i^2}{3^2-(2i)^2}$
$= \frac{15+7i-2(-1)}{9-4i^2}$
$= \frac{15+2+7i}{9+4}$
$= \frac{17+7i}{13}$
$= \frac{17}{13} + \frac{7}{13}i$

39. $(7i)(-9i) = -63i^2 = -63(-1) = 63$

41. $(6-3i)-(4-2i) = 6-3i-4+2i = 2-i$

43. $(6-2i)(3+i) = 2(3-i)(3+i)$
$= 2(3^2 - i^2)$
$= 2(9+1)$
$= 2(10)$
$= 20$

45. $(8-3i)+(2+3i) = 8-3i+2+3i$
$= 10$

47. $(1-i)(1+i) = 1^2 - i^2 = 1-(-1) = 1+1 = 2$

49. $\frac{16+15i}{-3i} = \frac{16+5i}{-3i} \cdot \frac{i}{i}$
$= \frac{16i+15i^2}{-3i^2}$
$= \frac{16i+15(-1)}{-3(-1)}$
$= \frac{-15+16i}{3}$
$= \frac{-15}{3} + \frac{16}{3}i$
$= -5 + \frac{16}{3}i$

51. $(9+8i)^2 = 9^2 + 2(9)(8i) + (8i)^2$
$= 81 + 144i + 64i^2$
$= 81 + 144i + 64(-1)$
$= 81 - 64 + 144i$
$= 17 + 144i$

53. $\frac{2}{3+i} = \frac{2}{3+i} \cdot \frac{3-i}{3-i}$
$= \frac{2(3-i)}{3^2 - i^2}$
$= \frac{6-2i}{9+1}$
$= \frac{6}{10} - \frac{2}{10}i$
$= \frac{3}{5} - \frac{1}{5}i$

55. $(5-6i) - 4i = 5 - 6i - 4i = 5 - 10i$

57. $\frac{2-3i}{2+i} = \frac{2-3i}{2+i} \cdot \frac{2-i}{2-i}$
$= \frac{4 - 2i - 6i + 3i^2}{2^2 - i^2}$
$= \frac{4 - 8i + 3(-1)}{4+1}$
$= \frac{4-3-8i}{5}$
$= \frac{1-8i}{5}$
$= \frac{1}{5} - \frac{8}{5}i$

59. $(2+4i) + (6-5i) = (2+6) + (4-5)i$
$= 8 - i$

61. $i^8 = (i^4)^2 = 1^2 = 1$

63. $i^{21} = i^{20}i = (i^4)^5 i = 1^5 i = 1i = i$

65. $i^{11} = i^{10}i = (i^2)^5 i = (-1)^5 i = (-1)i = -i$

67. $i^{-6} = (i^2)^{-3} = (-1)^{-3} = \frac{1}{(-1)^3} = \frac{1}{-1} = -1$

69. $i^3 + i^4 = -i + 1 = 1 - i$

71. $i^6 + i^8 = (i^4)(i^2) + (i^4)^2$
$= 1(-1) + 1^2$
$= -1 + 1$
$= 0$

73. $2 + \sqrt{-9} = 2 + 3i$

75. $\frac{6+\sqrt{-18}}{3} = \frac{6+3i\sqrt{2}}{3}$
$= \frac{6}{3} + \frac{3i\sqrt{2}}{3}$
$= 2 + i\sqrt{2}$

77. $\frac{5-\sqrt{-75}}{10} = \frac{5-5i\sqrt{3}}{10}$
$= \frac{5}{10} - \frac{5i\sqrt{3}}{10}$
$= \frac{1}{2} - \frac{\sqrt{3}}{2}i$

79. Answers may vary.

81. $\left(8-\sqrt{-4}\right) - \left(2+\sqrt{-16}\right)$
$= (8-2i) - (2+4i)$
$= (8-2) + (-2i-4i)$
$= 6 - 6i$

83. $x^2 + 2x \stackrel{?}{=} -2$
$(-1+i)^2 + 2(-1+i) \stackrel{?}{=} -2$
$1 - 2i + i^2 - 2 + 2i \stackrel{?}{=} -2$
$1 - 2i - 1 - 2 + 2i \stackrel{?}{=} -2$
$-2 \stackrel{?}{=} -2$

Yes, $-1 + i$ is a solution.

85. $x + 57° + 90° = 180°$
$x + 147° = 180°$
$x = 33°$

87.
$$\begin{array}{r|rrrrr} -2 & 5 & 0 & -3 & 0 & 2 \\ & & -10 & 20 & -34 & 68 \\ \hline & 5 & -10 & 17 & -34 & 70 \end{array}$$
$(5x^4 - 3x^2 + 2) \div (x+2)$
$= 5x^3 - 10x^2 + 17x - 34 + \dfrac{70}{x+2}$

89. 5 people reported an average checking balance of \$0 to \$100.

91. $6 + 2 + 3 = 11$
11 people reported an average checking balance of \$301 or more.

93. $\dfrac{5}{30} = 0.166\overline{6}$
16.7% of the people reprted an average checking balance of \$0 to \$100.

Chapter 9 Review Exercises

1. $\sqrt{81} = 9$ because $9^2 = 81$.

2. $\sqrt[4]{81} = 3$ because $3^4 = 81$

3. $\sqrt[3]{-8} = -2$ because $(-2)^3 = -8$.

4. $\sqrt[4]{-16}$ is not a real number since there is no real number whose 4th power is negative.

5. $-\sqrt{\dfrac{1}{49}} = -\dfrac{1}{7}$ because $\left(\dfrac{1}{7}\right)^2 = \dfrac{1}{49}$.

6. $\sqrt{x^{64}} = x^{32}$ because $(x^{32})^2 = x^{32\cdot 2} = x^{64}$.

7. $-\sqrt{36} = -6$ because $6^2 = 36$

8. $\sqrt[3]{64} = 4$ because $4^3 = 64$.

9. $\sqrt[3]{-a^6b^9} = \sqrt[3]{-1}\sqrt[3]{a^6}\sqrt[3]{b^9}$
$= (-1)a^2b^3$
$= -a^2b^3$

10. $\sqrt{16a^4b^{12}} = \sqrt{16}\sqrt{a^4}\sqrt{b^{12}} = 4a^2b^6$

11. $\sqrt[5]{32a^5b^{10}} = \sqrt[5]{32}\sqrt[5]{a^5}\sqrt[5]{b^{10}} = 2ab^2$

12. $\sqrt[5]{-32x^{15}y^{20}} = \sqrt[5]{-32}\sqrt[5]{x^{15}}\sqrt[5]{y^{20}}$
$= -2x^3y^4$

13. $\sqrt{\dfrac{x^{12}}{36y^2}} = \dfrac{\sqrt{x^{12}}}{\sqrt{36y^2}} = \dfrac{x^6}{6y}$

14. $\sqrt[3]{\dfrac{27y^3}{z^{12}}} = \dfrac{\sqrt[3]{27y^3}}{\sqrt[3]{z^{12}}} = \dfrac{3y}{z^4}$

15. $\sqrt{(-x)^2} = |-x|$

16. $\sqrt[4]{(x^2-4)^4} = |x^2-4|$

17. $\sqrt[3]{(-27)^3} = -27$

18. $\sqrt[5]{(-5)^5} = -5$

19. $-\sqrt[5]{x^5} = -x$

20. $\sqrt[4]{16(2y+z)^{12}} = \sqrt[4]{16}\sqrt[4]{[(2y+z)^3]^4}$
$= 2|2y+z|^3$

21. $\sqrt{25(x-y)^{10}} = \sqrt{25}\sqrt{[(x-y)^5]^2}$
$= 5|(x-y)^5|$

22. $\sqrt[5]{-y^5} = \sqrt[5]{-1}\sqrt[5]{y^5} = (-1)y = -y$

23. $\sqrt[9]{-x^9} = \sqrt[9]{-1}\sqrt[9]{x^9} = -x$

24. $f(x) = \sqrt{x} + 3$
Domain: $[0, \infty)$

x	$f(x) = \sqrt{x} + 3$
0	$\sqrt{0} + 3 = 3$
1	$\sqrt{1} + 3 = 4$
4	$\sqrt{4} + 3 = 5$
9	$\sqrt{9} + 3 = 6$

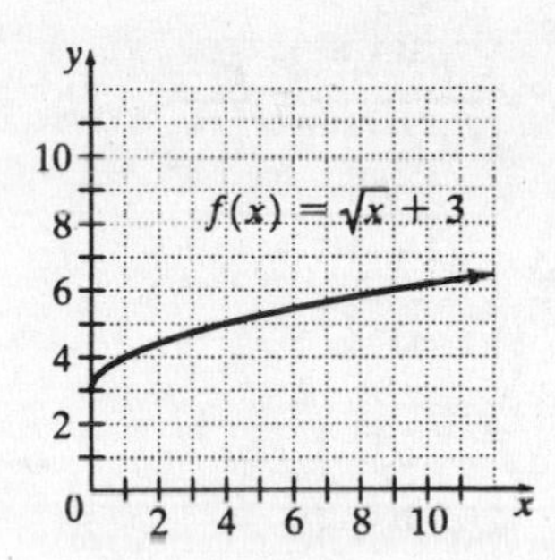

25. $g(x) = \sqrt[3]{x-3}$
Domain: $(-\infty, \infty)$

x	$g(x) = \sqrt[3]{x-3}$
-5	$\sqrt[3]{-5-3} = \sqrt[3]{-8} = -2$
2	$\sqrt[3]{2-3} = \sqrt[3]{-1} = -1$
3	$\sqrt[3]{3-3} = \sqrt[3]{0} = 0$
4	$\sqrt[3]{4-3} = \sqrt[3]{1} = 1$
11	$\sqrt[3]{11-3} = \sqrt[3]{8} = 2$

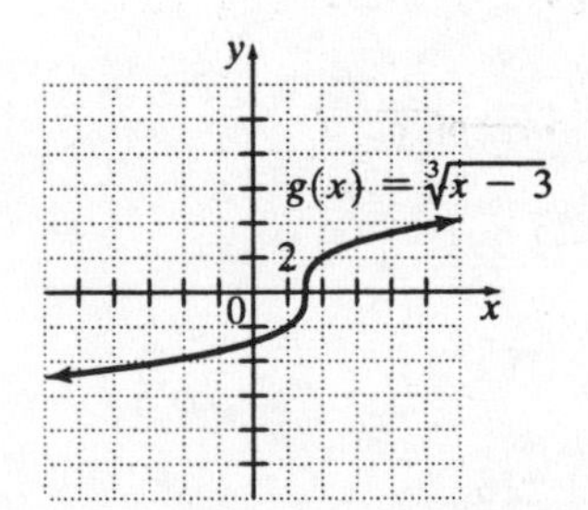

26. $\left(\frac{1}{81}\right)^{1/4} = \frac{1}{81^{1/4}} = \frac{1}{\sqrt[4]{81}} = \frac{1}{3}$

27. $\left(-\frac{1}{27}\right)^{1/3} = \frac{1}{(-27)^{1/3}} = \frac{1}{\sqrt[3]{-27}} = \frac{1}{-3} = -\frac{1}{3}$

28. $(-27)^{-1/3} = \frac{1}{(-27)^{1/3}} = \frac{1}{\sqrt[3]{-27}} = \frac{1}{-3} = -\frac{1}{3}$

29. $(-64)^{-1/3} = \frac{1}{(-64)^{1/3}}$

$= \frac{1}{\sqrt[3]{-64}}$

$= \frac{1}{-4}$

$= -\frac{1}{4}$

30. $-9^{3/2} = -(\sqrt{9})^3 = -3^3 = -27$

31. $64^{-1/3} = \frac{1}{(64)^{1/3}} = \frac{1}{\sqrt[3]{64}} = \frac{1}{4}$

32. $(-25)^{5/2} = [(-25)^{1/2}]^5$, which is not a real number, since there is no real number whose square is negative.

33. $\left(\frac{25}{49}\right)^{-3/2} = \frac{1}{\left(\frac{25}{49}\right)^{3/2}}$

$= \frac{1}{\left(\sqrt{\frac{25}{49}}\right)^3}$

$= \frac{1}{\left(\frac{5}{7}\right)^3}$

$= \frac{1}{\frac{125}{343}}$

$= \frac{343}{125}$

34. $\left(\frac{8}{27}\right)^{-2/3} = \frac{1}{\left(\frac{8}{27}\right)^{2/3}} = \frac{1}{\left(\sqrt[3]{\frac{8}{27}}\right)^2} = \frac{1}{\left(\frac{2}{3}\right)^2} = \frac{1}{\frac{4}{9}} = \frac{9}{4}$

35. $\left(-\frac{1}{36}\right)^{-1/4} = \frac{1}{\left(-\frac{1}{36}\right)^{1/4}}$, which is not a real number, since there is no real number whose 4th power is negative.

36. $\sqrt[3]{x^2} = x^{2/3}$

37. $\sqrt[5]{5x^2y^3} = (5x^2y^3)^{1/5} = 5^{1/5}(x^2)^{1/5}(y^3)^{1/5} = 5^{1/5}x^{2/5}y^{3/5}$

38. $y^{4/5} = \sqrt[5]{y^4}$

39. $5(xy^2z^5)^{1/3} = 5\sqrt[3]{xy^2z^5}$

40. $(x+2y)^{-1/2} = \frac{1}{(x+2y)^{1/2}} = \frac{1}{\sqrt{x+2y}}$

41. $a^{1/3}a^{4/3}a^{1/2} = a^{1/3+4/3+1/2} = a^{5/3+1/2} = a^{10/6+3/6} = a^{13/6}$

42. $\frac{b^{1/3}}{b^{4/3}} = b^{1/3-4/3} = b^{-1} = \frac{1}{b}$

43. $(a^{1/2}a^{-2})^3 = (a^{1/2-2})^3 = (a^{-3/2})^3 = a^{-9/2} = \frac{1}{a^{9/2}}$

44. $(x^{-3}y^6)^{1/3} = (x^{-3})^{1/3}(y^6)^{1/3} = x^{-1}y^2 = \frac{y^2}{x}$

45. $\left(\frac{b^{3/4}}{a^{-1/2}}\right)^8 = \frac{(b^{3/4})^8}{(a^{-1/2})^8} = \frac{b^6}{a^{-4}} = a^4b^6$

46. $\frac{x^{1/4}x^{-1/2}}{x^{2/3}} = \frac{x^{1/4-1/2}}{x^{2/3}} = \frac{x^{-1/4}}{x^{2/3}} = x^{-1/4-2/3} = x^{-3/12-8/12} = x^{-11/12} = \frac{1}{x^{11/12}}$

47. $\left(\frac{49c^{5/3}}{a^{-1/4}b^{5/6}}\right)^{-1} = \frac{49^{-1}c^{-5/3}}{a^{1/4}b^{-5/6}} = \frac{b^{5/6}}{49a^{1/4}c^{5/3}}$

48. $a^{-1/4}(a^{5/4} - a^{9/4}) = a^{-1/4+5/4} - a^{-1/4+9/4} = a - a^2$

49. $\sqrt{20} \approx 4.472$

50. $\sqrt[3]{-39} \approx -3.391$

51. $\sqrt[4]{726} \approx 5.191$

52. $56^{1/3} \approx 3.826$

53. $-78^{3/4} \approx -26.246$

54. $105^{-2/3} \approx 0.045$

55. $\sqrt[3]{2} \cdot \sqrt{7} = 2^{1/3} \cdot 7^{1/2}$
$= 2^{2/6} \cdot 7^{3/6}$
$= (2^2)^{1/6} \cdot (7^3)^{1/6}$
$= 4^{1/6} \cdot 343^{1/6}$
$= (4 \cdot 343)^{1/6}$
$= \sqrt[6]{1372}$

56. $\sqrt[3]{3} \cdot \sqrt[4]{x} = 3^{1/3} \cdot x^{1/4}$
$= 3^{4/12} \cdot x^{3/12}$
$= (3^4)^{1/12}(x^3)^{1/12}$
$= (81)^{1/12}(x^3)^{1/12}$
$= (81x^3)^{1/12}$
$= \sqrt[12]{81x^3}$

57. $\sqrt{3} \cdot \sqrt{8} = \sqrt{24} = \sqrt{4 \cdot 6} = 2\sqrt{6}$

58. $\sqrt[3]{7y} \cdot \sqrt[3]{x^2 z} = \sqrt[3]{7x^2 yz}$

59. $\dfrac{\sqrt{44x^3}}{\sqrt{11x}} = \sqrt{\dfrac{44x^3}{11x}} = \sqrt{4x^2} = 2x$

60. $\dfrac{\sqrt[4]{a^6 b^{13}}}{\sqrt[4]{a^2 b}} = \sqrt[4]{\dfrac{a^6 b^{13}}{a^2 b}}$
$= \sqrt[4]{a^{6-2} b^{13-1}}$
$= \sqrt[4]{a^4 b^{12}}$
$= ab^3$

61. $\sqrt{60} = \sqrt{4 \cdot 15} = 2\sqrt{15}$

62. $-\sqrt{75} = -\sqrt{25 \cdot 3} = -5\sqrt{3}$

63. $\sqrt[3]{162} = \sqrt[3]{27 \cdot 6} = 3\sqrt[3]{6}$

64. $\sqrt[3]{-32} = \sqrt[3]{(-8)(4)} = -2\sqrt[3]{4}$

65. $\sqrt{36x^7} = \sqrt{36x^6 \cdot x} = 6x^3\sqrt{x}$

66. $\sqrt[3]{24a^5 b^7} = \sqrt[3]{8a^3 b^6 \cdot 3a^2 b} = 2ab^2\sqrt[3]{3a^2 b}$

67. $\sqrt{\dfrac{p^{17}}{121}} = \dfrac{\sqrt{p^{16} \cdot p}}{\sqrt{121}} = \dfrac{p^8\sqrt{p}}{11}$

68. $\sqrt[3]{\dfrac{y^5}{27x^6}} = \dfrac{\sqrt[3]{y^3 \cdot y^2}}{\sqrt[3]{27x^6}} = \dfrac{y\sqrt[3]{y^2}}{3x^2}$

69. $\sqrt[4]{\dfrac{xy^6}{81}} = \dfrac{\sqrt[4]{y^4 \cdot xy^2}}{\sqrt[4]{81}} = \dfrac{y\sqrt[4]{xy^2}}{3}$

70. $\sqrt{\dfrac{2x^3}{49y^4}} = \dfrac{\sqrt{x^2 \cdot 2x}}{\sqrt{49y^4}} = \dfrac{x\sqrt{2x}}{7y^2}$

71. $r = \sqrt{\dfrac{A}{\pi}}$

a. $r = \sqrt{\dfrac{25}{\pi}} = \dfrac{\sqrt{25}}{\sqrt{\pi}} = \dfrac{5}{\sqrt{\pi}}$

$\dfrac{5}{\sqrt{\pi}}$ meters or $\dfrac{5\sqrt{\pi}}{\pi}$ meters

b. $r = \sqrt{\dfrac{104}{\pi}} \approx 5.75$ inches

72. $x\sqrt{75xy} - \sqrt{27x^3 y}$
$= x\sqrt{25}\sqrt{3xy} - \sqrt{9x^2}\sqrt{3xy}$
$= 5x\sqrt{3xy} - 3x\sqrt{3xy}$
$= 2x\sqrt{3xy}$

73. $2\sqrt{32x^2 y^3} - xy\sqrt{98y}$
$= 2\sqrt{16x^2 y^2 \cdot 2y} - xy\sqrt{49 \cdot 2y}$
$= 2\sqrt{16x^2 y^2}\sqrt{2y} - xy\sqrt{49}\sqrt{2y}$
$= 8xy\sqrt{2y} - 7xy\sqrt{2y}$
$= xy\sqrt{2y}$

74. $\sqrt[3]{128}+\sqrt[3]{250}=\sqrt[3]{64\cdot 2}+\sqrt[3]{125\cdot 2}$
$=\sqrt[3]{64}\sqrt[3]{2}+\sqrt[3]{125}\sqrt[3]{2}$
$=4\sqrt[3]{2}+5\sqrt[3]{2}$
$=9\sqrt[3]{2}$

75. $3\sqrt[4]{32a^5}-a\sqrt[4]{162a}$
$=3\sqrt[4]{16a^4\cdot 2a}-a\sqrt[4]{81\cdot 2a}$
$=3\sqrt[4]{16a^4}\sqrt[4]{2a}-a\sqrt[4]{81}\sqrt[4]{2a}$
$=6a\sqrt[4]{2a}-3a\sqrt[4]{2a}$
$=3a\sqrt[4]{2a}$

76. $\frac{5}{\sqrt{4}}+\frac{\sqrt{3}}{3}=\frac{5}{2}+\frac{\sqrt{3}}{3}$
$=\frac{5(3)+2\sqrt{3}}{2\cdot 3}=\frac{15+2\sqrt{3}}{6}$

77. $\sqrt{\frac{8}{x^2}}-\sqrt{\frac{50}{16x^2}}=\frac{2\sqrt{2}}{x}-\frac{5\sqrt{2}}{4x}$
$=\frac{8\sqrt{2}-5\sqrt{2}}{4x}$
$=\frac{3\sqrt{2}}{4x}$

78. $2\sqrt{50}-3\sqrt{125}+\sqrt{98}$
$=2\sqrt{25}\sqrt{2}-3\sqrt{25}\sqrt{5}+\sqrt{49}\sqrt{2}$
$=2(5)\sqrt{2}-3(5)\sqrt{5}+7\sqrt{2}$
$=10\sqrt{2}-15\sqrt{5}+7\sqrt{2}$
$=17\sqrt{2}-15\sqrt{5}$

79. $2a\sqrt[4]{32b^5}-3b\sqrt[4]{162a^4b}+\sqrt[4]{2a^4b^5}$
$=2a\sqrt[4]{16b^4}\sqrt[4]{2b}-3b\sqrt[4]{81a^4}\sqrt[4]{2b}$
$+\sqrt[4]{a^4b^4}\sqrt[4]{2b}$
$=[(2a)(2b)-(3b)(3a)+ab]\sqrt[4]{2b}$
$=(4ab-9ab+ab)\sqrt[4]{2b}$
$=-4ab\sqrt[4]{2b}$

80. $\sqrt{3}(\sqrt{27}-\sqrt{3})=\sqrt{3}(\sqrt{9}\sqrt{3}-\sqrt{3})$
$=\sqrt{3}(3\sqrt{3}-\sqrt{3})$
$=\sqrt{3}\left(2\sqrt{3}\right)$
$=2(3)$
$=6$

81. $(\sqrt{x}-3)^2=\left(\sqrt{x}\right)^2-2(3)\sqrt{x}+3^2$
$=x-6\sqrt{x}+9$

82. $(\sqrt{5}-5)(2\sqrt{5}+2)$
$=2\sqrt{25}+2\sqrt{5}-10\sqrt{5}-10$
$=10-8\sqrt{5}-10$
$=-8\sqrt{5}$

83. $(2\sqrt{x}-3\sqrt{y})(2\sqrt{x}+3\sqrt{y})$
$=(2\sqrt{x})^2-(3\sqrt{y})^2$
$=4x-9y$

84. $(\sqrt{a}+3)(\sqrt{a}-3)=\left(\sqrt{a}\right)^2-3^2$
$=a-9$

85. $\left(\sqrt[3]{a}+2\right)^2=(\sqrt[3]{a})^2+2(2)\sqrt[3]{a}+2^2$
$=\sqrt[3]{a^2}+4\sqrt[3]{a}+4$

86. $(\sqrt[3]{5x}+9)(\sqrt[3]{5x}-9)=(\sqrt[3]{5x})^2-9^2$
$=\sqrt[3]{25x^2}-81$

87. $(\sqrt[3]{a}+4)(\sqrt[3]{a^2}-4\sqrt[3]{a}+16)=(\sqrt[3]{a})^3+4^3$
$=a+64$

88. $\frac{3}{\sqrt{7}}=\frac{3\sqrt{7}}{\sqrt{7}\cdot\sqrt{7}}=\frac{3\sqrt{7}}{7}$

89. $\sqrt{\frac{x}{12}} = \frac{\sqrt{x}}{\sqrt{12}}$
$= \frac{\sqrt{x}\cdot\sqrt{12}}{\sqrt{12}\cdot\sqrt{12}}$
$= \frac{\sqrt{12x}}{12}$
$= \frac{2\sqrt{3x}}{12}$
$= \frac{\sqrt{3x}}{6}$

90. $\frac{5}{\sqrt[3]{4}} = \frac{5\cdot\sqrt[3]{16}}{\sqrt[3]{4}\cdot\sqrt[3]{16}}$
$= \frac{5\sqrt[3]{8\cdot 2}}{\sqrt[3]{64}}$
$= \frac{10\sqrt[3]{2}}{4}$
$= \frac{5\sqrt[3]{2}}{2}$

91. $\sqrt{\frac{24x^5}{3y^2}} = \sqrt{\frac{8x^5}{y^2}}$
$= \frac{\sqrt{4x^4\cdot 2x}}{\sqrt{y^2}}$
$= \frac{2x^2\sqrt{2x}}{y}$

92. $\sqrt[3]{\frac{15x^6y^7}{z^2}} = \frac{\sqrt[3]{15x^6y^7}}{\sqrt[3]{z^2}}$
$= \frac{\sqrt[3]{x^6y^6\cdot 15y}\sqrt[3]{z}}{\sqrt[3]{z^2}\cdot\sqrt[3]{z}}$
$= \frac{x^2y^2\sqrt[3]{15yz}}{z}$

93. $\frac{5}{2-\sqrt{7}} = \frac{5}{2-\sqrt{7}}\cdot\frac{2+\sqrt{7}}{2+\sqrt{7}}$
$= \frac{5(2+\sqrt{7})}{2^2-(\sqrt{7})^2}$
$= \frac{10+5\sqrt{7}}{4-7}$
$= \frac{10+5\sqrt{7}}{-3}$
$= -\frac{10+5\sqrt{7}}{3}$

94. $\frac{3}{\sqrt{y}-2}\cdot\frac{\sqrt{y}+2}{\sqrt{y}+2} = \frac{3\sqrt{y}+6}{(\sqrt{y})^2-2^2}$
$= \frac{3\sqrt{y}+6}{y-4}$

95. $\frac{\sqrt{2}-\sqrt{3}}{\sqrt{2}+\sqrt{3}} = \frac{\sqrt{2}-\sqrt{3}}{\sqrt{2}+\sqrt{3}}\cdot\frac{\sqrt{2}-\sqrt{3}}{\sqrt{2}-\sqrt{3}}$
$= \frac{(\sqrt{2})^2-2\sqrt{2}\sqrt{3}+(\sqrt{3})^2}{(\sqrt{2})^2-(\sqrt{3})^2}$
$= \frac{2-2\sqrt{6}+3}{2-3}$
$= \frac{5-2\sqrt{6}}{-1}$
$= -5+2\sqrt{6}$

96. $\frac{\sqrt{11}}{3} = \frac{\sqrt{11}\cdot\sqrt{11}}{3\cdot\sqrt{11}} = \frac{11}{3\sqrt{11}}$

97. $\sqrt{\frac{18}{y}} = \frac{\sqrt{18}}{\sqrt{y}} = \frac{3\sqrt{2}}{\sqrt{y}} = \frac{3\sqrt{2}\cdot\sqrt{2}}{\sqrt{y}\cdot\sqrt{2}} = \frac{3\cdot 2}{\sqrt{2y}}$
$= \frac{6}{\sqrt{2y}}$

98. $\frac{\sqrt[3]{9}}{7} = \frac{\sqrt[3]{9}\cdot\sqrt[3]{3}}{7\cdot\sqrt[3]{3}} = \frac{3}{7\sqrt[3]{3}}$

99. $\sqrt{\dfrac{24x^5}{3y^2}} = \dfrac{\sqrt{4x^4 \cdot 6x}}{\sqrt{y^2 \cdot 3}}$
$= \dfrac{2x^2\sqrt{6x}}{y\sqrt{3}}$
$= \dfrac{2x^2\sqrt{6x} \cdot \sqrt{6x}}{y\sqrt{3} \cdot \sqrt{6x}}$
$= \dfrac{2x^2(6x)}{y\sqrt{18x}}$
$= \dfrac{12x^3}{3y\sqrt{2x}}$
$= \dfrac{4x^3}{y\sqrt{2x}}$

100. $\sqrt[3]{\dfrac{xy^2}{10z}} = \dfrac{\sqrt[3]{xy^2} \cdot \sqrt[3]{x^2y}}{\sqrt[3]{10z} \cdot \sqrt[3]{x^2y}} = \dfrac{xy}{\sqrt[3]{10x^2yz}}$

101. $\dfrac{\sqrt{x}+5}{-3} = \dfrac{(\sqrt{x}+5)(\sqrt{x}-5)}{-3(\sqrt{x}-5)} = \dfrac{x-25}{-3\sqrt{x}+15}$

102. $\sqrt{y-7} = 5$
$y - 7 = 25$
$y = 32$
The solution is 32.

103. $\sqrt{2x} + 10 = -4$
$\sqrt{2x} = -14$
No solution exists since the principal square root of a number is not negative.

104. $\sqrt[3]{2x-6} = 4$
$(\sqrt[3]{2x-6})^3 = (4)^3$
$2x - 6 = 64$
$2x = 70$
$x = 35$
The solution is 35.

105. $\sqrt{x+6} = \sqrt{x+2}$
$x + 6 = x + 2$
$6 = 2$
Since the last equation is never satisfied, there is no solution.

106. $2x - 5\sqrt{x} = 3$
$2x - 3 = 5\sqrt{x}$
$(2x-3)^2 = (5\sqrt{x})^2$
$4x^2 - 12x + 9 = 25x$
$4x^2 - 37x + 9 = 0$
$(4x-1)(x-9) = 0$
$4x - 1 = 0$ or $x - 9 = 0$
$4x = 1$ or $x = 9$
$x = \dfrac{1}{4}$
When $x = \dfrac{1}{4}$,
$2x - 5\sqrt{x} = 2\left(\dfrac{1}{4}\right) - 5\sqrt{\dfrac{1}{4}}$
$= \dfrac{1}{2} - \dfrac{5}{2}$
$= -2 \neq 3.$
When $x = 9$,
$2x - 5\sqrt{x} = 2(9) - 5\sqrt{9}$
$= 18 - 15$
$= 3.$
The solution is 9.

107. $\sqrt{x+9} = 2 + \sqrt{x-7}$
$x + 9 = 4 + 4\sqrt{x-7} + x - 7$
$12 = 4\sqrt{x-7}$
$3 = \sqrt{x-7}$
$9 = x - 7$
$16 = x$
The solution is 16.

108. Let c = the length of the hypotenuse of the right triangle. By the Pythagorean theorem,
$c^2 = 3^2 + 3^2 = 9 + 9 = 9(2)$
so $c = \sqrt{9(2)} = \sqrt{9}\sqrt{2} = 3\sqrt{2}$

109. Let c = the length of the hypotenuse of the right triangle. By the Pythagorean theorem,
$c^2 = 7^2 + (8\sqrt{3})^2$
$= 49 + 192$
$= 241$
so $c = \sqrt{241}$

110. $c^2 = a^2 + b^2$
$(65)^2 = (40)^2 + b^2$
$4225 - 1600 = b^2$
$2625 = b^2$
$\pm\sqrt{2625} = b$
$\pm 51.235 = b$
51.2 feet

111. $c^2 = 3^2 + 3^2$
$c^2 = 9 + 9$
$c^2 = 18$
$c = \pm\sqrt{18}$
$c = \pm 4.243$
4.24 feet

112. $\sqrt{-8} = \sqrt{4}\sqrt{2}\sqrt{-1} = 2i\sqrt{2}$

113. $-\sqrt{-6} = -\sqrt{6}\sqrt{-1} = -i\sqrt{6}$

114. $\sqrt{-4} + \sqrt{-16} = 2i + 4i = 6i$

115. $\sqrt{-2} \cdot \sqrt{-5} = i\sqrt{2} \cdot i\sqrt{5} = i^2\sqrt{2 \cdot 5} = -\sqrt{10}$

116. $(12 - 6i) + (3 + 2i) = (12 + 3) + (-6 + 2)i$
$= 15 - 4i$

117. $(-8 - 7i) - (5 - 4i) = (-8 - 5) + (-7 + 4)i$
$= -13 - 3i$

118. $(\sqrt{3} + \sqrt{2}) + (3\sqrt{2} - \sqrt{-8})$
$= \sqrt{3} + \sqrt{2} + 3\sqrt{2} - \sqrt{4}\sqrt{2}\sqrt{-1}$
$= \sqrt{3} + 4\sqrt{2} - 2i\sqrt{2}$

119. $2i(2 - 5i) = 4i - 10i^2$
$= 4i - 10(-1)$
$= 10 + 4i$

120. $-3i(6 - 4i) = -18i + 12i^2$
$= -18i + 12(-1)$
$= -12 - 18i$

121. $(3 + 2i)(1 + i) = 3 + 3i + 2i + 2i^2$
$= 3 + 5i + 2(-1)$
$= (3 - 2) + 5i$
$= 1 + 5i$

122. $(2 - 3i)^2 = (2 - 3i)(2 - 3i)$
$= (2)^2 + 2(2)(-3i) + (-3i)^2$
$= 4 - 12i + 9i^2$
$= 4 - 12i - 9$
$= -5 - 12i$

123. $(\sqrt{6} - 9i)(\sqrt{6} + 9i) = (\sqrt{6})^2 - (9i)^2$
$= 6 - 81i^2$
$= 6 + 81$
$= 87$

124. $\dfrac{2 + 3i}{2i} = \dfrac{(2 + 3i)i}{2i^2}$
$= \dfrac{2i + 3i^2}{2(-1)}$
$= \dfrac{2i + 3(-1)}{-2}$
$= \dfrac{-3 + 2i}{-2}$
$= \dfrac{3}{2} - i$

125. $\dfrac{1 + i}{-3i} = \dfrac{1 + i}{-3i} \cdot \dfrac{3i}{3i}$
$= \dfrac{3i + 3i^2}{-9i^2}$
$= \dfrac{3i + 3(-1)}{-9(-1)}$
$= \dfrac{3(-1 + i)}{9}$
$= \dfrac{-1 + i}{3}$
$= -\dfrac{1}{3} + \dfrac{1}{3}i$

Chapter 9 Test

1. $\sqrt{216} = \sqrt{36 \cdot 6} = 6\sqrt{6}$

2. $-\sqrt[4]{x^{64}} = -(x^{64})^{1/4} = -x^{16}$

3. $\left(\frac{1}{125}\right)^{1/3} = \frac{1}{5}$

4. $\left(\frac{1}{125}\right)^{-1/3} = \left[\left(\frac{1}{125}\right)^{-1}\right]^{1/3} = 125^{1/3} = 5$

5. $\left(\frac{8x^3}{27}\right)^{2/3} = \frac{8^{2/3}(x^3)^{2/3}}{27^{2/3}}$
$= \frac{(8^{1/3})^2 x^2}{(27^{1/3})^2}$
$= \frac{2^2 x^2}{3^2}$
$= \frac{4x^2}{9}$

6. $\sqrt[3]{-a^{18}b^9} = \sqrt[3]{-1}\sqrt[3]{a^{18}}\sqrt[3]{b^9}$
$= (-1)a^6b^3$
$= -a^6b^3$

7. $\left(\frac{64c^{4/3}}{a^{-2/3}b^{5/6}}\right)^{1/2}$
$= \left(\frac{64a^{2/3}c^{4/3}}{b^{5/6}}\right)^{1/2}$
$= \frac{64^{1/2}\left(a^{2/3}\right)^{1/2}\left(c^{4/3}\right)^{1/2}}{\left(b^{5/6}\right)^{1/2}}$
$= \frac{8a^{1/3}c^{2/3}}{b^{5/12}}$

8. $a^{-2/3}(a^{5/4} - a^3)$
$= a^{-2/3+5/4} - a^{-2/3+3}$
$= a^{-8/12+15/12} - a^{-2/3+9/3}$
$= a^{7/12} - a^{7/3}$

9. $\sqrt[4]{(4xy)^4} = |4xy|$ or $4|xy|$

10. $\sqrt[3]{(-27)^3} = -27$

11. $\sqrt{\frac{9}{y}} = \frac{\sqrt{9}}{\sqrt{y}} = \frac{3}{\sqrt{y}} \cdot \frac{\sqrt{y}}{\sqrt{y}} = \frac{3\sqrt{y}}{y}$

12. $\frac{4-\sqrt{x}}{4+2\sqrt{x}} = \frac{(4-\sqrt{x})(2-\sqrt{x})}{2(2+\sqrt{x})(2-\sqrt{x})}$
$= \frac{8-4\sqrt{x}-2\sqrt{x}+x}{2(2^2-\sqrt{x}^2)}$
$= \frac{8-6\sqrt{x}+x}{2(4-x)}$
$= \frac{8-6\sqrt{x}+x}{8-2x}$

13. $\frac{\sqrt[3]{ab}}{\sqrt[3]{ab^2}} = \sqrt[3]{\frac{ab}{ab^2}}$
$= \sqrt[3]{\frac{1}{b^1}}$
$= \frac{1}{\sqrt[3]{b}}$
$= \frac{\sqrt[3]{b^2}}{\sqrt[3]{b^3}}$
$= \frac{\sqrt[3]{b^2}}{b}$

14. $\frac{\sqrt{6}+x}{8} = \frac{(\sqrt{6}+x)(\sqrt{6}-x)}{8(\sqrt{6}-x)}$
$= \frac{(\sqrt{6})^2 - x^2}{8(\sqrt{6}-x)}$
$= \frac{6-x^2}{8(\sqrt{6}-x)}$

15. $\sqrt{125x^3} - 3\sqrt{20x^3}$
$= \sqrt{25x^2}\sqrt{5x} - 3\sqrt{4x^2}\sqrt{5x}$
$= [5x - 3(2x)]\sqrt{5x}$
$= 5x - 6x\sqrt{5x}$
$= -x\sqrt{5x}$

16. $\sqrt{3}(\sqrt{16} - \sqrt{2}) = \sqrt{3}(4 - \sqrt{2})$
$= 4\sqrt{3} - \sqrt{3}\sqrt{2}$
$= 4\sqrt{3} - \sqrt{6}$

17. $(\sqrt{x} + 1)^2 = (\sqrt{x})^2 + 2\sqrt{x} + 1$
$= x + 2\sqrt{x} + 1$

18. $(\sqrt{2} - 4)(\sqrt{3} + 1) = \sqrt{2}\sqrt{3} + \sqrt{2} - 4\sqrt{3} - 4$
$= \sqrt{6} - 4\sqrt{3} + \sqrt{2} - 4$

19. $(\sqrt{5} + 5)(\sqrt{5} - 5) = (\sqrt{5})^2 - 5^2 = 5 - 25$
$= -20$

20. $\sqrt{561} = 23.685$

21. $386^{-2/3} = 0.019$

22.
$x = \sqrt{x-2} + 2$
$x - 2 = \sqrt{x-2}$
$(x-2)^2 = x - 2$
$x^2 - 4x + 4 = x - 2$
$x^2 - 5x + 6 = 0$
$(x-2)(x-3) = 0$
$x - 2 = 0$ or $x - 3 = 0$
$x = 2$ or $x = 3$
The solutions are 2 and 3.

23. $\sqrt{x^2 - 7} + 3 = 0$
$\sqrt{x^2 - 7} = -3$
No solution exists since a principal square root is not negatively valued.

24. $\sqrt{x+5} = \sqrt{2x-1}$
$x + 5 = 2x - 1$
$6 = x$
The solution is 6.

25. $\sqrt{-2} = i\sqrt{2}$

26. $-\sqrt{-8} = -\sqrt{4}\sqrt{2}\sqrt{-1} = -2i\sqrt{2}$

27. $(12 - 6i) - (12 - 3i) = 12 - 6i - 12 + 3i$
$= -3i$

28. $(6 - 2i)(6 + 2i) = 6^2 + 2^2 = 36 + 4 = 40$

29. $(4 + 3i)^2 = 16 + 24i + 9i^2$
$= 16 + 24i + 9(-1)$
$= (16 - 9) + 24i$
$= 7 + 24i$

30. $\frac{1+4i}{1-i} = \frac{(1+4i)(1+i)}{(1-i)(1+i)}$
$= \frac{1 + i + 4i + 4i^2}{1 - i^2}$
$= \frac{-3 + 5i}{2}$
$= \frac{3}{2} + \frac{5}{2}i$

31. Let x = the length of a leg of the isosceles right triangle. By the Pythagorean theorem,
$x^2 + x^2 = 5^2$
$2x^2 = 25$
$x^2 = \frac{25}{2}$
$x = \sqrt{\frac{25}{2}} = \frac{\sqrt{25}}{\sqrt{2}} = \frac{5}{\sqrt{2}} \cdot \frac{\sqrt{2}}{\sqrt{2}} = \frac{5\sqrt{2}}{2}$.

32. $g(x) = \sqrt{x+2}$

Domain: $[-2, \infty)$

x	$g(x) = \sqrt{x+2}$
-2	$\sqrt{-2+2} = 0$
-1	$\sqrt{-1+2} = 1$
2	$\sqrt{2+2} = 2$
7	$\sqrt{7+2} = 3$

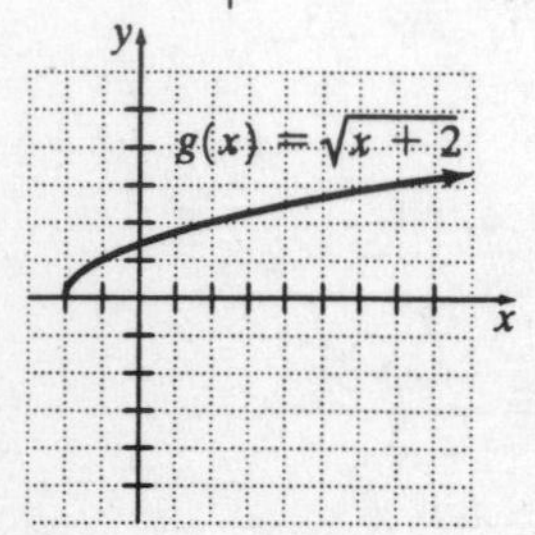

33. $V(r) = \sqrt{2.5r}$

$V(300) = \sqrt{2.5(300)} \approx 27$

The maximum safe speed is about 27 miles per hour.

34. $V(r) = \sqrt{2.5r}$

$30 = \sqrt{2.5r}$

$900 = 2.5r$

$360 = r$

The radius of curvature is 360 feet.

Chapter 10

Section 10.1

Graphing Calculator Explorations

1. −1.27, 6.27

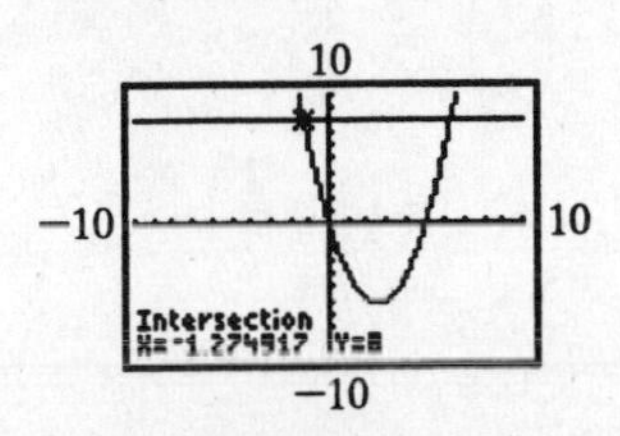

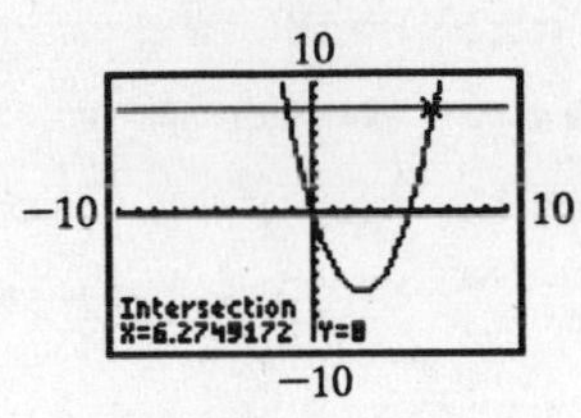

3. −1.10, 0.90

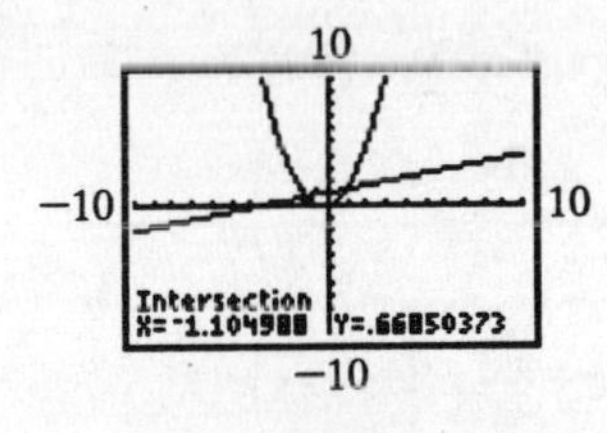

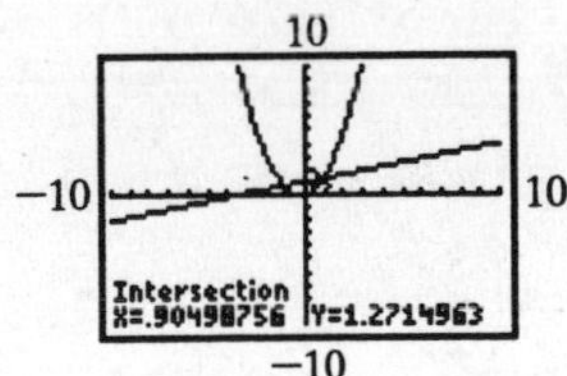

5. There is no real solution.

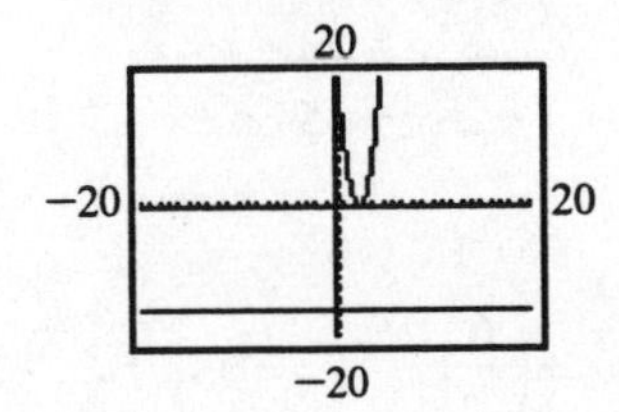

Exercise Set 10.1

1. $x^2 = 16$
$x = \pm\sqrt{16}$
$x = \pm 4$

3. $x^2 - 7 = 0$
$x^2 = 7$
$x = \pm\sqrt{7}$

5. $x^2 = 18$
$x = \pm\sqrt{18}$
$x = \pm\sqrt{9}\sqrt{2}$
$x = \pm 3\sqrt{2}$

7. $3z^2 - 30 = 0$
$3z^2 = 30$
$z^2 = 10$
$z = \pm\sqrt{10}$

9. $(x+5)^2 = 9$
$x + 5 = \pm\sqrt{9}$
$x + 5 = \pm 3$
$x = -5 \pm 3$
$x = -8$ or $x = -2$

11. $(z-6)^2 = 18$
$z - 6 = \pm\sqrt{18}$
$z - 6 = \pm\sqrt{9}\sqrt{2}$
$z - 6 = \pm 3\sqrt{2}$
$z = 6 \pm 3\sqrt{2}$

13. $(2x-3)^2 = 8$
$2x - 3 = \pm\sqrt{8}$
$2x - 3 = \pm\sqrt{4}\sqrt{2}$
$2x - 3 = \pm 2\sqrt{2}$
$2x = 3 \pm 2\sqrt{2}$
$x = \dfrac{3 \pm 2\sqrt{2}}{2}$

15. $x^2+9=0$
$x^2=-9$
$x=\pm\sqrt{-9}$
$x=\pm 3i$

17. $x^2-6=0$
$x^2=6$
$x=\pm\sqrt{6}$

19. $2z^2+16=0$
$2z^2=-16$
$z^2=-8$
$z=\pm\sqrt{-8}$
$z=\pm\sqrt{4}\sqrt{2}\sqrt{-1}$
$z=\pm 2i\sqrt{2}$

21. $(x-1)^2=-16$
$x-1=\pm\sqrt{-16}$
$x-1=\pm 4i$
$x=1\pm 4i$

23. $(z+7)^2=5$
$z+7=\pm\sqrt{5}$
$z=-7\pm\sqrt{5}$

25. $(x+3)^2=-8$
$x+3=\pm\sqrt{-8}$
$x+3=\pm\sqrt{4}\sqrt{2}\sqrt{-1}$
$x+3=\pm 2i\sqrt{2}$
$x=-3\pm 2i\sqrt{2}$

27. $x^2+16x+\left(\frac{16}{2}\right)^2=x^2+16x+64$
$=(x+8)^2$

29. $z^2-12z+\left(\frac{12}{2}\right)^2=z^2-12z+36$
$=(z-6)^2$

31. $p^2+9p+\left(\frac{9}{2}\right)^2=p^2+9p+\frac{81}{4}$
$=\left(p+\frac{9}{2}\right)^2$

33. $x^2+x+\left(\frac{1}{2}\right)^2=x^2+x+\frac{1}{4}$
$=\left(x+\frac{1}{2}\right)^2$

35. $x^2+\_\_+16$
$\pm 2\sqrt{16}=\pm 2(4)=\pm 8$
$x^2-8x+16=(x-4)^2$
$x^2+8x+16=(x+4)^2$
Answer: $\pm 8x$

37. $z^2+\_\_+\frac{25}{4}$
$\pm 2\sqrt{\frac{25}{4}}=\pm 2\left(\frac{5}{2}\right)=\pm 5$
$z^2-5z+\frac{25}{4}=\left(z-\frac{5}{2}\right)^2$
$z^2+5z+\frac{25}{4}=\left(z+\frac{5}{2}\right)^2$
Answer: $\pm 5z$

39. $x^2+8x=-15$
$x^2+8x+\left(\frac{8}{2}\right)^2=-15+16$
$(x+4)^2=1$
$x+4=\pm\sqrt{1}$
$x=-4\pm 1$
$x=-5$ or $x=-3$

41. $x^2+6x+2=0$
$x^2+6x+\left(\frac{6}{2}\right)^2=-2+9$
$(x+3)^2=7$
$x+3=\pm\sqrt{7}$
$x=-3\pm\sqrt{7}$

43.
$$x^2+x-1=0$$
$$x^2+x+\left(\frac{1}{2}\right)^2=1+\left(\frac{1}{2}\right)^2$$
$$x^2+x+\frac{1}{4}=1+\frac{1}{4}$$
$$\left(x+\frac{1}{2}\right)^2=\frac{5}{4}$$
$$x+\frac{1}{2}=\pm\sqrt{\frac{5}{4}}$$
$$x=-\frac{1}{2}\pm\frac{\sqrt{5}}{2}$$
$$x=\frac{-1\pm\sqrt{5}}{2}$$

45.
$$x^2+2x-5=0$$
$$x^2+2x+\left(\frac{2}{2}\right)^2=5+1$$
$$(x+1)^2=6$$
$$x+1=\pm\sqrt{6}$$
$$x=-1\pm\sqrt{6}$$

47.
$$3p^2-12p+2=0$$
$$3p^2-12p=-2$$
$$p^2-4p=-\frac{2}{3}$$
$$p^2-4p+\left(\frac{4}{2}\right)^2=-\frac{2}{3}+4$$
$$(p-2)^2=\frac{10}{3}$$
$$p-2=\pm\sqrt{\frac{10}{3}}$$
$$p-2=\pm\sqrt{\frac{10\cdot 3}{3^2}}$$
$$p-2=\pm\frac{\sqrt{30}}{3}$$
$$p=2\pm\frac{\sqrt{30}}{3}$$
$$p=\frac{6\pm\sqrt{30}}{3}$$

49.
$$4y^2-12y-2=0$$
$$4y^2-12y=2$$
$$y^2-3y=\frac{1}{2}$$
$$y^2-3y+\left(\frac{3}{2}\right)^2=\frac{1}{2}+\frac{9}{4}$$
$$\left(y-\frac{3}{2}\right)^2=\frac{11}{4}$$
$$y-\frac{3}{2}=\pm\sqrt{\frac{11}{4}}$$
$$y-\frac{3}{2}=\pm\frac{\sqrt{11}}{2}$$
$$y=\frac{3}{2}\pm\frac{\sqrt{11}}{2}$$

51.
$$2x^2+7x=4$$
$$x^2+\frac{7}{2}x=2$$
$$x^2+\frac{7}{2}x+\left(\frac{\frac{7}{2}}{2}\right)^2=2+\frac{49}{16}$$
$$\left(x+\frac{7}{4}\right)^2=\frac{81}{16}$$
$$x+\frac{7}{4}=\pm\sqrt{\frac{81}{16}}$$
$$x=-\frac{7}{4}\pm\frac{9}{4}$$
$$x=-4,\ \frac{1}{2}$$

53.
$$\begin{aligned} x^2 - 4x - 5 &= 0 \\ x^2 - 4x &= 5 \\ x^2 - 4x + \left(\frac{4}{2}\right)^2 &= 5 + 4 \\ (x-2)^2 &= 9 \\ x - 2 &= \pm\sqrt{9} \\ x - 2 &= \pm 3 \\ x &= 2 \pm 3 \\ x &= -1,\ 5 \end{aligned}$$

55.
$$\begin{aligned} x^2 + 8x + 1 &= 0 \\ x^2 + 8x &= -1 \\ x^2 + 8x + \left(\frac{8}{2}\right)^2 &= -1 + 16 \\ (x+4)^2 &= 15 \\ x + 4 &= \pm\sqrt{15} \\ x &= -4 \pm \sqrt{15} \end{aligned}$$

57.
$$\begin{aligned} 3y^2 + 6y - 4 &= 0 \\ 3y^2 + 6y &= 4 \\ y^2 + 2y &= \frac{4}{3} \\ y^2 + 2y + \left(\frac{2}{2}\right)^2 &= \frac{4}{3} + 1 \\ (y+1)^2 &= \frac{7}{3} \\ y + 1 &= \pm\sqrt{\frac{7}{3}} \\ y + 1 &= \pm\frac{\sqrt{7}}{\sqrt{3}} \cdot \frac{\sqrt{3}}{\sqrt{3}} \\ y + 1 &= \pm\frac{\sqrt{21}}{3} \\ y &= -1 \pm \frac{\sqrt{21}}{3} \end{aligned}$$

59.
$$\begin{aligned} 2x^2 - 3x - 5 &= 0 \\ 2x^2 - 3x &= 5 \\ x^2 - \frac{3}{2}x &= \frac{5}{2} \\ x^2 - \frac{3}{2}x + \left(\frac{\frac{3}{2}}{2}\right)^2 &= \frac{5}{2} + \frac{9}{16} \\ \left(x - \frac{3}{4}\right)^2 &= \frac{49}{16} \\ x - \frac{3}{4} &= \pm\frac{7}{4} \\ x &= \frac{3}{4} \pm \frac{7}{4} \\ x &= -1,\ \frac{5}{2} \end{aligned}$$

61.
$$\begin{aligned} y^2 + 2y + 2 &= 0 \\ y^2 + 2y + \left(\frac{2}{2}\right)^2 &= -2 + 1 \\ (y+1)^2 &= -1 \\ y + 1 &= \pm\sqrt{-1} \\ y + 1 &= \pm i \\ y &= -1 \pm i \end{aligned}$$

63.
$$\begin{aligned} x^2 - 6x + 3 &= 0 \\ x^2 - 6x + \left(\frac{6}{2}\right)^2 &= -3 + 9 \\ (x-3)^2 &= 6 \\ x - 3 &= \pm\sqrt{6} \\ x &= 3 \pm \sqrt{6} \end{aligned}$$

65.
$$2a^2+8a=-12$$
$$a^2+4a=-6$$
$$a^2+4a=-6$$
$$a^2+4a+\left(\frac{4}{2}\right)^2=-6+4$$
$$(a+2)^2=-2$$
$$a+2=\pm\sqrt{-2}$$
$$a+2=\pm i\sqrt{2}$$
$$a=-2\pm i\sqrt{2}$$

67.
$$5x^2+15x-1=0$$
$$5x^2+15x=1$$
$$x^2+3x=\frac{1}{5}$$
$$x^2+3x+\left(\frac{3}{2}\right)^2=\frac{1}{5}+\frac{9}{4}$$
$$\left(x+\frac{3}{2}\right)^2=\frac{49}{20}$$
$$x+\frac{3}{2}=\pm\sqrt{\frac{49}{20}}$$
$$=\frac{\pm\sqrt{49}}{\sqrt{4}\sqrt{5}}$$
$$=\pm\frac{7}{2\sqrt{5}}$$
$$=\pm\frac{7\sqrt{5}}{2\cdot\sqrt{5^2}}$$
$$=\pm\frac{7\sqrt{5}}{2\cdot 5}$$
$$x+\frac{3}{2}=\pm\frac{7\sqrt{5}}{10}\text{ so}$$
$$x=-\frac{3}{2}\pm\frac{7\sqrt{5}}{10}$$

69.
$$2x^2-x+6=0$$
$$2x^2-x=-6$$
$$x^2-\frac{1}{2}x=-3$$
$$x^2-\frac{1}{2}x+\left(\frac{\frac{1}{2}}{2}\right)^2=-3+\frac{1}{16}$$
$$\left(x-\frac{1}{4}\right)^2=-\frac{47}{16}$$
$$x-\frac{1}{4}=\pm\sqrt{-\frac{47}{16}}$$
$$=\pm\frac{\sqrt{47}\sqrt{-1}}{\sqrt{16}}$$
$$=\pm\frac{i\sqrt{47}}{4}$$
$$x=\frac{1\pm i\sqrt{47}}{4}$$

71.
$$x^2+10x+28=0$$
$$x^2+10x=-28$$
$$x^2+10x+\left(\frac{10}{2}\right)^2=-28+25$$
$$(x+5)^2=-3$$
$$x+5=\pm\sqrt{-3}$$
$$x+5=\pm i\sqrt{3}$$
$$x=-5\pm i\sqrt{3}$$

73.
$$z^2+3z-4=0$$
$$z^2+3z=4$$
$$z^2+3z+\left(\frac{3}{2}\right)^2=4+\frac{9}{4}$$
$$\left(z+\frac{3}{2}\right)^2=\frac{25}{4}$$
$$z+\frac{3}{2}=\pm\sqrt{\frac{25}{4}}$$
$$z+\frac{3}{2}=\pm\frac{5}{2}$$
$$z=-\frac{3}{2}\pm\frac{5}{2}$$
$$z=-4\text{ or }z=1$$

75.
$$2x^2 - 4x + 3 = 0$$
$$2x^2 - 4x = -3$$
$$x^2 - 2x = -\frac{3}{2}$$
$$x^2 - 2x + \left(\frac{2}{2}\right)^2 = -\frac{3}{2} + 1$$
$$(x-1)^2 = -\frac{1}{2}$$
$$x - 1 = \pm\sqrt{-\frac{1}{2}}$$
$$x - 1 = \pm\frac{\sqrt{-1}}{\sqrt{2}}$$
$$x - 1 = \pm\frac{i}{\sqrt{2}}\cdot\frac{\sqrt{2}}{\sqrt{2}}$$
$$x - 1 = \pm\frac{\sqrt{2}}{2}i$$
$$x = 1 \pm \frac{\sqrt{2}}{2}i$$
$$x = \frac{2 \pm i\sqrt{2}}{2}$$

77.
$$3x^2 + 3x = 5$$
$$x^2 + x = \frac{5}{3}$$
$$x^2 + x + \left(\frac{1}{2}\right)^2 = \frac{5}{3} + \frac{1}{4}$$
$$\left(x + \frac{1}{2}\right)^2 = \frac{23}{12}$$
$$x + \frac{1}{2} = \pm\sqrt{\frac{23}{12}}$$
$$= \pm\frac{\sqrt{23}}{2\sqrt{3}}$$
$$= \pm\frac{\sqrt{23}\sqrt{3}}{2\sqrt{3^2}}$$
$$= \pm\frac{\sqrt{69}}{2\cdot 3}$$
$$x + \frac{1}{2} = \pm\frac{\sqrt{69}}{6}$$
$$x = -\frac{1}{2} \pm \frac{\sqrt{69}}{6}$$
$$x = \frac{-3 \pm \sqrt{69}}{6}$$

79.
$$A = P(1+r)^t$$
$$4320 = 3000(1+r)^2$$
$$\frac{4320}{3000} = (1+r)^2$$
$$1.44 = (1+r)^2$$
$$\pm\sqrt{1.44} = 1 + r$$
$$\pm 1.2 = 1 + r$$
$$-1 \pm 1.2 = r$$
$r = -1 + 1.2$ or $r = -1 - 1.2$
$r = 0.2$ or $r = -2.2$
Rate cannot be negative, so the rate is $r = 0.2 = 20\%$.

81.
$$A = P(1+r)^t$$
$$1000 = 810(1+r)^2$$
$$\frac{1000}{810} = (1+r)^2$$
$$\frac{100}{81} = (1+r)^2$$
$$\pm\sqrt{\frac{100}{81}} = 1+r$$
$$\pm\frac{10}{9} = 1+r$$
$$-1\pm\frac{10}{9} = r$$
$$r = -1+\frac{10}{9} \text{ or } r = -1-\frac{10}{9}$$
$$r = \frac{1}{9} \text{ or } r = -\frac{19}{9}$$
Rate cannot be negative, so
$r = \frac{1}{9}$, or about 11%.

83. Answers may vary.

85. Simple; answers may vary.

87.
$$s(t) = 16t^2$$
$$1053 = 16t^2$$
$$t = \pm\sqrt{\frac{1053}{16}}$$
$t = 8.11$ or -8.11 (disregard)
It would take 8.11 seconds.

89.
$$s(t) = 16t^2$$
$$725 = 16t^2$$
$$t = \pm\sqrt{\frac{725}{16}}$$
$t = 6.73$ or -6.73 (disregard)
It would take 6.73 seconds.

91.
$$A = \pi r^2$$
$$\frac{36\pi}{\pi} = \frac{\pi r^2}{\pi}$$
$$\sqrt{36} = \sqrt{r^2}$$
$$\pm 6 = r$$
disregard -6
6 inches

93.
$$x^2 + x^2 = 27^2$$
$$\frac{2x^2}{2} = \frac{729}{2}$$
$$\sqrt{x^2} = \sqrt{\frac{729}{2}}$$
$$x = \pm\frac{27}{\sqrt{2}}\cdot\frac{\sqrt{2}}{\sqrt{2}}$$
$$x = \pm\frac{27\sqrt{2}}{\sqrt{4}}$$
$$x = \pm\frac{27\sqrt{2}}{2}$$
Disregard the negative.
$\frac{27\sqrt{2}}{2}$ inches

95.
$$p = -x^2 + 15$$
$$7 = -x^2 + 15$$
$$x^2 = 8$$
$$x = \pm\sqrt{8}$$
$$x \approx 2.828$$
Demand cannot be negative, therefore the demand is approximately 2.828 thousand units.

97. $\frac{3}{5} + \sqrt{\frac{16}{25}} = \frac{3}{5} + \frac{4}{5} = \frac{7}{5}$

99. $\frac{9}{10} - \sqrt{\frac{49}{100}} = \frac{9}{10} - \frac{7}{10} = \frac{2}{10} = \frac{1}{5}$

101. $\frac{10-20\sqrt{3}}{2} = \frac{10}{2} - \frac{20\sqrt{3}}{2} = 5 - 10\sqrt{3}$

103. $\frac{12-8\sqrt{7}}{16}=\frac{12}{16}-\frac{8\sqrt{7}}{16}$
$=\frac{3}{4}-\frac{\sqrt{7}}{2}$
$=\frac{3}{4}-\frac{2\sqrt{7}}{4}$
$=\frac{3-2\sqrt{7}}{4}$

105. $\sqrt{b^2-4ac}$
$a=1, b=6, c=2$
$\sqrt{(6)^2-4(1)(2)}=\sqrt{36-8}$
$=\sqrt{28}$
$=\sqrt{4}\sqrt{7}$
$=2\sqrt{7}$

107. $\sqrt{b^2-4ac}$
$a=1, b=-3, c=-1$
$\sqrt{(-3)^2-4(1)(-1)}=\sqrt{9+4}=\sqrt{13}$

Exercise Set 10.2

1. $m^2+5m-6=0$
$a=1, b=5, c=-6$
$m=\frac{-5\pm\sqrt{5^2-4(1)(-6)}}{2(1)}$
$m=\frac{-5\pm\sqrt{25+24}}{2}=\frac{-5\pm\sqrt{49}}{2}$
$m=\frac{-5\pm 7}{2}$
$m=-6$ or $m=1$
The solutions are –6 and 1.

3. $2y=5y^2-3$
$5y^2-2y-3=0$
$a=5, b=-2, c=-3$
$y=\frac{2\pm\sqrt{(-2)^2-4(5)(-3)}}{2(5)}$
$y=\frac{2\pm\sqrt{4+60}}{10}=\frac{2\pm\sqrt{64}}{10}$
$y=\frac{2\pm 8}{10}$
$y=-\frac{3}{5}$, or $y=1$
The solutions are $-\frac{3}{5}$ and 1.

5. $x^2-6x+9=0$
$a=1, b=-6, c=9$
$x=\frac{6\pm\sqrt{(-6)^2-4(1)(9)}}{2(1)}$
$x=\frac{6\pm\sqrt{36-36}}{2}=\frac{6\pm\sqrt{0}}{2}=\frac{6}{2}=3$
The solution is 3.

7. $x^2+7x+4=0$
$a=1, b=7, c=4$
$x=\frac{-7\pm\sqrt{7^2-4(1)(4)}}{2(1)}$
$x=\frac{-7\pm\sqrt{49-16}}{2}=\frac{-7\pm\sqrt{33}}{2}$
The solutions are $\frac{-7+\sqrt{33}}{2}$ and $\frac{-7-\sqrt{33}}{2}$.

9. $8m^2 - 2m = 7$

$8m^2 - 2m - 7 = 0$

$a = 8, b = -2, c = -7$

$m = \frac{2 \pm \sqrt{(-2)^2 - 4(8)(-7)}}{2(8)}$

$m = \frac{2 \pm \sqrt{4 + 224}}{16} = \frac{2 \pm \sqrt{228}}{16}$

$m = \frac{2 \pm \sqrt{4}\sqrt{57}}{16} = \frac{2 \pm 2\sqrt{57}}{16}$

$m = \frac{1 \pm \sqrt{57}}{8}$

The solutions are $\frac{1+\sqrt{57}}{8}$ and $\frac{1-\sqrt{57}}{8}$.

11. $3m^2 - 7m = 3$

$3m^2 - 7m - 3 = 0$

$a = 3, b = -7, c = -3$

$m = \frac{7 \pm \sqrt{(-7)^2 - 4(3)(-3)}}{2(3)}$

$m = \frac{7 \pm \sqrt{49 + 36}}{6} = \frac{7 \pm \sqrt{85}}{6}$

The solutions are $\frac{7+\sqrt{85}}{6}$ and $\frac{7-\sqrt{85}}{6}$.

13. $\frac{1}{2}x^2 - x - 1 = 0$

$x^2 - 2x - 2 = 0$

$a = 1, b = -2, c = -2$

$x = \frac{2 \pm \sqrt{(-2)^2 - 4(1)(-2)}}{2(1)}$

$x = \frac{2 \pm \sqrt{4 + 8}}{2} = \frac{2 \pm \sqrt{12}}{2}$

$x = \frac{2 \pm \sqrt{4}\sqrt{3}}{2} = \frac{2 \pm 2\sqrt{3}}{2}$

$x = 1 \pm \sqrt{3}$

The solutions are $1+\sqrt{3}$ and $1-\sqrt{3}$.

15. $\frac{2}{5}y^2 + \frac{1}{5}y = \frac{3}{5}$

$2y^2 + y = 3$

$2y^2 + y - 3 = 0$

$a = 2, b = 1, c = -3$

$y = \frac{-1 \pm \sqrt{1^2 - 4(2)(-3)}}{2(2)}$

$y = \frac{-1 \pm \sqrt{1 + 24}}{4} = \frac{-1 \pm \sqrt{25}}{4}$

$y = \frac{-1 \pm 5}{4}$

$y = -\frac{3}{2}$ or $y = 1$

The solutions are $-\frac{3}{2}$ and 1.

17. $\frac{1}{3}y^2 - y - \frac{1}{6} = 0$

$2y^2 - 6y - 1 = 0$

$a = 2, b = -6, c = -1$

$y = \frac{6 \pm \sqrt{(-6)^2 - 4(2)(-1)}}{2(2)}$

$y = \frac{6 \pm \sqrt{36 + 8}}{4} = \frac{6 \pm \sqrt{44}}{4}$

$y = \frac{6 \pm \sqrt{4}\sqrt{11}}{4} = \frac{6 \pm 2\sqrt{11}}{4}$

$y = \frac{3 \pm \sqrt{11}}{2}$

The solutions are $\frac{3+\sqrt{11}}{2}$ and $\frac{3-\sqrt{11}}{2}$.

19. $m^2 + 5m - 6 = 0$

$(m+6)(m-1) = 0$

$m + 6 = 0$ or $m - 1 = 0$

$m = -6$ or $m = 1$

The results are the same.

21. $6 = -4x^2 + 3x$

$4x^2 - 3x + 6 = 0$

$a = 4, b = -3, c = 6$

$$x = \frac{3 \pm \sqrt{(-3)^2 - 4(4)(6)}}{2(4)}$$

$$x = \frac{3 \pm \sqrt{9 - 96}}{8}$$

$$x = \frac{3 \pm \sqrt{-87}}{8}$$

$$x = \frac{3 \pm i\sqrt{87}}{8}$$

The solutions are $\frac{3 + i\sqrt{87}}{8}$ and $\frac{3 - i\sqrt{87}}{8}$.

23. $(x + 5)(x - 1) = 2$

$x^2 + 4x - 5 = 2$

$x^2 + 4x - 7 = 0$

$a = 1, b = 4, c = -7$

$$x = \frac{-4 \pm \sqrt{4^2 - 4(1)(-7)}}{2(1)}$$

$$x = \frac{-4 \pm \sqrt{16 + 28}}{2}$$

$$x = \frac{-4 + \sqrt{44}}{2}$$

$$x = \frac{-4 \pm \sqrt{4}\sqrt{11}}{2}$$

$$x = \frac{-4 \pm 2\sqrt{11}}{2}$$

$x = -2 \pm \sqrt{11}$

The solutions are $-2 + \sqrt{11}$ and $-2 - \sqrt{11}$.

25. $10y^2 + 10y + 3 = 0$

$a = 10, b = 10, c = 3$

$$y = \frac{-10 \pm \sqrt{10^2 - 4(10)(3)}}{2(10)}$$

$$y = \frac{-10 \pm \sqrt{100 - 120}}{20}$$

$$y = \frac{-10 \pm \sqrt{-20}}{20}$$

$$y = \frac{-10 \pm \sqrt{4}\sqrt{5}\sqrt{-1}}{20}$$

$$y = \frac{-10 \pm 2i\sqrt{5}}{20}$$

$$y = \frac{-5 \pm i\sqrt{5}}{10}$$

The solutions are $\frac{-5 + i\sqrt{5}}{10}$ and $\frac{-5 - i\sqrt{5}}{10}$.

27. $$\frac{-b + \sqrt{b^2 - 4ac}}{2a} + \frac{-b - \sqrt{b^2 - 4ac}}{2a}$$

$$= \frac{-b + \sqrt{b^2 - 4ac} - b - \sqrt{b^2 - 4ac}}{2a}$$

$$= \frac{-2b}{2a}$$

$$= -\frac{b}{a}$$

29. $9x - 2x^2 + 5 = 0$

$-2x^2 + 9x + 5 = 0$

$a = -2, b = 9, c = 5$

$b^2 - 4ac = 9^2 - 4(-2)(5)$

$b^2 - 4ac = 81 + 40$

$b^2 - 4ac = 121 > 0$

Therefore, there are two real solutions.

31. $4x^2 + 12x = -9$

$4x^2 + 12x + 9 = 0$

$a = 4, b = 12, c = 9$

$b^2 - 4ac = 12^2 - 4(4)(9)$

$b^2 - 4ac = 144 - 144 = 0$

Therefore, there is 1 real solution.

33. $3x = -2x^2 + 7$
$2x^2 + 3x - 7 = 0$
$a = 2, b = 3, c = -7$
$b^2 - 4ac = 3^2 - 4(2)(-7)$
$b^2 - 4ac = 9 + 56$
$b^2 - 4ac = 65 > 0$
Therefore, there are two real solutions.

35. $6 = 4x - 5x^2$
$5x^2 - 4x + 6 = 0$
$a = 5, b = -4, c = 6$
$b^2 - 4ac = (-4)^2 - 4(5)(6)$
$b^2 - 4ac = 16 - 120$
$b^2 - 4ac = -104 < 0$
Therefore, there are two complex but not real solutions.

37. $x^2 + 5x = -2$
$x^2 + 5x + 2 = 0$
$a = 1, b = 5, c = 2$
$$x = \frac{-5 \pm \sqrt{5^2 - 4(1)(2)}}{2(1)}$$
$$x = \frac{-5 \pm \sqrt{25 - 8}}{2} = \frac{-5 \pm \sqrt{17}}{2}$$
The solutions are $\frac{-5 + \sqrt{17}}{2}$ and $\frac{-5 - \sqrt{17}}{2}$.

39. $(m + 2)(2m - 6) = 5(m - 1) - 12$
$2m^2 - 2m - 12 = 5m - 5 - 12$
$2m^2 - 7m + 5 = 0$
$a = 2, b = -7, c = 5$
$$m = \frac{-(-7) \pm \sqrt{(-7)^2 - 4(2)(5)}}{2(2)}$$
$$m = \frac{7 \pm \sqrt{49 - 40}}{4}$$
$$m = \frac{7 \pm \sqrt{9}}{4}$$
$$m = \frac{7 \pm 3}{4}$$
$$m = \frac{10}{4} = \frac{5}{2}, \text{ or } m = \frac{4}{4} = 1$$
The solutions are $\frac{5}{2}$ and 1.

41. $\frac{x^2}{3} - x = \frac{5}{3}$
$x^2 - 3x = 5$
$x^2 - 3x - 5 = 0$
$a = 1, b = -3, c = -5$
$$x = \frac{3 \pm \sqrt{(-3)^2 - 4(1)(-5)}}{2(1)}$$
$$x = \frac{3 \pm \sqrt{9 + 20}}{2} = \frac{3 \pm \sqrt{29}}{2}$$
The solutions are $\frac{3 + \sqrt{29}}{2}$ and $\frac{3 - \sqrt{29}}{2}$.

43. $x(6x + 2) - 3 = 0$
$6x^2 + 2x - 3 = 0$
$a = 6, b = 2, c = -3$
$$x = \frac{-2 \pm \sqrt{2^2 - 4(6)(-3)}}{2(6)}$$
$$x = \frac{-2 \pm \sqrt{4 + 72}}{12} = \frac{-2 \pm \sqrt{76}}{12}$$
$$x = \frac{-2 \pm \sqrt{4}\sqrt{19}}{12} = \frac{-2 \pm 2\sqrt{19}}{12}$$
$$x = \frac{-1 \pm \sqrt{19}}{6}$$
The solutions are $\frac{-1 + \sqrt{19}}{6}$ and $\frac{-1 - \sqrt{19}}{6}$.

45. $x^2 + 6x + 13 = 0$
$a = 1, b = 6, c = 13$
$$x = \frac{-6 \pm \sqrt{6^2 - 4(1)(13)}}{2(1)}$$
$$x = \frac{-6 \pm \sqrt{36 - 52}}{2} = \frac{-6 \pm \sqrt{-16}}{2}$$
$$x = \frac{-6 \pm 4i}{2} = -3 \pm 2i$$
The solutions are $-3 + 2i$ and $-3 - 2i$.

47. $\frac{2}{5}y^2+\frac{1}{5}y+\frac{3}{5}=0$

$2y^2+y+3=0$

$a=2, b=1, c=3$

$y=\frac{-1\pm\sqrt{1^2-4(2)(3)}}{2(2)}$

$y=\frac{-1\pm\sqrt{1-24}}{4}=\frac{-1\pm\sqrt{-23}}{4}$

$y=\frac{-1\pm i\sqrt{23}}{4}$

The solutions are $\frac{-1+i\sqrt{23}}{4}$ and $\frac{-1-i\sqrt{23}}{4}$.

49. $\frac{1}{2}y^2=y-\frac{1}{2}$

$y^2=2y-1$

$y^2-2y+1=0$

$a=1, b=-2, c=1$

$y=\frac{2\pm\sqrt{(-2)^2-4(1)(1)}}{2(1)}$

$y=\frac{2\pm\sqrt{4-4}}{2}=\frac{2\pm\sqrt{0}}{2}=1$

The solution is 1.

51. $(n-2)^2=15n$

$n^2-4n+4=15n$

$n^2-19n+4=0$

$a=1, b=-19, c=4$

$n=\frac{19\pm\sqrt{(-19)^2-4(1)(4)}}{2(1)}$

$n=\frac{19\pm\sqrt{361-16}}{2}$

$n=\frac{19\pm\sqrt{345}}{2}$

$n=\frac{19}{2}\pm\frac{\sqrt{345}}{2}$

The solutions are $\frac{19+\sqrt{345}}{2}$ and $\frac{19-\sqrt{345}}{2}$.

53. $(x+8)^2+x^2=36^2$

$x^2+16x+64+x^2=1296$

$2x^2+16x-1232=0$

$a=2, b=16, c=-1232$

$x=\frac{-16\pm\sqrt{16^2-4(2)(-1232)}}{2(2)}$

$x=\frac{-16\pm\sqrt{10{,}112}}{4}$

$x\approx 21$ or $x\approx -29$ (disregard)

$x+x+8=2(21)+8=50$

$50-36=14$

They save about 14 feet of walking distance.

55. Let x = length of leg

$x+2$ = length of hypotenuse

$x^2+x^2=(x+2)^2$

$2x^2=x^2+4x+4$

$x^2-4x-4=0$

$a=1, b=-4, c=-4$

$x=\frac{4\pm\sqrt{(-4)^2-4(1)(-4)}}{2(1)}$

$x=\frac{4\pm\sqrt{32}}{2}$

$x=\frac{4\pm4\sqrt{2}}{2}$

$x=2\pm2\sqrt{2}$

(disregard a negative value)

The sides measure $2+2\sqrt{2}$ cm, $2+2\sqrt{2}$ cm, and $4+2\sqrt{2}$ cm.

57. Let x = width, then $x + 10$ = length.
Area = length · width

$$400 = (x+10)x$$
$$0 = x^2 + 10x - 400$$
$$a = 1, b = 10, c = -400$$
$$x = \frac{-10 \pm \sqrt{10^2 - 4(1)(-400)}}{2(1)}$$
$$x = \frac{-10 \pm \sqrt{1700}}{2}$$
$$x = \frac{-10 \pm 10\sqrt{17}}{2}$$
$$x = -5 \pm 5\sqrt{17}$$

Disregard a negative length. The width is $-5 + 5\sqrt{17}$ ft and the length is $5 + 5\sqrt{17}$ ft.

59. a. Let x = length.

$$x^2 + x^2 = 100^2$$
$$2x^2 - 10{,}000 = 0$$
$$a = 2, b = 0, c = -10{,}000$$
$$x = \frac{0 \pm \sqrt{0^2 - 4(2)(-10{,}000)}}{2(2)}$$
$$x = \frac{\pm\sqrt{80{,}000}}{4}$$
$$x = \frac{\pm 200\sqrt{2}}{4}$$
$$x = \pm 50\sqrt{2}$$

Disregard a negative length. The side measures $50\sqrt{2}$ meters.

b. Area $= s^2$

$$= \left(50\sqrt{2}\right)^2$$
$$= 50^2\left(\sqrt{2}\right)^2$$
$$= 2500 \cdot 2$$
$$= 5000$$

The area is 5000 square meters.

61.

$$\frac{x-1}{1} = \frac{1}{x}$$
$$x(x-1) = 1 \cdot 1$$
$$x^2 - x - 1 = 0$$
$$a = 1, b = -1, c = -1$$
$$x = \frac{1 \pm \sqrt{(-1)^2 - 4(1)(-1)}}{2(1)}$$
$$x = \frac{1 \pm \sqrt{5}}{2}$$

(disregard a negative length)

The value is $\frac{1+\sqrt{5}}{2}$.

63.

$$h = -16t^2 + 20t + 1100$$
$$0 = -16t^2 + 20t + 1100$$
$$a = -16, b = 20, c = 1100$$
$$t = \frac{-20 \pm \sqrt{20^2 - 4(-16)(1100)}}{2(-16)}$$
$$t = \frac{-20 \pm \sqrt{70{,}800}}{-32}$$

$t \approx 8.9$ or $t \approx -7.7$ (disregard)

It will take about 8.9 seconds.

65.

$$h(t) = -16t^2 - 20t + 180$$
$$0 = -16t^2 - 20t + 180$$
$$0 = -4(4t^2 + 5t - 45)$$
$$0 - 4t^2 + 5t - 45$$
$$t = \frac{-5 \pm \sqrt{5^2 - 4(4)(-45)}}{2(4)}$$
$$t = \frac{-5 \pm \sqrt{745}}{8}$$
$$t \approx \frac{-5 \pm 27.3}{8}$$

$t \approx 2.8$ or $t \approx -4.0$

Time cannot be negative, therefore the time until it strikes the ground is 2.8 seconds.

67. Sunday to Monday

69. Wednesday

71. $f(4) = 3(4)^2 - 18(4) + 56 = 32$

This answer appears to agree with the graph.

73. a. $f(x) = 128.5x^2 - 69.5x + 2681$
$x = 1997 - 1995 = 2$
$f(2) = 128.5(2)^2 - 69.5(2) + 2681$
$f(2) = 3056$
Their net income was \$3056 million.

b. $15{,}000 = 128.5x^2 - 69.5x + 2681$
$0 = 128.5x^2 - 69.5x - 12{,}319$
$a = 128.5, b = -69.5, c = -12{,}319$
$$x = \frac{69.5 \pm \sqrt{(-69.5)^2 - 4(128.5)(-12{,}319)}}{2(128.5)}$$
$x \approx 10$ or $x \approx -10$ (disregard)
$1995 + 10 = 2005$
10 years after 1995 or in 2005, the net income will be \$15,000 million.

75. Exercise 63:

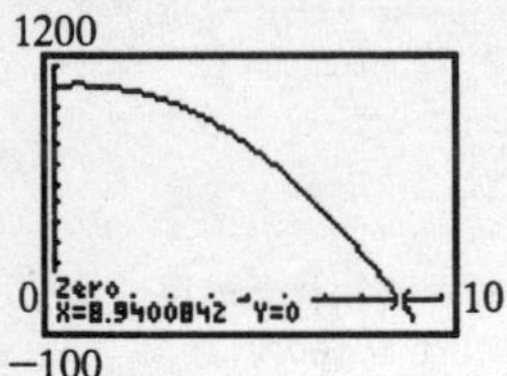

Exercise 65:

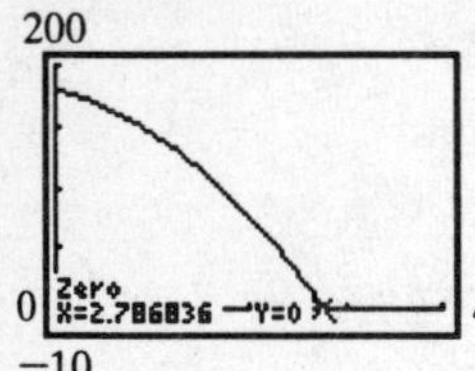

77.

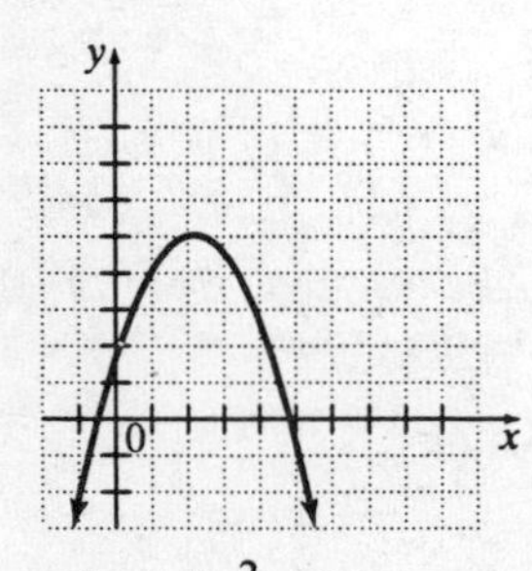

$y = 9x - 2x^2 + 5$
There are 2 intercepts.

79. $\sqrt{5x-2} = 3$
$(\sqrt{5x-2})^2 = 3^2$
$5x - 2 = 9$
$5x = 11$
$x = \frac{11}{5}$

Checking:
$\sqrt{5\left(\frac{11}{5}\right) - 2} \stackrel{?}{=} 3$
$\sqrt{11-2} \stackrel{?}{=} 3$
$\sqrt{9} \stackrel{?}{=} 3$
$3 = 3$
The solution is $\frac{11}{5}$.

81. $\frac{1}{x} + \frac{2}{5} = \frac{7}{x}$
$5x \cdot \frac{1}{x} + 5x \cdot \frac{2}{5} = 5x \cdot \frac{7}{x}$
$5 + 2x = 35$
$2x = 30$
$x = 15$

83. $x^4 + x^2 - 20 = (x^2 + 5)(x^2 - 4)$
$= (x^2 + 5)(x + 2)(x - 2)$

85. $z^4 - 13z^2 + 36 = (z^2 - 9)(z^2 - 4)$
$= (z + 3)(z - 3)(z + 2)(z - 2)$

87. $3x^2 - \sqrt{12}x + 1 = 0$

$a = 3,\ b = -\sqrt{12},\ c = 1$

$$x = \frac{\sqrt{12} \pm \sqrt{(-\sqrt{12})^2 - 4(3)(1)}}{2(3)}$$

$$x = \frac{\sqrt{12} \pm \sqrt{12 - 12}}{6}$$

$$x = \frac{\sqrt{4}\sqrt{3} \pm \sqrt{0}}{6}$$

$$x = \frac{2\sqrt{3}}{6}$$

$$x = \frac{\sqrt{3}}{3}$$

The solution is $\frac{\sqrt{3}}{3}$.

89. $x^2 + \sqrt{2}x + 1 = 0$

$a = 1,\ b = \sqrt{2},\ c = 1$

$$x = \frac{-\sqrt{2} \pm \sqrt{(\sqrt{2})^2\ \ 4(1)(1)}}{2(1)}$$

$$x = \frac{-\sqrt{2} \pm \sqrt{2 - 4}}{2}$$

$$x = \frac{-\sqrt{2} \pm \sqrt{-2}}{2}$$

$$x = \frac{-\sqrt{2} \pm i\sqrt{2}}{2}$$

The solutions are $\frac{-\sqrt{2} + i\sqrt{2}}{2}$ and $\frac{-\sqrt{2} - i\sqrt{2}}{2}$.

91. $2x^2 - \sqrt{3}x - 1 = 0$

$a = 2,\ b = -\sqrt{3},\ c = -1$

$$x = \frac{\sqrt{3} \pm \sqrt{(-\sqrt{3})^2 - 4(2)(-1)}}{2(2)}$$

$$x = \frac{\sqrt{3} \pm \sqrt{3 + 8}}{4}$$

$$x = \frac{\sqrt{3} \pm \sqrt{11}}{4}$$

The solutions are $\frac{\sqrt{3} + \sqrt{11}}{4}$ and $\frac{\sqrt{3} - \sqrt{11}}{4}$.

Exercise Set 10.3

1.
$$2x = \sqrt{10 + 3x}$$
$$4x^2 = 10 + 3x$$
$$4x^2 - 3x - 10 = 0$$
$$(4x + 5)(x - 2) = 0$$
$$4x + 5 = 0 \quad \text{or} \quad x - 2 = 0$$
$$x = -\frac{5}{4} \quad \text{or} \quad x = 2$$

Discard $x = -\frac{5}{4}$.

The solution is 2.

3. $x - 2\sqrt{x} = 8$
$$2\sqrt{x} = x - 8$$
$$4x = x^2 - 16x + 64$$
$$0 = x^2 - 20x + 64$$
$$0 = (x - 16)(x - 4)$$
$$x - 16 = 0 \quad \text{or} \quad x - 4 = 0$$
$$x = 16 \quad \text{or} \quad x = 4 \text{ (discard)}$$

The solution is 16.

5. $\sqrt{9x} = x + 2$
$$9x = x^2 + 4x + 4$$
$$0 = x^2 - 5x + 4$$
$$0 = (x - 4)(x - 1)$$
$$x - 4 = 0 \quad \text{or} \quad x - 1 = 0$$
$$x = 4 \quad \text{or} \quad x = 1$$

The solutions are 4 and 1.

7. $\frac{2}{x}+\frac{3}{x-1}=1$

$\frac{2(x-1)+3x}{x(x-1)}=1$

$\frac{2x-2+3x}{x^2-x}=1$

$5x-2=x^2-x$

$x^2-6x+2=0$

$x=\frac{6\pm\sqrt{(-6)^2-4(1)(2)}}{2(1)}=\frac{6\pm\sqrt{28}}{2}=\frac{6\pm2\sqrt{7}}{2}=3\pm\sqrt{7}$

The solutions are $3+\sqrt{7}$ and $3-\sqrt{7}$.

9. $\frac{3}{x}+\frac{4}{x+2}=2$

$\frac{3(x+2)+4x}{x(x+2)}=2$

$\frac{3x+6+4x}{x^2+2x}=2$

$7x+6=2(x^2+2x)$

$7x+6=2x^2+4x$

$2x^2-3x-6=0$

$x=\frac{3\pm\sqrt{(-3)^2-4(2)(-6)}}{2(2)}$

$x=\frac{3\pm\sqrt{57}}{4}$

The solutions are $\frac{3+\sqrt{57}}{4}$ and $\frac{3-\sqrt{57}}{4}$.

11. $\frac{7}{x^2-5x+6}=\frac{2x}{x-3}-\frac{x}{x-2}$

$\frac{7}{(x-3)(x-2)}=\frac{2x(x-2)-x(x-3)}{(x-3)(x-2)}$

$7=2x^2-4x-x^2+3x$

$7=x^2-x$

$0=x^2-x-7$

$x=\frac{1\pm\sqrt{(-1)^2-4(1)(-7)}}{2(1)}$

$x=\frac{1\pm\sqrt{29}}{2}$

The solutions are $\frac{1+\sqrt{29}}{2}$ and $\frac{1-\sqrt{29}}{2}$.

13.
$$p^4 - 16 = 0$$
$$(p^2 + 4)(p^2 - 4) = 0$$
$$(p + 2i)(p - 2i)(p + 2)(p - 2) = 0$$
$$p + 2i = 0 \text{ or } p - 2i = 0 \text{ or } p + 2 = 0 \text{ or } p - 2 = 0$$
$$p = -2i \quad \text{or} \quad p = 2i \quad \text{or} \quad p = -2 \quad \text{or} \quad p = 2$$
The solutions are $-2i$, $2i$, -2, and 2.

15.
$$4x^4 + 11x^2 = 3$$
$$4x^4 + 11x^2 - 3 = 0$$
$$(4x^2 - 1)(x^2 + 3) = 0$$
$$(2x + 1)(2x - 1)\left(x + i\sqrt{3}\right)\left(x - i\sqrt{3}\right) = 0$$
$$2x + 1 = 0 \quad \text{or} \quad 2x - 1 = 0 \quad \text{or} \quad x + i\sqrt{3} = 0 \quad \text{or} \quad x - i\sqrt{3} = 0$$
$$2x = -1 \quad \text{or} \quad 2x = 1 \quad \text{or} \quad x = -i\sqrt{3} \quad \text{or} \quad x = i\sqrt{3}$$
$$x = -\frac{1}{2} \quad \text{or} \quad x = \frac{1}{2}$$
The solutions are $-\frac{1}{2}, \frac{1}{2}, -i\sqrt{3}$, and $i\sqrt{3}$.

17.
$$z^4 - 13z^2 + 36 = 0$$
$$(z^2 - 9)(z^2 - 4) = 0$$
$$(z + 3)(z - 3)(z + 2)(z - 2) = 0$$
$$z + 3 = 0 \quad \text{or} \quad z - 3 = 0 \quad \text{or} \quad z + 2 = 0 \quad \text{or} \quad z - 2 = 0$$
$$z = -3 \quad \text{or} \quad z = 3 \quad \text{or} \quad z = -2 \quad \text{or} \quad z = 2$$
The solutions are -3, 3, -2, and 2.

19. $x^{2/3} - 3x^{1/3} - 10 = 0$

Let $y = x^{1/3}$.
$$y^2 - 3y - 10 = 0$$
$$(y - 5)(y + 2) = 0$$
$$y - 5 = 0 \quad \text{or} \quad y + 2 = 0$$
$$y = 5 \quad \text{or} \quad y = -2$$
$$x^{1/3} = 5 \quad \text{or} \quad x^{1/3} = -2$$
$$x = 125 \quad \text{or} \quad x = -8$$
The solutions are 125 and -8.

21. $(5n+1)^2+2(5n+1)-3=0$

Let $y = 5n + 1$.

$$y^2+2y-3=0$$
$$(y+3)(y-1)=0$$
$$y+3=0 \quad \text{or} \quad y-1=0$$
$$y=-3 \quad \text{or} \quad y=1$$
$$5n+1=-3 \quad \text{or} \quad 5n+1=1$$
$$5n=-4 \quad \text{or} \quad 5n=0$$
$$n=-\frac{4}{5} \quad \text{or} \quad n=0$$

The solutions are $-\frac{4}{5}$ and 0.

23. $2x^{2/3}-5x^{1/3}=3$

Let $y=x^{1/3}$.

$$2y^2-5y=3$$
$$2y^2-5y-3=0$$
$$(2y+1)(y-3)=0$$
$$2y+1=0 \quad \text{or} \quad y-3=0$$
$$y=-\frac{1}{2} \quad \text{or} \quad y=3$$
$$x^{1/3}=-\frac{1}{2} \quad \text{or} \quad x^{1/3}=3$$
$$x=-\frac{1}{8} \quad \text{or} \quad x=27$$

The solutions are $-\frac{1}{8}$ and 27.

25. $$1+\frac{2}{3t-2}=\frac{8}{(3t-2)^2}$$

$$(3t-2)^2+2(3t-2)-8=0$$

Let $y = 3t - 2$.

$$y^2+2y-8=0$$
$$(y+4)(y-2)=0$$
$$y+4=0 \quad \text{or} \quad y-2=0$$
$$y=-4 \quad \text{or} \quad y=2$$
$$3t-2=-4 \quad \text{or} \quad 3t-2=2$$
$$3t=-2 \quad \text{or} \quad 3t=4$$
$$t=-\frac{2}{3} \quad \text{or} \quad t=\frac{4}{3}$$

The solutions are $-\frac{2}{3}$ and $\frac{4}{3}$.

27. $20x^{2/3}-6x^{1/3}-2=0$

Let $y=x^{1/3}$.

$$20y^2-6y-2=0$$
$$(5y+1)(4y-2)=0$$
$$5y=-1 \quad \text{or} \quad 4y=2$$
$$y=-\frac{1}{5} \quad \text{or} \quad y=\frac{1}{2}$$
$$x^{1/3}=-\frac{1}{5} \quad \text{or} \quad x^{1/3}=\frac{1}{2}$$
$$x=-\frac{1}{125} \quad \text{or} \quad x=\frac{1}{8}$$

The solutions are $-\frac{1}{125}$ and $\frac{1}{8}$.

29. $$a^4-5a^2+6=0$$

$$(a^2-3)(a^2-2)=0$$
$$a^2-3=0 \quad \text{or} \quad a^2-2=0$$
$$a^2=3 \quad \text{or} \quad a^2=2$$
$$a=\pm\sqrt{3} \quad \text{or} \quad a=\pm\sqrt{2}$$

The solutions are $-\sqrt{3}$, $\sqrt{3}$, $-\sqrt{2}$, and $\sqrt{2}$.

31. $$\frac{2x}{x-2}+\frac{x}{x+3}=\frac{-5}{x+3}$$

$$\frac{2x}{x-2}=\frac{-x}{x+3}-\frac{5}{x+3}$$
$$\frac{2x}{x-2}=\frac{-x-5}{x+3}$$
$$2x(x+3)=(x-2)(-x-5)$$
$$2x^2+6x=-x^2+2x-5x+10$$
$$2x^2+6x=-x^2-3x+10$$
$$3x^2+9x-10=0$$
$$x=\frac{-9\pm\sqrt{9^2-4(3)(-10)}}{2(3)}$$
$$x=\frac{-9\pm\sqrt{201}}{6}$$

The solutions are $\frac{-9+\sqrt{201}}{6}$ and $\frac{-9-\sqrt{201}}{6}$.

33. $(p+2)^2 = 9(p+2) - 20$

$(p+2)^2 - 9(p+2) + 20 = 0$

Let $x = p + 2$.

$x^2 - 9x + 20 = 0$

$(x-5)(x-4) = 0$

$x = 5$ or $x = 4$

$p+2 = 5$ or $p+2 = 4$

$p = 3$ or $p = 2$

The solutions are 3 and 2.

35. $2x = \sqrt{11x+3}$

$4x^2 = 11x + 3$

$4x^2 - 11x - 3 = 0$

$(4x+1)(x-3) = 0$

$x = -\frac{1}{4}$ (discard) or $x = 3$

The solution is 3.

37. $x^{2/3} - 8x^{1/3} + 15 = 0$

Let $y = x^{1/3}$.

$y^2 - 8y + 15 = 0$

$(y-5)(y-3) = 0$

$y = 5$ or $y = 3$

$x^{1/3} = 5$ or $x^{1/3} = 3$

$x = 5^3$ or $x = 3^3$

$x = 125$ or $x = 27$

The solutions are 125 and 27.

39. $y^3 + 9y - y^2 - 9 = 0$

$y(y^2+9) - 1(y^2+9) = 0$

$(y^2+9)(y-1) = 0$

$(y+3i)(y-3i)(y-1) = 0$

$y + 3i = 0$ or $y - 3i = 0$ or $y - 1 = 0$

$y = -3i$ or $y = 3i$ or $y = 1$

The solutions are $-3i$, $3i$, and 1.

41. $2x^{2/3} + 3x^{1/3} - 2 = 0$

Let $m = x^{1/3}$.

$2m^2 + 3m - 2 = 0$

$(2m-1)(m+2) = 0$

$m = \frac{1}{2}$ or $m = -2$

$x^{1/3} = \frac{1}{2}$ or $x^{1/3} = -2$

$x = \frac{1}{8}$ or $x = -8$

The solutions are $\frac{1}{8}$ and -8.

43. $x^{-2} - x^{-1} - 6 = 0$

Let $y = x^{-1}$.

$y^2 - y - 6 = 0$

$(y-3)(y+2) = 0$

$y = 3$ or $y = -2$

$x^{-1} = 3$ or $x^{-1} = -2$

$x = \frac{1}{3}$ or $x = -\frac{1}{2}$

The solutions are $\frac{1}{3}$ and $-\frac{1}{2}$.

45. $x - \sqrt{x} = 2$

$x - 2 = \sqrt{x}$

$x^2 - 4x + 4 = x$

$x^2 - 5x + 4 = 0$

$(x-4)(x-1) = 0$

$x = 4$ or $x = 1$ (discard)

The solution is 4.

47. $\frac{x}{x-1} + \frac{1}{x+1} = \frac{2}{x^2-1}$

$\frac{x(x+1) + x - 1}{(x-1)(x+1)} = \frac{2}{(x+1)(x-1)}$

$x^2 + x + x - 1 = 2$

$x^2 + 2x - 3 = 0$

$(x+3)(x-1) = 0$

$x + 3 = 0$ or $x - 1 = 0$

$x = -3$ or $x = 1$ (discard)

The solution is -3.

49. $p^4 - p^2 - 20 = 0$
$(p^2 - 5)(p^2 + 4) = 0$
$p^2 - 5 = 0$ or $p^2 + 4 = 0$
$p^2 = 5$ or $(p + 2i)(p - 2i) = 0$
$p = \pm\sqrt{5}$ or $p + 2i = 0$ or $p - 2i = 0$
$p = -2i$ or $p = 2i$

The solutions are $-\sqrt{5}$, $\sqrt{5}$, $-2i$, and $2i$.

51. $2x^3 = -54$
$x^3 = -27$
$x^3 + 27 = 0$
$(x + 3)(x^2 - 3x + 9) = 0$
$x + 3 = 0$ or $x^2 - 3x + 9 = 0$
$x = -3$ or

$$x = \frac{3 \pm \sqrt{(-3)^2 - 4(1)(9)}}{2(1)}$$

$$x = \frac{3 \pm \sqrt{-27}}{2}$$

$$x = \frac{3 \pm 3i\sqrt{3}}{2}$$

The solutions are -3, $\frac{3 + 3i\sqrt{3}}{2}$, and $\frac{3 - 3i\sqrt{3}}{2}$.

53. $1 = \frac{4}{x - 7} + \frac{5}{(x - 7)^2}$
$(x - 7)^2 = 4(x - 7) + 5$
Let $y = x - 7$.
$y^2 - 4y - 5 = 0$
$(y - 5)(y + 1) = 0$
$y = 5$ or $y = -1$
$x - 7 = 5$ or $x - 7 = -1$
$x = 12$ or $x = 6$

The solutions are 12 and 6.

55. $27y^4 + 15y^2 = 2$
$27y^4 + 15y^2 - 2 = 0$
$(9y^2 - 1)(3y^2 + 2) = 0$
$(3y + 1)(3y - 1)(3y^2 + 2) = 0$
$3y + 1 = 0$ or $3y - 1 = 0$ or $3y^2 + 2 = 0$
$3y = -1$ or $3y = 1$ or $3y^2 = -2$
$y = -\frac{1}{3}$ or $y = \frac{1}{3}$ or $y^2 = -\frac{2}{3}$

$$y = \pm\sqrt{-\frac{2}{3}}$$

$$y = \pm\frac{i\sqrt{6}}{3}$$

The solutions are $-\frac{1}{3}$, $\frac{1}{3}$, $-\frac{i\sqrt{6}}{3}$, and $\frac{i\sqrt{6}}{3}$.

57. Let x = speed on first part.
$x - 1$ = speed on second part
$D = r \cdot t$ or $t = \frac{D}{r}$, $1\frac{3}{5} = \frac{8}{5}$

$$\frac{3}{x} + \frac{4}{x - 1} = \frac{8}{5}$$

$3 \cdot 5(x - 1) + 4 \cdot 5x = 8 \cdot x(x - 1)$
$15x - 15 + 20x = 8x^2 - 8x$
$0 = 8x^2 - 43x + 15$
$0 = (8x - 3)(x - 5)$
$8x - 3 = 0$ or $x - 5 = 0$
$x = \frac{3}{8}$ or $x = 5$
$x - 1 = 4$

Her speeds were 5 mph and 4 mph.

59. Let x = time for hose alone, then
$x - 1$ = time for inlet pipe alone.

$$\frac{1}{x} + \frac{1}{x-1} = \frac{1}{8}$$
$$8(x-1) + 8x = x(x-1)$$
$$8x - 8 + 8x = x^2 - x$$
$$0 = x^2 - 17x + 8$$

$$x = \frac{17 \pm \sqrt{(-17)^2 - 4(1)(8)}}{2(1)}$$
$$x = \frac{17 \pm \sqrt{257}}{2}$$
$x \approx 0.5$ (discard) or $x \approx 16.5$
$x - 1 \approx 15.5$

Inlet pipe: 15.5 hours
Hose: 16.5 hours

61. Let x = original speed.
$x + 11$ = return speed

$D = r \cdot t$ or $t = \frac{D}{r}$

$$\frac{330}{x} - \frac{330}{x+11} = 1$$
$$330(x+11) - 330x = x(x+11)$$
$$330x + 3630 - 330x = x^2 + 11x$$
$$0 = x^2 + 11x - 3630$$
$$0 = (x-55)(x+66)$$
$x = 55$ or $x = -66$ (discard)
$x + 11 = 66$
orginal speed: 55 mph
return speed: 66 mph

63.

	Hours	Part complete in one hour
Bill	$x-1$	$\frac{1}{x-1}$
Billy	x	$\frac{1}{x}$
Together	4	$\frac{1}{4}$

$$\frac{1}{x-1} + \frac{1}{x} = \frac{1}{4}$$
$$\frac{x + (x-1)}{x(x-1)} = \frac{1}{4}$$
$$4(2x-1) = x(x-1)$$
$$8x - 4 = x^2 - x$$
$$0 = x^2 - 9x + 4$$
$$x = \frac{-(-9) \pm \sqrt{(-9)^2 - 4(1)(4)}}{2(1)}$$
$$x = \frac{9 - \sqrt{65}}{2} \quad \text{or} \quad x = \frac{9 + \sqrt{65}}{2}$$
$x \approx 0.5$ (discard) or $x \approx 8.5$
It takes his son about 8.5 hours.

65. Let x = the number.
$$x(x-4) = 96$$
$$x^2 - 4x = 96$$
$$x^2 - 4x - 96 = 0$$
$$(x-12)(x+8) = 0$$
$x - 12 = 0$ or $x + 8 = 0$
$x = 12$ or $x = -8$
The number is 12 or –8.

67. **a.** $x - 3 - 3 = (x - 6)$ cm

b. $V = l \cdot w \cdot h$
300 cu. cm $= (x-6)(x-6) \cdot 3$ cu. cm

c. $(x-6)(x-6) = 100$
$$x^2 - 12x + 36 = 100$$
$$x^2 - 12x - 64 = 0$$
$$(x-16)(x+4) = 0$$
$x = 16$ or $x = -4$ (discard)
The sheet is 16 cm by 16 cm.

69. a. Let x = Papis's fastest lap speed and $x + 3.8$ = Montoya's fastest lap speed.
Using time = distance ÷ speed,

$$\frac{7920}{x} = \frac{7920}{x+3.8} + 0.376$$

$$7920(x+3.8) = 7920x + 0.376x(x+3.8)$$

$$7920x + 30096 = 7920x + 0.376x^2 + 1.4288x$$

$$0 = 0.376x^2 + 1.4288x - 30096$$

$$x = \frac{-1.4288 \pm \sqrt{(1.4288)^2 - 4(0.376)(-30096)}}{2(0.376)}$$

Using the positive square root, $x \approx 281.0$ feet per second.

b. $x + 3.8 = 281.0 + 3.8 = 284.8$ feet per second.

c. There are 5,280 feet in one mile, and 3,600 seconds in one hour.

Papis: $\frac{281 \text{ ft}}{\text{sec}} \cdot \frac{3600 \text{ sec}}{1 \text{ hour}} \cdot \frac{1 \text{ mile}}{5280 \text{ ft}} \approx 191.6$ mph

Montoya: $\frac{284.8 \text{ ft}}{\text{sec}} \cdot \frac{3600 \text{ sec}}{1 \text{ hour}} \cdot \frac{1 \text{ mile}}{5280 \text{ ft}} \approx 194.2$ mph

71. Answers may vary.

73. $\frac{5x}{3} + 2 \le 7$

$$\frac{5x}{3} \le 5$$

$$\frac{3}{5}\left(\frac{5x}{3}\right) \le \frac{3}{5}(5)$$

$$x \le 3$$

The solution is $(-\infty, 3]$.

75. $\frac{y-1}{15} > -\frac{2}{5}$

$$15\left(\frac{y-1}{15}\right) > 15\left(-\frac{2}{5}\right)$$

$$y - 1 > -6$$

$$y > -5$$

The solution is $(-5, \infty)$.

77. Domain: $\{x | x \text{ is a real number}\}$

Range: $\{y | y \text{ is a real number}\}$

It is a function.

79. Domain: $\{x|x \text{ is a real number}\}$
Range: $\{y|y \ge -1\}$
It is a function.

Exercise Set 10.4

1. $(x+1)(x+5) > 0$
$x+1=0$ or $x+5=0$
$x=-1$ or $x=-5$

Region	Interval	Test Point
A	$(-\infty, -5)$	-6
B	$(-5, -1)$	-2
C	$(-1, \infty)$	0

$x=-6$	$(-6+1)(-6+5) > 0$	True
$x=-2$	$(-2+1)(-2+5) > 0$	False
$x=0$	$(0+1)(0+5) > 0$	True

-5 -1

The solution is $(-\infty, -5) \cup (-1, \infty)$.

3. $(x-3)(x+4) \le 0$
$x-3=0$ or $x+4=0$
$x=3$ or $x=-4$

Region	Interval	Test Point
A	$(-\infty, -4)$	-5
B	$(-4, 3)$	0
C	$(3, \infty)$	4

$x=-5$	$(-5-3)(-5+4) \le 0$	False
$x=0$	$(0-3)(0+4) \le 0$	True
$x=4$	$(4-3)(4+4) \le 0$	False

-4 3

The solution is $[-4, 3]$.

5. $x^2-7x+10 \le 0$
$(x-5)(x-2) \le 0$
$(x-5)(x-2) = 0$
$x-5=0$ or $x-2=0$
$x=5$ or $x=2$

Region	Interval	Test Point
A	$(-\infty, 2)$	0
B	$(2, 5)$	3
C	$(5, \infty)$	6

$x=0$	$(0-5)(0-2) \le 0$	False
$x=3$	$(3-5)(3-2) \le 0$	True
$x=6$	$(6-5)(6-2) \le 0$	False

2 5

The solution is $[2, 5]$.

7. $3x^2+16x < -5$
$3x^2+16x+5 < 0$
$(3x+1)(x+5) < 0$
$(3x+1)(x+5) = 0$
$3x+1=0$ or $x+5=0$
$3x=-1$ or $x=-5$
$x=-\frac{1}{3}$

Region	Interval	Test Point
A	$(-\infty, -5)$	-6
B	$\left(-5, -\frac{1}{3}\right)$	-1
C	$\left(-\frac{1}{3}, \infty\right)$	0

$x=-6$	$[3(-6)+1](-6+5) < 0$	False
$x=-1$	$[3(-1)+1](-1+5) < 0$	True
$x=0$	$[3(0)+1](0+5) < 0$	False

-5 $-\frac{1}{3}$

The solution is $\left(-5, -\frac{1}{3}\right)$.

9. $(x-6)(x-4)(x-2) > 0$
$(x-6)(x-4)(x-2) = 0$
$x-6=0$ or $x-4=0$ or $x-2=0$
$x=6$ or $x=4$ or $x=2$

Region	Interval	Test Point
A	$(-\infty, 2)$	1
B	$(2, 4)$	3
C	$(4, 6)$	5
D	$(6, \infty)$	7

$x=1$	$(1-6)(1-4)(1-2) > 0$	False
$x=3$	$(3-6)(3-4)(3-2) > 0$	True
$x=5$	$(5-6)(5-4)(5-2) > 0$	False
$x=7$	$(7-6)(7-4)(7-2) > 0$	True

2 4 6

The solution is $(2, 4) \cup (6, \infty)$.

11. $x(x-1)(x+4) \le 0$
$x(x-1)(x+4) = 0$
$x=0$ or $x-1=0$ or $x+4=0$
$x=0$ or $x=1$ or $x=-4$

Region	Interval	Test Point
A	$(-\infty, -4)$	-5
B	$(-4, 0)$	-1
C	$(0, 1)$	$\frac{1}{2}$
D	$(1, \infty)$	2

$x=-5$	$-5(-5-1)(-5+4) \le 0$	True
$x=-1$	$-1(-1-1)(-1+4) \le 0$	False
$x=\frac{1}{2}$	$\frac{1}{2}\left(\frac{1}{2}-1\right)\left(\frac{1}{2}+4\right) \le 0$	True
$x=2$	$2(2-1)(2+4) \le 0$	False

−4 0 1

The solution is $(-\infty, -4] \cup [0, 1]$.

13. $(x^2-9)(x^2-4)>0$

$(x+3)(x-3)(x+2)(x-2)>0$

$(x+3)(x-3)(x+2)(x-2)=0$

$x+3=0$ or $x-3=0$ or $x+2=0$ or $x-2=0$

$x=-3$ or $x=3$ or $x=-2$ or $x=2$

Region	Interval	Test Point
A	$(-\infty, -3)$	-4
B	$(-3, -2)$	$-\frac{5}{2}$
C	$(-2, 2)$	0
D	$(2, 3)$	$\frac{5}{2}$
E	$(3, \infty)$	4

$x=-4$	$(-4+3)(-4-3)(-4+2)(-4-2)>0$	True
$x=-\frac{5}{2}$	$\left(-\frac{5}{2}+3\right)\left(-\frac{5}{2}-3\right)\left(-\frac{5}{2}+2\right)\left(-\frac{5}{2}-2\right)>0$	False
$x=0$	$(0+3)(0-3)(0+2)(0-2)>0$	True
$x=\frac{5}{2}$	$\left(\frac{5}{2}+3\right)\left(\frac{5}{2}-3\right)\left(\frac{5}{2}+2\right)\left(\frac{5}{2}-2\right)>0$	False
$x=4$	$(4+3)(4-3)(4+2)(4-2)>0$	True

−3 −2 2 3

The solution is $(-\infty, -3) \cup (-2, 2) \cup (3, \infty)$.

15. $\frac{x+7}{x-2}<0$

$x=-7$ solves the related equation and $x=2$ makes the denominator 0.

Region	Interval	Test Point
A	$(-\infty, -7)$	-8
B	$(-7, 2)$	0
C	$(2, \infty)$	3

$x=-8$	$\frac{-8+7}{-8-2}<0$	False
$x=0$	$\frac{0+7}{0-2}<0$	True
$x=3$	$\frac{3+7}{3-2}<0$	False

−7 2

The solution is $(-7, 2)$.

17. $\frac{5}{x+1} > 0$

$x = -1$ makes the denominator 0.

Region	Interval	Test Point
A	$(-\infty, -1)$	-2
B	$(-1, \infty)$	0
$x = -2$	$\frac{5}{-2+1} > 0$	False
$x = 0$	$\frac{5}{0+1} > 0$	True

-1

The solution is $(-1, \infty)$.

19. $\frac{x+1}{x-4} \geq 0$

$x = -1$ solves the related equation and $x = 4$ makes the denominator 0.

Region	Interval	Test Point
A	$(-\infty, -1)$	-2
B	$(-1, 4)$	0
C	$(4, \infty)$	5
$x = -2$	$\frac{-2+1}{-2-4} \geq 0$	True
$x = 0$	$\frac{0+1}{0-4} \geq 0$	False
$x = 5$	$\frac{5+1}{5-4} \geq 0$	True

-1 4

The solution is $(-\infty, -1] \cup (4, \infty)$.

21. Answers may vary.

23.

$$\frac{3}{x-2} < 4$$

$$\frac{3}{x-2} - 4 < 0$$

$$\frac{3-4(x-2)}{x-2} < 0$$

$$\frac{3-4x+8}{x-2} < 0$$

$$\frac{11-4x}{x-2} < 0$$

$11 - 4x = 0$ and $x - 2 = 0$

$11 = 4x$ and $x = 2$

$x = \frac{11}{4}$

Region	Interval	Test Point
A	$(-\infty, 2)$	1
B	$\left(2, \frac{11}{4}\right)$	$\frac{5}{2}$
C	$\left(\frac{11}{4}, \infty\right)$	3
$x = 1$	$\frac{11-4(1)}{1-2} < 0$	True
$x = \frac{5}{2}$	$\frac{11-4\left(\frac{5}{2}\right)}{\frac{5}{2}-2} < 0$	False
$x = 3$	$\frac{11-4(3)}{4-2} < 0$	True

2 $\frac{11}{4}$

The solution is $(-\infty, 2) \cup \left(\frac{11}{4}, \infty\right)$.

25. $\frac{x^2+6}{5x} \geq 1$

$\frac{x^2+6}{5x} - 1 \geq 0$

$\frac{x^2+6-5x}{5x} \geq 0$

$\frac{(x-2)(x-3)}{5x} \geq 0$

$x-2=0$ or $x-3=0$ or $5x=0$

$x=2$ or $x=3$ or $x=0$

Region	Interval	Test Point
A	$(-\infty, 0)$	-1
B	$(0, 2)$	1
C	$(2, 3)$	$\frac{5}{2}$
D	$(3, \infty)$	4

$x=-1$	$\frac{(-1-2)(-1-3)}{5(-1)} \geq 0$	False
$x=1$	$\frac{(1-2)(1-3)}{5(1)} \geq 0$	True
$x=\frac{5}{2}$	$\frac{\left(\frac{5}{2}-2\right)\left(\frac{5}{2}-3\right)}{5\left(\frac{5}{2}\right)} \geq 0$	False
$x=4$	$\frac{(4-2)(4-3)}{5(4)} \geq 0$	True

0 2 3

The solution is $(0, 2] \cup [3, \infty)$.

27. $(x-8)(x+7)>0$

$(x-8)(x+7)=0$

$x-8=0$ or $x+7=0$

$x=8$ or $x=-7$

Region	Interval	Test Point
A	$(-\infty, -7)$	-8
B	$(-7, 8)$	0
C	$(8, \infty)$	9

$x=-8$	$(-8-8)(-8+7)>0$	True
$x=0$	$(0-8)(0+7)>0$	False
$x=9$	$(9-8)(9+7)>0$	True

−7 8

The solution is $(-\infty, -7) \cup (8, \infty)$.

29. $(2x-3)(4x+5) \le 0$
$(2x-3)(4x+5) = 0$
$2x-3=0$ or $4x+5=0$
$2x=3$ or $4x=-5$
$x=\frac{3}{2}$ or $x=-\frac{5}{4}$

Region	Interval	Test Point
A	$\left(-\infty, -\frac{5}{4}\right)$	-2
B	$\left(-\frac{5}{4}, \frac{3}{2}\right)$	0
C	$\left(\frac{3}{2}, \infty\right)$	2

$x=-2$	$[2(-2)-3][4(-2)+5] \le 0$	False
$x=0$	$[2(0)-3][4(0)+5] \le 0$	True
$x=2$	$[2(2)-3][4(2)+5] \le 0$	False

$-\frac{5}{4}$ $\frac{3}{2}$

The solution is $\left[-\frac{5}{4}, \frac{3}{2}\right]$.

31. $x^2 > x$
$x^2 - x > 0$
$x(x-1) > 0$
$x(x-1) = 0$
$x=0$ or $x-1=0$
$x=1$

Region	Interval	Test Point
A	$(-\infty, 0)$	-1
B	$(0, 1)$	$\frac{1}{2}$
C	$(1, \infty)$	2

$x=-1$	$-1(-1-1) > 0$	True
$x=\frac{1}{2}$	$\frac{1}{2}\left(\frac{1}{2}-1\right) > 0$	False
$x=2$	$2(2-1) > 0$	True

0 1

The solution is $(-\infty, 0) \cup (1, \infty)$.

33. $(2x-8)(x+4)(x-6) \le 0$
$(2x-8)(x+4)(x-6) = 0$
$2x-8=0$ or $x+4=0$ or $x-6=0$
$2x=8$ or $x=-4$ or $x=6$
$x=4$

Region	Interval	Test Point
A	$(-\infty, -4)$	-5
B	$(-4, 4)$	0
C	$(4, 6)$	5
D	$(6, \infty)$	7

$x=-5$	$[2(-5)-8](-5+4)(-5-6) \le 0$	True
$x=0$	$[2(0)-8](0+4)(0-6) \le 0$	False
$x=5$	$[2(5)-8](5+4)(5-6) \le 0$	True
$x=7$	$[2(7)-8)](7+4)(7-6) \le 0$	False

−4 4 6

The solution is $(-\infty, -4] \cup [4, 6]$.

35. $6x^2 - 5x \ge 6$
$6x^2 - 5x - 6 \ge 0$
$(3x+2)(2x-3) \ge 0$
$(3x+2)(2x-3) = 0$
$3x+2=0$ or $2x-3=0$
$3x=-2$ or $2x=3$
$x=-\frac{2}{3}$ or $x=\frac{3}{2}$

Region	Interval	Test Point
A	$\left(-\infty, -\frac{2}{3}\right)$	-1
B	$\left(-\frac{2}{3}, \frac{3}{2}\right)$	0
C	$\left(\frac{3}{2}, \infty\right)$	2

$x=-1$	$[3(-1)+2][2(-1)-3] \ge 0$	True
$x=0$	$[3(0)+2][2(0)-3] \ge 0$	False
$x=2$	$[3(2)+2][2(2)-3] \ge 0$	True

$-\frac{2}{3}$ $\frac{3}{2}$

The solution is $\left(-\infty, -\frac{2}{3}\right] \cup \left[\frac{3}{2}, \infty\right)$.

37. $4x^3 + 16x^2 - 9x - 36 > 0$

$$4x^2(x+4) - 9(x+4) > 0$$
$$(4x^2 - 9)(x+4) > 0$$
$$(2x-3)(2x+3)(x+4) > 0$$
$$2x - 3 = 0 \quad \text{or} \quad 2x + 3 = 0 \quad \text{or} \quad x + 4 = 0$$
$$x = \frac{3}{2} \quad \text{or} \quad x = -\frac{3}{2} \quad \text{or} \quad x = -4$$

Region	Interval	Test Point
A	$(-\infty, -4)$	−5
B	$\left(-4, -\frac{3}{2}\right)$	−3
C	$\left(-\frac{3}{2}, \frac{3}{2}\right)$	0
D	$\left(\frac{3}{2}, \infty\right)$	2

$x = -5$	$4(-5)^3 + 16(-5)^2 - 9(-5) - 36 = -91$ $-91 > 0$	False
$x = -3$	$4(-3)^3 + 16(-3)^2 - 9(-3) - 36 = 27$ $27 > 0$	True
$x = 0$	$4(0)^3 + 16(0)^2 - 9(0) - 36 = -36$ $-36 > 0$	False
$x = 2$	$4(2)^3 + 16(2)^2 - 9(2) - 36 = 42$ $42 > 0$	True

−4 $\quad -\frac{3}{2} \quad \frac{3}{2}$

The solution is $\left(-4, -\frac{3}{2}\right) \cup \left(\frac{3}{2}, \infty\right)$.

39.

$$x^4 - 26x^2 + 25 \ge 0$$
$$(x^2 - 25)(x^2 - 1) \ge 0$$
$$(x-5)(x+5)(x-1)(x+1) \ge 0$$
$$(x-5)(x+5)(x-1)(x+1) = 0$$
$$x-5=0 \text{ or } x+5=0 \text{ or } x-1=0 \text{ or } x+1=0$$
$$x=5 \text{ or } x=-5 \text{ or } x=1 \text{ or } x=-1$$

Region	Interval	Test Point
A	$(-\infty, -5)$	-6
B	$(-5, -1)$	-3
C	$(-1, 1)$	0
D	$(1, 5)$	3
E	$(5, \infty)$	6
$x=-6$	$(-6-5)(-6+5)(-6-1)(-6+1) \ge 0$	True
$x=-3$	$(-3-5)(-3+5)(-3-1)(-3+1) \ge 0$	False
$x=0$	$(0-5)(0+5)(0-1)(0+1) \ge 0$	True
$x=3$	$(3-5)(3+5)(3-1)(3+1) \ge 0$	False
$x=6$	$(6-5)(6+5)(6-1)(6+1) \ge 0$	True

$-5 \quad -1 \quad 1 \quad 5$

The solution is $(-\infty, -5] \cup [-1, 1] \cup [5, \infty)$.

41. $(2x-7)(3x+5)>0$

$(2x-7)(3x+5)=0$

$2x-7=0$ or $3x+5=0$

$2x=7$ or $3x=-5$

$x=\frac{7}{2}$ or $x=-\frac{5}{3}$

Region	Interval	Test Point
A	$\left(-\infty, -\frac{5}{3}\right)$	-2
B	$\left(-\frac{5}{3}, \frac{7}{2}\right)$	0
C	$\left(\frac{7}{2}, \infty\right)$	4

$x=-2$	$[2(-2)-7][3(-2)+5]>0$	True
$x=0$	$[2(0)-7][3(0)+5]>0$	False
$x=4$	$[2(4)-7][3(4)+5]>0$	True

$-\frac{5}{3}$ $\frac{7}{2}$

The solution is $\left(-\infty, -\frac{5}{3}\right)\cup\left(\frac{7}{2}, \infty\right)$.

43. $\frac{x}{x-10}<0$

$x=0$ or $x-10=0$

$x=10$

Region	Interval	Test Point
A	$(-\infty, 0)$	-1
B	$(0, 10)$	1
C	$(10, \infty)$	11

$x=-1$	$\frac{-1}{-1-10}<0$	False
$x=1$	$\frac{1}{1-10}<0$	True
$x=11$	$\frac{11}{11-10}<0$	False

0 10

The solution is $(0, 10)$.

45. $\frac{x-5}{x+4}\geq 0$

$x-5=0$ or $x+4=0$

$x=5$ or $x=-4$

Region	Interval	Test Point
A	$(-\infty, -4)$	-5
B	$(-4, 5)$	0
C	$(5, \infty)$	6

$x=-5$	$\frac{-5-5}{-5+4}\geq 0$	True
$x=0$	$\frac{0-5}{0+4}\geq 0$	False
$x=6$	$\frac{6-5}{6+4}\geq 0$	True

-4 5

The solution is $(-\infty, -4)\cup[5, \infty)$.

47. $\frac{x(x+6)}{(x-7)(x+1)} \geq 0$

$x = 0$ or $x + 6 = 0$ or $x - 7 = 0$ or $x + 1 = 0$

$x = 0$ or $x = -6$ or $x = 7$ or $x = -1$

Region	Interval	Test Point
A	$(-\infty, -6)$	-7
B	$(-6, -1)$	-2
C	$(-1, 0)$	$-\frac{1}{2}$
D	$(0, 7)$	1
E	$(7, \infty)$	8

$x = -7$	$\frac{-7(-7+6)}{(-7-7)(-7+1)} \geq 0$	True
$x = -2$	$\frac{-2(-2+6)}{(-2-7)(-2+1)} \geq 0$	False
$x = -\frac{1}{2}$	$\frac{-\frac{1}{2}\left(-\frac{1}{2}+6\right)}{\left(-\frac{1}{2}-7\right)\left(-\frac{1}{2}+1\right)} \geq 0$	True
$x = 1$	$\frac{1(1+6)}{(1-7)(1+1)} \geq 0$	False
$x = 8$	$\frac{8(8+6)}{(8-7)(8+1)} \geq 0$	True

−6   −1   0   7

The solution is $(-\infty, -6] \cup (-1, 0] \cup (7, \infty)$.

49.

$$\frac{-1}{x-1} > -1$$
$$\frac{1}{x-1} < 1$$
$$\frac{1}{x-1} - 1 < 0$$
$$\frac{1-(x-1)}{x-1} < 0$$
$$\frac{1-x+1}{x-1} < 0$$
$$\frac{2-x}{x-1} < 0$$
$$2-x=0 \quad \text{or} \quad x-1=0$$
$$2=x \quad \text{or} \quad x=1$$

Region	Interval	Test Point
A	$(-\infty, 1)$	0
B	$(1, 2)$	$\frac{3}{2}$
C	$(2, \infty)$	3
$x=0$	$\frac{2-0}{0-1} < 0$	True
$x=\frac{3}{2}$	$\frac{2-\frac{3}{2}}{\frac{3}{2}-1} < 0$	False
$x=3$	$\frac{2-3}{3-1} < 0$	True

The solution is $(-\infty, 1) \cup (2, \infty)$.

51.

$$\frac{x}{x+4} \le 2$$
$$\frac{x}{x+4} - 2 \le 0$$
$$\frac{x-2(x+4)}{x+4} \le 0$$
$$\frac{-x-8}{x+4} \le 0$$
$$-x-8=0 \quad \text{or} \quad x+4=0$$
$$-x=8 \quad \text{or} \quad x=-4$$
$$x=-8$$

Region	Interval	Test Point
A	$(-\infty, -8)$	-9
B	$(-8, -4)$	-5
C	$(-4, \infty)$	0
$x=-9$	$\frac{-(-9)-8}{-9+4} \le 0$	True
$x=-5$	$\frac{-(-5)-8}{-5+4} \le 0$	False
$x=0$	$\frac{-0-8}{0+4} \le 0$	True

The solution is $(-\infty, -8] \cup (-4, \infty)$.

53.

$$\frac{z}{z-5} \geq 2z$$

$$\frac{z}{z-5} - 2z \geq 0$$

$$\frac{z - 2z(z-5)}{z-5} \geq 0$$

$$\frac{z - 2z^2 + 10z}{z-5} \geq 0$$

$$\frac{11z - 2z^2}{z-5} \geq 0$$

$$\frac{z(11-2z)}{z-5} \geq 0$$

$z = 0$ or $11 - 2z = 0$ or $z - 5 = 0$

$11 = 2z$ or $z = 5$

$\frac{11}{2} = z$

Region	Interval	Test Point
A	$(-\infty, 0)$	-1
B	$(0, 5)$	1
C	$\left(5, \frac{11}{2}\right)$	$\frac{21}{4}$
D	$\left(\frac{11}{2}, \infty\right)$	6

$z = -1$	$\frac{-1[11-2(-1)]}{-1-5} \geq 0$	True
$z = 1$	$\frac{1[11-2(1)]}{1-5} \geq 0$	False
$z = \frac{21}{4}$	$\frac{\frac{21}{4}\left[11-2\left(\frac{21}{4}\right)\right]}{\frac{21}{4}-5} \geq 0$	True
$z = 6$	$\frac{6[11-2(6)]}{6-5} \geq 0$	False

0 5 $\frac{11}{2}$

The solution is $(-\infty, 0] \cup \left(5, \frac{11}{2}\right]$.

55. $\frac{(x+1)^2}{5x} > 0$

$(x+1)^2 = 0$ or $5x = 0$

$x + 1 = 0$ or $x = 0$

$x = -1$

Region	Interval	Test Point
A	$(-\infty, -1)$	-2
B	$(-1, 0)$	$-\frac{1}{2}$
C	$(0, \infty)$	1

$x = -2$	$\frac{(-2+1)^2}{5(-2)} > 0$	False
$x = -\frac{1}{2}$	$\frac{\left(-\frac{1}{2}+1\right)^2}{5\left(-\frac{1}{2}\right)} > 0$	False
$x = 1$	$\frac{(1+1)^2}{5(1)} > 0$	True

0

The solution is $(0, \infty)$.

57. Let x = the number.

$\frac{1}{x}$ = the reciprocal of the number

$$x - \frac{1}{x} < 0$$

$$\frac{x^2 - 1}{x} < 0$$

$$\frac{(x+1)(x-1)}{x} < 0$$

$x + 1 = 0$ and $x - 1 = 0$ and $x = 0$

$x = -1$ and $x = 1$ and $x = 0$

Region	Interval	Test Point
A	$(-\infty, -1)$	-2
B	$(-1, 0)$	$-\frac{1}{2}$
C	$(0, 1)$	$\frac{1}{2}$
D	$(1, \infty)$	2

$x = -2$ $\quad \frac{(-2+1)(-2-1)}{-2} < 0$ $\quad$ True

$x = -\frac{1}{2}$ $\quad \frac{\left(-\frac{1}{2}+1\right)\left(-\frac{1}{2}-1\right)}{-\frac{1}{2}} < 0$ $\quad$ False

$x = \frac{1}{2}$ $\quad \frac{\left(\frac{1}{2}+1\right)\left(\frac{1}{2}-1\right)}{\frac{1}{2}} < 0$ $\quad$ True

$x = 2$ $\quad \frac{(2+1)(2-1)}{2} < 0$ $\quad$ False

−1 0 1

The solution is $(-\infty, -1) \cup (0, 1)$.

59. $P(x) = -2x^2 + 26x - 44$

$$-2x^2 + 26x - 44 > 0$$

$$-2(x^2 - 13x + 22) > 0$$

$$-2(x - 11)(x - 2) > 0$$

$x - 11 = 0$ or $x - 2 = 0$

$x = 11$ or $x = 2$

Region	Interval	Test Point
A	$(-\infty, 2)$	0
B	$(2, 11)$	3
C	$(11, \infty)$	12

$x = 0$ $\quad -2(0 - 11)(0 - 2) > 0$ $\quad$ False

$x = 3$ $\quad -2(3 - 11)(3 - 2) > 0$ $\quad$ True

$x = 12$ $\quad -2(12 - 11)(12 - 2) > 0$ $\quad$ False

2 11

The solution is $(2, 11)$.

61.

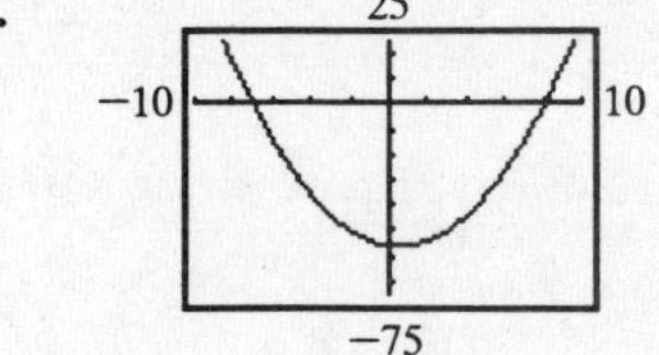

63.

150

−10 10

−150

65. $g(x) = |x| + 2$

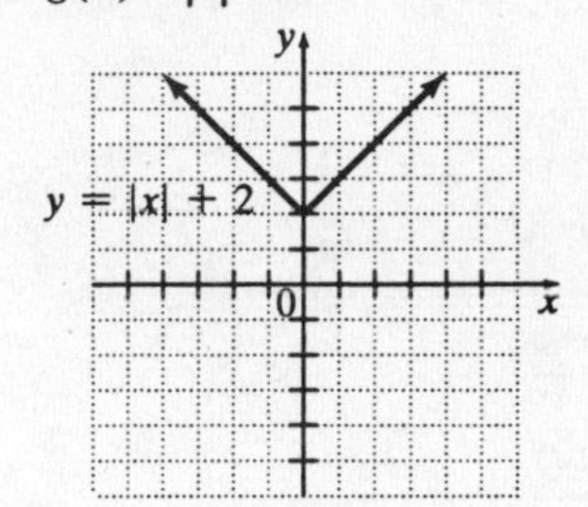

67. $F(x) = |x| - 1$

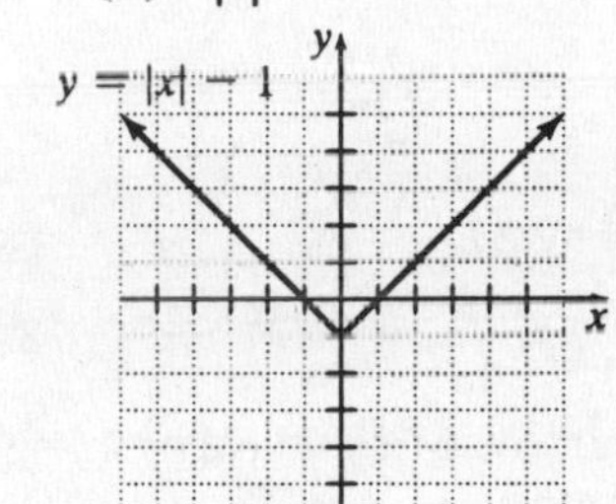

69. $F(x) = x^2 - 3$

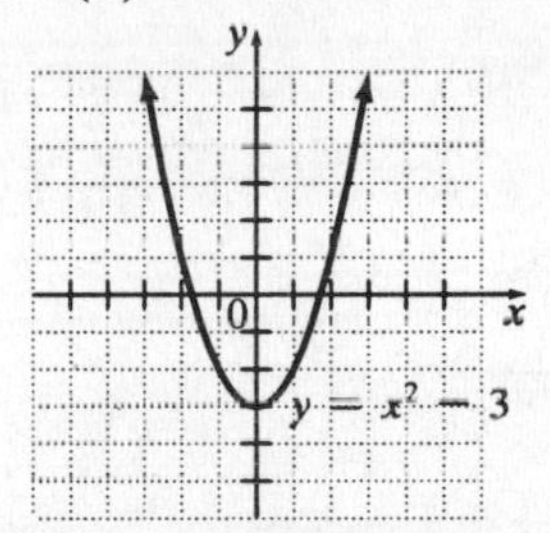

71. $H(x) = x^2 + 1$

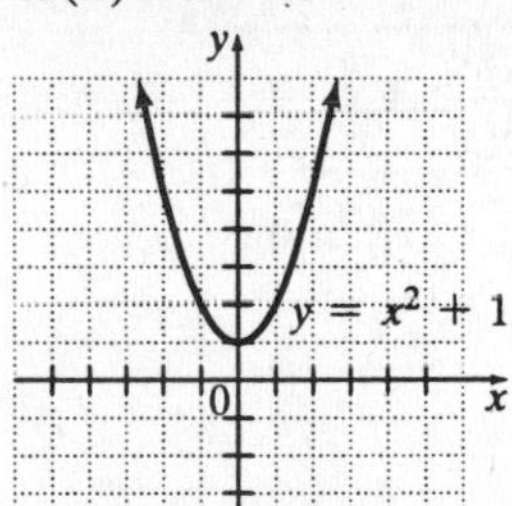

Exercise Set 10.5

1. $y = kx$
$4 = k(20)$
$k = \frac{1}{5}$

$y = \frac{1}{5}x$

3. $y = kx$
$6 = k(4)$
$k = \frac{3}{2}$

$y = \frac{3}{2}x$

5. $y = kx$
$7 = k\left(\frac{1}{2}\right)$
$k = 14$

$y = 14x$

7. $y = kx$
$0.2 = k(0.8)$
$k = 0.25$

$y = 0.25x$

9. $W = kr^3$
$1.2 = k \cdot 2^3$
$k = \frac{1.2}{8} = 0.15$

$W = 0.15r^3 = 0.15(3)^3$
$W = 0.15(27) = 4.05$ lb

11. $P = kN$
$260{,}000 = k(450{,}000)$
$k = \frac{26}{45}$

$P = \frac{26}{45}N = \frac{26}{45}(980{,}000)$
$P \approx 566{,}222$ tons

13. $y = \frac{k}{x}$

$6 = \frac{k}{5}$

$k = 30$

$y = \frac{30}{x}$

15. $y = \frac{k}{x}$

$100 = \frac{k}{7}$

$k = 700$

$y = \frac{700}{x}$

17. $y = \frac{k}{x}$

$\frac{1}{8} = \frac{k}{16}$

$k = 2$

$y = \frac{2}{x}$

19. $y = \frac{k}{x}$

$0.2 = \frac{k}{0.7}$

$k = 0.14$

$y = \frac{0.14}{x}$

21. $R = \frac{k}{T}$

$45 = \frac{k}{6}$

$k = 270$

$R = \frac{270}{T} = \frac{270}{5}$

$R = 54$ mph

23. $I = \frac{k}{R}$

$40 = \frac{k}{270}$

$k = 10,800$

$I = \frac{10,800}{R}$

$= \frac{10,800}{150}$

$= 72$ amps

25. $I_1 = \frac{k}{d^2}$

Replace d by $2d$.

$I_2 = \frac{k}{(2d)^2} = \frac{k}{4d^2} = \frac{1}{4}I_1$

Thus, the intensity is divided by 4.

27. $x = kyz$

29. $r = kst^3$

31. $W = \frac{kwh^2}{l}$

$12 = \frac{k\left(\frac{1}{2}\right)\left(\frac{1}{3}\right)^2}{10}$

$k = 2160$

$W = \frac{2160wh^2}{l}$

$= \frac{2160\left(\frac{2}{3}\right)\left(\frac{1}{2}\right)^2}{16}$

$= \frac{45}{2}$ tons or 22.5 tons

33. $V = kr^2h$

$32\pi = k(4)^2(6)$

$k = \dfrac{\pi}{3}$

$V = \dfrac{\pi}{3}r^2h$

$= \dfrac{\pi}{3}(3)^2(5)$

$= 15\pi$ cu. in.

35. $H = ksd^3$

$40 = k(120)(2)^3$

$k = \dfrac{1}{24}$

$H = \dfrac{1}{24}sd^3$

$= \dfrac{1}{24}(80)(3)^3$

$= 90$ hp

37. $y = \dfrac{k}{x}$

$400 = \dfrac{k}{8}$

$k = 3200$

$y = \dfrac{3200}{x} = \dfrac{3200}{4} = 800$ millibars

39. $V_1 = khr^2$

$V_2 = k\left(\dfrac{1}{2}h\right)(2r)^2$

$V_2 = 2khr^2 = 2V_1$

It is multiplied by 2.

41. $y_1 = kx^2$

$y_2 = k(2x)^2$

$y_2 = 4kx^2 = 4y_1$

It is multiplied by 4.

43.

x	$\frac{1}{4}$	$\frac{1}{2}$	1	2	4
$y = \frac{3}{x}$	12	6	3	$\frac{3}{2}$	$\frac{3}{4}$

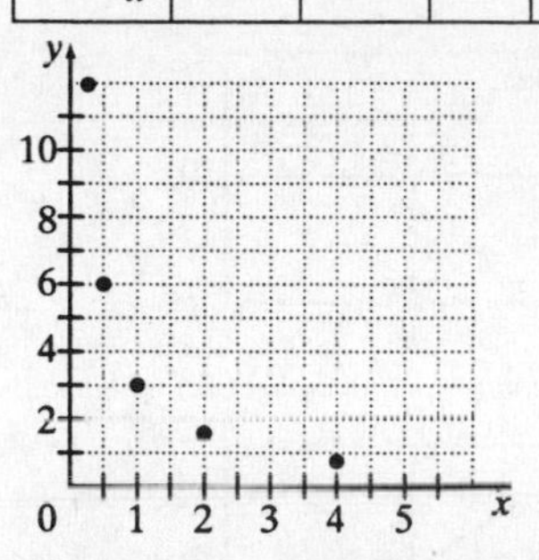

45.

x	$\frac{1}{4}$	$\frac{1}{2}$	1	2	4
$y = \frac{1}{2x}$	2	1	$\frac{1}{2}$	$\frac{1}{4}$	$\frac{1}{8}$

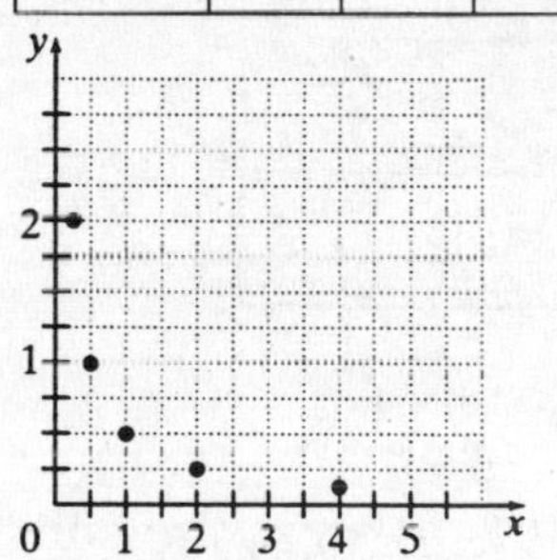

47. $r = 6$ cm

$C = 2\pi r = 2\pi(6) = 12\pi$ cm

$A = \pi r^2 = \pi(6^2) = 36\pi$ sq. cm

49. $r = 7$ m

$C = 2\pi r = 2\pi(7) = 14\pi$ m

$A = \pi r^2 = \pi(7^2) = 49\pi$ sq. m

51. $\sqrt{36} = 6$

53. $\sqrt{4} = 2$

55. $\sqrt{\dfrac{1}{25}} = \dfrac{1}{5}$

57. $\sqrt{\dfrac{25}{121}} = \dfrac{5}{11}$

Section 10.6

Graphing Calculator Explorations

1.

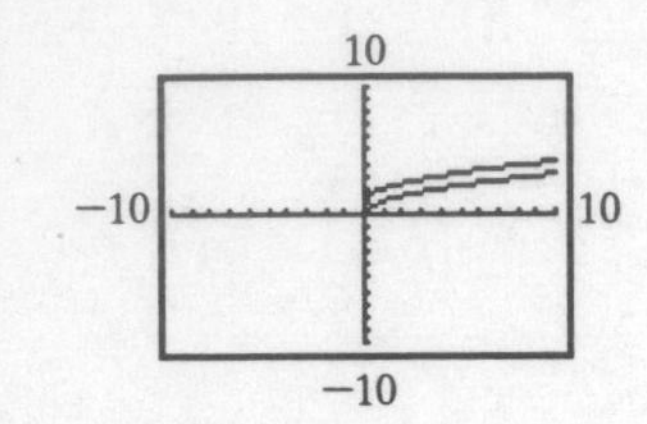

3.

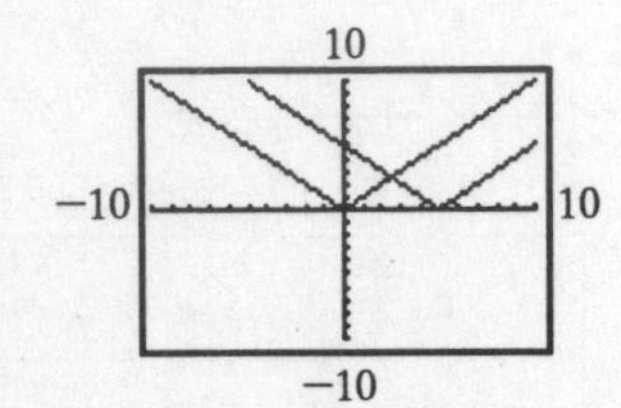

5.

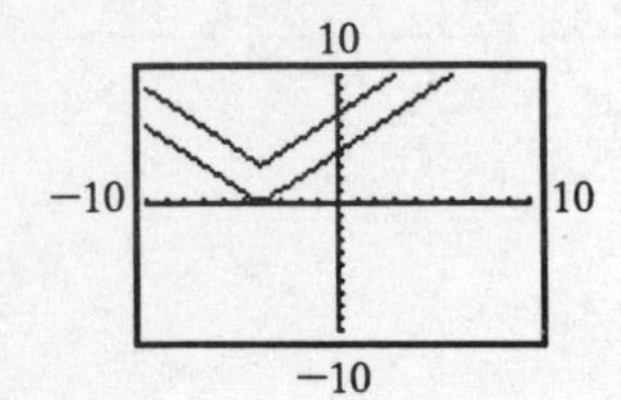

Mental Math

1. $f(x) = x^2$; vertex: (0, 0)

2. $f(x) = -5x^2$; vertex: (0, 0)

3. $g(x) = (x-2)^2$; vertex: (2, 0)

4. $g(x) = (x+5)^2$; vertex (–5, 0)

5. $f(x) = 2x^2 + 3$; vertex: (0, 3)

6. $h(x) = x^2 - 1$; vertex: (0, –1)

7. $g(x) = (x+1)^2 + 5$; vertex: (–1, 5)

8. $h(x) = (x-10)^2 - 7$; vertex: (10, –7)

Exercise Set 10.6

1. $f(x) = x^2 - 1$

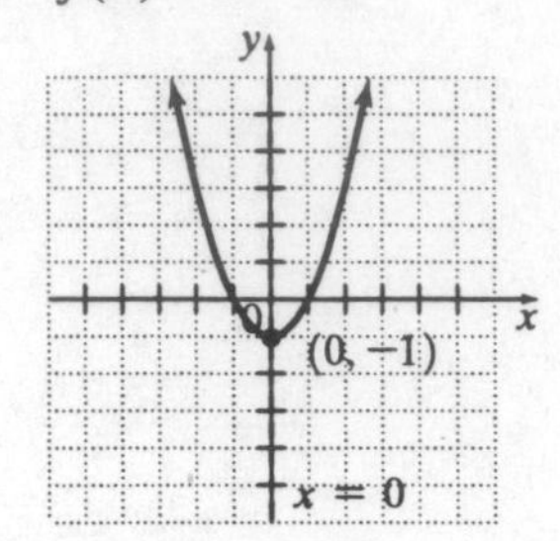

3. $h(x) = x^2 + 5$

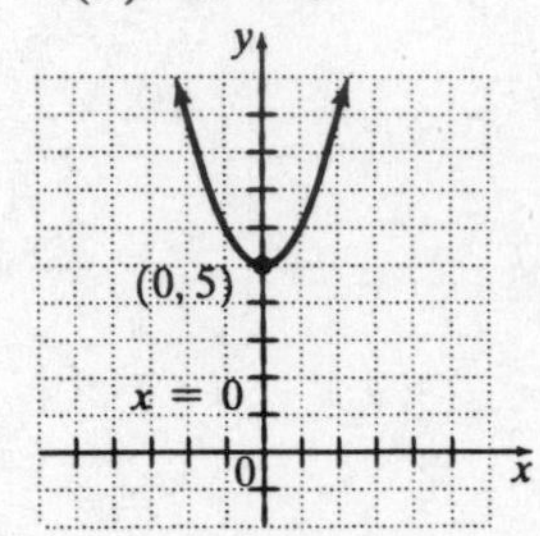

5. $g(x) = x^2 + 7$

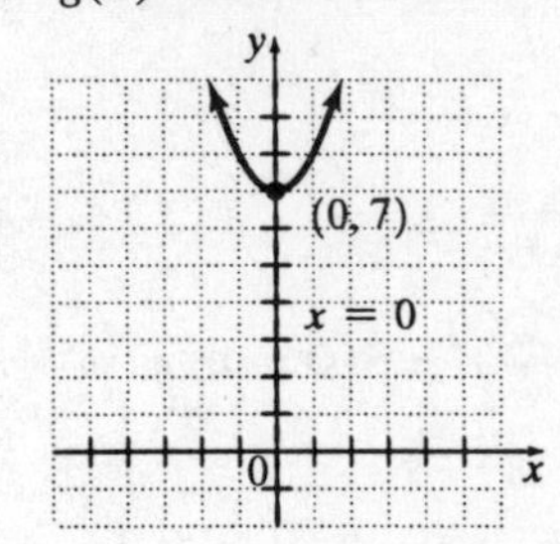

7. $f(x) = (x-5)^2$

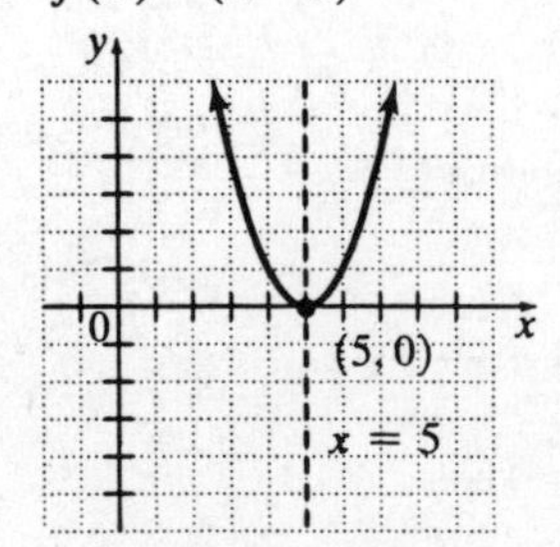

9. $h(x) = (x+2)^2$

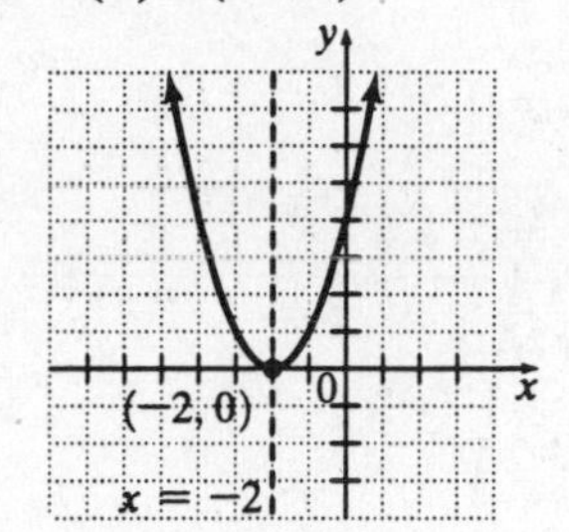

11. $G(x) = (x+3)^2$

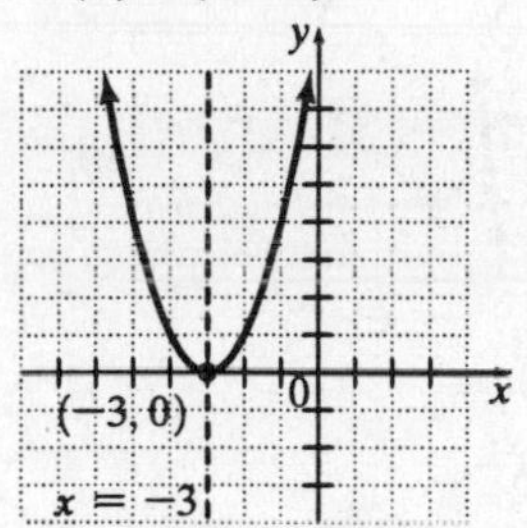

13. $f(x) = (x-2)^2 + 5$

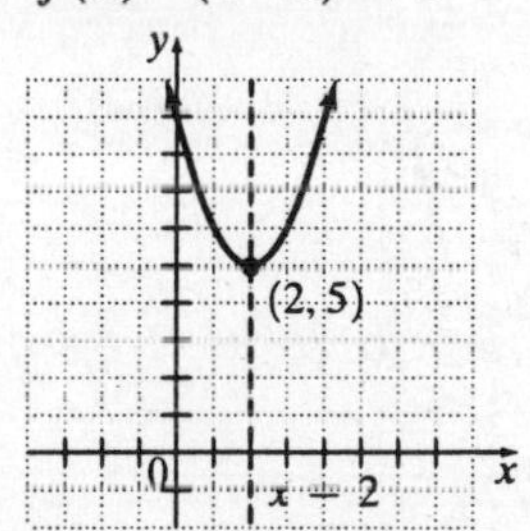

15. $h(x) = (x+1)^2 + 4$

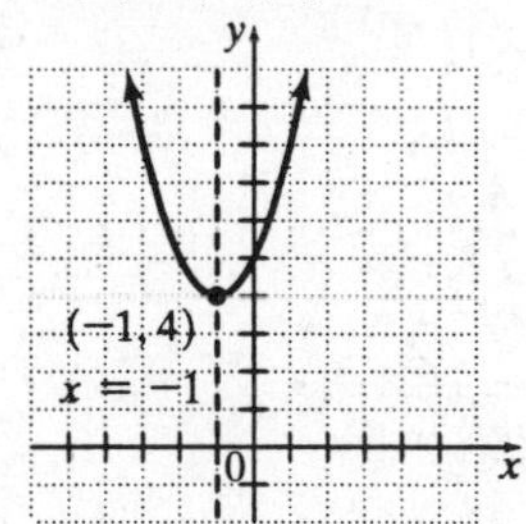

17. $g(x) = (x+2)^2 - 5$

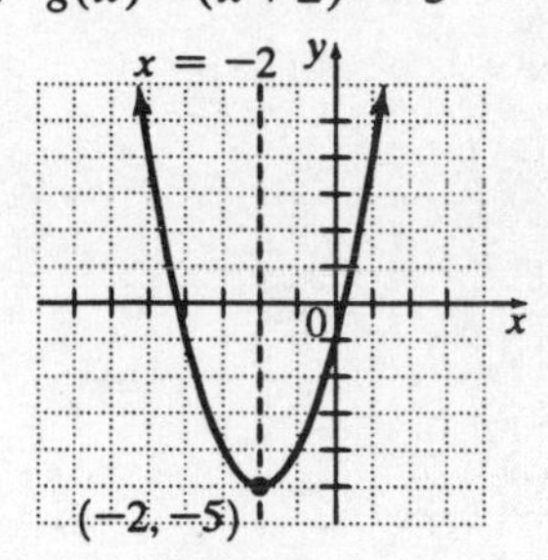

19. $g(x) = -x^2$

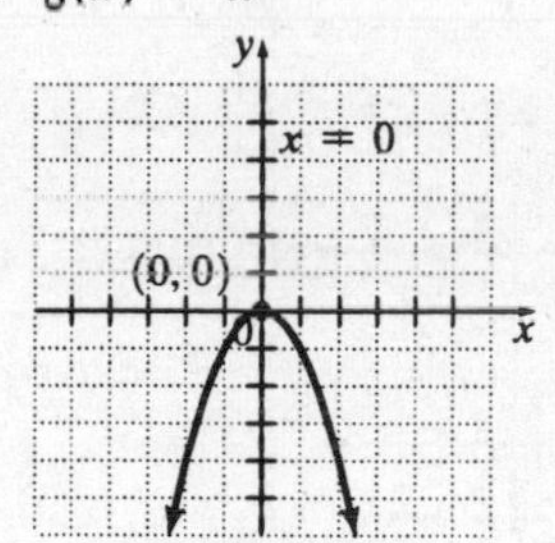

21. $h(x) = \frac{1}{3}x^2$

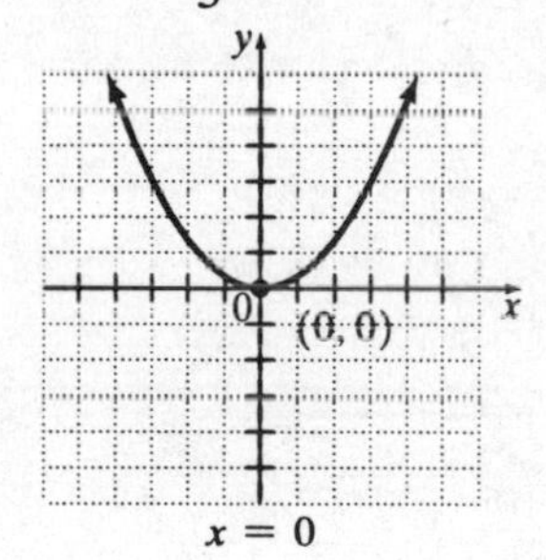

23. $H(x) = 2x^2$

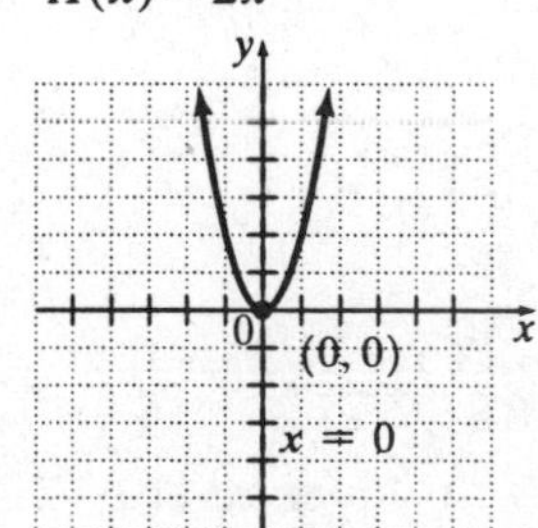

25. $f(x) = 2(x-1)^2 + 3$

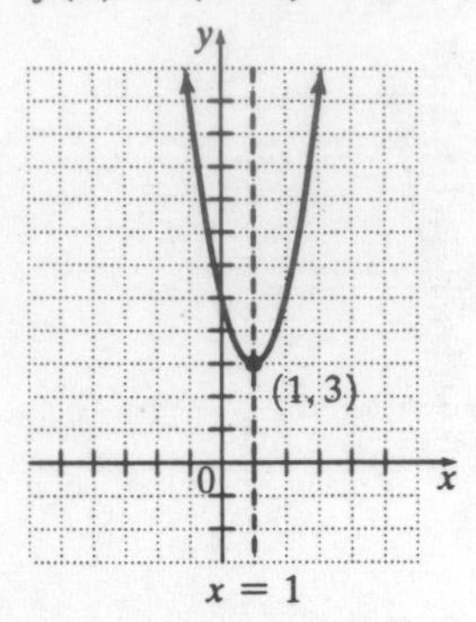

27. $h(x) = -3(x+3)^2 + 1$

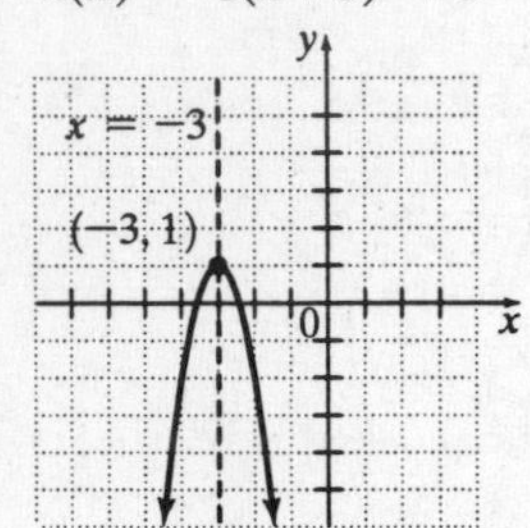

29. $H(x) = \frac{1}{2}(x-6)^2 - 3$

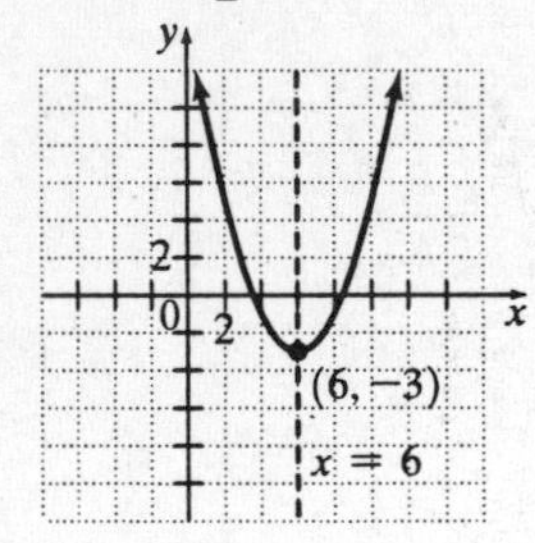

31. $f(x) = -(x-2)^2$

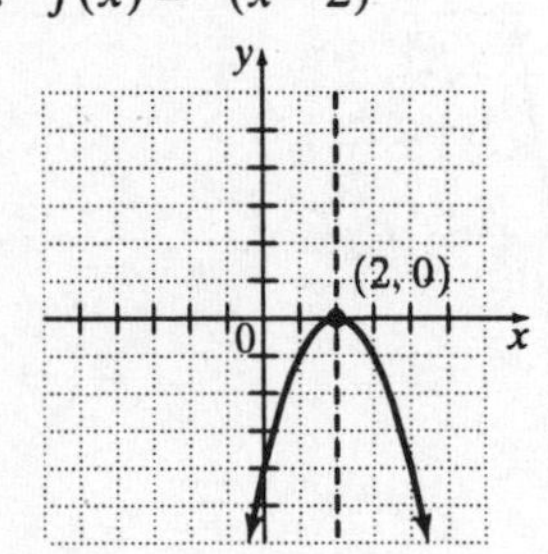

33. $F(x) = -x^2 + 4$

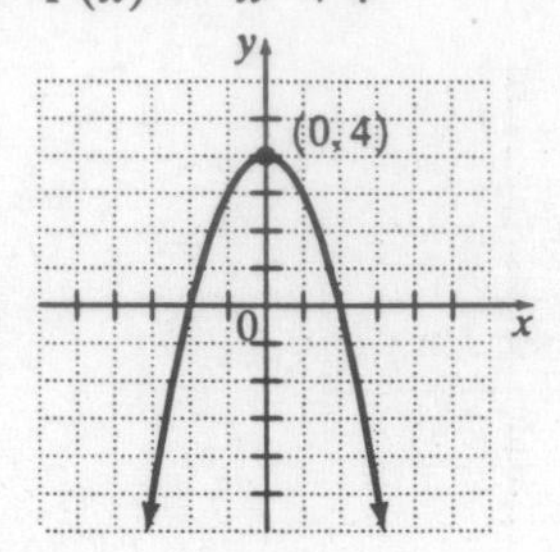

35. $F(x) = 2x^2 - 5$

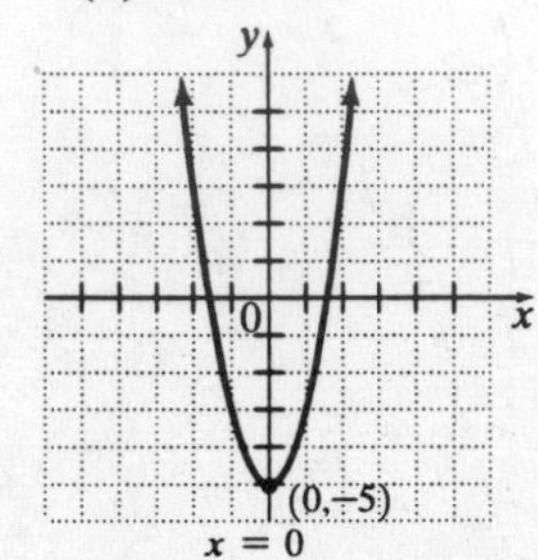

37. $h(x) = (x-6)^2 + 4$

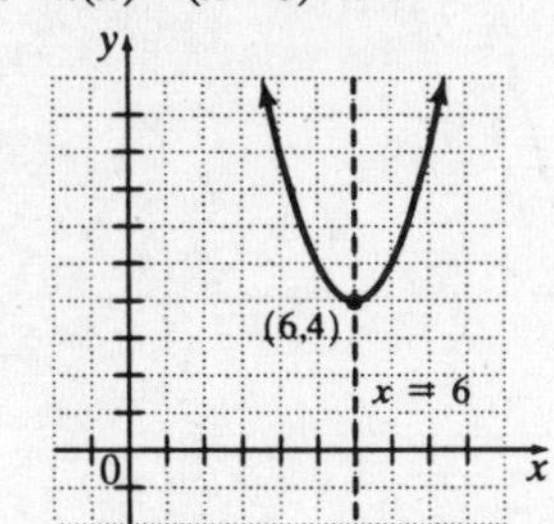

39. $F(x) = \left(x + \frac{1}{2}\right)^2 - 2$

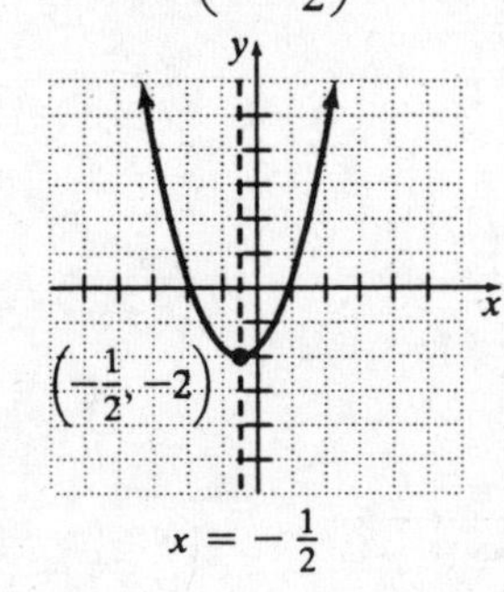

41. $F(x) = \frac{3}{2}(x+7)^2 + 1$

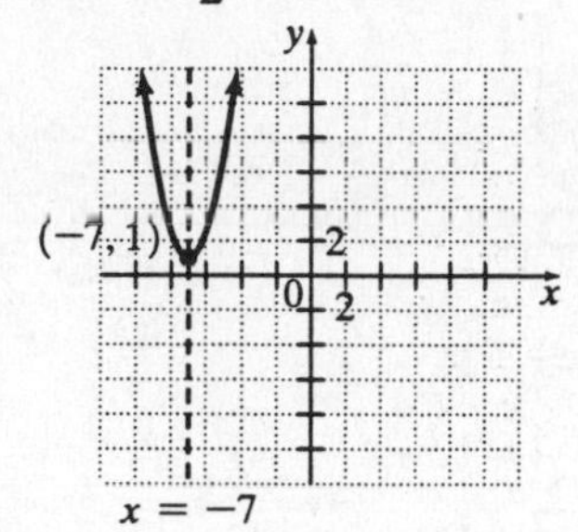

43. $f(x) = \frac{1}{4}x^2 - 9$

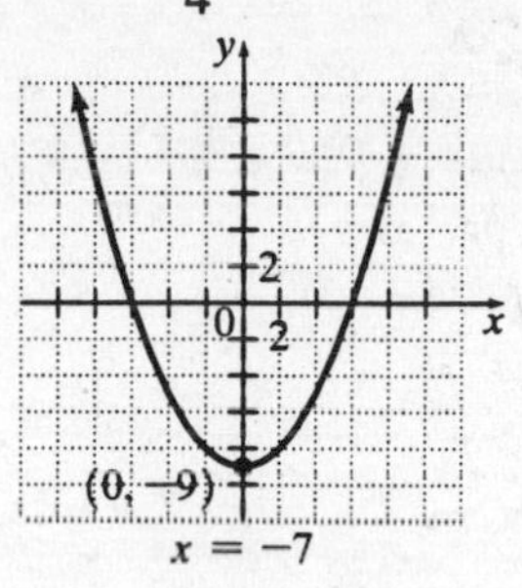

45. $G(x) = 5\left(x + \frac{1}{2}\right)^2$

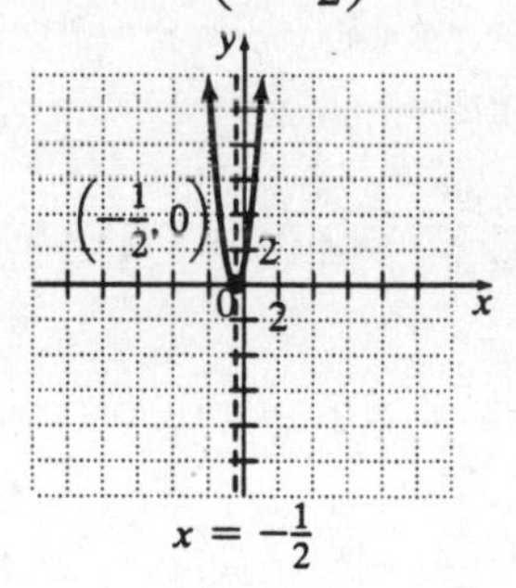

47. $f(x) = -(x-1)^2 - 1$

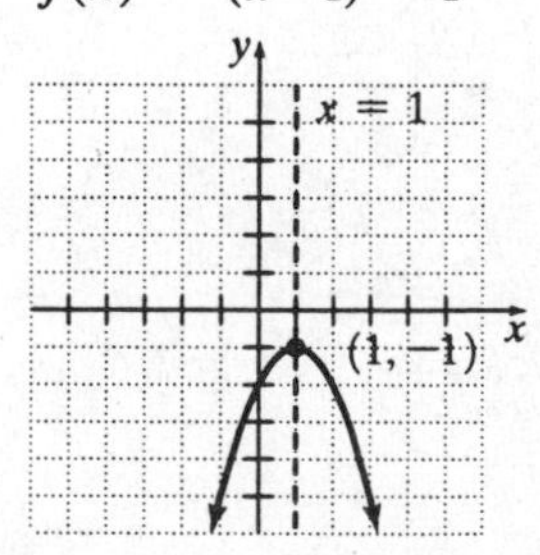

49. $g(x) = \sqrt{3}(x+5)^2 + \frac{3}{4}$

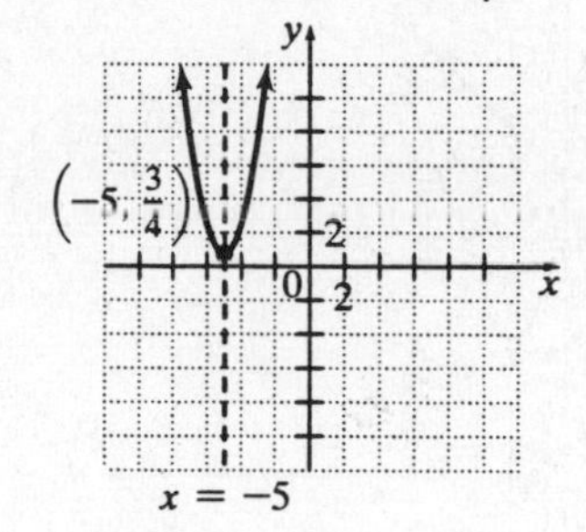

51. $h(x) = 10(x+4)^2 - 6$

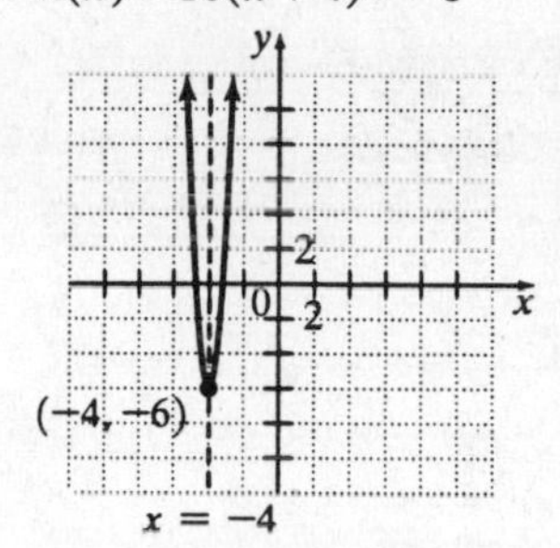

53. $f(x) = -2(x-4)^2 + 5$

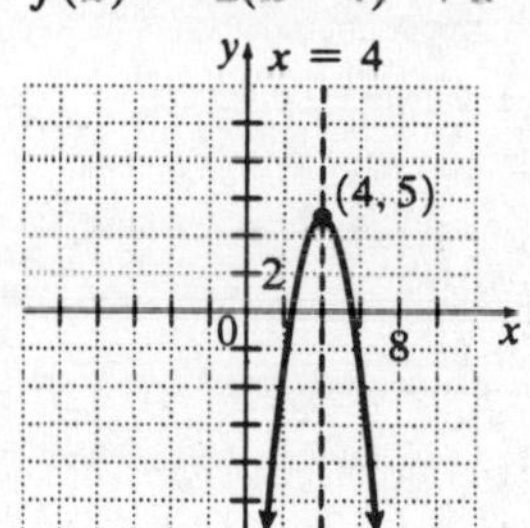

55. $f(x) = 5(x-2)^2 + 3$

57. $f(x) = 5[x-(-3)]^2 + 6$
$f(x) = 5(x+3)^2 + 6$

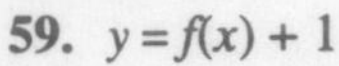

59. $y = f(x) + 1$

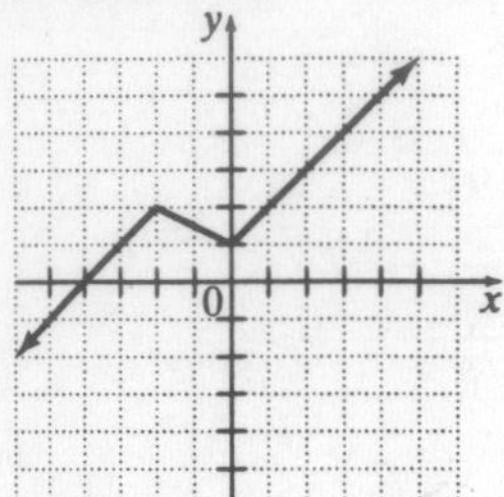

61. $y = f(x - 3)$

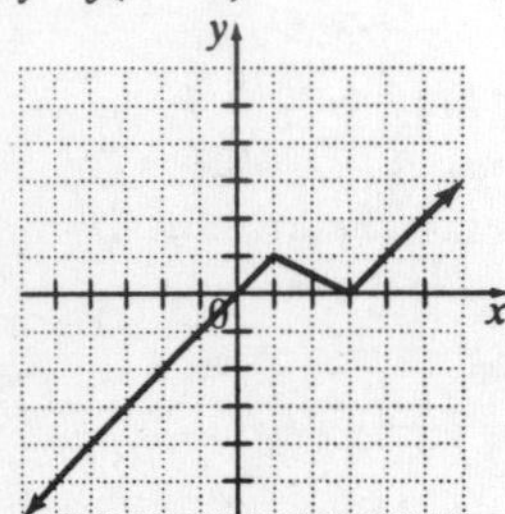

63. $y = f(x + 2) + 2$

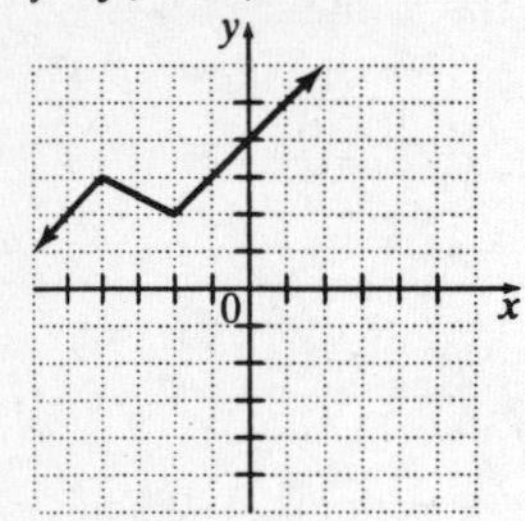

65. $x^2 + 8x$

$\left[\frac{1}{2}(8)\right]^2 = [4]^2 = 16$

$x^2 + 8x + 16$

67. $z^2 - 16z$

$\left[\frac{1}{2}(-16)\right]^2 = [-8]^2 = 64$

$z^2 - 16z + 64$

69. $y^2 + y$

$\left[\frac{1}{2}(1)\right]^2 = \left[\frac{1}{2}\right]^2 = \frac{1}{4}$

$y^2 + y + \frac{1}{4}$

71.

$$x^2 + 4x = 12$$
$$x^2 + 4x + 4 = 12 + 4$$
$$(x+2)^2 = 16$$
$$x + 2 = \pm\sqrt{16}$$
$$x + 2 = \pm 4$$
$$x = -2 \pm 4$$
$$x = -2 + 4 \quad \text{or} \quad x = -2 - 4$$
$$x = 2 \quad \text{or} \quad x = -6$$

The solutions are 2 and –6.

73.

$$z^2 + 10z - 1 = 0$$
$$z^2 + 10z = 1$$
$$z^2 + 10z + 25 = 1 + 25$$
$$(z+5)^2 = 26$$
$$z + 5 = \pm\sqrt{26}$$
$$z = -5 \pm \sqrt{26}$$
$$z = -5 + \sqrt{26} \text{ or } z = -5 - \sqrt{26}$$

The solutions are $-5 + \sqrt{26}$ and $-5 - \sqrt{26}$.

75.

$$z^2 - 8z = 2$$
$$z^2 - 8z + 16 = 2 + 16$$
$$(z-4)^2 = 18$$
$$z - 4 = \pm\sqrt{18}$$
$$z - 4 = \pm 3\sqrt{2}$$
$$z = 4 \pm 3\sqrt{2}$$
$$z = 4 + 3\sqrt{2} \text{ or } z = 4 - 3\sqrt{2}$$

The solutions are $4 + 3\sqrt{2}$ and $4 - 3\sqrt{2}$.

Exercise Set 10.7

1. $f(x) = x^2 + 8x + 7$

$-\frac{b}{2a} = \frac{-8}{2(1)} = -4$ and

$f(-4) = (-4)^2 + 8(-4) + 7$

$f(-4) = 16 - 32 + 7 = -9$

Thus, the vertex is (–4, –9).

3. $f(x) = -x^2 + 10x + 5$

$\frac{-b}{2a} = \frac{-10}{2(-1)} = 5$ and

$f(5) = -5^2 + 10(5) + 5$

$f(5) = -25 + 50 + 5 = 30$

Thus, the vertex is (5, 30).

5. $f(x) = 5x^2 - 10x + 3$

$\frac{-b}{2a} = \frac{-(-10)}{2(5)} = 1$ and

$f(1) = 5(1)^2 - 10(1) + 3$

$f(1) = 5 - 10 + 3 = -2$

Thus, the vertex is (1, –2).

7. $f(x) = -x^2 + x + 1$

$\frac{-b}{2a} = \frac{-1}{2(-1)} = \frac{1}{2}$ and

$f\left(\frac{1}{2}\right) = -\left(\frac{1}{2}\right)^2 + \frac{1}{2} + 1$

$f\left(\frac{1}{2}\right) = -\frac{1}{4} + \frac{1}{2} + 1 = \frac{5}{4}$

Thus, the vertex is $\left(\frac{1}{2}, \frac{5}{4}\right)$.

9. $f(x) = x^2 - 4x + 3$

$\frac{-b}{2a} = \frac{-(-4)}{2(1)} = 2$ and

$f(2) = 2^2 - 4(2) + 3 = -1$

The vertex is (2, –1), so Graph D.

11. $f(x) = x^2 - 2x - 3$

$\frac{-b}{2a} = \frac{-(-2)}{2(1)} = 1$ and

$f(1) = 1^2 - 2(1) - 3 = -4$

The vertex is (1, –4), so Graph B.

13. $f(x) = x^2 + 4x - 5$

$\frac{-b}{2a} = \frac{-4}{2(1)} = -2$ and

$f(-2) = (-2)^2 + 4(-2) - 5$

$f(-2) = 4 - 8 - 5 = -9$

Thus, the vertex is (–2, –9).

The graph opens upward since $a = 1 > 0$.

$x^2 + 4x - 5 = 0$

$(x + 5)(x - 1) = 0$

$x + 5 = 0$ or $x - 1 = 0$

$x = -5$ or $x = 1$

The x-intercepts are (–5, 0) and (1, 0).

$f(0) = -5$, so the y-intercept is (0, –5).

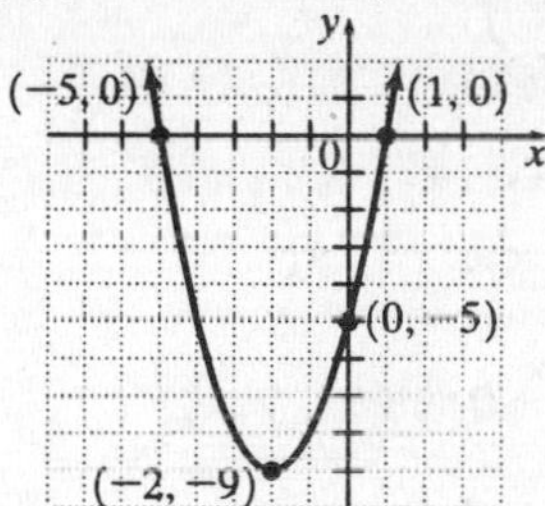

15. $f(x) = -x^2 + 2x - 1$

$\frac{-b}{2a} = \frac{-2}{2(-1)} = 1$ and

$f(1) = -1^2 + 2(1) - 1$

$f(1) = -1 + 2 - 1 = 0$

Thus, the vertex is (1, 0).

The graph opens downward since $a = -1 < 0$.

$$-x^2 + 2x - 1 = 0$$
$$-(x^2 - 2x + 1) = 0$$
$$-(x-1)^2 = 0$$
$$x - 1 = 0$$
$$x = 1$$

The x-intercept is (1, 0).

$f(0) = -1$, so the y-intercept is (0, –1).

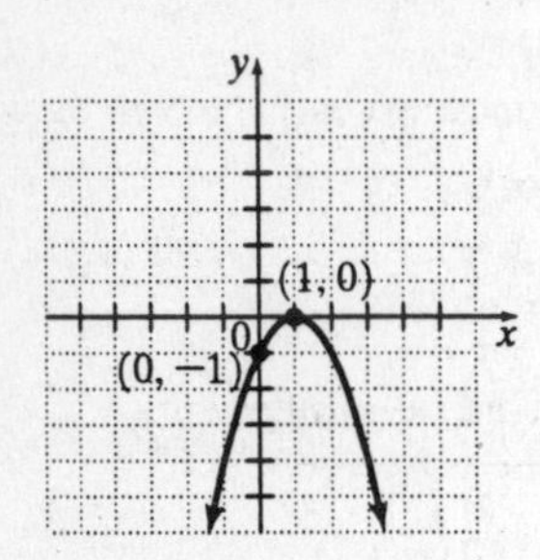

17. $f(x) = x^2 - 4$

$\frac{-b}{2a} = \frac{-0}{2(1)} = 0$ and

$f(0) = (0)^2 - 4$

$f(0) = 0 - 4 = -4$

Thus, the vertex is (0, –4).

The graph opens upward since $a = 1 > 0$.

$$x^2 - 4 = 0$$
$$(x+2)(x-2) = 0$$
$$x + 2 = 0 \quad \text{or} \quad x - 2 = 0$$
$$x = -2 \quad \text{or} \quad x = 2$$

The x-intercepts are (–2, 0) and (2, 0).

$f(0) = -4$, so the y-intercept is (0, –4).

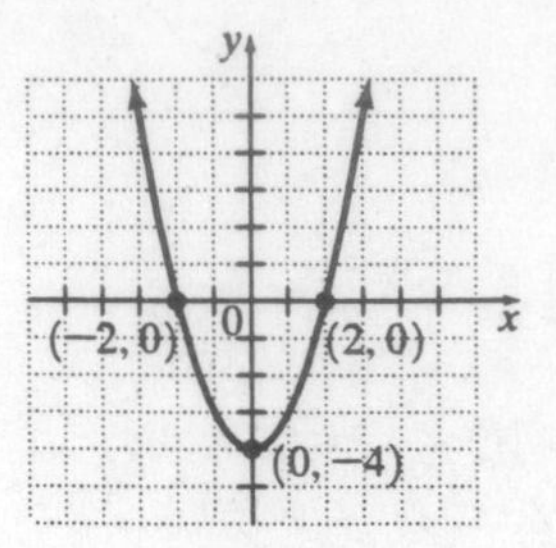

19. $f(x) = 4x^2 + 4x - 3$

$\frac{-b}{2a} = \frac{-4}{2(4)} = -\frac{1}{2}$ and

$f\left(-\frac{1}{2}\right) = 4\left(-\frac{1}{2}\right)^2 + 4\left(-\frac{1}{2}\right) - 3$

$f\left(-\frac{1}{2}\right) = 1 - 2 - 3 = -4$

Thus, the vertex is $\left(-\frac{1}{2}, -4\right)$.

The graph opens upward since $a = 4 > 0$.

$$4x^2 + 4x - 3 = 0$$
$$(2x-1)(2x+3) = 0$$
$$2x - 1 = 0 \quad \text{or} \quad 2x + 3 = 0$$
$$2x = 1 \quad \text{or} \quad 2x = -3$$
$$x = \frac{1}{2} \quad \text{or} \quad x = -\frac{3}{2}$$

The x-intercepts are $\left(\frac{1}{2}, 0\right)$ and $\left(-\frac{3}{2}, 0\right)$.

$f(0) = -3$, so the y-intercept is (0, –3).

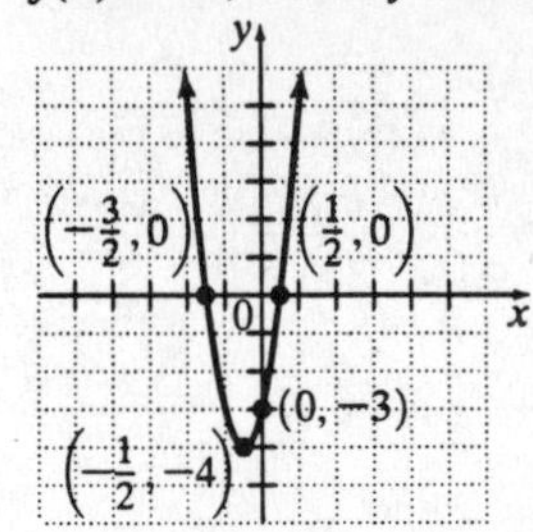

21. $f(x) = x^2 + 8x + 15$

$\frac{-b}{2a} = \frac{-8}{2(1)} = -4$

$f(-4) = (-4)^2 + 8(-4) + 15$

$f(-4) = 16 - 32 + 15 = -1$

Thus, the vertex is $(-4, -1)$.

The graph opens upward since $a = 1 > 0$.

$x^2 + 8x + 15 = 0$

$(x+3)(x+5) = 0$

$x + 3 = 0$ or $x + 5 = 0$

$x = -3$ or $x = -5$

The x-intercepts are $(-3, 0)$ and $(-5, 0)$.

$f(0) = 15$, so the y-intercept is $(0, 15)$.

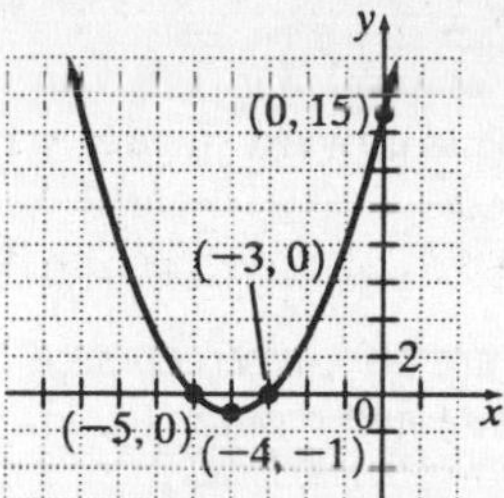

23.
$$y = x^2 - 6x + 5$$
$$y - 5 = x^2 - 6x$$
$$y - 5 + 9 = x^2 - 6x + 9$$
$$y + 4 = (x-3)^2$$
$$f(x) = (x-3)^2 - 4$$

Thus, the vertex is $(3, -4)$.

The graph opens upward since $a = 1 > 0$.

$x^2 - 6x + 5 = 0$

$(x-1)(x-5) = 0$

$x - 1 = 0$ or $x - 5 = 0$

$x = 1$ or $x = 5$

The x-intercepts are $(1, 0)$ and $(5, 0)$.

$f(0) = 5$, so the y-intercept is $(0, 5)$.

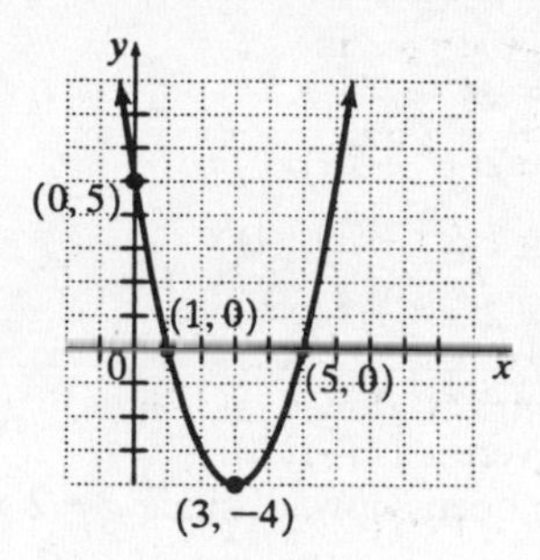

25.
$$y = x^2 - 4x + 5$$
$$y - 5 = x^2 - 4x$$
$$y - 5 + 4 = x^2 - 4x + 4$$
$$y - 1 = (x-2)^2$$
$$f(x) = (x-2)^2 + 1$$

Thus, the vertex is $(2, 1)$.

The graph opens upward since $a = 1 > 0$.

$x^2 - 4x + 5 = 0$

$(x-2)^2 + 1 = 0$

$(x-2)^2 = -1$

Hence, there are no x-intercepts.

$f(0) = 5$, so the y-intercept is $(0, 5)$.

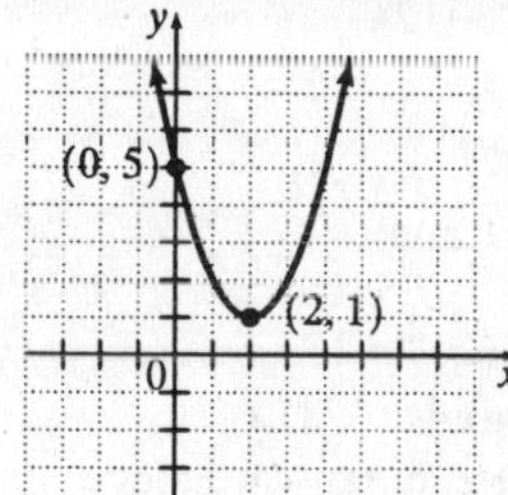

27.
$$y = 2x^2 + 4x + 5$$
$$y - 5 = 2(x^2 + 2x)$$
$$y - 5 + 2 = 2(x^2 + 2x + 1)$$
$$y - 3 = 2(x + 1)^2$$
$$f(x) = 2(x + 1)^2 + 3$$
Thus, the vertex is (–1, 3).
The graph opens upward since $a = 2 > 0$.
$$2x^2 + 4x + 5 = 0$$
$$2(x + 1)^2 + 3 = 0$$
$$2(x + 1)^2 = -3$$
Hence, there are no x-intercepts.
$f(0) = 5$, so the y-intercept is (0, 5).

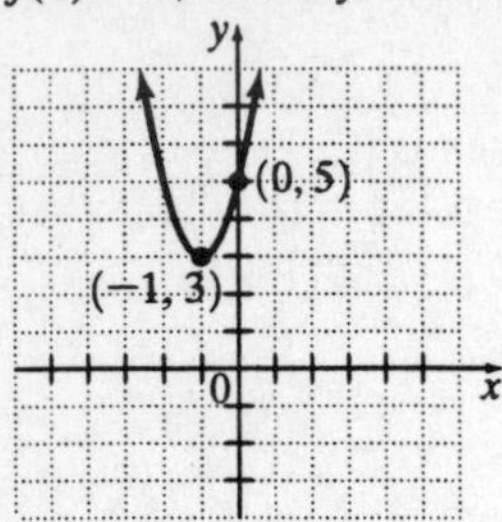

29.
$$y = -2x^2 + 12x$$
$$y = -2(x^2 - 6x)$$
$$y - 18 = -2(x^2 - 6x + 9)$$
$$y - 18 = -2(x - 3)^2$$
$$f(x) = -2(x - 3)^2 + 18$$
Thus, the vertex is (3, 18).
The graph opens downward since $a = -2 < 0$.
$$-2x^2 + 12x = 0$$
$$-2x(x - 6) = 0$$
$$-2x = 0 \quad \text{or} \quad x - 6 = 0$$
$$x = 0 \quad \text{or} \quad x = 6$$
The x-intercepts are (0, 0) and (6, 0).
$f(0) = 0$, so the y-intercept is (0, 0).

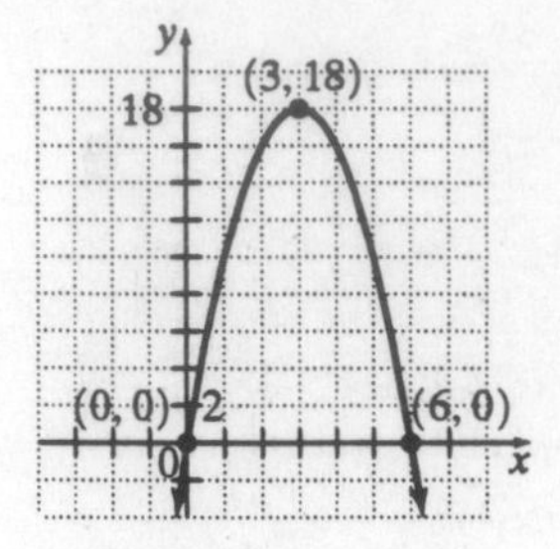

31. $f(x) = x^2 + 1$
$$\frac{-b}{2a} = \frac{-0}{2(1)} = 0 \text{ and}$$
$$f(0) = 0^2 + 1$$
$$f(0) = 0 + 1 = 1$$
Thus, the vertex is (0, 1).
The graph opens upward since $a = 1 > 0$.
$$x^2 + 1 = 0$$
$$x^2 = -1$$
Hence, there are no x-intercepts.
$f(0) = 1$, so the y-intercept is (0, 1).

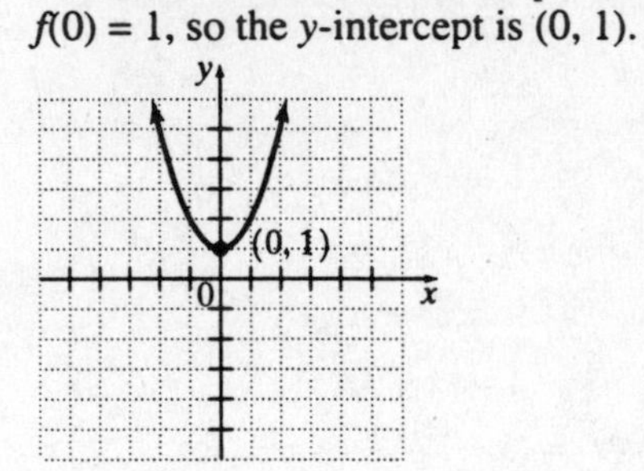

33. $f(x) = x^2 - 2x - 15$

$\frac{-b}{2a} = \frac{-(-2)}{2(1)} = 1$ and

$f(1) = 1^2 - 2(1) - 15$

$f(1) = 1 - 2 - 15 = -16$

Thus, the vertex is (1, –16).

The graph opens upward since $a = 1 > 0$.

$x^2 - 2x - 15 = 0$

$(x-5)(x+3) = 0$

$x - 5 = 0$ or $x + 3 = 0$

$x = 5$ or $x = -3$

The x-intercepts are (5, 0) and (–3, 0).

$f(0) = -15$, so the y-intercept is (0, –15).

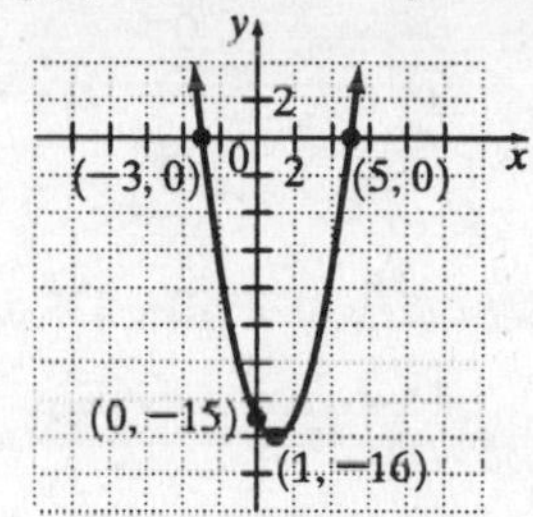

35. $f(x) = -5x^2 + 5x$

$\frac{-b}{2a} = \frac{-5}{2(-5)} = \frac{1}{2}$ and

$f\left(\frac{1}{2}\right) = -5\left(\frac{1}{2}\right)^2 + 5\left(\frac{1}{2}\right)$

$f\left(\frac{1}{2}\right) = -\frac{5}{4} + \frac{5}{2} = \frac{5}{4}$

Thus, the vertex is $\left(\frac{1}{2}, \frac{5}{4}\right)$.

The graph opens downward since $a = -5 < 0$.

$-5x^2 + 5x = 0$

$-5x(x-1) = 0$

$-5x = 0$ or $x - 1 = 0$

$x = 0$ or $x = 1$

The x-intercepts are (0, 0) and (1, 0).

$f(0) = 0$, so the y-intercept is (0, 0).

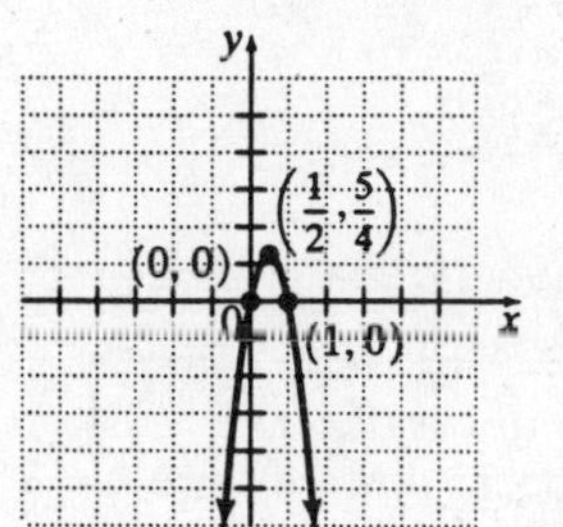

37. $f(x) = -x^2 + 2x - 12$

$\frac{-b}{2a} = \frac{-2}{2(-1)} = 1$ and

$f(1) = -1^2 + 2(1) - 12$

$f(1) = -1 + 2 - 12 = -11$

Thus, the vertex is (1, –11).

The graph opens downward since $a = -1 < 0$.

$-x^2 + 2x - 12 = 0$

$x^2 - 2x = -12$

$x^2 - 2x + 1 = -12 + 1$

$(x-1)^2 = -11$

Hence, there are no x-intercepts.

$f(0) = -12$, so the y-intercept is (0, –12).

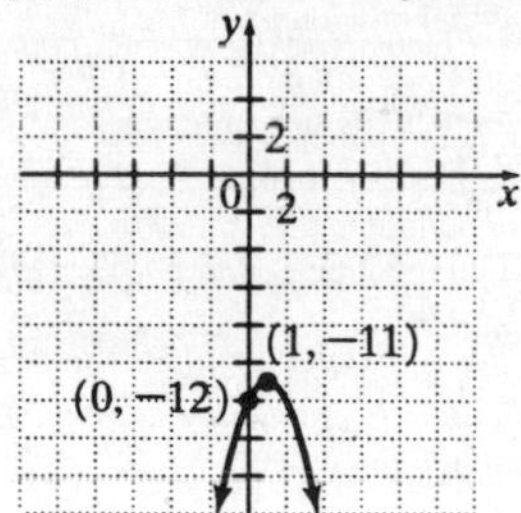

39. $f(x) = 3x^2 - 12x + 15$

$\frac{-b}{2a} = \frac{-(-12)}{2(3)} = 2$ and

$f(2) = 3(2)^2 - 12(2) + 15$

$f(2) = 12 - 24 + 15 = 3$

Thus, the vertex is (2, 3).

The graph opens upward since $a = 3 > 0$.

$3x^2 - 12x + 15 = 0$

$x^2 - 4x + 5 = 0$

$x^2 - 4x + 4 = -5 + 4$

$(x-2)^2 = -1$

Hence, there are no x-intercepts.

$f(0) = 15$, so the y-intercept is (0, 15).

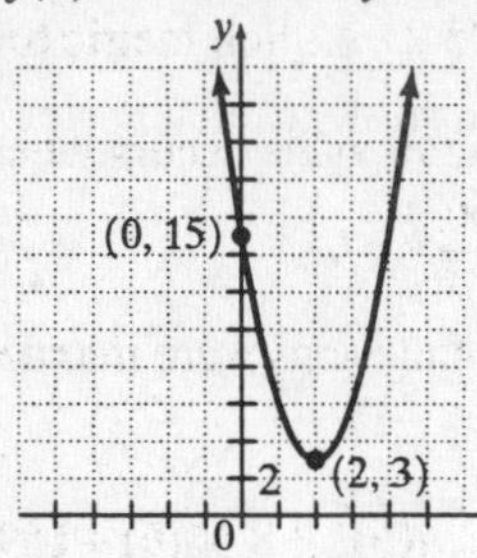

41. $f(x) = x^2 + x - 6$

$\frac{-b}{2a} = \frac{-1}{2(1)} = -\frac{1}{2}$ and

$f\left(-\frac{1}{2}\right) = \left(-\frac{1}{2}\right)^2 + \left(-\frac{1}{2}\right) - 6$

$f\left(-\frac{1}{2}\right) = \frac{1}{4} - \frac{1}{2} - 6 = -\frac{25}{4}$

Thus, the vertex is $\left(-\frac{1}{2}, -\frac{25}{4}\right)$.

The graph opens upward since $a = 1 > 0$.

$x^2 + x - 6 = 0$

$(x+3)(x-2) = 0$

$x + 3 = 0$ or $x - 2 = 0$

$x = -3$ or $x = 2$

The x-intercepts are (–3, 0) and (2, 0).

$f(0) = -6$, so the y-intercept is (0, –6).

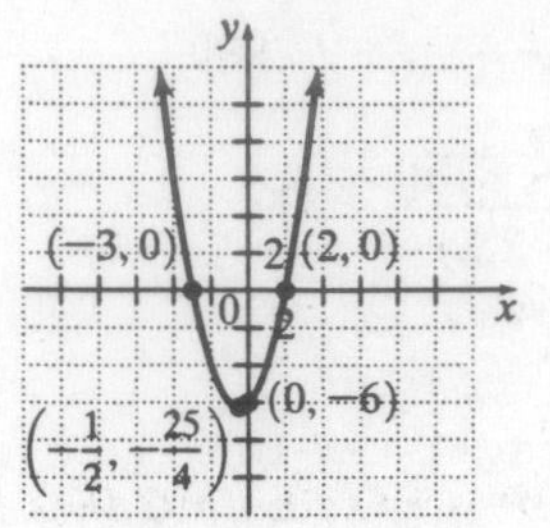

43. $f(x) = -2x^2 - 3x + 35$

$\frac{-b}{2a} = \frac{-(-3)}{2(-2)} = -\frac{3}{4}$ and

$f\left(-\frac{3}{4}\right) = -2\left(-\frac{3}{4}\right)^2 - 3\left(-\frac{3}{4}\right) + 35$

$f\left(-\frac{3}{4}\right) = -\frac{9}{8} + \frac{9}{4} + 35 = \frac{289}{8}$

Thus, the vertex is $\left(-\frac{3}{4}, \frac{289}{8}\right)$.

The graph opens downward since $a = -2 < 0$.

$-2x^2 - 3x + 35 = 0$

$2x^2 + 3x - 35 = 0$

$(2x-7)(x+5) = 0$

$2x - 7 = 0$ or $x + 5 = 0$

$2x = 7$ or $x = -5$

$x = \frac{7}{2}$

The x-intercepts are $\left(\frac{7}{2}, 0\right)$ and (–5, 0).

$f(0) = 35$, so the y-intercept is (0, 35).

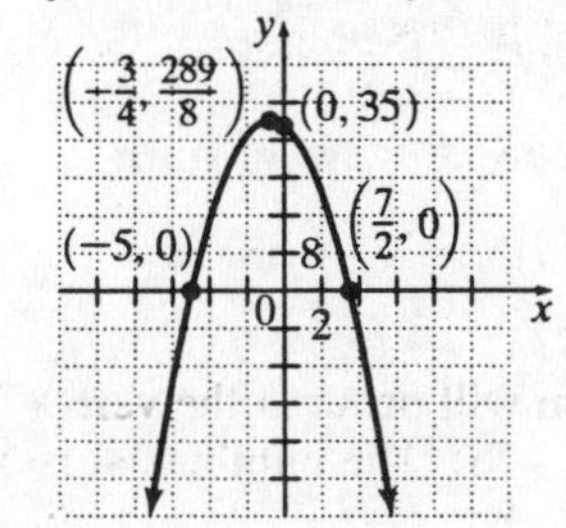

45. $C(x) = 2x^2 - 800x + 92{,}000$

a. $\frac{-b}{2a} = \frac{-(-800)}{2(2)} = 200$
200 bicycles

b. $C(200)$
$= 2(200)^2 - 800(200) + 92{,}000$
$= 12{,}000$
The minimum cost is $12,000.

47. $h(t) = -16t^2 + 32t$
$\frac{-b}{2a} = \frac{-32}{2(-16)} = 1$
$h(1) = -16(1)^2 + 32(1)$
$h(1) = 16$ feet
The maximum height of the ball is 16 feet.

49. Let x = one number.
$60 - x$ = other number
$f(x) = x(60 - x)$
$f(x) = 60x - x^2$
$f(x) = -x^2 + 60x$
$f(x) = -1(x^2 - 60x)$
$f(x) = -1(x^2 - 60x + 900) + 900$
$f(x) = -(x - 30)^2 + 900$
The maximum will occur at the vertex which is (30, 900). The numbers are 30 and 30.

51. Let x = one number.
$10 + x$ = other number
$f(x) = x(10 + x)$
$f(x) = 10x + x^2$
$f(x) = (x^2 + 10x + 25) - 25$
$f(x) = (x + 5)^2 - 25$
The minimum will occur at the vertex which is (–5, –25). The numbers are –5 and 5.

53. Let x = width.
$40 - x$ = the length
$f(x) = x(40 - x)$
$f(x) = 40x - x^2$
$f(x) = -x^2 + 40x$
$f(x) = -1(x^2 - 40x)$
$f(x) = -1(x^2 - 40x + 400) + 400$
$f(x) = -(x - 20)^2 + 400$
The maximum will occur at the vertex which is (20, 400). The width is 20 units and the length is 20 units.

55. a. $f(14) = -0.74(14)^2 + 8.66(14) + 159.07$
≈ 135.27 million metric tons

b. The maximum value occurs when
$x = \frac{-8.66}{2(-0.74)} \approx 6.$
Therefore emissions were maximized in 1996.

c. $f(6) = -0.74(6)^2 + 8.66(6) + 159.07$
≈ 184.39 million metric tons

57. $f(x) = x^2 + 10x + 15$

$\frac{-b}{2a} = \frac{-10}{2(1)} = -5$ and

$f(-5) = (-5)^2 + 10(-5) + 15 = -10$

Thus, the vertex is (–5, –10).

The graph opens upward since $a = 1 > 0$.

$x^2 + 10x + 15 = 0$

$$x = \frac{-10 \pm \sqrt{(10)^2 - 4(1)(15)}}{2(1)}$$

The x-intercepts are approximately (–8.2, 0) and (–1.8, 0).

$f(0) = 15$, so the y-intercept is (0, 15).

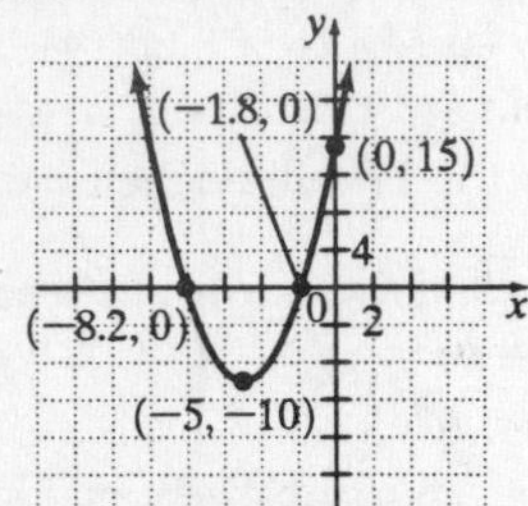

59. $f(x) = 3x^2 - 6x + 7$

$\frac{-b}{2a} = \frac{-(-6)}{2(3)} = 1$ and

$f(1) = 3(1)^2 - 6(1) + 7 = 4$

Thus the vertex is (1, 4).

The graph opens upward since $a = 3 > 0$.

$3x^2 - 6x + 7 = 0$

$$x = \frac{-(-6) \pm \sqrt{(-6)^2 - 4(3)(7)}}{2(3)}$$

There are no x-intercepts.

$f(0) = 7$, so the y-intercept is (0, 7).

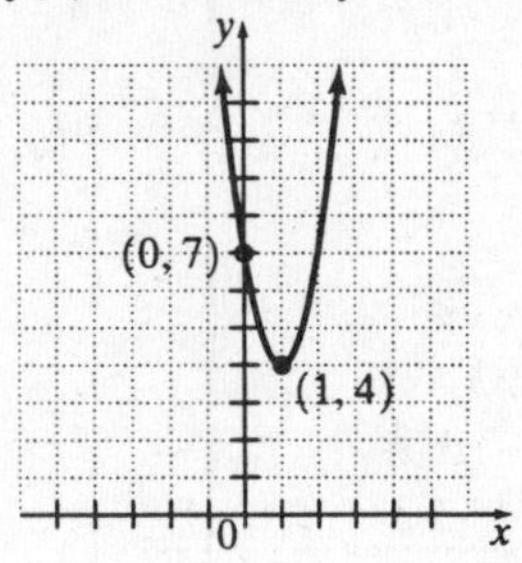

61.

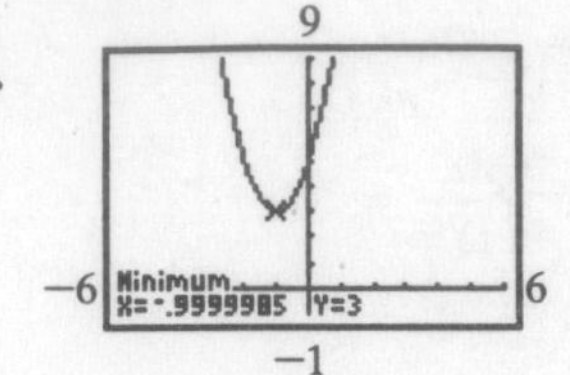

63.

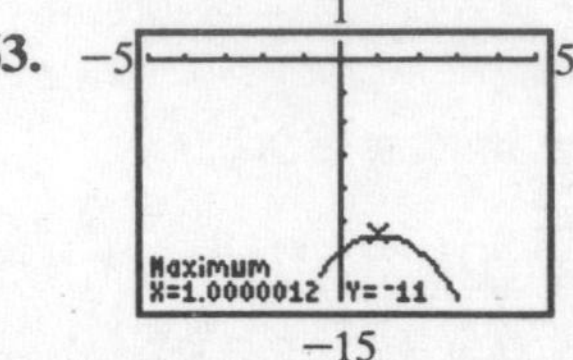

65. $f(x) = 2.3x^2 - 6.1x + 3.2$

minimum ≈ -0.84

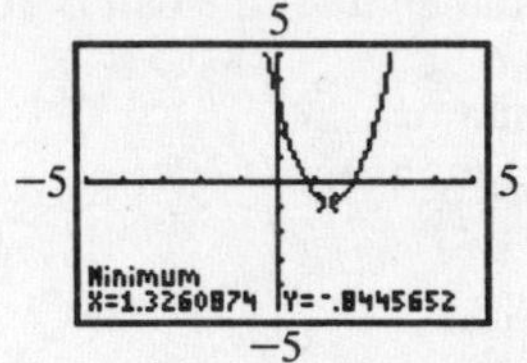

67. $f(x) = -1.9x^2 + 5.6x - 2.7$

maximum ≈ 1.43

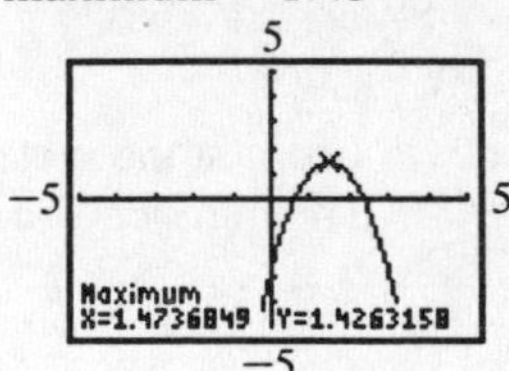

69.

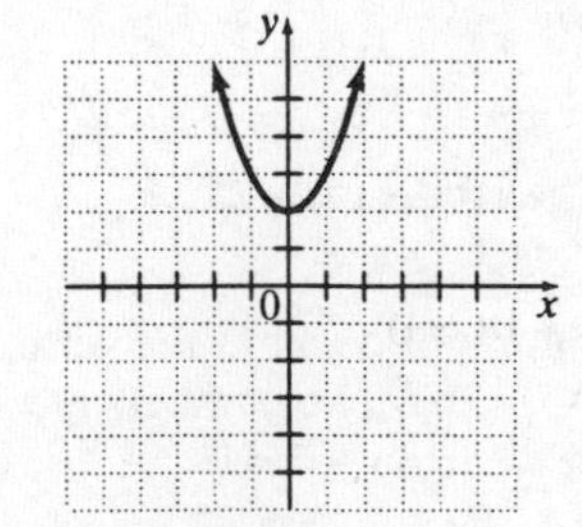

71.

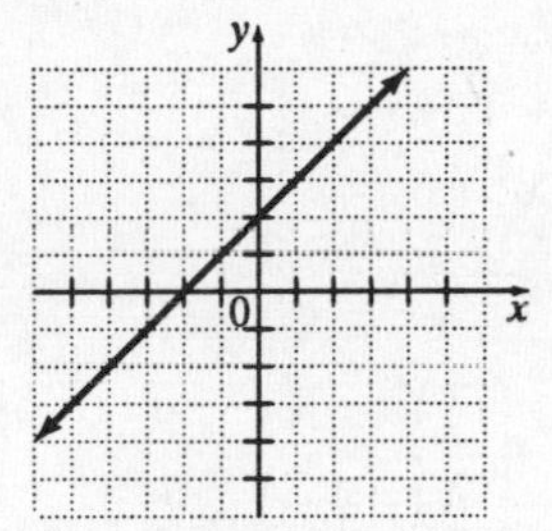

73.

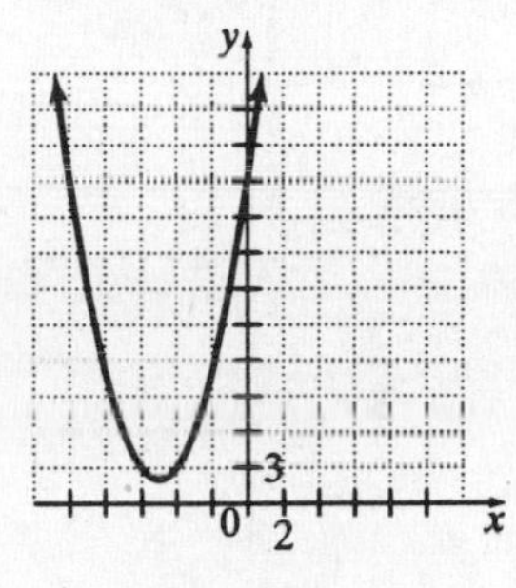

75.

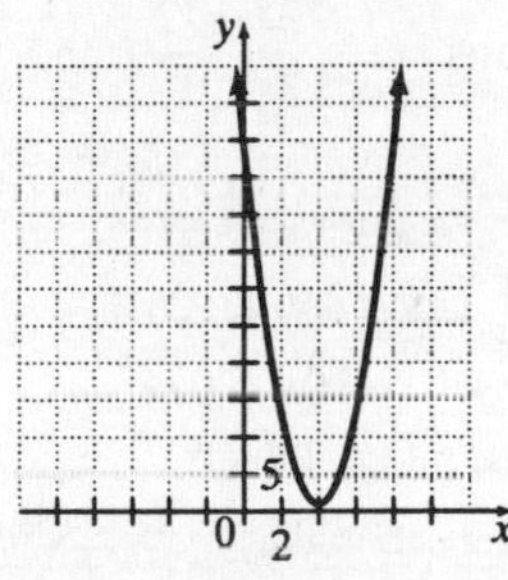

77.

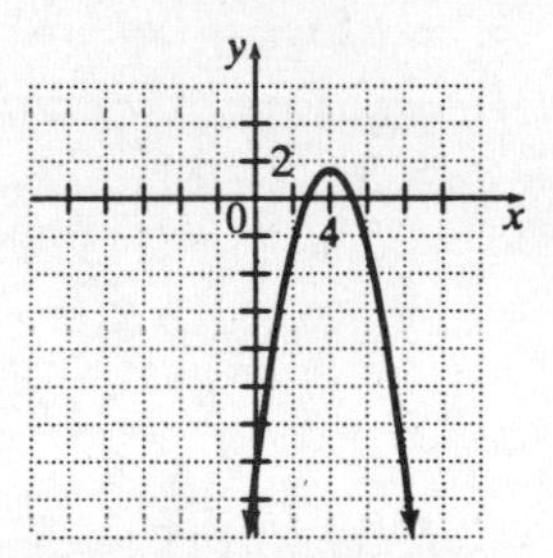

Chapter 10 Review Exercises

1. $x^2 - 15x + 14 = 0$
$(x-14)(x-1) = 0$
$x - 14 = 0$ or $x - 1 = 0$
$x = 14$ or $x = 1$
The solutions are 14 and 1.

2. $x^2 - x - 30 = 0$
$(x+5)(x-6) = 0$
$x + 5 = 0$ or $x - 6 = 0$
$x = -5$ or $x = 6$
The solutions are –5 and 6.

3. $10x^2 = 3x + 4$
$10x^2 - 3x - 4 = 0$
$(5x-4)(2x+1) = 0$
$5x - 4 = 0$ or $2x + 1 = 0$
$5x = 4$ or $2x = -1$
$x = \frac{4}{5}$ or $x = -\frac{1}{2}$
The solutions are $\frac{4}{5}$ and $-\frac{1}{2}$.

4. $7a^2 = 29a + 30$
$7a^2 - 29a - 30 = 0$
$(7a+6)(a-5) = 0$
$7a + 6 = 0$ or $a - 5 = 0$
$a = -\frac{6}{7}$ or $a = 5$
The solutions are $-\frac{6}{7}$ and 5.

5. $4m^2 = 196$
$m^2 = 49$
$m = \pm\sqrt{49}$
$m = \pm 7$
The solutions are 7 and –7.

6. $9y^2 = 36$
$y^2 = 4$
$y = \pm 2$
The solutions are 2 and –2.

7. $(9n+1)^2 = 9$
$9n + 1 = \pm\sqrt{9}$
$9n + 1 = \pm 3$
$9n = -1 \pm 3$
$n = \frac{-1 \pm 3}{9} = \frac{2}{9}, -\frac{4}{9}$
The solutions are $\frac{2}{9}$ and $-\frac{4}{9}$.

8. $(5x-2)^2=2$
$$5x-2=\pm\sqrt{2}$$
$$5x=2\pm\sqrt{2}$$
$$x=\frac{2\pm\sqrt{2}}{5}$$

The solutions are $\frac{2+\sqrt{2}}{5}$ and $\frac{2-\sqrt{2}}{5}$.

9. $z^2+3z+1=0$
$$z^2+3z=-1$$
$$z^2+3z+\left(\frac{3}{2}\right)^2=-1+\frac{9}{4}$$
$$\left(z+\frac{3}{2}\right)^2=\frac{5}{4}$$
$$z+\frac{3}{2}=\pm\sqrt{\frac{5}{4}}$$
$$z+\frac{3}{2}=\frac{\pm\sqrt{5}}{\sqrt{4}}$$
$$z=-\frac{3}{2}\pm\frac{\sqrt{5}}{2}$$

The solutions are $-\frac{3}{2}+\frac{\sqrt{5}}{2}$ and $-\frac{3}{2}-\frac{\sqrt{5}}{2}$.

10. $x^2+x+7=0$
$$x^2+x+\frac{1}{4}=-7+\frac{1}{4}$$
$$\left(x+\frac{1}{2}\right)^2=-\frac{27}{4}$$
$$x+\frac{1}{2}=\pm\sqrt{\frac{-27}{4}}$$
$$x=-\frac{1}{2}\pm\frac{3i\sqrt{3}}{2}$$
$$x=\frac{-1\pm 3i\sqrt{3}}{2}$$

The solutions are $\frac{-1+3i\sqrt{3}}{2}$ and $\frac{-1-3i\sqrt{3}}{2}$.

11. $(2x+1)^2=x$
$$4x^2+4x+1=x$$
$$4x^2+3x=-1$$
$$x^2+\frac{3}{4}x=-\frac{1}{4}$$
$$x^2+\frac{3}{4}x+\left(\frac{\frac{3}{4}}{2}\right)^2=-\frac{1}{4}+\frac{9}{64}$$
$$\left(x+\frac{3}{8}\right)^2=-\frac{7}{64}$$
$$x+\frac{3}{8}=\pm\sqrt{\frac{-7}{64}}$$
$$x+\frac{3}{8}=\pm\frac{\sqrt{7}i}{8}$$
$$x=-\frac{3}{8}\pm\frac{\sqrt{7}}{8}i$$

The solutions are $-\frac{3}{8}+\frac{\sqrt{7}}{8}i$ and $-\frac{3}{8}-\frac{\sqrt{7}}{8}i$.

12. $(3x-4)^2=10x$
$$9x^2-24x+16-10x=0$$
$$9x^2-34x+16=0$$
$$x^2-\frac{34}{9}x+\frac{289}{81}=-\frac{16}{9}+\frac{289}{81}$$
$$\left(x-\frac{17}{9}\right)^2=\frac{145}{81}$$
$$x-\frac{17}{9}=\pm\sqrt{\frac{145}{81}}$$
$$x=\frac{17\pm\sqrt{145}}{9}$$

The solutions are $\frac{17+\sqrt{145}}{9}$ and $\frac{17-\sqrt{145}}{9}$.

13. In this problem,
$P = 2500, A = 2717$

$$A = P(1+r)^2$$
$$2717 = 2500(1+r)^2$$
$$\frac{2717}{2500} = (1+r)^2$$
$$1.0868 = (1+r)^2$$
$$1.0425 = 1+r$$
$$0.0425 = r$$

The interest rate is 4.25%.

14. $c^2 = a^2 + b^2$ where $a = b$

$$c^2 = a^2 + a^2$$
$$(150)^2 = 2a^2$$
$$11,250 = a^2$$
$$\pm 75\sqrt{2} = a$$

Disregard the negative.
$75\sqrt{2}$ or 106.1 miles

15. Two complex but not real solutions exist.

16. Two real solutions exist.

17. Two real solutions exist.

18. One real solution exists.

19. $x^2 - 16x + 64 = 0$
$a = 1$, $b = -16$, and $c = 64$

$$x = \frac{16 \pm \sqrt{(-16)^2 - 4(1)(64)}}{2(1)}$$
$$x = \frac{16 \pm \sqrt{256-256}}{2} = \frac{16 \pm \sqrt{0}}{2} = 8$$

The solution is 8.

20. $x^2 + 5x = 0$

$$x = \frac{-5 \pm \sqrt{(5)^2 - 4(1)(0)}}{2(1)}$$
$$x = \frac{-5 \pm \sqrt{25}}{2} = \frac{-5 \pm 5}{2}$$
$$x = 0 \text{ or } x = -5$$

The solutions are 0 and –5.

21. $x^2 + 11 = 0$
$a = 1$, $b = 0$, and $c = 11$

$$x = \frac{-0 \pm \sqrt{0^2 - 4(1)(11)}}{2(1)}$$
$$= \frac{\pm\sqrt{-44}}{2}$$
$$= \frac{\pm 2\sqrt{11}i}{2}$$
$$= \pm\sqrt{11}i$$

The solutions are $\sqrt{11}i$ and $-\sqrt{11}i$.

22. $2x^2 + 3x = 5$

$$2x^2 + 3x - 5 = 0$$
$$x = \frac{-3 \pm \sqrt{9 - 4(2)(-5)}}{2(2)}$$
$$x = \frac{-3 \pm \sqrt{49}}{4} = \frac{-3 \pm 7}{4}$$
$$x = 1 \text{ or } x = -\frac{5}{2}$$

The solutions are 1 and $-\frac{5}{2}$.

23. $6x^2 + 7 = 5x$

$$6x^2 - 5x + 7 = 0$$

$a = 6$, $b = -5$, and $c = 7$

$$x = \frac{5 \pm \sqrt{(-5)^2 - 4(6)(7)}}{2(6)}$$
$$= \frac{5 \pm \sqrt{25 - 168}}{12}$$
$$= \frac{5 \pm \sqrt{-143}}{12}$$
$$= \frac{5 \pm \sqrt{143}i}{12}$$
$$= \frac{5}{12} \pm \frac{\sqrt{143}}{12}i$$

The solutions are $\frac{5}{12} + \frac{\sqrt{143}}{12}i$ and $\frac{5}{12} - \frac{\sqrt{143}}{12}i$.

24. $9a^2 + 4 = 2a$

$9a^2 - 2a + 4 = 0$

$a = \frac{2 \pm \sqrt{4 - 4(9)(4)}}{2(9)}$

$= \frac{2 \pm \sqrt{-140}}{18}$

$= \frac{2 \pm 2i\sqrt{35}}{18}$

$= \frac{1 \pm i\sqrt{35}}{9}$

The solutions are $\frac{1 + i\sqrt{35}}{9}$ and $\frac{1 - i\sqrt{35}}{9}$.

25. $(5a - 2)^2 - a = 0$

$25a^2 - 20a + 4 - a = 0$

$25a^2 - 21a + 4 = 0$

$a = 25$, $b = -21$ and $c = 4$

$a = \frac{21 \pm \sqrt{(-21)^2 - 4(25)(4)}}{2(25)}$

$= \frac{21 \pm \sqrt{441 - 400}}{50}$

$= \frac{21 \pm \sqrt{41}}{50}$

The solutions are $\frac{21 + \sqrt{41}}{50}$ and $\frac{21 - \sqrt{41}}{50}$.

26. $(2x - 3)^2 = x$

$4x^2 - 12x + 9 - x = 0$

$4x^2 - 13x + 9 = 0$

$x = \frac{13 \pm \sqrt{169 - 4(4)(9)}}{2(4)}$

$x = \frac{13 \pm \sqrt{25}}{8} = \frac{13 \pm 5}{8}$

$x = \frac{9}{4}$ or $x = 1$

The solutions are $\frac{9}{4}$ and 1.

27. $d(t) = -16t^2 + 30t + 6$

a. $d(1) = -16(1)^2 + 30(1) + 6$

$= -16 + 30 + 6$

$= 20$ feet

b. $-16t^2 + 30t + 6 = 0$

$8t^2 - 15t - 3 = 0$

$a = 8$, $b = -15$, and $c = -3$

$x = \frac{15 \pm \sqrt{(-15)^2 - 4(8)(-3)}}{2(8)}$

$= \frac{15 \pm \sqrt{225 + 96}}{16}$

$= \frac{15 \pm \sqrt{321}}{16}$

Discarding the negative value as extraneous, we find that

$x = \frac{15 + \sqrt{321}}{16} \approx 2.1$ seconds.

28. Let x = length of leg.

$x + 6$ = length of hypotenuse

$x^2 + x^2 = (x + 6)^2$

$2x^2 = x^2 + 12x + 36$

$x^2 - 12x - 36 = 0$

$x = \frac{12 \pm \sqrt{144 - 4(1)(-36)}}{2}$

$x = \frac{12 \pm \sqrt{288}}{2}$

$x = \frac{12 \pm 12\sqrt{2}}{2} = 6 \pm 6\sqrt{2}$

The length of each leg is $(6 + 6\sqrt{2})$ cm.

29.
$$x^3 = 27$$
$$x^3 - 27 = 0$$
$$(x-3)(x^2+3x+9) = 0$$
$$x-3=0 \quad \text{or} \quad x^2+3x+9=0$$
$$x = 3$$
$$a = 1,\ b = 3, \text{ and } c = 9$$
$$x = \frac{-3 \pm \sqrt{3^2 - 4(1)(9)}}{2(1)}$$
$$= \frac{-3 \pm \sqrt{9-36}}{2}$$
$$= \frac{-3 \pm \sqrt{-27}}{2}$$
$$= \frac{-3 \pm 3\sqrt{3}i}{2}$$
$$= -\frac{3}{2} \pm \frac{3\sqrt{3}i}{2}$$

The solutions are $3, -\frac{3}{2} + \frac{3\sqrt{3}i}{2}$ and $-\frac{3}{2} - \frac{3\sqrt{3}i}{2}$.

30.
$$y^3 = -64$$
$$y^3 + 64 = 0$$
$$(y+4)(y^2 - 4y + 16) = 0$$
$$y+4=0 \quad \text{or} \quad y^2 - 4y + 16 = 0$$
$$y = -4 \quad \text{or} \quad y = \frac{4 \pm \sqrt{16 - 4(1)(16)}}{2}$$
$$y = \frac{4 \pm \sqrt{-48}}{2}$$
$$y = \frac{4 \pm 4i\sqrt{3}}{2}$$
$$y = 2 \pm 2i\sqrt{3}$$

The solutions are -4, $2 + 2i\sqrt{3}$, and $2 - 2i\sqrt{3}$.

31.
$$\frac{5}{x} + \frac{6}{x-2} = 3$$
$$\frac{5(x-2)+6x}{x(x-2)} = 3$$
$$5x - 10 + 6x = 3x(x-2)$$
$$11x - 10 = 3x^2 - 6x$$
$$3x^2 - 17x + 10 = 0$$
$$(3x-2)(x-5) = 0$$
$$3x-2=0 \quad \text{or} \quad x-5=0$$
$$3x = 2 \quad \text{or} \quad x = 5$$
$$x = \frac{2}{3}$$

The solutions are $\frac{2}{3}$ and 5.

32.
$$\frac{7}{8} = \frac{8}{x^2}$$
$$7x^2 = 64$$
$$x^2 = \frac{64}{7}$$
$$x = \pm\sqrt{\frac{64}{7}} = \frac{\pm 8\sqrt{7}}{7}$$

The solutions are $\frac{8\sqrt{7}}{7}$ and $-\frac{8\sqrt{7}}{7}$.

33.
$$x^4 - 21x^2 - 100 = 0$$
$$(x^2 - 25)(x^2 + 4) = 0$$
$$x^2 - 25 = 0 \quad \text{or} \quad x^2 + 4 = 0$$
$$x^2 = 25 \quad \text{or} \quad x^2 = -4$$
$$x = \pm\sqrt{25} \quad \text{or} \quad x = \pm\sqrt{-4}$$
$$x = \pm 5 \quad \text{or} \quad x = \pm 2i$$

The solutions are 5, -5, $2i$, and $-2i$.

34. $5(x+3)^2 - 19(x+3) = 4$
Let $u = x + 3$. Then
$$5u^2 - 19u - 4 = 0$$
$$(5u+1)(u-4) = 0$$
$$5u + 1 = 0 \quad \text{or} \quad u - 4 = 0$$
$$u = -\frac{1}{5} \quad \text{or} \quad u = 4$$
$$x + 3 = -\frac{1}{5} \quad \text{or} \quad x + 3 = 4$$
$$x = -\frac{16}{5} \quad \text{or} \quad x = 1$$
The solutions are $-\frac{16}{5}$ and 1.

35. $x^{2/3} - 6x^{1/3} + 5 = 0$
$$(x^{1/3} - 1)(x^{1/3} - 5) = 0$$
$$x^{1/3} - 1 = 0 \quad \text{or} \quad x^{1/3} - 5 = 0$$
$$x^{1/3} = 1 \quad \text{or} \quad x^{1/3} = 5$$
$$x = 1^3 = 1 \quad \text{or} \quad x = 5^3 = 125$$
The solutions are 1 and 125.

36. $x^{2/3} - 6x^{1/3} + 8 = 0$
Let $m = x^{1/3}$. Then
$$m^2 - 6m + 8 = 0$$
$$(m-2)(m-4) = 0$$
$$m - 2 = 0 \quad \text{or} \quad m - 4 = 0$$
$$m = 2 \quad \text{or} \quad m = 4$$
$$x^{1/3} = 2 \quad \text{or} \quad x^{1/3} = 4$$
$$x = 2^3 = 8 \quad \text{or} \quad x = 4^3 = 64$$
The solutions are 8 and 64.

37. $a^6 - a^2 = a^4 - 1$
$$a^2(a^4 - 1) - (a^4 - 1) = 0$$
$$(a^2 - 1)(a^4 - 1) = 0$$
$$(a^2 - 1)(a^2 - 1)(a^2 + 1) = 0$$
$$(a^2 - 1)^2(a^2 + 1) = 0$$
$$[(a+1)(a-1)]^2(a^2 + 1) = 0$$
$$(a+1)^2(a-1)^2(a^2 + 1) = 0$$
$$(a+1)^2 = 0 \quad \text{or} \quad (a-1)^2 = 0 \quad \text{or} \quad a^2 + 1 = 0$$
$$a + 1 = 0 \quad \text{or} \quad a - 1 = 0 \quad \text{or} \quad a^2 = -1$$
$$a = -1 \quad \text{or} \quad a = 1 \quad \text{or} \quad a = \pm\sqrt{-1} = \pm i$$
The solutions are 1, −1, i and $-i$.

38. $y^{-2}+y^{-1}=20$

$$\frac{1}{y^2}+\frac{1}{y}=20$$
$$1+y=20y^2$$
$$0=20y^2-y-1$$
$$0=(5y+1)(4y-1)$$
$$5y+1=0 \quad \text{or} \quad 4y-1=0$$
$$y=-\frac{1}{5} \quad \text{or} \quad y=\frac{1}{4}$$

The solutions are $-\frac{1}{5}$ and $\frac{1}{4}$.

39.

	Hours	Job complete in one hour
Jerome	x	$\frac{1}{x}$
Tim	$x-1$	$\frac{1}{x-1}$
Together	5	$\frac{1}{5}$

$$\frac{1}{x}+\frac{1}{x-1}=\frac{1}{5}$$
$$\frac{(x-1)+x}{x(x-1)}=\frac{1}{5}$$
$$\frac{2x-1}{x^2-x}=\frac{1}{5}$$
$$10x-5=x^2-x$$
$$0=x^2-11x+5$$

$a=1$, $b=-11$, and $c=5$

$$x=\frac{-(-11)\pm\sqrt{(-11)^2-4(1)(5)}}{2(1)}$$
$$x=\frac{11\pm\sqrt{101}}{2}$$

$x\approx 0.475$ or 10.525

$x\approx 10.5,\ x-1\approx 9.5$

Jerome, 10.5 hours

Tim, 9.5 hours

40. Let x = the number so $\frac{1}{x}$ = the reciprocal of the number. Then

$$x-\frac{1}{x}=-\frac{24}{5}$$
$$\frac{x^2-1}{x}=-\frac{24}{5}$$
$$5(x^2-1)=-24x$$
$$5x^2-5=-24x$$
$$5x^2+24x-5=0$$
$$(5x-1)(x+5)=0$$
$$5x-1=0 \quad \text{or} \quad x+5=0$$
$$5x=1 \quad \text{or} \quad x=-5$$
$$x=\frac{1}{5}$$

Discarding the positive value as extraneous, we find the number to be -5.

41. $2x^2-50\le 0$

$$2x^2\le 50$$
$$x^2\le 25$$
$$-5\le x\le 5$$

$[-5, 5]$

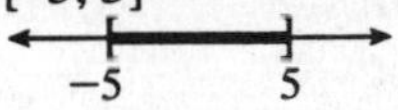

42. $\frac{1}{4}x^2 < \frac{1}{16}$

$x^2 < \frac{1}{4}$

$x^2 - \frac{1}{4} < 0$

$\left(x - \frac{1}{2}\right)\left(x + \frac{1}{2}\right) < 0$

A	B	C

$-\frac{1}{2}$, $\frac{1}{2}$

Region	Test Point	$\left(x-\frac{1}{2}\right)\left(x+\frac{1}{2}\right) < 0$
A	−1	$\left(-\frac{3}{2}\right)\left(-\frac{1}{2}\right) > 0$
B	0	$\left(-\frac{1}{2}\right)\left(\frac{1}{2}\right) < 0$
C	1	$\left(\frac{1}{2}\right)\left(\frac{3}{2}\right) > 0$

The solution is $\left(-\frac{1}{2}, \frac{1}{2}\right)$.

$-\frac{1}{2}$ $\frac{1}{2}$

43. $(2x - 3)(4x + 5) \geq 0$

First solve

$(2x - 3)(4x + 5) = 0$.

$2x - 3 = 0$ or $4x + 5 = 0$

$2x = 3$ or $4x = -5$

$x = \frac{3}{2}$ or $x = -\frac{5}{4}$

Now select three appropriate test points.

$x = -2$: $[2(-2) - 3][4(-2) + 5] = (-)(-) = +$

$x = 2$: $[2 \cdot 2 - 3][4 \cdot 2 + 5] = (+)(+) = +$

$x = 0$: $[2(0) - 3][4(0) + 5] = (-)(+) = -$

Thus, the solutions are $x \geq \frac{3}{2}$ and

$x \leq -\frac{5}{4}$.

$\left(-\infty, -\frac{5}{4}\right] \cup \left[\frac{3}{2}, \infty\right)$

$-\frac{5}{4}$ $\frac{3}{2}$

44. $(x^2 - 16)(x^2 - 1) > 0$

$(x + 4)(x - 4)(x + 1)(x - 1) > 0$

A	B	C	D	E

−4 −1 1 4

Region	Test Point	$(x^2 - 16)(x^2 - 1) > 0$
A	−5	$(9)(24) > 0$
B	−2	$(-12)(3) < 0$
C	0	$(-16)(-1) > 0$
D	2	$(-12)(3) < 0$
E	5	$(9)(24) > 0$

$(-\infty, -4) \cup (-1, 1) \cup (4, \infty)$ satisfies the inequality.

−4 −1 1 4

45. $\frac{x - 5}{x - 6} < 0$

First solve both

$x - 5 = 0$ and $x - 6 = 0$

$x = 5$ and $x = 6$.

Now select three appropriate test points.

$x = 4$: $\frac{4 - 5}{4 - 6} = \frac{-}{-} = +$

$x = \frac{11}{2}$: $\frac{\frac{11}{2} - 5}{\frac{11}{2} - 6} = \frac{+}{-} = -$

$x = 7$: $\frac{7 - 5}{7 - 6} = \frac{+}{+} = +$

Thus, the solution is $5 < x < 6$.

$(5, 6)$

5 6

46. $\dfrac{x(x+5)}{4x-3} \geq 0$

A	B	C	D

$-5 \quad 0 \quad \frac{3}{4}$

Region	Test Point	$\dfrac{x(x+5)}{4x-3} \geq 0$
A	–6	$\dfrac{(-6)(-1)}{-27} \leq 0$
B	–1	$\dfrac{(-1)(4)}{-7} \geq 0$
C	$\frac{1}{2}$	$\dfrac{\frac{1}{2}\left(5\frac{1}{2}\right)}{(-1)} \leq 0$
D	1	$\dfrac{1(6)}{1} \geq 0$

$[-5,\ 0] \cup \left(\dfrac{3}{4},\ \infty\right)$ satisfies the inequality.

$-5 \quad 0 \quad \frac{3}{4}$

47. $\dfrac{(4x+3)(x-5)}{x(x+6)} > 0$

First solve both:

$(4x + 3)(x - 5) = 0$ and $x(x + 6) = 0$

$4x + 3 = 0$ or $x - 5 = 0$ and $x = 0$ or $x + 6 =$
0

$4x = -3$ or $x = 5$ and $x = 0$ or $x = -6$

$x = -\dfrac{3}{4}$

Now select five appropriate test points.

$x = -7$:

$\dfrac{[4(-7)+3][-7-5]}{-7(-7+6)} = \dfrac{(-)(-)}{(-)(-)} = \dfrac{+}{+} = +$

$x = -1$:

$\dfrac{[4(-1)+3](-1-5)}{-1(-1+6)} = \dfrac{(-)(-)}{(-)(+)} = \dfrac{+}{-} = -$

$x = -\dfrac{1}{2}$: $\dfrac{\left[4\left(-\frac{1}{2}\right)+3\right]\left[-\frac{1}{2}-5\right]}{-\frac{1}{2}\left(-\frac{1}{2}+6\right)} = \dfrac{(+)(-)}{(-)(+)}$

$= \dfrac{-}{-}$

$= +$

$x = 1$: $\dfrac{[4(1)+3][1-5]}{1(1+6)} = \dfrac{(+)(-)}{(+)(+)} = \dfrac{-}{+} = -$

$x = 6$: $\dfrac{[4(6)+3][6-5]}{6(6+6)} = \dfrac{(+)(+)}{(+)(+)} = \dfrac{+}{+} = +$

Thus, the solutions are $x < -6$,

$-\dfrac{3}{4} < x < 0$, and

$x > 5$.

$(-\infty,\ -6) \cup \left(-\dfrac{3}{4},\ 0\right) \cup (5,\ \infty)$

$-6 \quad -\frac{3}{4} \quad 0 \quad 5$

48. $(x + 5)(x - 6)(x + 2) \leq 0$

A	B	C	D

$-5 \quad -2 \quad 6$

Region	Test Point	$(x + 5)(x - 6)(x + 2) \leq 0$
A	–6	$(-1)(-12)(-4) \leq 0$
B	–3	$(2)(-9)(-1) \geq 0$
C	0	$(5)(-6)(2) \leq 0$
D	7	$(12)(1)(9) \geq 0$

The solution is $(-\infty,\ -5] \cup [-2,\ 6]$.

$-5 \quad -2 \quad 6$

49. $x^3 + 3x^2 - 25x - 75 > 0$

$x^2(x+3) - 25(x+3) > 0$

$(x+3)(x^2 - 25) > 0$

$(x+3)(x+5)(x-5) > 0$

Region	A	B	C	D
x-values	$(-\infty, -5)$	$(-5, -3)$	$(-3, 5)$	$(5, \infty)$

Region	Test Point	$(x+3)(x+5)(x-5) > 0$
A	−6	$(-3)(-1)(-11) < 0$
B	−4	$(-1)(1)(-9) > 0$
C	0	$(3)(5)(-5) < 0$
D	6	$(9)(11)(1) > 0$

$(-5, -3) \cup (5, \infty)$ satisfies the inequality.

−5 −3 5

50. $\dfrac{x^2+4}{3x} \le 1$

First solve both

$3x = 0$ and $\dfrac{x^2+4}{3x} = 1.$

$x = 0$

$x^2 + 4 = 3x$

$x^2 - 3x + 4 = 0$

has no real solution

Now select two appropriate test points.

$x = -5$: $\dfrac{(-5)^2+4}{3(-5)} = \dfrac{+}{-} = -$

$x = 1$: $\dfrac{1^2+4}{3(1)} = \dfrac{+}{+} = +$

Thus the solution is $x < 0$.

$(-\infty, 0)$

0

51. $\dfrac{(5x+6)(x-3)}{x(6x-5)} < 0$

A	B	C	D	E

$-\frac{6}{5}$ 0 $\frac{5}{6}$ 3

Region	Test Point	$\dfrac{(5x+6)(x-3)}{x(6x-5)} < 0$
A	−2	$\dfrac{(-4)(-5)}{(-2)(-17)} > 0$
B	−1	$\dfrac{(1)(-4)}{(-1)(-11)} < 0$
C	$\frac{1}{2}$	$\dfrac{\left(8\frac{1}{2}\right)\left(-2\frac{1}{2}\right)}{\frac{1}{2}(-2)} > 0$
D	1	$\dfrac{(11)(-2)}{1(1)} < 0$
E	4	$\dfrac{(26)(1)}{4(19)} > 0$

The solution is $\left(-\dfrac{6}{5}, 0\right) \cup \left(\dfrac{5}{6}, 3\right)$.

$-\frac{6}{5}$ 0 $\frac{5}{6}$ 3

52. $\frac{3}{x-2} > 2$

$3 > 2x - 4$

$0 > 2x - 7$

$x = 2$ (undefined)

$x = \frac{7}{2}$

Region	A	B	C
x-values	$(-\infty, 2)$	$\left(2, \frac{7}{2}\right)$	$\left(\frac{7}{2}, \infty\right)$

Region	Test Point	$\frac{3}{x-2} > 2$
A	0	$-\frac{3}{2} < 2$
B	3	$3 > 2$
C	4	$\frac{3}{2} < 2$

$\left(2, \frac{7}{2}\right)$ satisfies the inequality.

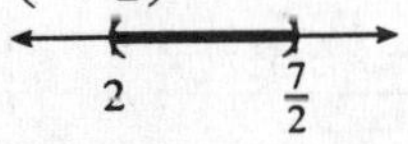

53. $A = kB$

$6 = k(14)$

$k = \frac{6}{14} = \frac{3}{7}$

$A = \frac{3}{7}B = \frac{3}{7}(21) = 3(3) = 9$

54. $C = \frac{k}{D}$

$12 = \frac{k}{8}$

$96 = k$

$C = \frac{96}{D}$

$C = \frac{96}{24} = 4$

55. $P = \frac{k}{V}$

$1250 = \frac{k}{2}$

$k = 2500$

$P = \frac{2500}{V}$

$800 = \frac{2500}{V}$

$V = \frac{2500}{800} = 3.125$ cubic feet

56. $A = kr^2$

$36\pi = k(3)^2$

$4\pi = k$

$A = 4\pi r^2$

$A = 4\pi(4)^2 = 64\pi$ sq. in.

57. $f(x) = x^2 - 4$

$\frac{-b}{2a} = \frac{0}{2(1)} = 0$

$f(0) = 0^2 - 4 = -4$

Vertex: (0, –4)

Axis of symmetry: $x = 0$

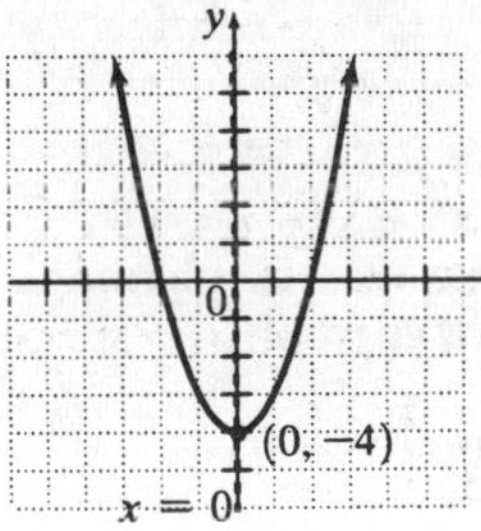

58. $g(x) = x^2 + 7$

$\frac{-b}{2a} = \frac{0}{2(1)} = 0$

$g(0) = 0^2 + 7$

Vertex: (0, 7)

Axis of symmetry: $x = 0$

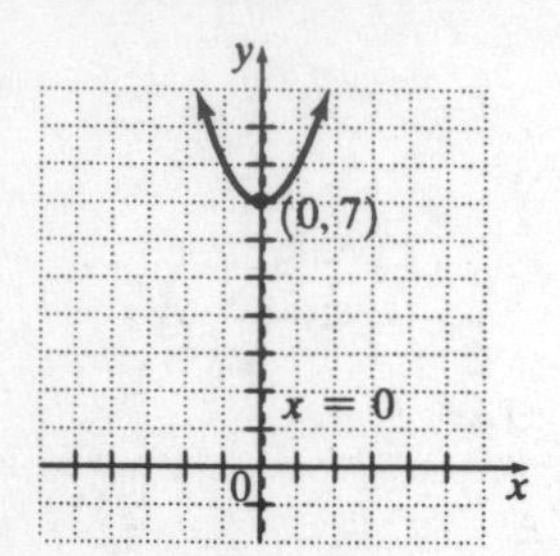

59. $H(x) = 2x^2$

Vertex: (0, 0)

Axis of symmetry: $x = 0$

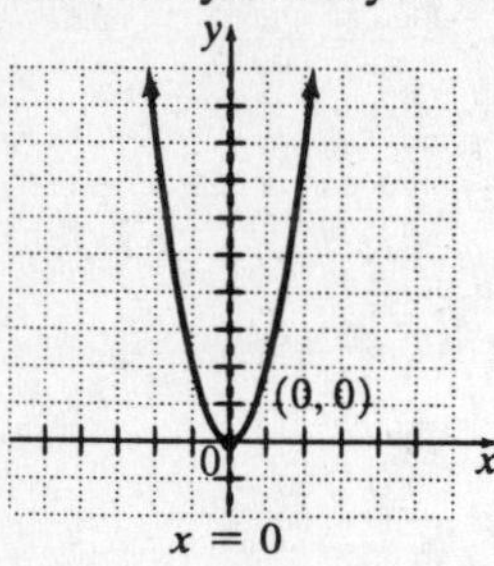

60. $h(x) = -\frac{1}{3}x^2$

Vertex: (0, 0)

Axis of symmetry: $x = 0$

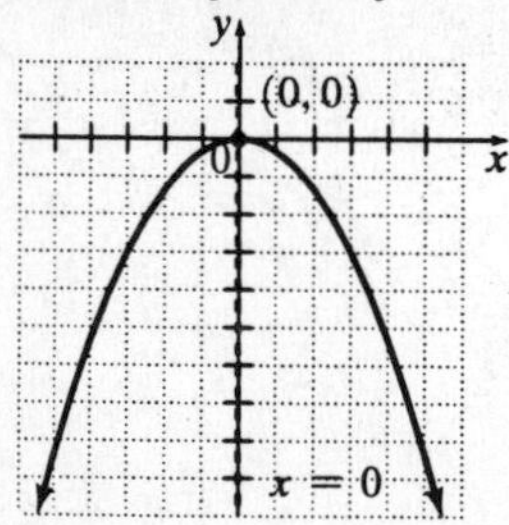

61. $F(x) = (x-1)^2$

$F(x) = x^2 - 2x + 1$

$\frac{-b}{2a} = \frac{-(-2)}{2(1)} = 1$

$F(1) = 0$

Vertex: (1, 0)

Axis of symmetry: $x = 1$

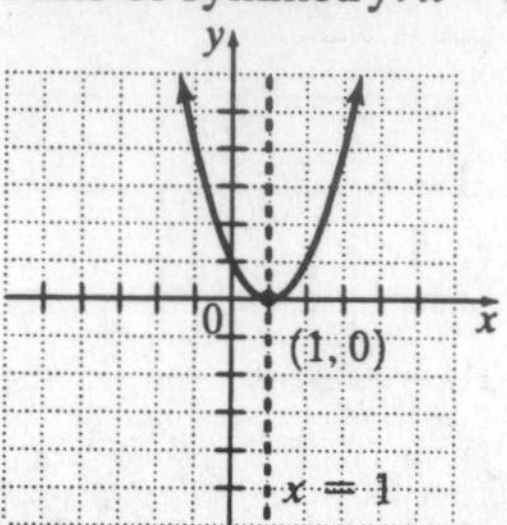

62. $G(x) = (x+5)^2$

$G(x) = x^2 + 10x + 25$

$\frac{-b}{2a} = \frac{-10}{2(1)} = -5$

$G(-5) = 0$

Vertex: (–5, 0)

Axis of symmetry: $x = -5$

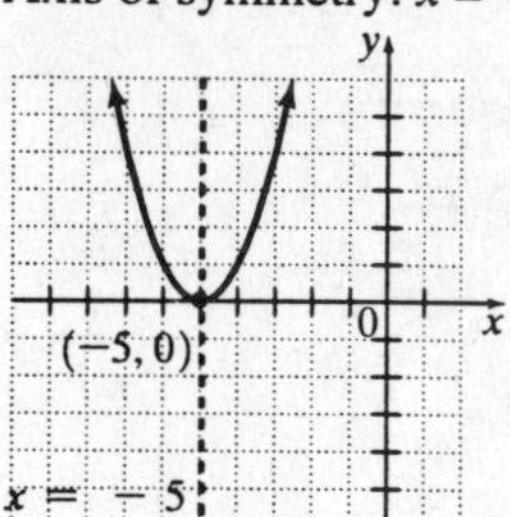

63. $f(x) = (x-4)^2 - 2$

Vertex: (4, –2)

Axis of symmetry: $x = 4$

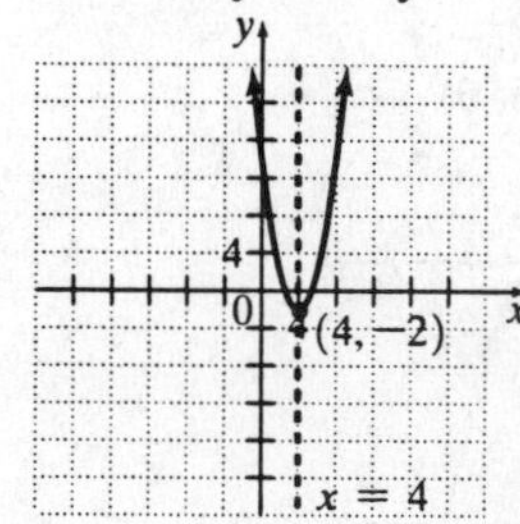

64. $y = -3(x-1)^2 + 1$

Vertex: (1, 1)

Axis of symmetry: $x = 1$

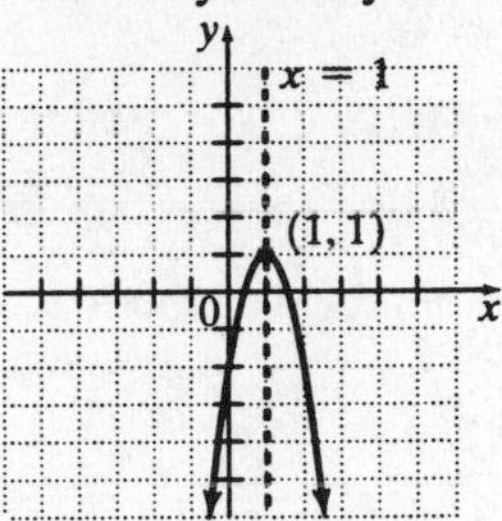

65. $f(x) = x^2 + 10x + 25$ or

$f(x) = (x+5)^2 + 0$

Vertex: (–5, 0).

$(x+5)^2 = 0 \Rightarrow x + 5 = 0$

x-intercept: (–5, 0).

y-intercept: (0, 25).

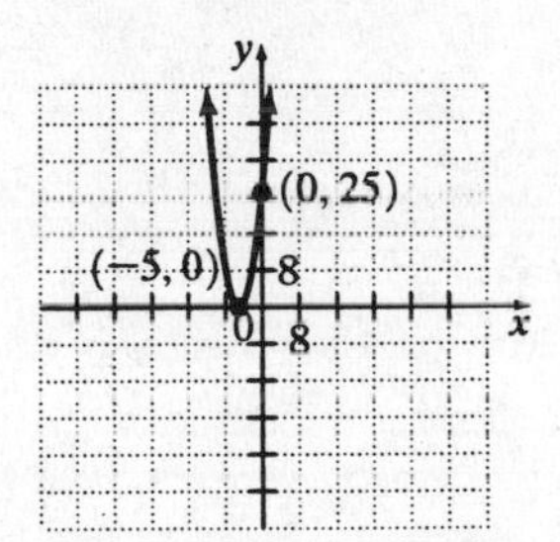

66. $f(x) = -x^2 + 6x - 9$

$\frac{-b}{2a} = \frac{-6}{2(-1)} = 3$

$f(x) = -(3)^2 + 6(3) - 9 = 0$

Vertex: (3, 0)

$y(0) = -9$

y-intercept: (0, –9)

x-intercept: (3, 0)

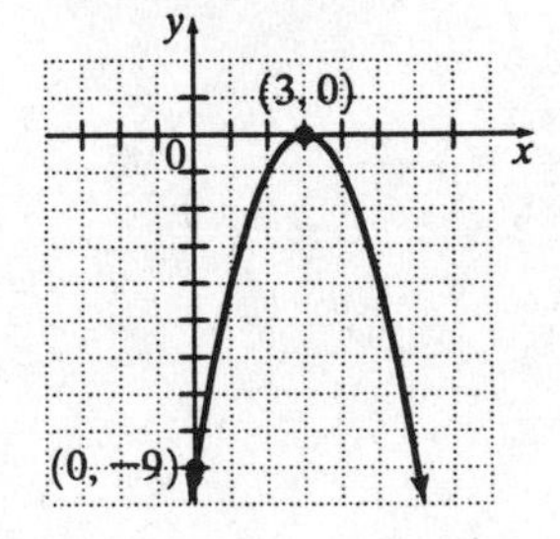

67. $f(x) = 4x^2 - 1$ or

$f(x) = 4(x-0)^2 - 1$

Vertex: (0, –1)

$4x^2 - 1 = 0$

$4x^2 = 1$

$x^2 = \frac{1}{4}$

$x = \pm\frac{1}{2}$

x-intercepts: $\left(\frac{1}{2}, 0\right), \left(-\frac{1}{2}, 0\right)$

y-intercept: (0, –1)

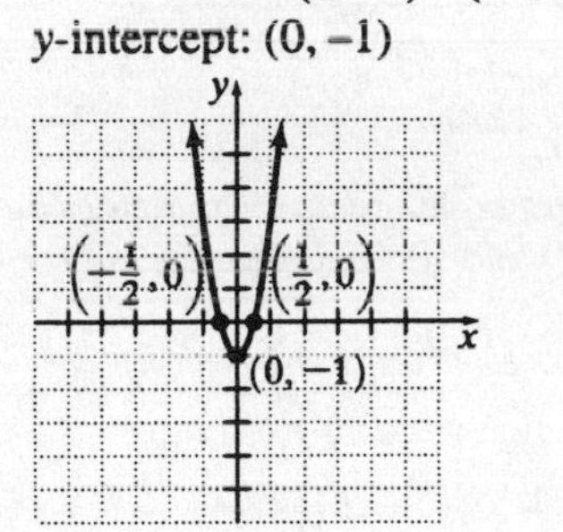

68. $f(x) = -5x^2 + 5$

$\frac{-b}{2a} = 0$

$y = 0 + 5$

Vertex: (0, 5)

y-intercept: (0, 5)

x-intercepts: (–1, 0), (1, 0)

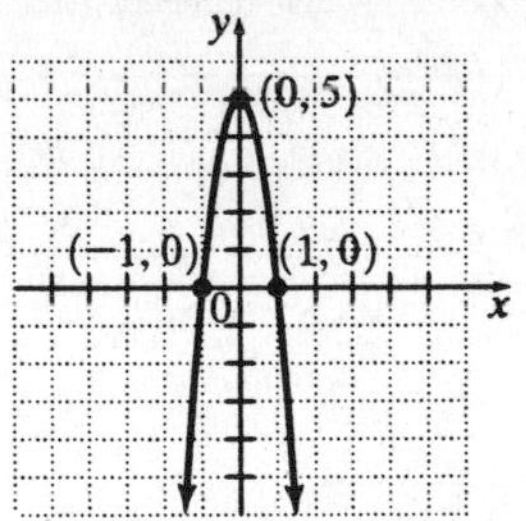

69. $f(x) = -3x^2 - 5x + 4$

$\frac{-b}{2a} = \frac{-(-5)}{2(-3)} = -\frac{5}{6}$

$f\left(-\frac{5}{6}\right) = -3\left(-\frac{5}{6}\right)^2 - 5\left(-\frac{5}{6}\right) + 4 = \frac{73}{12}$

Vertex: $\left(-\frac{5}{6}, \frac{73}{12}\right)$

The graph opens down because $a = -3 < 0$.

y-intercept: (0, 4).

$$x = \frac{-(-5) \pm \sqrt{(-5)^2 - 4(-3)(4)}}{2(-3)}$$

$$x = \frac{5 \pm \sqrt{73}}{-6}$$

The x-intercepts are approximately (–2.26, 0) and (0.59, 0).

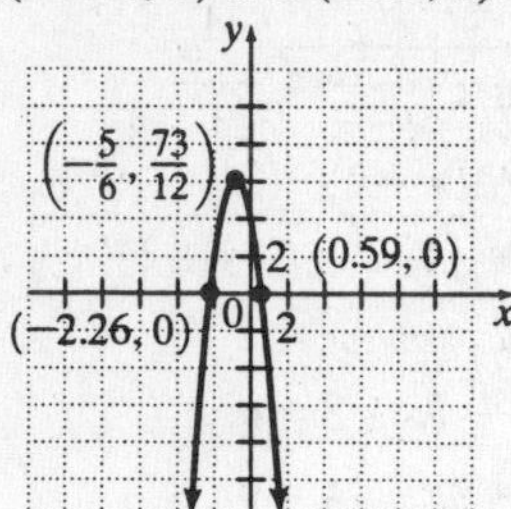

70. $h(t) = -16t^2 + 120t + 300$

a. $350 = -16t^2 + 120t + 300$

$0 = -16t^2 + 120t - 50$

$0 = -8t^2 + 60t - 25$

$a = -8, b = 60, c = -25$

$$t = \frac{-60 \pm \sqrt{(60)^2 - 4(-8)(-25)}}{2(-8)}$$

$$t = \frac{-60 \pm \sqrt{2800}}{-16}$$

$t \approx 0.4$ seconds and 7.1 seconds

b. The object will be at 350 feet on the way up and on the way down.

71. Let x = one number so $420 - x$ = the other number. Let $f(x)$ represent their product. Thus,

$f(x) = x(420 - x)$

$f(x) = -x^2 + 420x$

$\frac{-b}{2a} = \frac{-420}{2(-1)} = 210$

Therefore, the numbers are both 210.

72. $y = a(x - h)^2 + k$

$y = a(x + 3)^2 + 7$

$0 = a(0 + 3)^2 + 7$

$-7 = 9a$

$-\frac{7}{9} = a$

$y = -\frac{7}{9}(x + 3)^2 + 7$

Chapter 10 Test

1. $5x^2 - 2x = 7$

$5x^2 - 2x - 7 = 0$

$(5x - 7)(x + 1) = 0$

$5x - 7 = 0$ or $x + 1 = 0$

$5x = 7$ or $x = -1$

$x = \frac{7}{5}$

The solutions are $\frac{7}{5}$ and –1.

2. $(x + 1)^2 = 10$

$x + 1 = \pm\sqrt{10}$

$x = -1 \pm \sqrt{10}$

The solutions are $-1 + \sqrt{10}$ and $-1 - \sqrt{10}$.

3. $m^2 - m + 8 = 0$

$a = 1$, $b = -1$, and $c = 8$

$$m = \frac{1 \pm \sqrt{(-1)^2 - 4(1)(8)}}{2(1)} = \frac{1 \pm \sqrt{1-32}}{2} = \frac{1 \pm \sqrt{-31}}{2} = \frac{1 \pm \sqrt{31}i}{2} = \frac{1}{2} \pm \frac{\sqrt{31}}{2}i$$

The solutions are $\frac{1}{2} + \frac{\sqrt{31}}{2}i$ and $\frac{1}{2} - \frac{\sqrt{31}}{2}i$.

4. $u^2 - 6u + 2 = 0$

$a = 1$, $b = -6$, $c = 2$

$$u = \frac{6 \pm \sqrt{(-6)^2 - 4(1)(2)}}{2(1)} = \frac{6 \pm \sqrt{36-8}}{2} = \frac{6 \pm \sqrt{28}}{2} = \frac{6 \pm 2\sqrt{7}}{2} = 3 \pm \sqrt{7}$$

The solutions are $3 + \sqrt{7}$ and $3 - \sqrt{7}$.

5. $7x^2 + 8x + 1 = 0$

$(7x + 1)(x + 1) = 0$

$7x + 1 = 0$ or $x + 1 = 0$

$7x = -1$ or $x = -1$

$x = -\frac{1}{7}$

The solutions are $-\frac{1}{7}$ and -1.

6. $a^2 - 3a = 5$

$a^2 - 3a - 5 = 0$

$a = 1$, $b = -3$, and $c = -5$

$$a = \frac{3 \pm \sqrt{(-3)^2 - 4(1)(-5)}}{2(1)} = \frac{3 \pm \sqrt{9+20}}{2} = \frac{3 \pm \sqrt{29}}{2} = \frac{3}{2} \pm \frac{\sqrt{29}}{2}$$

The solutions are $\frac{3}{2} + \frac{\sqrt{29}}{2}$ and $\frac{3}{2} - \frac{\sqrt{29}}{2}$.

7. $$\frac{4}{x+2} + \frac{2x}{x-2} = \frac{6}{x^2-4}$$

$$\frac{4(x-2) + 2x(x+2)}{(x+2)(x-2)} = \frac{6}{(x+2)(x-2)}$$

$4x - 8 + 2x^2 + 4x = 6$

$2x^2 + 8x - 8 = 6$

$2x^2 + 8x - 14 = 0$

$x^2 + 4x - 7 = 0$

$a = 1$, $b = 4$, and $c = -7$

$$x = \frac{-4 + \sqrt{4^2 - 4(1)(-7)}}{2(1)} = \frac{-4 \pm \sqrt{16+28}}{2} = \frac{-4 \pm \sqrt{44}}{2} = \frac{-4 \pm 2\sqrt{11}}{2} = -2 \pm \sqrt{11}$$

The solutions are $-2 + \sqrt{11}$ and $-2 - \sqrt{11}$.

8. $x^4 - 8x^2 - 9 = 0$

$(x^2 - 9)(x^2 + 1) = 0$

$x^2 - 9 = 0$ or $x^2 + 1 = 0$

$x^2 = 9$ or $x^2 = -1$

$x = \pm\sqrt{9}$ or $x = \pm\sqrt{-1}$

$x = \pm 3$ or $x = \pm i$

The solutions are 3, –3, i and $-i$.

9. $x^6 + 1 = x^4 + x^2$

$x^6 - x^4 - x^2 + 1 = 0$

$x^4(x^2 - 1) - (x^2 - 1) = 0$

$(x^4 - 1)(x^2 - 1) = 0$

$(x^2 + 1)(x^2 - 1)(x^2 - 1) = 0$

$(x^2 + 1)(x^2 - 1)^2 = 0$

$(x^2 + 1)[(x + 1)(x - 1)]^2 = 0$

$(x^2 + 1)(x + 1)^2(x - 1)^2 = 0$

$x^2 + 1 = 0$ or $(x + 1)^2 = 0$ or $(x - 1)^2 = 0$

$x^2 = -1$ or $x + 1 = 0$ or $x - 1 = 0$

$x = \pm\sqrt{-1}$ or $x = -1$ or $x = 1$

The solutions are 1, –1, i, and $-i$.

10. $(x + 1)^2 - 15(x + 1) + 56 = 0$

Let $u = x + 1$.

$u^2 - 15u + 56 = 0$

$(u - 7)(u - 8) = 0$

$u - 7 = 0$ or $u - 8 = 0$

$u = 7$ or $u = 8$

Substitute $x + 1$ for u.

$x + 1 = 7$ or $x + 1 = 8$

$x = 6$ or $x = 7$

The solutions are 6 and 7.

11. $x^2 - 6x = -2$

$$x^2 - 6x + \left(\frac{6}{2}\right)^2 = -2 + 9$$

$(x - 3)^2 = 7$

$x - 3 = \pm\sqrt{7}$

$x = 3 \pm \sqrt{7}$

The solutions are $3 + \sqrt{7}$ and $3 - \sqrt{7}$.

12. $2a^2 + 5 = 4a$

$2a^2 - 4a = -5$

$$a^2 - 2a = -\frac{5}{2}$$

$$a^2 - 2a + \left(\frac{2}{2}\right)^2 = -\frac{5}{2} + 1$$

$$(a - 1)^2 = -\frac{3}{2}$$

$$a - 1 = \pm\sqrt{\frac{-3}{2}} = \pm\frac{\sqrt{3}i}{\sqrt{2}}$$

$$a - 1 = \frac{\pm\sqrt{6}i}{2}$$

$$a = 1 \pm \frac{\sqrt{6}}{2}i$$

The solutions are $1 + \frac{\sqrt{6}}{2}i$ and $1 - \frac{\sqrt{6}}{2}i$.

13. $2x^2 - 7x > 15$

$2x^2 - 7x - 15 > 0$

$(2x + 3)(x - 5) > 0$

$2x + 3 = 0$ or $x - 5 = 0$

$x = -\frac{3}{2}$ or $x = 5$

Now select three appropriate test points.

$x = -2$: $2(-2)^2 - 7(-2) - 15 = 7 > 0$

$x = 0$: $2(0)^2 - 7(0) - 15 = -15 < 0$

$x = 6$: $2(6)^2 - 7(6) - 15 = 15 > 0$

Thus, the solutions are

$$\left(-\infty, -\frac{3}{2}\right) \cup (5, \infty)$$

$-\frac{3}{2}$ 5

14. $(x^2 - 16)(x^2 - 25) > 0$

Let $u = x^2$.

$(u - 16)(u - 25) > 0$

First solve $(u - 16)(u - 25) = 0$

$u - 16 = 0$ or $u - 25 = 0$

$u = 16$ or $u = 25$

Now select three appropriate test points.

$u = 15$: $(15 - 16)(15 - 25) = (-)(-) = +$

$u = 20$: $(20 - 16)(20 - 25) = (+)(-) = -$

$u = 30$: $(30 - 16)(30 - 25) = (+)(+) = +$

Thus, the solutions are $u < 16$ or $u > 25$.

Now substitute back.

$x^2 < 16$ or $x^2 > 25$

$-4 < x < 4$ or $x < -5$ or $x > 5$

$(-\infty, -5) \cup (-4, 4) \cup (5, \infty)$

15. $\frac{5}{x+3} < 1$

$x = -3$ (undefined)

$5 = x + 3$

$x = 2$

Now select three appropriate test points.

$x = -4$: $\frac{5}{-4+3} < 1$

$x = 0$: $\frac{5}{0+3} > 1$

$x = 3$: $\frac{5}{3+3} < 1$

$(-\infty, -3) \cup (2, \infty)$

16. $\frac{7x - 14}{x^2 - 9} \le 0$

First solve both

$7x - 14 = 0$ and $x^2 - 9 = 0$.

$7x = 14$ and $x^2 = 9$

$x = 2$ and $x = \pm\sqrt{9} = \pm 3$

Now select four appropriate test points.

$x = -4$: $\frac{7(-4) - 14}{(-4)^2 - 9} = \frac{-}{+} = -$

$x = 0$: $\frac{7(0) - 14}{0^2 - 9} = \frac{-}{-} = +$

$x = \frac{5}{2}$: $\frac{7\left(\frac{5}{2}\right) - 14}{\left(\frac{5}{2}\right)^2 - 9} = \frac{+}{-} = -$

$x = 4$: $\frac{7(4) - 14}{4^2 - 9} = \frac{+}{+} = +$

Thus, the solutions are $x < -3$ and $2 \le x < 3$.

$(-\infty, -3) \cup [2, 3)$

17. $f(x) = 3x^2$

vertex: (0, 0)

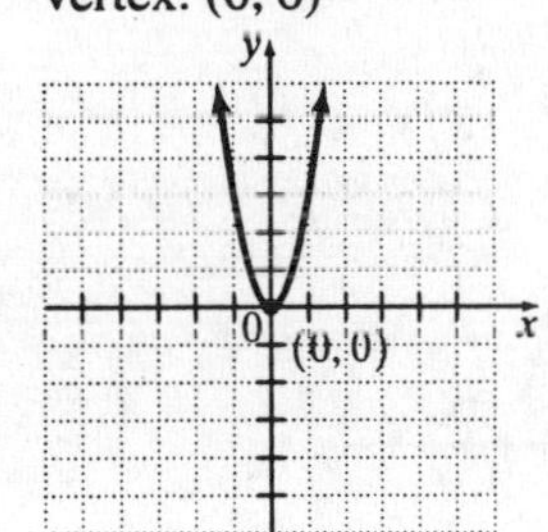

18. $G(x) = -2(x - 1)^2 + 5$

vertex: (1, 5)

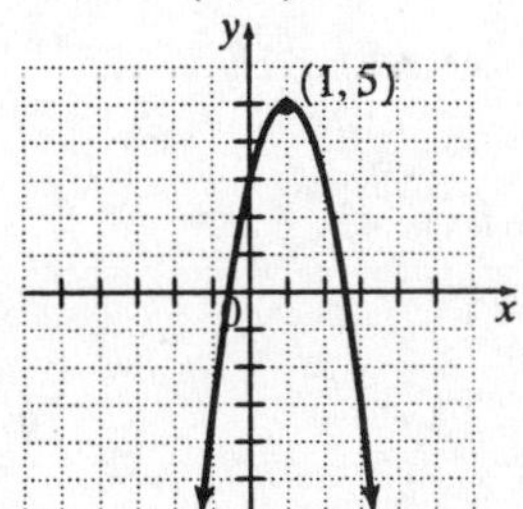

19. $h(x) = x^2 - 4x + 4$

$\frac{-b}{2a} = \frac{-(-4)}{2(1)} = 2$

$h(2) = 2^2 - 4(2) + 4 = 0$

vertex: (2, 0)

y-intercept: (0, 4).

$0 = x^2 - 4x + 4 = (x-2)^2$

x-intercept: (2, 0)

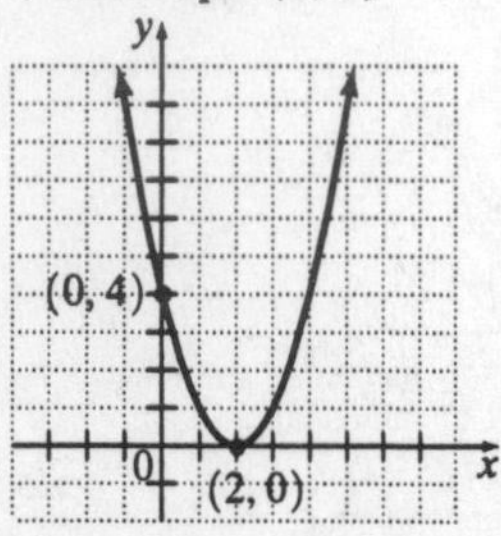

20. $F(x) = 2x^2 - 8x + 9$

$\frac{-b}{2a} = \frac{-(-8)}{2(2)} = 2$

$F(2) = 2(2)^2 - 8(2) + 9 = 1$

Vertex: (2, 1)

y-intercept: (0, 9)

There are no x-intercepts.

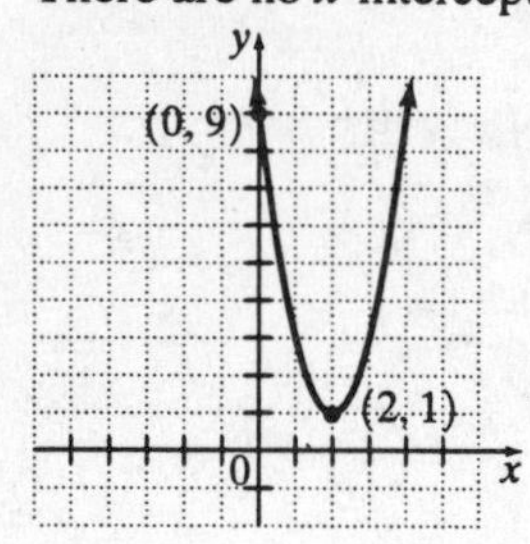

21. $c^2 = a^2 + b^2$

$(10)^2 = x^2 + (x-4)^2$

$100 = x^2 + x^2 - 8x + 16$

$0 = 2x^2 - 8x - 84$

$0 = x^2 - 4x - 42$

$a = 1, b = -4, c = -42$

$x = \frac{-(-4) \pm \sqrt{(-4)^2 - 4(1)(-42)}}{2(1)}$

$x = \frac{4 \pm \sqrt{16 + 168}}{2}$

$x = \frac{4 \pm \sqrt{184}}{2}$

$x = \frac{4 \pm 2\sqrt{46}}{2}$

$x = 2 \pm \sqrt{46}$

Disregard the negative result.

$2 + \sqrt{46}$ or 8.8 feet

22.

	Hours	Job complete in 1 hr
Dave	$x-2$	$\frac{1}{x-2}$
Sandy	x	$\frac{1}{x}$
Together	4	$\frac{1}{4}$

$\frac{1}{x-2} + \frac{1}{x} = \frac{1}{4}$

$\frac{x + (x-2)}{x(x-2)} = \frac{1}{4}$

$\frac{2x-2}{x^2 - 2x} = \frac{1}{4}$

$8x - 8 = x^2 - 2x$

$0 = x^2 - 10x + 8$

$a = 1, b = -10, c = 8$

$x = \frac{-(-10) \pm \sqrt{(-10)^2 - 4(1)(8)}}{2(1)}$

$x = \frac{10 \pm \sqrt{68}}{2}$

$x = \frac{10 \pm 2\sqrt{17}}{2}$

$x = 5 \pm \sqrt{17}$

Since $x - 2$ must be positive,

$x = 5 + \sqrt{17}$.

$5 + \sqrt{17}$ or 9.12 hours

23. $s(t) = -16t^2 + 32t + 256$

a. $\frac{-b}{2a} = \frac{-32}{2(-16)} = 1$

$s(1) = -16(1)^2 + 32(1) + 256 = 272$

Vertex: (1, 272)

Maximum height = 272 feet

b. $0 = -16t^2 + 32t + 256$

$0 = -t^2 + 2t + 16$

$a = -1, b = 2, c = 16$

$x = \frac{-2 \pm \sqrt{2^2 - 4(-1)(16)}}{2(-1)}$

$x = \frac{-2 \pm \sqrt{68}}{-2}$

$x = \frac{-2 \pm 2\sqrt{17}}{-2}$

$x = 1 \pm \sqrt{17}$

$x \approx -3.12$ and 5.12

Disregard the negative.

5.12 seconds

24.

$(x+8)^2 + x^2 = 20^2$

$x^2 + 16x + 64 + x^2 = 400$

$2x^2 + 16x - 336 = 0$

$x^2 + 8x - 168 = 0$

$x = \frac{-8 \pm \sqrt{8^2 - 4(1)(-168)}}{2(1)}$

$x = \frac{-8 \pm \sqrt{736}}{2}$

$x = -4 \pm 2\sqrt{46}$

$x + 8 + x = 2x + 8$

$= 2(-4 + 2\sqrt{46}) + 8$

$= -8 + 4\sqrt{46} + 8$

$= 4\sqrt{46}$

≈ 27 feet

$27 - 20 = 7$ feet

25. $W = \frac{k}{V}$

$20 = \frac{k}{12}$

$k = 240$

so $W = \frac{240}{V} = \frac{240}{15} = 16$

26. $Q = kRS^2$

$24 = k(3)4^2$

$24 = 48k$

$k = \frac{1}{2}$

so $Q = \frac{1}{2}RS^2 = \frac{1}{2}(2)3^2 = 9$

27. $S = k\sqrt{d}$

$160 = k\sqrt{400}$

$k = \frac{160}{20} = 8$

so

$S = 8\sqrt{d}$

$128 = 8\sqrt{d}$

$\sqrt{d} = 16$

$d = 256$

The height of the cliff is 256 feet.

Chapter 11

Exercise Set 11.1

1. a. $(f+g)(x) = x-7+2x+1 = 3x-6$

b. $(f-g)(x) = x-7-(2x+1)$
$= x-7-2x-1$
$= -x-8$

c. $(f \cdot g)(x) = (x-7)(2x+1)$
$= 2x^2-13x-7$

d. $\left(\dfrac{f}{g}\right)(x) = \dfrac{x-7}{2x+1}$, where $x \neq -\dfrac{1}{2}$

3. a. $(f+g)(x) = x^2+1+5x = x^2+5x+1$

b. $(f-g)(x) = x^2+1-5x = x^2-5x+1$

c. $(f \cdot g)(x) = (x^2+1)(5x) = 5x^3+5x$

d. $\left(\dfrac{f}{g}\right)(x) = \dfrac{x^2+1}{5x}$, where $x \neq 0$

5. a. $(f+g)(x) = \sqrt{x}+x+5$

b. $(f-g)(x) = \sqrt{x}-(x+5)$
$= \sqrt{x}-x-5$

c. $(f \cdot g)(x) = \sqrt{x}(x+5) = x\sqrt{x}+5\sqrt{x}$

d. $\left(\dfrac{f}{g}\right)(x) = \dfrac{\sqrt{x}}{x+5}$, where $x \neq -5$

7. a. $(f+g)(x) = -3x+5x^2 = 5x^2-3x$

b. $(f-g)(x) = -3x-5x^2 = -5x^2-3x$

c. $(f \cdot g)(x) = -3x(5x^2) = -15x^3$

d. $\left(\dfrac{f}{g}\right)(x) = \dfrac{-3x}{5x^2} = -\dfrac{3}{5x}$, where $x \neq 0$.

9. $(f \circ g)(2) = f(g(2))$
$= f(-2 \cdot 2)$
$= f(-4)$
$= (-4)^2-6(-4)+2$
$= 16+24+2$
$= 42$

11. $(g \circ f)(-1) = g(f(-1))$
$= g((-1)^2-6(-1)+2)$
$= g(9)$
$= -2(9)$
$= -18$

13. $(g \circ h)(0) = g(h(0))$
$= g(\sqrt{0})$
$= g(0)$
$= -2(0)$
$= 0$

15. $(f \circ g)(x) = f(g(x))$
$= f(5x)$
$= (5x)^2+1$
$= 25x^2+1$

$(g \circ f)(x) = g(f(x))$
$= g(x^2+1)$
$= 5(x^2+1)$
$= 5x^2+5$

17. $(f \circ g)(x) = f(g(x))$
$= f(x+7)$
$= 2(x+7)-3$
$= 2x+14-3$
$= 2x+11$

$(g \circ f)(x) = g(f(x))$
$= g(2x-3)$
$= (2x-3)+7$
$= 2x+4$

19. $(f \circ g)(x) = f(g(x))$
$= f(-2x)$
$= (-2x)^3 + (-2x) - 2$
$= -8x^3 - 2x - 2$

$(g \circ f)(x) = g(f(x))$
$= g(x^3 + x - 2)$
$= -2(x^3 + x - 2)$
$= -2x^3 - 2x + 4$

21. $(f \circ g)(x) = f(g(x))$
$= f(-5x + 2)$
$= \sqrt{-5x + 2}$

$(g \circ f)(x) = g(f(x)) = g\left(\sqrt{x}\right) = -5\sqrt{x} + 2$

23. $H(x) = (g \circ h)(x)$
$= g(h(x))$
$= g(x^2 + 2)$
$= \sqrt{x^2 + 2}$

25. $F(x) = (h \circ f)(x)$
$= h(f(x))$
$= h(3x)$
$= (3x)^2 + 2$
$= 9x^2 + 2$

27. $G(x) = (f \circ g)(x)$
$= f(g(x))$
$= f\left(\sqrt{x}\right)$
$= 3\sqrt{x}$

29. Answers may vary. For example, $g(x) = x + 2$ and $f(x) = x^2$.

31. Answers may vary. For example, $g(x) = x + 5$ and $f(x) = \sqrt{x} + 2$.

33. Answers may vary. For example, $g(x) = 2x - 3$ and $f(x) = \frac{1}{x}$.

35. $(f + g)(2) = f(2) + g(2) = 7 + (-1) = 6$

37. $(f \circ g)(2) = f(g(2)) = f(-1) = 4$

39. $(f \cdot g)(7) = f(7) \cdot g(7) = 1 \cdot 4 = 4$

41. $\left(\frac{f}{g}\right)(-1) = \frac{f(-1)}{g(-1)} = \frac{4}{-4} = -1$

43. $P(x) = R(x) - C(x)$

45. $x = y + 2$
$x - 2 = y + 2 - 2$
$x - 2 = y$
$y = x - 2$

47. $x = 3y$
$\frac{x}{3} = \frac{3y}{3}$
$\frac{x}{3} = y$
$y = \frac{x}{3}$

49. $x = -2y - 7$
$x + 7 = -2y$
$\frac{x + 7}{-2} = y$
$y = -\frac{x + 7}{2}$

Exercise Set 11.2

1. $f = \{(-1, -1), (1, 1), (0, 2), (2, 0)\}$ is a one-to-one function.
$f^{-1} = \{(-1, -1), (1, 1), (2, 0), (0, 2)\}$

3. $h = \{(10, 10)\}$ is a one-to-one function.
$h^{-1} = \{(10, 10)\}$

5. $f = \{(11, 12), (4, 3), (3, 4), (6, 6)\}$ is a one-to-one function.
$f^{-1} = \{(12, 11), (3, 4), (4, 3), (6, 6)\}$

7. This function is not one-to-one because there are two months with the same output: (January, 282) and (May, 282).

9. This function is one-to-one. The inverse function is:

Rank in population (Input)	1	49	12	2	45
State (Output)	California	Vermont	Virginia	Texas	South Dakota

11. $f(x) = x^3 + 2$

a. $f(1) = 1^3 + 2 = 3$

b. $f^{-1}(3) = 1$

13. $f(x) = x^3 + 2$

a. $f(-1) = (-1)^3 + 2 = 1$

b. $f^{-1}(1) = -1$

15. The graph represents a one-to-one function because it passes the horizontal line test.

17. The graph does not represent a one-to-one function because it does not pass the horizontal line test.

19. The graph represents a one-to-one function because it passes the horizontal line test.

21. The graph does not represent a one-to-one function because it does not pass the horizontal line test.

23. $f(x) = x + 4$
$y = x + 4$
$x = y + 4$
$y = x - 4$
$f^{-1}(x) = x - 4$

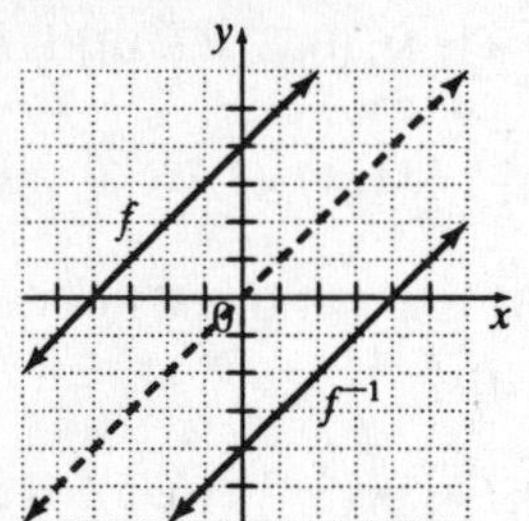

25. $f(x) = 2x - 3$
$$y = 2x - 3$$
$$x = 2y - 3$$
$$2y = x + 3$$
$$y = \frac{x+3}{2}$$
$$f^{-1}(x) = \frac{x+3}{2}$$

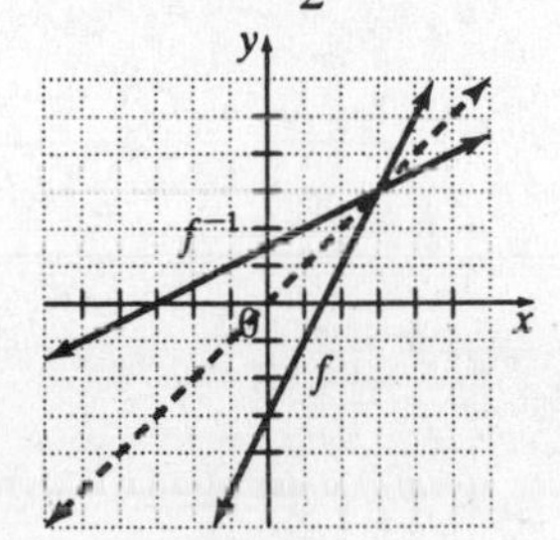

27. $f(x) = \frac{1}{2}x - 1$
$$y = \frac{1}{2}x - 1$$
$$x = \frac{1}{2}y - 1$$
$$\frac{1}{2}y = x + 1$$
$$y = 2x + 2$$
$$f^{-1}(x) = 2x + 2$$

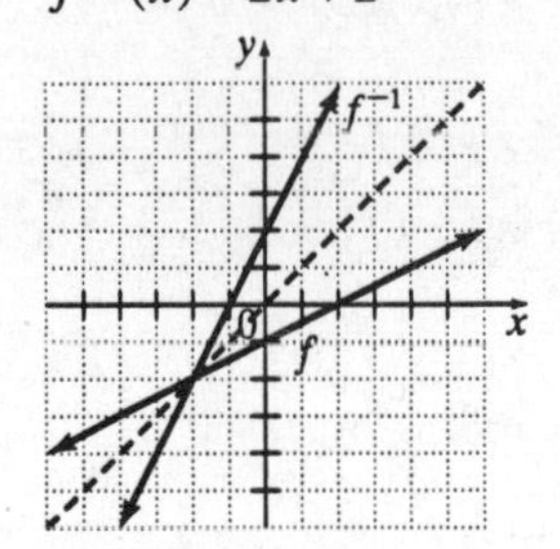

29. $f(x) = x^3$
$$y = x^3$$
$$x = y^3$$
$$y = \sqrt[3]{x}$$
$$f^{-1}(x) = \sqrt[3]{x}$$

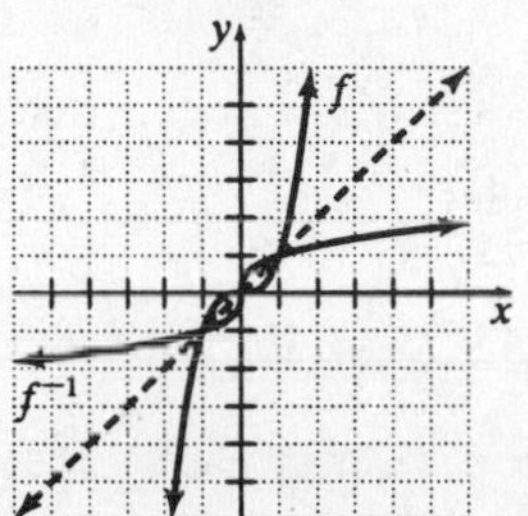

31. $f(x) = 5x + 2$
$$y = 5x + 2$$
$$x = 5y + 2$$
$$5y = x - 2$$
$$y = \frac{x-2}{5}$$
$$f^{-1}(x) = \frac{x-2}{5}$$

33. $f(x) = \frac{x-2}{5}$
$$y = \frac{x-2}{5}$$
$$x = \frac{y-2}{5}$$
$$5x = y - 2$$
$$y = 5x + 2$$
$$f^{-1}(x) = 5x + 2$$

35. $f(x) = \sqrt[3]{x}$
$$y = \sqrt[3]{x}$$
$$x = \sqrt[3]{y}$$
$$x^3 = y$$
$$f^{-1}(x) = x^3$$

37. $f(x)=\dfrac{5}{3x+1}$

$$y=\frac{5}{3x+1}$$
$$x=\frac{5}{3y+1}$$
$$3y+1=\frac{5}{x}$$
$$3y=\frac{5}{x}-1$$
$$3y=\frac{5-x}{x}$$
$$y=\frac{5-x}{3x}$$
$$f^{-1}(x)=\frac{5-x}{3x}$$

39. $f(x)=(x+2)^3$

$$y=(x+2)^3$$
$$x=(y+2)^3$$
$$\sqrt[3]{x}=y+2$$
$$\sqrt[3]{x}-2=y$$
$$f^{-1}(x)=\sqrt[3]{x}-2$$

41.

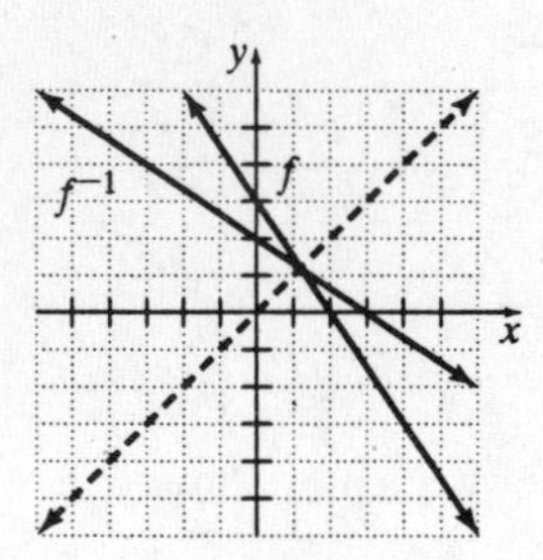

43.

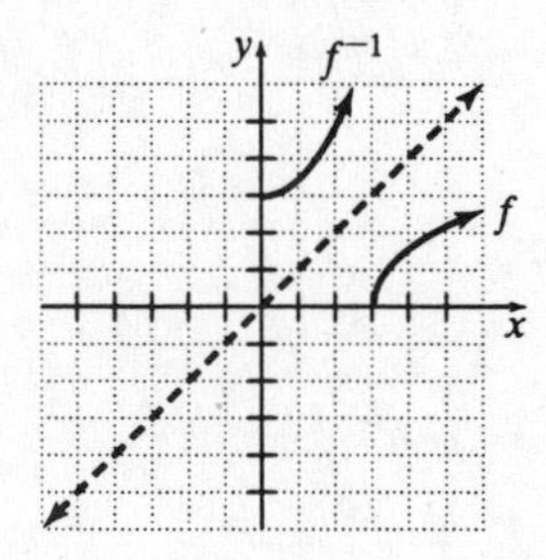

45.

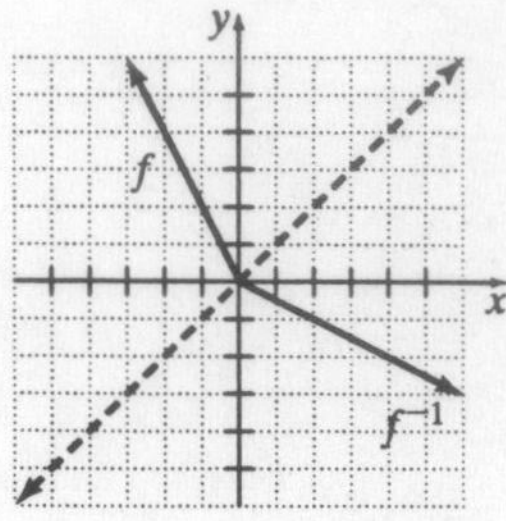

47. $(f\circ f^{-1})(x)=f(f^{-1}(x))$

$$=f\left(\frac{x-1}{2}\right)$$
$$=2\left(\frac{x-1}{2}\right)+1$$
$$=x-1+1$$
$$=x$$

$(f^{-1}\circ f)(x)=f^{-1}(f(x))$

$$=f^{-1}(2x+1)$$
$$=\frac{2x+1-1}{2}$$
$$=x$$

49. $(f\circ f^{-1})(x)=f(f^{-1}(x))$

$$=f(\sqrt[3]{x-6})$$
$$=(\sqrt[3]{x-6})^3+6$$
$$=x-6+6$$
$$=x$$

$(f^{-1}\circ f)(x)=f^{-1}(f(x))$

$$=f^{-1}(x^3+6)$$
$$=\sqrt[3]{(x^3+6)-6}$$
$$=\sqrt[3]{x^3}$$
$$=x$$

51. a. $\left(-2,\ \frac{1}{4}\right),\left(-1,\ \frac{1}{2}\right)$, (0, 1), (1, 2), (2, 5)

b. Reverse the coordinates: $\left(\frac{1}{4},\ -2\right),\left(\frac{1}{2},\ -1\right)$, (1, 0), (2, 1), (5, 2)

c.

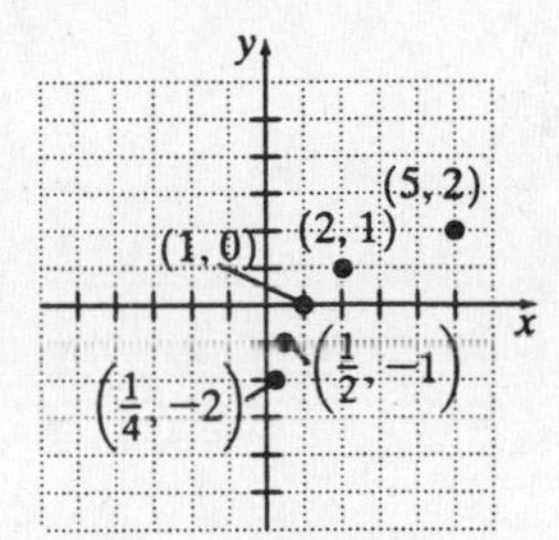

d.

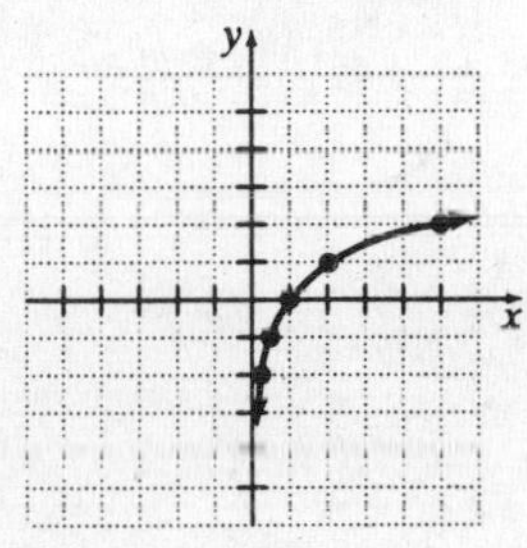

53. $f(x) = 3x + 1$ or $y = 3x + 1$
Find inverse

$$x = 3y + 1$$
$$3y = x - 1$$
$$y = \frac{x-1}{3}$$
$$f^{-1}(x) = \frac{x-1}{3}$$

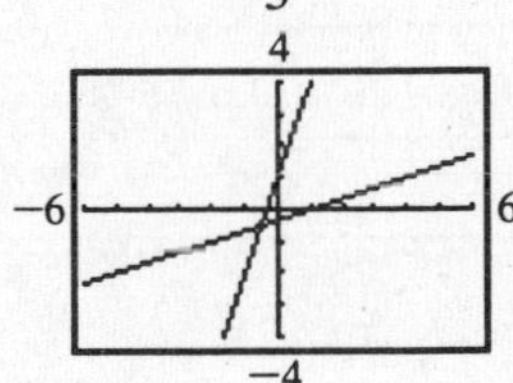

55. $f(x) = \sqrt[3]{x+1}$ or $y = \sqrt[3]{x+1}$
Find inverse

$$x = \sqrt[3]{y+1}$$
$$x^3 = y + 1$$
$$y = x^3 - 1$$
$$f^{-1}(x) = x^3 - 1$$

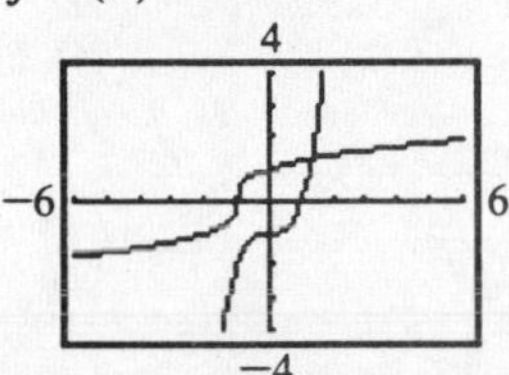

57. $25^{1/2} = \sqrt{25} = 5$

59. $16^{3/4} = (\sqrt[4]{16})^3 = 2^3 = 8$

61. $9^{-3/2} = \frac{1}{9^{3/2}} = \frac{1}{(\sqrt{9})^3} = \frac{1}{3^3} = \frac{1}{27}$

63. $f(x) = 3^x$
$f(2) = 3^2 = 9$

65. $f(x) = 3^x$
$f\left(\frac{1}{2}\right) = 3^{1/2} = \sqrt{3} \approx 1.73$

Section 11.3

Graphing Calculator Explorations

1.

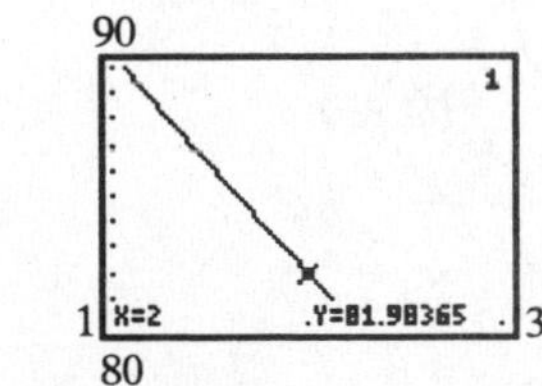

81.98%

3. 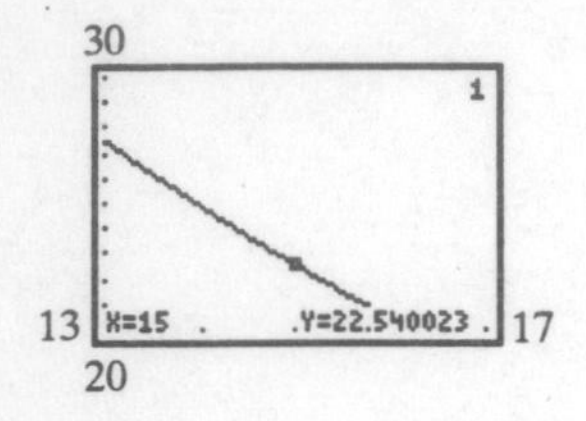

22.54%

Exercise Set 11.3

1. $y = 4^x$

x	−2	−1	0	1	2
y	$\frac{1}{16}$	$\frac{1}{4}$	1	4	16

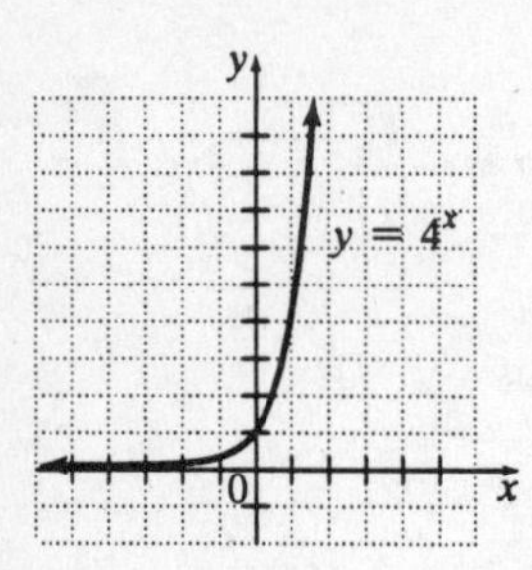

3. $y = 2^x + 1$

x	−2	−1	0	1	2
y	$\frac{5}{4}$	$\frac{3}{2}$	2	3	5

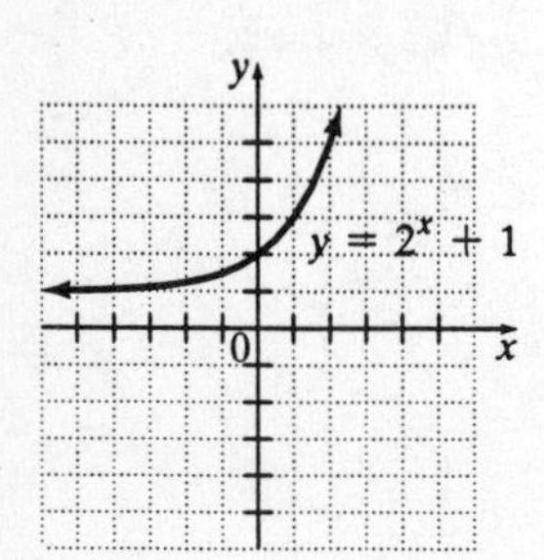

5. $y = \left(\frac{1}{4}\right)^x$

x	−2	−1	0	1	2
y	16	4	1	$\frac{1}{4}$	$\frac{1}{16}$

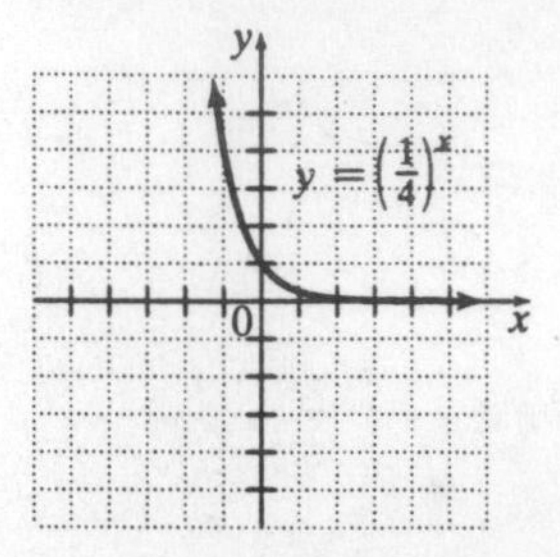

7. $y = \left(\frac{1}{2}\right)^x - 2$

x	−2	−1	0	1	2
y	2	0	−1	$-\frac{3}{2}$	$-\frac{7}{4}$

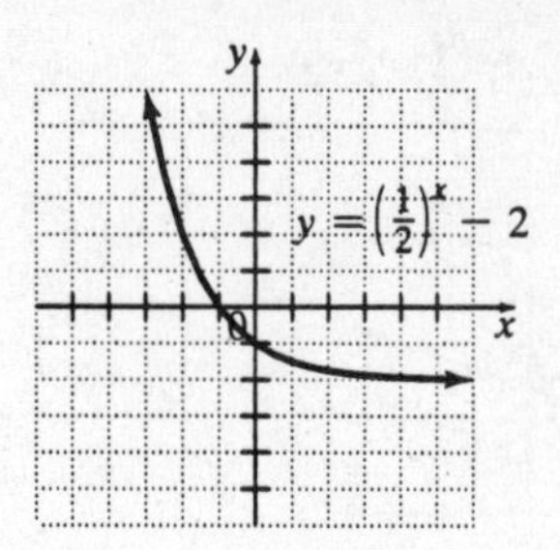

9. $y = -2^x$

x	−2	−1	0	1	2
y	$-\frac{1}{4}$	$-\frac{1}{2}$	−1	−2	−4

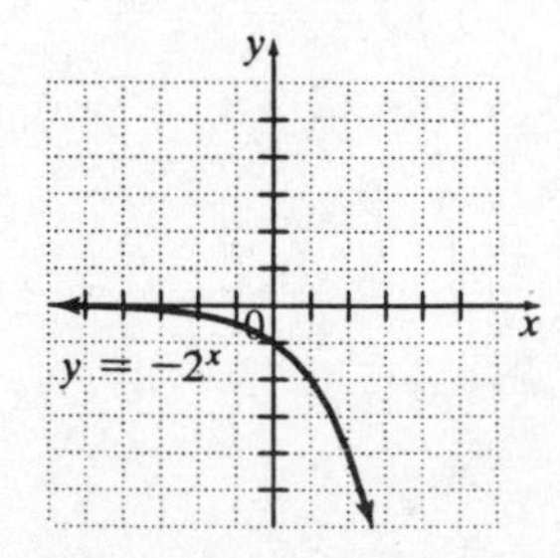

11. $y = -\left(\frac{1}{4}\right)^x$

x	-2	-1	0	1	2
y	-16	-4	-1	$-\frac{1}{4}$	$-\frac{1}{16}$

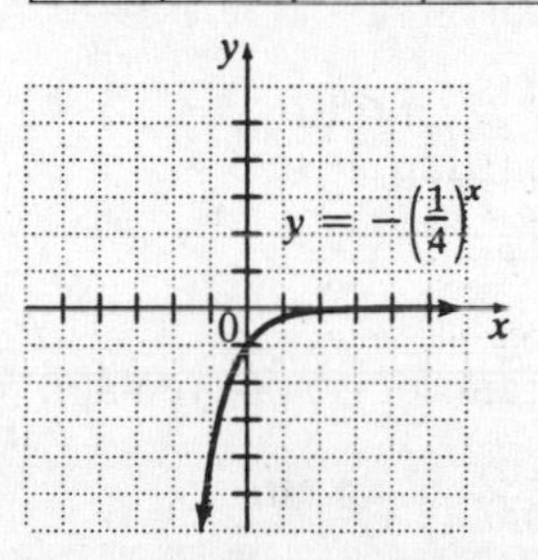

13. $f(x) = 2^{x+1}$

x	-2	-1	0	1	2
y	$\frac{1}{2}$	1	2	4	8

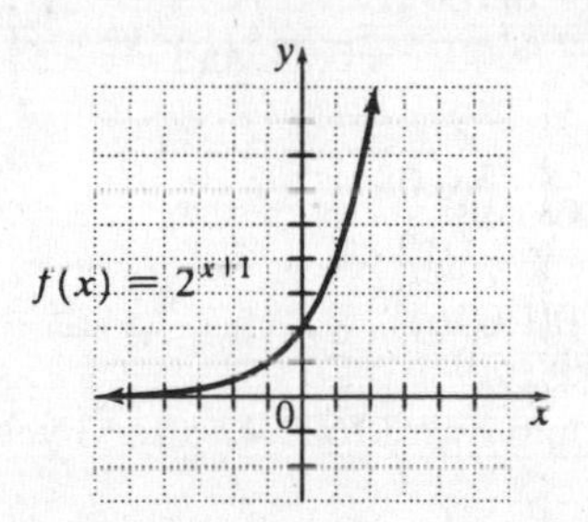

15. $f(x) = 4^{x-2}$

x	-1	0	1	2	3
y	$\frac{1}{64}$	$\frac{1}{16}$	$\frac{1}{4}$	1	4

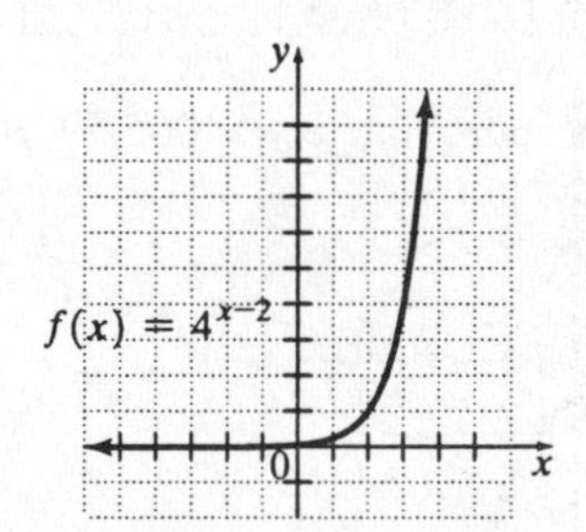

17. Answers may vary.

19. $3^x = 27$
$3^x = 3^3$
$x = 3$
The solution is 3.

21. $16^x = 8$
$2^{4x} = 2^3$
$4x = 3$
$x = \frac{3}{4}$
The solution is $\frac{3}{4}$.

23. $32^{2x-3} = 2$
$2^{5(2x-3)} = 2^1$
$10x - 15 = 1$
$10x = 16$
$x = \frac{8}{5}$
The solution is $\frac{8}{5}$.

25. $\frac{1}{4} = 2^{3x}$
$2^{-2} = 2^{3x}$
$3x = -2$
$x = -\frac{2}{3}$
The solution is $-\frac{2}{3}$.

27. $5^x = 625$
$5^x = 5^4$
$x = 4$
The solution is 4.

29. $4^x = 8$
$2^{2x} = 2^3$
$2x = 3$
$x = \frac{3}{2}$
The solution is $\frac{3}{2}$.

31. $27^{x+1} = 9$
$3^{3(x+1)} = 3^2$
$3^{3x+3} = 3^2$
$3x + 3 = 2$
$3x = -1$
$x = -\frac{1}{3}$
The solution is $-\frac{1}{3}$.

33. $81^{x-1} = 27^{2x}$
$3^{4(x-1)} = 3^{6x}$
$3^{4x-4} = 3^{6x}$
$4x - 4 = 6x$
$-4 = 2x$
$x = -2$
The solution is –2.

35. Decreasing, includes point $\left(1, \frac{1}{2}\right)$:
C

37. Decreasing, includes point $\left(1, \frac{1}{4}\right)$:
D

39. $y = 30(2.7)^{-0.004t}$, $t = 50$
$= 30(2.7)^{-(0.004)(50)}$
$= 30(2.7)^{-0.2}$
≈ 24.6
Approximately 24.6 pounds of uranium will remain after 50 days.

41. $y = 200(2.7)^{0.08t}$, $t = 12$
$= 200(2.7)^{0.08(12)}$
$= 200(2.7)^{0.96}$
≈ 519
There should be approximately 519 rats by next January.

43. $y = 5(2.7)^{-0.15t}$, $t = 10$
$= 5(2.7)^{-0.15(10)}$
$= 5(2.7)^{-1.5}$
≈ 1.1
About 1.1 grams of isotope remain.

45. $y = 15{,}525{,}000(2.7)^{0.007t}$, $t = 6$
$= 15{,}525{,}000(2.7)^{0.007(6)}$
$= 15{,}525{,}000(2.7)^{0.042}$
$\approx 16{,}190{,}000$
Approximately 16,190,000 residents.

47. $A = P\left(1 + \frac{r}{n}\right)^{nt}$
$t = 3$, $P = 6000$, $r = 0.08$, and $n = 12$.
Then
$A = 6000\left(1 + \frac{0.08}{12}\right)^{12(3)}$
$= 6000(1.00\overline{6})^{36}$
≈ 7621.42
Erica would owe \$7621.42 after 3 years.

49. $A = P\left(1 + \frac{r}{n}\right)^{nt}$
$t = 12$, $P = 2000$, $r = 0.06$, and $n = 2$.
Then
$A = 2000\left(1 + \frac{0.06}{2}\right)^{2(12)}$
$= 2000(1.03)^{24}$
≈ 4065.59
Janina has approximately \$4065.59 in her savings account.

51. 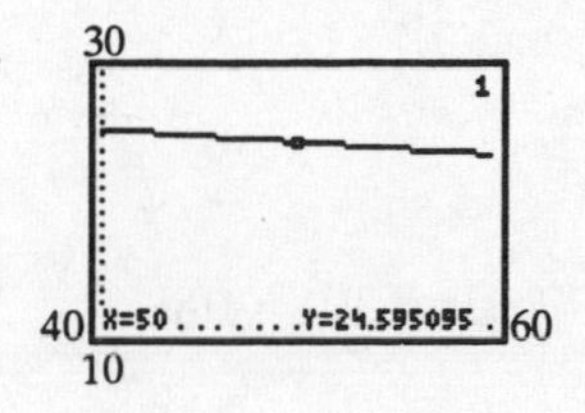

24.60 pounds

53. 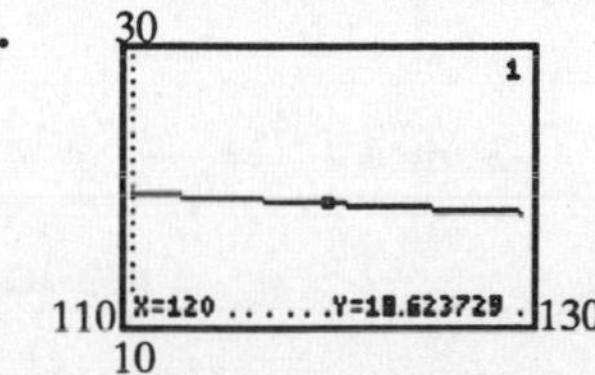

18.62 pounds

55. 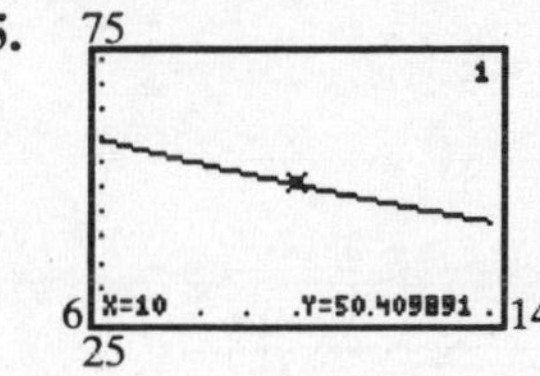

50.41 grams

57. $y = 5,926,466,814(2.7)^{0.0132t}, t = 7$
$y = 5,926,466,814(2.7)^{0.0132(7)}$
$\approx 6,496,117,457$ people
Approximately 6,496,000 people.

59. $y = 358(1.004)^t$

a. $t = 2004 - 1994 = 10$
$y = 358(1.004)^{10}$
≈ 372.6 parts per million by volume

b. $t = 2025 - 1994 = 31$
$y = 358(1.004)^{31}$
≈ 405.2 parts per million by volume

61. $y = 6.052(1.378)^x$
$t = 2008 - 1991 = 17$
$y = 6.052(1.378)^{17} \approx 1410$
There will be approximately 1410 million cellular phone users in 2008.

63. $3x - 7 = 11$
$3x = 18$
$x = 6$
The solution is 6.

65. $2 - 6x = 6(1 - x)$
$2 - 6x = 6 - 6x$
$2 = 6$
There is no solution.

67. $18 = 11x - x^2$
$x^2 - 11x + 18 = 0$
$(x - 2)(x - 9) = 0$
$x - 2 = 0$ or $x - 9 = 0$
$x = 2$ or $x = 9$
The solution is 2 or 9.

69. $3^x = 9$
$3^2 = 9$
$x = 2$
The solution is 2.

71. $4^x = 1$
$4^0 = 1$
$x = 0$
The solution is 0.

Exercise Set 11.4

1. $\log_6 36 = 2$
$6^2 = 36$

3. $\log_3 \frac{1}{27} = -3$
$3^{-3} = \frac{1}{27}$

5. $\log_{10} 1000 = 3$
$10^3 = 1000$

7. $\log_e x = 4$
$e^4 = x$

9. $\log_e \frac{1}{e^2} = -2$
$e^{-2} = \frac{1}{e^2}$

11. $\log_7 \sqrt{7} = \frac{1}{2}$
$7^{1/2} = \sqrt{7}$

13. $2^4 = 16$
$\log_2 16 = 4$

15. $10^2 = 100$
$\log_{10} 100 = 2$

17. $e^3 = x$
$\log_e x = 3$

19. $10^{-1} = \frac{1}{10}$
$\log_{10} \frac{1}{10} = -1$

21. $4^{-2} = \frac{1}{16}$
$\log_4 \frac{1}{16} = -2$

23. $5^{1/2} = \sqrt{5}$
$\log_5 \sqrt{5} = \frac{1}{2}$

25. $\log_2 8 = 3$ since $2^3 = 8$.

27. $\log_3 \frac{1}{9} = -2$ since $3^{-2} = \frac{1}{9}$.

29. $\log_{25} 5 = \frac{1}{2}$ since $25^{1/2} = 5$.

31. $\log_{1/2} 2 = -1$ since $\left(\frac{1}{2}\right)^{-1} = 2$.

33. $\log_7 1 = 0$ since $7^0 = 1$.

35. $\log_2 2^4 = 4$

37. $\log_{10} 100 = 2$ since $10^2 = 100$.

39. $3^{\log_3 5} = 5$

41. $\log_3 81 = 4$ since $3^4 = 81$.

43. $\log_4\left(\frac{1}{64}\right) = -3$ since $4^{-3} = \frac{1}{64}$.

45. Answers may vary.

47. $\log_3 9 = x$
$\log_3(3^2) = x$
$2 = x$
The solution is 2.

49. $\log_3 x = 4$
$x = 3^4 = 81$
The solution is 81.

51. $\log_x 49 = 2$
$x^2 = 49$
$x = \pm 7$
We discard the negative base.
The solution is 7.

53. $\log_2 \frac{1}{8} = x$
$2^x = \frac{1}{8}$
$2^x = 2^{-3}$
$x = -3$
The solution is –3.

55. $\log_3\left(\frac{1}{27}\right) = x$
$\log_3 3^{-3} = x$
$-3 = x$
The solution is –3.

57. $\log_8 x = \frac{1}{3}$

$x = 8^{1/3} = 2$

The solution is 2.

59. $\log_4 16 = x$

$\log_4 4^2 = x$

$2 = x$

The solution is 2.

61. $\log_{3/4} x = 3$

$x = \left(\frac{3}{4}\right)^3 = \frac{27}{64}$

The solution is $\frac{27}{64}$.

63. $\log_x 100 = 2$

$x^2 = 100$

$x = \pm 10$

We discard the negative base. The solution is 10.

65. $\log_5 5^3 = 3$

67. $2^{\log_2 3} = 3$

69. $\log_9 9 = 1$

71. $y = \log_3 x$

$y = 0$:

$\log_3 x = 0$

$x = 3^0 = 1$ is the only x-intercept.

$x = 0$: $y = \log_3 0$ which is not defined. No y-intercept exists.

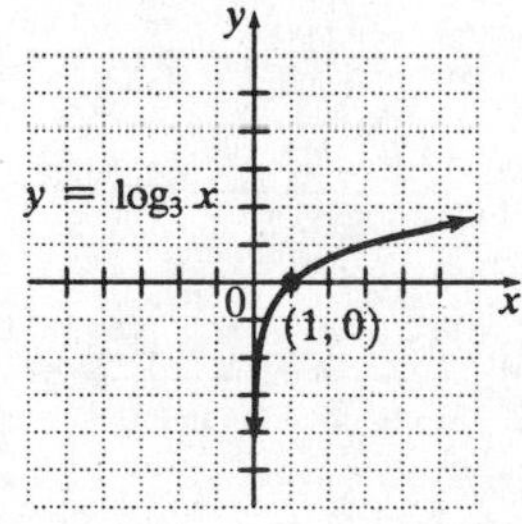

73. $f(x) = \log_{1/4} x$ or $y = \log_{1/4} x$

$y = 0$:

$0 = \log_{1/4} x$

$x = \left(\frac{1}{4}\right)^0 = 1$ is the x-intercept.

$x = 0$: $y = \log_{1/4} 0$ which is not defined.

No y-intercept exists.

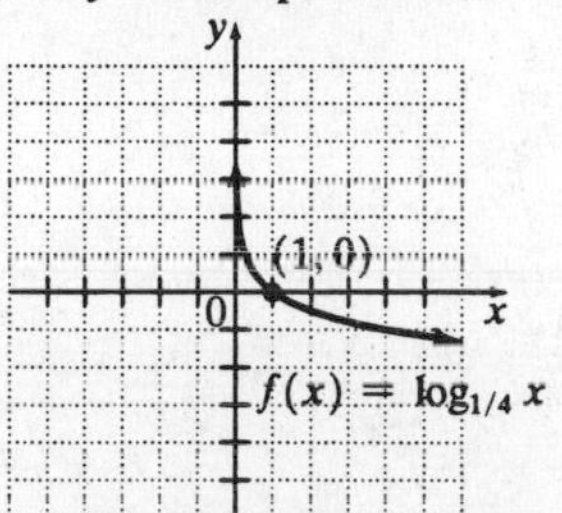

75. $f(x) = \log_5 x$ or $y = \log_5 x$

$y = 0$:

$0 = \log_5 x$

$x = 5^0 = 1$ is the x-intercept.

$x = 0$:

$y = \log_5 0$ is not defined so there is no y-intercept.

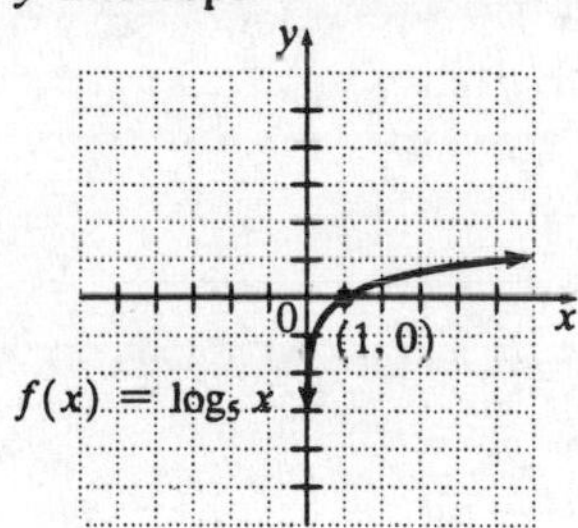

77. $f(x) = \log_{1/6} x$ or $y = \log_{1/6} x$

$y = 0$:

$0 = \log_{1/6} x$

$x = \left(\frac{1}{6}\right)^0 = 1$ is the x-intercept.

$x = 0$:

$y = \log_{1/6} 0$ is not defined so there is no y-intercept.

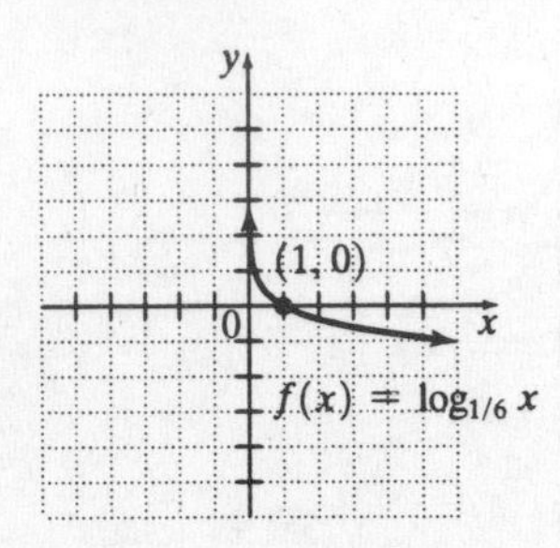

79. $y = 4^x$; $y = \log_4 x$

$x = 0$: $y = 4^0 = 1$ is the y-intercept of $y = 4^x$, hence the x-intercept of $y = \log_4 x$.

$y = 0$: $4^x = 0$ has no solution so $y = 4^x$ has no x-intercept; hence $y = \log_4 x$ has no y-intercept.

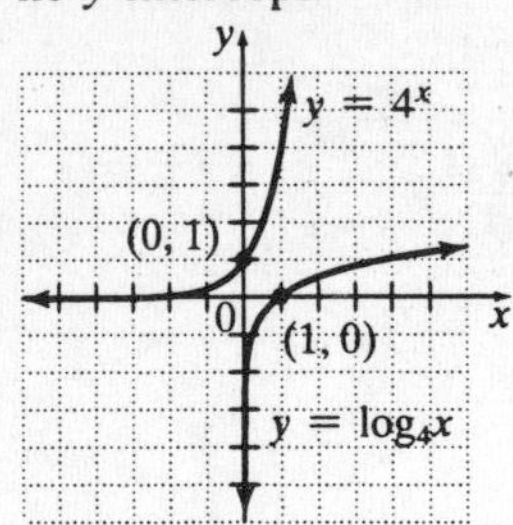

81. $y = \left(\frac{1}{3}\right)^x$; $y = \log_{1/3} x$

$x = 0$: $y = \left(\frac{1}{3}\right)^0 = 1$ is the y-intercept of $y = \left(\frac{1}{3}\right)^x$, and hence the x-intercept of $y = \log_{1/3} x$.

$y = 0$: $0 = \left(\frac{1}{3}\right)^x$ has no solution so $y = \left(\frac{1}{3}\right)^x$ has no x-intercept; hence $y = \log_{1/3} x$ has no y-intercept.

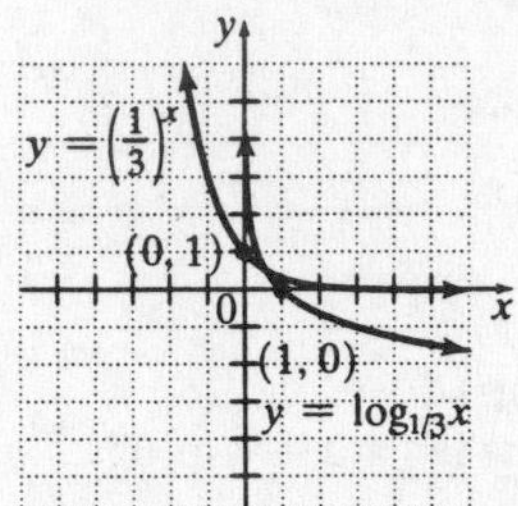

83. $\log_{10}(1-k) = \frac{-0.3}{H}$, $H = 8$

$$\log_{10}(1-k) = \frac{-0.3}{8} = -0.0375$$

$$1 - k = 10^{-0.0375}$$

$$1 - 10^{-0.0375} = k$$

$$k \approx 0.0827$$

85. $\log_3 10$ is between 2 and 3 because $3^2 = 9$ and $3^3 = 27$

87. $\frac{x-5}{5-x} = \frac{x-5}{-(x-5)} = \frac{1}{-1} = -1$

89. $\frac{x^2 - 3x - 10}{2 + x} = \frac{(x+2)(x-5)}{x+2} = x - 5$

91. $\frac{3x}{x+3} + \frac{9}{x+3} = \frac{3x+9}{x+3} = \frac{3(x+3)}{x+3} = 3$

93. $\frac{5}{y+1}-\frac{4}{y-1}=\frac{5(y-1)-4(y+1)}{(y+1)(y-1)}$
$=\frac{5y-5-4y-4}{y^2-1}$
$=\frac{y-9}{y^2-1}$

Exercise Set 11.5

1. $\log_5 2+\log_5 7=\log_5(2\cdot 7)=\log_5 14$

3. $\log_4 9+\log_4 x=\log_4 9\cdot x=\log_4 9x$

5. $\log_{10} 5+\log_{10} 2+\log_{10}(x^2+2)$
$=\log_{10}[5\cdot 2(x^2+2)]$
$=\log_{10}(10x^2+20)$

7. $\log_5 12-\log_5 4=\log_5\frac{12}{4}=\log_5 3$

9. $\log_2 x-\log_2 y=\log_2\frac{x}{y}$

11. $\log_4 2+\log_4 10-\log_4 5$
$=\log_4(2\cdot 10)-\log_4 5$
$=\log_4\frac{20}{5}$
$=\log_4 4=1$

13. $\log_3 x^2=2\log_3 x$

15. $\log_4 5^{-1}=-1\log_4 5=-\log_4 5$

17. $\log_5\sqrt{y}=\log_5 y^{1/2}=\frac{1}{2}\log_5 y$

19. $2\log_2 5=\log_2 5^2=\log_2 25$

21. $3\log_5 x+6\log_5 z$
$=\log_5 x^3+\log_5 z^6$
$=\log_5 x^3z^6$

23. $\log_{10} x-\log_{10}(x+1)+\log_{10}(x^2-2)$
$=\log_{10}\frac{x}{x+1}+\log_{10}(x^2-2)$
$=\log_{10}\frac{x(x^2-2)}{x+1}$
$=\log_{10}\frac{x^3-2x}{x+1}$

25. $\log_4 5+\log_4 7=\log_4(5\cdot 7)=\log_4 35$

27. $\log_3 8-\log_3 2=\log_3\left(\frac{8}{2}\right)=\log_3 4$

29. $\log_7 6+\log_7 3-\log_7 4$
$=\log_7(6\cdot 3)-\log_7 4$
$=\log_7\left(\frac{18}{4}\right)$
$=\log_7\frac{9}{2}$

31. $3\log_4 2+\log_4 6=\log_4 2^3+\log_4 6$
$=\log_4 8+\log_4 6$
$=\log_4(8\cdot 6)$
$=\log_4 48$

33. $3\log_2 x+\frac{1}{2}\log_2 x-2\log_2(x+1)$
$=\log_2 x^3+\log_2 x^{1/2}-\log_2(x+1)^2$
$=\log_2(x^3\cdot x^{1/2})-\log_2(x+1)^2$
$=\log_2 x^{7/2}-\log_2(x+1)^2$
$=\log_2\frac{x^{7/2}}{(x+1)^2}$

35. $2\log_8 x-\frac{2}{3}\log_8 x+4\log_8 x$
$=\left(2-\frac{2}{3}+4\right)\log_8 x$
$=\frac{16}{3}\log_8 x$
$=\log_8 x^{16/3}$

37. $\log_2 \frac{7\cdot 11}{3} = \log_2(7\cdot 11) - \log_2 3$
$= \log_2 7 + \log_2 11 - \log_2 3$

39. $\log_3\left(\frac{4y}{5}\right) = \log_3 4y - \log_3 5$
$= \log_3 4 + \log_3 y - \log_3 5$

41. $\log_2\left(\frac{x^3}{y}\right) = \log_2 x^3 - \log_2 y$
$= 3\log_2 x - \log_2 y$

43. $\log_b \sqrt{7x} = \log_b(7x)^{1/2}$
$= \frac{1}{2}\log_b(7x)$
$= \frac{1}{2}[\log_b 7 + \log_b x]$
$= \frac{1}{2}\log_b 7 + \frac{1}{2}\log_b x$

45. $\log_7\left(\frac{5x}{4}\right) = \log_7 5x - \log_7 4$
$= \log_7 5 + \log_7 x - \log_7 4$

47. $\log_5 x^3(x+1) = \log_5 x^3 + \log_5(x+1)$
$= 3\log_5 x + \log_5(x+1)$

49. $\log_6 \frac{x^2}{x+3} = \log_6 x^2 - \log_6(x+3)$
$= 2\log_6 x - \log_6(x+3)$

51. $\log_b \frac{5}{3} = \log_b 5 - \log_b 3$
$= 0.7 - 0.5 = 0.2$

53. $\log_b 15 = \log_b(5\cdot 3)$
$= \log_b 5 + \log_b 3$
$= 0.7 + 0.5 = 1.2$

55. $\log_b \sqrt[3]{5} = \log_b 5^{1/3}$
$= \frac{1}{3}\log_b 5$
$= \frac{1}{3}(0.7)$
≈ 0.233

57. $\log_2 x^3 = 3\log_2 x$
True

59. $\frac{\log_7 10}{\log_7 5} = \log_7 2$
False

61. $\frac{\log_7 x}{\log_7 y} = (\log_7 x) - (\log_7 y)$
False

63. $\log_b 8 = \log_b 2^3$
$= 3\log_b 2$
$= 3(0.43)$
$= 1.29$

65. $\log_b \frac{3}{9} = \log_b \frac{1}{3}$
$= \log_b 3^{-1}$
$= (-1)\log_b 3$
$= -(0.68)$
$= -0.68$

67. $\log_b \sqrt{\frac{2}{3}} = \log_b\left(\frac{2}{3}\right)^{1/2}$
$= \frac{1}{2}\log_b \frac{2}{3}$
$= \frac{1}{2}(\log_b 2 - \log_b 3)$
$= \frac{1}{2}(0.43 - 0.68)$
$= \frac{1}{2}(-0.25)$
$= -0.125$

69. $y = 10^x$ and $y = \log_{10} x$

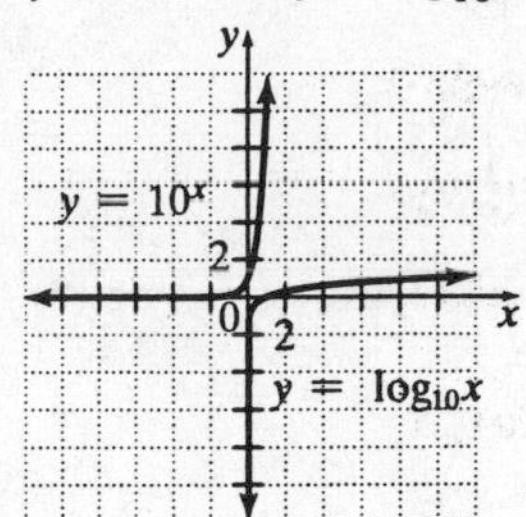

71. $\log_{10} \frac{1}{10} = \log_{10} 10^{-1} = -1$

73. $\log_7 \sqrt{7} = \log_7 7^{1/2} = \frac{1}{2}$

Exercise Set 11.6

1. $\log 8 \approx 0.9031$

3. $\log 2.31 \approx 0.3636$

5. $\ln 2 \approx 0.6931$

7. $\ln 0.0716 \approx -2.6367$

9. $\log 12.6 \approx 1.1004$

11. $\ln 5 \approx 1.6094$

13. $\log 41.5 \approx 1.6180$

15. Answers may vary.

17. $\log 100 = \log 10^2 = 2$

19. $\log \frac{1}{1000} = \log 10^{-3} = -3$

21. $\ln e^2 = 2$

23. $\ln \sqrt[4]{e} = \ln e^{1/4} = \frac{1}{4}$

25. $\log 10^3 = 3$

27. $\ln e^9 = 9$

29. $\log 0.0001 = \log 10^{-4} = -4$

31. $\ln \sqrt{e} = \ln e^{1/2} = \frac{1}{2}$

33. ln 50 is larger, because $e < 10$, and so e must be raised to a greater power to equal 50.

35. $\log x = 1.3$

$x = 10^{1.3}$

The solution is $10^{1.3} \approx 19.9526$.

37. $\log 2x = 1.1$

$2x = 10^{1.1}$

$x = \frac{10^{1.1}}{2}$

The solution is $\frac{10^{1.1}}{2} \approx 6.2946$.

39. $\ln x = 1.4$

$x = e^{1.4}$

The solution is $e^{1.4} \approx 4.0552$.

41. $\ln(3x - 4) = 2.3$

$3x - 4 = e^{2.3}$

$3x = 4 + e^{2.3}$

$x = \frac{4 + e^{2.3}}{3}$

The solution is $\frac{4 + e^{2.3}}{3} \approx 4.6581$.

43. $\log x = 2.3$

$x = 10^{2.3}$

The solution is $10^{2.3} \approx 199.5262$.

45. $\ln x = -2.3$

$x = e^{-2.3}$

The solution is $e^{-2.3} \approx 0.1003$.

47. $\log(2x+1) = -0.5$

$$2x+1 = 10^{-0.5}$$
$$2x = 10^{-0.5} - 1$$
$$x = \frac{10^{-0.5}-1}{2}$$

The solution is $\frac{10^{-0.5}-1}{2} \approx -0.3419$.

49. $\ln 4x = 0.18$

$$4x = e^{0.18}$$
$$x = \frac{e^{0.18}}{4}$$

The solution is $\frac{e^{0.18}}{4} \approx 0.2993$.

51. $\log_2 3 = \frac{\ln 3}{\ln 2} \approx 1.5850$

53. $\log_{1/2} 5 = \frac{\ln 5}{\ln\left(\frac{1}{2}\right)} \approx -2.3219$

55. $\log_4 9 = \frac{\ln 9}{\ln 4} \approx 1.5850$

57. $\log_3 \frac{1}{6} = \log_3 6^{-1}$

$$= (-1)\log_3 6$$
$$= -\frac{\ln 6}{\ln 3} \approx -1.6309$$

59. $\log_8 6 = \frac{\ln 6}{\ln 8} \approx 0.8617$

61. $R = \log\left(\frac{a}{T}\right) + B$

$$= \log\left(\frac{200}{1.6}\right) + 2.1$$
$$\approx 4.2$$

The earthquake measures 4.2 on the Richter scale.

63. $R = \log\left(\frac{a}{T}\right) + B$

$$= \log\left(\frac{400}{2.6}\right) + 3.1$$
$$\approx 5.3$$

The earthquake measures 5.3 on the Richter scale.

65. $A = Pe^{rt}$

$$= 1400e^{(0.08)12}$$
$$\approx 3656.38$$

Dana has $3656.38 after 12 years.

67. $A = Pe^{rt}$

$$= 2000e^{(0.06)4}$$
$$\approx 2542.50$$

Barbara owes $2542.50 at the end of 4 years.

69.

x	$f(x)$
-2	$e^{-2} \approx 0.14$
-1	$e^{-1} \approx 0.37$
0	$e^{0} = 1$
1	$e^{1} \approx 2.72$
2	$e^{2} \approx 7.39$

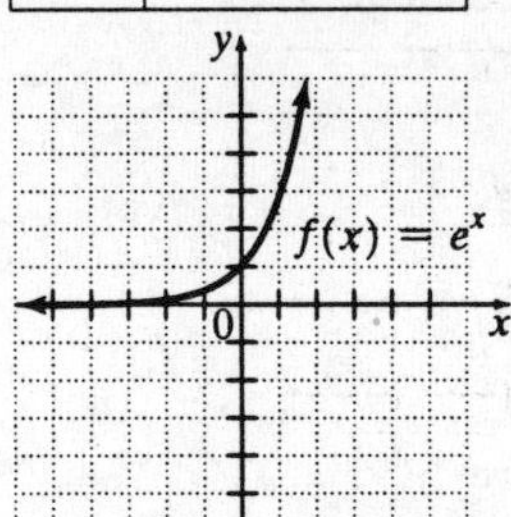

71.

x	$f(x)$
–2	$e^6 \approx 403.43$
–1	$e^3 \approx 20.09$
0	$e^0 = 1$
1	$e^{-3} \approx 0.05$
2	$e^{-6} \approx 0.002$

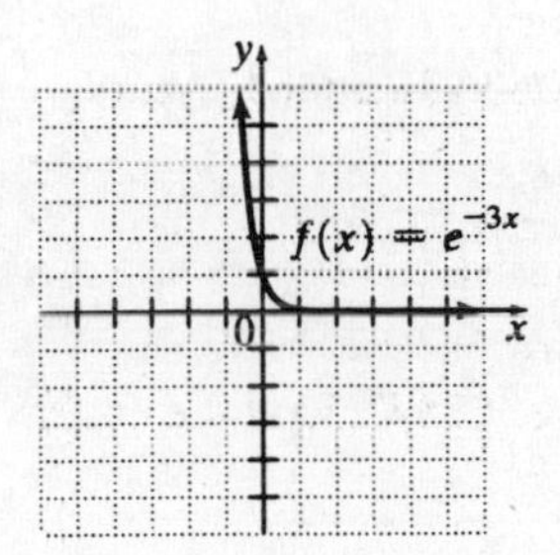

75.

x	$f(x)$
–2	$e^{-3} \approx 0.05$
–1	$e^{-2} \approx 0.14$
0	$e^{-1} \approx 0.37$
1	$e^0 = 1$
2	$e^1 \approx 2.72$

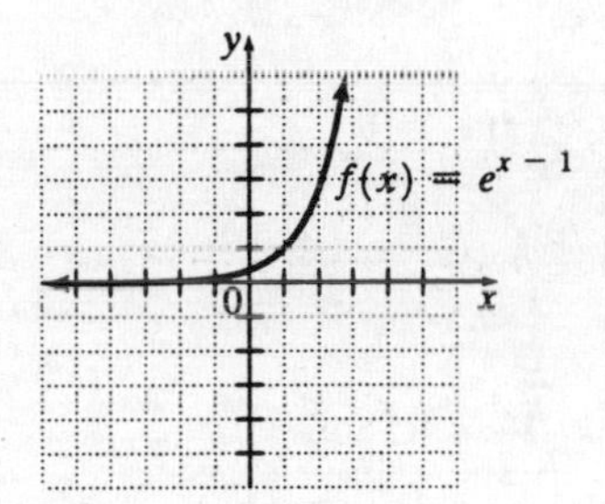

73.

x	$f(x)$
–2	$e^{-2} + 2 \approx 2.14$
–1	$e^{-1} + 2 \approx 2.37$
0	$e^0 + 2 = 3$
1	$e^1 + 2 \approx 4.72$
2	$e^2 + 2 \approx 9.39$

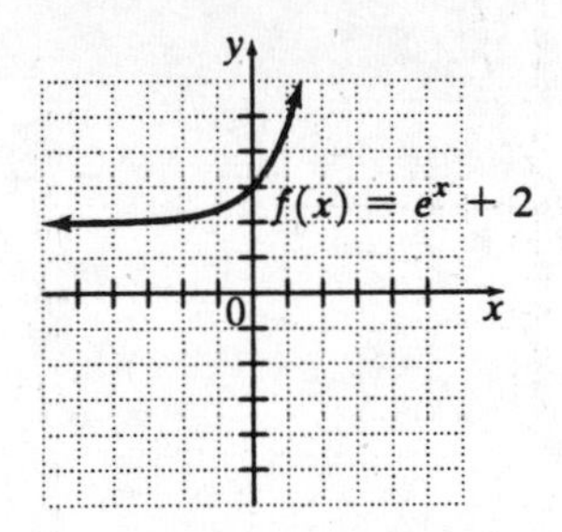

77.

x	$f(x)$
–2	$3e^{-2} \approx 0.41$
–1	$3e^{-1} \approx 1.10$
0	$3e^0 = 3$
1	$3e^1 \approx 8.15$
2	$3e^2 \approx 22.17$

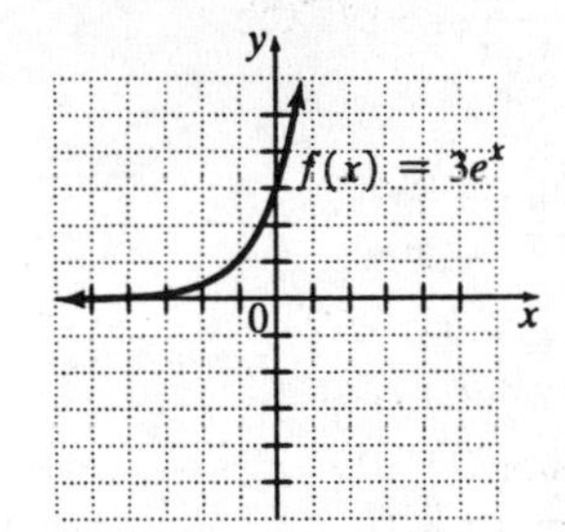

79.

x	$f(x)$
1	$\ln 1 = 0$
2	$\ln 2 \approx 0.69$
3	$\ln 3 \approx 1.10$
4	$\ln 4 \approx 1.39$
5	$\ln 5 \approx 1.61$

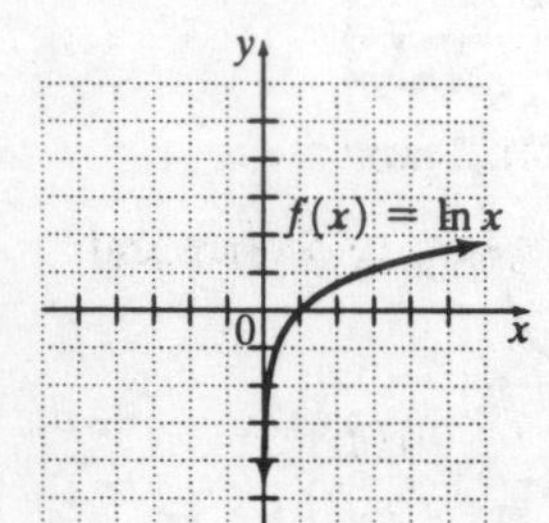

83.

x	$f(x)$
–1	$\log 1 = 0$
0	$\log 2 \approx 0.30$
1	$\log 3 \approx 0.48$
2	$\log 4 \approx 0.60$
3	$\log 5 \approx 0.70$

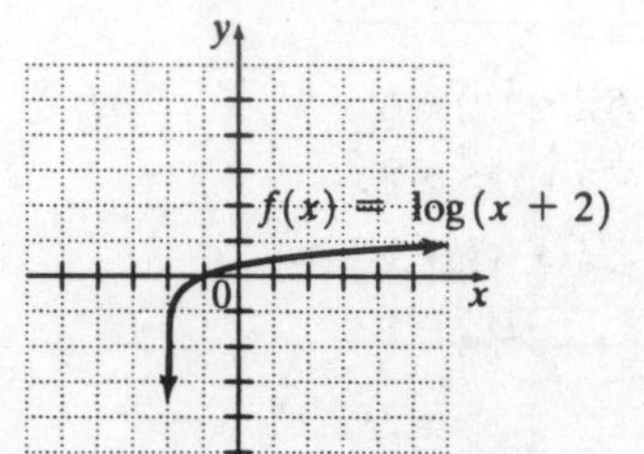

81.

x	$f(x)$
1	$-2\log 1 = 0$
2	$-2\log 2 \approx -0.60$
3	$-2\log 3 \approx -0.95$
4	$-2\log 4 \approx -1.20$
5	$-2\log 5 \approx -1.40$

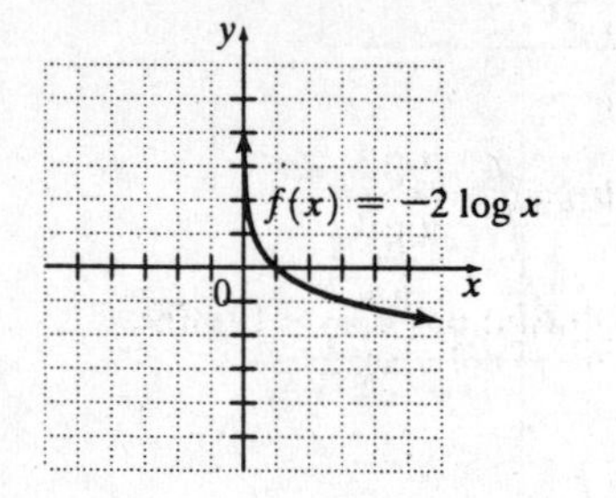

85.

x	$f(x)$
1	$\ln 1 - 3 = -3$
2	$\ln 2 - 3 \approx -2.31$
3	$\ln 3 - 3 \approx -1.90$
4	$\ln 4 - 3 \approx -1.61$
5	$\ln 5 - 3 \approx -1.39$

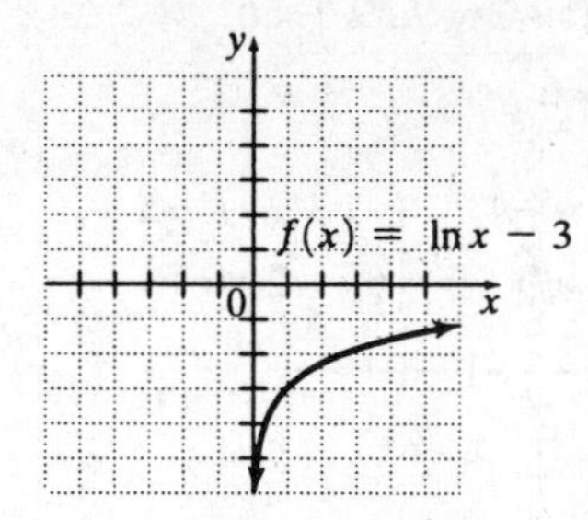

87. $f(x) = e^x + 2$
$f(x) = e^x$
$f(x) = e^x - 3$

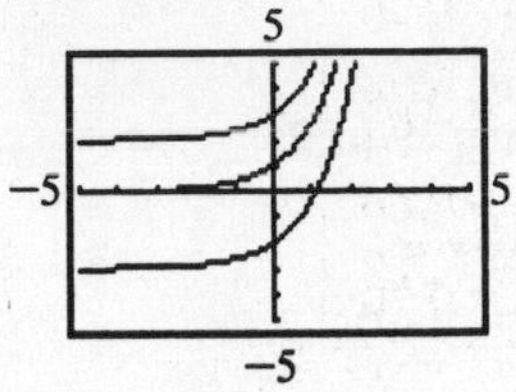

89. $6x - 3(2 - 5x) = 6$
$6x - 6 + 15x = 6$
$21x - 6 = 6$
$21x = 12$
$x = \frac{12}{21} = \frac{4}{7}$

91. $2x + 3y = 6x$
$3y = 4x$
$x = \frac{3y}{4}$

93. $x^2 + 7x = -6$
$x^2 + 7x + 6 = 0$
$(x + 6)(x + 1) = 0$
$x + 6 = 0$ or $x + 1 = 0$
$x = -6$ or $x = -1$

95. $\begin{cases} x + 2y = -4 \\ 3x - y = 9 \end{cases} \rightarrow \begin{cases} x + 2y = -4 \\ 6x - 2y = 18 \end{cases}$
Add.
$7x = 14$
$x = 2$
Substitute back.
$2 + 2y = -4$
$2y = -6$
$y = -3$
The solution is (2, –3).

Section 11.7

Graphing Calculator Explorations

1. $Y_1 = 5000\left(1 + \frac{0.05}{4}\right)^{4x}$
$Y_2 = 6000$

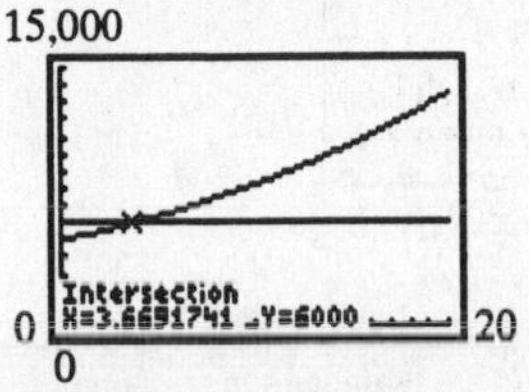

It takes 3.67 years, or 3 years and 8 months.

3. $Y_1 = 10{,}000\left(1 + \frac{0.06}{12}\right)^{12x}$
$Y_2 = 40{,}000$

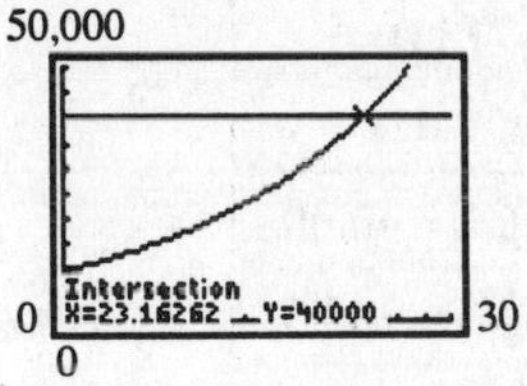

It takes 23.16 years, or 23 years and 2 months.

Exercise Set 11.7

1. $3^x = 6$
$x = \log_3 6 = \frac{\log 6}{\log 3}$
The solution is $\frac{\log 6}{\log 3} \approx 1.6309$.

3. $3^{2x} = 3.8$
$2x = \log_3 3.8 = \frac{\log 3.8}{\log 3}$ so
$x = \frac{\log 3.8}{2\log 3}$
The solution is $\frac{\log 3.8}{2\log 3} \approx 0.6076$.

5. $2^{x-3} = 5$

$x - 3 = \log_2 5$

$x = 3 + \log_2 5 = 3 + \dfrac{\log 5}{\log 2}$

The solution is $3 + \dfrac{\log 5}{\log 2} \approx 5.3219$.

7. $9^x = 5$

$x = \log_9 5 = \dfrac{\log 5}{\log 9}$

The solution is $\dfrac{\log 5}{\log 9} \approx 0.7325$.

9. $4^{x+7} = 3$

$x + 7 = \log_4 3$

$x = \log_4 3 - 7$

$= \dfrac{\log 3}{\log 4} - 7$

The solution is $\dfrac{\log 3}{\log 4} - 7 \approx -6.2075$.

11. $7^{3x-4} = 11$

$3x - 4 = \log_7 11$

$3x = 4 + \log_7 11$

$x = \dfrac{1}{3}\left(4 + \dfrac{\log 11}{\log 7}\right)$

The solution is $\dfrac{1}{3}\left(4 + \dfrac{\log 11}{\log 7}\right) \approx 1.7441$.

13. $e^{6x} = 5$

$6x = \ln 5$

$x = \dfrac{\ln 5}{6}$

The solution is $\dfrac{\ln 5}{6} \approx 0.2682$.

15. $\log_2(x+5) = 4$

$x + 5 = 2^4$

$x + 5 = 16$

$x = 11$

The solution is 11.

17. $\log_3 x^2 = 4$

$x^2 = 3^4$

$x^2 = 81$

$x = \pm 9$

The solution is –9 or 9.

19. $\log_4 2 + \log_4 x = 0$

$\log_4(2x) = 0$

$2x = 4^0$

$2x = 1$

$x = \dfrac{1}{2}$

The solution is $\dfrac{1}{2}$.

21. $\log_2 6 - \log_2 x = 3$

$\log_2\left(\dfrac{6}{x}\right) = 3$

$\dfrac{6}{x} = 2^3$

$\dfrac{6}{x} = 8$

$8x = 6$

$x = \dfrac{6}{8} = \dfrac{3}{4}$

The solution is $\dfrac{3}{4}$.

23. $\log_4 x + \log_4(x+6) = 2$

$\log_4 x(x+6) = 2$

$x(x+6) = 4^2$

$x^2 + 6x = 16$

$x^2 + 6x - 16 = 0$

$(x+8)(x-2) = 0$

$x + 8 = 0$ or $x - 2 = 0$

$x = -8$ or $x = 2$

We discard –8 as extraneous. The solution is 2.

25. $\log_5(x+3)-\log_5 x=2$

$$\log_5\left(\frac{x+3}{x}\right)=2$$
$$\frac{x+3}{x}=5^2$$
$$\frac{x+3}{x}=25$$
$$x+3=25x$$
$$3=24x$$
$$x=\frac{3}{24}=\frac{1}{8}$$

The solution is $\frac{1}{8}$.

27. $\log_3(x-2)=2$

$$x-2=3^2$$
$$x-2=9$$
$$x=11$$

The solution is 11.

29. $\log_4(x^2-3x)=1$

$$x^2-3x=4^1$$
$$x^2-3x=4$$
$$x^2-3x-4=0$$
$$(x-4)(x+1)=0$$
$$x-4=0 \quad \text{or} \quad x+1=0$$
$$x=4 \quad \text{or} \quad x=-1$$

The solution is 4 or −1.

31. $\ln 5+\ln x=0$

$$\ln(5x)=0$$
$$5x=e^0$$
$$5x=1$$
$$x=\frac{1}{5}$$

The solution is $\frac{1}{5}$.

33. $3\log x-\log x^2=2$

$$3\log x-2\log x=2$$
$$\log x=2$$
$$x=10^2$$
$$x=100$$

The solution is 100.

35. $\log_2 x+\log_2(x+5)=1$

$$\log_2 x(x+5)=1$$
$$x(x+5)=2^1$$
$$x^2+5x-2=0$$

$a=1$, $b=5$, and $c=-2$

$$x=\frac{-5\pm\sqrt{5^2-4(1)(-2)}}{2(1)}$$
$$=\frac{-5\pm\sqrt{25+8}}{2}$$
$$=\frac{-5\pm\sqrt{33}}{2}$$

Discard $\frac{-5-\sqrt{33}}{2}$. The solution is $\frac{-5+\sqrt{33}}{2}$.

37. $\log_4 x-\log_4(2x-3)=3$

$$\log_4\frac{x}{2x-3}=3$$
$$\frac{x}{2x-3}=4^3$$
$$\frac{x}{2x-3}=64$$
$$x=64(2x-3)$$
$$x=128x-192$$
$$192=127x$$
$$x=\frac{192}{127}$$

The solution is $\frac{192}{127}$.

39. $\log_2 x + \log_2(3x+1) = 1$
$\log_2 x(3x+1) = 1$
$x(3x+1) = 2^1$
$3x^2 + x - 2 = 0$
$(3x-2)(x+1) = 0$
$3x - 2 = 0$ or $x + 1 = 0$
$3x = 2$ or $x = -1$
$x = \frac{2}{3}$
We discard −1 as extraneous. The solution is $\frac{2}{3}$.

41. $y = y_0 e^{0.043t}$, $y_0 = 83$ and $t = 5$
$y = 83e^{0.043(5)} = 83e^{0.215} \approx 103$
There should be 103 wolves in 5 years.

43. $y = y_0 e^{0.026t}$, $y_0 = 10{,}052{,}000$ and $t = 6$
$y = 10{,}052{,}000e^{0.026(6)} \approx 11{,}750{,}000$
There will be approximately 11,750,000 inhabitants in 2005.

45. $y = y_0 e^{-0.005t}$,
$y_0 = 146{,}394$, and $y = 120{,}000$
$120{,}000 = 146{,}394e^{-0.005t}$
$\frac{120{,}000}{146{,}394} = e^{-0.005t}$
$$t = \frac{\ln\left(\frac{120{,}000}{146{,}394}\right)}{-0.005} \approx 39.8$$
It will take approximately 39.8 years to reach 120,000 thousand.

47. $A = P\left(1 + \frac{r}{n}\right)^{nt}$, $P = 600$,
$A = 2(600) = 1200$, $r = 0.07$, and $n = 12$
$1200 = 600\left(1 + \frac{0.07}{12}\right)^{12t}$
$2 = (1.0058\overline{3})^{12t}$
$12t = \log_{1.0058\overline{3}}(2)$
$t = \frac{1}{12}\log_{1.0058\overline{3}}(2)$
$= \frac{1}{12}\frac{\ln 2}{\ln(1.0058\overline{3})} \approx 10$
It would take approximately 10 years for the \$600 to double.

49. $A = P\left(1 + \frac{r}{n}\right)^{nt}$, $P = 1200$,
$A = P + I = 1200 + 200 = 1400$, $r = 0.09$, and $n = 4$
$1400 = 1200\left(1 + \frac{0.09}{4}\right)^{4t}$
$\frac{7}{6} = (1.0225)^{4t}$
$4t = \log_{1.0225}\left(\frac{7}{6}\right)$
$t = \frac{1}{4}\log_{1.0225}\left(\frac{7}{6}\right)$
$= \frac{1}{4}\frac{\ln\left(\frac{7}{6}\right)}{\ln 1.0225} \approx 1.73$
It would take the investment approximately 1.7 years to earn \$200.

51. $A = P\left(1 + \frac{r}{n}\right)^{nt}$, $P = 1000$,
$A = 2(1000) = 2000$, $r = 0.08$, and $n = 2$.
$2000 = 1000\left(1 + \frac{0.08}{2}\right)^{2t}$
$2 = (1.04)^{2t}$
$2t = \log_{1.04} 2$
$t = \frac{1}{2}\frac{\ln 2}{\ln 1.04} \approx 8.8$
It would take approximately 8.8 years for the \$1000 to double.

53. $w = 0.00185h^{2.67}$, and $h = 35$

$w = 0.00185(35)^{2.67} \approx 24.5$

The expected weight of a boy 35 inches tall is 24.5 pounds.

55. $w = 0.00185h^{2.67}$, $w = 85$

$85 = 0.00185h^{2.67}$

$\frac{85}{0.00185} = h^{2.67}$

$h = \left(\frac{85}{0.00185}\right)^{1/2.67} \approx 55.7$

The expected height of the boy is approximately 55.7 inches.

57. $P = 14.7e^{-0.21x}$, $x = 1$

$= 14.7e^{-0.21(1)}$

$= 14.7e^{-0.21}$

≈ 11.9

The average atmospheric pressure of Denver is approximately 11.9 pounds per square inch.

59. $P = 14.7e^{-0.21x}$, $P = 7.5$

$7.5 = 14.7e^{-0.21x}$

$\frac{7.5}{14.7} = e^{-0.21x}$

$-0.21x = \ln\left(\frac{7.5}{14.7}\right)$

$x = -\frac{1}{0.21}\ln\left(\frac{7.5}{14.7}\right) \approx 3.2$

The elevation of the jet is approximately 3.2 miles.

61. $t = \frac{1}{c}\ln\left(\frac{A}{A-N}\right)$

$t = \frac{1}{0.09}\ln\left(\frac{75}{75-50}\right)$

$t = \frac{1}{0.09}\ln(3)$

$t \approx 12.21$

It will take 12 weeks.

63. $t = \frac{1}{c}\ln\left(\frac{A}{A-N}\right)$

$t = \frac{1}{0.07}\ln\left(\frac{210}{210-150}\right)$

$t = \frac{1}{0.07}\ln(3.5)$

$t \approx 17.9$

It will take 18 weeks.

65. $Y_1 = e^{0.3x}$ and $Y_2 = 8$

Intersection X=6.9314718 Y=8

$x \approx 6.93$

67. $Y_1 = 2\log(-5.6x + 1.3)$

$Y_2 = -x - 1$

Intersection X=-3.681549 Y=2.6815494

Intersection X=.18658985 Y=-1.18659

$x \approx -3.68$ or $x \approx 0.19$

69. $Y_1 = 7^{3x-4}$

$Y_2 = 11$

Intersection X=1.7440915 Y=11

$x \approx 1.74$

71. $Y_1 = \ln 5 + \ln x$
$Y_2 = 0$

Intersection X=.2 Y=0

$x = 0.2$

73. $\frac{x^2 - y + 2z}{3x} = \frac{(-2)^2 - (0) + 2(3)}{3(-2)}$
$= \frac{4+6}{-6}$
$= \frac{10}{-6}$
$= -\frac{5}{3}$

75. $\frac{3z - 4x + y}{x + 2z} = \frac{3(3) - 4(-2) + 0}{-2 + 2(3)}$
$= \frac{9+8}{-2+6}$
$= \frac{17}{4}$

77. $f(x) = 5x + 2$
$y = 5x + 2$
$x = 5y + 2$
$x - 2 = 5y$
$y = \frac{x-2}{5}$
$f^{-1}(x) = \frac{x-2}{5}$

Chapter 11 Review

1. $(f + g)(x) = f(x) + g(x)$
$= (x - 5) + (2x + 1)$
$= x - 5 + 2x + 1$
$= 3x - 4$

2. $(f - g)(x) = f(x) - g(x)$
$= (x - 5) - (2x + 1)$
$= x - 5 - 2x - 1$
$= -x - 6$

3. $(f \cdot g)(x) = f(x) \cdot g(x)$
$= (x - 5)(2x + 1)$
$= 2x^2 + x - 10x - 5$
$= 2x^2 - 9x - 5$

4. $\left(\frac{g}{f}\right)(x) = \frac{g(x)}{f(x)} = \frac{2x+1}{x-5},\ x \neq 5$

5. $(f \circ g)(x) = f(g(x))$
$= f(x + 1)$
$= (x + 1)^2 - 2$
$= x^2 + 2x + 1 - 2$
$= x^2 + 2x - 1$

6. $(g \circ f)(x) = g(f(x))$
$= g(x^2 - 2)$
$= x^2 - 2 + 1$
$= x^2 - 1$

7. $(h \circ g)(2) = h(g(2))$
$= h(2 + 1)$
$= h(3)$
$= 3^3 - 3^2$
$= 18$

8. $(f \circ f)(x) = f(f(x))$
$= f(x^2 - 2)$
$= (x^2 - 2)^2 - 2$
$= x^4 - 4x^2 + 4 - 2$
$= x^4 - 4x^2 + 2$

9. $(f \circ g)(-1) = f(g(-1))$
$= f(-1 + 1)$
$= f(0)$
$= 0^2 - 2$
$= -2$

10. $(h \circ h)(2) = h(h(2))$
$= h(2^3 - 2^2)$
$= h(4)$
$= 4^3 - 4^2$
$= 48$

11. The function is one-to-one. To obtain the inverse, switch the x- and the y-values:
$h^{-1} = \{(14, -9), (8, 6), (12, -11), (15, 15)\}$

12. The function is not one-to-one. The y-value 5 is assigned to two x-values.

13. The function is one-to-one.

Rank in Automobile Thefts (Input)	2	4	1	3
U.S. Region (Output)	West	Midwest	South	Northeast

14. The function is not one-to-one. The output 4 is assigned to three inputs.

15. $f(x) = \sqrt{x+2}$

a. $f(7) = \sqrt{7+2} = \sqrt{9} = 3$

b. $f^{-1}(3) = 7$

16. $f(x) = \sqrt{x+2}$

a. $f(-1) = \sqrt{-1+2} = \sqrt{1} = 1$

b. $f^{-1}(1) = -1$

17. The graph is not a one-to-one function since it fails the horizontal line test.

18. The graph is not a one-to-one function since it fails the horizontal line test.

19. The graph is not a one-to-one function since it fails the horizontal line test.

20. The graph is a one-to-one function since it passes the horizontal line test.

21.
$$\begin{aligned} f(x) &= x-9 \\ y &= x-9 \\ x &= y-9 \\ y &= x+9 \\ f^{-1}(x) &= x+9 \end{aligned}$$

22.
$$\begin{aligned} f(x) &= x+8 \\ y &= x+8 \\ x &= y+8 \\ y &= x-8 \\ f^{-1}(x) &= x-8 \end{aligned}$$

23.
$$\begin{aligned} f(x) &= 6x+11 \\ y &= 6x+11 \\ x &= 6y+11 \\ 6y &= x-11 \\ y &= \frac{x-11}{6} \\ f^{-1}(x) &= \frac{x-11}{6} \end{aligned}$$

24.
$$\begin{aligned} f(x) &= 12x \\ y &= 12x \\ x &= 12y \\ y &= \frac{x}{12} \\ f^{-1}(x) &= \frac{x}{12} \end{aligned}$$

25.
$$\begin{aligned} f(x) &= x^3-5 \\ y &= x^3-5 \\ x &= y^3-5 \\ y^3 &= x+5 \\ y &= \sqrt[3]{x+5} \\ f^{-1}(x) &= \sqrt[3]{x+5} \end{aligned}$$

26. $f(x)=\sqrt[3]{x+2}$

$y=\sqrt[3]{x+2}$

$x=\sqrt[3]{y+2}$

$x^3=y+2$

$y=x^3-2$

$f^{-1}(x)=x^3-2$

27. $g(x)=\dfrac{12x-7}{6}$

$y=\dfrac{12x-7}{6}$

$x=\dfrac{12y-7}{6}$

$6x=12y-7$

$6x+7=12y$

$\dfrac{6x+7}{12}=y$

$g^{-1}(x)=\dfrac{6x+7}{12}$

28. $r(x)=\dfrac{13}{2}x-4$

$y=\dfrac{13}{2}x-4$

$x=\dfrac{13}{2}y-4$

$x+4=\dfrac{13}{2}y$

$2(x+4)=13y$

$\dfrac{2(x+4)}{13}=y$

$r^{-1}(x)=\dfrac{2(x+4)}{13}$

29. $g(x)=\sqrt{x}$

$y=\sqrt{x}$

$x=\sqrt{y}$

$x^2=y$

$g^{-1}(x)=x^2,\ x\geq 0$

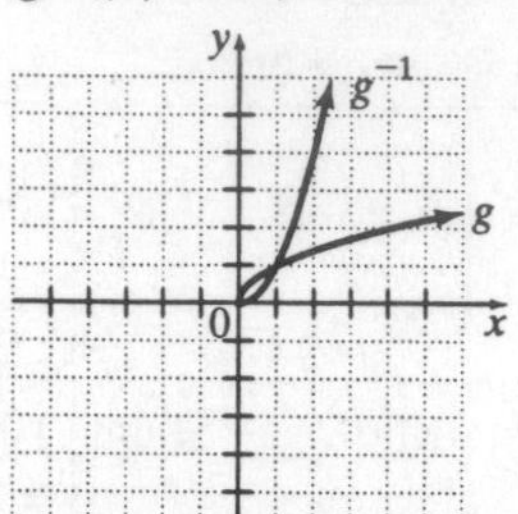

30. $h(x)=5x-5$

$y=5x-5$

$x=5y-5$

$5y=x+5$

$y=\dfrac{x+5}{5}$

$h^{-1}(x)=\dfrac{x+5}{5}$

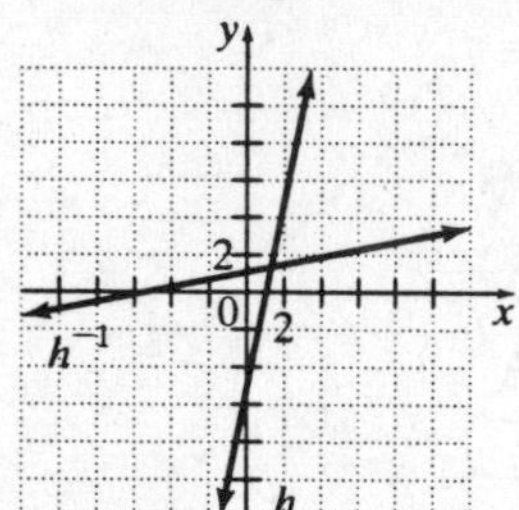

31. $f(x)=2x-3$ or $y=2x-3$

Find inverse

$x=2y-3$

$y=\dfrac{x+3}{2}$

$f^{-1}(x)=\dfrac{x+3}{2}$

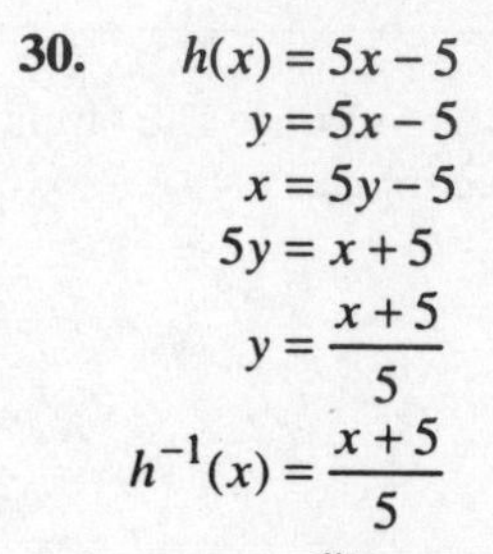

32. $4^x = 64$

$4^x = 4^3$

$x = 3$

33. $3^x = \frac{1}{9}$

$3^x = 3^{-2}$

$x = -2$

34. $2^{3x} = \frac{1}{16}$

$2^{3x} = 2^{-4}$

$3x = -4$

$x = -\frac{4}{3}$

35. $5^{2x} = 125$

$5^{2x} = 5^3$

$2x = 3$

$x = \frac{3}{2}$

36. $9^{x+1} = 243$

$(3^2)^{x+1} = 3^5$

$3^{2x+2} = 3^5$

$2x + 2 = 5$

$2x = 3$

$x = \frac{3}{2}$

37. $8^{3x-2} = 4$

$(2^3)^{3x-2} = 2^2$

$2^{9x-6} = 2^2$

$9x - 6 = 2$

$9x = 8$

$x = \frac{8}{9}$

38.

x	y
−2	$3^{-2} = \frac{1}{9}$
−1	$3^{-1} = \frac{1}{3}$
0	$3^0 = 1$
1	$3^1 = 3$
2	$3^2 = 9$

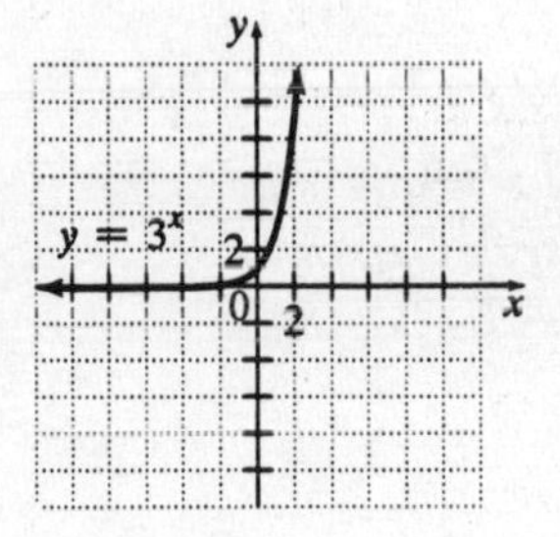

39.

x	y
−2	$\left(\frac{1}{3}\right)^{-2} = 9$
−1	$\left(\frac{1}{3}\right)^{-1} = 3$
0	$\left(\frac{1}{3}\right)^{0} = 1$
1	$\left(\frac{1}{3}\right)^{1} = \frac{1}{3}$
2	$\left(\frac{1}{3}\right)^{2} = \frac{1}{9}$

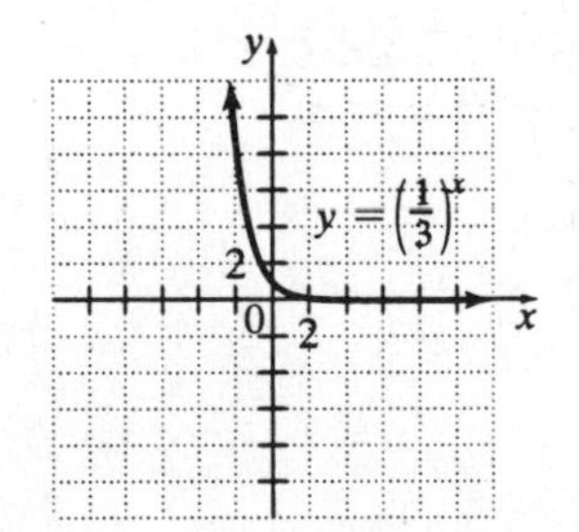

40.

x	y
-2	$4 \cdot 2^{-2} = 1$
-1	$4 \cdot 2^{-1} = 2$
0	$4 \cdot 2^{0} = 4$
1	$4 \cdot 2^{1} = 8$
2	$4 \cdot 2^{2} = 16$

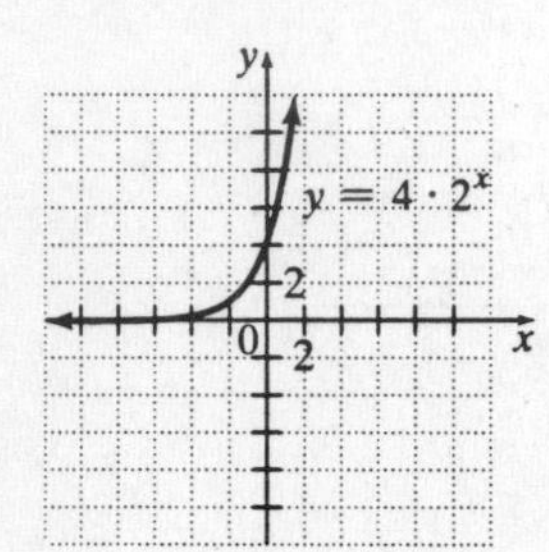

41.

x	y
-2	$2^{-2} + 4 = \frac{17}{4}$
-1	$2^{-1} + 4 = \frac{9}{2}$
0	$2^{0} + 4 = 5$
1	$2^{1} + 4 = 6$
2	$2^{2} + 4 = 8$

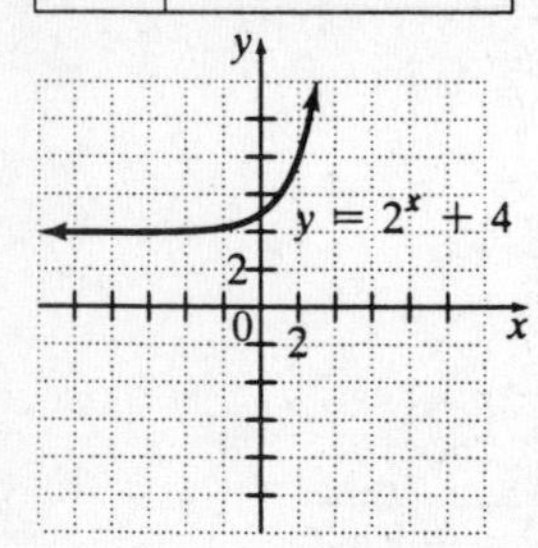

42. $A = P\left(1 + \frac{r}{n}\right)^{nt}$

$A = 1600\left(1 + \frac{0.09}{2}\right)^{(2)(7)}$

$A = \$2963.11$

After 7 years the investment is worth \$2963.11.

43. $A = P\left(1 + \frac{r}{n}\right)^{nt}$

$A = 800\left(1 + \frac{0.07}{4}\right)^{4(5)}$

$= 800(1.0175)^{20}$

$= 1131.82$

The certificate is worth \$1131.82 at the end of 5 years.

44. $y = 4 \cdot 2^x$

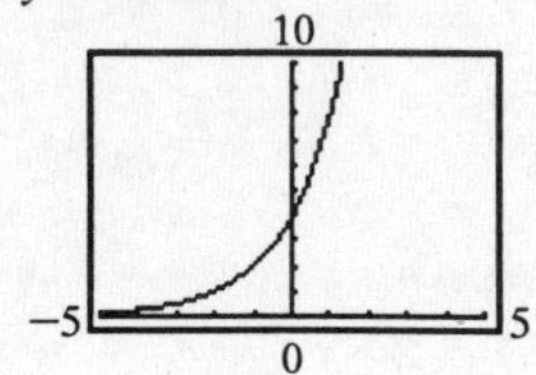

45. $7^2 = 49$

$\log_7 49 = 2$

46. $2^{-4} = \frac{1}{16}$

$\log_2 \frac{1}{16} = -4$

47. $\log_{1/2} 16 = -4$

$\left(\frac{1}{2}\right)^{-4} = 16$

48. $\log_{0.4} 0.064 = 3$

$0.4^3 = 0.064$

49. $\log_4 x = -3$

$x = 4^{-3} = \frac{1}{64}$

50. $\log_3 x = 2$

$x = 3^2 = 9$

51. $\log_3 1 = x$

$3^x = 1$

$3^x = 3^0$

$x = 0$

52. $\log_4 64 = x$

$4^x = 64$

$4^x = 4^3$

$x = 3$

53. $\log_x 64 = 2$

$x^2 = 64$

$x = \pm\sqrt{64} = \pm 8$

$x = 8$ since base > 0

54. $\log_x 81 = 4$

$x^4 = 81$

$x = \sqrt[4]{81} = \pm 3$

We discard the negative base -3, so $x = 3$.

55. $\log_4 4^5 = x$

$x = 5$

56. $\log_7 7^{-2} = x$

$x = -2$

57. $5^{\log_5 4} = x$

$x = 4$

58. $2^{\log_2 9} = x$

$9 = x$

59. $\log_2(3x - 1) = 4$

$3x - 1 = 2^4 = 16$

$3x = 17$

$x = \dfrac{17}{3}$

60. $\log_3(2x + 5) = 2$

$2x + 5 = 3^2$

$2x + 5 = 9$

$2x = 4$

$x = 2$

61. $\log_4(x^2 - 3x) = 1$

$x^2 - 3x = 4^1 = 4$

$x^2 - 3x - 4 = 0$

$(x + 1)(x - 4) = 0$

$x + 1 = 0$ or $x - 4 = 0$

$x = -1$ or $x = 4$

62. $\log_8(x^2 + 7x) = 1$

$x^2 + 7x = 8^1 = 8$

$x^2 + 7x - 8 = 0$

$(x + 8)(x - 1) = 0$

$x + 8 = 0$ or $x - 1 = 0$

$x = -8$ or $x = 1$

63.

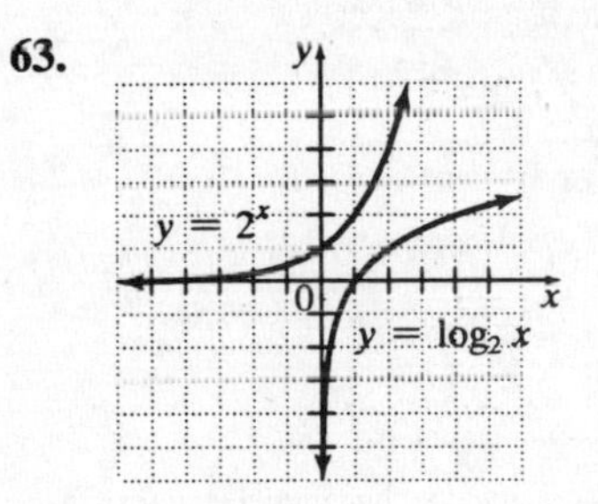

64.

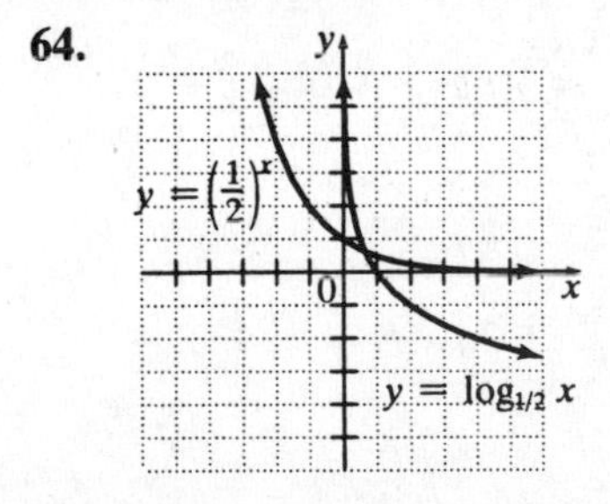

65. $\log_3 8 + \log_3 4 = \log_3(8 \cdot 4) = \log_3 32$

66. $\log_2 6 + \log_2 3 = \log_2(6 \cdot 3) = \log_2 18$

67. $\log_7 15 - \log_7 20 = \log_7 \frac{15}{20} = \log_7 \frac{3}{4}$

68. $\log 18 - \log 12 = \log \frac{18}{12} = \log \frac{3}{2}$

69. $\log_{11} 8 + \log_{11} 3 - \log_{11} 6 = \log_{11} \frac{(8 \cdot 3)}{6}$
$= \log_{11} 4$

70. $\log_5 14 + \log_5 3 - \log_5 21$
$= \log_5(14 \cdot 3) - \log_5 21$
$= \log_5 \frac{42}{21} = \log_5 2$

71. $2\log_5 x - 2\log_5(x+1) + \log_5 x$
$= \log_5 x^2 - \log_5(x+1)^2 + \log_5 x$
$= \log_5 \frac{(x^2)(x)}{(x+1)^2}$
$= \log_5 \frac{x^3}{(x+1)^2}$

72. $4\log_3 x - \log_3 x + \log_3(x+2)$
$= 3\log_3 x + \log_3(x+2)$
$= \log_3 x^3 + \log_3(x+2)$
$= \log_3[x^3(x+2)]$
$= \log_3(x^4 + 2x^3)$

73. $\log_3 \frac{x^3}{x+2} = \log_3 x^3 - \log_3(x+2)$
$= 3\log_3 x - \log_3(x+2)$

74. $\log_4 \frac{x+5}{x^2} = \log_4(x+5) - \log_4 x^2$
$= \log_4(x+5) - 2\log_4 x$

75. $\log_2 \frac{3x^2y}{z}$
$= \log_2 3 + \log_2 x^2 + \log_2 y - \log_2 z$
$= \log_2 3 + 2\log_2 x + \log_2 y - \log_2 z$

76. $\log_7 \frac{yz^3}{x} = \log_7(yz^3) - \log_7 x$
$= \log_7 y + \log_y z^3 - \log_7 x$
$= \log_7 y + 3\log_7 z - \log_7 x$

77. $\log_b 50 = \log_b(5)(5)(2)$
$= \log_b(5) + \log_b(5) + \log_b(2)$
$= 0.83 + 0.83 + 0.36$
$= 2.02$

78. $\log_b \frac{4}{5} = \log_b 4 - \log_b 5$
$= \log_b 2^2 - \log_b 5$
$= 2\log_b 2 - \log_b 5$
$= 2(0.36) - 0.83$
$= 0.72 - 0.83$
$= -0.11$

79. $\log 3.6 \approx 0.5563$

80. $\log 0.15 \approx -0.8239$

81. $\ln 1.25 \approx 0.2231$

82. $\ln 4.63 \approx 1.5326$

83. $\log 1000 = \log 10^3 = 3$

84. $\log \frac{1}{10} = \log 10^{-1} = -1$

85. $\ln \frac{1}{e} = \ln 1 - \ln e = 0 - 1 = -1$

86. $\ln e^4 = 4$

87. $\ln(2x) = 2$
$2x = e^2$
$x = \frac{e^2}{2}$

88. $\ln(3x) = 1.6$
$$3x = e^{1.6}$$
$$x = \frac{e^{1.6}}{3}$$

89. $\ln(2x - 3) = -1$
$$2x - 3 = e^{-1}$$
$$2x = e^{-1} + 3$$
$$x = \frac{e^{-1} + 3}{2}$$

90. $\ln(3x + 1) = 2$
$$3x + 1 = e^2$$
$$3x = e^2 - 1$$
$$x = \frac{e^2 - 1}{3}$$

91. $\ln\frac{I}{I_0} = -kx$
$$\ln\frac{0.03I_0}{I_0} = -2.1x$$
$$\ln 0.03 = -2.1x$$
$$\frac{\ln 0.03}{-2.1} = x$$
$$x \approx 1.67 \text{ mm}$$
The radiation is reduced to 3% at a depth of approximately 1.67 mm.

92. $\ln\frac{I}{I_0} = -kx$
$$\ln\frac{0.02I_0}{I_0} = -3.2x$$
$$\ln 0.02 = -3.2x$$
$$x = \frac{\ln 0.02}{-3.2} \approx 1.22$$
2% of the original radioactivity will penetrate at a depth of approximately 1.22 millimeters.

93. $\log_5 1.6 = \frac{\log 1.6}{\log 5} = 0.2920$

94. $\log_3 4 = \frac{\log 4}{\log 3} \approx 1.2619$

95. $A = Pe^{rt}$
$$A = 1450e^{(0.06)(5)}$$
$$A = \$1957.30$$
The investment grows to $1957.30.

96. $A = Pe^{rt}$
$$A = 940e^{0.11(3)} = 940e^{0.33} \approx 1307.51$$
The $940 investment grows to $1307.51 in 3 years.

97. $3^{2x} = 7$
$$2x \log 3 = \log 7$$
$$x = \frac{\log 7}{2\log 3}$$
The solution is $\frac{\log 7}{2\log 3} \approx 0.8856$.

98. $6^{3x} = 5$
$$3x = \log_6 5$$
$$x = \frac{1}{3}\log_6 5 = \frac{\log 5}{3\log 6}$$
The solution is $\frac{\log 5}{3\log 6} \approx 0.2994$.

99. $3^{2x+1} = 6$
$$(2x + 1)\log 3 = \log 6$$
$$2x = \frac{\log 6}{\log 3} - 1$$
$$x = \frac{1}{2}\left(\frac{\log 6}{\log 3} - 1\right)$$
The solution is $\frac{1}{2}\left(\frac{\log 6}{\log 3} - 1\right) \approx 0.3155$.

100. $4^{3x+2} = 9$
$$3x + 2 = \log_4 9$$
$$3x = \log_4 9 - 2$$
$$x = \frac{1}{3}\left(\frac{\log 9}{\log 4} - 2\right)$$
The solution is $\frac{1}{3}\left(\frac{\log 9}{\log 4} - 2\right) \approx -0.1383$.

101. $5^{3x-5} = 4$

$(3x-5)\log 5 = \log 4$

$3x = \frac{\log 4}{\log 5} + 5$

$x = \frac{1}{3}\left(\frac{\log 4}{\log 5} + 5\right)$

The solution is $\frac{1}{3}\left(\frac{\log 4}{\log 5} + 5\right) \approx 1.9538$.

102. $8^{4x-2} = 3$

$4x - 2 = \log_8 3$

$4x = \log_8 3 + 2$

$4x = \frac{\log 3}{\log 8} + 2$

$x = \frac{1}{4}\left(\frac{\log 3}{\log 8} + 2\right)$

The solution is $\frac{1}{4}\left(\frac{\log 3}{\log 8} + 2\right) \approx 0.6321$.

103. $2 \cdot 5^{x-1} = 1$

$\log 2 + (x-1)\log 5 = \log 1$

$(x-1)\log 5 = -\log 2$

$x = -\frac{\log 2}{\log 5} + 1$

The solution is $-\frac{\log 2}{\log 5} + 1 \approx 0.5693$.

104. $3 \cdot 4^{x+5} = 2$

$4^{x+5} = \frac{2}{3}$

$x + 5 = \log_4\left(\frac{2}{3}\right)$

$x = \log_4\left(\frac{2}{3}\right) - 5$

$x = \frac{\log \frac{2}{3}}{\log 4} - 5$

The solution is $\frac{\log \frac{2}{3}}{\log 4} - 5 \approx -5.2925$.

105. $\log_5 2 + \log_5 x = 2$

$\log_5 2x = 2$

$2x = 5^2 = 25$

$x = \frac{25}{2}$

The solution is $\frac{25}{2}$.

106. $\log_3 x + \log_3 10 = 2$

$\log_3(10x) = 2$

$10x = 3^2$

$10x = 9$

$x = \frac{9}{10}$

The solution is $\frac{9}{10}$.

107. $\log(5x) - \log(x+1) = 4$

$\log \frac{5x}{x+1} = 4$

$\frac{5x}{x+1} = 10^4 = 10{,}000$

$5x = 10{,}000x + 10{,}000$

$-10{,}000 = 9995x$

$-1.0005 \approx x$

This is outside the domains of log $(5x)$ and log $(x + 1)$, so there is no solution.

108. $\ln(3x) - \ln(x-3) = 2$

$\ln \frac{3x}{x-3} = 2$

$\frac{3x}{x-3} = e^2$

$3x = e^2(x-3)$

$3x = e^2x - 3e^2$

$3x - e^2x = -3e^2$

$(3 - e^2)x = -3e^2$

$x = \frac{-3e^2}{3 - e^2} = \frac{3e^2}{e^2 - 3}$

The solution is $\frac{3e^2}{e^2 - 3}$.

109. $\log_2 x + \log_2 2x - 3 = 1$
$$\log_2(x)(2x) = 4$$
$$2x^2 = 2^4 = 16$$
$$x^2 = 8$$
$$x = \pm\sqrt{8} = \pm 2\sqrt{2}$$
Discard $-2\sqrt{2}$ as extraneous. The solution is $2\sqrt{2}$.

110. $-\log_6(4x+7) + \log_6 x = 1$
$$\log_6 \frac{x}{4x+7} = 1$$
$$\frac{x}{4x+7} = 6^1$$
$$x = 6(4x+7)$$
$$x = 24x + 42$$
$$-42 = 23x$$
$$x = -\frac{42}{23}$$
We discard $x = -\frac{42}{23}$ as extraneous leaving no solution.

111. $y = y_0 e^{kt}$
$$y = 155{,}000e^{(0.06)(4)}$$
$$y = 197{,}044$$
197,044 ducks are expected in 4 weeks.

112. $y = y_0 e^{kt}$
$$y = 212{,}942{,}000e^{0.015(8)}$$
$$= 212{,}942{,}000e^{0.12}$$
$$\approx 240{,}091{,}435$$
The expected population of Indonesia by the year 2006 is approximately 240,091,435.

113. $y = y_0 e^{kt}$
$$140{,}000{,}000 = 125{,}932{,}000e^{0.002t}$$
$$t = \frac{\ln \frac{140{,}000{,}000}{125{,}932{,}000}}{0.002}$$
$$t \approx 53$$
It will take approximately 53 years.

114.
$$y = y_0 e^{kt}$$
$$2(30{,}675{,}000) = 30{,}675{,}000e^{0.011t}$$
$$2 = e^{0.011t}$$
$$t = \frac{\ln 2}{0.011}$$
$$t \approx 63$$
It will take approximately 63 years.

115. $2(66{,}050{,}000) = 66{,}050{,}000e^{(0.019)t}$
$$2 = e^{0.019t}$$
$$t = \frac{\ln 2}{0.019} \approx 36$$
It will take approximately 36 years.

116.
$$A = P\left(1 + \frac{r}{n}\right)^{nt}$$
$$10{,}000 = 5000\left(1 + \frac{0.08}{4}\right)^{4t}$$
$$2 = (1.02)^{4t}$$
$$\log 2 = 4t \log 1.02$$
$$t = \frac{\log 2}{4 \log 1.02} \approx 8.8$$
It will take about 8.8 years.

117. $A = P\left(1 + \frac{r}{n}\right)^{nt}$
$$10{,}000 = 6{,}000\left(1 + \frac{0.06}{12}\right)^{12t}$$
$$\frac{5}{3} = (1.005)^{12t}$$
$$12t = \log_{1.005}\left(\frac{5}{3}\right)$$
$$t = \frac{1}{12}\log_{1.005}\left(\frac{5}{3}\right)$$
$$= \frac{1}{12}\frac{\log\left(\frac{5}{3}\right)}{\log(1.005)}$$
$$\approx 8.5$$
It was invested for approximately 8.5 years.

118.

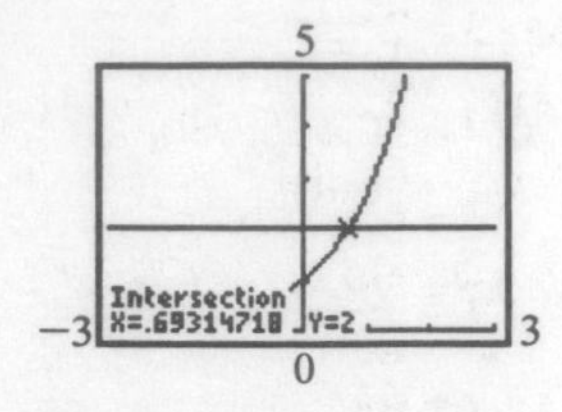

The solution is ≈ 0.69.

119.

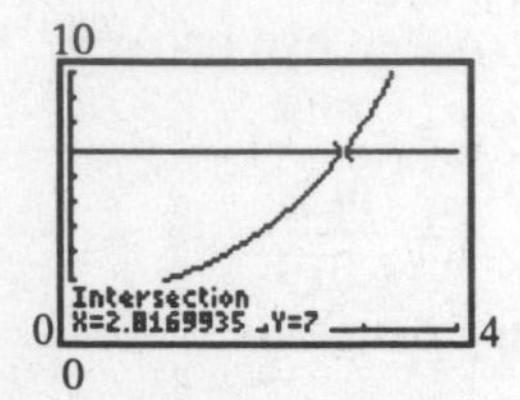

The solution is ≈ 2.82.

Chapter 11 Test

1. $(f \circ h)(0) = f(h(0)) = f(5) = 5$

2. $(g \circ f)(x) = g(f(x)) = g(x) = x - 7$

3.
$$\begin{aligned}(g \circ h)(x) &= g(h(x))\\ &= g(x^2 - 6x + 5)\\ &= x^2 - 6x + 5 - 7\\ &= x^2 - 6x - 2\end{aligned}$$

4.
$$\begin{aligned}f(x) &= 7x - 14\\ y &= 7x - 14\\ x &= 7y - 14\\ x + 14 &= 7y\\ y &= \frac{x+14}{7}\\ f^{-1}(x) &= \frac{x+14}{7}\end{aligned}$$

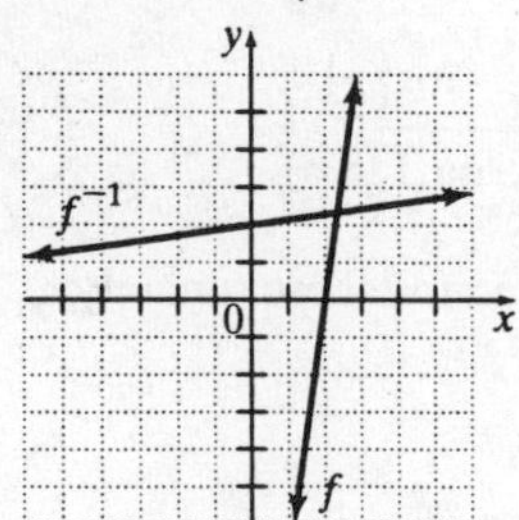

5. The graph does represent a one-to-one function since it passes the horizontal line test.

6. The graph does not represent a function since it fails the vertical line test, hence does not represent a one-to-one function.

7. $y = 6 - 2x$ is one-to-one.
$$\begin{aligned}x &= 6 - 2y\\ 2y &= -x + 6\\ y &= \frac{-x+6}{2}\\ f^{-1}(x) &= \frac{-x+6}{2}\end{aligned}$$

8. $f = \{(0, 0), (2, 3), (-1, 5)\}$ is one-to-one. Switch the ordered pairs to obtain the inverse:
$$f^{-1} = \{(0, 0), (3, 2), (5, -1)\}$$

9. This function is not one-to-one, since Dog and Desk have the same output.

10. $\log_3 6 + \log_3 4 = \log_3(6 \cdot 4) = \log_3 24$

11.
$$\begin{aligned}&\log_5 x + 3\log_5 x - \log_5(x+1)\\ &= 4\log_5 x - \log_5(x+1)\\ &= \log_5 x^4 - \log_5(x+1)\\ &= \log_5 \frac{x^4}{x+1}\end{aligned}$$

12.
$$\begin{aligned}\log_6 \frac{2x}{y^3} &= \log_6 2x - \log_6 y^3\\ &= \log_6 2 + \log_6 x - 3\log_6 y\end{aligned}$$

13.
$$\begin{aligned}\log_b\left(\frac{3}{25}\right) &= \log_b 3 - \log_b 25\\ &= \log_b 3 - \log_b 5^2\\ &= \log_b 3 - 2\log_b 5\\ &= 0.79 - 2(1.16)\\ &= -1.53\end{aligned}$$

14. $\log_7 8 = \frac{\ln 8}{\ln 7} \approx 1.0686$

15. $8^{x-1} = \frac{1}{64}$

$8^{x-1} = 8^{-2}$

$x - 1 = -2$

$x = -1$

The solution is –1.

16. $3^{2x+5} = 4$

$2x + 5 = \log_3 4$

$2x + 5 = \frac{\log 4}{\log 3}$

$2x = \frac{\log 4}{\log 3} - 5$

$x = \frac{1}{2}\left(\frac{\log 4}{\log 3} - 5\right)$

The solution is $\frac{1}{2}\left(\frac{\log 4}{\log 3} - 5\right) \approx -1.8691$.

17. $\log_3 x = -2$

$x = 3^{-2}$

$x = \frac{1}{9}$

The solution is $\frac{1}{9}$.

18. $\ln \sqrt{e} = x$

$\ln e^{1/2} = x$

$\frac{1}{2} = x$

The solution is $\frac{1}{2}$.

19. $\log_8(3x - 2) = 2$

$3x - 2 = 8^2$

$3x - 2 = 64$

$3x = 66$

$x = \frac{66}{3} = 22$

The solution is 22.

20. $\log_5 x + \log_5 3 = 2$

$\log_5(3x) = 2$

$3x = 5^2$

$3x = 25$

$x = \frac{25}{3}$

The solution is $\frac{25}{3}$.

21. $\log_4(x+1) - \log_4(x-2) = 3$

$\log_4 \frac{x+1}{x-2} = 3$

$\frac{x+1}{x-2} = 4^3$

$x + 1 = 64(x - 2)$

$x + 1 = 64x - 128$

$129 = 63x$

$x = \frac{129}{63} = \frac{43}{21}$

The solution is $\frac{43}{21}$.

22. $\ln(3x + 7) = 1.31$

$3x + 7 = e^{1.31}$

$3x = e^{1.31} - 7$

$x = \frac{e^{1.31} - 7}{3}$

The solution is $\frac{e^{1.31} - 7}{3} \approx -1.0979$.

23. $y = \left(\frac{1}{2}\right)^x + 1$

x	y
−2	$\left(\frac{1}{2}\right)^{-2} + 1 = 5$
−1	$\left(\frac{1}{2}\right)^{-1} + 1 = 3$
0	$\left(\frac{1}{2}\right)^{0} + 1 = 2$
1	$\left(\frac{1}{2}\right)^{1} + 1 = \frac{3}{2}$
2	$\left(\frac{1}{2}\right)^{2} + 1 = \frac{5}{4}$

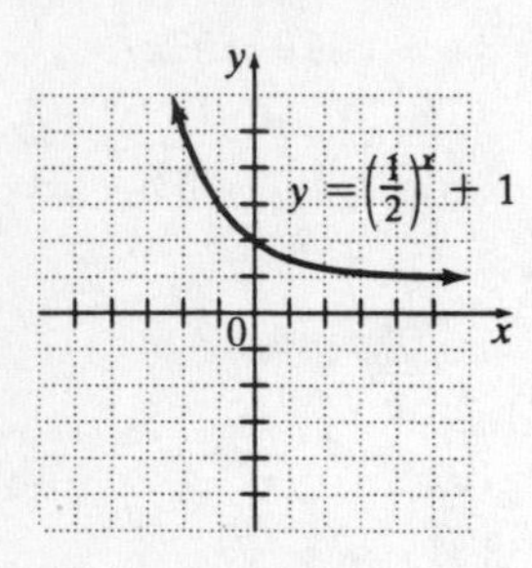

24. $y = 3^x$ and $y = \log_3 x$

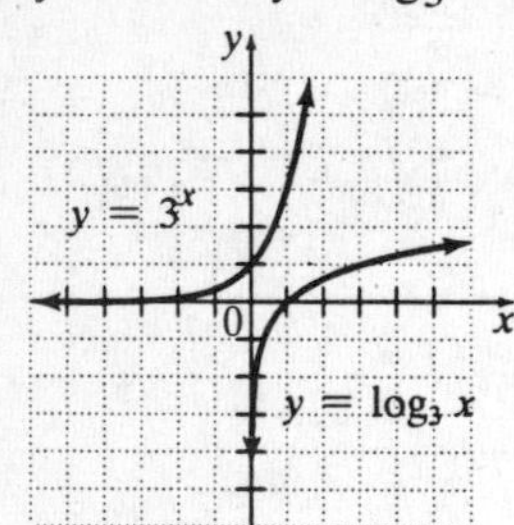

25. $A = P\left(1 + \frac{r}{n}\right)^{nt}$, $P = 4000$, $t = 3$, $r = 0.09$, and $n = 12$

$$A = 4000\left(1 + \frac{0.09}{12}\right)^{12(3)}$$
$$= 4000(1.0075)^{36}$$
$$\approx 5234.58$$

$5234.58 will be in the account.

26. $A = P\left(1 + \frac{r}{n}\right)^{nt}$, $P = 2000$, $A = 3000$, $r = 0.07$, and $n = 2$

$$3000 = 2000\left(1 + \frac{0.07}{2}\right)^{2t}$$
$$1.5 = (1.035)^{2t}$$
$$2t = \log_{1.035} 1.5$$
$$t = \frac{1}{2}\log_{1.035} 1.5$$
$$= \frac{1}{2}\frac{\ln 1.5}{\ln 1.035}$$
$$\approx 5.9$$

It would take 6 years for the investment to reach $3000.

27. $y = y_0 e^{kt}$, $y_0 = 57{,}000$, $k = 0.026$, and $t = 5$

$$y = 57{,}000e^{0.026(5)}$$
$$= 57{,}000e^{0.13}$$
$$\approx 64{,}913$$

There will be approximately 64,913 prairie dogs 5 years from now.

28. $y = y_0 e^{kt}$, $y_0 = 400$, $y = 1000$, and $k = 0.062$

$$1000 = 400e^{0.062t}$$
$$2.5 = e^{0.062t}$$
$$0.062t = \ln 2.5$$
$$t = \frac{\ln 2.5}{0.062}$$
$$\approx 14.8$$

It will take the naturalists approximately 15 years to reach their goal.

29. $\log(1+k) = \frac{0.3}{D}$, $D = 56$

$\log(1+k) = \frac{0.3}{56}$

$1+k = 10^{\frac{0.3}{56}}$

$k = -1 + 10^{\frac{0.3}{56}}$

$k \approx .012$

The rate of population increase is approximately 1.2%.

30. $Y_1 = e^{0.2x}$

$Y_2 = e^{-0.4x} + 2$

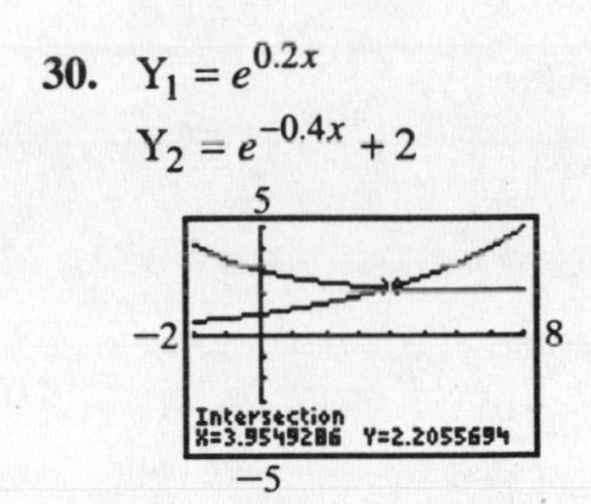

The solution is ≈ 3.95.

Chapter 12

Exercise Set 12.1

1. $a_n = n + 4$

$a_1 = 1 + 4 = 5$

$a_2 = 2 + 4 = 6$

$a_3 = 3 + 4 = 7$

$a_4 = 4 + 4 = 8$

$a_5 = 5 + 4 = 9$

Thus, the first five terms of the sequence $a_n = n + 4$ are or 5, 6, 7, 8, 9.

3. $a_n = (-1)^n$

$a_1 = (-1)^1 = -1$

$a_2 = (-1)^2 = 1$

$a_3 = (-1)^3 = -1$

$a_4 = (-1)^4 = 1$

$a_5 = (-1)^5 = -1$

Thus, the first five terms of the sequence $a_n = (-1)^n$ are −1, 1, −1, 1, −1.

5. $a_n = \frac{1}{n+3}$

$a_1 = \frac{1}{1+3} = \frac{1}{4}$

$a_2 = \frac{1}{2+3} = \frac{1}{5}$

$a_3 = \frac{1}{3+3} = \frac{1}{6}$

$a_4 = \frac{1}{4+3} = \frac{1}{7}$

$a_5 = \frac{1}{5+3} = \frac{1}{8}$

Thus, the first five terms of the sequence $a_n = \frac{1}{n+3}$ are $\frac{1}{4}, \frac{1}{5}, \frac{1}{6}, \frac{1}{7}, \frac{1}{8}$.

7. $a_n = 2n$

$a_1 = 2(1) = 2$

$a_2 = 2(2) = 4$

$a_3 = 2(3) = 6$

$a_4 = 2(4) = 8$

$a_5 = 2(5) = 10$

or 2, 4, 6, 8, 10

9. $a_n = -n^2$

$a_1 = -1^2 = -1$

$a_2 = -2^2 = -4$

$a_3 = -3^2 = -9$

$a_4 = -4_2 = -16$

$a_5 = -5^2 = -25$

Thus, the first five terms of the sequence $a_n = -n^2$ are −1, −4, −8, −16, −25.

11. $a_n = 2^n$

$a_1 = 2^1 = 2$

$a_2 = 2^2 = 4$

$a_3 = 2^3 = 8$

$a_4 = 2^4 = 16$

$a_5 = 2^5 = 32$

Thus, the first five terms of the sequence $a_n = 2^n$ are 2, 4, 8, 16, 32.

13. $a_n = 2n + 5$

$a_1 = 2(1) + 5 = 2 + 5 = 7$

$a_2 = 2(2) + 5 = 4 + 5 = 9$

$a_3 = 2(3) + 5 = 6 + 5 = 11$

$a_4 = 2(4) + 5 = 8 + 5 = 13$

$a_5 = 2(5) + 5 = 10 + 5 = 15$

Thus, the first five terms of the sequence $a_n = 2n + 5$ are 7, 9, 11, 13, 15.

15. $a_n = (-1)^n n^2$

$a_1 = (-1)^1(1^2) = -1(1) = -1$

$a_2 = (-1)^2(2)^2 = 1(4) = 4$

$a_3 = (-1)^3(3)^2 = -1(9) = -9$

$a_4 = (-1)^4(4)^2 = 1(16) = 16$

$a_5 = (-1)^5(5)^2 = -1(25) = -25$

Thus, the first five terms of the sequence $a_n = (-1)^n n^2$ are −1, 4, −9, 16, −25.

17. $a_n = 3n^2$

$a_5 = 3(5)^2 = 3(25) = 75$

19. $a_n = 6n - 2$

$a_{20} = 6(20) - 2 = 120 - 2 = 118$

21. $a_n = \dfrac{n+3}{n}$

$a_{15} = \dfrac{15+3}{15} = \dfrac{18}{15} = \dfrac{6}{5}$

23. $a_n = (-3)^n$

$a_6 = (-3)^6 = 729$

25. $a_n = \dfrac{n-2}{n+1}$

$a_6 = \dfrac{6-2}{6+1} = \dfrac{4}{7}$

27. $a_n = \dfrac{(-1)^n}{n}$

$a_8 = \dfrac{(-1)^8}{8} = \dfrac{1}{8}$

29. $a_n = -n^2 + 5$

$a_{10} = -10^2 + 5 = -100 + 5 = -95$

31. $a_n = \dfrac{(-1)^n}{n+6}$

$a_{19} = \dfrac{(-1)^{19}}{19+6} = -\dfrac{1}{25}$

33. 3, 7, 11, 15 or
4(1) − 1, 4(2) − 1, 4(3) − 1, 4(4) − 1
In general, $a_n = 4n - 1$

35. −2, −4, −8, −16, or $-2, -2^2, -2^3, -2^4$
In general, $a_n = -2^n$

37. $\dfrac{1}{3}, \dfrac{1}{9}, \dfrac{1}{27}, \dfrac{1}{81}$, or

$\dfrac{1}{3}, \dfrac{1}{3^2}, \dfrac{1}{3^3}, \dfrac{1}{3^4}$

In general, $a_n = \dfrac{1}{3^n}$

39. $a_n = 32n - 16$

$a_2 = 32(2) - 16 = 64 - 16 = 48$ ft

$a_3 = 32(3) - 16 = 96 - 16 = 80$ ft

$a_4 = 32(4) - 16 = 128 - 16 = 112$ ft

41. 0.10, 0.20, 0.40 or
$0.10, 0.10(2), 0.10(2)^2$
In general, $a_n = 0.10(2)^{n-1}$

$a_{14} = 0.10(2)^{13} \approx \$819.20$

43. $a_n = 75(2)^{n-1}$

$a_6 = 75(2)^5 = 75(32) = 2400$ cases

$a_1 = 75(2)^0 = 75(1) = 75$ cases

45. $a_n = \frac{1}{2}a_{n-1}$ for $n > 1$, $a_1 = 800$

In 2000, $n = 1$ and $a_1 = 800$.

In 2001, $n = 2$ and $a_2 = \frac{1}{2}(800) = 400$.

In 2002, $n = 3$ and $a_3 = \frac{1}{2}(400) = 200$.

In 2003, $n = 4$ and $a_4 = \frac{1}{2}(200) = 100$.

In 2004, $n = 5$ and $a_5 = \frac{1}{2}(100) = 50$.

The population estimate for 2004 is 50 sparrows.

Continuing the sequence:

in 2005, $n = 6$ and $a_6 = \frac{1}{2}(50) = 25$;

in 2006, $n = 7$ and $a_7 = \frac{1}{2}(25) = 12$;

in 2007, $n = 8$ and $a_8 = \frac{1}{2}(12) = 6$;

in 2008, $n = 9$ and $a_9 = \frac{1}{2}(6) = 3$;

in 2009, $n = 10$ and $a_{10} = \frac{1}{2}(3) \approx 1$;

in 2010, $n = 11$ and $a_{11} = \frac{1}{2}(1) \approx 0$.

The population is estimated to become extinct in 2010.

47. $a_n = \frac{1}{\sqrt{n}}$

$a_1 = \frac{1}{\sqrt{1}} = \frac{1}{1} = 1$

$a_2 = \frac{1}{\sqrt{2}} \approx 0.7071$

$a_3 = \frac{1}{\sqrt{3}} \approx 0.5774$

$a_4 = \frac{1}{\sqrt{4}} = \frac{1}{2} = 0.5$

$a_5 = \frac{1}{\sqrt{5}} \approx 0.4472$

Thus, the first five terms of the sequence $a_n = \frac{1}{\sqrt{n}}$ are $1, 0.7071, 0.5774, 0.5, 0.4472$.

49. $a_n = \left(1+\frac{1}{n}\right)^n$

$a_1 = \left(1+\frac{1}{1}\right)^1 = (2)^1 = 2$

$a_2 = \left(1+\frac{1}{2}\right)^2 = \left(\frac{3}{2}\right)^2 = 2.25$

$a_3 = \left(1+\frac{1}{3}\right)^3 = \left(\frac{4}{3}\right)^3 \approx 2.3704$

$a_4 = \left(1+\frac{1}{4}\right)^4 = \left(\frac{5}{4}\right)^4 \approx 2.4414$

$a_5 = \left(1+\frac{1}{5}\right)^5 = \left(\frac{6}{5}\right)^5 \approx 2.4883$

Thus, the first five terms of the sequence $a_n = \left(1+\frac{1}{n}\right)^n$ are 2, 2.25, 2.3704, 2.4414, 2.4883.

51. $f(x) = (x-1)^2 + 3$

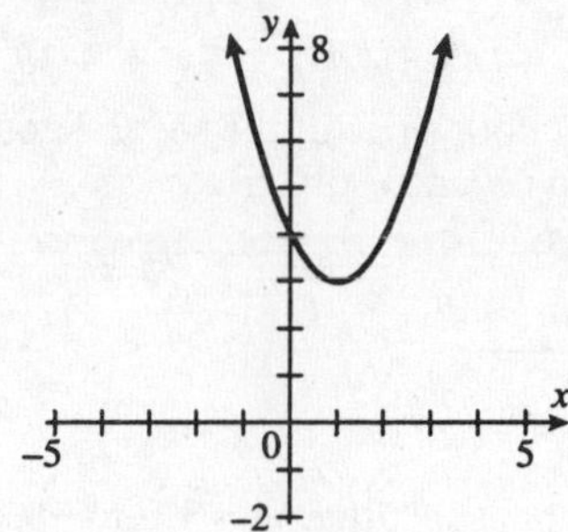

53. $f(x) = 2(x+4)^2 + 2$

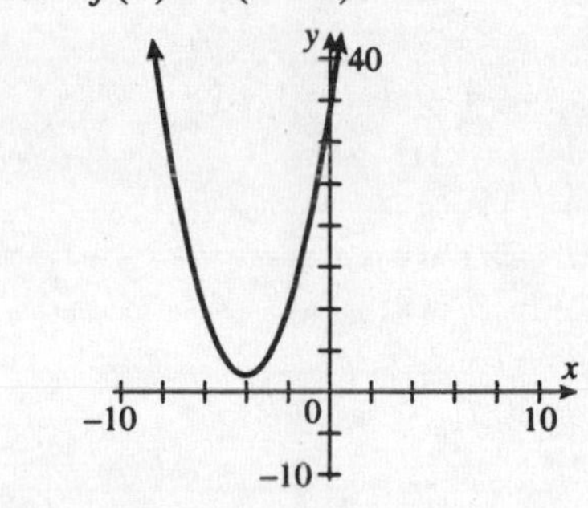

Exercise Set 12.2

1. $a_n = a_1 + (n-1)d$

$a_1 = 4;\ d = 2$

$a_1 = 4$

$a_2 = 4 + (2-1)2 = 6$

$a_3 = 4 + (3-1)2 = 8$

$a_4 = 4 + (4-1)2 = 10$

$a_5 = 4 + (5-1)2 = 12$

The first five terms are 4, 6, 8, 10, 12.

3. $a_n = a_1 + (n-1)d$

$a_1 = 6,\ d = -2$

$a_1 = 6$

$a_2 = 6 + (2-1)(-2) = 4$

$a_3 = 6 + (3-1)(-2) = 2$

$a_4 = 6 + (4-1)(-2) = 0$

$a_5 = 6 + (5-1)(-2) = -2$

The first five terms are 6, 4, 2, 0, –2.

5. $a_n = a_1 r^{n-1}$

$a_1 = 1,\ r = 3$

$a_1 = 1(3)^{1-1} = 1$

$a_2 = 1(3)^{2-1} = 3$

$a_3 = 1(3)^{3-1} = 9$

$a_4 = 1(3)^{4-1} = 27$

$a_5 = 1(3)^{5-1} = 81$

The first five terms are 1, 3, 9, 27, 81.

7. $a_n = a_1 r^{n-1}$

$a_1 = 48,\ r = \frac{1}{2}$

$a_1 = 48\left(\frac{1}{2}\right)^{1-1} = 48$

$a_2 = 48\left(\frac{1}{2}\right)^{2-1} = 24$

$a_3 = 48\left(\frac{1}{2}\right)^{3-1} = 12$

$a_4 = 48\left(\frac{1}{2}\right)^{4-1} = 6$

$a_5 = 48\left(\frac{1}{2}\right)^{5-1} = 3$

The first five terms are 48, 24, 12, 6, 3.

9. $a_n = a_1 + (n-1)d$

$a_1 = 12,\ d = 3$

$a_n = 12 + (n-1)3$

$a_8 = 12 + 7(3) = 12 + 21 = 33$

11. $a_n = a_1 r^{n-1}$

$a_1 = 7,\ r = -5$

$a_n = a_1 r^{n-1}$

$a_4 = 7(-5)^3 = 7(-125) = -875$

13. $a_n = a_1 + (n-1)d$

$a_1 = -4,\ d = -4$

$a_n = -4 + (n-1)(-4)$

$a_{15} = -4 + 14(-4) = -4 - 56 = -60$

15. 0, 12, 24

$a_1 = 0$ and $d = 12$

$a_n = 0 + (n-1)12$

$a_9 = 8(12) = 96$

17. 20, 18, 16

$a_1 = 20$ and $d = -2$

$a_n = 20 + (n-1)(-2)$

$a_{25} = 20 + 24(-2) = 20 - 48 = -28$

19. 2, −10, 50

$a_1 = 2$ and $r = -5$

$a_n = 2(-5)^{n-1}$

$a_5 = 2(-5)^4 = 2(625) = 1250$

21. $a_4 = 19,\ a_{15} = 52$

$$\begin{cases} a_4 = a_1 + (4-1)d \\ a_{15} = a_1 + (15-1)d \end{cases} \text{ or } \begin{cases} 19 = a_1 + 3d \\ 52 = a_1 + 14d \end{cases}$$

Multiply both sides of the second equation by -1 in order to solve for d.

$$\begin{aligned} 19 &= a_1 + 3d \\ -52 &= -a_1 - 14d \\ \hline -33 &= \quad -11d \end{aligned}$$

$3 = d$

To find a_1, substitute $d = 3$ into the first equation.

$19 = a_1 + 3(3)$

$19 = a_1 + 9$

$10 = a_1$

Thus, $a_1 = 10$ and $d = 3$, so

$$\begin{aligned} a_n &= 10 + (n-1)3 \\ &= 10 + 3n - 3 \\ &= 7 + 3n \end{aligned}$$

and
$$\begin{aligned} a_8 &= 7 + 3(8) \\ &= 7 + 24 \\ a_8 &= 31 \end{aligned}$$

23. $a_2 = -1$ and $a_4 = 5$

$$\begin{cases} a_2 = a_1 + (2-1)d \\ a_4 = a_1 + (4-1)d \end{cases} \text{ or } \begin{cases} -1 = a_1 + d \\ 5 = a_1 + 3d \end{cases}$$

Multiply both sides of the second equation by -1 in order to solve for d.

$$\begin{aligned} -1 &= a_1 + d \\ -5 &= -a_1 - 3d \\ \hline -6 &= \quad -2d \end{aligned}$$

$3 = d$

To find a_1, substitute $d = 3$ into the first equation.

$-1 = a_1 + 3$

$-4 = a_1$

Thus, $a_1 = -4$ and $d = 3$, so

$$\begin{aligned} a_n &= -4 + (n-1)3 \\ &= -4 + 3n - 3 \\ &= -7 + 3n \end{aligned}$$

and
$$\begin{aligned} a_9 &= -7 + 3(9) \\ &= -7 + 27 \\ a_9 &= 20 \end{aligned}$$

25. $a_2 = -\frac{4}{3}$ and $a_3 = \frac{8}{3}$

Notice that $\frac{8}{3} \div \frac{-4}{3} = \frac{8}{3} \cdot -\frac{3}{4} = -2$, so $r = -2$. Then

$a_2 = a_1(-2)^{2-1}$

$-\frac{4}{3} = a_1(-2)$

$\frac{2}{3} = a,$

The first term is $\frac{2}{3}$ and the common ratio is -2.

27. Answers may vary.

29. 2, 4, 6 is an arithmetic sequence.

$a_1 = 2$ and $d = 2$

31. 5, 10, 20 is a geometric sequence.

$a_1 = 5$ and $r = 2$

33. $\frac{1}{2}, \frac{1}{10}, \frac{1}{50}$ is a geometric sequence.
$a_1 = \frac{1}{2}$ and $r = \frac{1}{5}$

35. $x, 5x, 25x$ is a geometric sequence.
$a_1 = x$ and $r = 5$

37. $p, p+4, p+8$ is an arithmetic sequence.
$a_1 = p$ and $d = 4$

39. $a_1 = 14$ and $d = \frac{1}{4}$
$a_n = 14 + (n-1)\frac{1}{4}$
$a_{21} = 14 + 20\left(\frac{1}{4}\right) = 14 + 5 = 19$

41. $a_1 = 3$ and $r = -\frac{2}{3}$
$a_n = 3\left(-\frac{2}{3}\right)^{n-1}$
$a_4 = 3\left(-\frac{2}{3}\right)^3 = 3\left(-\frac{8}{27}\right) = -\frac{8}{9}$

43. $\frac{3}{2}, 2, \frac{5}{2}, \ldots$
$a_1 = \frac{3}{2}$ and $d = \frac{1}{2}$
$a_n = \frac{3}{2} + (n-1)\frac{1}{2}$
$a_{15} = \frac{3}{2} + 14\left(\frac{1}{2}\right) = \frac{17}{2}$

45. $24, 8, \frac{8}{3}, \ldots$
$a_1 = 24$ and $r = \frac{1}{3}$
$a_n = 24\left(\frac{1}{3}\right)^{n-1}$
$a_6 = 24\left(\frac{1}{3}\right)^5 = 24\left(\frac{1}{243}\right) = \frac{8}{81}$

47. $a_3 = 2$ and $a_{17} = -40$
$\begin{cases} a_3 = a_1 + (3-1)d \\ a_{17} = a_1 + (17-1)d \end{cases}$ or $\begin{array}{l} 2 = a_1 + 2d \\ -40 = a_1 + 16d \end{array}$
Multiply both sides of the second equation by -1 in order to solve for d.

$$\begin{array}{rl} 2 = & a_1 + 2d \\ 40 = & -a_1 - 16d \\ \hline 42 = & -14d \end{array}$$

$-3 = d$
To find a_1, substitute $d = -3$ into the first equation.
$2 = a_1 + 2(-3)$
$2 = a_1 - 6$
$8 = a_1$
Thus, $a_1 = 8$ and $d = -3$, so
$a_n = 8 + (n-1)(-3)$
$= 8 - 3n + 3$
$= 11 - 3n$
and $a_{10} = 11 - 3(10)$
$= 11 - 30$
$a_{10} = -19$

49. 54, 58, 62
$a_1 = 54$ and $d = 4$
$a_n = 54 + (n-1)4$
$a_{20} = 54 + 19(4) = 54 + 76 = 130$
The general term of the sequence is $a_n = 4n + 50$.
There are 130 seats in the twentieth row.

51. $a_1 = 6$ and $r = 3$
$a_n = 6(3)^{n-1} = 2 \cdot 3 \cdot 3^{n-1} = 2(3)^n$
The general term of the sequence is $a_n = 6(3)^{n-1}$ or $a_n = 2(3)^n$.

53. $a_1 = 486$ and $r = \frac{1}{3}$

Initial Height $= a_1 = 486\left(\frac{1}{3}\right)^{1-1} = 486$

Rebound 1 $= a_2 = 486\left(\frac{1}{3}\right)^{2-1} = 162$

Rebound 2 $= a_3 = 486\left(\frac{1}{3}\right)^{3-1} = 54$

Rebound 3 $= a_4 = 486\left(\frac{1}{3}\right)^{4-1} = 18$

Rebound 4 $= a_5 = 486\left(\frac{1}{3}\right)^{5-1} = 6$

The first five terms of the sequence are 486, 162, 54, 18, 6.

$a_n = 486\left(\frac{1}{3}\right)^{n-1}$

Solve $486\left(\frac{1}{3}\right)^{n-1} = 1$

$$\left(\frac{1}{3}\right)^{n-1} = \frac{1}{486}$$
$$\frac{1}{3^{n-1}} = \frac{1}{486}$$
$$3^{n-1} = 486$$
$$3^n = 1458$$
$$n = \log_3 1458$$
$$n = \frac{\ln 1458}{\ln 3}$$
$$\approx 6.6$$

Since a_7 is less than a foot and a_7 corresponds to the 6th rebound, the ball will rebound less than a foot on the 6th bounce.

55. $a_1 = 4000$ and $d = 125$

$a_n = 4000 + (n-1)125$ or

$a_n = 3875 + 125n$

$a_{12} = 4000 + 11(125)$

$a_{12} = 5375$

His salary for his last month of training is $5375.

57. $a_1 = 400$ and $r = \frac{1}{2}$

12 hrs = 4(3 hrs), so we seek the fourth term after a_1, namely a_5.

$a_n = a_1 r^{n-1}$

$a_5 = 400\left(\frac{1}{2}\right)^4 = \frac{400}{16} = 25$

25 grams of the radioactive material remains after 12 hours.

59. $a_1 = \$11,782.40$

$r = 0.5$

$a_2 = (11,782.40)(0.5) = \$5891.20$

$a_3 = (5891.20)(0.5) = \$2945.60$

$a_4 = (2945.60)(0.5) = \$1472.80$

The first four terms of the sequence are $11,782.40, $5891.20, $2945.60, $1472.80.

61. $a_1 = 19.652$ and $d = -0.034$

$a_2 = 19.652 - 0.034 = 19.618$

$a_3 = 19.618 - 0.034 = 19.584$

$a_4 = 19.584 - 0.034 = 19.550$

The first four terms of the sequence are 19.652, 19.618, 19.584, 19.550.

63. Answers may vary.

65. $\frac{1}{3(1)} + \frac{1}{3(2)} + \frac{1}{3(3)}$

$= \frac{1}{3}\left(\frac{1}{1} + \frac{1}{2} + \frac{1}{3}\right)$

$= \frac{1}{3}\left(\frac{6}{6} + \frac{3}{6} + \frac{2}{6}\right)$

$= \frac{1}{3}\left(\frac{11}{6}\right) = \frac{11}{18}$

67. $3^0 + 3^1 + 3^2 + 3^3 = 1 + 3 + 9 + 27 = 40$

69. $\frac{8-1}{8+1} + \frac{8-2}{8+2} + \frac{8-3}{8+3}$

$= \frac{7}{9} + \frac{6}{10} + \frac{5}{11}$

$= \frac{770}{990} + \frac{594}{990} + \frac{450}{990}$

$= \frac{1814}{990} = \frac{907}{495}$

Exercise Set 12.3

1. $\sum_{i=1}^{4}(i-3)$
$= (1-3)+(2-3)+(3-3)+(4-3)$
$= -2+(-1)+0+1=-2$

3. $\sum_{i=4}^{7}(2i+4)$
$= [2(4)+4]+[2(5)+4]+[2(6)+4] + [2(7)+4]$
$= 12+14+16+18=60$

5. $\sum_{i=2}^{4}(i^2-3)$
$= (2^2-3)+(3^2-3)+(4^2-3)$
$= 1+6+13=20$

7. $\sum_{i-1}^{3}\frac{1}{i+5}$
$= \frac{1}{1+5}+\frac{1}{2+5}+\frac{1}{3+5}$
$= \frac{1}{6}+\frac{1}{7}+\frac{1}{8}=\frac{28}{168}+\frac{24}{168}+\frac{21}{168}=\frac{73}{168}$

9. $\sum_{i=1}^{3}\frac{1}{6i}=\frac{1}{6(1)}+\frac{1}{6(2)}+\frac{1}{6(3)}$
$= \frac{1}{6}+\frac{1}{12}+\frac{1}{18}=\frac{6+3+2}{36}=\frac{11}{36}$

11. $\sum_{i=2}^{6}3i$
$= 3(2)+3(3)+3(4)+3(5)+3(6)$
$= 6+9+12+15+18=60$

13. $\sum_{i=3}^{5}i(i+2)$
$= 3(3+2)+4(4+2)+5(5+2)$
$= 15+24+35=74$

15. $\sum_{i=1}^{5}2^i=2^1+2^2+2^3+2^4+2^5$
$= 2+4+8+16+32=62$

17. $\sum_{i=1}^{4}\frac{4i}{i+3}=\frac{4(1)}{1+3}+\frac{4(2)}{2+3}+\frac{4(3)}{3+3}+\frac{4(4)}{4+3}$
$= 1+\frac{8}{5}+2+\frac{16}{7}=\frac{105}{35}+\frac{56}{35}+\frac{80}{35}=\frac{241}{35}$

19. $1+3+5+7+9$
$= [(2)-1]+[2(2)-1]+[2(3)-1] + [2(4)-1]+[2(5)-1]$
$= \sum_{i=1}^{5}(2i-1)$

21. $4+12+36+108$
$= 4+4(3)+4(3^2)+4(3^3)$
$= \sum_{i=1}^{4}4(3)^{i-1}$

23. $12+9+6+3+0+(-3)$
$= [-3(1)+15]+[-3(2)+15] + [-3(3)+15]+[-3(4)+15] + [-3(5)+15]+[-3(6)+15]$
$= \sum_{i=1}^{6}(-3i+15)$

25. $12+4+\frac{4}{3}+\frac{4}{9}=\frac{4}{3^{-1}}+\frac{4}{3^0}+\frac{4}{3^1}+\frac{4}{3^2}$
$= \sum_{i=1}^{4}\frac{4}{3^{i-2}}$

27. $1+4+9+16+25+36+49$
$= 1^2+2^2+3^2+4^2+5^2+6^2+7^2$
$= \sum_{i=1}^{7}i^2$

29. $a_n=(n+2)(n-5)$
$a_1=(1+2)(1-5)=3(-4)=-12$
$a_2=(2+2)(2-5)=4(-3)=-12$
$a_1+a_2=-12+(-12)=-24$

31. $a_n = n(n-6)$
$= a_1 + a_2 = 1(1-6) + 2(2-6)$
$= 1(-5) + 2(-4) = -13$

33. $a_n = (n+3)(n+1)$
$a_1 = (1+3)(1+1) = 4(2) = 8$
$a_2 = (2+3)(2+1) = 5(3) = 15$
$a_3 = (3+3)(3+1) = 6(4) = 24$
$a_4 = (4+3)(4+1) = 7(5) = 35$
$\sum_{i=1}^{4} a_i = 8 + 15 + 24 + 35 = 82$

35. $a_n = -2n$
$\sum_{i=1}^{4} (-2i)$
$= -2(1) + (-2)(2) + (-2)(3) + (-2)(4)$
$= -2 - 4 - 6 - 8 = -20$

37. $a_n = -\frac{n}{3}$
$a_1 + a_2 + a_3 = -\frac{1}{3} - \frac{2}{3} - \frac{3}{3} = -2$

39. 1, 2, 3, …, 10
$a_n = n$
$\sum_{i=1}^{10} i = 1 + 2 + 3 + \cdots + 10$
$= \frac{10(11)}{2} = 55$
A total of 55 trees were planted.

41. $a_1 = 6$ and $r = 2$
$a_n = 6 \cdot 2^{n-1}$
$a_5 = 6 \cdot 2^4 = 6 \cdot 16 = 96$
There will be 96 fungus units at the beginning of the 5th day.

43. $a_1 = 50$ and $r = 2$
Since 48 = 4(12), we seek the fourth term after a_1, namely a_5.
The general term of the sequence is $a_n = 50(2)^{n-1}$, where n represents the number of 12-hr periods.
$a_5 = 50(2)^4 = 50(16) = 800$
There are 800 bacteria after 48 hours.

45. $a_n = (n+1)(n+2)$
$a_4 = (4+1)(4+2) = 5(6) = 30$ opossums
$a_1 = (1+1)(1+2) = 2(3) = 6$
$a_2 = (2+1)(2+2) = 3(4) = 12$
$a_3 = (3+1)(3+2) = 4(5) = 20$
$\sum_{i=1}^{4} a_i = 6 + 12 + 20 + 30 = 68$ opossums

47. $a_n = 100(0.5)^n$
$a_4 = 100(0.5)^4 = 6.25$ lbs of decay.
$a_1 = 100(0.5)^1 = 50$
$a_2 = 100(0.5)^2 = 25$
$a_3 = 100(0.5)^3 = 12.5$
$\sum_{i=1}^{4} a_i = 50 + 25 + 12.5 + 6.25$
$= 93.75$ lbs of decay

49. $a_1 = 40$ and $r = \frac{4}{5}$
$a_5 = 40\left(\frac{4}{5}\right)^4 = 16.384$ or 16.4 in.
$a_2 = 40\left(\frac{4}{5}\right)^1 = 32$
$a_3 = 40\left(\frac{4}{5}\right)^2 = 25.6$
$a_4 = 40\left(\frac{4}{5}\right)^3 = 20.48$
$\sum_{i=1}^{5} a_i = 40 + 32 + 25.6 + 20.48 + 16.384$
$= 134.464$ or 134.5 in.

51. a. $\sum_{i=1}^{7} i+i^2$

$= (1+1^2)+(2+2^2)+(3+3^2)$
$+(4+4^2)+(5+5^2)+(6+6^2)$
$+(7+7^2)$
$= 2 + 6 + 12 + 20 + 30 + 42 + 56$

b. $\sum_{i=1}^{7} i + \sum_{i=1}^{7} i^2$

$= 1 + 2 + 3 + 4 + 5 + 6 + 7 + 1 + 4$
$+ 9 + 16 + 25 + 36 + 49$

c. Answers may vary.

d. True; answers may vary.

53. $\frac{5}{1-\frac{1}{2}} = \frac{5}{\frac{1}{2}} = 5 \cdot \frac{2}{1} = 10$

55. $\frac{\frac{1}{3}}{1-\frac{1}{10}} = \frac{\frac{1}{3}}{\frac{9}{10}} = \frac{1}{3} \cdot \frac{10}{9} = \frac{10}{27}$

57. $\frac{3(1-2^4)}{1-2} = \frac{3(1-16)}{-1}$

$= \frac{3(-15)}{-1} = \frac{-45}{-1} = 45$

59. $\frac{10}{2}(3+15) = \frac{10}{2}(18) = \frac{180}{2} = 90$

Exercise Set 12.4

1. 1, 3, 5, 7, ...
The first term of this arithmetic sequence is $a_1 = 1$ and the sixth term is $a_6 = 11$.

$S_6 = \frac{6}{2}(a_1 + a_6)$
$= 3(1+11)$
$= 3(12)$
$S_6 = 36$

3. 4, 12, 36, ...
$a_1 = 4,\ r = 3, n = 5$

$S_5 = \frac{4(1-3^5)}{1-3} = -2(1-243)$
$S_5 = -2(-242) = 484$

5. 3, 6, 9, ...
The first term of this arithmetic sequence is $a_1 = 3$ and the sixth term is $a_6 = 18$.

$S_6 = \frac{6}{2}(a_1 + a_6)$
$= 3(3+18)$
$= 3(21)$
$S_6 = 63$

7. $2, \frac{2}{5}, \frac{2}{25}, \ldots$
$a_1 = 2,\ r = \frac{1}{5},\ n = 4$

$S_4 = \frac{2\left[1-\left(\frac{1}{5}\right)^4\right]}{1-\frac{1}{5}} = \frac{5}{2}\left(1-\frac{1}{625}\right)$

$S_4 = \frac{5}{2}\left(\frac{624}{625}\right) = \frac{312}{125} = 2.496$

9. 1, 2, 3, ..., 10
The first term of this arithmetic sequence is $a_1 = 1$ and the tenth term is $a_{10} = 10$.

$S_{10} = \frac{10}{2}(a_1 + a_{10})$
$= 5(1+10)$
$= 5(11)$
$S_{10} = 55$

11. 1, 3, 5, 7
The first term of this arithmetic sequence is $a_1 = 1$ and the fourth term is $a_4 = 7$.

$S_4 = \frac{4}{2}(a_1 + a_4)$
$= 2(1+7)$
$= 2(8)$
$S_4 = 16$

13. 12, 6, 3, ...

$a_1 = 12$ and $r = \frac{1}{2}$

$$S_\infty = \frac{12}{1-\frac{1}{2}} = \frac{12}{\frac{1}{2}} = 24$$

15. $\frac{1}{10}, \frac{1}{100}, \frac{1}{1000}, \ldots$

$a_1 = \frac{1}{10}$ and $r = \frac{1}{10}$

$$S_\infty = \frac{\frac{1}{10}}{1-\frac{1}{10}} = \frac{\frac{1}{10}}{\frac{9}{10}} = \frac{1}{9}$$

17. $-10, -5, -\frac{5}{2}, \ldots$

$a_1 = -10$ and $r = \frac{1}{2}$

$$S_\infty = \frac{-10}{1-\frac{1}{2}} = \frac{-10}{\frac{1}{2}} = -20$$

19. $2, -\frac{1}{4}, \frac{1}{32}, \ldots$

$a_1 = 2$ and $r = -\frac{1}{8}$

$$S_\infty = \frac{2}{1-\left(-\frac{1}{8}\right)} = \frac{2}{\frac{9}{8}} = \frac{16}{9}$$

21. $\frac{2}{3}, -\frac{1}{3}, \frac{1}{6}, \ldots$

$a_1 = \frac{2}{3}$ and $r = -\frac{1}{2}$

$$S_\infty = \frac{\frac{2}{3}}{1-\left(-\frac{1}{2}\right)} = \frac{\frac{2}{3}}{\frac{3}{2}} = \frac{4}{9}$$

23. –4, 1, 6, ..., 41

The first term of this arithmetic sequence is $a_1 = -4$ and the tenth term is $a_{10} = 41$.

$$S_{10} = \frac{10}{2}(a_1 + a_{10})$$
$$= 5(-4+41)$$
$$= 5(37)$$
$$S_{10} = 185$$

25. $3, \frac{3}{2}, \frac{3}{4}, \ldots$

$a_1 = 3$, $r = \frac{1}{2}$, $n = 7$

$$S_7 = \frac{3\left[1-\left(\frac{1}{2}\right)^7\right]}{1-\frac{1}{2}} = 6\left(1-\frac{1}{128}\right)$$

$$S_7 = 6\left(\frac{127}{128}\right) = \frac{381}{64}$$

27. –12, 6, –3, ...

$a_1 = -12$, $r = -\frac{1}{2}$, $n = 5$

$$S_5 = \frac{-12\left[1-\left(-\frac{1}{2}\right)^5\right]}{1-\left(-\frac{1}{2}\right)}$$

$$S_5 = -8\left(1+\frac{1}{32}\right) = -8\left(\frac{33}{32}\right)$$

$$S_5 = -\frac{33}{4} \text{ or } -8.25$$

29. $\frac{1}{2}, \frac{1}{4}, 0, \ldots, -\frac{17}{4}$

The first term of this arithmetic sequence is $a_1 = \frac{1}{2}$ and the twentieth term is $a_{20} = -\frac{17}{4}$.

$$S_{20} = \frac{20}{2}(a_1 + a_{20})$$
$$= 10\left(\frac{1}{2} - \frac{17}{4}\right)$$
$$= 10\left(-\frac{15}{4}\right)$$
$$S_{20} = -\frac{75}{2}$$

31. $a_1 = 8,\ r = -\frac{2}{3},\ n = 3$

$$S_3 = \frac{8\left[1-\left(-\frac{2}{3}\right)^3\right]}{1-\left(-\frac{2}{3}\right)} = \frac{24}{5}\left(1+\frac{8}{27}\right)$$

$$S_3 = \frac{24}{5}\left(\frac{35}{27}\right) = \frac{8}{1}\left(\frac{7}{9}\right) = \frac{56}{9}$$

33. The first five terms of the sequence are 4000, 3950, 3900, 3850, and 3800.

$a_1 = 4000,\ d = -50,\ n = 12$

$a_{12} = 4000 + 11(-50)$

$a_{12} = 4000 - 550$

$a_{12} = 3450$ cars sold in month 12.

$S_{12} = \frac{12}{2}[4000 + 3450]$

$S_{12} = 6(7450)$

$S_{12} = 44,700$ cars sold in the first 12 months.

35. Firm A:

The first term of this arithmetic sequence is $a_1 = 22,000$ and the tenth term is $a_{10} = 31,000$.

$$S_{10} = \frac{10}{2}(a_1 + a_{10})$$
$$= 5(22,000 + 31,000)$$
$$= 5(53,000)$$
$$S_{10} = \$265,000$$

Firm B:

The first term of this arithmetic sequence is $a_1 = 20,000$ and the tenth term is $a_{10} = 30,800$.

$$S_{10} = \frac{10}{2}(a_1 + a_{10})$$
$$= \frac{10}{2}(20,000 + 30,800)$$
$$= 5(50,800)$$
$$S_{10} = \$254,000$$

Thus, Firm A is making the more profitable offer.

37. $a_1 = 30,000,\ r = 1.10,\ n = 4$

$a_4 = 30,000(1.10)^{4-1}$

$a_4 = \$39,930$ made during her fourth year of business.

$$S_4 = \frac{30,000(1-1.10^4)}{1-1.10}$$

$S_4 = \$139,230$ made during the first four years.

39. $a_1 = 30,\ r = 0.9,\ n = 5$

$a_5 = 30(0.9)^{5-1} = 19.683$ or approximately 20 minutes to assemble the fifth computer.

$S_5 = \frac{30(1-0.9^5)}{1-0.9} = 122.853$, or approximately 123 minutes to assemble the first 5 computers.

41. $a_1 = 20$ and $r = \frac{4}{5}$

$$S_\infty = \frac{20}{1-\frac{4}{5}} = \frac{20}{\frac{1}{5}} = 100$$

We double the number (to account for the flight up as well as down) and subtract 20 (since the first bounce was preceded by only a downward flight). Thus, the ball traveled a total distance of $2(100) - 20 = 180$ feet.

43. Player A:
The first term of this arithmetic sequence is $a_1 = 1$ and the ninth term is $a_9 = 9$.

$$S_9 = \frac{9}{2}(a_1 + a_9)$$
$$S_9 = \frac{9}{2}(1+9)$$
$$= \frac{9}{2}(10)$$
$$S_{10} = 45 \text{ points}$$

Player B:
The first term of this arithmetic sequence is $a_1 = 10$ and the sixth term is $a_6 = 15$.

$$S_6 = \frac{6}{2}(a_1 + a_6)$$
$$= 3(10+15)$$
$$= 3(25)$$
$$S_6 = 75 \text{ points}$$

45. The first term of this arithmetic sequence is $a_1 = 200$ and the twentieth term is $a_{20} = 105$.

$$S_{20} = \frac{20}{2}(a_1 + a_{20})$$
$$S_9 = 10(200+105)$$
$$= 10(305)$$
$$S_{20} = 3050$$

Thus, \$3050 rent is paid for 20 days during the holiday rush.

47. $a_1 = 0.01,\ r = 2, n = 30$

$$S_{30} = \frac{0.01(1-2^{30})}{1-2} = 10,737,418.23$$

He would pay \$10,737,418.23 in room and board for the 30 days.

49. $0.888\overline{8} = 0.8 + 0.08 + 0.008 + \cdots$
$$= \frac{8}{10} + \frac{8}{100} + \frac{8}{1000} + \cdots$$

This is a geometric series with $a_1 = \frac{8}{10}$ and $r = \frac{1}{10}$

$$S_\infty = \frac{\frac{8}{10}}{1-\frac{1}{10}} = \frac{\frac{8}{10}}{\frac{9}{10}} = \frac{8}{10} \cdot \frac{10}{9} = \frac{8}{9}$$

51. Answers may vary.

53. $6 \cdot 5 \cdot 4 \cdot 3 \cdot 2 \cdot 1 = (30)(12)(2) = 720$

55. $\frac{3 \cdot 2 \cdot 1}{2 \cdot 1} = 3$

57. $(x+5)^2 = x^2 + 2(5x) + 25$
$$= x^2 + 10x + 25$$

59. $(2x-1)^3$
$$= (2x-1)(2x-1)^2$$
$$= (2x-1)(4x^2 - 4x + 1)$$
$$= 8x^3 - 8x^2 + 2x - 4x^2 + 4x - 1$$
$$= 8x^3 - 12x^2 + 6x - 1$$

Exercise Set 12.5

1. $(m+n)^3$
$$= 1 \cdot m^3 + 3 \cdot m^2 n + 3 \cdot mn^2 + 1 \cdot n^3$$
$$= m^3 + 3m^2 n + 3mn^2 + n^3$$

3. $(c+d)^5$
$= 1\cdot c^5 + 5\cdot c^4 d + 10\cdot c^3 d^2 + 10\cdot c^2 d^3 + 5\cdot cd^4 + 1\cdot d^5$
$= c^5 + 5c^4 d + 10c^3 d^2 + 10c^2 d^3 + 5cd^4 + d^5$

5. $(y-x)^5 = [y+(-x)]^5$
$= 1y^5 + 5y^4(-x) + 10y^3(-x)^2 + 10y^2(-x)^3 + 5y(-x)^4 + (-x)^5$
$= y^5 - 5y^4 x + 10y^3 x^2 - 10y^2 x^3 + 5yx^4 - x^5$

7. Answers may vary.

9. $\frac{8!}{7!} = \frac{8\cdot 7!}{7!} = 8$

11. $\frac{7!}{5!} = \frac{7\cdot 6\cdot 5!}{5!} = 7\cdot 6 = 42$

13. $\frac{10!}{7!2!} = \frac{10\cdot 9\cdot 8\cdot 7!}{7!2!} = \frac{10\cdot 9\cdot 8}{2\cdot 1} = \frac{720}{2} = 360$

15. $\frac{8!}{6!0!} = \frac{8\cdot 7\cdot 6!}{6!\cdot 1} = 8\cdot 7 = 56$

17. $(a+b)^7$
$= a^7 + 7a^6 b + \frac{7\cdot 6}{2!}a^5 b^2 + \frac{7\cdot 6\cdot 5}{3!}a^4 b^3 + \frac{7\cdot 6\cdot 5\cdot 4}{4!}a^3 b^4 + \frac{7\cdot 6\cdot 5\cdot 4\cdot 3}{5!}a^2 b^5$
$+ \frac{7\cdot 6\cdot 5\cdot 4\cdot 3\cdot 2}{6!}ab^6 + b^7$
$= a^7 + 7a^6 b + 21a^5 b^2 + 35a^4 b^3 + 35a^3 b^4 + 21a^2 b^5 + 7ab^6 + b^7$

19. $(a+2b)^5$
$= a^5 + 5a^4(2b) + \frac{5\cdot 4}{2!}a^3(2b)^2 + \frac{5\cdot 4\cdot 3}{3!}a^2(2b)^3 + \frac{5\cdot 4\cdot 3\cdot 2}{4!}a(2b)^4 + (2b)^5$
$= a^5 + 10a^4 b + 40a^3 b^2 + 80a^2 b^3 + 80ab^4 + 32b^5$

21. $(q+r)^9$
$= q^9 + 9q^8 r + \frac{9\cdot 8}{2!}q^7 r^2 + \frac{9\cdot 8\cdot 7}{3!}q^6 r^3 + \frac{9\cdot 8\cdot 7\cdot 6}{4!}q^5 r^4 + \frac{9\cdot 8\cdot 7\cdot 6\cdot 5}{5!}q^4 r^5 + \frac{9\cdot 8\cdot 7\cdot 6\cdot 5\cdot 4}{6!}q^3 r^6$
$+ \frac{9\cdot 8\cdot 7\cdot 6\cdot 5\cdot 4\cdot 3}{7!}q^2 r^7 + \frac{9\cdot 8\cdot 7\cdot 6\cdot 5\cdot 4\cdot 3\cdot 2}{8!}qr^8 + r^9$
$= q^9 + 9q^8 r + 36q^7 r^2 + 84q^6 r^3 + 126q^5 r^4 + 126q^4 r^5 + 84q^3 r^6 + 36q^2 r^7 + 9qr^8 + r^9$

23. $(4a+b)^5$
$= (4a)^5 + 5(4a)^4 b + \frac{5\cdot 4}{2!}(4a)^3 b^2 + \frac{5\cdot 4\cdot 3}{3!}(4a)^2 b^3 + \frac{5\cdot 4\cdot 3\cdot 2}{4!}(4a)b^4 + b^5$
$= 1024a^5 + 1280a^4 b + 640a^3 b^2 + 160a^2 b^3 + 20ab^4 + b^5$

25. $(5a-2b)^4$
$= (5a)^4 + 4(5a)^3(-2b) + \frac{4\cdot 3}{2!}(5a)^2(-2b)^2 + \frac{4\cdot 3\cdot 2}{3!}(5a)(-2b)^3 + (-2b)^4$
$= 625a^4 - 1000a^3b + 600a^2b^2 - 160ab^3 + 16b^4$

27. $(2a+3b)^3$
$= (2a)^3 + 3(2a)^2(3b) + \frac{3\cdot 2}{2!}(2a)(3b)^2 + (3b)^3$
$= 8a^3 + 36a^2b + 54ab^2 + 27b^3$

29. $(x+2)^5$
$= x^5 + 5x^4(2) + \frac{5\cdot 4}{2!}x^3(2^2) + \frac{5\cdot 4\cdot 3}{3!}x^2(2^3) + \frac{5\cdot 4\cdot 3\cdot 2}{4!}x(2^4) + 2^5$
$= x^5 + 10x^4 + 40x^3 + 80x^2 + 80x + 32$

31. 5th term of $(c-d)^5$ corresponds to $r=4$: $\frac{5!}{4!(5-4)!}c^{5-4}(-d)^4 = 5cd^4$

33. 8th term of $(2c+d)^7$ corresponds to $r=7$: $\frac{7!}{7!(7-7)!}(2c)^{7-7}d^7 = d^7$

35. 4th term of $(2r-s)^5$ corresponds to $r=3$: $\frac{5!}{3!(5-3)!}(2r)^{5-3}(-s)^3 = -40r^2s^3$

37. 3rd term of $(x+y)^4$ corresponds to $r=2$: $\frac{4!}{2!(4-2)!}x^{4-2}y^2 = 6x^2y^2$

39. 2nd term of $(a+3b)^{10}$ corresponds to $r=1$: $\frac{10!}{1!(10-1)!}a^{10-1}(3b)^1 = 30a^9b$

41. $f(x) = |x|$

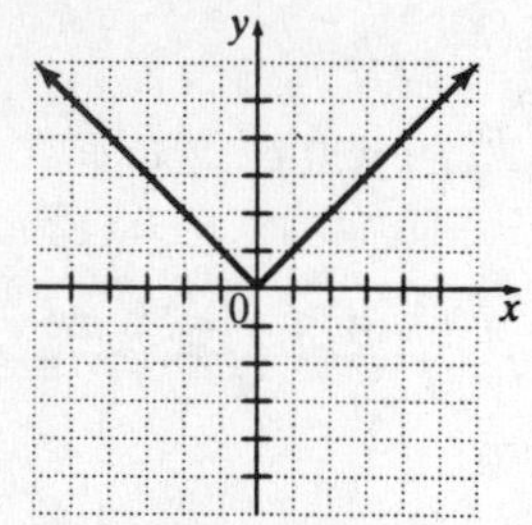

Not one-to-one

43. 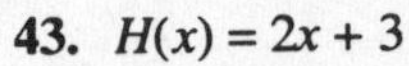
$H(x) = 2x + 3$

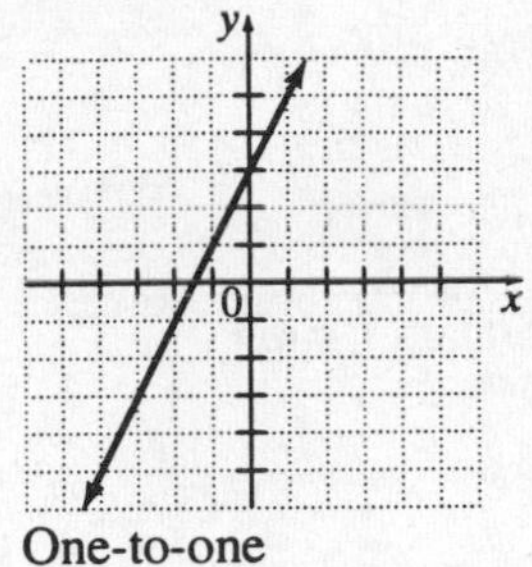

One-to-one

45. $f(x) = x^2 + 3$

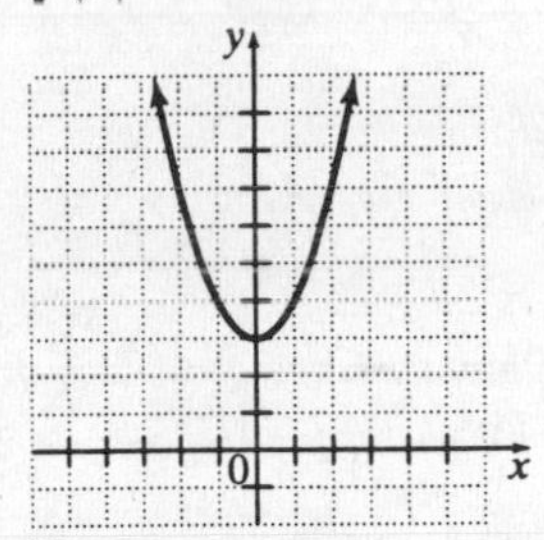

Not one-to-one

Chapter 12 Review

1. $a_n = -3n^2$

$a_1 = -3(1)^2 = -3$

$a_2 = -3(2)^2 = -12$

$a_3 = -3(3)^2 = -27$

$a_4 = -3(4)^2 = -48$

$a_5 = -3(5)^2 = -75$

The first five terms of the sequence are –3, –12, –27, –48, –75.

2. $a_n = n^2 + 2n$

$a_1 = 1^2 + 2(1) = 1 + 2 = 3$

$a_2 = 2^2 + 2(2) = 4 + 4 = 8$

$a_3 = 3^2 + 2(3) = 9 + 6 = 15$

$a_4 = 4^2 + 2(4) = 16 + 8 = 24$

$a_5 = 5^2 + 2(5) = 25 + 10 = 35$

The first five terms of the sequence are 3, 8, 15, 24, 35.

3. $a_n = \dfrac{(-1)^n}{100}$

$a_{100} = \dfrac{(-1)^{100}}{100} = \dfrac{1}{100}$

4. $a_n = \dfrac{2n}{(-1)^2}$

$a_{50} = \dfrac{2(50)}{(-1)^2} = 100$

5. $\dfrac{1}{6}, \dfrac{1}{12}, \dfrac{1}{18}, \ldots$

or $\dfrac{1}{6 \cdot 1}, \dfrac{1}{6 \cdot 2}, \dfrac{1}{6 \cdot 3}, \ldots$

In general, $a_n = \dfrac{1}{6n}$

6. $-1, 4, -9, 16, \ldots$

$a_n = (-1)^n n^2$

7. $a_n = 32n - 16$

$a_5 = 32(5) - 16 = 160 - 16 = 144$ ft

$a_6 = 32(6) - 16 = 192 - 16 = 176$ ft

$a_7 = 32(7) - 16 = 224 - 16 = 208$ ft

8. $a_n = 100(2)^{n-1}$

$10{,}000 = 100(2)^{n-1}$

$100 = 2^{n-1}$

$\log 100 = (n - 1)\log 2$

$n = \dfrac{\log 100}{\log 2} + 1 \approx 7.6$

Eighth day culture will be at least 10,000.

Originally, $a_0 = 100(2)^{-1} = 50$.

9. $a_1 = 450$

$a_2 = 3(450) = 1350$

$a_3 = 3(1350) = 4050$

$a_4 = 3(4050) = 12{,}150$

$a_5 = 3(12{,}150) = 36{,}450$

This predicts that in 2003 (when $n = 5$) the number of infected people should be 36,450.

10. $a_n = 50 + (n-1)8$
$a_1 = 50$
$a_2 = 50 + 8 = 58$
$a_3 = 50 + 2(8) = 66$
$a_4 = 50 + 3(8) = 74$
$a_5 = 50 + 4(8) = 82$
$a_6 = 50 + 5(8) = 90$
$a_7 = 50 + 6(8) = 98$
$a_8 = 50 + 7(8) = 106$
$a_9 = 50 + 8(8) = 114$
$a_{10} = 50 + 9(8) = 122$
There are 122 seats in the tenth row.

11. $a_1 = -2$ and $r = \frac{2}{3}$
The first 5 terms of the sequence are
$-2, -\frac{4}{3}, -\frac{8}{9}, -\frac{16}{27}, -\frac{32}{81}$.

12. $a_n = 12 + (n-1)(-1.5)$
$a_1 = 12$
$a_2 = 12 + 1(-1.5) = 10.5$
$a_3 = 12 + 2(-1.5) = 9$
$a_4 = 12 + 3(-1.5) = 7.5$
$a_5 = 12 + 4(-1.5) = 6$
The first 5 terms of the sequence are 12, 10.5, 9, 7.5, 6.

13. $a_1 = -5$, $d = 4$, and $n = 30$
$a_{30} = 5 + (30-1)4 = -5 + 116 = 111$

14. $a_n = 2 + (n-1)\frac{3}{4}$
$a_{11} = 2 + 10\left(\frac{3}{4}\right) = \frac{38}{4} = \frac{19}{2}$

15. 12, 7, 2, ...
$a_1 = 12$, $d = -5$, and $n = 20$
$a_{20} = 12 + (20-1)(-5) = 12 - 95 = -83$

16. $a_n = a_1 r^{n-1}$
$a_n = 4\left(\frac{3}{2}\right)^{n-1}$
$a_6 = 4\left(\frac{3}{2}\right)^{6-1} = 4\left(\frac{3}{2}\right)^5 = \frac{4(243)}{32} = \frac{243}{8}$

17. $a_4 = 18$ and $a_{20} = 98$
Use the relationship:
$a_4 + 16d = a_{20}$
$18 + 16d = 98$
$16d = 80$
$d = 5$
Now use the relationship:
$a_4 = a_1 + 3d$
$18 = a_1 + 3(5)$
$18 = a_1 + 15$
$a_1 = 3$

18. $-48 = a_3 = a_1 r^{3-1}$
$192 = a_4 = a_1 r^{4-1}$
$-48 = a_1 r^2$
$192 = a_1 r^3$
Dividing we have $r = -4$.
Substitution gives $-48 = 16a_1$ or $a_1 = -3$.

19. $\frac{3}{10}, \frac{3}{100}, \frac{3}{1000}, \ldots$ or
$\frac{3}{10}, \frac{3}{10^2}, \frac{3}{10^3}, \ldots$
In general, $a_n = \frac{3}{10^n}$

20. 50, 58, 66, ...
$a_n = 50 + (n-1)8$ or $a_n = 42 + 8n$

21. $\frac{8}{3}$, 4, 6, ...
Geometric, since $\frac{4}{\frac{8}{3}} = \frac{6}{4} = \frac{3}{2}$
$a_1 = \frac{8}{3}$ and $r = \frac{3}{2}$

22. arithmetic; $a_1 = -10.5$, $d = 4.4$

23. $7x, -14x, 28x$
Geometric, since
$\frac{-14x}{7x} = \frac{28x}{-14x} = -2$
$a_1 = 7x$ and $r = -2$

24. neither

25. $a_1 = 8$ and $r = 0.75$
8, 6, 4.5, 3.4, 2.5, 1.9
Yes, a ball that rebounds to a height of 2.5 feet after the fifth bounce is good, since $2.5 \geq 1.9$.

26. $a_n = 25 + (n-1)(-4)$
$a_n = 25 + 6(-4) = 1$
Continuing the progression as far as possible leaves 1 can in the top row.

27. $a_1 = 1$ and $r = 2$
$a_n = 1 \cdot 2^{n-1}$ or
$a_n = 2^{n-1}$
$a_{10} = 2^{10-1} = 2^9 = \$512$
$a_{30} = 2^{30-1} = 2^{29} = \$536,870,912$

28. $a_n = a_1 r^{n-1}$
$a_5 = 30(0.7)^4 = 7.203$ in.

29. $a_1 = 900$ and $d = 150$
$a_n = 900 + (n-1)150$
$a_6 = 900 + (6-1)150$
$= 900 + 750 = \$1650$/month

30. $\frac{1}{512}, \frac{1}{256}, \frac{1}{128}, \ldots$
first fold $a_1 = \frac{1}{256}$, $r = 2$
$a_n = a_1 r^{n-1}$
$a_{15} = \frac{1}{256}(2)^{15-1} = 64$ inches

31. $\sum_{i=1}^{5}(2i-1)$
$= [2(1)-1] + [2(2)-1] + [2(3)-1] + [2(4)-1] + [2(5)-1]$
$= 1 + 3 + 5 + 7 + 9 = 25$

32. $\sum_{i=1}^{5} i(i+2)$
$= 1(1+2) + 2(2+2) + 3(3+2) + 4(4+2) + 5(5+2)$
$= 3 + 8 + 15 + 24 + 35 = 85$

33. $\sum_{i=2}^{4} \frac{(-1)^i}{2i}$
$= \frac{(-1)^2}{2(2)} + \frac{(-1)^3}{2(3)} + \frac{(-1)^4}{2(4)}$
$= \frac{1}{4} - \frac{1}{6} + \frac{1}{8}$
$= \frac{6-4+3}{24} = \frac{5}{24}$

34. $\sum_{i=3}^{5} 5(-1)^{i-1}$
$= 5(-1)^{3-1} + 5(-1)^{4-1} + 5(-1)^{5-1}$
$= 5(1) + 5(-1) + 5(1) = 5$

35. $a_n = (n-3)(n+2)$
$S_4 = (1-3)(1+2) + (2-3)(2+2) + (3-3)(3+2) + (4-3)(4+2)$
$= -6 - 4 + 0 + 6 = -4$

36. $a_n = n^2$
$S_6 = (1)^2 + (2)^2 + (3)^2 + (4)^2 + (5)^2 + (6)^2$
$= 1 + 4 + 9 + 16 + 25 + 36 = 91$

37. $a_n = -8 + (n-1)3$
$a_1 = -8 + (1-1)3 = -8 + 0 = -8$
$a_2 = -8 + (2-1)3 = -8 + 3 = -5$
$a_3 = -8 + (3-1)3 = -8 + 6 = -2$
$a_4 = -8 + (4-1)3 = -8 + 9 = 1$
$a_5 = -8 + (5-1)3 = -8 + 12 = 4$
So $S_5 = -8 + (-5) + (-2) + 1 + 4 = -10$.

38. $a_n = 5(4)^{n-1}$
$S_3 = 5(4)^0 + 5(4)^1 + 5(4)^2$
$= 5 + 20 + 80 = 105$

39. $1 + 3 + 9 + 27 + 81 + 243$
$= 3^0 + 3^1 + 3^2 + 3^3 + 3^4 + 3^5$
$= \sum_{i=1}^{6} 3^{i-1}$

40. $6+2+(-2)+(-6)+(-10)+(-14)+(-18)$
$a_n = 6+(n-1)(-4)$
$\sum_{i=1}^{7} 6+(i-1)(-4)$

41. $\frac{1}{4}+\frac{1}{16}+\frac{1}{64}+\frac{1}{256}=\frac{1}{4^1}+\frac{1}{4^2}+\frac{1}{4^3}+\frac{1}{4^4}$
$\sum_{i=1}^{4} \frac{1}{4^i}$

42. $1+\left(-\frac{3}{2}\right)+\frac{9}{4}$
$a_n = 1\left(-\frac{3}{2}\right)^{n-1}$
$\sum_{i=1}^{3} \left(-\frac{3}{2}\right)^{i-1}$

43. $a_1 = 20$ and $r = 2$
$a_n = 20(2)^n$ represents the number of yeast, where n represents the number of 8-hr periods. Since $48 = 6(8)$ here, $n = 6$.
$a_6 = 20(2)^6 = 1280$ yeast

44. $a_n = n^2 + 2n - 1$
$a_4 = (4)^2 + 2(4) - 1 = 23$ cranes
$\sum_{i=1}^{4} i^2 + 2i - 1$
$= (1 + 2 - 1) + (4 + 4 - 1) + (9 + 6 - 1) + (16 + 8 - 1)$
$= 2 + 7 + 14 + 23 = 46$ cranes

45. For Job A: $a_1 = 39{,}500$ and $d = 2200$;
$a_5 = 39{,}500 + (5-1)2200 = \$48{,}300$
For Job B: $a_1 = 41{,}000$ and $d = 1400$;
$a_5 = 41{,}000 + (5-1)1400 = \$46{,}600$
For the fifth year, Job A has a higher salary.

46. $a_n = 200(0.5)^n$
$a_3 = 200(0.5)^3 = 25$ kg
$\sum_{i=1}^{3} 200(0.5)^i$
$= 200(0.5) + 200(0.5)^2 + 200(0.5)^3$
$= 100 + 50 + 25 = 175$ kg

47. 15, 19, 23, ...
$a_1 = 15$ and $d = 4$
$S_6 = \frac{6}{2}[2(15) + (6-1)4]$
$= 3[30 + 20] = 3(50) = 150$

48. 5, −10, 20, ⋯
$a_1 = 5$, $r = -2$
$S_n = \frac{a_1(1-r^n)}{1-r}$
$S_9 = \frac{5(1-(-2)^9)}{1-(-2)} = 855$

49. $a_1 = 1$, $d = 2$, and $n = 30$
$S_{30} = \frac{30}{2}[2(1) + (30-1)2]$
$= 15[2 + 58] = 15(60) = 900$

50. 7, 14, 21, 28, ...
$a_n = 7 + (n-1)7$
$a_{20} = 7 + (20-1)7 = 140$
$S_n = \frac{n}{2}(a_1 + a_n)$
$S_{20} = \frac{20}{2}(7 + 140) = 1470$

51. 8, 5, 2, ...
$a_1 = 8$, $d = -3$, and $n = 20$
$S_{20} = \frac{20}{2}[2(8) + (20-1)(-3)]$
$= 10[16 - 57] = 10(-41) = -410$

52. $\frac{3}{4}, \frac{9}{4}, \frac{27}{4}, \ldots$

$a_1 = \frac{3}{4}, r = 3$

$S_n = \frac{a_1(1-r^n)}{1-r}$

$S_8 = \frac{\frac{3}{4}(1-3^8)}{1-3} = 2460$

53. $a_1 = 6$ and $r = 5$

$S_4 = \frac{6(1-5^4)}{1-5} = \frac{-3}{2}(1-625)$

$= \frac{3}{2}(624) = 936$

54. $a_1 = -3, d = -6$

$a_n = -3 + (n-1)(-6)$

$a_{100} = -3 + (100-1)(-6) = -597$

$S_n = \frac{n}{2}(a_1 + a_n)$

$S_{100} = \frac{100}{2}(-3 + (-597))$

$S_{100} = -30,000$

55. $5, \frac{5}{2}, \frac{5}{4}, \ldots$

$a_1 = 5$ and $r = \frac{1}{2}$

$S_\infty = \frac{5}{1-\frac{1}{2}} = \frac{5}{\frac{1}{2}} = 10$

56. $18, -2, \frac{2}{9}, \ldots$

$a_1 = 18, r = -\frac{1}{9}$

$S_\infty = \frac{a_1}{1-r} = \frac{18}{1+\frac{1}{9}} = \frac{81}{5}$

57. $-20, -4, -\frac{4}{5}, \ldots$

$a_1 = -20$ and $r = \frac{1}{5}$

$S_\infty = \frac{-20}{1-\frac{1}{5}} = \frac{-20}{\frac{4}{5}} = -25$

58. $0.2, 0.02, 0.002, \ldots$

$a_1 = 0.2, r = \frac{1}{10}$

$S_\infty = \frac{a_1}{1-r} = \frac{0.2}{1-\frac{1}{10}} = \frac{2}{9}$

59. $a_1 = 20,000$, $r = 1.15$, and $n = 4$.

$a_4 = 20,000(1.15)^{4-1} = \$30,418$ earned in his fourth year.

$S_4 = \frac{20,000(1-1.15^4)}{1-1.15} = \$99,868$ earned in his first four years.

60. $a_n = 40(0.8)^{n-1}$

$a_4 = 40(0.8)^{4-1} \approx 20$ min

$S_n = \frac{a_1(1-r^n)}{1-r} = \frac{40(1-0.8^4)}{1-0.8} = 118$ min

61. $a_1 = 100$, $d = -7$, and $n = 7$

$a_7 = 100 + (7-1)(-7) = 100 - 42$

$= \$58$ rent paid for the seventh day.

$S_7 = \frac{7}{2}[100 + 58]$

$= \frac{7}{2}(158)$

$= \$553$ rent paid for the first seven days.

62. $a_1 = 15, r = 0.8$

$S_\infty = \frac{a_1}{1-r} = \frac{15}{1-0.8} = 75$ feet downward

$a_1 = 12, r = 0.8$

$S_\infty = \frac{a_1}{1-r} = \frac{12}{1-0.8} = 60$ feet upward

The total is 135 feet.

63. 1800, 600, 200, ...

$a_1 = 1800,\ r = \frac{1}{3},$ and $n = 6$

$$S_6 = 1800\frac{\left[1-\left(\frac{1}{3}\right)^6\right]}{1-\frac{1}{3}} = 2700\left(1-\frac{1}{729}\right) = 2700\left(\frac{728}{729}\right) = \frac{72800}{27}$$

≈ 2696 mosquitoes killed during the first six days after the spraying

64. 1800, 600, 200, ...

For which n is $a_n > 1$?

$a_n = a_1 r^{n-1} = 1800\left(\frac{1}{3}\right)^{n-1} > 1$

$(n-1)\log\left(\frac{1}{3}\right) > \log\frac{1}{1800}$

$(n-1)(-0.4771213) > (-3.2552725)$

$n - 1 < 6.8$

$n < 7.8$

No longer effective on the 8th day. About 2700 mosquitos were killed.

65. $0.55\overline{5} = 0.5 + 0.05 + 0.005 + \cdots$

$a_1 = 0.5$ and $r = 0.1$

$S_\infty = \frac{0.5}{1-0.1} = \frac{0.5}{0.9} = \frac{5}{9}$

$0.55\overline{5} = \frac{5}{9}$

66. 27, 30, 33, $\cdots$

$a_n = 27 + (n-1)(3)$

$a_{20} = 27 + (20-1)(3) = 84$

$S_n = \frac{n}{2}(a_1 + a_n)$

$S_{20} = \frac{20}{2}(27 + 84) = 1110$

There are 1110 seats.

67. $(x+z)^5 = 1\cdot x^5 + 5\cdot x^4 z + 10\cdot x^3 z^2 + 10\cdot x^2 z^3 + 5\cdot xz^4 + 1\cdot z^5$
$= x^5 + 5x^4 z + 10x^3 z^2 + 10x^2 z^3 + 5xz^4 + z^5$

68. $(y-r)^6 = y^6 - 6y^5 r + 15y^4 r^2 - 20y^3 r^3 + 15y^2 r^4 - 6yr^5 + r^6$

69. $(2x+y)^4 = 1\cdot(2x)^4 + 4\cdot(2x)^3 y + 6\cdot(2x)^2 y^2 + 4\cdot(2x)y^3 + 1\cdot y^4$
$= 16x^4 + 32x^3 y + 24x^2 y^2 + 8xy^3 + y^4$

70. $(3y-z)^4 = (3y)^4 + 4(3y)^3(-z) + 6(3y)^2(-z)^2 + 4(3y)(-z)^3 + (-z)^4$
$= 81y^4 - 108y^3 z + 54y^2 z^2 - 12yz^3 + z^4$

71. $(b+c)^8$

$$= b^8 + 8b^7c + \frac{8\cdot7}{2!}b^6c^2 + \frac{8\cdot7\cdot6}{3!}b^5c^3 + \frac{8\cdot7\cdot6\cdot5}{4!}b^4c^4 + \frac{8\cdot7\cdot6\cdot5\cdot4}{5!}b^3c^5 + \frac{8\cdot7\cdot6\cdot5\cdot4\cdot3}{6!}b^2c^6 + \frac{8\cdot7\cdot6\cdot5\cdot4\cdot3\cdot2}{7!}bc^7 + c^8$$

$$= b^8 + 8b^7c + 28b^6c^2 + 56b^5c^3 + 70b^4c^4 + 56b^3c^5 + 28b^2c^6 + 8bc^7 + c^8$$

72. $(x-w)^7 = x^7 - 7x^6w + \frac{7\cdot6}{2!}x^5w^2 - \frac{7\cdot6\cdot5}{3!}x^4w^3 + \frac{7\cdot6\cdot5\cdot4}{4!}x^3w^4 - \frac{7\cdot6\cdot5\cdot4\cdot3}{5!}x^2w^5 + \frac{7\cdot6\cdot5\cdot4\cdot3\cdot2}{6!}xw^6 - w^7$

$$= x^7 - 7x^6w + 21x^5w^2 - 35x^4w^3 + 35x^3w^4 - 21x^2w^5 + 7xw^6 - w^7$$

73. $(4m-n)^4 = [4m+(-n)]^4$

$$= (4m)^4 + 4(4m)^3(-n) + \frac{4\cdot3}{2!}(4m)^2(-n)^2 + \frac{4\cdot3\cdot2}{3!}(4m)(-n)^3 + (-n)^4$$

$$= 256m^4 - 256m^3n + 96m^2n^2 - 16mn^3 + n^4$$

74. $(p-2r)^5 = p^5 + 5p^4(-2r) + \frac{5\cdot4}{2!}p^3(-2r)^2 + \frac{5\cdot4\cdot3}{3!}p^2(-2r)^3 + \frac{5\cdot4\cdot3\cdot2}{4!}p(-2r)^4 + (-2r)^5$

$$= p^5 - 10p^4r + 40p^3r^2 - 80p^2r^3 + 80pr^4 - 32r^5$$

75. 4th term of $(a+b)^7$ corresponds to $r = 3$.

$$\frac{7!}{3!(7-3)!}a^{7-3}b^3 = 35a^4b^3$$

76. The 11th term is $\frac{n!}{r!(n-r)!}a^{n-r}b^r$ where $n = 10, r = 10, a = y, b = 2z$.

$$\frac{10!}{10!0!}y^{10-10}(2z)^{10} = 1024z^{10}$$

Chapter 12 Test

1. $a_n = \frac{(-1)^n}{n+4}$

$$a_1 = \frac{(-1)^1}{1+4} = -\frac{1}{5}$$

$$a_2 = \frac{(-1)^2}{2+4} = \frac{1}{6}$$

$$a_3 = \frac{(-1)^3}{3+4} = -\frac{1}{7}$$

$$a_4 = \frac{(-1)^4}{4+4} = \frac{1}{8}$$

$$a_5 = \frac{(-1)^5}{5+4} = -\frac{1}{9}$$

or $-\frac{1}{5}, \frac{1}{6}, -\frac{1}{7}, \frac{1}{8}, -\frac{1}{9}$

2. $a_n = \frac{3}{(-1)^n}$

$a_1 = \frac{3}{(-1)^1} = -3$

$a_2 = \frac{3}{(-1)^2} = 3$

$a_3 = \frac{3}{(-1)^3} = -3$

$a_4 = \frac{3}{(-1)^4} = 3$

$a_5 = \frac{3}{(-1)^5} = -3$

or $-3, 3, -3, 3, -3$

3. $a_n = 10 + 3(n-1)$

$a_{80} = 10 + 3(80-1) = 10 + 237 = 247$

4. $a_n = (n+1)(n-1)(-1)^n$

$a_{200} = (200+1)(200-1)(-1)^{200}$

$= 200^2 - 1^2 = 40000 - 1 = 39,999$

5. $\frac{2}{5}, \frac{2}{25}, \frac{2}{125}, \cdots$ or $\frac{2}{5}, \frac{2}{5^2}, \frac{2}{5^3}, \cdots$

In general, $a_n = \frac{2}{5}\left(\frac{1}{5}\right)^{n-1}$

6. $-9, 18, -27, 36, \cdots$

or $(-1)^1 9 \cdot 1, (-1)^2 9 \cdot 2,$

$(-1)^3 9 \cdot 3, (-1)^4 9 \cdot 4, \cdots$

In general, $a_n = (-1)^n 9n$

7. $a_n = 5(2)^{n-1}$

Geometric sequence with $a_1 = 5$ and $r = 2$.

$S_5 = \frac{5(1-2^5)}{1-2} = -5(1-32) = 5(31) = 155$

8. $a_n = 18 + (n-1)(-2)$

Arithmetic sequence with $a_1 = 18$ and $d = -2$.

$S_{30} = \frac{30}{2}[2(18) + (30-1)(-2)]$

$= 15[36 - 58] = 15(-22) = -330$

9. $a_1 = 24$ and $r = \frac{1}{6}$

$S_\infty = \frac{24}{1-\frac{1}{6}} = \frac{24}{\frac{5}{6}} = \frac{144}{5}$

10. $\frac{3}{2}, -\frac{3}{4}, \frac{3}{8}, \cdots$

$a_1 = \frac{3}{2}$ and $r = -\frac{1}{2}$

$S_\infty = \frac{\frac{3}{2}}{1-\left(-\frac{1}{2}\right)} = \frac{\frac{3}{2}}{\frac{3}{2}} = 1$

11. $\sum_{i=1}^{4} i(i-2)$

$= 1(1-2) + 2(2-2) + 3(3-2) + 4(4-2)$

$= -1 + 0 + 3 + 8 = 10$

12. $\sum_{i=2}^{4} 5(2)^i(-1)^{i-1}$

$= 5(2)^2(-1)^{2-1} + 5(2)^3(-1)^{3-1} + 5(2)^4(-1)^{4-1} = -20 + 40 - 80 = -60$

13. $(a-b)^6 = (a+(-b))^6$

$= 1 \cdot a^6 + 6 \cdot a^5(-b) + 15 \cdot a^4(-b)^2 + 20a^3(-b)^3 + 15 \cdot a^2(-b)^4 + 6 \cdot a(-b)^5 + 1 \cdot (-b)^6$

$= a^6 - 6a^5b + 15a^4b^2 - 20a^3b^3 + 15a^2b^4 - 6ab^5 + b^6$

14. $(2x+y)^5 = 1 \cdot (2x)^5 + 5 \cdot (2x)^4 y + 10 \cdot (2x)^3 y^2 + 10 \cdot (2x)^2 y^3 + 5 \cdot (2x) y^4 + y^5$

$= 32x^5 + 80x^4y + 80x^3y^2 + 40x^2y^3 + 10xy^4 + y^5$

15. $(y+z)^8 = y^8 + 8y^7z + \frac{8 \cdot 7}{2!} y^6z^2 + \frac{8 \cdot 7 \cdot 6}{3!} y^5z^3 + \frac{8 \cdot 7 \cdot 6 \cdot 5}{4!} y^4z^4 + \frac{8 \cdot 7 \cdot 6 \cdot 5 \cdot 4}{5!} y^3z^5$

$+ \frac{8 \cdot 7 \cdot 6 \cdot 5 \cdot 4 \cdot 3}{6!} y^2z^6 + \frac{8 \cdot 7 \cdot 6 \cdot 5 \cdot 4 \cdot 3 \cdot 2}{7!} yz^7 + z^8$

$= y^8 + 8y^7z + 28y^6z^2 + 56y^5z^3 + 70y^4z^4 + 56y^3z^5 + 28y^2z^6 + 8yz^7 + z^8$

16. $(2p+r)^7 = (2p)^7 + 7(2p)^6r + \frac{7 \cdot 6}{2!}(2p)^5r^2 + \frac{7 \cdot 6 \cdot 5}{3!}(2p)^4r^3 + \frac{7 \cdot 6 \cdot 5 \cdot 4}{4!}(2p)^3r^4$

$+ \frac{7 \cdot 6 \cdot 5 \cdot 4 \cdot 3}{5!}(2p)^2r^5 + \frac{7 \cdot 6 \cdot 5 \cdot 4 \cdot 3 \cdot 2}{6!}(2p)r^6 + r^7$

$= 128p^7 + 448p^6r + 672p^5r^2 + 560p^4r^3 + 280p^3r^4 + 84p^2r^5 + 14pr^6 + r^7$

17. $a_n = 250 + 75(n-1)$

$a_{10} = 250 + 75(10-1) = 250 + 675 = 925$

There were 925 people in the town at the beginning of the tenth year.

$a_1 = 250 + 75(1-1) = 250$

There were 250 people in the town at the beginning of the first year.

18. 1, 3, 5, ...

$a_1 = 1$, $d = 2$, and $n = 8$

$a_8 = 1 + (8-1)2 = 1 + 14 = 15$

We want $1 + 3 + 5 + \ldots + 15$

$S_8 = \frac{8}{2}[1+15] = 4(16) = 64$

There were 64 shrubs planted in the 8 rows.

19. $a_1 = 80,\ r = \frac{3}{4}$, and $n = 4$

$a_4 = 80\left(\frac{3}{4}\right)^{4-1} = 80\left(\frac{27}{64}\right) = \frac{135}{4}$ or 33.75

The arc length is 33.75 cm on the 4th swing.

$$S_4 = \frac{80\left(1-\left(\frac{3}{4}\right)^4\right)}{1-\frac{3}{4}} = 320\left(1-\frac{81}{256}\right) = 320\left(\frac{175}{256}\right) = \frac{875}{4} = 218.75$$

The total of the arc lengths is 218.75 cm for the first 4 swings.

20. $a_1 = 80$ and $r = \frac{3}{4}$

$$S_\infty = \frac{80}{1-\frac{3}{4}} = \frac{80}{\frac{1}{4}} = 320$$

The total of the arc lengths is 320 cm before the pendulum comes to rest.

21. 16, 48, 80, ...

$a_1 = 16,\ d = 32$, and $n = 10$

$a_{10} = 16 + (10-1)32 = 16 + 288 = 304$

He falls 304 feet during the 10th second.

$S_{10} = \frac{10}{2}[16 + 304] = 5(320) = 1600$

He falls 1600 feet during the first 10 seconds.

22. $0.42\overline{42} = 0.42 + 0.0042 + 0.000042 + \cdots$

$a_1 = 0.42$ and $r = 0.01$

$$S_\infty = \frac{0.42}{1-0.01} = \frac{0.42}{0.99} = \frac{42}{99} = \frac{14}{33}$$

Thus, $0.42\overline{42} = \frac{14}{33}$.

Chapter 13

Section 13.1

Graphing Calculator Explorations

1. $x^2 + y^2 = 55$

$$y^2 = 55 - x^2$$

$$y = \pm\sqrt{55 - x^2}$$

10

−15 15

−10

3. $5x^2 + 5y^2 = 50$

$$5y^2 = 50 - 5x^2$$

$$y^2 = 10 - x^2$$

$$y = \pm\sqrt{10 - x^2}$$

10

−15 15

−10

5. $2x^2 + 2y^2 - 34 = 0$

$$2y^2 = 34 - 2x^2$$

$$y^2 = 17 - x^2$$

$$y = \pm\sqrt{17 - x^2}$$

10

−15 15

−10

7. $7x^2 + 7y^2 - 89 = 0$

$$7y^2 = 89 - 7x^2$$

$$y^2 = \frac{89 - 7x^2}{7}$$

$$y = \pm\sqrt{\frac{89 - 7x^2}{7}}$$

10

−15 15

−10

Mental Math

1. $y = x^2 - 7x + 5$; upward

2. $y = -x^2 + 16$; downward

3. $x = -y^2 - y + 2$; to the left

4. $x = 3y^2 + 2y - 5$; to the right

5. $y = -x^2 + 2x + 1$; downward

6. $x = -y^2 + 2y - 6$; to the left

Exercise Set 13.1

1. $x = 3y^2$

$x = 3(y - 0)^2$

The vertex is (0, 0).

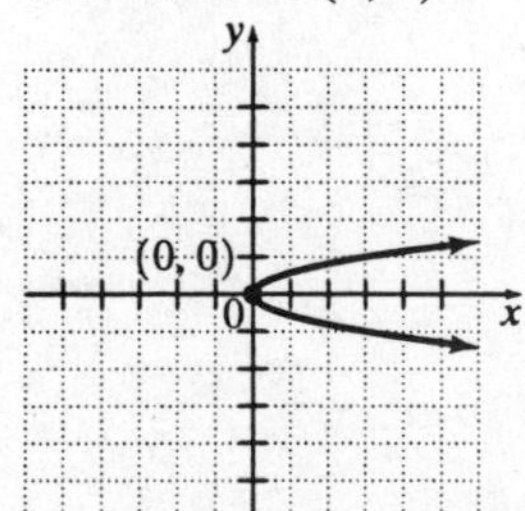

3. $x = (y-2)^2 + 3$

The vertex is (3, 2).

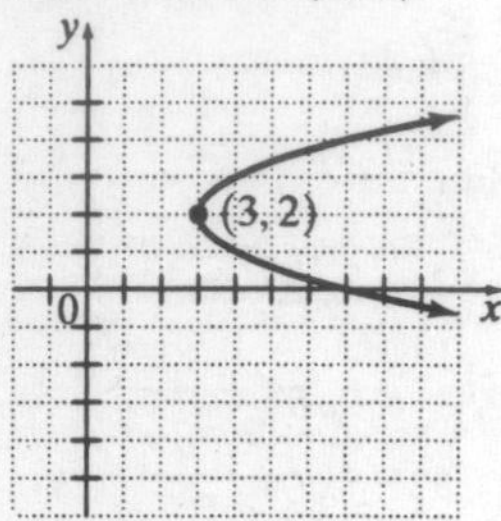

5. $y = 3(x-1)^2 + 5$

The vertex is (1, 5).

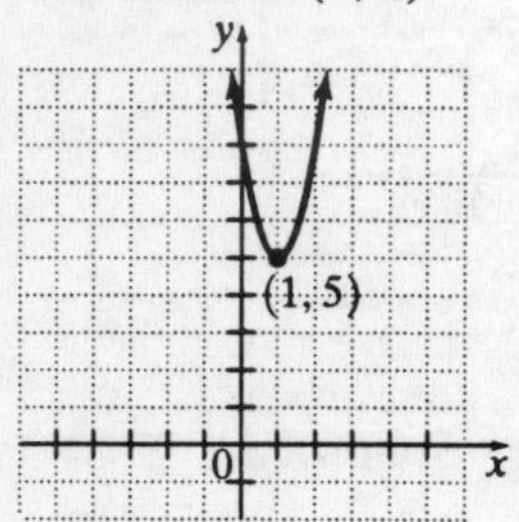

7. $x = y^2 + 6y + 8$

$x = y^2 + 6y + 9 + 8 - 9$

$x = (y+3)^2 - 1$

The vertex is (–1, –3).

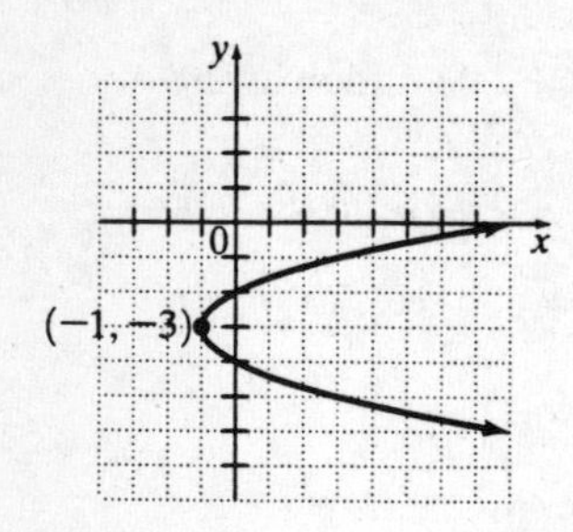

9. $y = x^2 + 10x + 20$

$y = x^2 + 10x + 25 + 20 - 25$

$y = (x+5)^2 - 5$

The vertex is (–5, –5).

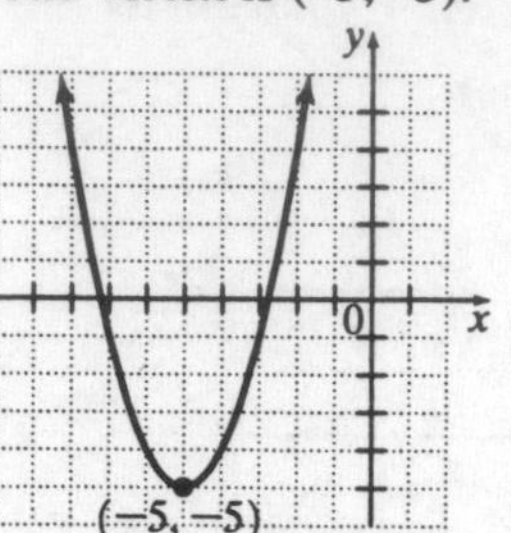

11. $x = -2y^2 + 4y + 6$

$x = -2(y^2 - 2y) + 6$

$x = -2(y^2 - 2y + 1) + 6 + 2$

$x = -2(y-1)^2 + 8$

Thus, the vertex is (8, 1).

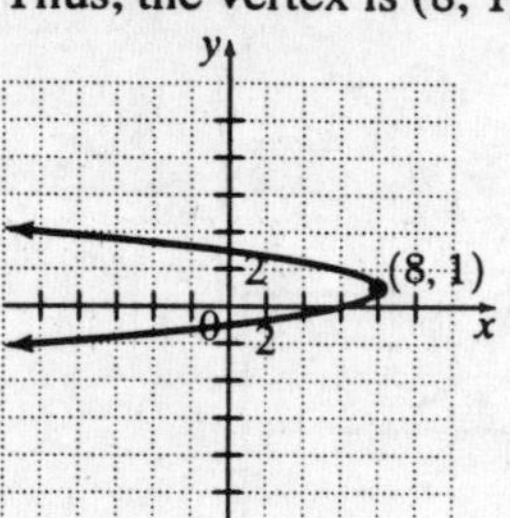

13. (5, 1), (8, 5)

$d = \sqrt{(8-5)^2 + (5-1)^2}$

$d = \sqrt{9+16}$

$d = \sqrt{25}$

$d = 5$ units

15. (–3, 2), (1, –3)

$d = \sqrt{[1-(-3)]^2 + (-3-2)^2}$

$d = \sqrt{16+25}$

$d = \sqrt{41}$ units

17. $(-9, 4), (-8, 1)$

$d = \sqrt{[-8-(-9)]^2 + (1-4)^2}$

$d = \sqrt{(-8+9)^2 + (-3)^2}$

$d = \sqrt{1^2 + 9}$

$d = \sqrt{10}$ units

19. $(0, -\sqrt{2}), (\sqrt{3}, 0)$

$d = \sqrt{(\sqrt{3}-0)^2 + [0-(-\sqrt{2})]^2}$

$d = \sqrt{(\sqrt{3})^2 + (\sqrt{2})^2}$

$d = \sqrt{3+2}$

$d = \sqrt{5}$ units

21. $(1.7, -3.6), (-8.6, 5.7)$

$d = \sqrt{(-8.6-1.7)^2 + [5.7-(-3.6)]^2}$

$d = \sqrt{(-10.3)^2 + (9.3)^2}$

$d = \sqrt{192.58}$

$d = 13.88$ units

23. $(2\sqrt{3}, \sqrt{6}), (-\sqrt{3}, 4\sqrt{6})$

$d = \sqrt{(-\sqrt{3}) - 2\sqrt{3})^2 + (4\sqrt{6} - \sqrt{6})^2}$

$d = \sqrt{(-3\sqrt{3})^2 + (3\sqrt{6})^2}$

$d = \sqrt{27+54}$

$d = \sqrt{81}$

$d = 9$ units

25. $(6, -8), (2, 4)$

$\left(\frac{6+2}{2}, \frac{-8+4}{2}\right)$

The midpoint of the segment is $(4, -2)$.

27. $(-2, -1), (-8, 6)$

$\left(\frac{-2+(-8)}{2}, \frac{-1+6}{2}\right)$

The midpoint of the segment is $\left(-5, \frac{5}{2}\right)$.

29. $(7, 3), (-1, -3)$

$\left(\frac{7+(-1)}{2}, \frac{3+(-3)}{2}\right)$

The midpoint of the segment is $(3, 0)$.

31. $\left(\frac{1}{2}, \frac{3}{8}\right), \left(-\frac{3}{2}, \frac{5}{8}\right)$

$\left(\frac{\frac{1}{2}+\left(-\frac{3}{2}\right)}{2}, \frac{\frac{3}{8}+\frac{5}{8}}{2}\right)$

The midpoint of the segment is $\left(-\frac{1}{2}, \frac{1}{2}\right)$.

33. $(\sqrt{2}, 3\sqrt{5}), (\sqrt{2}, -2\sqrt{5})$

$\left(\frac{\sqrt{2}+\sqrt{2}}{2}, \frac{3\sqrt{5}-2\sqrt{5}}{2}\right)$

The midpoint of the segment is $\left(\sqrt{2}, \frac{\sqrt{5}}{2}\right)$.

35. $(4.6, -3.5), (7.8, -9.8)$

$\left(\frac{4.6+7.8}{2}, \frac{-3.5+(-9.8)}{2}\right)$

The midpoint of the segment is $(6.2, -6.65)$.

37. $x^2 + y^2 = 9$

$(x-0)^2 + (y-0)^2 = 3^2$

The center is $(0, 0)$ and the radius is 3.

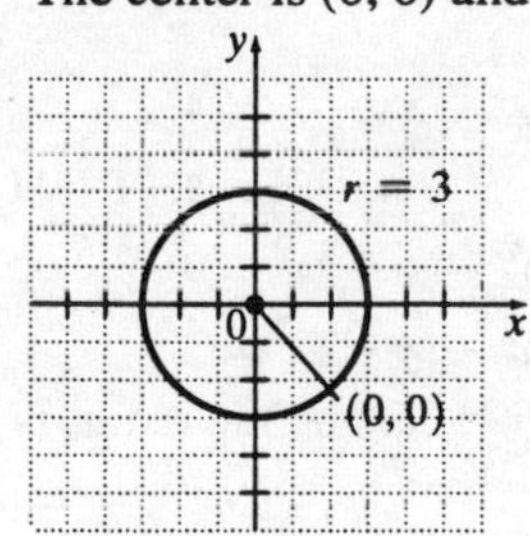

39. $x^2 + (y-2)^2 = 1$

$(x-0)^2 + (y-2)^2 = 1^2$

The center is $(0, 2)$ and the radius is 1.

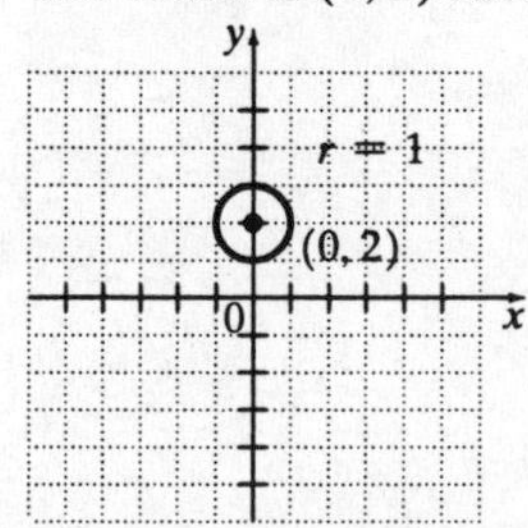

41. $(x-5)^2+(y+2)^2=1$
$(x-5)^2+(y+2)^2=1^2$
The center is (5, –2) and the radius is 1.

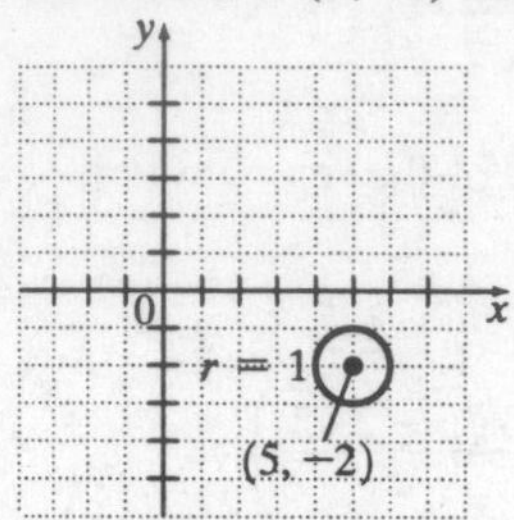

43. $x^2+y^2+6y=0$
$x^2+y^2+6y+9=9$
$(x-0)^2+(y+3)^2=3^2$
The center is (0, –3) and the radius is 3.

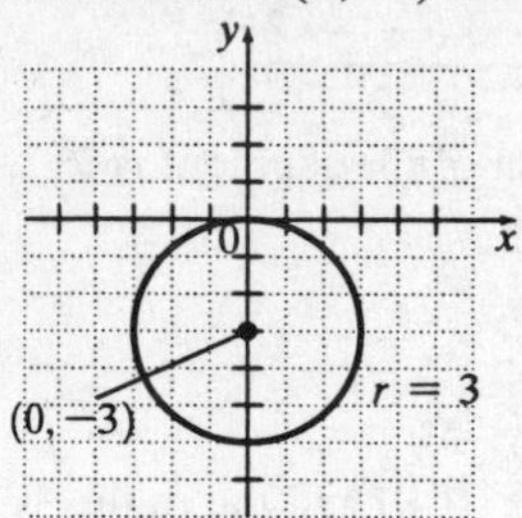

45. $x^2+y^2+2x-4y=4$
$x^2+2x+1+y^2-4y+4=4+1+4$
$(x+1)^2+(y-2)^2=9$
$(x+1)^2+(y-2)^2=3^2$
The center is (–1, 2) and the radius is 3.

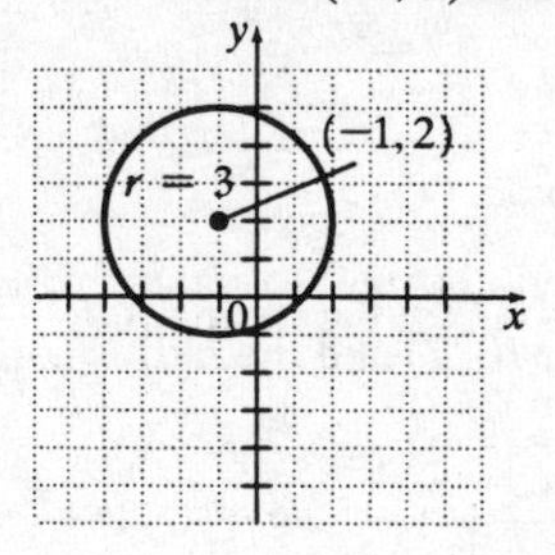

47. $x^2+y^2-4x-8y-2=0$
$(x^2-4x+4)+(y^2-8y+16)=2+4+16$
$(x-2)^2+(y-4)^2=22$
The center is (2, 4) and the radius is $\sqrt{22}$.

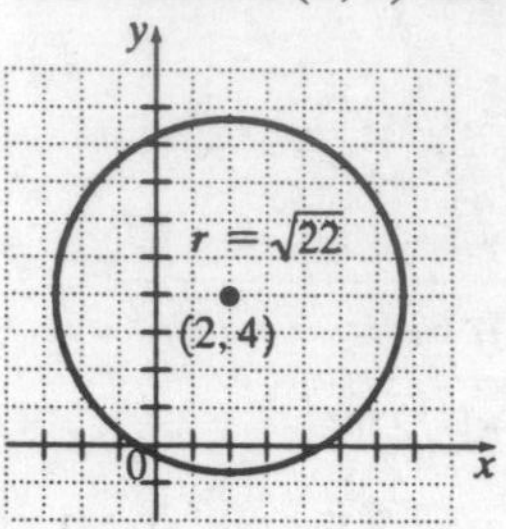

49. The center is (2, 3); the radius is 6.
$(x-2)^2+(y-3)^2=6^2$
$(x-2)^2+(y-3)^2=36$

51. The center is (0, 0); the radius is $\sqrt{3}$.
$(x-0)^2+(y-0)^2=(\sqrt{3})^2$
$x^2+y^2=3$

53. The center is (–5, 4); the radius is $3\sqrt{5}$.
$[x-(-5)]^2+(y-4)^2=(3\sqrt{5})^2$
$(x+5)^2+(y-4)^2=45$

55. Answers may vary.

57. $x=y^2-3$
The vertex is (–3, 0).

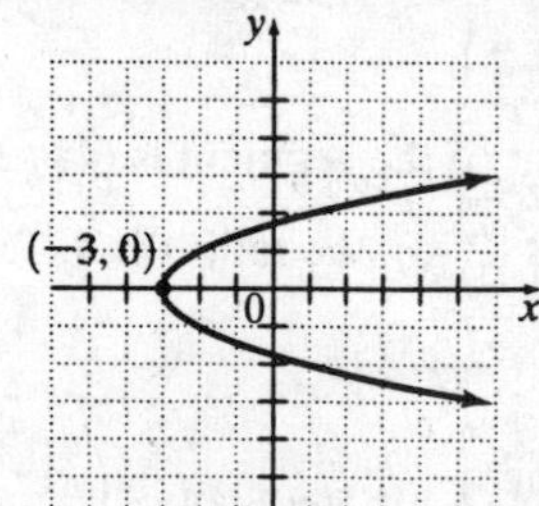

59. $y=(x-2)^2-2$
The vertex is (2, –2).

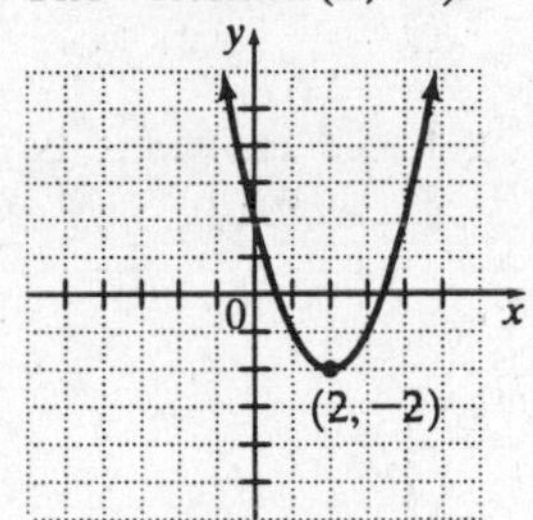

61. $x^2+y^2=1$
The vertex is (0, 0); the radius is 1.

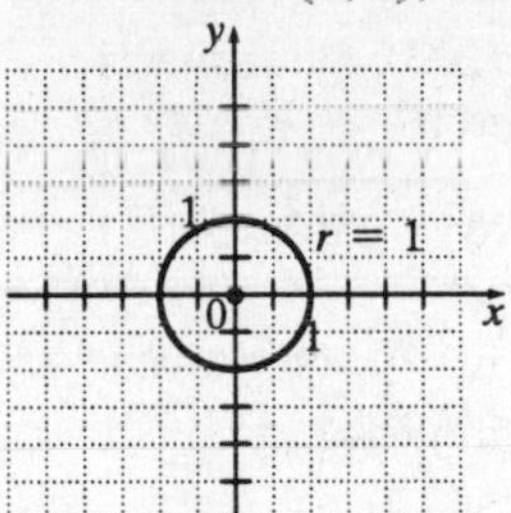

63. $x=(y+3)^2-1$
The vertex is (–1, –3).

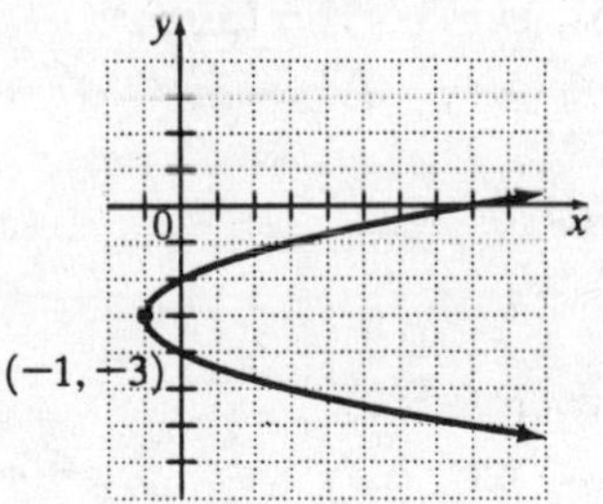

65. $(x-2)^2+(y-2)^2=16$
The center is (2, 2); the radius is 4.

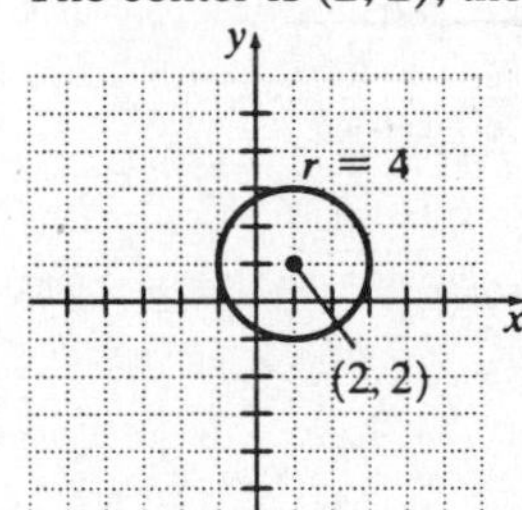

67. $x=-(y-1)^2$
The vertex is (0, 1).

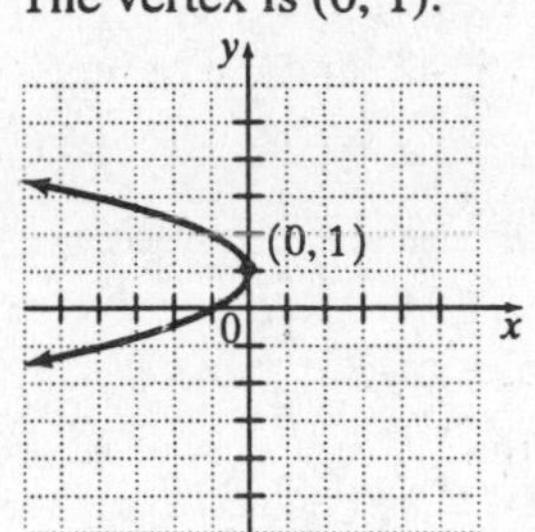

69. $(x-4)^2+y^2=7$
The center is (4, 0); the radius is $\sqrt{7}$.

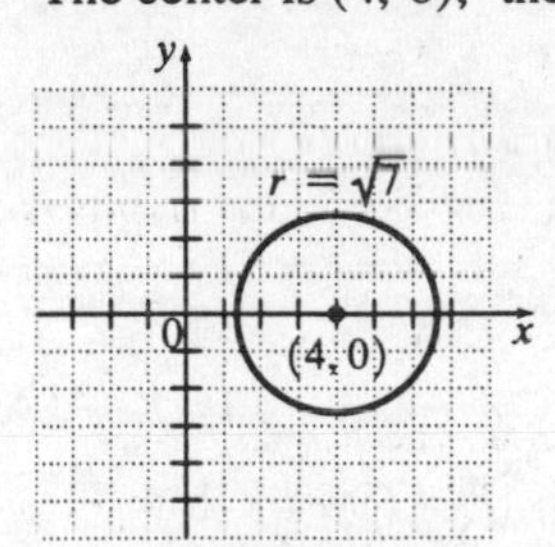

71. $y=5(x+5)^2+3$
The vertex is (–5, 3).

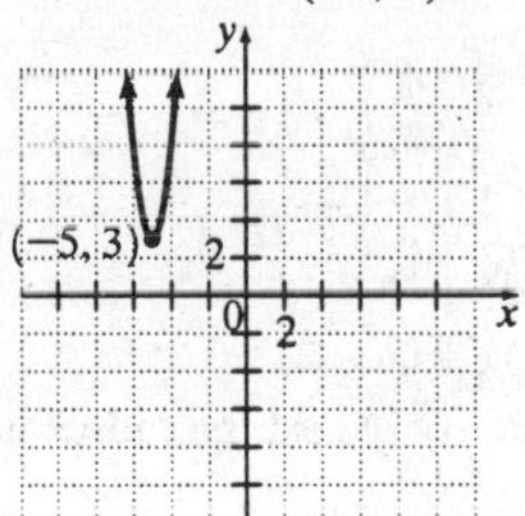

73. $\frac{x^2}{8}+\frac{y^2}{8}=2$
The center is (0, 0); the radius is 4.

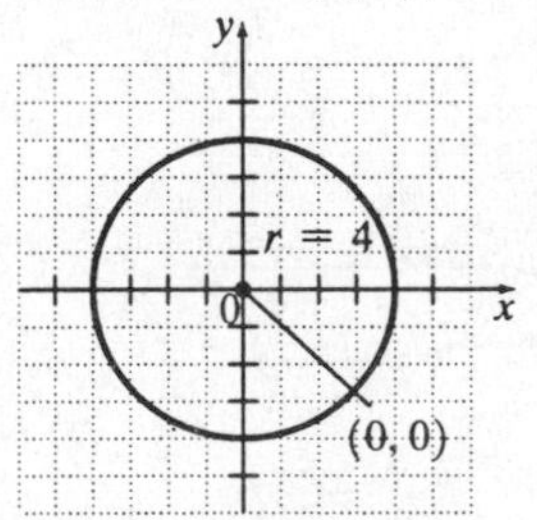

75. $y = x^2 + 7x + 6$

$y = x^2 + 7x + \frac{49}{4} + 6 - \frac{49}{4}$

$y = \left(x + \frac{7}{2}\right)^2 - \frac{25}{4}$

The vertex is $\left(-\frac{7}{2}, -\frac{25}{4}\right)$.

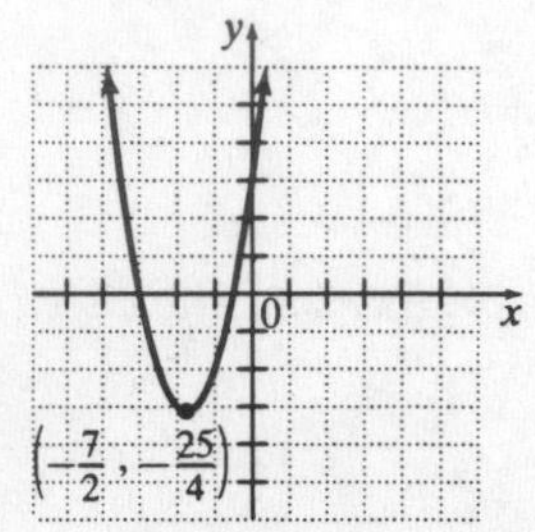

77. $x^2 + y^2 + 2x + 12y - 12 = 0$

$x^2 + 2x + 1 + y^2 + 12y + 36 = 12 + 1 + 36$

$(x+1)^2 + (y+6)^2 = 49$

The center is (–1, –6); the radius is 7.

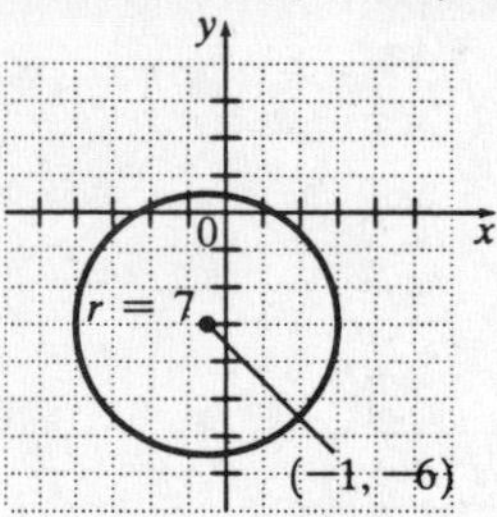

79. $x = y^2 + 8y - 4$

$x = y^2 + 8y + 16 - 4 - 16$

$x = (y+4)^2 - 20$

The vertex is (–20, –4).

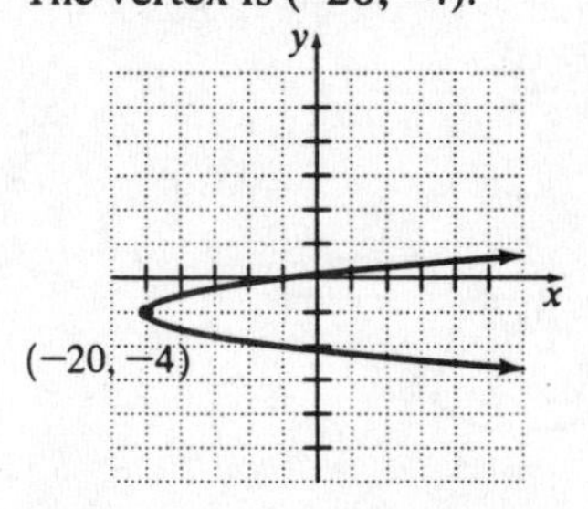

81. $x^2 - 10y + y^2 + 4 = 0$

$x^2 + y^2 - 10y + 25 = -4 + 25$

$x^2 + (y-5)^2 = 21$

The center is (0, 5); the radius is $\sqrt{21}$.

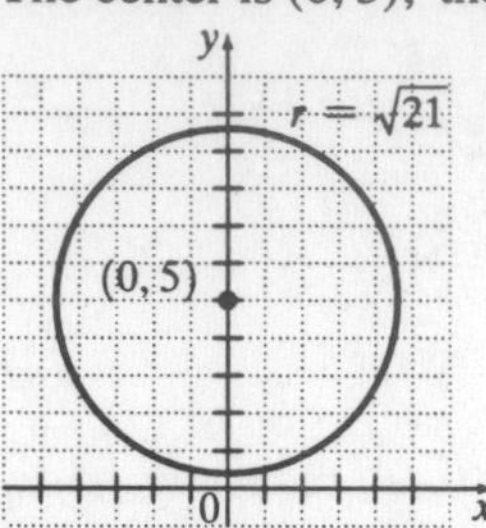

83. $x = -3y^2 + 30y$

$x = -3(y^2 - 10y + 25) + 75$

$x = -3(y-5)^2 + 75$

The vertex is (75, 5).

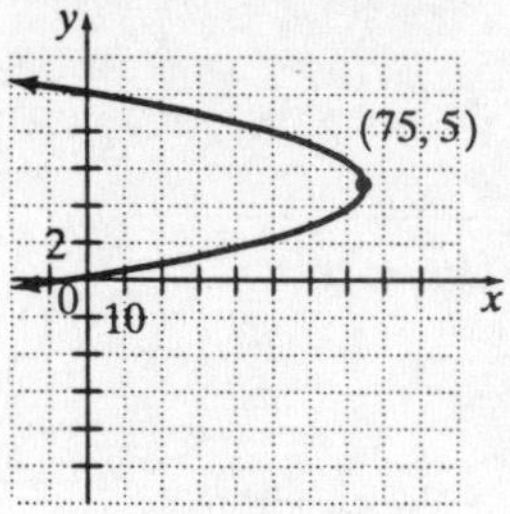

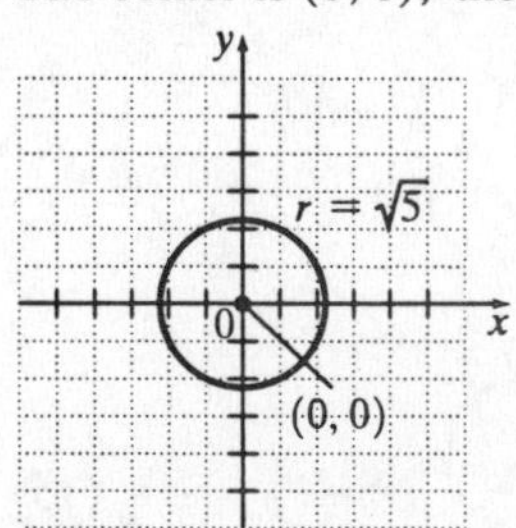

85. $5x^2 + 5y^2 = 25$

$x^2 + y^2 = 5$

The center is (0, 0); the radius is $\sqrt{5}$.

y
r = √5
0
x
(0, 0)

87. $y = 5x^2 - 20x + 16$

$y = 5(x^2 - 4x + 4) + 16 - 20$

$y = 5(x - 2)^2 - 4$

The vertex is (2, –4).

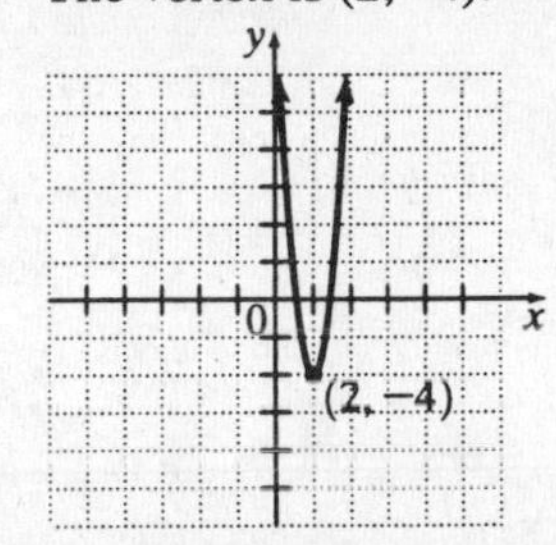

89. The distance between the points (5, 1) and (2, 6) is

$d = \sqrt{(5-2)^2 + (1-6)^2}$

$d = \sqrt{3^2 + (-5)^2}$

$d = \sqrt{9 + 25}$

$d = \sqrt{34}$ units.

The distance between the points (5, 1) and (0, –2) is

$d = \sqrt{(5-0)^2 + [1-(-2)]^2}$

$d = \sqrt{5^2 + 3^2}$

$d = \sqrt{25 + 9}$

$d = \sqrt{34}$ units.

Therefore, the triangle with vertices (2, 6), (0, –2) and (5, 1) is an isosceles triangle.

91. Setting up a coordinate system with the axis of symmetry as the y-axis and the base as the x-axis. The parabola would pass through the points (–50, 0), (0, 40), and (50, 0).

Use the equation for a parabola:

$y = ax^2 + bx + c$.

Substituting the x and y coordinates for the known points yields a system of equations.

$$\begin{cases} 0 = a(-50)^2 + b(-50) + c \\ 40 = a(0)^2 + b(0) + c \\ 0 = a(50)^2 + b(50) + c \end{cases}$$

$$\begin{cases} 0 = 2500a - 50b + c \\ 40 = c \\ 0 = 2500a + 50b + c \end{cases}$$

Solving $c = 40$:

$$\begin{cases} 0 = 2500a - 50b + 40 \\ 0 = 2500a + 50b + 40 \end{cases}$$

Adding the two equations:

$0 = 5000a + 80$

$-80 = 5000a$

$\frac{-80}{5000} = a$

$-\frac{2}{125} = a$

$0 = 2500\left(-\frac{2}{125}\right) - 50b + 40$

$0 = -40 - 50b + 40$

$0 = -50b$

$0 = b$

Thus, the equation of the parabola is

$y = -\frac{2}{125}x^2 + 40.$

93. $5x^2 + 5y^2 = 25$

$$5y^2 = 25 - 5x^2$$
$$y^2 = 5 - x^2$$
$$y = \pm\sqrt{5 - x^2}$$

Answers are the same.

95. $y = 5x^2 - 20x + 16$

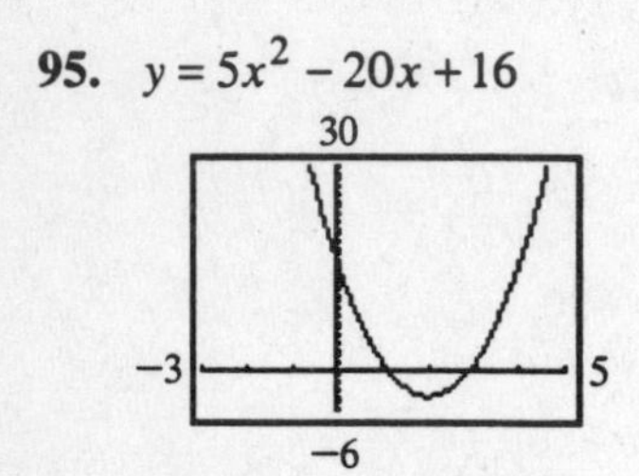

Answers are the same.

97. $y = -3x + 3$

99. $x = -2$

101. $\frac{\sqrt{5}}{\sqrt{8}} = \frac{\sqrt{5}}{2\sqrt{2}}$

$$= \frac{\sqrt{5}}{2\sqrt{2}} \cdot \frac{\sqrt{2}}{\sqrt{2}}$$
$$= \frac{\sqrt{10}}{2\sqrt{4}}$$
$$= \frac{\sqrt{10}}{2(2)}$$
$$= \frac{\sqrt{10}}{4}$$

103. $\frac{10}{\sqrt{5}} = \frac{10}{\sqrt{5}} \cdot \frac{\sqrt{5}}{\sqrt{5}}$

$$= \frac{10\sqrt{5}}{\sqrt{25}}$$
$$= \frac{10\sqrt{5}}{5}$$
$$= 2\sqrt{5}$$

Section 13.2

Graphing Calculator Explorations

1. $10x^2 + y^2 = 32$

$$y^2 = 32 - 10x^2$$
$$y = \pm\sqrt{32 - 10x^2}$$

3. $20x^2 + 5y^2 = 100$

$$5y^2 = 100 - 20x^2$$
$$y^2 = 20 - 4x^2$$
$$y = \pm\sqrt{20 - 4x^2}$$

6

−9 9

−6

5. $7.3x^2 + 15.5y^2 = 95.2$

$$15.5y^2 = 95.2 - 7.3x^2$$
$$y^2 = \frac{95.2 - 7.3x^2}{15.5}$$
$$y = \pm\sqrt{\frac{95.2 - 7.3x^2}{15.5}}$$

4

−6 6

−4

Exercise Set 13.2

1. $\frac{x^2}{4} + \frac{y^2}{25} = 1$

$\frac{x^2}{2^2} + \frac{y^2}{5^2} = 1$

The center is (0, 0).
x-intercepts: $(-2, 0)$ and $(2, 0)$
y-intercepts: $(0, -5)$ and $(0, 5)$

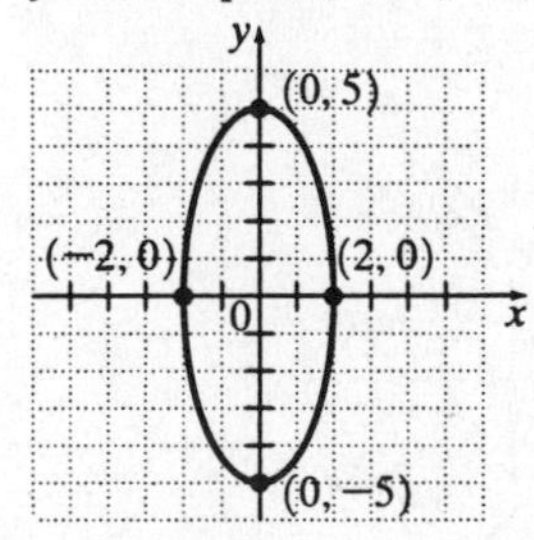

3. $\frac{x^2}{16} + \frac{y^2}{9} = 1$

$\frac{x^2}{4^2} + \frac{y^2}{3^2} = 1$

The center is (0, 0).
x-intercepts: $(-4, 0)$ and $(4, 0)$
y-intercepts: $(0, -3)$ and $(0, 3)$

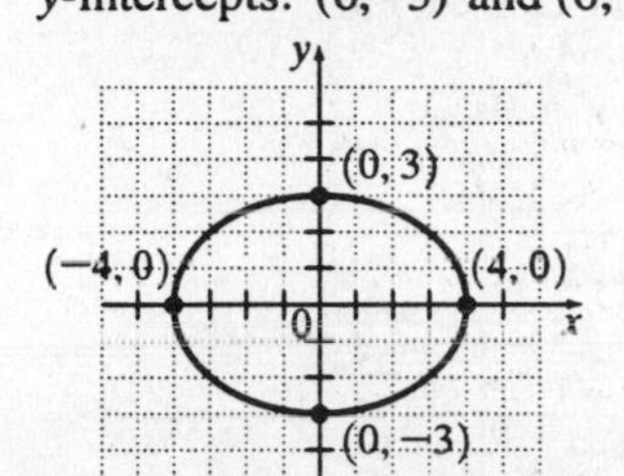

5. $9x^2 + 4y^2 = 36$

$\frac{x^2}{4} + \frac{y^2}{9} = 1$

$\frac{x^2}{2^2} + \frac{y^2}{3^2} = 1$

The center is (0, 0).
x-intercepts: $(-2, 0)$ and $(2, 0)$
y-intercepts: $(0, -3)$ and $(0, 3)$

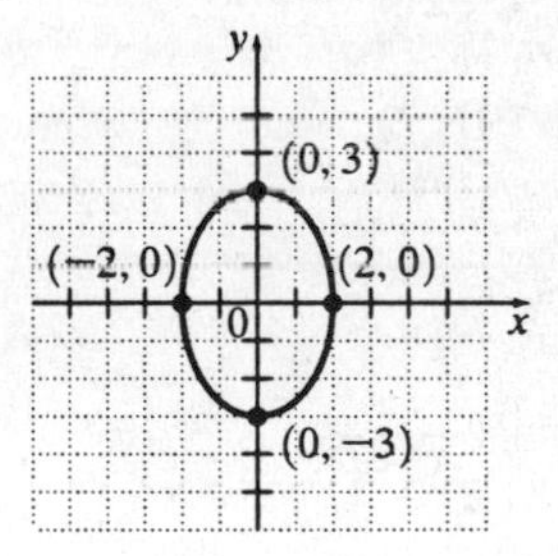

7. $4x^2 + 25y^2 = 100$

$\frac{x^2}{25} + \frac{y^2}{4} = 1$

$\frac{x^2}{5^2} + \frac{y^2}{2^2} = 1$

x-intercepts: $(-5, 0)$ and $(5, 0)$
y-intercepts: $(0, -2)$ and $(0, 2)$
The center is (0, 0).

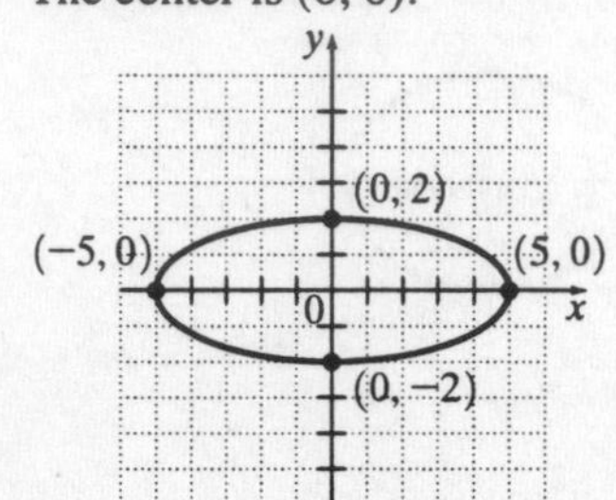

9. $\frac{(x+1)^2}{36} + \frac{(y-2)^2}{49} = 1$

$\frac{(x+1)^2}{6^2} + \frac{(y-2)^2}{7^2} = 1$

The center is (−1, 2).
Other points:
(−1 − 6, 2) or (−7, 2)
(−1 + 6, 2) or (5, 2)
(−1, 2 − 7) or (−1, −5)
(−1, 2 + 7) or (−1, 9)

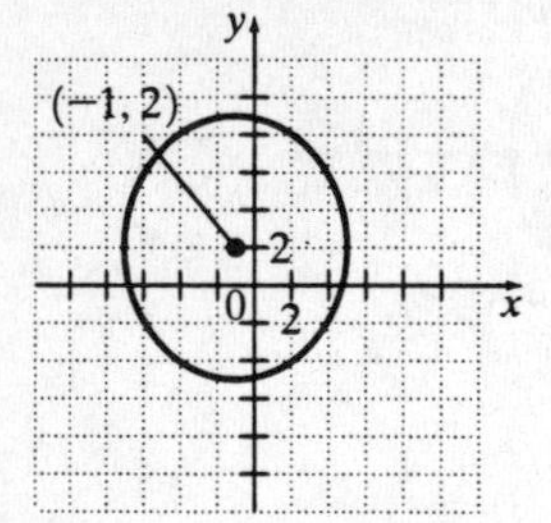

11. $\frac{(x-1)^2}{4} + \frac{(y-1)^2}{25} = 1$

$\frac{(x-1)^2}{2^2} + \frac{(y-1)^2}{5^2} = 1$

The center is (1, 1).
Other points:
(1 − 2, 1) or (−1, 1)
(1 + 2, 1) or (3, 1)
(1, 1 − 5) or (1, −4)
(1, 1 + 5) or (1, 6)

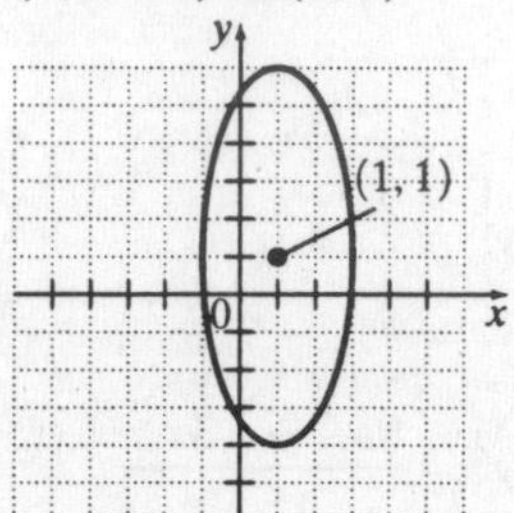

13. $\frac{x^2}{4} - \frac{y^2}{9} = 1$

$\frac{x^2}{2^2} - \frac{y^2}{3^2} = 1$

$a = 2, b = 3$

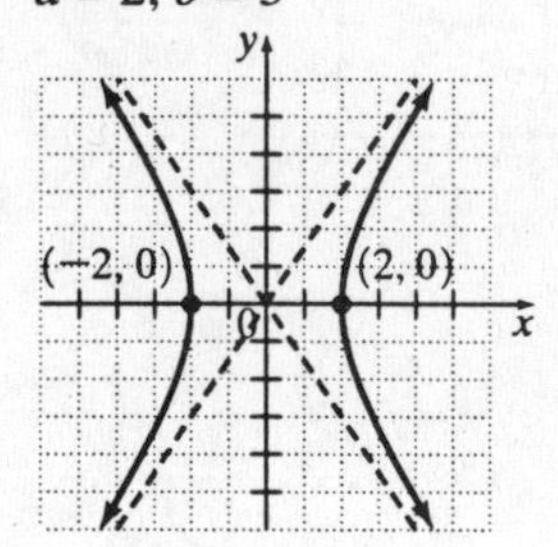

15. $\frac{y^2}{25} - \frac{x^2}{16} = 1$

$\frac{y^2}{5^2} - \frac{x^2}{4^2} = 1$

$a = 4, b = 5$

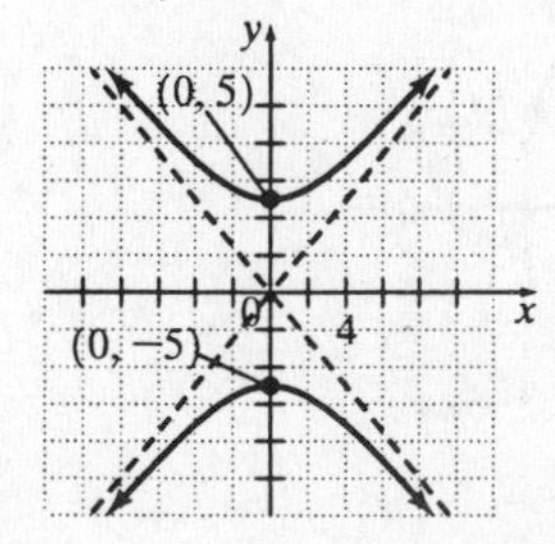

17. $x^2 - 4y^2 = 16$

$\frac{x^2}{16} - \frac{y^2}{4} = 1$

$\frac{x^2}{4^2} - \frac{y^2}{2^2} = 1$

$a = 4, b = 2$

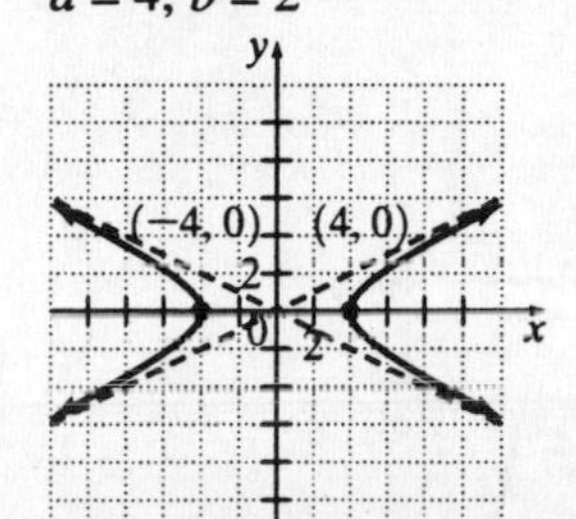

19. $16y^2 - x^2 = 16$

$\frac{y^2}{1} - \frac{x^2}{16} = 1$

$\frac{y^2}{1^2} - \frac{x^2}{4^2} = 1$

$a = 4, b = 1$

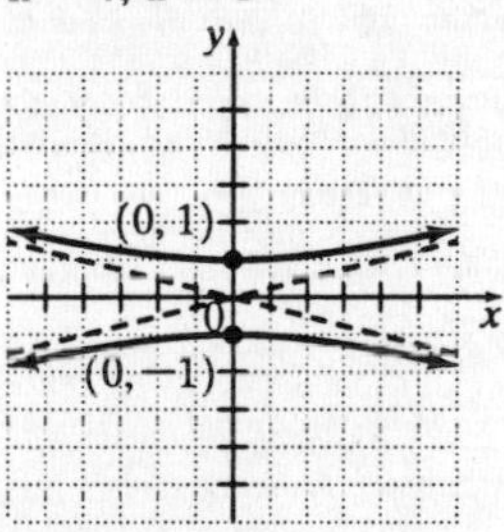

21. Answers may vary.

23. $y = x^2 + 4$

parabola: The vertex is (0, 4).
opens upward

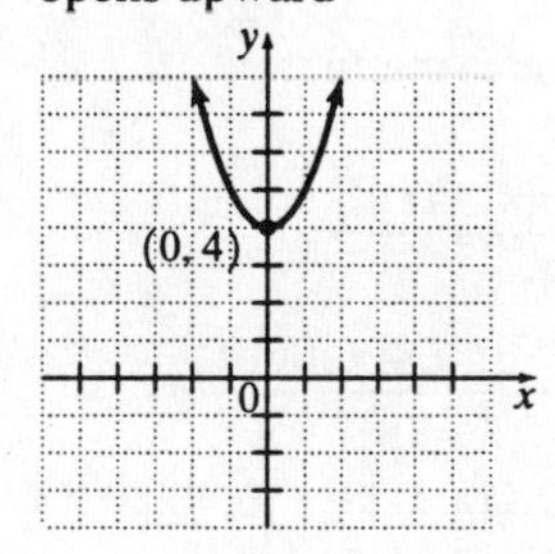

25. $\frac{x^2}{4} + \frac{y^2}{9} = 1$

ellipse: center (0, 0)
$a = 2, b = 3$
y-intercepts:(0, 3) and (0, −3)
x-intercepts: (2, 0) and (−2, 0)

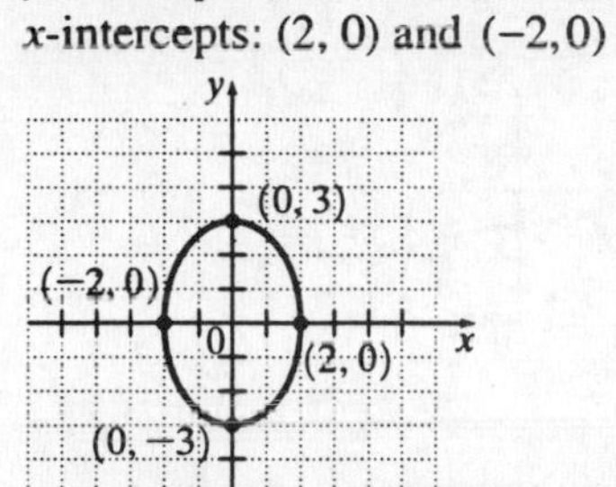

27. $\frac{x^2}{16} - \frac{y^2}{4} = 1$

hyperbola: center (0, 0)
$a = 4, b = 2$
x-intercepts: (4, 0) and (−4, 0)

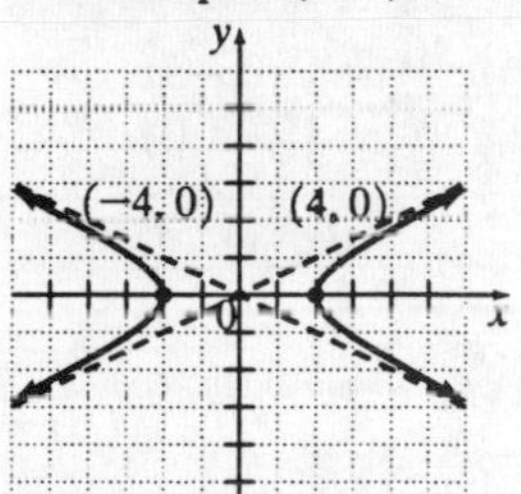

29. $x^2 + y^2 = 16$

circle: center (0, 0)
radius = 4

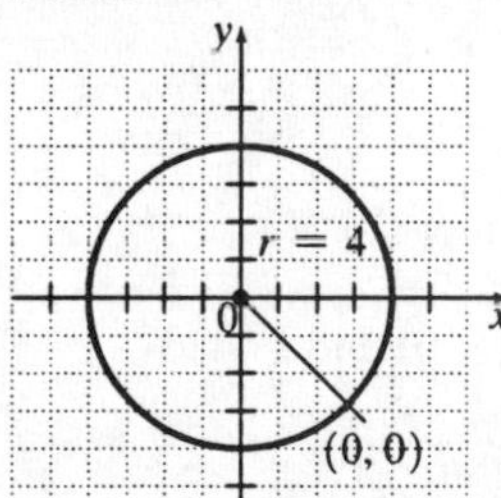

31. $x = -y^2 + 6y$

parabola: $\frac{-b}{2a} = \frac{-6}{2(-1)} = 3$

$x = -9 + 18 = 9$

The vertex is (9, 3).

opens to the left

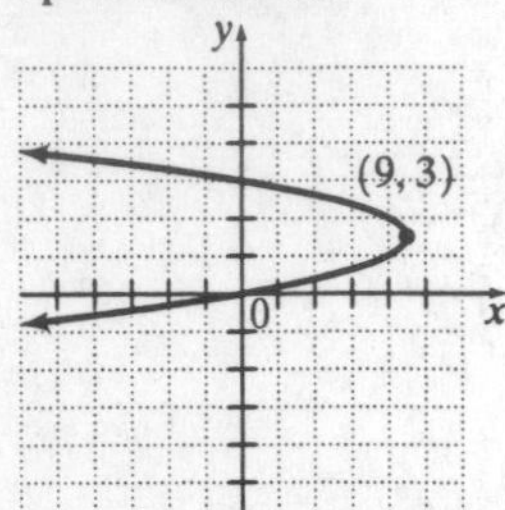

33. $9x^2 + 4y^2 = 36$

$\frac{x^2}{4} + \frac{y^2}{9} = 1$

ellipse: center (0, 0)

$a = 2, b = 3$

y-intercepts: (0, 3) and (0, −3)

x-intercepts: (2, 0) and (−2, 0)

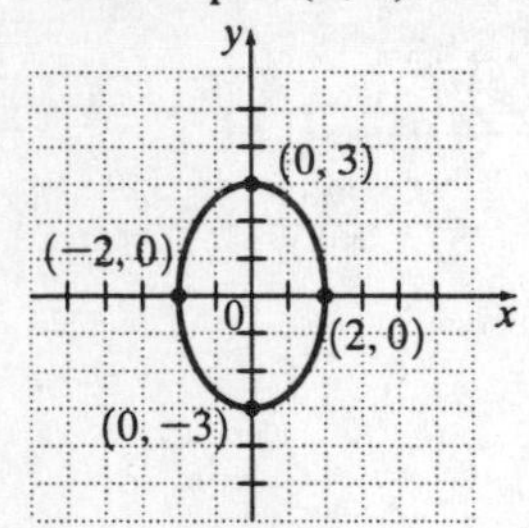

35. $y^2 = x^2 + 16$

$\frac{y^2}{16} - \frac{x^2}{16} = 1$

hyperbola: center (0, 0)

$a = 4, b = 4$

y-intercepts: (4, 0) and (−4, 0)

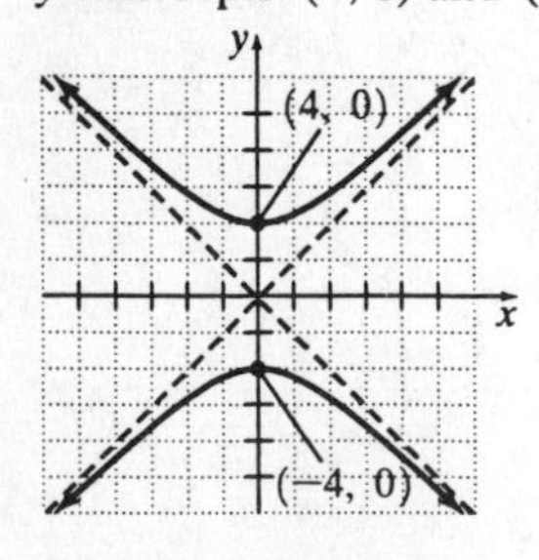

37. $y = -2x^2 + 4x - 3$

parabola: $\frac{-b}{2a} = \frac{-4}{2(-2)} = 1$

$y = -2 + 4 - 3 = -1$

The vertex is (1, −1).

opens downward

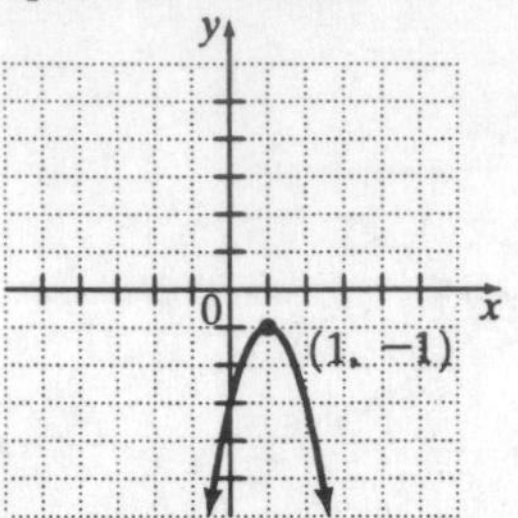

39. A: $a^2 = 36, b^2 = 13$

B: $a^2 = 4, b^2 = 4$

C: $a^2 = 25, b^2 = 16$

D: $a^2 = 39, b^2 = 25$

E: $a^2 = 17, b^2 = 81$

F: $a^2 = 36, b^2 = 36$

G: $a^2 = 16, b^2 = 65$

H: $a^2 = 144, b^2 = 140$

41. A: $d = 6$

B: $d = 2$

C: $d = 5$

D: $d = 5$

E: $d = 9$

F: $d = 6$

G: $d = 4$

H: $d = 12$

43. They are greater than 0 and less than 1.

45. They are greater than 1.

47. $$\frac{(x - 1{,}782{,}000{,}000)^2}{(3.42)(10^{23})} + \frac{(y - 356{,}400{,}000)^2}{(1.368)(10^{22})} = 1$$

center (1,782,000,000, 356,400,000)

49. $x^2 + 4y^2 = 16$

$$4y^2 = 16 - x^2$$

$$y^2 = \frac{16 - x^2}{4}$$

$$y = \pm\sqrt{\frac{16 - x^2}{4}}$$

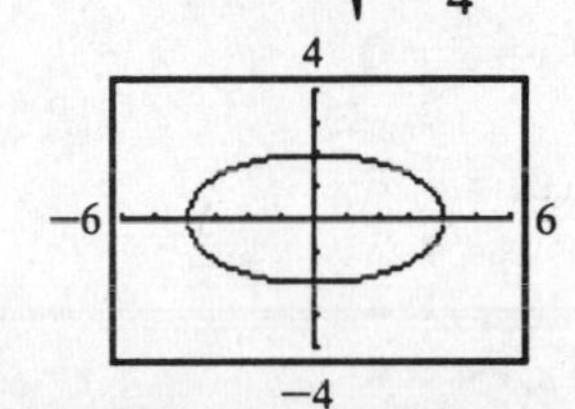

Answers are the same.

51. $x < 5$ or $x < 1$
The solution is $(-\infty, 5)$.

53. $2x - 1 \ge 7$ and $-3x \le -6$
$x \ge 4$ and $x \ge 2$
The solution is $[4, \infty)$.

55. $2x^3 - 4x^3 = -2x^3$

57. $(-5x^2)(x^2) = -5x^4$

59. $\frac{(x-1)^2}{4} - \frac{(y+1)^2}{25} = 1$
center $(1, -1)$
$a = 2$, $b = 5$

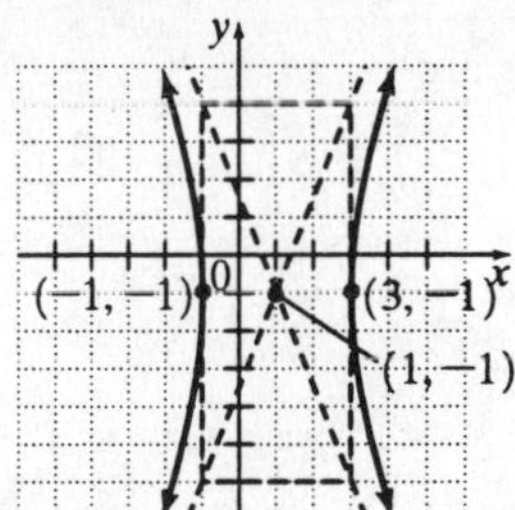

61. $\frac{y^2}{16} - \frac{(x+3)^2}{9} = 1$
center $(-3, 0)$
$a = 3$, $b = 4$

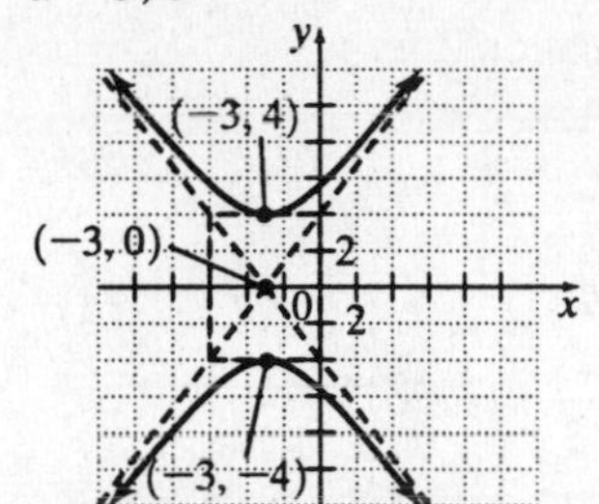

63. $\frac{(x+5)^2}{16} - \frac{(y+2)^2}{25} = 1$
center $(-5, -2)$
$a = 4$, $b = 5$

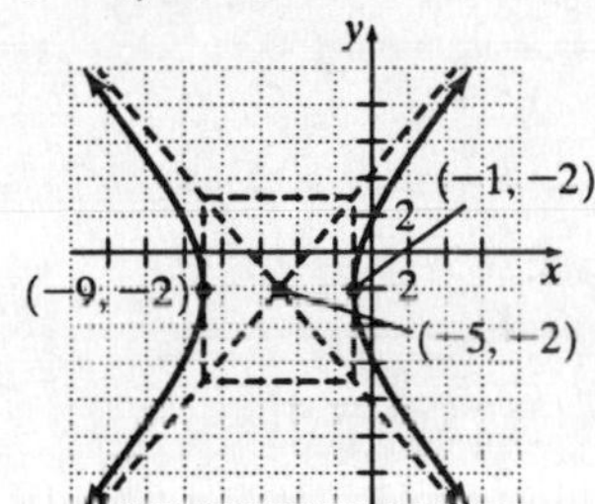

Exercise Set 13.3

1. $\begin{cases} x^2 + y^2 = 25 \\ 4x + 3y = 0 \end{cases}$

Solve equation 2 for *y*.

$3y = -4x$

$y = \frac{-4x}{3}$

Substitute.

$x^2 + \left(-\frac{4x}{3}\right)^2 = 25$

$x^2 + \frac{16x^2}{9} = 25$

$\frac{25}{9}x^2 = 25$

$\frac{x^2}{9} = 1$

$x^2 = 9$

$x = \pm\sqrt{9} = \pm 3$

$x = 3\text{: } y = -\frac{4}{3}(3) = -4$

$x = -3\text{: } y = -\frac{4}{3}(-3) = 4$

The solutions are (3, –4) and (–3, 4).

3. $\begin{cases} x^2 + 4y^2 = 10 \\ y = x \end{cases}$

Substitute.

$x^2 + 4x^2 = 10$

$5x^2 = 10$

$x^2 = 2$

$x = \pm\sqrt{2}$

$x = \sqrt{2};\ y = \sqrt{2}$

$x = -\sqrt{2};\ y = -\sqrt{2}$

The solutions are

$\left(\sqrt{2}, \sqrt{2}\right)$ and $\left(-\sqrt{2}, -\sqrt{2}\right)$.

5. $\begin{cases} y^2 = 4 - x \\ x - 2y = 4 \end{cases}$

$-2y = 4 - x$

Substitute.

$y^2 = -2y$

$y^2 + 2y = 0$

$y(y + 2) = 0$

$y = 0$ or $y + 2 = 0$

$y = -2$

$y = 0\text{: } x - 2(0) = 4$

$x = 4$

$y = -2\text{: } x - 2(-2) = 4$

$x + 4 = 4$

$x = 0$

The solutions are (4, 0) and (0, –2).

7. $\begin{cases} x^2 + y^2 = 9 \\ 16x^2 - 4y^2 = 64 \end{cases}$

Divide equation two by 4.

$\begin{cases} x^2 + y^2 = 9 \\ 4x^2 - y^2 = 16 \end{cases}$

Add.

$5x^2 = 25$

$x^2 = 5$

$x = \pm\sqrt{5}$

Substitute back.

$5 + y^2 = 9$

$y^2 = 4$

$y = \pm 2$

The solutions are

$\left(\sqrt{5}, 2\right)$, $\left(\sqrt{5}, -2\right)$, $\left(-\sqrt{5}, 2\right)$, and

$\left(-\sqrt{5}, -2\right)$.

9. $\begin{cases} x^2 + 2y^2 = 2 \\ x - y = 2 \end{cases}$

$x = y + 2$

Substitute.

$$(y+2)^2 + 2y^2 = 2$$
$$y^2 + 4y + 4 + 2y^2 = 2$$
$$3y^2 + 4y + 4 = 2$$
$$3y^2 + 4y + 2 = 0$$

$b^2 - 4ac = 4^2 - 4(3)(2) = 16 - 24 = -8 < 0$

There is no solution.

11. $\begin{cases} y = x^2 - 3 \\ 4x - y = 6 \end{cases}$

Substitute.

$$4x - (x^2 - 3) = 6$$
$$4x - x^2 + 3 = 6$$
$$0 = x^2 - 4x + 3$$
$$0 = (x-3)(x-1)$$

$x - 3 = 0$ or $x - 1 = 0$

$x = 3$ or $x = 1$

$x = 3$: $y = 3^2 - 3 = 9 - 3 = 6$

$x = 1$: $y = 1^2 - 3 = 1 - 3 = -2$

The solutions are (3, 6) and (1, −2).

13. $\begin{cases} y = x^2 \\ 3x + y = 10 \end{cases}$

Substitute.

$$3x + x^2 = 10$$
$$x^2 + 3x - 10 = 0$$
$$(x+5)(x-2) = 0$$

$x + 5 = 0$ or $x - 2 = 0$

$x = -5$ or $x = 2$

$x = -5$: $y = (-5)^2 = 25$

$x = 2$: $y = 2^2 = 4$

The solutions are (−5, 25) and (2, 4).

15. $\begin{cases} y = 2x^2 + 1 \\ x + y = -1 \end{cases}$

Substitute.

$$x + 2x^2 + 1 = -1$$
$$2x^2 + x + 1 = -1$$
$$2x^2 + x + 2 = 0$$

$b^2 - 4ac = 1^2 - 4(2)(2) = -15 < 0$

There is no solution.

17. $\begin{cases} y = x^2 - 4 \\ y = x^2 - 4x \end{cases}$

Substitute.

$$x^2 - 4 = x^2 - 4x$$
$$-4 = -4x$$
$$x = 1$$

$y = 1^2 - 4 = -3$

The solution is (1, −3).

19. $\begin{cases} 2x^2 + 3y^2 = 14 \\ -x^2 + y^2 = 3 \end{cases}$

$y^2 = x^2 + 3$

Substitute.

$$2x^2 + 3(x^2 + 3) = 14$$
$$2x^2 + 3x^2 + 9 = 14$$
$$5x^2 + 9 = 14$$
$$5x^2 = 5$$
$$x^2 = 1$$
$$x = \pm 1$$

Substitute back.

$$y^2 = 1 + 3$$
$$y^2 = 4$$
$$y = \pm 2$$

The solutions are (1, −2), (1, 2), (−1, −2), and (−1, 2).

21. $\begin{cases} x^2+y^2=1 \\ x^2+(y+3)^2=4 \end{cases}$

Subtract equation 1 from equation 2.

$$(y+3)^2-y^2=3$$
$$y^2+6y+9-y^2=3$$
$$6y+9=3$$
$$6y=-6$$
$$y=-1$$

Substitute back.

$$x^2+(-1)^2=1$$
$$x^2+1=1$$
$$x^2=0$$
$$x=0$$

The solution is (0, –1).

23. $\begin{cases} y=x^2+2 \\ y=-x^2+4 \end{cases}$

Substitute.

$$x^2+2=-x^2+4$$
$$2x^2=2$$
$$x^2=1$$
$$x=\pm 1$$

Substitute back.

$y=1+2=3$

The solutions are (1, 3) and (–1, 3).

25. $\begin{cases} 3x^2+y^2=9 \\ 3x^2-y^2=9 \end{cases}$

Subtract.

$$2y^2=0$$
$$y^2=0$$
$$y=0$$

Substitute back.

$$3x^2+0=9$$
$$3x^2=9$$
$$x^2=3$$
$$x=\pm\sqrt{3}$$

The solutions are $\left(\sqrt{3}, 0\right)$ and $\left(-\sqrt{3}, 0\right)$.

27. $\begin{cases} x^2+3y^2=6 \\ x^2-3y^2=10 \end{cases}$

Add.

$$2x^2=16$$
$$x^2=8$$
$$x=\pm\sqrt{8}$$
$$x=\pm 2\sqrt{2}$$

Substitute back.

$$8+3y^2=6$$
$$3y^2=-2$$
$$y^2=-\frac{2}{3}$$

There is no solution.

29. $\begin{cases} x^2+y^2=36 \\ y=\frac{1}{6}x^2-6 \end{cases}$

$$y+6=\frac{1}{6}x^2$$
$$x^2=6(y+6)$$

Substitute.

$$6(y+6)+y^2=36$$
$$6y+36+y^2=36$$
$$6y+y^2=0$$
$$y(6+y)=0$$
$$y=0 \quad \text{or} \quad 6+y=0$$
$$y=-6$$

$y=0$: $x^2+0^2=36$
$$x^2=36$$
$$x=\pm 6$$

$y=-6$: $x^2+(-6)^2=36$
$$x^2+36=36$$
$$x^2=0$$
$$x=0$$

The solutions are (6, 0), (–6, 0), and (0, –6).

31. There can be 0, 1, 2, 3, or 4 real solutions. For the circle $x^2 + y^2 = 9$:

a. The parabola $y = -x^2 - 4$ does not intersect it;

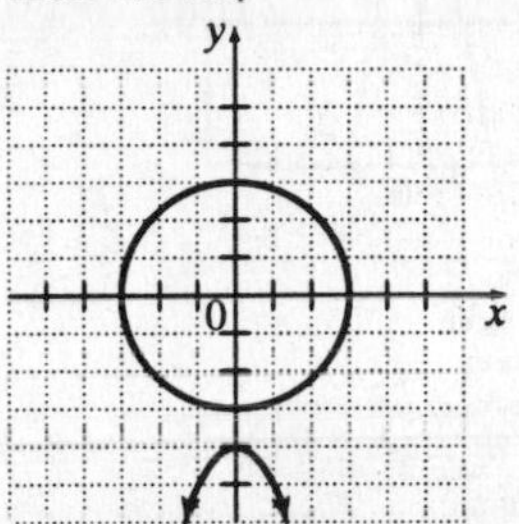

b. The parabola $y = x^2 + 3$ intersects it once;

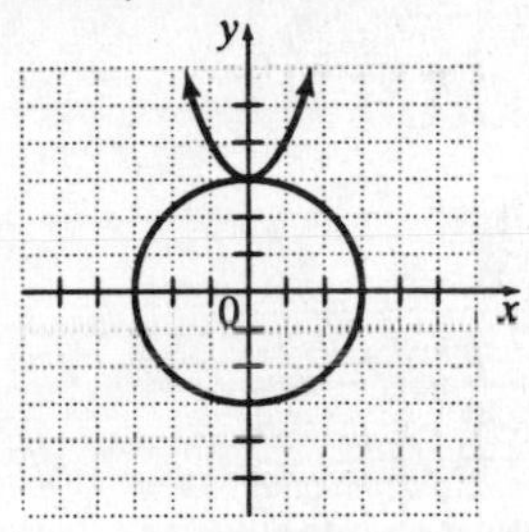

c. The parabola $y = x^2$ intersects it twice;

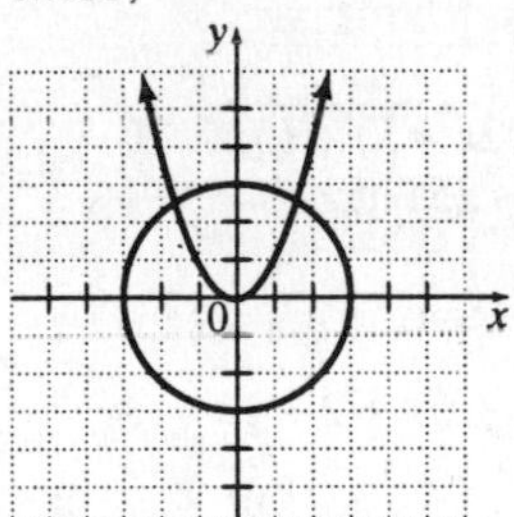

d. The parabola $y = x^2 - 3$ intersects it 3 times;

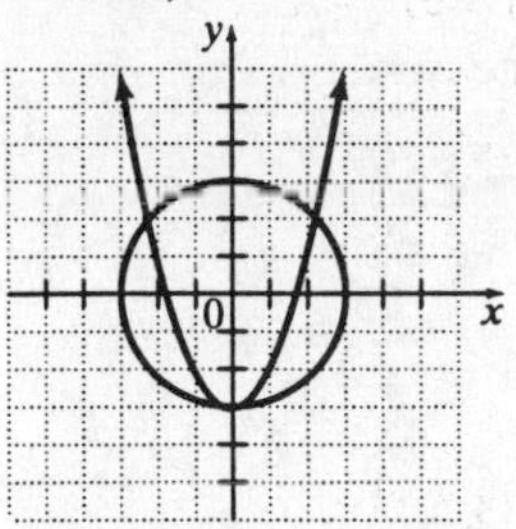

e. The parabola $y = x^2 - 4$ intersects it 4 times.

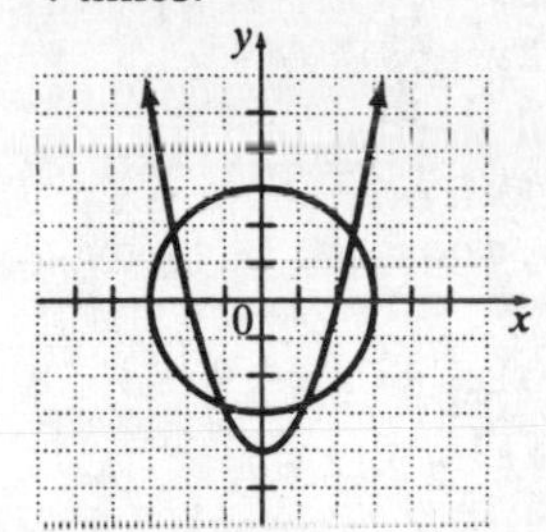

33. $\begin{cases} x^2 + y^2 = 130 \\ x^2 - y^2 = 32 \end{cases}$

Add.

$2x^2 = 162$

$x^2 = 81$

$x = \pm 9$

Substitute back.

$9^2 + y^2 = 130$ $\quad (-9)^2 + y^2 = 130$

$y^2 = 49$ $\quad y^2 = 49$

$y = \pm 7$ $\quad y = \pm 7$

Answer:

The possibilities are 9 and 7, 9 and –7, –9 and 7, –9 and –7.

35. $\begin{cases} 2x + 2y = 68 \\ xy = 285 \end{cases}$

Substitute $y = \frac{285}{x}$.

$$2x + 2\left(\frac{285}{x}\right) = 68$$
$$x + \frac{285}{x} = 34$$
$$x^2 - 34x + 285 = 0$$
$$(x - 19)(x - 15) = 0$$

$x = 19$ or $x = 15$

Substitute back.

$$2(19) + 2y = 68$$
$$38 + 2y = 68$$
$$2y = 30$$
$$y = 15$$

The dimensions are 19 cm by 15 cm.

37. $p = -0.01x^2 - 0.2x + 9$

$p = 0.01x^2 - 0.1x + 3$

$$-0.01x^2 - 0.2x + 9 = 0.01x^2 - 0.1x + 3$$
$$0 = 0.02x^2 + 0.1x - 6$$
$$0 = x^2 + 5x - 300$$
$$0 = (x + 20)(x - 15)$$

$x + 20 = 0$ or $x - 15 = 0$

$x = -20$ or $x = 15$

Disregard the negative.

$p = -0.01(15)^2 - 0.2(15) + 9$

$p = 3.75$

15 thousand compact discs; price: $3.75

39. $\begin{cases} x^2 + 4y^2 = 10 \\ y = x \end{cases}$

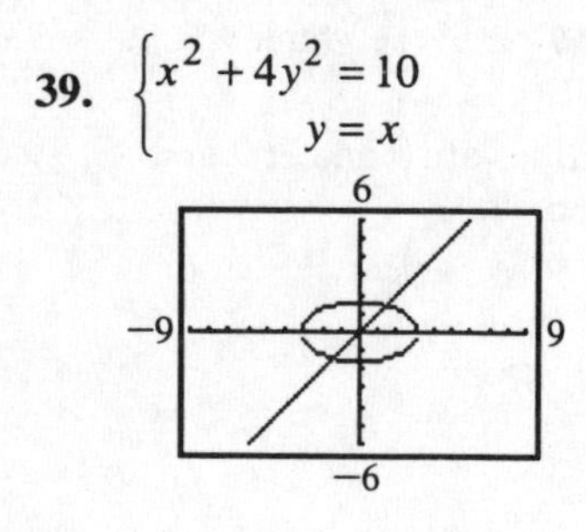

41. $\begin{cases} y = x^2 + 2 \\ y = -x^2 + 4 \end{cases}$

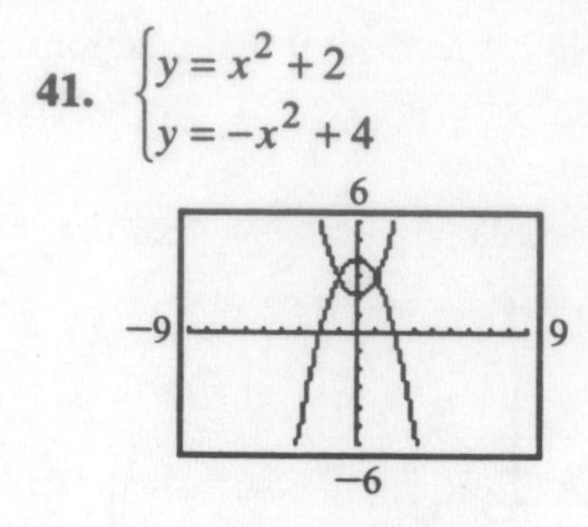

43. $x > -3$

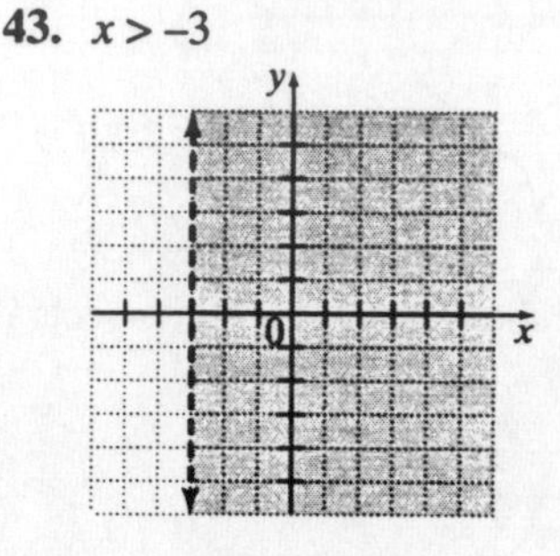

45. $y < 2x - 1$

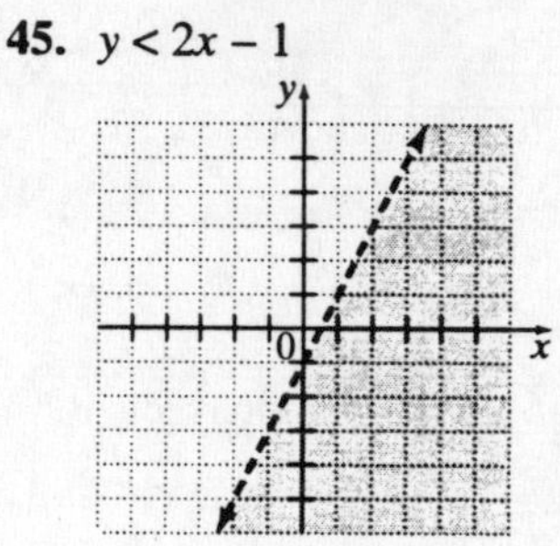

47. $P = x + (2x - 5) + (5x - 20)$

$P = (8x - 25)$ in.

49. $P = 2(x^2 + 3x + 1) + 2x^2$

$P = 2x^2 + 6x + 2 + 2x^2$

$P = (4x^2 + 6x + 2)$ m

Exercise Set 13.4

1. $y < x^2$
First, graph the parabola with a dashed line.
Does (0, 1) satisfy the inequality?
$1 < 0^2$
$1 < 0$ false
Shade portion of the graph which does not contain (0, 1).

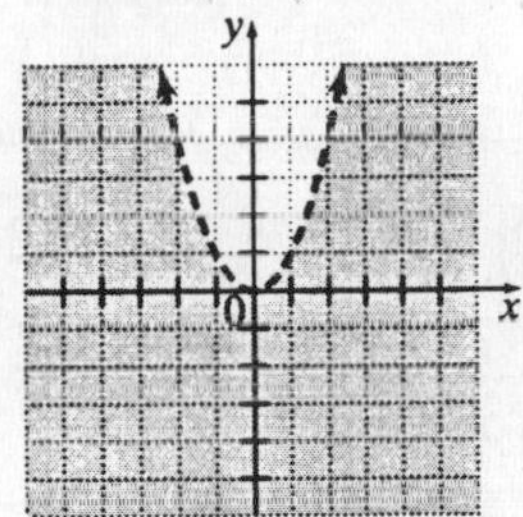

3. $x^2 + y^2 \geq 16$
First graph the circle with a solid line.
Does (0, 0) satisfy the inequality?
$0^2 + 0^2 \geq 16$
$0 \geq 16$ false
Shade the portion of the graph which does not contain (0, 0).

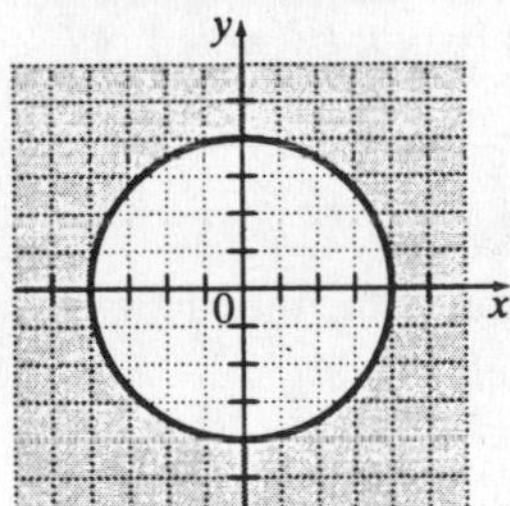

5. $\frac{x^2}{4} - y^2 < 1$
First graph the hyperbola with a dashed line.
Does (–4, 0) satisfy the inequality?
$\frac{(-4)^2}{4} - 0^2 < 1$
$4 < 1$ false
Does (0, 0) satisfy the inequality?
$\frac{0^2}{4} - 0^2 < 1$
$0 < 1$ true
Does (4, 0) satisfy the inequality?
$\frac{4^2}{4} - 0^2 < 1$
$4 < 1$ false
Shade the portion of the graph containing (0, 0).

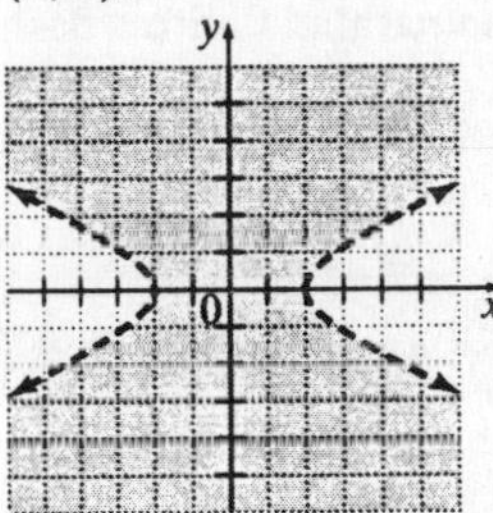

7. $y > (x-1)^2 - 3$
First graph the parabola with a dashed line.
Does (0, 0) satisfy the inequality?
$0 > (0-1)^2 - 3$
$0 > -2$ true
Shade the portion of the graph which contains (0, 0).

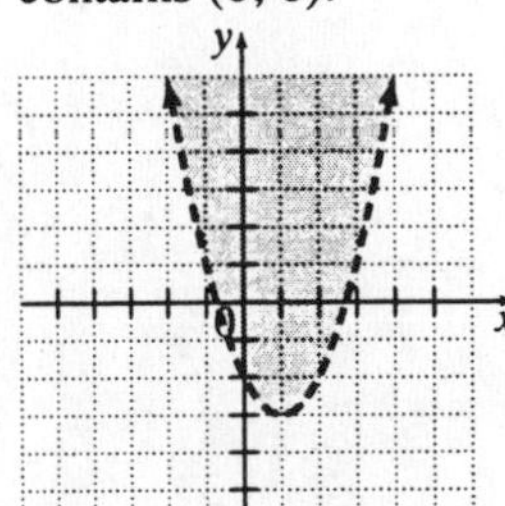

9. $x^2 + y^2 \le 9$

First graph the circle with a solid line.
Does (0, 0) satisfy the inequality?

$0^2 + 0^2 \le 9$

$0 \le 9$ true

Shade the portion of the graph which contains (0, 0).

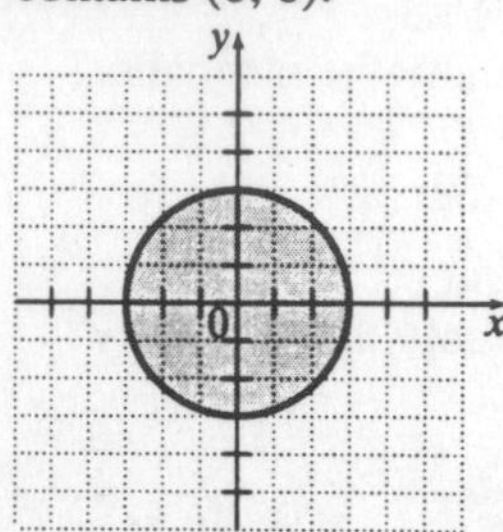

11. $y > -x^2 + 5$

First, graph the parabola with a dashed line.
Does (0, 0) satisfy the inequality?

$0 > -0^2 + 5$

$0 > 5$ false

Shade the portion of the graph which does not contain (0, 0).

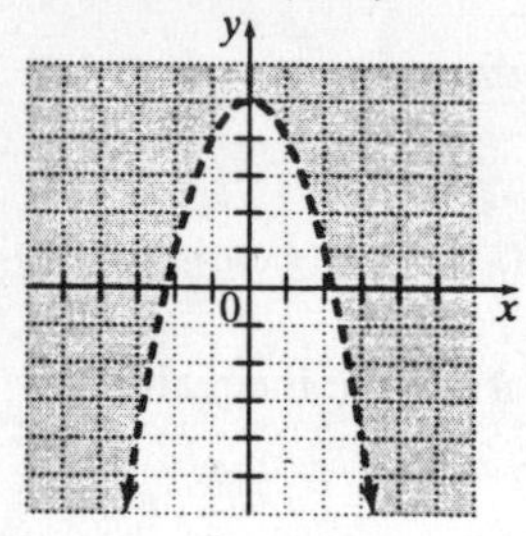

13. $\frac{x^2}{4} + \frac{y^2}{9} \le 1$

First, graph the ellipse with a solid line.
Does (0, 0) satisfy the inequality?

$\frac{0^2}{4} + \frac{0^2}{9} \le 1$

$0 \le 1$ true

Shade the portion of the graph which contains (0, 0).

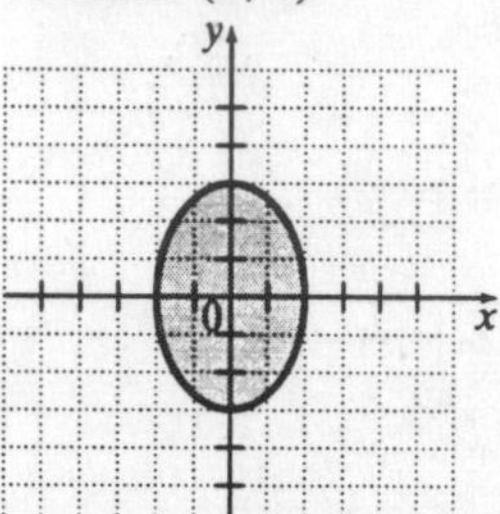

15. $\frac{y^2}{4} - x^2 \le 1$

First, graph the hyperbola with a solid line.
Does (0, −4) satisfy the inequality?

$\frac{(-4)^2}{4} - 0 \le 1$

$4 \le 1$ false

Does (0, 0) satisfy the inequality?

$\frac{0^2}{4} - 0 \le 1$

$0 \le 1$ true

Does (0, 4) satisfy the inequality?

$\frac{4^2}{4} - 0 \le 1$

$16 \le 1$ false

Shade the portion of the graph containing (0, 0).

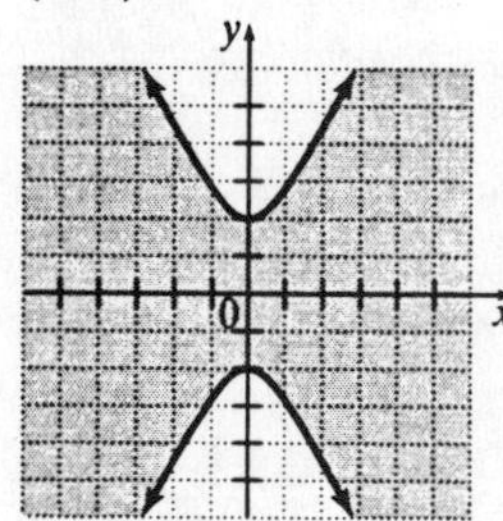

17. $y < (x-2)^2 + 1$
First, graph the parabola with a dashed line.
Does (2, 0) satisfy the inequality?
$0 < (2-2)^2 + 1$
$0 < 1$ true
Shade the portion of the graph which contains (2, 0).

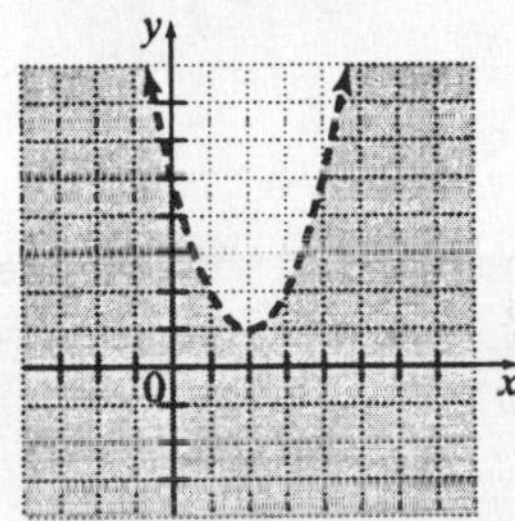

19. $y \le x^2 + x - 2$
First, graph the parabola with a solid line.
Does (0, 0) satisfy the inequality?
$0 \le 0^2 + 0 - 2$
$0 \le -2$ false
Shade the portion of the graph which does not contain (0, 0).

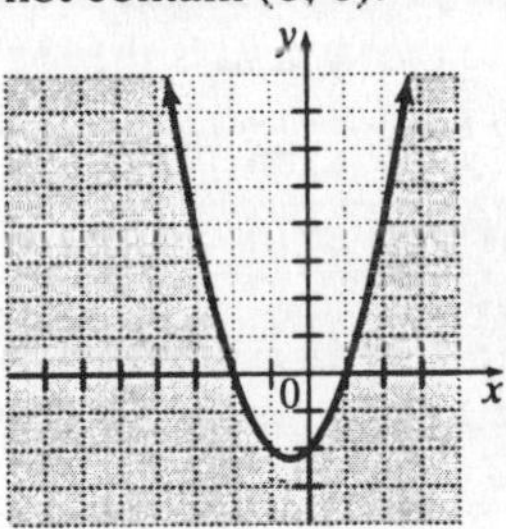

21. Answers may vary.

23. $\begin{cases} 2x - y < 2 \\ y \le -x^2 \end{cases}$
First, graph $2x - y = 2$ with a dotted line by solving for y.
$-y < -2x + 2$
$y > 2x - 2$
Does (0, 0) satisfy $2x - y < 2$?
$2(0) - 0 < 2$
$0 < 2$ true
Shade the portion of the graph which contains (0, 0).
Next, graph the parabola $y = -x^2$ with a solid line.
Does (0, 1) satisfy $y \le -x^2$?
$1 \le -0^2$
$1 \le 0$ false
Shade the portion of the graph which does not contain (0, 1).
The solution to the system is the overlapping region.

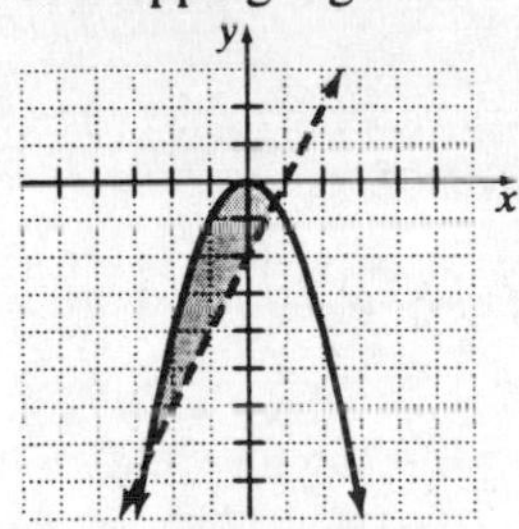

25. $\begin{cases} 4x+3y \geq 12 \\ x^2+y^2 < 16 \end{cases}$

First, graph the circle with a dashed line.
Does (0, 0) satisfy $x^2+y^2 < 16$?
$0^2+0^2 < 16$
$0 < 16$ true
Shade the portion of the graph which contains (0, 0).
Next, graph the line with a solid line.
Does (0, 0) satisfy $4x + 3y \geq 12$?
$4(0)+3(0) \geq 12$
$0 \geq 12$ false
Shade the portion of the graph which does not contain (0, 0).
The solution to the system is the overlapping region.

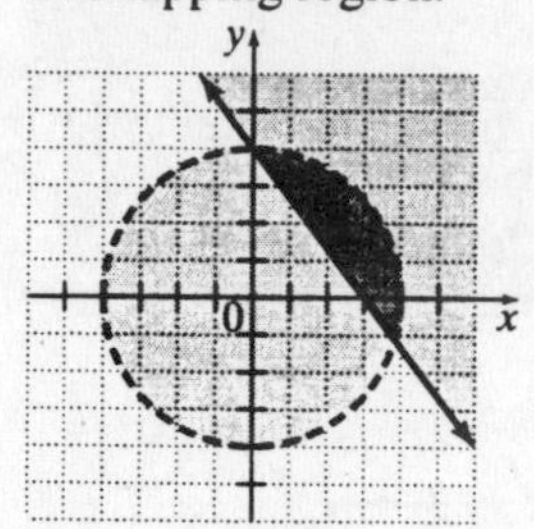

27. $\begin{cases} x^2+y^2 \leq 9 \\ x^2+y^2 \geq 1 \end{cases}$

First, graph the circle with radius 1.
Does (0, 0) satisfy $x^2+y^2 \geq 1$?
$0^2+0^2 \geq 1$
$0 \geq 1$ false
Shade the portion of the graph which does not contain (0, 0).
Next, graph the circle with radius 3.
Does (0, 0) satisfy $x^2+y^2 \leq 9$?
$0^2+0^2 \leq 9$
$0 \leq 9$ true
Shade the portion of the graph which contains (0, 0).
The solution to the system is the overlapping region.

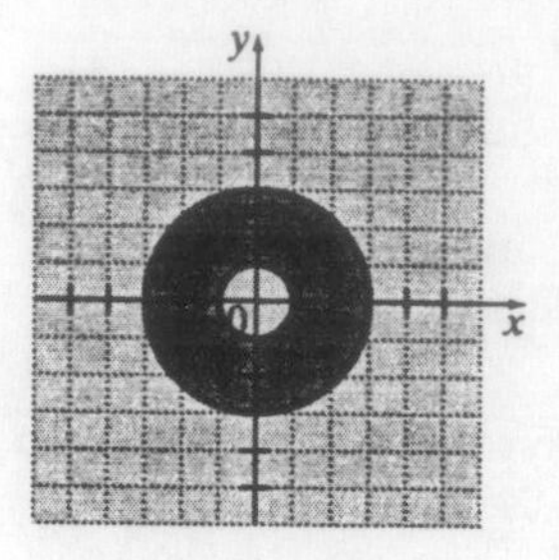

29. $\begin{cases} y > x^2 \\ y \geq 2x+1 \end{cases}$

First, graph the parabola with a dashed line.
Does (0, 2) satisfy $y > x^2$?
$2 > 0^2$
$2 > 0$ true
Shade the portion of the graph which contains (0, 2).
Next, graph the line with a solid line.
Does (0, 0) satisfy $y \geq 2x + 1$?
$0 \geq 2(0) + 1$
$0 \geq 1$ false
Shade the portion of the graph which does not contain (0, 0).
The solution to the system is the overlapping region.

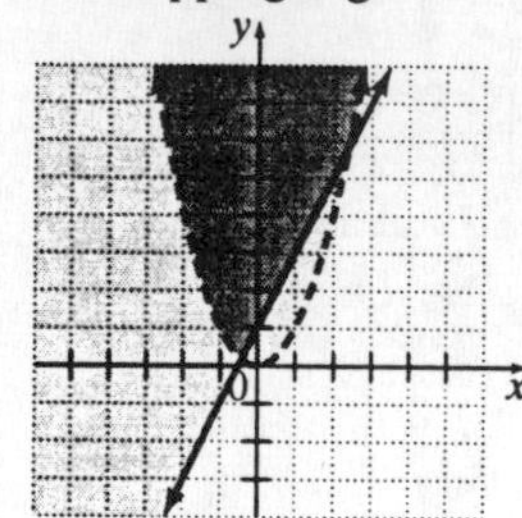

31. $\begin{cases} x > y^2 \\ y > 0 \end{cases}$

First graph the parabola with a dashed line.

Does (2, 0) satisfy $x > y^2$?

$2 > 0^2$

$2 > 0$ true

Shade the portion of the graph which contains (2, 0).

Next, graph the line with a dashed line.

Does (0, 1) satisfy $y > 0$?

$1 > 0$ true

Shade the portion of the graph which contains (0, 1).

The solution to the system is the overlapping region.

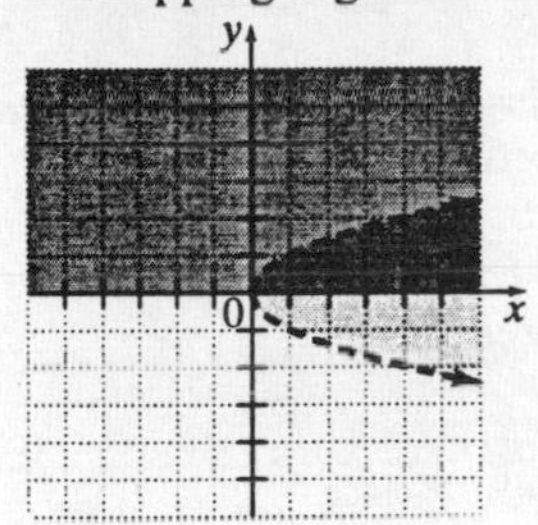

33. $\begin{cases} x^2 + y^2 > 9 \\ y > x^2 \end{cases}$

First, graph the circle with a dashed line.

Does (0, 0) satisfy $x^2 + y^2 > 9$?

$0^2 + 0^2 > 9$

$0 > 9$ false

Shade the portion of the graph which does not contain (0, 0).

Next, graph the parabola with a dashed line.

Does (0, 2) satisfy $y > x^2$?

$2 > 0^2$

$2 > 0$ true

Shade the portion of the graph which contains (0, 2).

The solution to the system is the overlapping region.

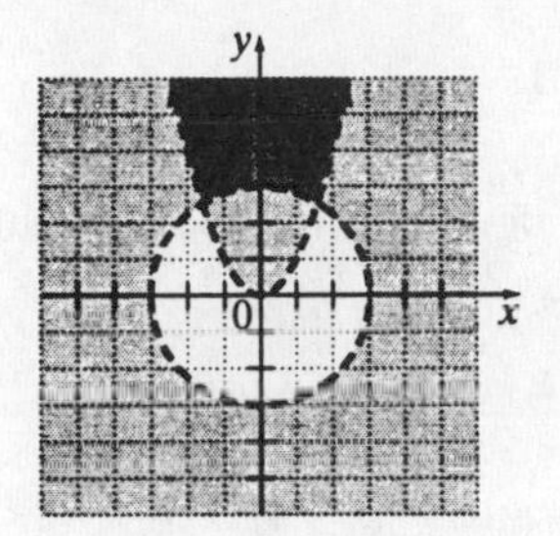

35. $\begin{cases} \frac{x^2}{4} + \frac{y^2}{9} \geq 1 \\ x^2 + y^2 \geq 4 \end{cases}$

First, graph the ellipse. Does (0, 0) satisfy

$\frac{x^2}{4} + \frac{y^2}{9} \geq 1$?

$\frac{0^2}{4} + \frac{0^2}{9} \geq 1$

$0 \geq 1$ false

Shade the portion of the graph which does not contain (0, 0).

Next, graph the circle. Does (0, 0) satisfy

$x^2 + y^2 \geq 4$?

$0^2 + 0^2 \geq 4$

$0 \geq 4$ false

Shade the portion of the graph which does not contain (0, 0).

The solution to the system is the overlapping region.

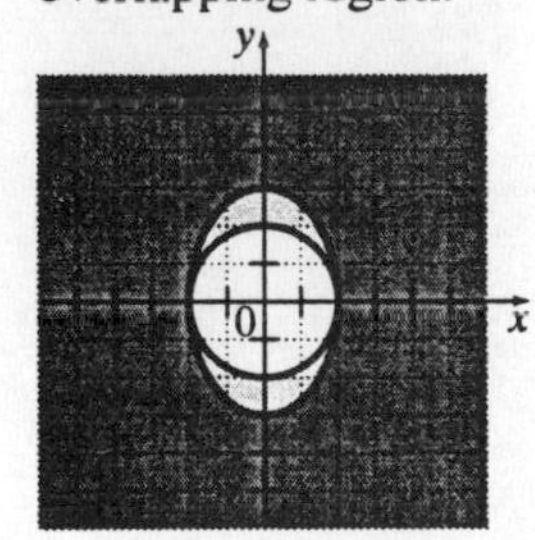

37. $\begin{cases} x^2 - y^2 \geq 1 \\ y \geq 0 \end{cases}$

First, graph the hyperbola. Does (–2, 0) satisfy $x^2 - y^2 \geq 1$?

$(-2)^2 - 0^2 \geq 1$

$4 \geq 1$ true

Does (0, 0) satisfy $x^2 - y^2 \geq 1$?

$0^2 + 0^2 \geq 1$

$0 \geq 1$ false

Does (2, 0) satisfy $x^2 + y^2 \geq 1$?

$2^2 + 0^2 \geq 1$

$4 \geq 1$ true

Shade the portion of the graph containing (–2, 0) and the portion of the graph containing (2, 0).

Next, graph the line. Does (0, 2) satisfy $y \geq 0$?

$2 \geq 0$ true

Shade the portion of the graph containing (0, 2).

The solution to the system is the overlapping region.

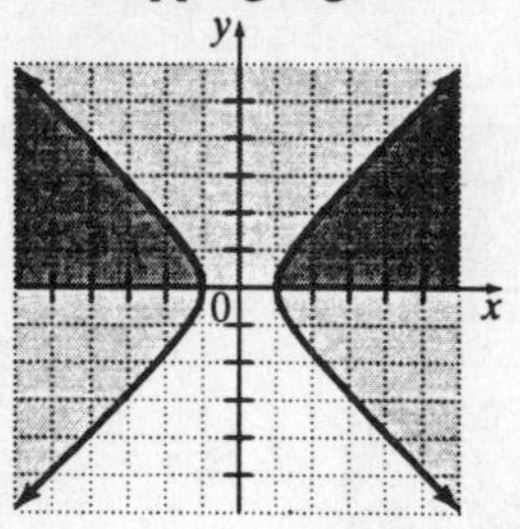

39. $\begin{cases} x + y \geq 1 \\ 2x + 3y < 1 \\ x > -3 \end{cases}$

First, graph $x = -3$ with a dashed line.

Does (0, 0) satisfy $x > -3$?

$0 > -3$ true

Shade the portion of the graph containing (0, 0).

Next, graph $2x + 3y = 1$ with a dashed line.

Does (0, 0) satisfy $2x + 3y < 1$?

$2(0) + 3(0) < 1$

$0 < 1$ true

Shade the portion of the graph containing (0, 0).

Next, graph $x + y = 1$ with a solid line.

Does (0, 0) satisfy $x + y \geq 1$?

$0 + 0 \geq 1$

$0 \geq 1$ false

Shade the portion of the graph which does not contain (0, 0).

The solution to the system is the overlapping region.

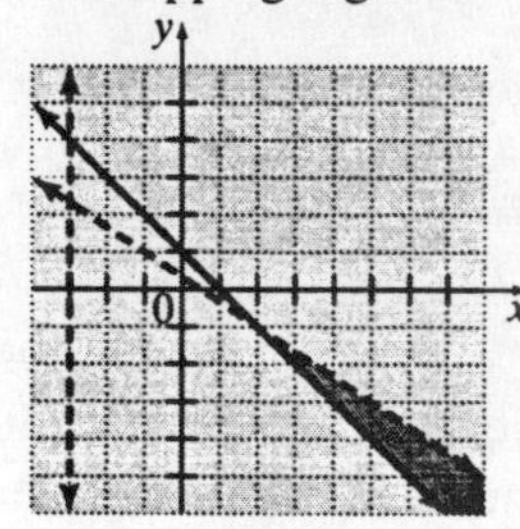

41. $\begin{cases} x^2 - y^2 < 1 \\ \frac{x^2}{16} + y^2 \le 1 \\ x \ge -2 \end{cases}$

First graph the hyperbola with a dashed line.

Does (–2, 0) satisfy $x^2 - y^2 < 1$?

$(-2)^2 - 0^2 < 1$

$4 < 1$ false

Does (0, 0) satisfy $x^2 - y^2 < 1$?

$0^2 - 0^2 < 1$

$0 < 1$ true

Does (2, 0) satisfy $x^2 - y^2 < 1$?

$2^2 - 0^2 < 1$

$4 < 1$ false

Shade the portion of the graph containing (0, 0).

Next, graph the ellipse with a solid line.

Does (0, 0) satisfy $\frac{x^2}{16} + y^2 \le 1$?

$\frac{0^2}{16} + 0^2 \le 1$

$0 \le 1$ true

Shade the portion of the graph containing (0, 0).

Next, graph the line.

Does (0, 0) satisfy $x \ge -2$?

$0 \ge -2$ true

Shade the portion of the graph containing (0, 0).

The solution to the system is the overlapping region.

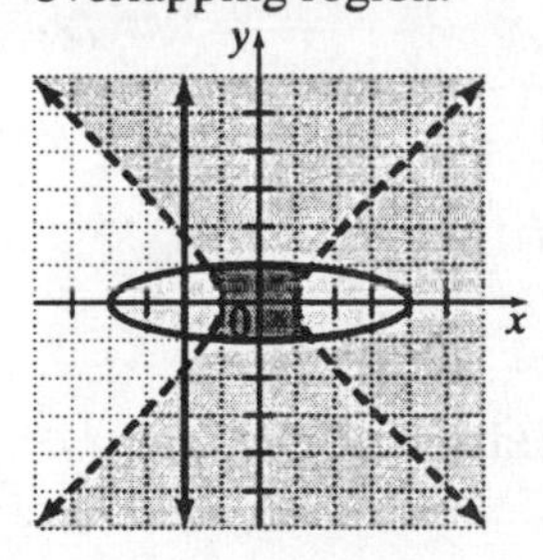

43. $\begin{cases} y \le x^2 \\ y \ge x + 2 \\ x \ge 0 \\ y \ge 0 \end{cases}$

First, graph the parabola with a solid line.

Does (0, 2) satisfy $y \le x^2$?

$2 \le 0^2$

$2 \le 0$ false

Shade the portion of the graph which does not contain (0, 2).

Next, graph $y = x + 2$ with a solid line.

Does (0, 0) satisfy $y \ge x + 2$?

$0 \ge 0 + 2$

$0 \ge 2$ false

Shade the portion of the graph which does not contain (0, 0).

Next, graph $x = 0$ with a solid line.

Does (1, 1) satisfy $x \ge 0$?

$1 \ge 0$ true

Shade the portion of the graph which contains (1, 1).

Next, graph $y = 0$ with a solid line.

Does (1, 1) satisfy $y \ge 0$?

$1 \ge 0$ true

Shade the portion of the graph which contains (1, 1).

The solution to the system is the overlapping region.

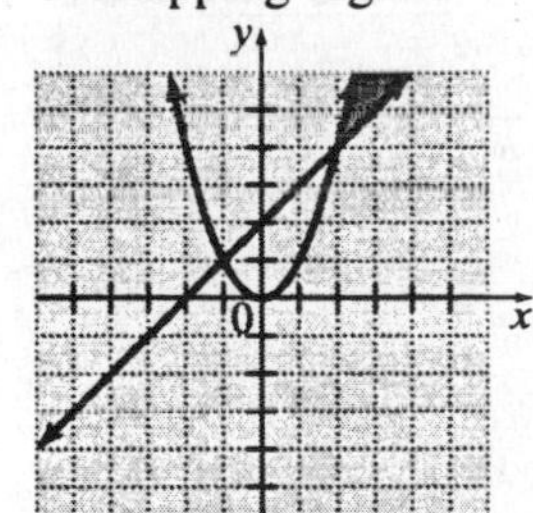

45. This is a function because any vertical line will intersect the graph only once.

47. This is not a function because a vertical line can cross the graph at two places.

49. $f(x) = 3x^2 - 2$

$f(-3) = 3(-3)^2 - 2$

$= 3(9) - 2$

$f(-3) = 25$

51. $f(x)=3x^2-2$
$f(b)=3b^2-2$

Chapter 13 Review

1. (–6, 3) and (8, 4)
$d=\sqrt{[8-(-6)]^2+(4-3)^2}$
$d=\sqrt{(14)^2+1^2}$
$d=\sqrt{197}$

2. (3, 5) and (8, 9)
$d=\sqrt{(8-3)^2+(9-5)^2}$
$d=\sqrt{5^2+4^2}$
$d=\sqrt{25+16}$
$d=\sqrt{41}$

3. (–4, –6) and (–1, 5)
$d=\sqrt{[-1-(-4)]^2+[5-(-6)]^2}$
$d=\sqrt{3^2+11^2}$
$d=\sqrt{9+121}$
$d=\sqrt{130}$

4. (–1, 5) and (2, –3)
$d=\sqrt{[2-(-1)]^2+(-3-5)^2}$
$d=\sqrt{3^2+(-8)^2}$
$d=\sqrt{9+64}$
$d=\sqrt{73}$

5. $(-\sqrt{2},\ 0)$ and $(0,\ -4\sqrt{6})$
$d=\sqrt{[0-(-\sqrt{2})]^2+(-4\sqrt{6}-0)^2}$
$d=\sqrt{(\sqrt{2})^2+(-4\sqrt{6})^2}$
$d=\sqrt{2+96}$
$d=\sqrt{98}$
$d=7\sqrt{2}$

6. $(-\sqrt{5},\ -\sqrt{11})$ and $(-\sqrt{5},\ -3\sqrt{11})$
$d=\sqrt{[-\sqrt{5}-(-\sqrt{5})]^2+[-3\sqrt{11}-(-\sqrt{11})]^2}$
$d=\sqrt{0+(-2\sqrt{11})^2}$
$d=\sqrt{44}$
$d=2\sqrt{11}$

7. (7.4, –8.6) and (–1.2, 5.6)
$d=\sqrt{(-1.2-7.4)^2+[5.6-(-8.6)]^2}$
$d=\sqrt{(-8.6)^2+(14.2)^2}$
$d=\sqrt{275.6}$
$d\approx 16.60$

8. (2.3, 1.8) and (10.7, –9.2)
$d=\sqrt{(10.7-2.3)^2+(-9.2-1.8)^2}$
$d=\sqrt{(8.4)^2+(-11)^2}$
$d=\sqrt{191.56}$
$d\approx 13.84$

9. (2, 6) and (–12, 4)
$\left(\frac{2-12}{2},\ \frac{6+4}{2}\right)$
The midpoint is (–5, 5).

10. (–3, 8) and (11, 24)
$\left(\frac{-3+11}{2},\ \frac{8+24}{2}\right)$
The midpoint is (4, 16).

11. (–6, –5) and (–9, 7)
$\left(\frac{-6-9}{2},\ \frac{-5+7}{2}\right)$
The midpoint is $\left(-\frac{15}{2},\ 1\right)$.

12. (4, –6) and (–15, 2)
$\left(\frac{4-15}{2},\ \frac{-6+2}{2}\right)$
The midpoint is $\left(-\frac{11}{2},\ -2\right)$.

13. $\left(0, -\frac{3}{8}\right)$ and $\left(\frac{1}{10}, 0\right)$

$$\left(\frac{0+\frac{1}{10}}{2}, \frac{-\frac{3}{8}+0}{2}\right)$$

The midpoint is $\left(\frac{1}{20}, -\frac{3}{16}\right)$.

14. $\left(\frac{3}{4}, -\frac{1}{7}\right)$ and $\left(-\frac{1}{4}, -\frac{3}{7}\right)$

$$\left(\frac{\frac{3}{4}-\frac{1}{4}}{2}, \frac{-\frac{1}{7}-\frac{3}{7}}{2}\right)$$

The midpoint is $\left(\frac{1}{4}, -\frac{2}{7}\right)$.

15. $(\sqrt{3}, -2\sqrt{6})$ and $(\sqrt{3}, -4\sqrt{6})$

$$\left(\frac{\sqrt{3}+\sqrt{3}}{2}, \frac{-2\sqrt{6}-4\sqrt{6}}{2}\right)$$

The midpoint is $(\sqrt{3}, -3\sqrt{6})$.

16. $(-5\sqrt{3}, 2\sqrt{7})$ and $(-3\sqrt{3}, 10\sqrt{7})$

$$\left(\frac{-5\sqrt{3}-3\sqrt{3}}{2}, \frac{2\sqrt{7}+10\sqrt{7}}{2}\right)$$

The midpoint is $(-4\sqrt{3}, 6\sqrt{7})$.

17. center (–4, 4), radius 3

$$(x-(-4))^2+(y-4)^2=3^2$$
$$(x+4)^2+(y-4)^2=9$$

18. center (5, 0), radius 5

$$(x-5)^2+(y-0)^2=5^2$$
$$(x-5)^2+y^2=25$$

19. center (–7, –9), radius $\sqrt{11}$

$$(x-(-7))^2+(y-(-9))^2=\left(\sqrt{11}\right)^2$$
$$(x+7)^2+(y+9)^2=11$$

20. center (0, 0), radius $\frac{7}{2}$

$$(x-0)^2+(y-0)^2=\left(\frac{7}{2}\right)^2$$
$$x^2+y^2=\frac{49}{4}$$

21. $x^2+y^2=7$

or $(x-0)^2+(y-0)^2=\sqrt{7}^2$

The center is (0, 0).

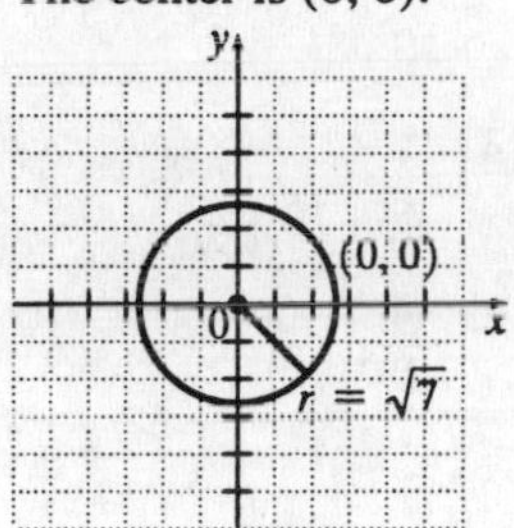

22. $x=2(y-5)^2+4$

The vertex is (4, 5).

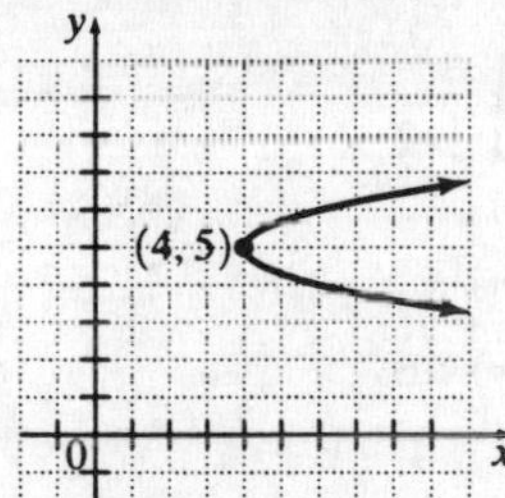

23. $x=-(y+2)^2+3$

The vertex is (3, –2).

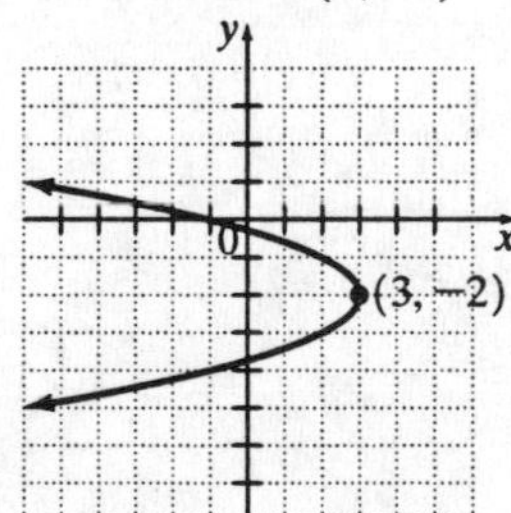

24. 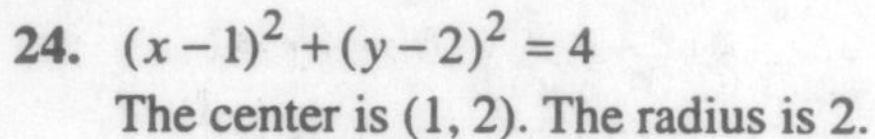
$(x-1)^2+(y-2)^2=4$

The center is (1, 2). The radius is 2.

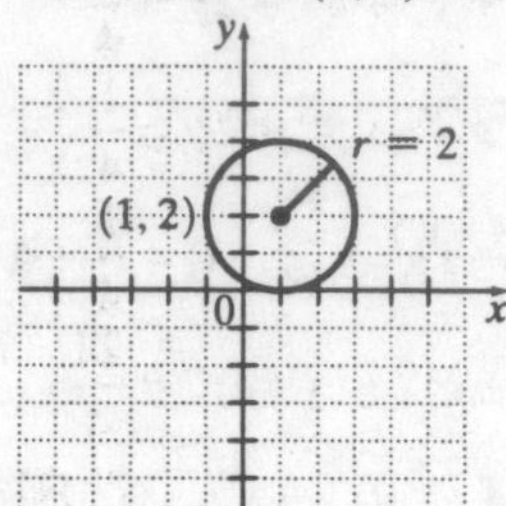

25. $y=-x^2+4x+10$

$y=-(x^2-4x)+10$

$y=-(x^2-4x+4)+10+4$

$y=-(x-2)^2+14$

The vertex is (2, 14).

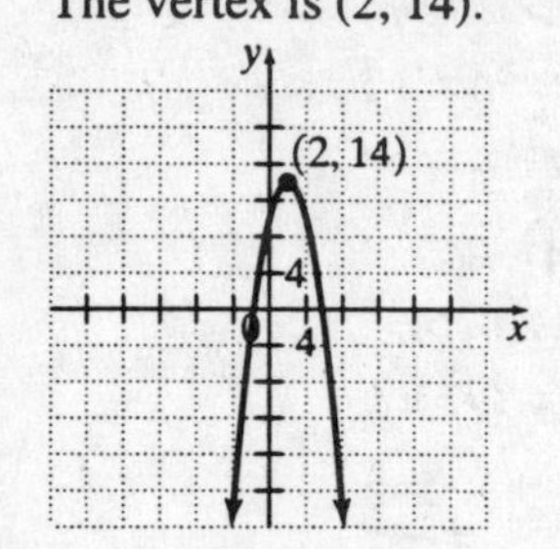

26. $x=-y^2-4y+6$

$x=-(y^2+4y)+6$

$x=-(y^2+4y+4)+6+4$

$x=-(y+2)^2+10$

The vertex is (10, –2).

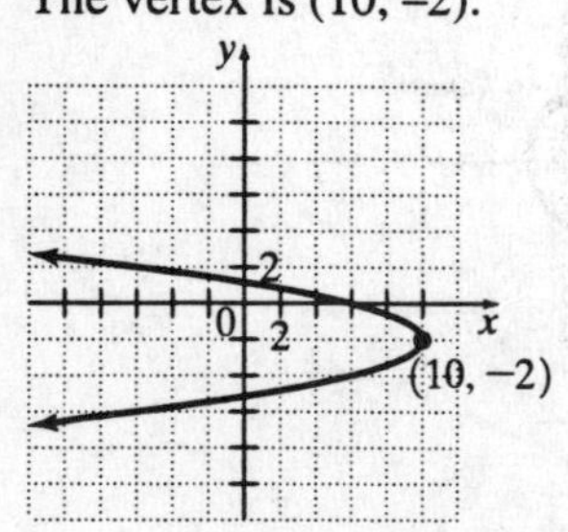

27. $x=\frac{1}{2}y^2+2y+1$

$x=\frac{1}{2}(y^2+4y)+1$

$x=\frac{1}{2}(y^2+4y+4)+1-2$

$x=\frac{1}{2}(y+2)^2-1$

The vertex is (–1, –2).

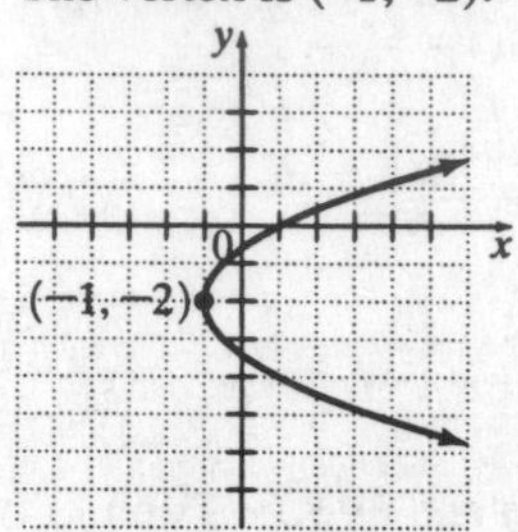

28. $y=-3x^2+\frac{1}{2}x+4$

$y=-3\left(x^2-\frac{1}{6}x\right)+4$

$y=-3\left(x^2-\frac{1}{6}x+\frac{1}{144}\right)+4+\frac{3}{144}$

$y=-3\left(x-\frac{1}{12}\right)^2+\frac{193}{48}$

The vertex is $\left(\frac{1}{12}, \frac{193}{48}\right)$.

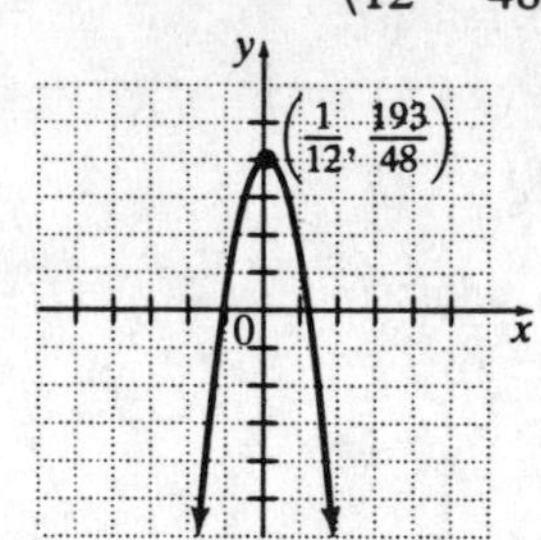

29.

$$x^2 + y^2 + 2x + y = \frac{3}{4}$$
$$(x^2 + 2x + 1) + \left(y^2 + y + \frac{1}{4}\right) = \frac{3}{4} + 1 + \frac{1}{4}$$
$$(x+1)^2 + \left(y + \frac{1}{2}\right)^2 = 2$$
$$(x+1)^2 + \left(y + \frac{1}{2}\right)^2 = \left(\sqrt{2}\right)^2$$

The center is $\left(-1, -\frac{1}{2}\right)$.

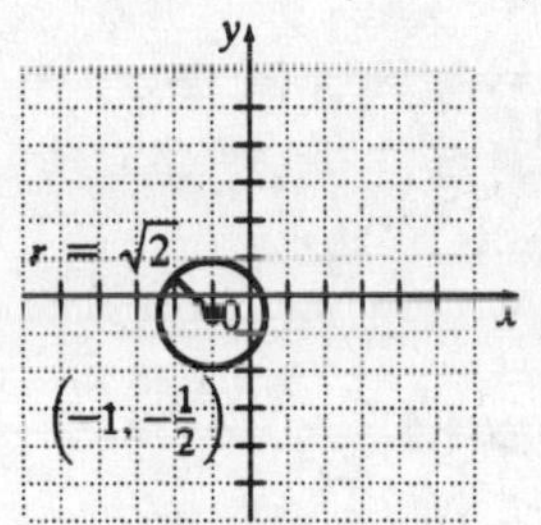

30.

$$x^2 + y^2 + 3y = \frac{7}{4}$$
$$x^2 + \left(y^2 + 3y + \frac{9}{4}\right) = \frac{7}{4} + \frac{9}{4}$$
$$x^2 + \left(y + \frac{3}{2}\right)^2 = 4$$
$$x^2 + \left(y + \frac{3}{2}\right)^2 = 2^2$$

The center is $\left(0, -\frac{3}{2}\right)$.

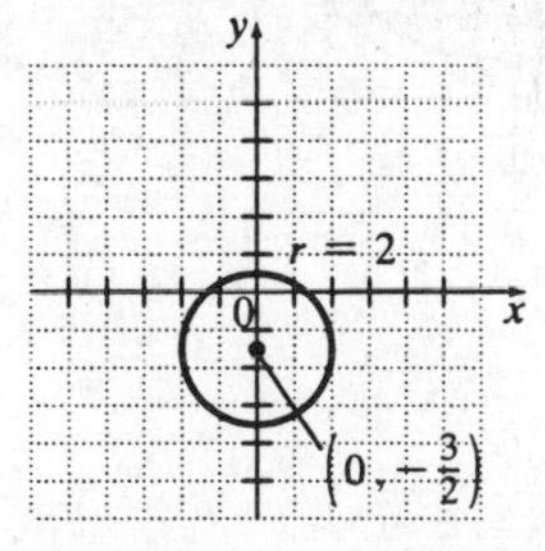

31.

$$4x^2 + 4y^2 + 16x + 8y = 1$$
$$x^2 + y^2 + 4x + 2y = \frac{1}{4}$$
$$x^2 + 4x + y^2 + 2y = \frac{1}{4}$$
$$(x^2 + 4x + 4) + (y^2 + 2y + 1) = \frac{1}{4} + 4 + 1$$
$$(x+2)^2 + (y+1)^2 = \frac{21}{4}$$
$$(x+2)^2 + (y+1)^2 = \left(\frac{\sqrt{21}}{2}\right)^2$$

The center is (–2, –1).

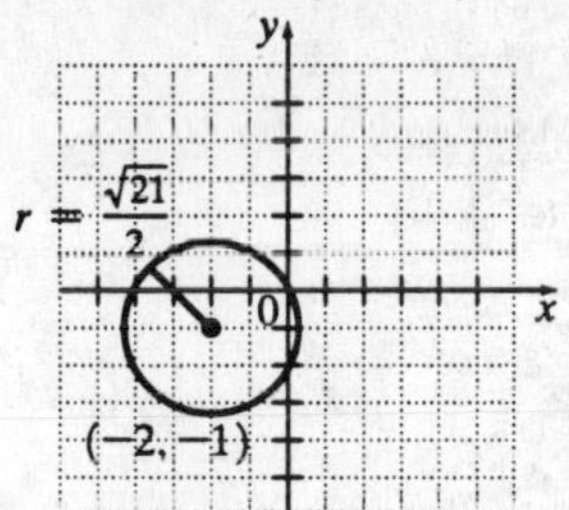

32.

$$3x^2 + 6x + 3y^2 = 9$$
$$x^2 + 2x + y^2 = 3$$
$$(x^2 + 2x + 1) + y^2 = 3 + 1$$
$$(x+1)^2 + y^2 = 4$$
$$(x+1)^2 + y^2 = 2^2$$

The center is (–1, 0).

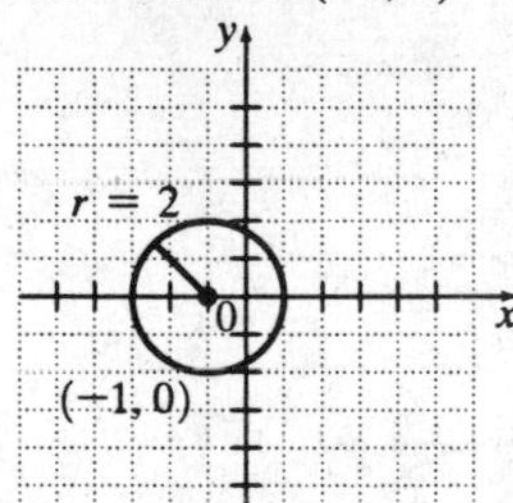

33. $y = x^2 + 6x + 9$
$y = (x+3)^2 + 0$
The vertex is (–3, 0).

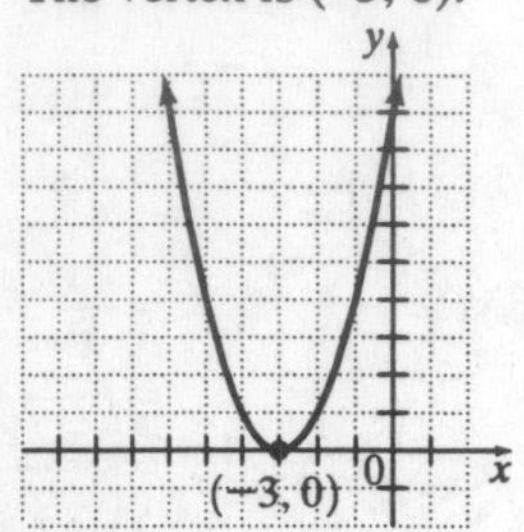

34. $x = y^2 + 6y + 9$
$x = (y+3)^2 + 0$
The vertex is (0, –3).

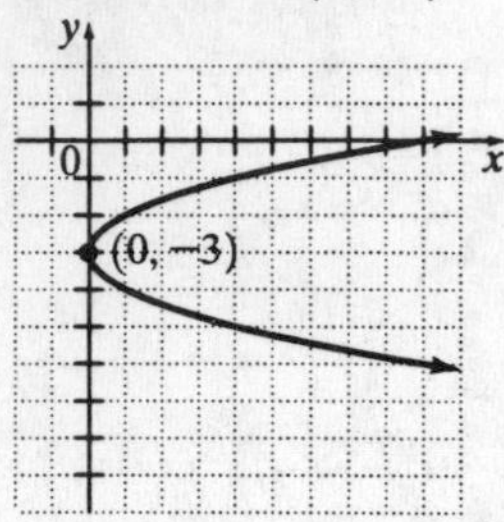

35. The center is (5.6, –2.4).
$\text{radius} = \frac{6.2}{2} = 3.1$
$(x-5.6)^2 + [y-(-2.4)]^2 = 3.1^2$
or $(x-5.6)^2 + (y+2.4)^2 = 9.61$

36. $x^2 + \frac{y^2}{4} = 1$
$a = 1, b = 2$

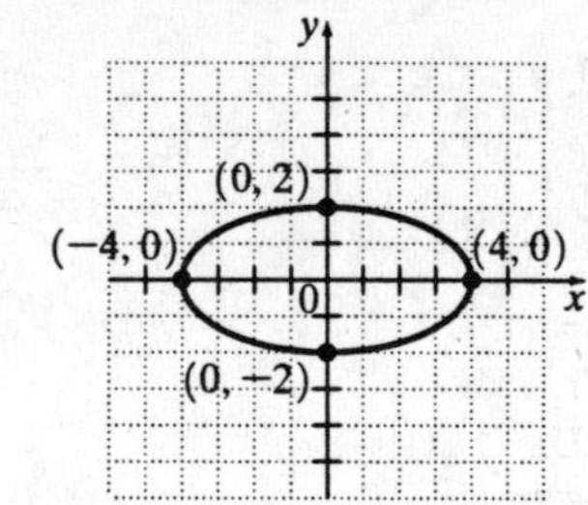

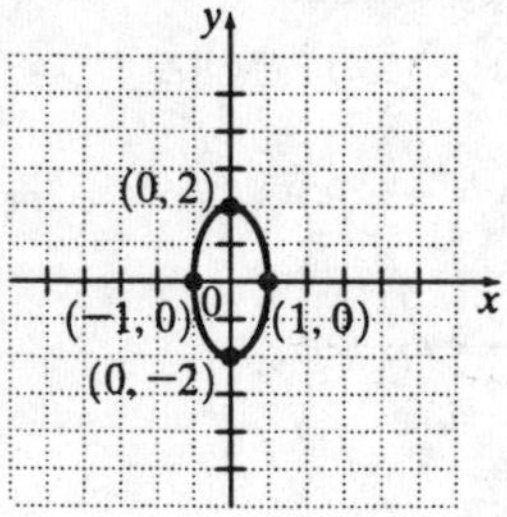

37. $x^2 - \frac{y^2}{4} = 1$
$a = 1, b = 2$

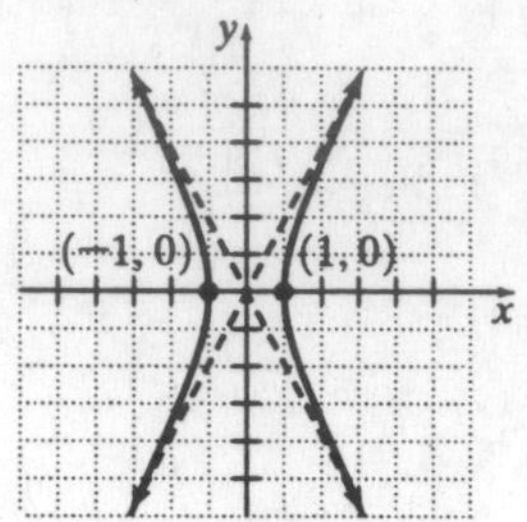

38. $\frac{y^2}{4} - \frac{x^2}{16} = 1$
$a = 4, b = 2$

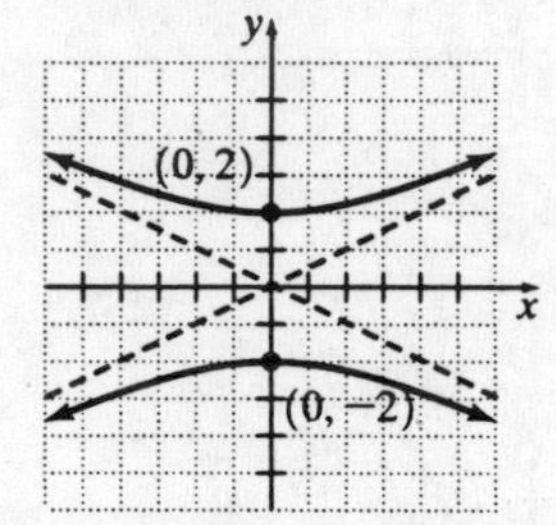

39. $\frac{y^2}{4} + \frac{x^2}{16} = 1$
$a = 4, b = 2$

40. $$\frac{x^2}{5}+\frac{y^2}{5}=1$$
$$x^2+y^2=5$$
$$(x-0)^2+(y-0)^2=(\sqrt{5})^2$$
center (0, 0), radius = $\sqrt{5}$

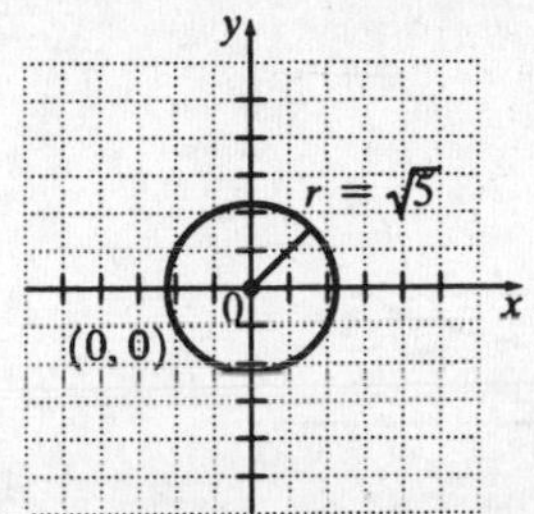

41. $$\frac{x^2}{5}-\frac{y^2}{5}=1$$
$a=\sqrt{5},\ b=\sqrt{5}$

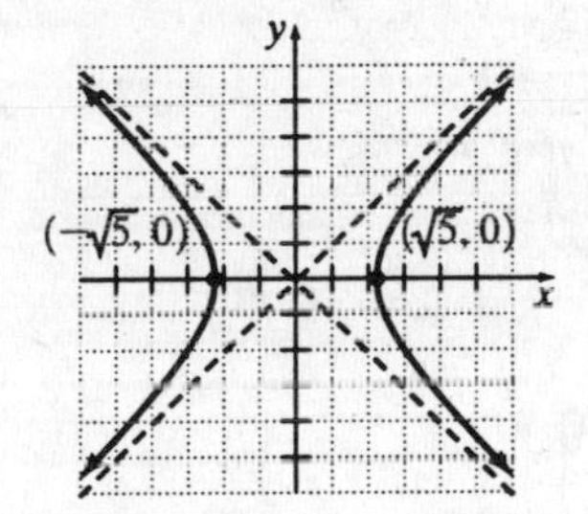

42. $$-5x^2+25y^2=125$$
$$\frac{y^2}{5}-\frac{x^2}{25}=1$$
$a=5,\ b=\sqrt{5}$

43. $$4y^2+9x^2=36$$
$$\frac{y^2}{9}+\frac{x^2}{4}=1$$
$a=2,\ b=3$

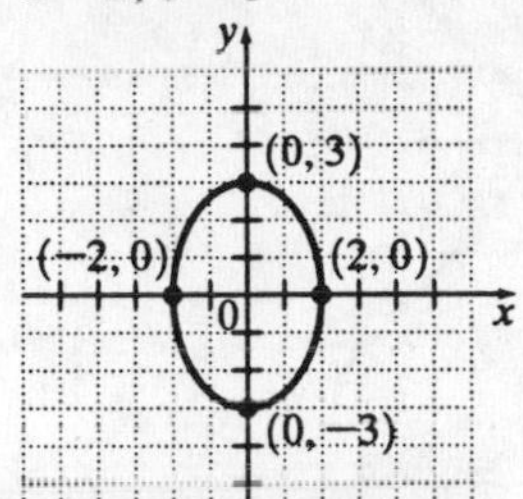

44. $$\frac{(x-2)^2}{4}+(y-1)^2=1$$
$$\frac{(x-2)^2}{2^2}+\frac{(y-1)^2}{1^2}=1$$
$a=2,\ b=1$, center (2, 1)

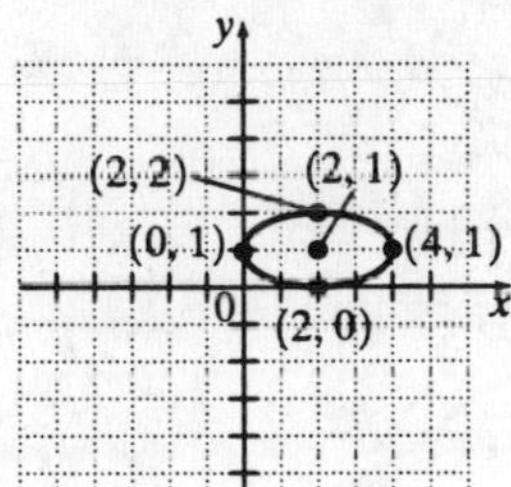

45. $$\frac{(x+3)^2}{9}+\frac{(y-4)^2}{25}=1$$
$$\frac{(x+3)^2}{3^2}+\frac{(y-4)^2}{5^2}=1$$
$a=3,\ b=5$, center (−3, 4)

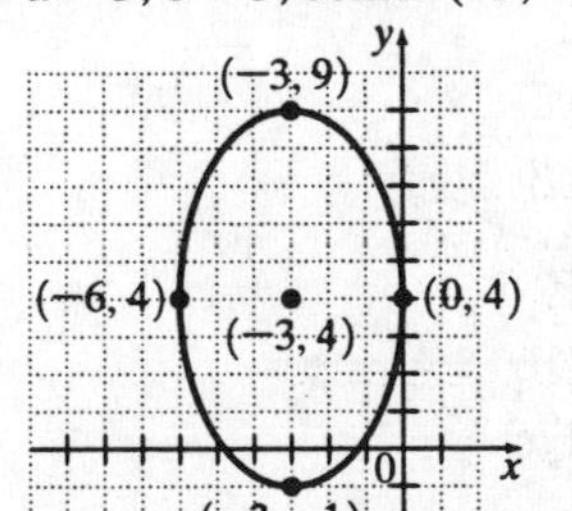

46. $x^2 - y^2 = 1$

$a = 1, b = 1$

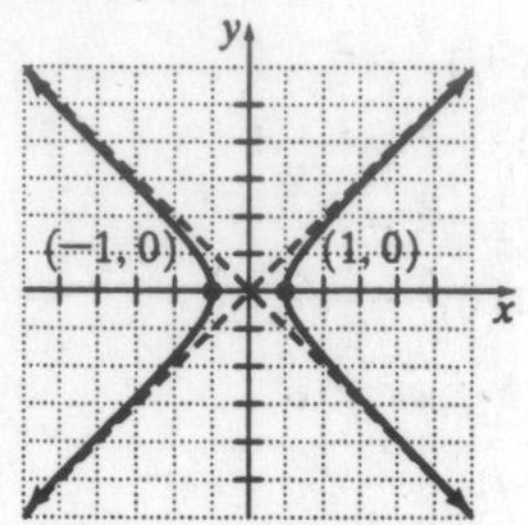

47. $36y^2 - 49x^2 = 1764$

$$\frac{y^2}{49} - \frac{x^2}{36} = 1$$

$$\frac{y^2}{7^2} - \frac{x^2}{6^2} = 1$$

$a = 6, b = 7$

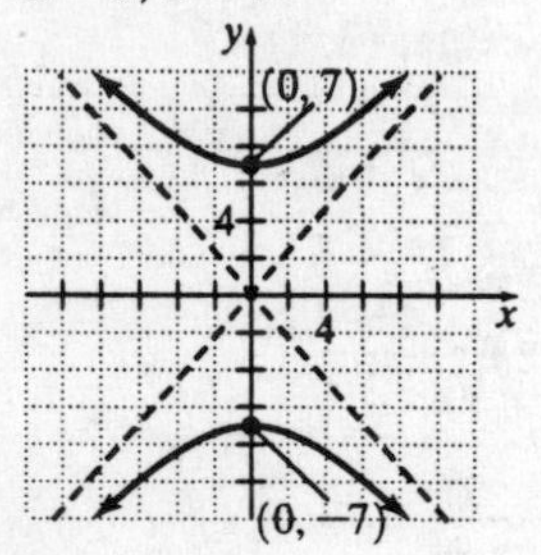

48. $y^2 = x^2 + 9$

$$\frac{y^2}{9} - \frac{x^2}{9} = 1$$

$$\frac{y^2}{3^2} - \frac{x^2}{3^2} = 1$$

$a = 3, b = 3$

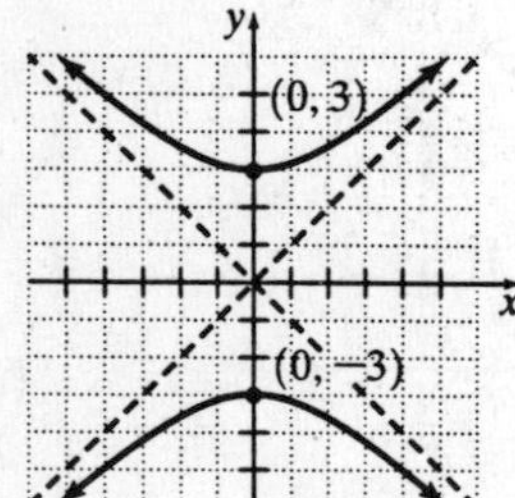

49. $x^2 = 4y^2 - 16$

$$16 = 4y^2 - x^2$$

$$1 = \frac{y^2}{4} - \frac{x^2}{16}$$

$$\frac{y^2}{2^2} - \frac{x^2}{4^2} = 1$$

$a = 4, b = 2$

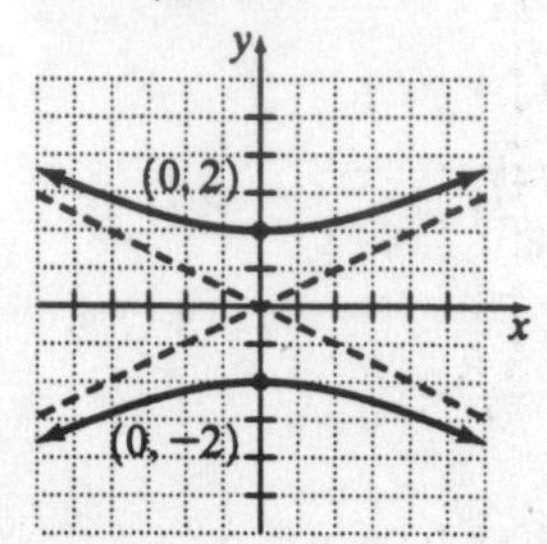

50. $100 - 25x^2 = 4y^2$

$$\frac{x^2}{4} + \frac{y^2}{25} = 1$$

$$\frac{x^2}{2^2} + \frac{y^2}{5^2} = 1$$

$a = 2, b = 5$

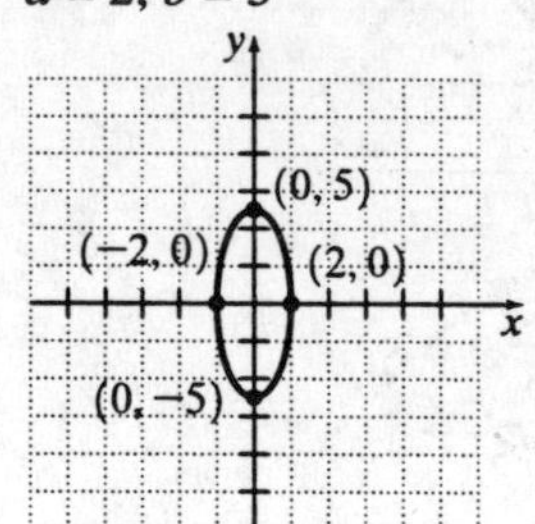

51. parabola

$$y = x^2 + 4x + 6$$
$$y = (x^2 + 4x + 4) + 6 - 4$$
$$y = (x+2)^2 + 2$$

vertex (−2, 2)

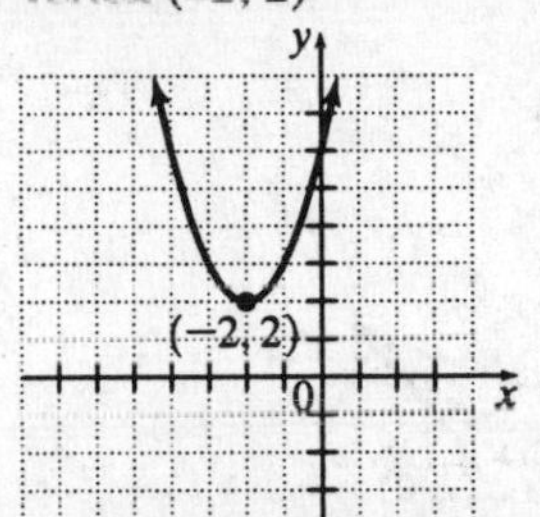

52. hyperbola

$$y^2 = x^2 + 6$$
$$\frac{y^2}{6} - \frac{x^2}{6} = 1$$
$$\frac{y^2}{(\sqrt{6})^2} - \frac{x^2}{(\sqrt{6})^2} = 1$$

$a = \sqrt{6},\ b = \sqrt{6}$

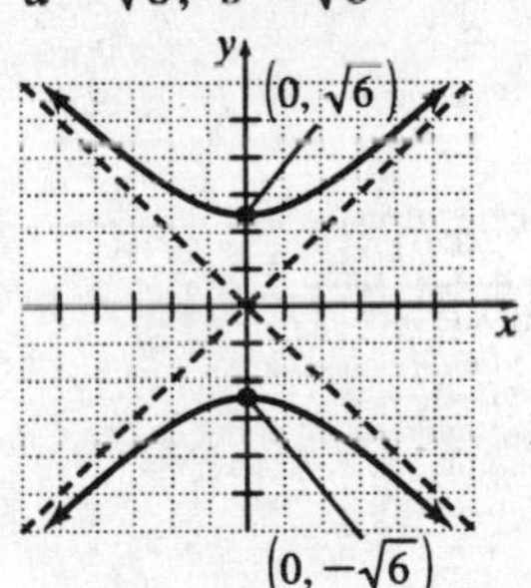

53. circle

$$y^2 + x^2 = 4x + 6$$
$$x^2 - 4x + y^2 = 6$$
$$(x^2 - 4x + 4) + y^2 = 6 + 4$$
$$(x-2)^2 + y^2 = 10$$
$$(x-2)^2 + (y-0)^2 = (\sqrt{10})^2$$

center (2, 0), radius $\sqrt{10}$

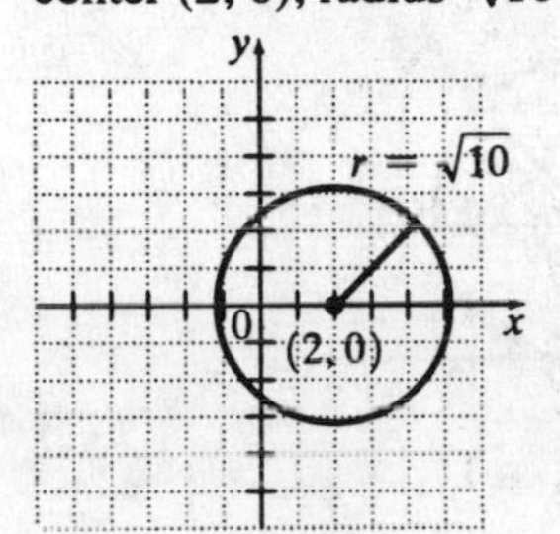

54. ellipse

$$y^2 + 2x^2 = 4x + 6$$
$$y^2 + 2(x^2 - 2x + 1) = 6 + 2$$
$$y^2 + 2(x-1)^2 = 8$$
$$\frac{y^2}{8} + \frac{(x-1)^2}{4} = 1$$
$$\frac{y^2}{(2\sqrt{2})^2} + \frac{(x-1)^2}{(2)^2} = 1$$

$a = 2,\ b = 2\sqrt{2}$

center (1, 0)

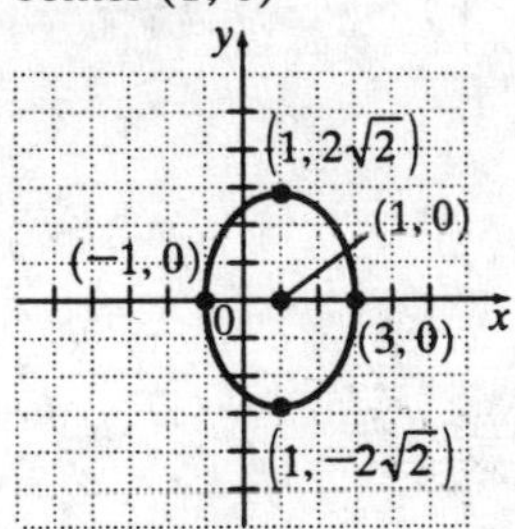

55. circle

$$x^2 + y^2 - 8y = 0$$
$$x^2 + (y^2 - 8y + 16) = 0 + 16$$
$$x^2 + (y - 4)^2 = 16$$
$$(x - 0)^2 + (y - 4)^2 = 4^2$$

center (0, 4), radius 4

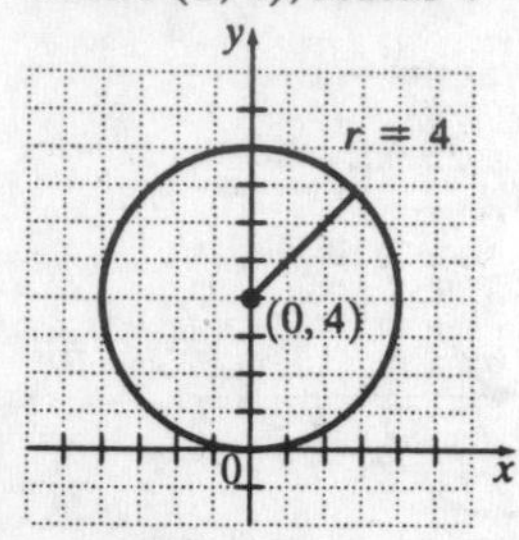

56. parabola

$$x - 4y = y^2$$
$$x = (y^2 + 4y + 4) - 4$$
$$x = (y + 2)^2 - 4$$

vertex (–4, –2)

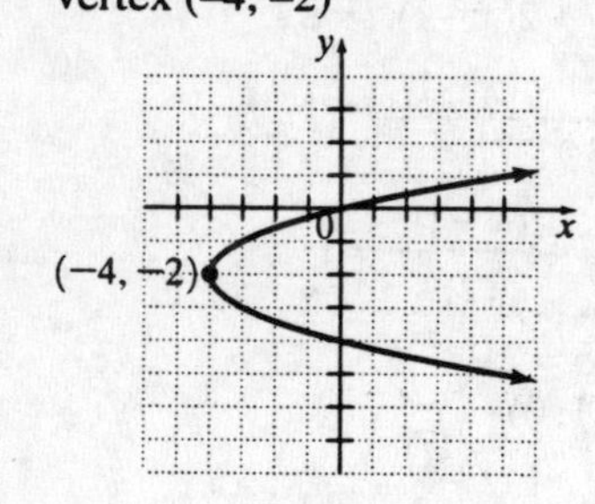

57. hyperbola

$$x^2 - 4 = y^2$$
$$x^2 - y^2 = 4$$
$$\frac{x^2}{4} - \frac{y^2}{4} = 1$$
$$\frac{x^2}{2^2} - \frac{y^2}{2^2} = 1$$

$a = 2, b = 2$

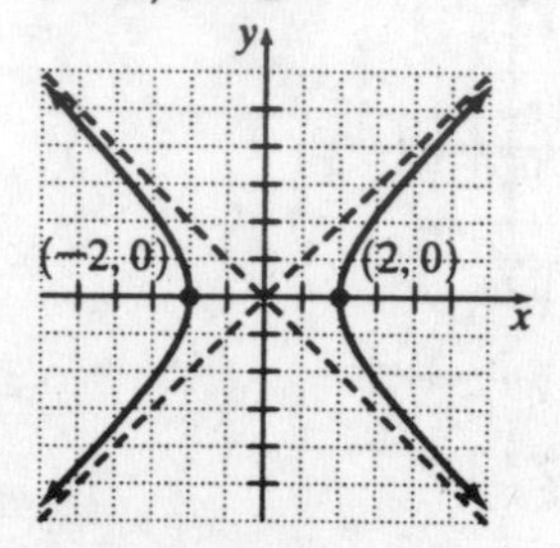

58. circle

$$x^2 = 4 - y^2$$
$$x^2 + y^2 = 4$$

center (0, 0), radius 2

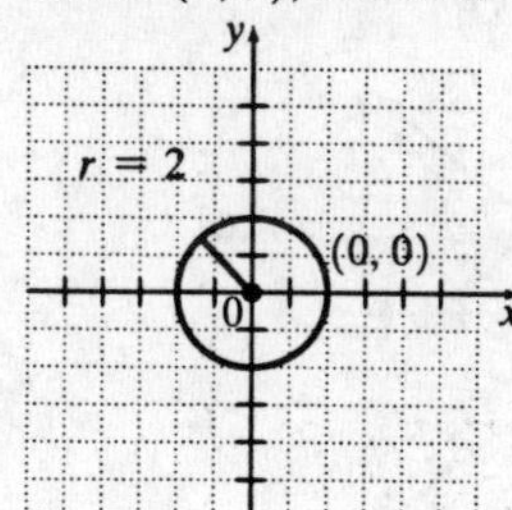

59. ellipse

$6(x-2)^2+9(y+5)^2=36$

$\frac{(x-2)^2}{6}+\frac{(y+5)^2}{4}=1$

$\frac{(x-2)^2}{\sqrt{6}^2}+\frac{(y+5)^2}{2^2}=1$

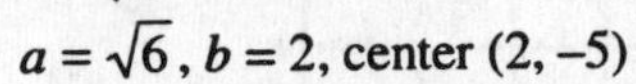

$a=\sqrt{6}$, $b=2$, center $(2,-5)$

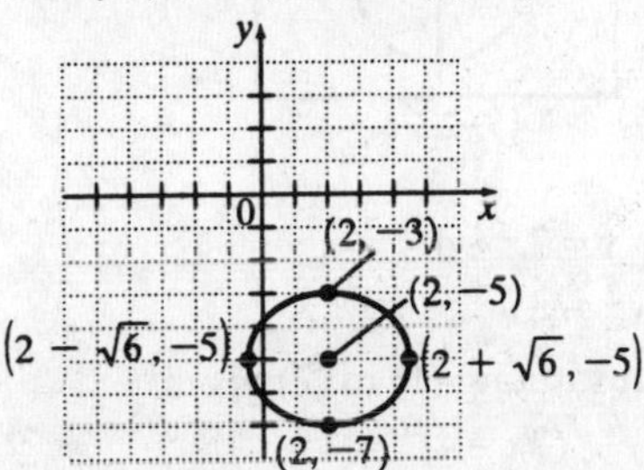

60. hyperbola

$36y^2=576+16x^2$

$\frac{y^2}{16}-\frac{x^2}{36}=1$

$\frac{y^2}{4^2}-\frac{x^2}{6^2}=1$

$a=6$, $b=4$

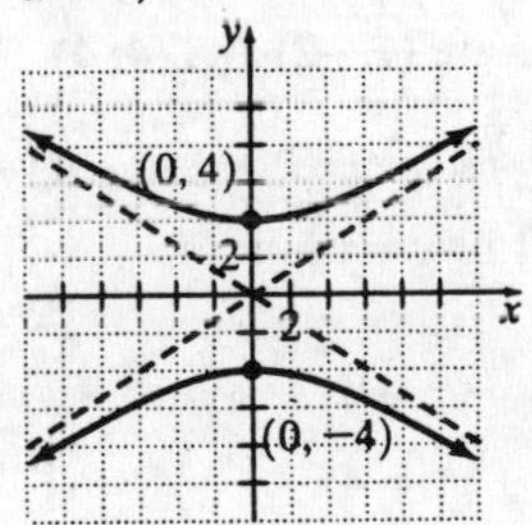

61. hyperbola

$\frac{x^2}{16}-\frac{y^2}{25}=1$

$\frac{x^2}{4^2}-\frac{y^2}{5^2}=1$

$a=4$, $b=5$

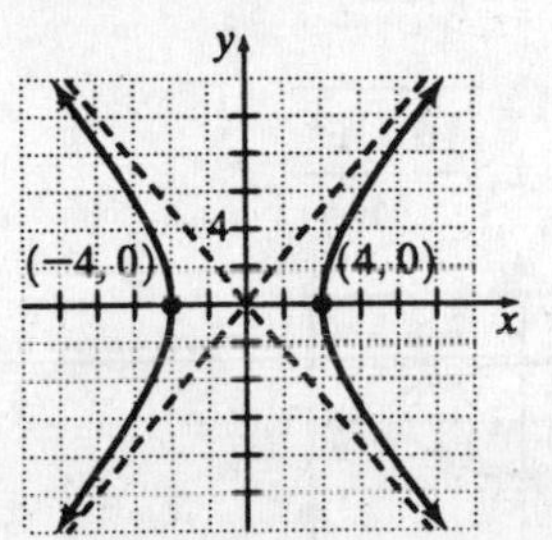

62. circle

$3(x-7)^2+3(y+4)^2=1$

$(x-7)^2+(y+4)^2=\frac{1}{3}$

$(x-7)^2+(y+4)^2=\left(\frac{\sqrt{3}}{3}\right)^2$

center $(7,-4)$, radius $=\frac{\sqrt{3}}{3}$

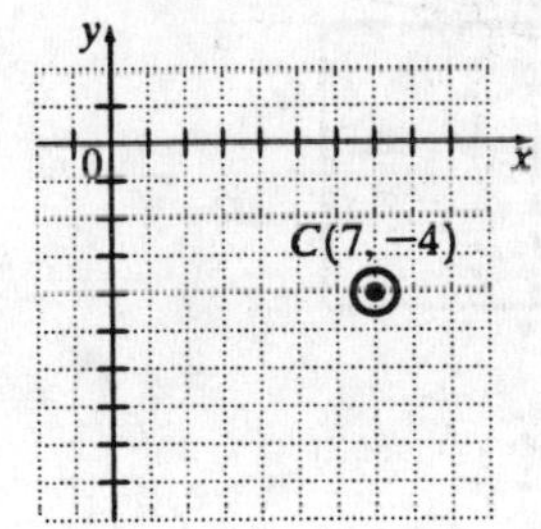

63. $\frac{y^2}{4}+\frac{x^2}{16}=1$

$4y^2+x^2=16$

$4y^2=16-x^2$

$y^2=\frac{16-x^2}{4}$

$y=\pm\sqrt{\frac{16-x^2}{4}}$

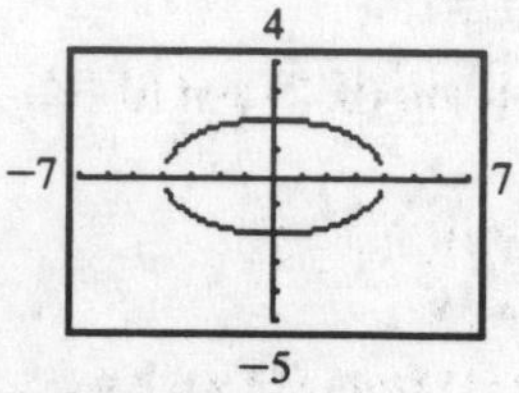

64. $\frac{x^2}{5}+\frac{y^2}{5}=1$

$x^2+y^2=5$

$y^2=5-x^2$

$y=\pm\sqrt{5-x^2}$

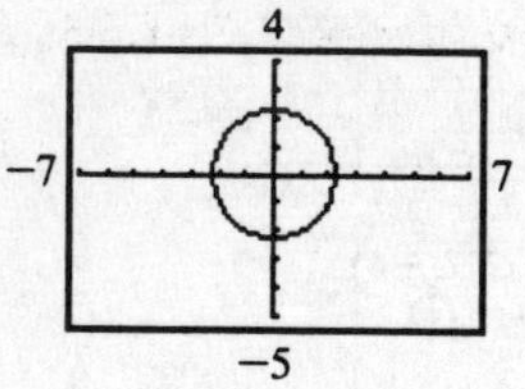

65. $y=x^2+4x+6$

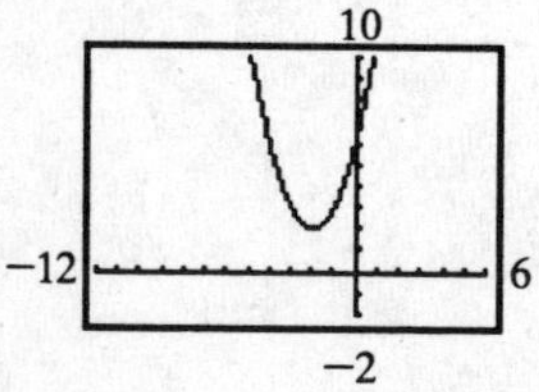

66. $x^2=4-y^2$

$y^2=4-x^2$

$y=\pm\sqrt{4-x^2}$

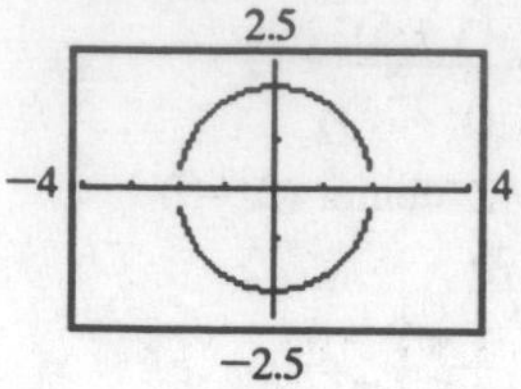

67. $\begin{cases} y=2x-4 \\ y^2=4x \end{cases}$

Substituting (1) in (2) gives

$(2x-4)^2=4x$

$4x^2-16x+16=4x$

$4x^2-20x+16=0$

$x^2-5x+4=0$

$(x-1)(x-4)=0$

$x-1=0$ or $x-4=0$

$x=1$ or $x=4$

$x=1: y=2(1)-4=-2$

$x=4: y=2(4)-4=4$

The solutions are (1, –2) and (4, 4).

68. $\begin{cases} x^2+y^2=4 \\ x-y=4 \text{ or } x=y+4 \end{cases}$

Substitute.

$(y+4)^2+y^2=4$

$y^2+8y+16+y^2=4$

$2y^2+8y+12=0$

$y^2+4y+6=0$

$b^2-4ac=4^2-4\cdot 1\cdot 6=16-24=-8$

Therefore, no real solutions exist.

The system has no solution.

69. $\begin{cases} y = x+2 \\ y = x^2 \end{cases}$

Substituting (1) in (2) gives

$x+2 = x^2$

$0 = x^2 - x - 2$

$0 = (x+1)(x-2)$

$x+1=0$ or $x-2=0$

$x=-1$ or $x=2$

$x=-1: y=-1+2=1$

$x=2: y=2+2=4$

The solutions are (–1, 1) and (2, 4).

70. $\begin{cases} y = x^2 - 5x + 1 \\ y = -x + 6 \end{cases}$

Substitute.

$-x+6 = x^2 - 5x + 1$

$0 = x^2 - 4x - 5$

$0 = (x-5)(x+1)$

$x-5=0$ or $x+1=0$

$x=5$ or $x=-1$

$x=5: y=-5+6=1$

$x=-1: y=-(-1)+6=1+6=7$

The solutions are (5, 1) and (–1, 7).

71. $\begin{cases} 4x - y^2 = 0 \\ 2x^2 + y^2 = 16 \end{cases}$

From (1) we have $y^2 = 4x$. Substituting in (2) gives

$2x^2 + 4x = 16$

$2x^2 + 4x - 16 = 0$

$x^2 + 2x - 8 = 0$

$(x-2)(x+4) = 0$

$x-2=0$ or $x+4=0$

$x=2$ or $x=-4$

$x=2: 4(2) - y^2 = 0$

$y^2 = 8$

$y = \pm 2\sqrt{2}$

$x=-4: 4(-4) - y^2 = 0$

$y^2 = -16$ non-real

The solutions are $(2, 2\sqrt{2})$ and $(2, -2\sqrt{2})$.

72. $\begin{cases} x^2 + 4y^2 = 16 \\ x^2 + y^2 = 4 \end{cases}$

Subtract.

$3y^2 = 12$

$y^2 = 4$

$y = \pm 2$

Substitute back.

$x^2 + 4 = 4$

$x^2 = 0$

$x = 0$

The solutions are (0, 2) and (0, –2).

73. $\begin{cases} x^2 + y^2 = 10 \\ 9x^2 + y^2 = 18 \end{cases}$

Subtracting (1) from (2) we have

$8x^2 = 8$

$x^2 = 1$

$x = \pm 1$

Substitution gives

$1 + y^2 = 10$

$y^2 = 9$

$y = \pm 3$

The solutions are (–1, 3), (–1, –3), (1, 3), and (1, –3).

74. $\begin{cases} x^2 + 2y = 9 \\ 5x - 2y = 5 \end{cases}$

Add.

$$x^2 + 5x = 14$$
$$x^2 + 5x - 14 = 0$$
$$(x+7)(x-2) = 0$$
$$x + 7 = 0 \quad \text{or} \quad x - 2 = 0$$
$$x = -7 \quad \text{or} \quad x = 2$$

$x = -7$: $(-7)^2 + 2y = 9$
$$49 + 2y = 9$$
$$2y = -40$$
$$y = -20$$

$x = 2$: $2^2 + 2y = 9$
$$4 + 2y = 9$$
$$2y = 5$$
$$y = \frac{5}{2}$$

The solutions are $(-7, -20)$ and $\left(2, \frac{5}{2}\right)$.

75. $\begin{cases} y = 3x^2 + 5x - 4 \\ y = 3x^2 - x + 2 \end{cases}$

Subtracting (2) from (1) gives
$$0 = 6x - 6$$
$$6x = 6$$
$$x = 1$$

Substitution gives
$y = 3(1)^2 - 1 + 2 = 4.$
The solution is $(1, 4)$.

76. $\begin{cases} x^2 - 3y^2 = 1 \text{ or } x^2 = 3y^2 + 1 \\ 4x^2 + 5y^2 = 21 \end{cases}$

Substitute.
$$4(3y^2 + 1) + 5y^2 = 21$$
$$12y^2 + 4 + 5y^2 = 21$$
$$17y^2 + 4 = 21$$
$$17y^2 = 17$$
$$y^2 = 1$$
$$y = \pm 1$$

Substitute back.
$$x^2 - 3(1) = 1$$
$$x^2 - 3 = 1$$
$$x^2 = 4$$
$$x = \pm 2$$

The solutions are $(2, 1)$, $(2, -1)$, $(-2, 1)$, and $(-2, -1)$.

77. $\begin{cases} xy = 150 \\ 2x + 2y = 50 \end{cases}$

$$x = \frac{150}{y}$$
$$2\left(\frac{150}{y}\right) + 2y = 50$$
$$\frac{150}{y} + y = 25$$
$$150 + y^2 = 25y$$
$$y^2 - 25y + 150 = 0$$
$$(y - 15)(y - 10) = 0$$
$$y = 15 \text{ or } y = 10$$

$$2x + 2(15) = 50 \qquad 2x + 2(10) = 50$$
$$2x = 20 \qquad 2x = 30$$
$$x = 10 \qquad x = 15$$

The room is 15 feet by 10 feet.

78. The graphs of an ellipse and a hyperbola may intersect at up to four points. Therefore, the greatest possible number of real solutions is four.

79. $y \le -x^2 + 3$

Graph $y = -x^2 + 3$. Does (0, 0) satisfy $y \le x^2 + 3$?

$0 \le -0^2 + 3$

$0 \le 3$ true

Shade the portion of the graph which contains (0, 0).

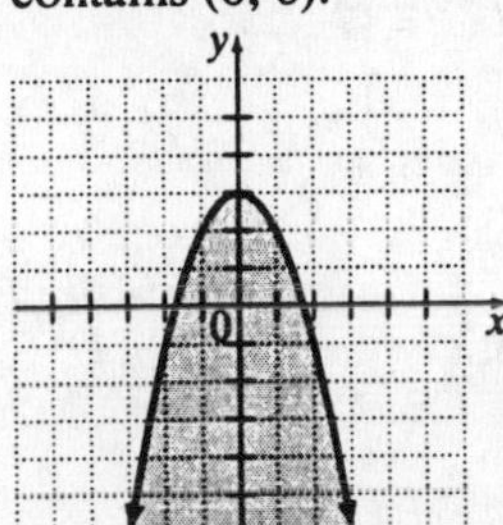

80. $x^2 + y^2 < 9$

Graph $x^2 + y^2 = 9$ with a dashed line.

Does (0, 0) satisfy $x^2 + y^2 < 9$?

$0 + 0 < 9$

$0 < 9$ true

Shade the portion of the graph which contains (0, 0).

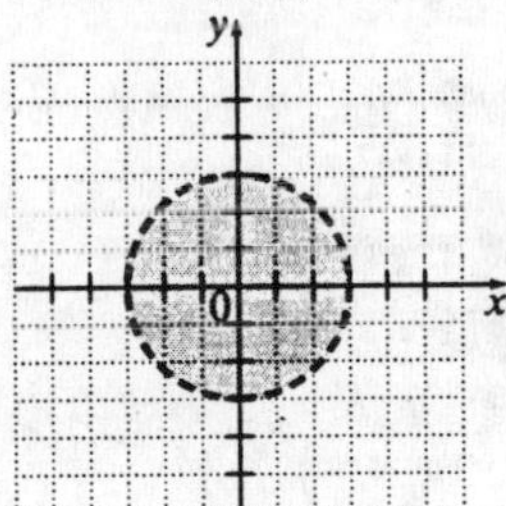

81. $x^2 - y^2 < 1$

Graph $x^2 - y^2 = 1$ with a dashed line.

Does (0, 0) satisfy the inequality?

$0 - 0 < 1$

$0 < 1$ true

Does (2, 0) satisfy the inequality?

$2^2 - 0^2 < 1$

$4 < 1$ false

Does $(-2, 0)$ satisfy the inequality?

$(-2)^2 - 0^2 < 1$

$4 < 1$ false

Shade the portion of the graph containing (0, 0).

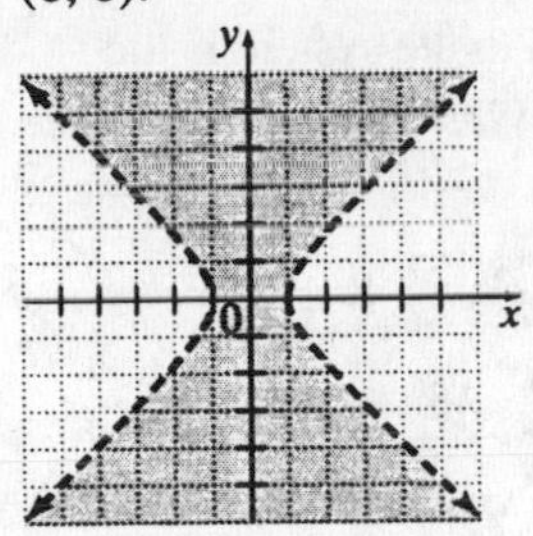

82. $\frac{x^2}{4} + \frac{y^2}{9} \ge 1$

Graph $\frac{x^2}{4} + \frac{y^2}{9} = 1$.

Does (0, 0) satisfy the inequality?

$\frac{0^2}{4} + \frac{0^2}{9} \ge 1$

$0 \ge 1$ false

Shade the portion of the graph that does not contain (0, 0).

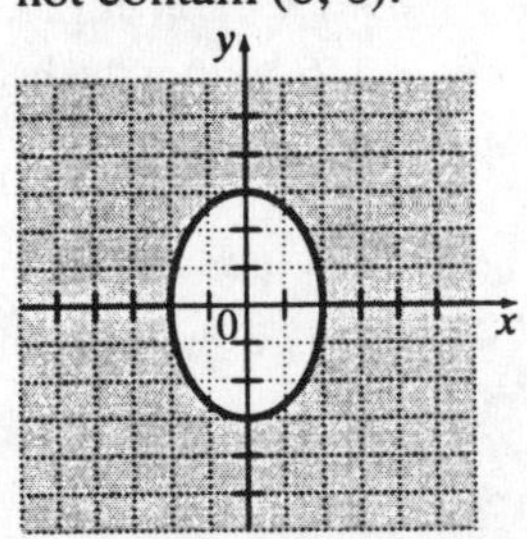

83. $\begin{cases} 2x \le 4 \\ x + y \ge 1 \end{cases}$

First, graph $2x = 4$.

Does (0, 0) satisfy the inequality?

$2 \cdot 0 \le 4$

$0 \le 4$ true

Shade the portion of the graph that contains (0, 0).

Next, graph $x + y = 1$.

Does (0, 0) satisfy the inequality?

$0 + 0 \ge 1$

$0 \ge 1$ false

Shade the portion of the graph that does not contain (0, 0).

The solution to the system is the overlapping region.

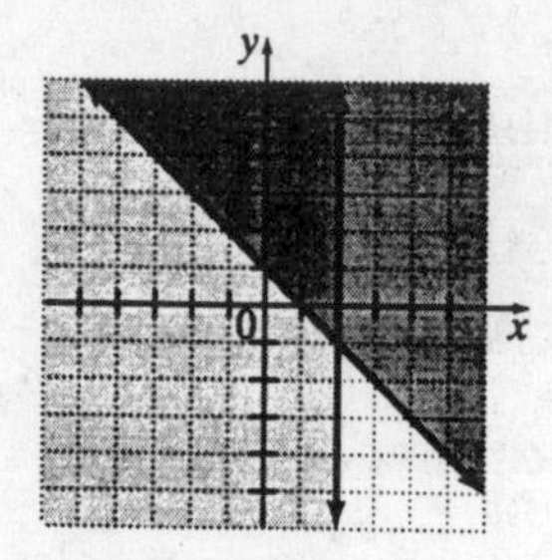

84. $\begin{cases} 3x + 4y \le 12 \\ x - 2y > 6 \end{cases}$

First, graph $3x + 4y = 12$.

Does (0, 0) satisfy the inequality?

$3(0) + 4(0) \le 12$

$0 \le 12$ true

Shade the portion of the graph containing (0, 0).

Next, graph $x - 2y = 6$ with a dashed line.

Does (0, 0) satisfy the inequality?

$0 - 2(0) > 6$

$0 > 6$ false

Shade the portion of the graph that does not contain (0, 0).

The solution to the system is the overlapping region.

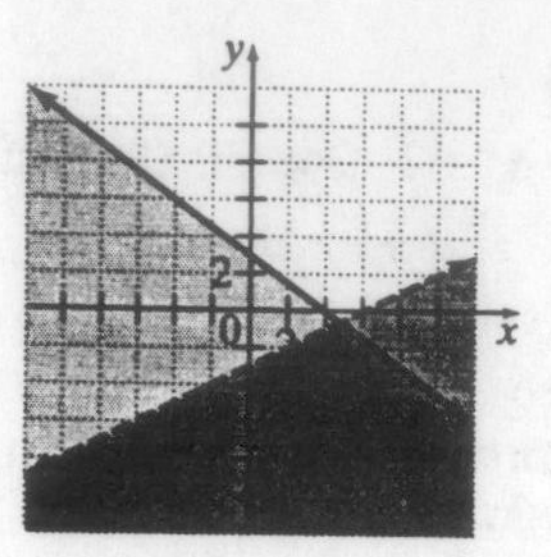

85. $\begin{cases} y > x^2 \\ x + y \ge 3 \end{cases}$

First, graph $y = x^2$ with a dashed line.

Does (0, 1) satisfy the inequality?

$0 > 1^2$

$0 > 1$ false

Shade the portion of the graph that does not contain (0, 0).

Next, graph $x + y = 3$.

Does (0, 0) satisfy the inequality?

$0 + 0 \ge 3$

$0 \ge 3$ false

Shade the portion of the graph that does not contain (0, 0).

The solution to the system is the overlapping region.

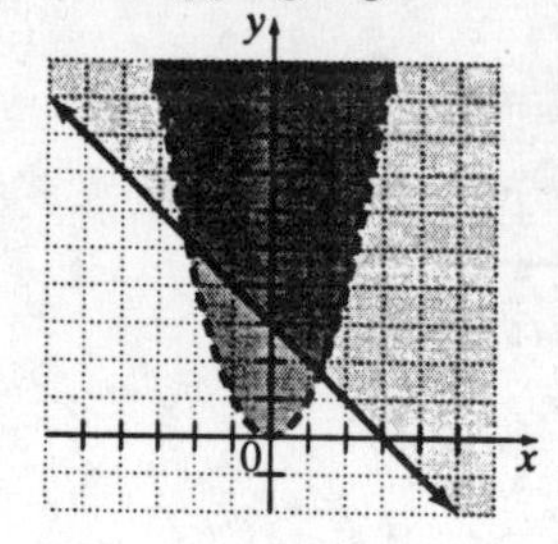

86. $\begin{cases} x^2 + y^2 \le 16 \\ x^2 + y^2 \ge 4 \end{cases}$

First, graph $x^2 + y^2 = 16$.
Does (0, 0) satisfy the inequality?
$0^2 + 0^2 \le 16$
$0 \le 16$ true
Shade the portion of the graph that contains (0, 0).
Next, graph $x^2 + y^2 = 4$.
Does (0, 0) satisfy the inequality?
$0^2 + 0^2 \ge 4$
$0 \ge 4$ false
Shade the portion of the graph that does not contain (0, 0).
The solution to the system is the overlapping region.

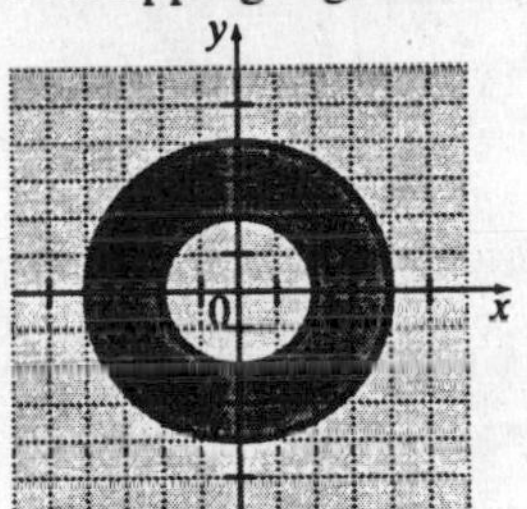

87. $\begin{cases} x^2 + y^2 < 4 \\ x^2 - y^2 \le 1 \end{cases}$

First, graph $x^2 + y^2 = 4$ with a dashed line.
Does (0, 0) satisfy the inequality?
$0^2 + 0^2 < 4$
$0 < 4$ true
Shade the portion of the graph that contains (0, 0).
Next, graph $x^2 - y^2 = 1$.
Does (0, 0) satisfy the inequality?
$0^2 - 0^2 \le 1$
$0 \le 1$ true
Does (2, 0) satisfy the inequality?
$2^2 - 0^2 \le 1$
$4 - 0 \le 1$
$4 \le 1$ false
Does $(-2, 0)$ satisfy the inequality?
$(-2)^2 - 0^2 \le 1$
$4 - 0 \le 1$
$4 \le 1$ false
Shade the portion of the graph that contains (0, 0).
The solution to the system is the overlapping region.

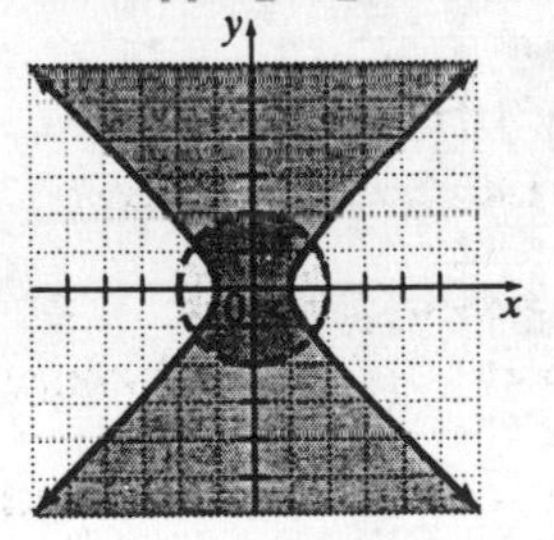

88. $\begin{cases} x^2 + y^2 < 4 \\ y \ge x^2 - 1 \\ x \ge 0 \end{cases}$

First, graph $x^2 + y^2 = 4$ with a dashed line.
Does (0, 0) satisfy the inequality?
$0^2 + 0^2 < 4$
$0 < 4$ true
Shade the portion of the graph that contains (0, 0).
Next, graph $y = x^2 - 1$.
Does (0, 0) satisfy the inequality?
$0 \ge 0^2 - 1$
$0 \ge -1$ true
Shade the portion of the graph that contains (0, 0).
Next, graph $x = 0$.
Does (1, 0) satisfy the inequality?
$1 \ge 0$ true
Shade the portion of the graph that contains (1, 0).
The solution to the system is the overlapping region.

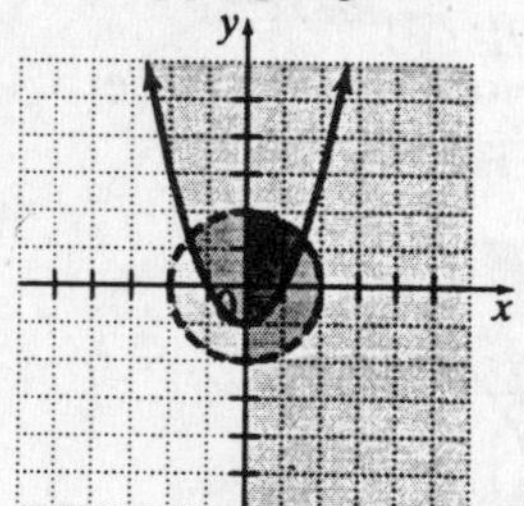

Chapter 13 Test

1. (–6, 3) and (–8, –7)
$d = \sqrt{(-8+6)^2 + (-7-3)^2}$
$d = \sqrt{(-2)^2 + (-10)^2}$
$d = \sqrt{4+100}$
$d = \sqrt{104}$
$d = 2\sqrt{26}$

2. $(-2\sqrt{5}, \sqrt{10})$ and $(-\sqrt{5}, 4\sqrt{10})$
$d = \sqrt{(-\sqrt{5}+2\sqrt{5})^2 + (4\sqrt{10}-\sqrt{10})^2}$
$d = \sqrt{(\sqrt{5})^2 + (3\sqrt{10})^2}$
$d = \sqrt{5+90}$
$d = \sqrt{95}$

3. (–2, –5) and (–6, 12)
$\left(\frac{-2-6}{2}, \frac{-5+12}{2}\right)$
The midpoint of the line segment is $\left(-4, \frac{7}{2}\right)$.

4. $\left(-\frac{2}{3}, -\frac{1}{5}\right)$ and $\left(-\frac{1}{3}, \frac{4}{5}\right)$
$\left(\frac{-\frac{2}{3}-\frac{1}{3}}{2}, \frac{-\frac{1}{5}+\frac{4}{5}}{2}\right)$
The midpoint of the line segment is $\left(-\frac{1}{2}, \frac{3}{10}\right)$.

5.
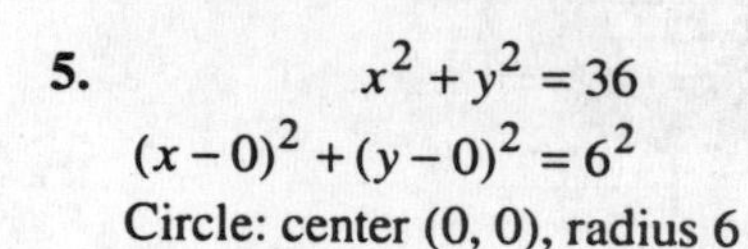
$x^2 + y^2 = 36$
$(x-0)^2 + (y-0)^2 = 6^2$
Circle: center (0, 0), radius 6

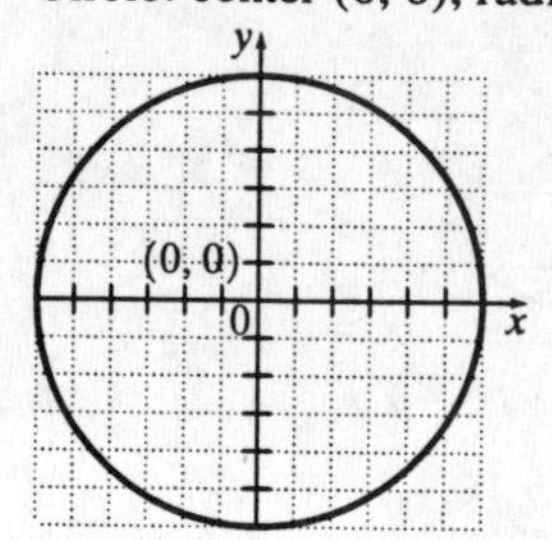

6. $x^2 - y^2 = 36$

$$\frac{x^2}{36} - \frac{y^2}{36} = 1$$

$$\frac{x^2}{6^2} - \frac{y^2}{6^2} = 1$$

Hyperbola:
$a = 6, b = 6$

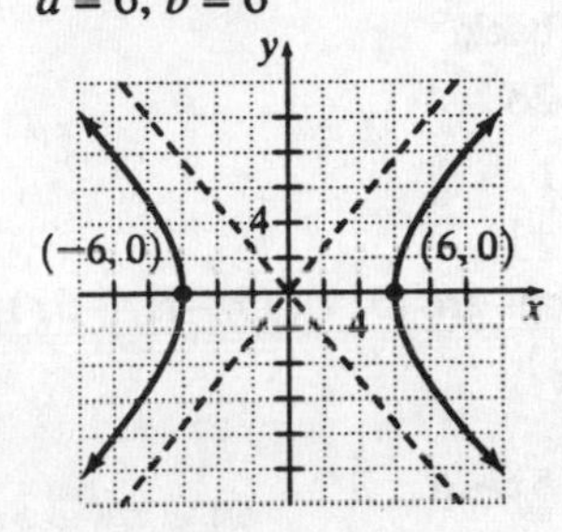

7. $16x^2 + 9y^2 = 144$

$$\frac{x^2}{9} + \frac{y^2}{16} = 1$$

$$\frac{x^2}{3^2} + \frac{y^2}{4^2} = 1$$

Ellipse: $a = 3$, $b = 4$, center (0, 0)

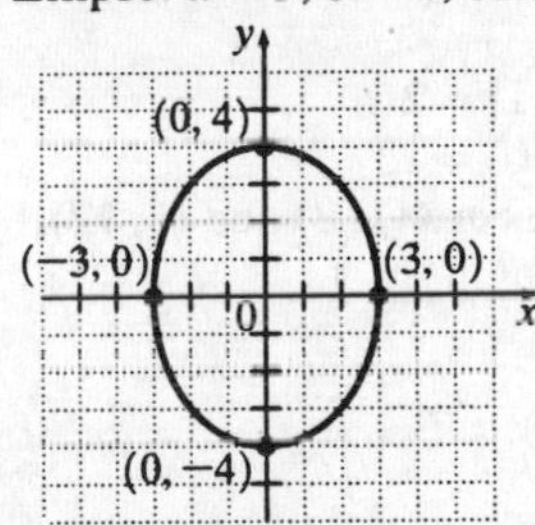

8. $y = x^2 - 8x + 16$

$y = (x - 4)^2 + 0$

Parabola: vertex (4, 0)

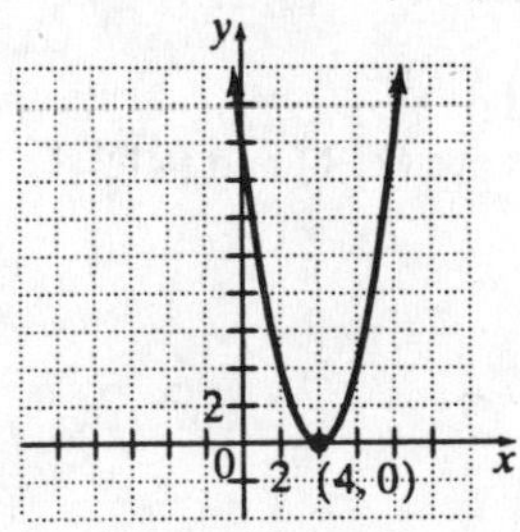

9.

$$x^2 + y^2 + 6x = 16$$

$$x^2 + 6x + y^2 = 16$$

$$(x^2 + 6x + 9) + y^2 = 16 + 9$$

$$(x + 3)^2 + y^2 = 5^2$$

Circle: center(−3, 0), radius 5

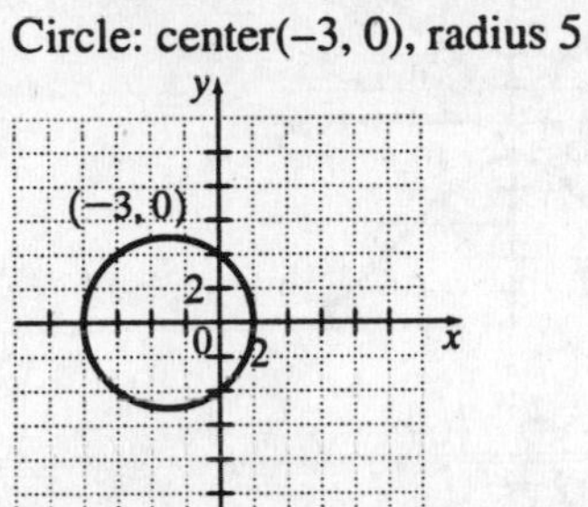

10. $x = y^2 + 8y - 3$

$x = (y^2 + 8y + 16) - 3 - 16$

$x = (y + 4)^2 - 19$

Parabola: vertex (−19, −4)

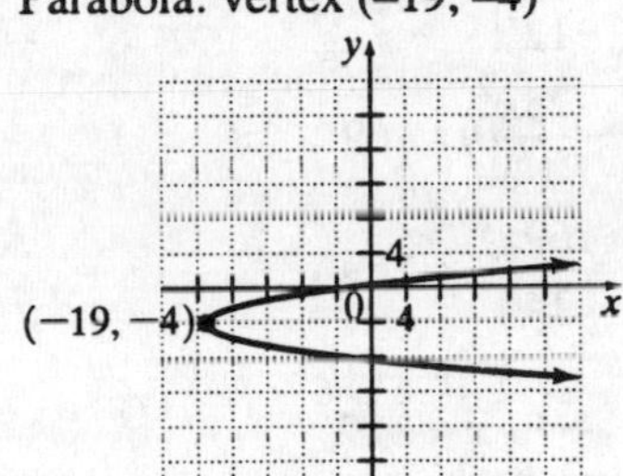

11.

$$\frac{(x-4)^2}{16} + \frac{(y-3)^2}{9} = 1$$

$$\frac{(x-4)^2}{4^2} + \frac{(y-3)^2}{3^2} = 1$$

Ellipse: $a = 4$, $b = 3$, center (4, 3)

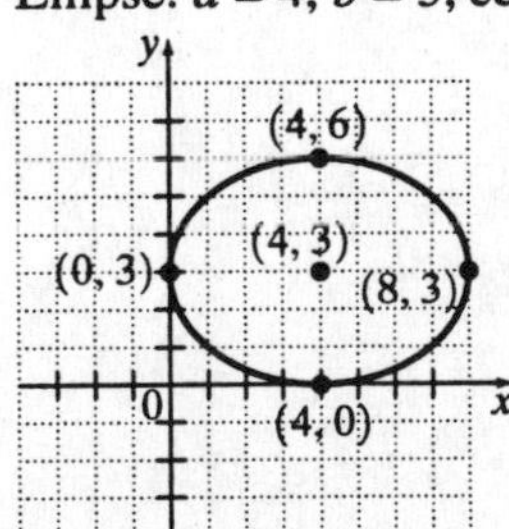

12. $y^2 - x^2 = 1$

Hyperbola: $a = 1, b = 1$

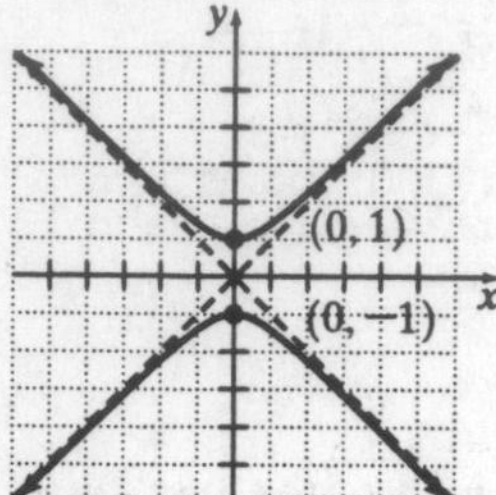

13. $\begin{cases} x^2 + y^2 = 169 \\ 5x + 12y = 0 \end{cases}$

$12y = -5x$

$y = -\frac{5x}{12}$

Substitute.

$x^2 + \left(-\frac{5x}{12}\right)^2 = 169$

$x^2 + \frac{25x^2}{144} = 169$

$\frac{169x^2}{144} = 169$

$x^2 = 144$

$x = \pm 12$

Substitute back.

$x = 12$: $y = -\frac{5}{12}(12) = -5$

$x = -12$: $y = -\frac{5}{12}(-12) = 5$

The solutions are (12, –5), (–12, 5).

14. $\begin{cases} x^2 + y^2 = 26 \\ x^2 - y^2 = 24 \end{cases}$

Add.

$2x^2 = 50$

$x^2 = 25$

$x = \pm 5$

Substitute back.

$25 + y^2 = 26$

$y^2 = 1$

$y = \pm 1$

The solutions are (5, 1), (5, –1), (–5, 1), and (–5, –1).

15. $\begin{cases} y = x^2 - 5x + 6 \\ y = 2x \end{cases}$

Substitute.

$2x = x^2 - 5x + 6$

$0 = x^2 - 7x + 6$

$0 = (x-1)(x-6)$

$x - 1 = 0$ or $x - 6 = 0$

$x = 1$ or $x = 6$

$x = 1$: $y = 2(1) = 2$

$x = 6$: $y = 2(6) = 12$

The solutions are (1, 2) and (6, 12).

16. $\begin{cases} x^2 + 4y^2 = 5 \\ y = x \end{cases}$

Substitute.

$x^2 + 4x^2 = 5$

$5x^2 = 5$

$x^2 = 1$

$x = \pm 1$

$x = 1$: $y = 1$

$x = -1$: $y = -1$

The solutions are (1, 1) and (–1, –1).

17. $\begin{cases} 2x + 5y \ge 10 \\ y \ge x^2 + 1 \end{cases}$

First, graph $2x + 5y = 10$.
Does (0, 0) satisfy the inequality?
$2(0) + 5(0) \ge 10$
$0 \ge 10$ false
Shade the portion of the graph that does not contain (0, 0).
Next, graph $y = x^2 + 1$.
Does (0, 0) satisfy the inequality?
$0 \ge 0^2 + 1$
$0 \ge 1$ false
Shade the portion of the graph that does not contain (0, 0).
The solution to the system is the overlapping region.

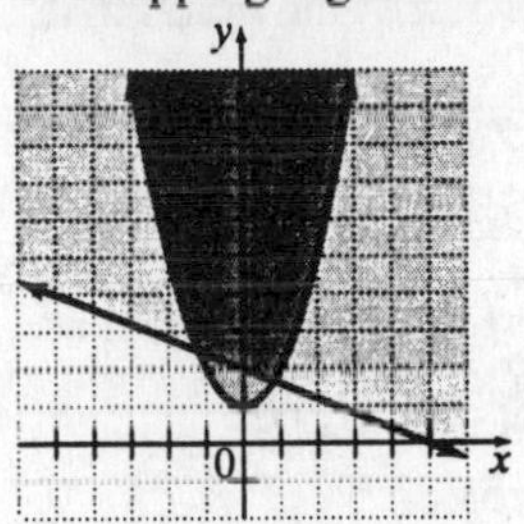

18. $\begin{cases} \frac{x^2}{4} + y^2 \le 1 \\ x + y > 1 \end{cases}$

First, graph $\frac{x^2}{4} + y^2 = 1$.
Does (0, 0) satisfy the inequality?
$\frac{0^2}{4} + 0^2 \le 1$
$0 \le 1$ true
Shade the portion of the graph that contains (0, 0).
Next, graph $x + y = 1$ with a dashed line.
Does (0, 0) satisfy the inequality?
$0 + 0 > 1$
$0 > 1$ false
Shade the portion of the graph that does not contain (0, 0).
The solution to the system is the overlapping region.

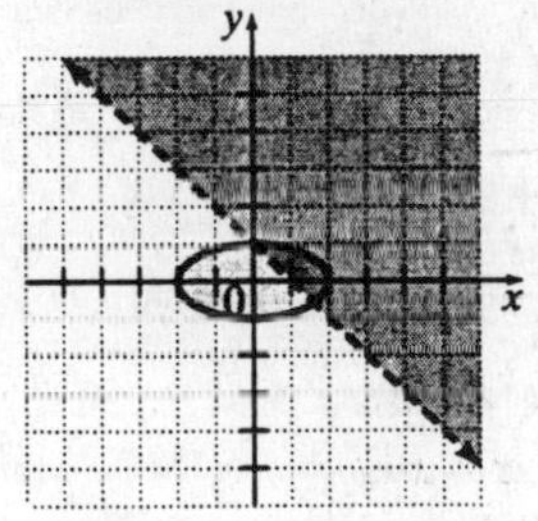

19. $\begin{cases} x^2 + y^2 > 1 \\ \dfrac{x^2}{4} - y^2 \ge 1 \end{cases}$

First, graph $x^2 + y^2 = 1$ with a dashed line.

Does (0, 0) satisfy the inequality?

$0^2 + 0^2 > 0$

$0 > 1$ false

Shade the portion of the graph that does not contain (0, 0).

Next, graph $\dfrac{x^2}{4} - y^2 = 1$.

Does (0, 0) satisfy the inequality?

$\dfrac{0^2}{4} - 0^2 \ge 1$

$0 \ge 1$ false

Does (3, 0) satisfy the inequality?

$\dfrac{3^2}{4} - 0^2 \ge 1$

$\dfrac{9}{4} \ge 1$ true

Does $(-3, 0)$ satisfy the inequality?

$\dfrac{(-3)^2}{4} - 0^2 \ge 1$

$\dfrac{9}{4} \ge 1$ true

Shade the portions of the graph that contain $(3, 0)$ and $(-3, 0)$.

The solution to the system is the overlapping region.

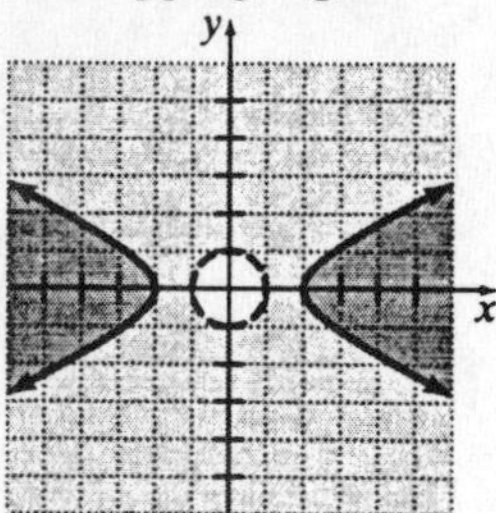

20. $\begin{cases} x^2 + y^2 \ge 4 \\ x^2 + y^2 < 16 \\ y \ge 0 \end{cases}$

First, graph $x^2 + y^2 = 4$.

Does (0, 0) satisfy the inequality?

$0^2 + 0^2 \ge 4$

$0 \ge 4$ false

Shade the portion of the graph that does not contain (0, 0).

Next, graph $x^2 + y^2 = 16$ with a dashed line.

Does (0, 0) satisfy the inequality?

$0^2 + 0^2 < 16$

$0 < 16$ true

Shade the portion of the graph that contains (0, 0).

Next, graph $y = 0$.

Does (0, 2) satisfy the inequality?

$2 \ge 0$ true

Shade the portion of the graph that contains (0, 2).

The solution to the system is the overlapping region.

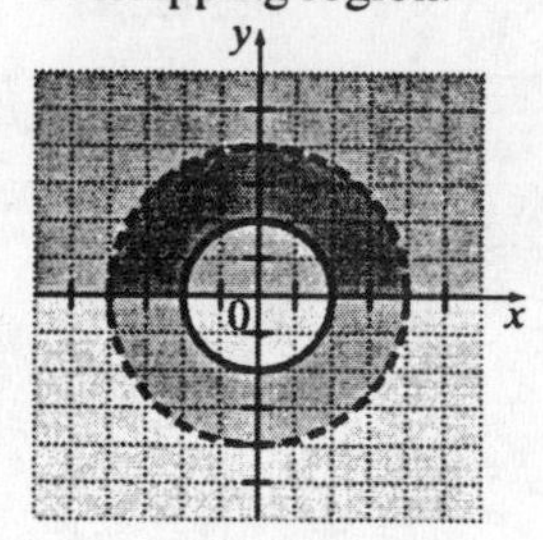

21. Since $a > 0$, the parabola opens either up or to the right. Since $h < 0$ and $k > 0$, (h, k) is in quadrant II. Since the squared term is $(y - k)$, the parabola opens to the left or to the right. Therefore, graph B best represents the equation.

22. $100x^2 + 225y^2 = 22,500$

$$\frac{x^2}{225} + \frac{y^2}{100} = 0$$

$a = \sqrt{225} = 15$

$b = \sqrt{100} = 10$

Width = 15 + 15 = 30 feet

Height = 10 feet

Appendix A

1.
```
   9.076
 + 8.004
  17.080
```

3.
```
    6 10
  27.004
 − 14.200
  12.804
```

5.
```
 107.92
 + 3.04
 110.96
```

7.
```
 0 9 10
  10.0
 − 7.6
   2.4
```

9.
```
 011151212
  126.32
 − 97.89
   28.43
```

11.
```
   3.25
 ×   70
 227.50
```

13.
```
      2.7
 3) 8.1
   −6
    21
   −21
```

15.
```
 4 15 310410
  55.4050
 − 6.1711
  49.2339
```

17.
```
          80
 75.) 6000.
      −600
        00
        −0
```

19.
```
        0.07612
 100) 7.61200
       −700
         612
        −600
         120
        −100
          200
         −200
```

21.
```
         4.56
 27.) 123.12
      −108
        151
       −135
        162
       −162
```

23.
```
  569.2
   71.25
 +  8.01
  648.46
```

25.
```
    7 910
  768.00
 −  0.17
  767.83
```

27.
```
   12.
 + 0.062
  12.062
```

29.
```
   5 910
  76.00
 − 14.52
  61.48
```

31.
```
        7.7
 43.) 331.1
      −301
       30 1
      −30 1
```

33.
$$\begin{array}{r} 762.12 \\ 89.7 \\ +\ 11.55 \\ \hline 863.37 \end{array}$$

35.
$$\begin{array}{r} {\scriptstyle 2\,13\,9\,10} \\ 23.400 \\ -\ 0.821 \\ \hline 22.579 \end{array}$$

37.
$$\begin{array}{r} {\scriptstyle 5\,10\,12} \\ 476.12 \\ -\ 112.97 \\ \hline 363.15 \end{array}$$

39.
$$\begin{array}{r} 0.007 \\ +\ 7. \\ \hline 7.007 \end{array}$$

Appendix B

1. $90° - 19° = 71°$

3. $90° - 70.8° = 19.2°$

5. $90° - 11\frac{1}{4}° = 78\frac{3}{4}°$

7. $180° - 150° = 30°$

9. $180° - 30.2° = 149.8°$

11. $180° - 79\frac{1}{2}° = 100\frac{1}{2}°$

13. $m\angle 1 = 110°$
$m\angle 2 = 180° - 110° = 70°$
$m\angle 3 = m\angle 2 = 70°$
$m\angle 4 = m\angle 2 = 70°$
$m\angle 5 = m\angle 1 = 110°$
$m\angle 6 = m\angle 4 = 70°$
$m\angle 7 = m\angle 5 = 110°$

15. $180° - 11° - 79° = 90°$

17. $180° - 25° - 65° = 90°$

19. $180° - 30° - 60° = 90°$

21. $90° - 45° = 45°$
The other angles are 45° and 90°.

23. $90° - 17° = 73°$
The other angles are 73° and 90°.

25. $90° - 39\frac{3}{4}° = 50\frac{1}{4}°$

The other angles are $50\frac{1}{4}°$ and 90°.

27.
$$\frac{12}{4} = \frac{18}{x}$$
$$4x\left(\frac{12}{4}\right) = 4x\left(\frac{18}{x}\right)$$
$$12x = 72$$
$$x = 6$$

29.
$$\frac{6}{9} = \frac{3}{x}$$
$$9x\left(\frac{6}{9}\right) = 9x\left(\frac{3}{x}\right)$$
$$6x = 27$$
$$x = 4.5$$

31.
$$a^2 + b^2 = c^2$$
$$6^2 + 8^2 = c^2$$
$$36 + 64 = c^2$$
$$100 = c^2$$
$$10 = c$$

33.
$$a^2 + b^2 = c^2$$
$$5^2 + b^2 = 13^2$$
$$25 + b^2 = 169$$
$$b^2 = 144$$
$$b = 12$$

Appendix D

1. Volume $= lwh = 6(4)(3) = 72$

$$\begin{aligned}\text{Surface area} &= 2lh + 2wh + 2lw \\ &= 2(6)(3) + 2(4)(3) + 2(6)(4) \\ &= 36 + 24 + 48 \\ &= 108\end{aligned}$$

The volume is 72 cubic inches and the surface area is 108 square inches.

3. Volume $= s^3 = 8^3 = 512$

Surface area $= 6s^2 = 6(8)^2 = 384$

The volume is 512 cubic centimeters and the surface area is 384 square centimeters.

5. $$\begin{aligned}\text{Volume} &= \frac{1}{3}\pi r^2 h \\ &= \frac{1}{3}\pi(2)^2(3) \\ &= 4\pi \\ &\approx 4\left(\frac{22}{7}\right) \\ &= 12\frac{4}{7}\end{aligned}$$

$$\begin{aligned}\text{Surface area} &= \pi r\sqrt{r^2 + h^2} + \pi r^2 \\ &= \pi(2)\sqrt{2^2 + 3^2} + \pi(2)^2 \\ &= (2\sqrt{13}\pi + 4\pi) \\ &\approx 2\sqrt{13}(3.14) + 4(3.14) \\ &\approx 35.20\end{aligned}$$

The volume is 4π cubic yards, or approximately $12\frac{4}{7}$ cubic yards, and the surface area is $(2\sqrt{13}\pi + 4\pi)$ square yards, or approximately 35.20 square yards.

7. $$\begin{aligned}\text{Volume} &= \frac{4}{3}\pi r^3 \\ &= \frac{4}{3}\pi(5)^3 \\ &= \frac{500}{3}\pi \\ &\approx \frac{500}{3}\left(\frac{22}{7}\right) \\ &= 523\frac{17}{21}\end{aligned}$$

$$\begin{aligned}\text{Surface area} &= 4\pi r^2 \\ &= 4\pi(5)^2 \\ &= 100\pi \\ &\approx 100\left(\frac{22}{7}\right) \\ &= 314\frac{2}{7}\end{aligned}$$

The volume is $\frac{500}{3}\pi$ cubic inches, or approximately $523\frac{17}{21}$ cubic inches, and the surface area is 100π square inches, or approximately $314\frac{2}{7}$ square inches.

9. Volume $= \frac{1}{3}s^2h = \frac{1}{3}(6)^2(4) = 48$

$$\begin{aligned}\text{Surface area} &= B + \frac{1}{2}pl \\ &= 36 + \frac{1}{2}(24)(5) \\ &= 96\end{aligned}$$

The volume is 48 cubic centimeters and the surface area is 96 square centimeters.

11. Volume $= s^3 = \left(1\frac{1}{3}\right)^3 = 2\frac{10}{27}$

The volume is $2\frac{10}{27}$ cubic inches.

13. Surface area $= 2lh + 2wh + 2lw$
$= 2(2)(1.4) + 2(3)(1.4) + 2(2)(3)$
$= 5.6 + 8.4 + 12$
$= 26$
The surface area is 26 square feet.

15. Volume $= \frac{1}{3}s^2h = \frac{1}{3}(5)^2(1.3) \approx 10.83$
The volume is approximately 10.83 cubic inches.

17. Volume $= \frac{1}{3}s^2h = \frac{1}{3}(12)^2(20) = 960$
The volume is 960 cubic centimeters.

19. Surface area $= 4\pi r^2 = 4\pi(7)^2 = 196\pi$
The surface area is 196π square inches.

21. Volume $= lwh = 2\left(2\frac{1}{2}\right)\left(1\frac{1}{2}\right) = 7\frac{1}{2}$
The volume is $7\frac{1}{2}$ cubic feet.

23. Volume $= \frac{1}{3}\pi r^2h \approx \frac{1}{3}\left(\frac{22}{7}\right)(2)^2(3) = 12\frac{4}{7}$
The volume is $12\frac{4}{7}$ cubic centimeters.

Appendix E

1. 21, 28, 16, 42, 38

$$\bar{x} = \frac{21+28+16+42+38}{5} = \frac{145}{5} = 29$$

16, 21, *28*, 38, 42
median = 28
no mode

3. 7.6, 8.2, 8.2, 9.6, 5.7, 9.1

$$\bar{x} = \frac{7.6+8.2+8.2+9.6+5.7+9.1}{6} = \frac{48.4}{6} = 8.1$$

5.7, 7.6, *8.2*, *8.2*, 9.1, 9.6

$$\text{median} = \frac{8.2+8.2}{2} = 8.2$$

mode = 8.2

5. 0.2, 0.3, 0.5, 0.6, 0.6, 0.9, 0.2, 0.7, 1.1

$$\bar{x} = \frac{0.2+0.3+0.5+0.6+0.6+0.9+0.2+0.7+1.1}{9} = \frac{5.1}{9} = 0.6$$

0.2, 0.2, 0.3, 0.5, *0.6*, 0.6, 0.7, 0.9, 1.1
median = 0.6
mode = 0.2 and 0.6

7. 231, 543, 601, 293, 588, 109, 334, 268

$$\bar{x} = \frac{231+543+601+293+588+109+334+268}{8} = \frac{2967}{8} = 370.9$$

109, 231, 268, *293*, *334*, 543, 588, 601

$$\text{median} = \frac{293+334}{2} = 313.5$$

no mode

9. $$\bar{x} = \frac{1454+1368+1362+1250+1136}{5} = \frac{5670}{5} = 1314$$

The mean height is 1314 feet.

11. 1454, 1368, 1362, 1250, *1136*, *1127*, 1107, 1046, 1023, 1002

$$\text{median} = \frac{1136+1127}{2} = 1131.5 \text{ feet}$$

The median height is 1131.5

13. $$\bar{x} = \frac{7.8+6.9+7.5+4.7+6.9+7.0}{6} = \frac{40.8}{6} = 6.8$$

The mean time is 6.8 seconds.

15. 4.7, 6.9, 6.9, 7.0, 7.5, 7.8
The mode is 6.9 seconds.

17. 74, 77, *85*, *86*, 91, 95

$\text{median} = \dfrac{85+86}{2} = 85.5$

The median score is 85.5.

19. $\bar{x} = \dfrac{78+80+66+68+71+64+82+71+70+65+70+75+77+86+72}{15} = \dfrac{1095}{15} = 73$

The mean pulse rate is 73.

21. 64, 65, 66, 68, 70, 70, 71, 71, 72, 75, 77, 78, 80, 82, 86

This set of pulse rates has two modes: 70 and 71.

23. 64, 65, 66, 68, 70, 70, 71, 71, 72, 75, 77, 78, 80, 82, 86

↑ mean = 73

9 rates were lower than the mean.

25. _, _, 16, 18, _;

Since the mode is 21, at least two of the missing numbers must be 21. The mean is 20. Let the one unknown number be x.

$$\bar{x} = \frac{21+21+16+18+x}{5} = 20$$

$$\frac{76+x}{5} = 20$$

$$76+x = 100$$

$$x = 24$$

The missing numbers are 21, 21, 24.

Appendix F

Viewing Window and Interpreting Window Settings Exercise Set

1. Yes, since every coordinate is between –10 and 10.

3. No, since –11 is less than –10.

5. Any values such that Xmin < –90, Ymin < –80, Xmax > 55, and Ymax > 80.

7. Any values such that Xmin < –11, Ymin < –5, Xmax > 7, and Ymax > 2.

9. Any values such that Xmin < 50, Ymin < –50, Xmax > 200, and Ymax > 200.

11. Xmin = –12 Ymin = –12
Xmax = 12 Ymax = 12
Xscl = 3 Yscl = 3

13. Xmin = –9 Ymin = –12
Xmax = 9 Ymax = 12
Xscl = 1 Yscl = 2

15. Xmin = –10 Ymin = –25
Xmax = 10 Ymax = 25
Xscl = 2 Yscl = 5

17. Xmin = –10 Ymin = –30
Xmax = 10 Ymax = 30
Xscl = 1 Yscl = 3

19. Xmin = –20 Ymin = –30
Xmax = 30 Ymax = 50
Xscl = 5 Yscl = 10

Graphing Equations and Square Viewing Window Exercise Set

1. Setting A:

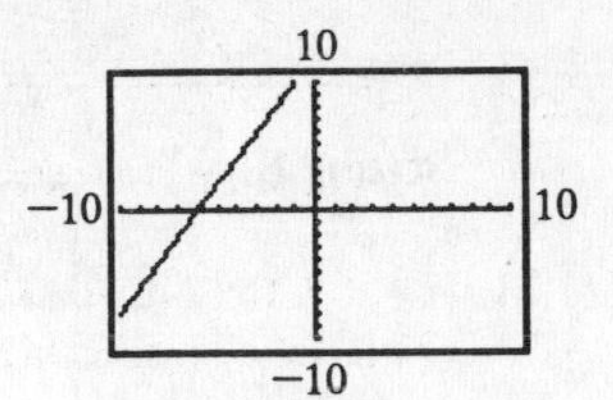

Setting B:

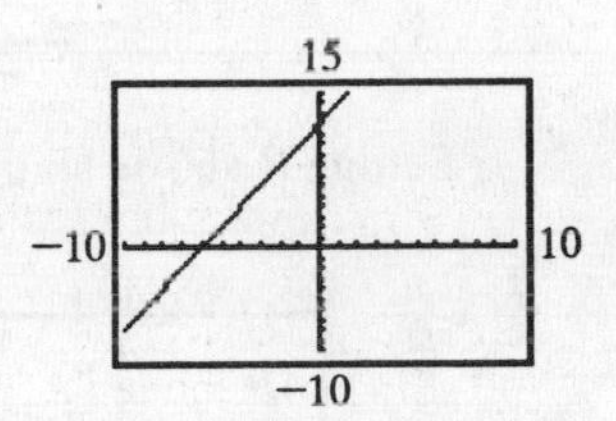

Setting B shows all the intercepts.

3. Setting A:

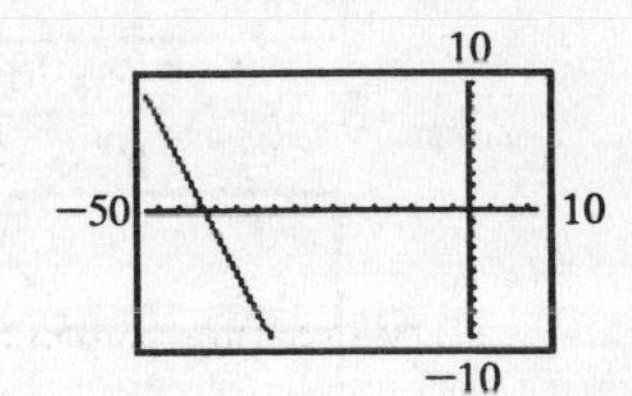

Setting B:

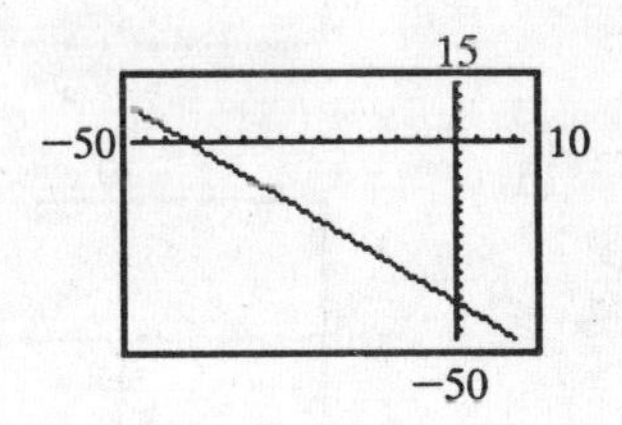

Setting B shows all the intercepts.

5. Setting A:

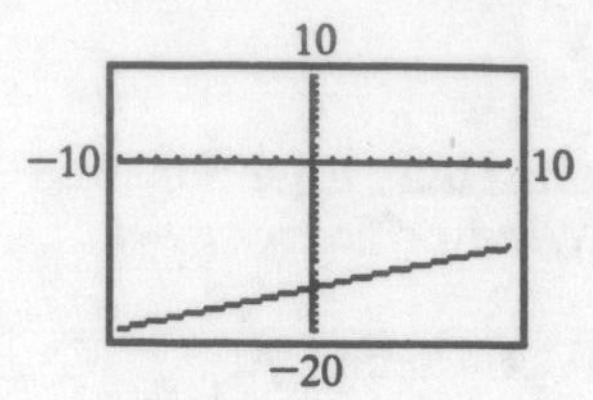

Setting B:

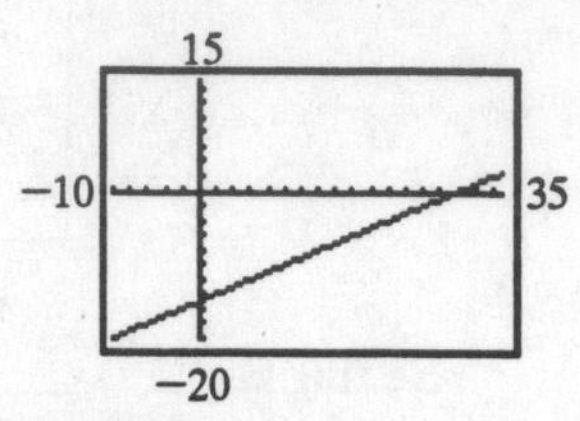

Setting B shows all the intercepts.

7.

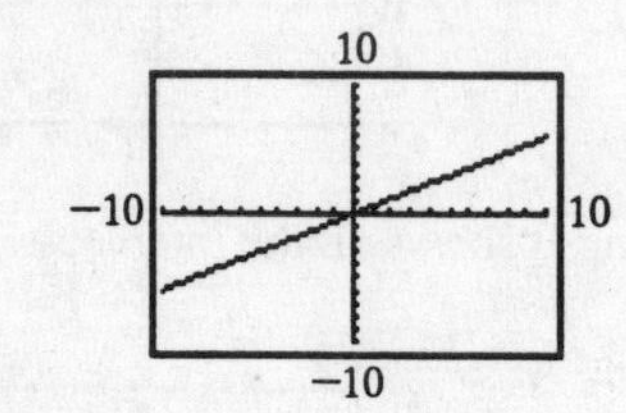

9.

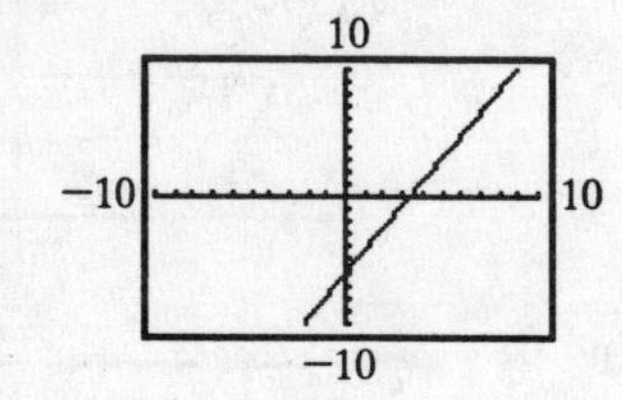

11.

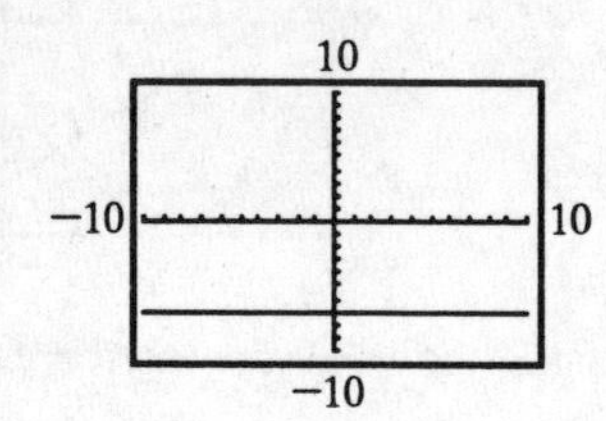

13.

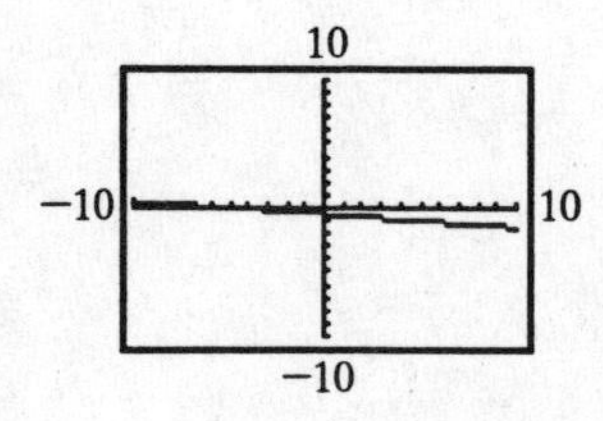

15.

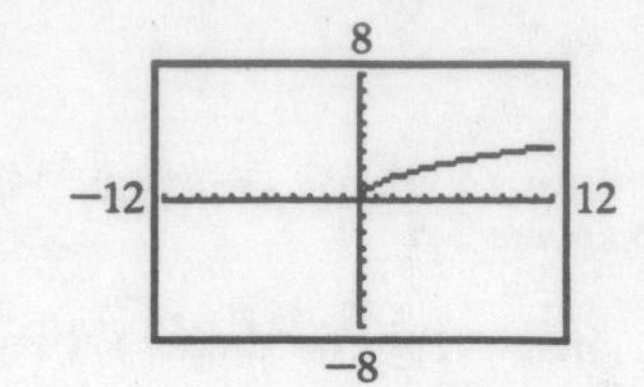

17.

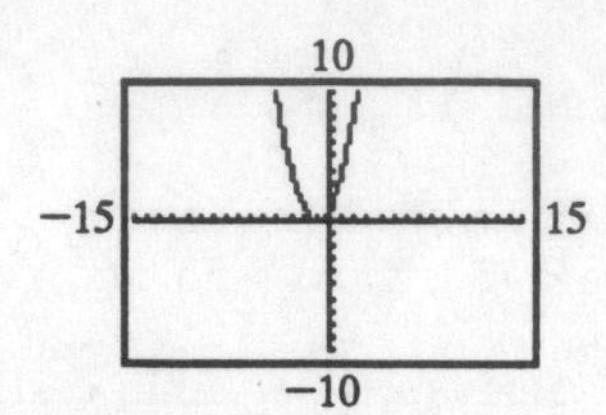

19.

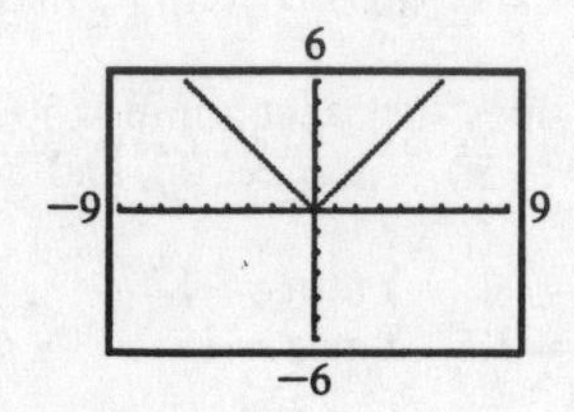

21. Standard window:

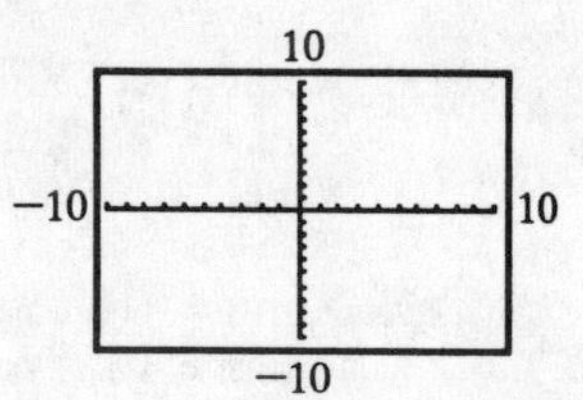

Adjusted window:

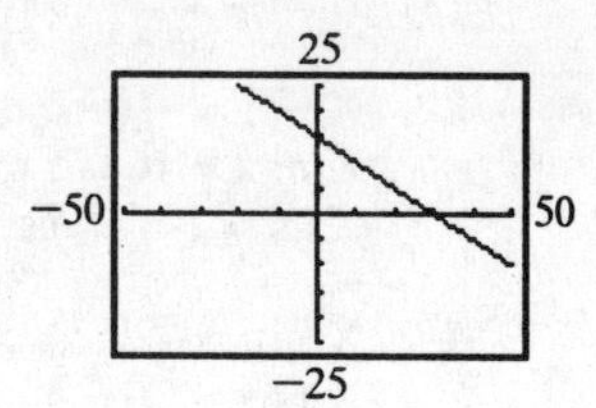